*Werner Vogel and
Dirk-Gunnar Welsch*
Quantum Optics

Related Titles

Audretsch, J. (ed.)
Entangled World
The Fascination of Quantum Information and Computation
approx. 312 pages with approx. 66 figures
Hardcover
ISBN 3-527-40470-8

Bachor, H.-A., Ralph, T. C.
A Guide to Experiments in Quantum Optics
434 pages with 195 figures
2004
Softcover
ISBN 3-527-40393-0

Weidemüller, M., Zimmermann, C. (eds.)
Interactions in Ultracold Gases
From Atoms to Molecules
519 pages with 241 figures and 16 tables
2003
Hardcover
ISBN 3-527-40389-2

Saleh, B.E.A.
Fundamentals of Photonics
Second Edition
2006
Hardcover
ISBN 0-471-35832-0

Schleich, W. P.
Quantum Optics in Phase Space
716 pages with approx. 220 figures
2001
Hardcover
ISBN 3-527-29435-X

Yariv, A.
Quantum Electronics
696 pages
1989
Hardcover
ISBN 0-471-60997-8

Werner Vogel and Dirk-Gunnar Welsch

Quantum Optics

Third, revised and extended edition

WILEY-VCH Verlag GmbH & Co. KGaA

The Authors

Prof. Dr. Werner Vogel
Universität Rostock
werner.vogel@physik.uni-rostock.de

Prof. Dr. Dirk-Gunnar Welsch
Friedrich-Schiller-Universität Jena
D.-G.Welsch@tpi.uni-jena.de

All books published by Wiley-VCH are carefully produced. Nevertheless, authors, editors, and publisher do not warrant the information contained in these books, including this book, to be free of errors. Readers are advised to keep in mind that statements, data, illustrations, procedural details or other items may inadvertently be inaccurate.

Library of Congress Card No.: applied for.

British Library Cataloging-in-Publication Data:
A catalogue record for this book is available from the British Library.

Bibliographic information published by Die Deutsche Bibliothek
Die Deutsche Bibliothek lists this publication in the Deutsche Nationalbibliografie; detailed bibliographic data is available in the Internet at <http://dnb.ddb.de>.

© 2006 WILEY-VCH Verlag GmbH & Co. KGaA, Weinheim

All rights reserved (including those of translation into other languages). No part of this book may be reproduced in any form – nor transmitted or translated into machine language without written permission from the publishers. Registered names, trademarks, etc. used in this book, even when not specifically marked as such, are not to be considered unprotected by law.

Printed in the Federal Republic of Germany
Printed on acid-free paper

Typesetting   Uwe Krieg, Berlin
Printing   betz-druck GmbH, Darmstadt
Binding   J. Schäffer GmbH i.G., Grünstadt

ISBN-13: 978-3-527-40507-7
ISBN-10: 3-527-40507-0

Contents

Preface *XI*

1 Introduction *1*
1.1 From Einstein's hypothesis to photon anti-bunching *2*
1.2 Nonclassical phenomena *5*
1.3 Source-attributed light *6*
1.4 Medium-assisted electromagnetic fields *7*
1.5 Measurement of light statistics *9*
1.6 Determination and preparation of quantum states *10*
1.7 Quantized motion of cold atoms *11*

2 Elements of quantum electrodynamics *15*
2.1 Basic classical equations *16*
2.2 The free electromagnetic field *20*
2.2.1 Canonical quantization *21*
2.2.2 Monochromatic-mode expansion *22*
2.2.3 Nonmonochromatic modes *28*
2.3 Interaction with charged particles *30*
2.3.1 Minimal coupling *31*
2.3.2 Multipolar coupling *33*
2.4 Dielectric background media *39*
2.4.1 Nondispersing and nonabsorbing media *41*
2.4.2 Dispersing and absorbing media *44*
2.5 Approximate interaction Hamiltonians *50*
2.5.1 The electric-dipole approximation *51*
2.5.2 The rotating-wave approximation *53*
2.5.3 Effective Hamiltonians *56*
2.6 Source-quantity representation *60*
2.7 Time-dependent commutation relations *65*
2.8 Correlation functions of field operators *69*

Quantum Optics, Third, revised and extended edition. Werner Vogel and Dirk-Gunnar Welsch
Copyright © 2006 WILEY-VCH Verlag GmbH & Co. KGaA, Weinheim
ISBN: 3-527-40507-0

3 Quantum states of bosonic systems 73
3.1 Number states 73
3.1.1 Statistics of the number states 77
3.1.2 Multi-mode number states 78
3.2 Coherent states 79
3.2.1 Statistics of the coherent states 84
3.2.2 Multi-mode coherent states 85
3.2.3 Displaced number states 87
3.3 Squeezed states 88
3.3.1 Statistics of the squeezed states 92
3.3.2 Multi-mode squeezed states 98
3.4 Quadrature eigenstates 102
3.5 Phase states 104
3.5.1 The eigenvalue problem of $\hat{V}$ 105
3.5.2 Cosine and sine phase states 109

4 Bosonic systems in phase space 113
4.1 The statistical density operator 113
4.2 Phase-space functions 116
4.2.1 Normal ordering: The P function 117
4.2.2 Anti-normal and symmetric ordering: The Q and the W function 120
4.2.3 Parameterized phase-space functions 121
4.3 Operator expansion in phase space 124
4.3.1 Orthogonalization relations 125
4.3.2 The density operator in phase space 126
4.3.3 Some elementary examples 129

5 Quantum theory of damping 135
5.1 Quantum Langevin equations and one-time averages 137
5.1.1 Hamiltonian 137
5.1.2 Heisenberg equations of motion 139
5.1.3 Born and Markov approximations 141
5.1.4 Quantum Langevin equations 142
5.2 Master equations and related equations 146
5.2.1 Master equations 147
5.2.2 Fokker–Planck equations 148
5.3 Damped harmonic oscillator 151
5.3.1 Langevin equations 151
5.3.2 Master equations 155
5.3.3 Fokker–Planck equations 156
5.3.4 Radiationless dephasing 158

5.4	Damped two-level system	*161*
5.4.1	Basic equations	*161*
5.4.2	Optical Bloch equations	*164*
5.5	Quantum regression theorem	*169*

6	**Photoelectric detection of light**	*173*
6.1	Photoelectric counting	*173*
6.1.1	Quantum-mechanical transition probabilities	*174*
6.1.2	Photoelectric counting probabilities	*179*
6.1.3	Counting moments and correlations	*183*
6.2	Photoelectric counts and photons	*187*
6.2.1	Detection scheme	*187*
6.2.2	Mode expansion	*189*
6.2.3	Photon-number statistics	*191*
6.3	Nonperturbative corrections	*195*
6.4	Spectral detection	*197*
6.4.1	Radiation-field modes	*198*
6.4.2	Input-output relations	*200*
6.4.3	Spectral correlation functions	*202*
6.5	Homodyne detection	*205*
6.5.1	Fields combining through a nonabsorbing beam splitter	*205*
6.5.2	Fields combining through an absorbing beam splitter	*210*
6.5.3	Unbalanced four-port homodyning	*213*
6.5.4	Balanced four-port homodyning	*217*
6.5.5	Balanced eight-port homodyning	*223*
6.5.6	Homodyne correlation measurement	*228*
6.5.7	Normally ordered moments	*231*

7	**Quantum-state reconstruction**	*237*
7.1	Optical homodyne tomography	*239*
7.1.1	Quantum state and phase-rotated quadratures	*240*
7.1.2	Wigner function	*244*
7.2	Density matrix in phase-rotated quadrature basis	*247*
7.3	Density matrix in the number basis	*250*
7.3.1	Sampling from quadrature components	*250*
7.3.2	Reconstruction from displaced number states	*254*
7.4	Local reconstruction of phase-space functions	*256*
7.5	Normally ordered moments	*257*
7.6	Canonical phase statistics	*260*

8	**Nonclassicality and entanglement of bosonic systems**	*265*
8.1	Quantum states with classical counterparts	*266*

8.2	Nonclassical light	270
8.2.1	Photon anti-bunching	270
8.2.2	Sub-Poissonian light	273
8.2.3	Squeezed light	276
8.3	Nonclassical characteristic functions	281
8.3.1	The Bochner theorem	282
8.3.2	First-order nonclassicality	283
8.3.3	Higher-order nonclassicality	285
8.4	Nonclassical moments	287
8.4.1	Reformulation of the Bochner condition	287
8.4.2	Criteria based on moments	288
8.5	Entanglement	290
8.5.1	Separable and nonseparable quantum states	290
8.5.2	Partial transposition and entanglement criteria	292
9	**Leaky optical cavities**	**299**
9.1	Radiation-field modes	301
9.1.1	Solution of the Helmholtz equation	301
9.1.2	Cavity-response function	303
9.2	Source-quantity representation	305
9.3	Internal field	308
9.3.1	Coarse-grained averaging	308
9.3.2	Nonmonochromatic modes and Langevin equations	311
9.4	External field	313
9.4.1	Source-quantity representation	314
9.4.2	Input-output relations	316
9.5	Commutation relations	317
9.5.1	Internal field	318
9.5.2	External field	321
9.6	Field correlation functions	323
9.7	Unwanted losses	327
9.8	Quantum-state extraction	329
10	**Medium-assisted electromagnetic vacuum effects**	**337**
10.1	Spontaneous emission	338
10.1.1	Weak atom–field coupling	341
10.1.2	Strong atom–field coupling	348
10.2	Vacuum forces	352
10.2.1	Force on an atom	353
10.2.2	The Casimir force	360

11	**Resonance fluorescence** *367*
11.1	Basic equations *367*
11.2	Two-level systems *370*
11.2.1	Intensity *372*
11.2.2	Intensity correlation and photon anti-bunching *375*
11.2.3	Squeezing *379*
11.2.4	Spectral properties *383*
11.3	Multi-level effects *391*
11.3.1	Dark resonances *391*
11.3.2	Intermittent fluorescence *394*
11.3.3	Vibronic coupling *398*

12	**A single atom in a high-Q cavity** *407*
12.1	The Jaynes–Cummings model *408*
12.2	Electronic-state dynamics *413*
12.2.1	Reduced density matrix *413*
12.2.2	Collapse and revival *415*
12.2.3	Quantum nature of the revivals *421*
12.2.4	Coherent preparation *422*
12.3	Field dynamics *424*
12.3.1	Reduced density matrix *424*
12.3.2	Photon statistics *425*
12.4	The Micromaser *428*
12.5	Quantum-state preparation *433*
12.5.1	Schrödinger-cat states *433*
12.5.2	Einstein–Podolsky–Rosen pairs of atoms *434*
12.6	Measurements of the cavity field *435*
12.6.1	Quantum state endoscopy *436*
12.6.2	QND measurement of the photon number *437*
12.6.3	Determining arbitrary quantum states *438*

13	**Laser-driven quantized motion of a trapped atom** *443*
13.1	Quantized motion of an ion in a Paul trap *444*
13.2	Interaction of a moving atom with light *446*
13.2.1	Radio-frequency radiation *447*
13.2.2	Optical radiation *448*
13.3	Dynamics in the resolved sideband regime *449*
13.3.1	Nonlinear Jaynes–Cummings model *449*
13.3.2	Decoherence effects *454*
13.3.3	Nonlinear motional dynamics *456*
13.4	Preparing motional quantum states *461*
13.4.1	Sideband laser-cooling *461*

13.4.2　Coherent, number and squeezed states　*463*
13.4.3　Schrödinger-cat states　*464*
13.4.4　Motional dark states　*466*
13.5　Measuring the quantum state　*472*
13.5.1　Tomographic methods　*472*
13.5.2　Local methods　*475*
13.5.3　Determination of entangled states　*478*

Appendix

A　**The medium-assisted Green tensor**　*481*
A.1　Basic relations　*481*
A.2　Asymptotic behavior　*482*

B　**Equal-time commutation relations**　*485*

C　**Algebra of bosonic operators**　*487*
C.1　Exponential-operator disentangling　*487*
C.2　Normal and anti-normal ordering　*490*

D　**Sampling function for the density matrix in the number basis**　*493*

Index　*497*

Preface

The refinement of experimental techniques has greatly stimulated progress in quantum optics. Understanding of the quantum nature of matter and light has been significantly widened and new insights have been gained. A number of fundamental predictions arising from the concepts of quantum physics have been proved by means of optical methods.

In our book *Quantum Optics*, which arose from lectures that we have given for many years in Jena, Güstrow and Rostock, an attempt is made to develop the theoretical concepts of modern quantum optics, with emphasis on current research trends. It is based on our book, *Lectures on Quantum Optics* (Akademie Verlag/VCH Publishers, Berlin/New York, 1994) and its revised and enlarged second edition, *Quantum Optics – An Introduction* (Wiley-VCH, Berlin, 2001), which we wrote together with S. Wallentowitz. Taking into account representative developments in the field, in the second edition we have included new topics such as quantization of radiation in dispersing and absorbing media, quantum-state measurement and reconstruction, and quantized motion of laser-driven trapped atoms. Following this line, in the present edition we have again included new topics. The new Chapter 10 is devoted to medium-assisted electromagnetic vacuum effects, with special emphasis on spontaneous emission and van der Waals and Casimir forces. In the substantially revised and extended Chapter 8, a unified concept of measurement-based nonclassicality and entanglement criteria for bosonic systems is presented. The new measurement principles needed in this context are explained in Chapter 6. Two sections are added to Chapter 9 in which the problem of unwanted losses in quantum-state extraction from leaky optical cavities is studied. A consideration of decoherence effects in the motion of trapped atoms is added to Chapter 13.

Quantum Optics should be useful for graduate students in physics as well as for research workers who want to become familiar with the ideas of quantum optics. A basic knowledge of quantum mechanics, electrodynamics and classical statistics is assumed.

Quantum Optics, Third, revised and extended edition. Werner Vogel and Dirk-Gunnar Welsch
Copyright © 2006 WILEY-VCH Verlag GmbH & Co. KGaA, Weinheim
ISBN: 3-527-40507-0

We are grateful to colleagues and students for their contributions to the research and for valuable comments on the manuscript. In particular we would like to thank S.Y. Buhmann, C. Di Fidio, T.D. Ho, T. Kampf, M. Khanbekyan, L. Knöll, C. Raabe, Th. Richter, S. Scheel, E. Shchukin, D. Vasylyev, S. Wallentowitz. Cordial thanks are due to the *Wiley-VCH* team for their helpful attitude and patience. Last but not least, we are greatly indebted to our wives for their patience with us during the period of preparing the manuscript.

W. Vogel and D.-G. Welsch   Rostock and Jena, March 2006

1
Introduction

Since the first experimental demonstration of nonclassical light in 1977, quantum optics has been a very rapidly developing and growing field of modern physics. There are a number of books on the subject [e.g., Agarwal (1974); Allen and Eberly (1975); Carmichael (1993, 1998); Cohen-Tannoudji, Dupont-Roc and Grynberg (1989, 1992); Gardiner (1991); Gerry and Knight (2004); Haken (1985); Klauder and Sudarshan (1968); Loudon (1983); Louisell (1973); Mandel and Wolf (1995); Meystre and Sargent (1990); Orszag (2000); Peřina (1985, 1991); Schleich (2001); Scully and Zubairy (1997); Shore (1990); Vogel and Welsch (1994); Vogel, Welsch and Wallentowitz (2001); Walls and Milburn (1994)], and it is covered in many journals.[1] Presently, in one journal alone (Physical Review A) hundreds of articles on a broad spectrum of quantum-optical and related topics appear every year. Moreover, there are close connections to other traditional fields, such as nonlinear optics, laser spectroscopy and optoelectronics, and the boundaries have often been flexible. The recent improvements in experimental techniques allow one to control the quantum states of various systems with increasing precision. These possibilities have also stimulated the development of rapidly increasing new fields of research such as atom optics and quantum information.

The aim of this book is to describe the fundamentals of quantum optics, and to introduce the basic theoretical concepts to a depth sufficient to apply them practically and to understand and treat specialized problems which have arisen in recent research. On the basis of a general quantum-field-theoretical approach, important topics are presented in a unified manner. Keeping in mind that any real light field is due to sources, time-dependent commutation rules are considered carefully. Nonclassical light is studied and a detailed analysis of measurement schemes is given, including the effect of passive optical instruments, such as beam splitters, spectral filters and leaky cavities. From this background, the basic concepts are developed that allow one to de-

---

[1] For example, see Europhysics Letters, European Physical Journal D, Journal of Modern Optics, Journal of Optics B, Journal of Physics A and B, Journal of the Optical Society of America B, Nature, Optics Communications, Optics Letters, Physical Review A, Physical Review Letters, Physics Letters A, Science.

Quantum Optics, Third, revised and extended edition. Werner Vogel and Dirk-Gunnar Welsch
Copyright © 2006 WILEY-VCH Verlag GmbH & Co. KGaA, Weinheim
ISBN: 3-527-40507-0

termine the quantum states of various systems from measured data. Methods of quantum-state preparation are outlined for particular systems, such as propagating light fields, cavity fields and the quantized motion of a trapped atom.

Any attempt to give a complete overview on the present state of the field, together with a complete list of references, would be a hopeless venture. We have therefore decided to refer to selected work that may be useful in the context of particular topics, with special emphasis on textbooks, review articles and research-stimulating original articles. Before giving a guide to the topics covered, we mention two important fields that, apart from some basic ideas, are not considered, although they are closely related to quantum optics. These are the large fields of nonlinear optics [see, e. g., Bloembergen (1965); Boyd (1991); Peřina (1991); Schubert and Wilhelmi (1986); Shen (1984)] and laser physics and laser spectroscopy [see, e. g., Sargent, Scully and Lamb (1977); Haken (1970); Levenson and Kano (1988); Milonni and Eberly (1988); Stenholm (1984)].

1.1
From Einstein's hypothesis to photon anti-bunching

At the beginning of the last century, one of the unresolved problems in physics was the photoelectric effect. When light falls on a metallic surface, photoelectrons may be ejected (Fig. 1.1), whose energy is insensitive to the intensity,

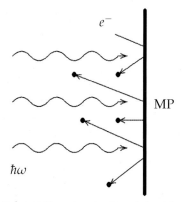

Fig. 1.1 Photoelectric effect: light of frequency ω falls on a metallic plate (MP) and ejects electrons (e$^-$).

but increases with the frequency of the incident light. This result is obviously in contradiction to the concepts of classical physics. From a classical point of view, one would expect the energy of the emitted electrons to increase with the light intensity. Einstein's explanation of the photoelectric effect in 1905, by postulating the existence of light quanta, photons, may be regarded as the birth of quantum optics. He assumed that light is composed of quanta of energy

$$E = \hbar\omega \tag{1.1}$$

and momentum

$$p = \hbar k = \frac{h}{\lambda}. \tag{1.2}$$

In this way, quantities that typically describe the wave aspects of light are related to those that describe particle aspects with the "coupling constant" between wave and particle features being given by the Planck constant $\hbar$. Hence the kinetic energy of an emitted electron, E_{kin}, is given by the difference between the energy of the absorbed photon, $\hbar\omega$, and the binding energy of the electron in the metal, E_b:

$$E_{kin} = \hbar\omega - E_b, \tag{1.3}$$

which implies that, in agreement with observations, the energy of the photoelectrons increases with the frequency of the incident light. Increasing the intensity of the light corresponds to increasing the number of light quanta falling on the metal surface, which gives rise to an increasing number of photoelectrons.

The photoelectric effect plays an important role in the photoelectric detection of light, the theory of which (Chapter 6) was developed at the end of the 1950s for classical radiation and extended to quantized radiation in the 1960s. Its experimental application has led to a deeper understanding of the statistics of light.

The invention of the laser at the beginning of the 1960s allowed qualitatively new developments in optical research and the growth of new fields such as nonlinear optics and laser spectroscopy. Intensive studies of lasers have stimulated the introduction of a series of basic theoretical concepts in quantum optics: coherent states (Chapter 3), the theory of phase-space functions (Chapter 4) and the quantum theory of damping (Chapter 5).

Modern quantum optics would be unthinkable without the availability of measurement techniques, such as the Hanbury Brown–Twiss experiment, which was first performed in 1956. By using a beam splitter and two photodetectors, the coincidences of photoelectric events were recorded and compared with the product of independently measured events (for the experimental setup see Fig. 8.1, p. 271). In the case of thermal light an excess of coincidences was observed. That is, the measured intensity correlation $G^{(2)}(\tau)$ as a function of the time delay τ, decays from its initial value at $\tau = 0$ towards a stationary value, cf. Fig. 1.2. This effect, which is called photon bunching, can

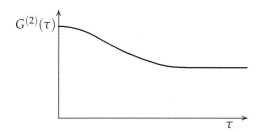

Fig. 1.2 Delay-time dependence of the intensity correlation as typically observed in a Hanbury Brown–Twiss experiment performed with light from a thermal source.

be understood by assuming that the light quanta arrive in bunches, so that the joint probability of events exceeds the product of the two probabilities measured independently of each other. Although this explanation is reasonable, it affords no proof of the existence of photons, since an intensity correlation behavior of the type observed can also be understood classically. It should be emphasized that, in the opposite case, where the measured intensity correlation has a positive initial slope (photon anti-bunching) there is no classical explanation (Chapter 8).

Notwithstanding the success of Einstein's hypothesis, the existence of photons was still a matter of discussion in the 1970s,[2] and the demonstration of photon anti-bunching in 1977 may be regarded as the first direct proof of their existence. The experimental apparatus was of the Hanbury Brown–Twiss type and the detected light was the resonance fluorescence (Chapter 11) from an atomic beam with such a low mean number of atoms that at most one atom contributed to the emitted light. Let us suppose that at a certain instant a single two-level atom that is (resonantly) driven by a laser pump is in the upper quantum state and ready to emit a photon. If the atom emits a photon, it undergoes a transition from the upper to the lower quantum state, which implies that it cannot emit a second photon simultaneously with the first one. The atom can emit a second photon only when it is again excited by the pump field. In other words, the measured intensity correlation vanishes for zero delay, $G^{(2)}(\tau \to 0) = 0$, and in the detection scheme considered there are no equal-time coincidences of photoelectric events. Note that any classical wave or wavepacket is divided by a 50%:50% beam splitter into two parts of equal intensity, which never leads to a vanishing intensity correlation at zero time delay. Photon anti-bunching is essentially a nonclassical property of light and its detection stimulated the formation of quantum optics as a specific field of research.

[2] See, e. g., the paper by Karp (1976), "Test for the non-existence of photons", and the response by Mandel (1977), "Photoelectric counting measurements as a test for the existence of photons".

1.2
Nonclassical phenomena

Nonclassical phenomena, that is, phenomena that are basically quantum mechanical, have been studied intensively in quantum optics and related fields. Nonclassical light has been considered and a number of (nonlinear-optical) methods have been developed to generate it (Chapter 8). Roughly speaking, in many cases the noise in nonclassical light is reduced below some standard quantum limit (e.g., the vacuum noise level), which is usually observed in the case of ideal laser light. As already mentioned, anti-bunched light shows an intensity anti-correlation at zero time delay. Another example of nonclassical light is sub-Poissonian light, which gives rise to a photocounting distribution narrower than a Poissonian one. Sub-Poissonian light was first observed in 1983, in resonance fluorescence from a low-intensity atomic beam. If the noise of a phase-sensitive field quantity, such as the electric-field strength, is reduced (as a function of the phase parameter) below the vacuum level, then the light is called squeezed light. This was first generated in 1985 by means of four-wave mixing.

A number of specific quantum states of radiation and other bosonic systems have been studied, which can be used to define various quantum-mechanical representations of observables (Chapter 3). They may also serve as examples of typical nonclassical effects. For example, photon-number states may be regarded as reflecting particle-like features of radiation rather than wave-like features. On the contrary, when a radiation field is prepared in a coherent state, then its properties, apart from the vacuum noise, become close to those of a classical, nonfluctuating wave.

An old and troublesome problem in quantum mechanics is the description of amplitude and phase and their measurement (Chapters 3 and 7). Since the 1920s, a number of attempts have been made to introduce phase operators and phase states in the quantum theory of light. Concepts based on quantum-mechanical first-principle definitions as well as measurement-assisted definitions have been considered.

In general, a radiation field is not prepared in a pure quantum state but in a mixture of states. In this case, information on the quantum statistics of the field is contained in the density operator. Rather than representations in an orthogonal Hilbert-space basis, representations in terms of phase-space functions are frequently preferred. The concept of phase-space functions (Chapter 4) bears a formal resemblance to classical statistics and allows, to some extent, the application of methods of classical probability theory.

Generation of nonclassical states on demand offers novel possibilities of exploiting quantum features in various fields of applied physics such as measurement technology and information processing. In particular, the increasing number of experimental realizations of nonclassical states of radiation and matter requires methods for characterizing the variety of nonclassical effects to be expected (Chapter 8). In this context, the question of measurable nonclassicality criteria arises, i.e., criteria that are directly applicable to experiments. Similarly, the question of measurable criteria for entangled states must be answered – states which play a key role in quantum communication such as quantum cryptography, quantum-state teleportation and quantum computation.

The quantum nature of radiation and matter becomes obvious both in their resonant and off-resonant interaction. Whereas spontaneous emission (Chapter 10) and resonance fluorescence (Chapter 11) from a single atom and the Jaynes–Cummings-type interaction of a single atom with a high-quality cavity field (Chapter 12) are examples of resonant interaction, van der Waals and Casimir forces (Chapter 10) are typical examples of virtual-photon-assisted off-resonant interaction.

1.3
Source-attributed light

Any real radiation field may be thought of as being due to sources, which essentially determine the quantum statistics of the radiation. Quantization of the radiation field requires, in principle, quantization of the matter and the radiation-matter interaction as well (Chapter 2). As is well known, commutation relations play an important role in quantum physics. Whereas commutation relations at equal times are given from quantum-mechanical first principles, determination of the time-dependent commutation relations requires knowledge of the dynamics of the coupled light–matter system. Therefore, to study general aspects of the generation, detection and processing of quantized light (such as quantum-optical correlation functions observed in the photoelectric detection of light), it is helpful to introduce appropriate source-quantity representations of field commutators (Chapter 2).

Light detection and processing are frequently performed in a source-free region of space, and the question arises as to the conditions under which it is possible to treat a quantized radiation field as being effectively free, that is to ignore the sources when considering the radiation. A criterion for an effectively free field may be seen in the agreement of the commutation relations of the field quantities at different times with the free-field commutation relations (Chapter 2). It is worth noting that the question of whether or not the commutation relations of field quantities at different times reduce to the free-field commutation relations, can be answered by means of their source-quantity representations, that is, by expressing them in terms of free-field commutators and so-called time-delayed terms. The latter can give rise to a nonvanishing contribution when the space-time arguments of the two field quantities under consideration can be connected to each other by the propagation of light from one of the space-time points to the other through the sources. Clearly, a light field may be regarded as being effectively free when the distances of the relevant points of observation from the light source are large enough and the considered time intervals are small enough to suppress the time-delayed terms. This rule can be established not only for the full field but also for appropriately chosen (multi-mode) parts of the field, such as the incoming and outgoing fields frequently introduced in connection with experimental apparatus. For example, if a (multi-mode) part of an optical light field propagates away from the sources and cannot return to them, then it may be regarded in many cases as being effectively free, independently of the chosen space-time points.

In practice, various optical instruments, which may substantially modify the propagation of light compared with that in free space, are used and a careful consideration of the time-dependent commutation relations is necessary to actually specify the free-field conditions and the correlation functions measurable by means of standard photodetectors. Typical examples are the theory of spectral filtering of quantized light (Chapter 6) and the treatment of an optical cavity with output coupling (Chapter 9).

To derive tractable equations of motions for a coupled light–matter system, various approximation schemes have been developed and applied, such as the dipole approximation and the rotating-wave approximation (Chapter 2). Furthermore, in nonlinear optics the concept of effective Hamiltonians is widely used, for example in the treatment of multi-photon absorption and emission, parametric optical processes (e. g., the optical parametric oscillator) and multi-wave mixing.

For gaining deeper insight into the quantum nature of light–matter interactions, models that are almost exactly solvable play an important role. In particular, there have been detailed studies of the resonant interaction of a single two-level atom with a (multi-mode) radiation field in free space within the framework of optical Bloch equations (Chapter 5) to describe resonance fluorescence (Chapter 11), and of the resonant interaction of a single two-level atom with a single-mode field in a high-quality cavity on the basis of the Jaynes–Cummings model (Chapter 12).

1.4
Medium-assisted electromagnetic fields

As is already known from classical optics, the use of instruments in optical experiments needs careful examination with regard to their action on the light under study [see, e. g., Born and Wolf (1980)]. In quantum optics an additional consideration is the influence of the presence of instruments on the quantum statistics of the light. For example, let us consider a 50%:50% beam splitter oriented at 45° to an incident light beam (Fig. 1.3). In classical optics the beam splitter divides the incoming beam into two (apart from a phase shift) equal outgoing parts propagating perpendicular to each other (Fig. 1.3a), and with the same scaling factor the classical noise of the incident field is transferred to the two fields in the output channels of the beam splitter. It is intuitively clear that, in quantum optics, the noise of the vacuum in the unused input port of the beam splitter introduces additional noise in the two output beams (Fig. 1.3b). Therefore, the quantum statistics of the output fields may differ significantly from that of the input field.

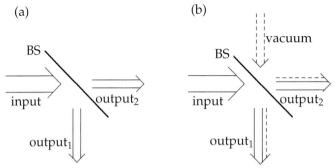

Fig. 1.3 Outline of the action of a 50%:50% beam splitter (BS). In classical optics (a) an incident light beam (input) is divided into two (apart from a phase shift) equal output beams (output$_1$, output$_2$). In quantum optics (b) the incident light beam and the quantum noise of the vacuum in the unused input port are combined to yield vacuum-noise-assisted output beams.

The above example shows that it is necessary to take into account the presence of optical instruments when considering the quantization of the radiation field. In principle, optical instruments could be included as part of the matter to which the radiation field is coupled and treated microscopically. However, in many cases, passive instruments are linearly responding macroscopic bodies that can be treated phenomenologically by introducing a spatially varying permittivity. In general, light propagation through such bodies is accompanied by dispersion and absorption, so that the permittivity is a complex function of frequency. Since in quantum physics any type of loss is unavoidably connected with fluctuations, for treatment of the effect of material absorption, quantization in an extended Hilbert space is required (Chapter 2). In some cases, in particular when the spectral range of the radiation is effectively limited to an appropriately chosen narrow interval, the effects of absorption and dispersion may become negligibly small and the description of the instruments considerably simplified. They can be modeled by bodies with real permittivities that may vary only in space. Obviously, both the time-dependent commutation relations and the quantum-statistical correlation functions of the field under study depend on the specific bodies used.

Typical examples of optical instruments whose action can be treated in this way are beam splitters and spectral filters of the Fabry–Perot type. Their main features can already be described by means of the simple model of a dielectric plate (Chapter 6). Moreover, progress in quantum optics would be unimaginable without the use of resonator-like equipment. A typical example is an optical cavity filled with an active medium and bounded by dielectric walls to allow for input and output coupling (Chapter 9). In particular, in the case of a high-quality cavity, applying the formalism of electromagnetic-field quantization in linear media naturally yields a description of the radiation field inside and outside the cavity in terms of quantum damping theory.

The formalism of electromagnetic-field quantization in linear media can also be used to treat body-assisted electromagnetic vacuum effects in a unified way (Chapter 10). Whereas the classical vacuum is the trivial state where the electromagnetic field identically vanishes, the quantum vacuum is very active and its interaction with atomic systems gives rise to a number of observable effects that are purely nonclassical. Since in the presence of macroscopic bodies the structure of the electromagnetic field is changed compared with that in free space, the electromagnetic vacuum is changed also, which can lead, e. g., to inhibition or enhancement of spontaneous emission. Moreover, forces of the van der Waals type in micro- and nano-structures can be controlled in this way.

1.5
Measurement of light statistics

To gain information on the quantum statistics of light from measured data, a careful consideration of the employed measurement scheme is needed (Chapter 6). In standard photoelectric detection of light the detection process is based on the internal photoelectric effect. In the spirit of Einstein's hypothesis, by absorbing a photon, a detector atom can undergo a transition from an initial state to a continuum of final states, ejecting a photoelectron. A combination of quantum mechanics (to treat the elementary acts of light absorption) and classical statistics (to deal with the macroscopic sample of photoelectrons produced in a chosen time interval of detection) yields the observed counting statistics in terms of either normally and time-ordered field correlation functions or the photon-number statistics.

There are various kinds of detection schemes that can be used to measure statistical properties of light that are not accessible from the photon-number statistics. On combining (single-mode) light fields by means of beam splitters before measuring the counting statistics (of the combined field), the quantum statistics of phase-sensitive light properties can be obtained. In particular, four-port homodyne detection (see Fig. 6.6, p. 206) and eight-port homodyne detection (see Fig. 6.8, p. 224) are typical examples of this measurement strategy.

If in a four-port homodyne detection one of the (single-mode) input fields is prepared in a coherent state with a sufficiently large mean number of photons, then the measured difference-count probabilities can be related to the phase-rotated quadrature probability distributions of the second input field. Using an eight-port scheme renders it possible to relate the measured joint difference-count probability to the Q function of the second input field, that is the phase-space function that applies directly to the calculation of expectation values of anti-normally ordered operator functions. Also, unbalanced homodyning is of interest since it leads to simple reconstruction methods for the quantum states.

Homodyne correlation measurements are of particular interest when a weak local oscillator is used. In this case new types of correlation properties can be observed. In principle, one may determine all normally ordered moments, including those containing unequal numbers of creation and annihilation operators. Such moments, which are not accessible by direct detection methods, are required, e.g., for implementing nonclassicality and entanglement criteria (Chapter 8).

1.6
Determination and preparation of quantum states

It is well known that the density matrix of a quantum system contains all the information necessary to completely determine its properties. Hence the determination of the density matrix from measured data is therefore an important problem (Chapter 7). The first reconstruction of a light-field density matrix from measured data was reported in 1993. Clearly, the density matrix can only be obtained from quantities that also contain the complete information on the system. For example, this information is contained in any phase-space function. Since the Q function of a (single-mode) field can be obtained from the data measured in eight-port homodyne detection (Chapter 6), the density matrix of the field can be obtained, in principle, from these data also. Moreover, knowledge of the phase-rotated quadrature probability distributions for every phase parameter in a π interval is equivalent to knowledge of any phase-space function, which implies that the density matrix can also be obtained from the phase-rotated quadrature distributions with the phase parameter varying in a π interval. Since these probability distributions can be obtained from the data measured in a succession of four-port homodyne detections (Chapter 6), the four-port homodyne detection scheme can also be used for the experimental determination of the density matrix.

An alternative way of determining the quantum state from measured data consists of a method that is local in phase space. For a radiation mode, the measurement scheme consists of unbalanced homodyning. By use of a local oscillator, the field to be measured is displaced in phase space, with the complex displacement amplitude being controlled by the phase and amplitude of the local oscillator. The resulting displacement amplitude defines the point in phase space where a chosen phase-space function can be determined locally. The phase-space function of interest is obtained in a simple manner as an appropriately weighted sum of the photon-number statistics of the displaced light field.

The basic concepts of determining the quantum state can also be modified to allow the determination of the quantum state of a high-Q cavity-field by transmission of probe atoms (Chapter 12). Moreover, methods have been developed for determining the motional quantum state of a trapped atom (Chapter 13) and the entangled state for the combined vibronic (vibrational-electronic) quantum state of an atom undergoing a quantized center-of-mass motion in a trap potential.

Appropriate methods of quantum-state preparation are needed for generating nonclassical states. The improvements of experimental techniques allowed one to prepare sophisticated quantum states such as entangled states of the Schrödinger-cat type or Einstein–Podolsky–Rosen states. Experiments of this type can be performed, for example, by using interactions of single atoms with high-Q cavity fields (Chapter 12) or by using the vibronic dynamics of trapped atoms (Chapter 13).

1.7
Quantized motion of cold atoms

The progress in developing techniques for cooling trapped atoms to extremely low temperatures has rendered it possible to visualize the quantum nature of the atomic center-of-mass motion, which is no longer hidden by thermal background noise. Control of the quantized atomic center-of-mass motion allows one to realize, e. g., atom interferometry, Bose–Einstein condensation and atom-laser like devices.[3]

If an atom is confined in a harmonic trap potential (Chapter 13), the laser-driven vibronic interaction shows some resemblance to the atom–field inter-

---

3) In atom lasers, the wavy nature of the atomic motion plays the role of the electromagnetic waves in conventional lasers and it is interesting to generate coherent (atomic) matter waves.

action in a high-Q cavity. An exactly solvable, nonlinear Jaynes–Cummings model is suited to describing the dynamics of the laser-driven trapped ion in the resolved sideband regime, where individual vibronic transitions are addressed by the laser. Besides the multi-quantum generalization of the standard Jaynes–Cummings model of cavity QED, there appears an additional nonlinear dependence of the interaction Hamiltonian on the vibrational excitation of the atom in the trap potential. The nonlinearity gives rise to interesting effects in the atomic dynamics, which can be employed to measure motional quantum states and prepare specific ones. In particular, it is possible to drive the motional quantum state of the atom in a nonlinear manner without affecting the electronic one.

In the first experimental realization of the nonlinear Jaynes–Cummings dynamics with a Raman-driven trapped ion, significant decoherence effects had already been observed. A detailed understanding of the underlying mechanisms is of great importance for any practical application of trapped atoms, e. g., in quantum information processing. In particular, in the case of a Raman-driven atom being cooled down to its motional ground state the decoherence is dominated by the, rarely occurring, excitation of an auxiliary electronic state used for the enhancement of the Raman coupling strength.

References

Agarwal, G.S. (1974) in *Quantum Statistical Theories of Spontaneous Emission and their Relation to Other Approaches*, ed. G. Höhler (Springer-Verlag, Berlin).

Allen, L. and J.H. Eberly (1975) *Optical Resonance and Two-Level Atoms* (Wiley, New York).

Bloembergen, N. (1965) *Nonlinear Optics* (Benjamin, New York).

Born, M. and E. Wolf (1980) *Principles of Optics* (Pergamon Press, Oxford).

Boyd, R.W. (1991) *Nonlinear Optics* (Academic Press, London).

Carmichael, H. (1993) *Lecture Notes in Physics: An Open Systems Approach to Quantum Optics* (Springer-Verlag, Berlin).

Carmichael, H. (1998) *Statistical Methods in Quantum Optics 1: Master Equations and Fokker-Planck Equations* (Springer-Verlag, Berlin).

Cohen-Tannoudji, C., J. Dupont-Roc and G. Grynberg (1989) *Photons and Atoms* (Wiley, New York).

Cohen-Tannoudji, C., J. Dupont-Roc and G. Grynberg (1992) *Atom-Photon Interactions* (Wiley, New York).

Gardiner, C.W. (1991) *Quantum Noise* (Springer-Verlag, Berlin).

Gerry, C. and P. Knight (2004) *Introductory Quantum Optics* (Cambridge University Press).

Haken, H. (1970) in *Light and Matter Ic*, ed. L. Genzel: Vol. XXV/2c of *Encyclopedia of Physics*, chief ed. S. Flügge (Springer-Verlag, Berlin).

Haken, H. (1985) *Light* (North-Holland, Amsterdam).

Karp, S. (1976) *J. Opt. Soc. Am.* **66**, 1421.

Klauder, J.R. and E.C.G. Sudarshan (1968) *Fundamentals of Quantum Optics* (Benjamin, New York).

Levenson, M.D. and S.S. Kano (1988) *Introduction to Nonlinear Laser Spectroscopy* (Academic Press, New York).

Loudon, R. (1983) *The Quantum Theory of Light* (Clarendon Press, Oxford).

Louisell, W.H. (1973) *Quantum Statistical Properties of Radiation* (Wiley, New York).

Mandel, L. (1977) *J. Opt. Soc. Am.* **67**, 1101.

Mandel, L. and E. Wolf (1995) *Optical Coherence and Quantum Optics* (Cambridge University Press, Cambridge).

Meystre, P. and M. Sargent III (1990) *Elements of Quantum Optics* (Springer-Verlag, Berlin).

Milonni, P.W. and J. H. Eberly (1988) *Lasers* (Wiley, New York).

Orszag, M. (2000) *Quantum Optics* (Springer-Verlag, Berlin).

Peřina, J. (1985) *Coherence of Light* (Reidel, Dordrecht).

Peřina, J. (1991) *Quantum Statistics of Linear and Nonlinear Optical Phenomena* (Reidel, Dordrecht).

Sargent III, M., M.O. Scully and W.E. Lamb, Jr. (1977) *Laser Physics* (Addison-Wesley, Reading).

Schleich, W.P. (2001) *Quantum Optics in Phase Space* (Wiley-VCH, Berlin).

Schubert, M. and B. Wilhelmi (1986) *Nonlinear Optics and Quantum Electronics* (Wiley, New York).

Scully, M.O. and M.S. Zubairy (1997) *Quantum Optics* (Cambridge University Press, Cambridge).

Shen, Y.R. (1984) *Principles of Nonlinear Optics* (Wiley, New York).

Shore, B.W. (1990) *Theory of Coherent Atomic Excitations* (Wiley, New York).

Stenholm, S. (1984) *Foundations of Laser Spectroscopy* (Wiley, New York).

Vogel, W. and D.–G. Welsch (1994) *Lectures on Quantum Optics* (Akademie-Verlag, Berlin).

Vogel, W., D.–G. Welsch and S. Wallentowitz (2001) *Quantum Optics – An Introduction* (Wiley-VCH, Berlin).

Walls, D.F. and G.J. Milburn (1994) *Quantum Optics* (Springer-Verlag, Berlin).

2
Elements of quantum electrodynamics

In order to arrive at the basic concepts for describing the quantum effects of radiation, it is necessary to consider the quantization of the electromagnetic field attributed to atomic sources in the presence of macroscopic bodies. For example, in many cases of practical interest the (passive) optical instruments through which radiation passes can be regarded as being more or less complicated dielectric bodies. In the quantization scheme developed here we will therefore allow for the presence of a dielectric medium with space- and frequency-dependent complex permittivity satisfying the Kramers–Kronig relations. For example, a standard situation is the spectral filtering of light pro-

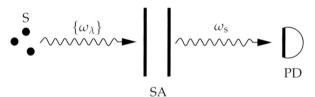

Fig. 2.1 Spectral photodetection scheme. After passing through a (Fabry–Perot-type) spectral apparatus (SA), whose spectral response function discriminates against values of the frequency ω_λ not equal to a given setting frequency ω_s, the light produced by the sources (S) falls on a photoelectric detection device (PD).

duced by some types of source, cf. Fig. 2.1. In homodyne detection a signal field is combined with a (local oscillator) reference field through a beam splitter. By means of photoelectric detection of the mixed output fields, phase information on the signal field becomes accessible. Further, resonators such as leaky optical cavities filled with optically active (nonlinear) matter are frequently used in quantum optics for generating and/or amplifying light, cf. Fig. 2.2.

Starting from the well-known classical equations of motion of microscopic electrodynamics (Section 2.1), canonical quantization of both the free electromagnetic field (Section 2.2) and the electromagnetic field with sources (Section 2.3) is performed. The theory is then extended to electrodynamics in dielectric media, transferring the powerful concepts of phenomenological clas-

Quantum Optics, Third, revised and extended edition. Werner Vogel and Dirk-Gunnar Welsch
Copyright © 2006 WILEY-VCH Verlag GmbH & Co. KGaA, Weinheim
ISBN: 3-527-40507-0

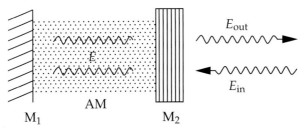

Fig. 2.2 Scheme of a resonator-like cavity bounded by a perfectly reflecting mirror M_1 and a fractionally transparent mirror M_2, the cavity being filled with optically active matter (AM). The mirror M_2 guarantees that the intra-cavity field (electric field strength E) is in contact with the incoming field (E_{in}) and the outgoing field (E_{out}), which may be utilized for subsequent optical processing.

sical electrodynamics to quantum theory (Section 2.4). With regard to the description of specific processes, frequently used concepts of approximate interaction Hamiltonians are discussed (Section 2.5). By formal solution of the Heisenberg equations of motion (Section 2.6), fundamental time-dependent commutation relations are derived (Section 2.7). This makes it possible to express observable field correlation functions in terms of source-quantity correlation functions (Section 2.8).

The standard concepts of canonical quantization are considered, for example, in the books of Cohen-Tannoudji, Dupont-Roc and Grynberg (1989), Haken (1985), Loudon (1983), Louisell (1973), Meystre and Sargent III (1990), Milonni (1994), Peřina (1991) and Schubert and Wilhelmi (1986). The concepts of inclusion in the quantization of dielectric media, are based on original work [Knöll, Vogel and Welsch (1987); Glauber and Lewenstein (1991); Huttner and Barnett (1992); Gruner and Welsch (1996); Scheel, Knöll and Welsch (1998); Ho, Buhmann, Knöll, Welsch, Scheel and Kästel (2003)].

2.1
Basic classical equations

In classical physics the electromagnetic field obeys Maxwell's equations[1]

$$\nabla \mathbf{B}(\mathbf{r}) = 0, \tag{2.1}$$

$$\nabla \times \mathbf{E}(\mathbf{r}) = -\dot{\mathbf{B}}(\mathbf{r}), \tag{2.2}$$

$$\nabla \mathbf{E}(\mathbf{r}) = \varepsilon_0^{-1} \rho(\mathbf{r}), \tag{2.3}$$

$$\nabla \times \mathbf{B}(\mathbf{r}) = \mu_0 \mathbf{j}(\mathbf{r}) + \mu_0 \varepsilon_0 \dot{\mathbf{E}}(\mathbf{r}) \tag{2.4}$$

---

1) Here, and in the following, for notational convenience we denote the scalar product of two vectors simply by $\mathbf{ab}$, the vector product by $\mathbf{a} \times \mathbf{b}$ and the tensor product by $\mathbf{a} \otimes \mathbf{b}$.

$[(\mu_0\varepsilon_0)^{-1/2}=c$, velocity of light in vacuum], with ρ and $\mathbf{j}$, respectively, being the total charge and current densities, which, according to Eqs (2.3) and (2.4), obey the continuity equation

$$\dot{\rho}(\mathbf{r}) + \nabla \mathbf{j}(\mathbf{r}) = 0 \tag{2.5}$$

as the local form of the charge conservation law. Other local balance equations that result from Maxwell's equations are the energy balance

$$\dot{w}(\mathbf{r}) + \nabla \mathbf{S}(\mathbf{r}) + \mathbf{j}(\mathbf{r})\mathbf{E}(\mathbf{r}) = 0 \tag{2.6}$$

and the momentum balance

$$\dot{\mathbf{p}}(\mathbf{r}) - \nabla T(\mathbf{r}) + \mathbf{f}(\mathbf{r}) = 0. \tag{2.7}$$

Here,

$$w(\mathbf{r}) = \frac{\varepsilon_0}{2}\mathbf{E}^2(\mathbf{r}) + \frac{1}{2\mu_0}\mathbf{B}^2(\mathbf{r}) \tag{2.8}$$

and

$$\mathbf{p}(\mathbf{r}) = \varepsilon_0 \mathbf{E}(\mathbf{r}) \times \mathbf{B}(\mathbf{r}) \tag{2.9}$$

are the electromagnetic energy and momentum densities, respectively, and

$$\mathbf{f}_\mathrm{L}(\mathbf{r}) = \rho(\mathbf{r})\mathbf{E}(\mathbf{r}) + \mathbf{j}(\mathbf{r}) \times \mathbf{B}(\mathbf{r}) \tag{2.10}$$

and

$$T(\mathbf{r}) = \varepsilon_0 \mathbf{E}(\mathbf{r}) \otimes \mathbf{E}(\mathbf{r}) + \mu_0^{-1}\mathbf{B}(\mathbf{r}) \otimes \mathbf{B}(\mathbf{r}) - w(\mathbf{r})I \tag{2.11}$$

are the Lorentz force density and the Maxwell stress tensor, respectively (I is the unity tensor). Further, the angular momentum density of the electromagnetic field,

$$\mathbf{l}(\mathbf{r}) = \mathbf{r} \times \mathbf{p}(\mathbf{r}) = \varepsilon_0 \mathbf{r} \times [\mathbf{E}(\mathbf{r}) \times \mathbf{B}(\mathbf{r})] \tag{2.12}$$

can be introduced, and from Eq. (2.7) it follows that

$$\dot{\mathbf{l}}(\mathbf{r}) + \nabla[T(\mathbf{r}) \times \mathbf{r}] + \mathbf{r} \times \mathbf{f}(\mathbf{r}) = 0. \tag{2.13}$$

Note that T is a symmetric tensor.

So far we have considered the equations for the electromagnetic-field variables. Provided that the charge and current densities are known, the field strengths can be calculated from the equations given above. In practice, both the charge positions and velocities depend sensitively on the field quantities (and vice versa), because of the field–matter interaction. Thus for a complete description of electrodynamics we also need the equations of motion for the charged particles (electrons, ions, etc.). Let us consider a collection of non-relativistic, point-like particles. Denoting the charge and the mass of the ath particle by Q_a and m_a respectively, we may write[2]

$$\rho(\mathbf{r}) = \sum_a Q_a \delta(\mathbf{r} - \mathbf{r}_a), \tag{2.14}$$

$$\mathbf{j}(\mathbf{r}) = \sum_a Q_a \dot{\mathbf{r}}_a \delta(\mathbf{r} - \mathbf{r}_a), \tag{2.15}$$

where the position vectors $\mathbf{r}_a$ of the particles obey the well-known Newtonian equations of motion

$$m_a \ddot{\mathbf{r}}_a = Q_a \left[\mathbf{E}(\mathbf{r}_a) + \dot{\mathbf{r}}_a \times \mathbf{B}(\mathbf{r}_a)\right]. \tag{2.16}$$

Introducing the scalar potential V and the vector potential $\mathbf{A}$ by

$$\mathbf{B}(\mathbf{r}) = \nabla \times \mathbf{A}(\mathbf{r}), \tag{2.17}$$

$$\mathbf{E}(\mathbf{r}) = -\dot{\mathbf{A}}(\mathbf{r}) - \nabla V(\mathbf{r}), \tag{2.18}$$

the Maxwell equations (2.1) and (2.2) are seen to be satisfied identically. Inserting Eqs (2.17) and (2.18) into the remaining Maxwell equations (2.3) and (2.4) yields

$$\nabla \times \nabla \times \mathbf{A}(\mathbf{r}) + \mu_0 \varepsilon_0 \ddot{\mathbf{A}}(\mathbf{r}) + \mu_0 \varepsilon_0 \nabla \dot{V}(\mathbf{r}) = \mu_0 \mathbf{j}(\mathbf{r}) \tag{2.19}$$

and

$$-\varepsilon_0 \left[\Delta V(\mathbf{r}) + \nabla \dot{\mathbf{A}}(\mathbf{r})\right] = \rho(\mathbf{r}). \tag{2.20}$$

Accordingly, we may rewrite Eq. (2.16) as

$$m_a \ddot{\mathbf{r}}_a = -Q_a [\nabla V(\mathbf{r}_a) + \dot{\mathbf{A}}(\mathbf{r}_a)] + Q_a \dot{\mathbf{r}}_a \times [\nabla \times \mathbf{A}(\mathbf{r}_a)]. \tag{2.21}$$

The field equations (2.19) and (2.20) and the Newtonian equations of motion (2.21) can be derived from the Lagrangian

$$L = \frac{1}{2} \int d^3 r \left\{ \varepsilon_0 [\dot{\mathbf{A}}(\mathbf{r}) + \nabla V(\mathbf{r})]^2 - \mu_0^{-1} [\nabla \times \mathbf{A}(\mathbf{r})]^2 \right\}$$
$$+ \frac{1}{2} \sum_a m_a \dot{\mathbf{r}}_a^2 + \int d^3 r \left[\mathbf{j}(\mathbf{r}) \mathbf{A}(\mathbf{r}) - \rho(\mathbf{r}) V(\mathbf{r}) \right]. \tag{2.22}$$

Hamilton's principle of least action

$$\delta \int dt\, L = 0 \tag{2.23}$$

[2] Note that Eqs (2.14) and (2.15) satisfy the equation of continuity (2.5).

then leads to the field equations (2.19) and (2.20) and the Newtonian equations of motion (2.21) via the Euler–Lagrange equations

$$\frac{d}{dt}\frac{\delta L}{\delta \dot{\mathbf{A}}(\mathbf{r})} - \frac{\delta L}{\delta \mathbf{A}(\mathbf{r})} = 0, \quad \frac{d}{dt}\frac{\delta L}{\delta \dot{V}(\mathbf{r})} - \frac{\delta L}{\delta V(\mathbf{r})} = 0, \tag{2.24}$$

$$\frac{d}{dt}\frac{\partial L}{\partial \dot{r}_a} - \frac{\partial L}{\partial r_a} = 0, \tag{2.25}$$

as may be proved by straightforward calculation.

The gauge freedom enables us to choose the potentials in a different way. In particular, in the Coulomb gauge, the vector potential is chosen to be a transverse vector function,

$$\nabla \mathbf{A}(\mathbf{r}) = 0, \tag{2.26}$$

which leads to the following simplifications of Eqs (2.19) and (2.20):

$$\Delta \mathbf{A}(\mathbf{r}) - \frac{1}{c^2}\ddot{\mathbf{A}}(\mathbf{r}) = -\mu_0 \mathbf{j}^\perp(\mathbf{r}), \tag{2.27}$$

$$-\varepsilon_0 \Delta V(\mathbf{r}) = \rho(\mathbf{r}), \tag{2.28}$$

where

$$\mathbf{j}^\perp(\mathbf{r}) = \mathbf{j}(\mathbf{r}) - \varepsilon_0 \nabla \dot{V}(\mathbf{r}) \tag{2.29}$$

is the transverse current density ($\nabla \mathbf{j}^\perp = 0$). From Eq. (2.28), the scalar potential V is simply the instantaneous Coulomb potential of the charged particles:

$$V(\mathbf{r}) = \frac{1}{4\pi\varepsilon_0}\int d^3r' \frac{\rho(\mathbf{r}')}{|\mathbf{r}-\mathbf{r}'|} = \frac{1}{4\pi\varepsilon_0}\sum_a \frac{Q_a}{|\mathbf{r}-\mathbf{r}_a|}. \tag{2.30}$$

Recalling the continuity equation (2.5), then from Eqs (2.29) and (2.30) we see that $\mathbf{j}^\perp$ can be expressed in term of $\mathbf{j}$:[3]

$$\mathbf{j}^\perp(\mathbf{r}) = \mathbf{j}(\mathbf{r}) + \nabla \otimes \int d^3r' \frac{\nabla \mathbf{j}(\mathbf{r}')}{4\pi|\mathbf{r}-\mathbf{r}'|}. \tag{2.31}$$

In the Coulomb gauge the first term in Eq. (2.18) is the transverse part of the electric field and the second term the longitudinal part, i. e.,

$$\mathbf{E}(\mathbf{r}) = \mathbf{E}^\perp(\mathbf{r}) + \mathbf{E}^\|(\mathbf{r}), \tag{2.32}$$

$$\mathbf{E}^\perp(\mathbf{r}) = -\dot{\mathbf{A}}(\mathbf{r}), \tag{2.33}$$

$$\mathbf{E}^\|(\mathbf{r}) = -\nabla V(\mathbf{r}). \tag{2.34}$$

---

[3] Note that the longitudinal ($\mathbf{F}^\|$) and transverse ($\mathbf{F}^\perp$) parts of a vector field $\mathbf{F}$ are given by $\mathbf{F}^{\|(\perp)}(\mathbf{r}) = \int d^3r'\, \delta^{\|(\perp)}(\mathbf{r}-\mathbf{r}')\mathbf{F}(\mathbf{r}')$, with $\delta^\|(\mathbf{r}) = -\nabla \otimes \nabla(4\pi|\mathbf{r}|)^{-1}$ and $\delta^\perp(\mathbf{r}) = \delta(\mathbf{r}) - \delta^\|(\mathbf{r})$ being respectively the longitudinal and transverse tensor-valued δ-functions.

Integrating by parts and using Eq. (2.28), we may rewrite Eq. (2.22) to obtain the Lagrangian in the Coulomb gauge

$$L = \tfrac{1}{2} \int d^3r \left\{ \varepsilon_0 \dot{\mathbf{A}}^2(\mathbf{r}) - \mu_0^{-1}[\nabla \times \mathbf{A}(\mathbf{r})]^2 \right\}$$
$$+ \tfrac{1}{2} \sum_a m_a \dot{\mathbf{r}}_a^2 - W_{\text{Coul}} + \int d^3r\, \mathbf{j}(\mathbf{r}) \mathbf{A}(\mathbf{r}), \tag{2.35}$$

where

$$W_{\text{Coul}} = \tfrac{1}{2} \int d^3r\, \rho(\mathbf{r}) V(\mathbf{r}) \tag{2.36}$$

is the Coulomb energy of the charged particles. It can be rewritten, on recalling Eqs (2.28), (2.30) and (2.34), as[4]

$$W_{\text{Coul}} = \frac{\varepsilon_0}{2} \int d^3r\, [\nabla V(\mathbf{r})]^2 = \frac{\varepsilon_0}{2} \int d^3r\, [\mathbf{E}^\parallel(\mathbf{r})]^2 = \frac{1}{8\pi\varepsilon_0} \sum_{a,a'}{}' \frac{Q_a Q_{a'}}{|\mathbf{r}_a - \mathbf{r}_{a'}|}. \tag{2.37}$$

Since the scalar potential is no longer an independent field variable, the Lagrangian L defined by Eq. (2.35) depends only on $\mathbf{r}_a$, $\dot{\mathbf{r}}_a$, $\mathbf{A}(\mathbf{r})$ and $\dot{\mathbf{A}}(\mathbf{r})$. In deriving the equations of motion from this Lagrangian, the principle of least action should be modified such that

$$\delta \int dt\, L = 0, \tag{2.38}$$

where the δ notation indicates that the vector potential $\mathbf{A}$ is varied in the space of transverse vector functions $\mathbf{g}(\mathbf{r})$ satisfying the condition $\nabla \mathbf{g} = 0$. The inhomogeneous wave equation (2.27) together with the gauge condition (2.26) may then be represented in the form of the following Euler–Lagrange equation:

$$\frac{d}{dt} \frac{\delta L}{\delta \dot{\mathbf{A}}(\mathbf{r})} - \frac{\delta L}{\delta \mathbf{A}(\mathbf{r})} = 0. \tag{2.39}$$

Here, the functional derivative $\delta / \delta \mathbf{g}(\mathbf{r})$ of a functional $F[\mathbf{g}(\mathbf{r})]$ is defined by

$$\frac{\delta F}{\delta \mathbf{g}(\mathbf{r})} = \lim_{\epsilon \to 0} \frac{1}{\epsilon} \{ F[\mathbf{g}(\mathbf{r}') + \epsilon \delta^\perp(\mathbf{r} - \mathbf{r}')] - F[\mathbf{g}(\mathbf{r}')] \}. \tag{2.40}$$

2.2
The free electromagnetic field

In the absence of charged particles the Lagrangian (2.35) reduces to the Lagrangian of the free electromagnetic field (in the Coulomb gauge)

$$L = \tfrac{1}{2} \int d^3r \left\{ \varepsilon_0 \dot{\mathbf{A}}^2(\mathbf{r}) - \mu_0^{-1}[\nabla \times \mathbf{A}(\mathbf{r})]^2 \right\}, \tag{2.41}$$

[4] Note that the infinite self-energies of the charges (terms with $a = a'$) must be removed from the Coulomb energy.

from which the Hamiltonian

$$H = \int d^3r\, \Pi(\mathbf{r})\dot{\mathbf{A}}(\mathbf{r}) - L \tag{2.42}$$

is obtained, with

$$\Pi(\mathbf{r}) = \frac{\bar{\delta} L}{\bar{\delta}\dot{\mathbf{A}}(\mathbf{r})} = \varepsilon_0 \dot{\mathbf{A}}(\mathbf{r}) \tag{2.43}$$

being the canonical momentum field. Note that the canonical momentum also belongs to the space of transverse vector functions, $\nabla\Pi(\mathbf{r})=0$.

2.2.1
Canonical quantization

Applying the canonical quantization scheme, we regard the canonically conjugate fields $\mathbf{A}$ and Π as Hilbert-space operators $\hat{\mathbf{A}}$ and $\hat{\Pi}$ and replace the classical (transverse) Poisson brackets

$$\{A_k(\mathbf{r}), \Pi_{k'}(\mathbf{r}')\} = \delta^{\perp}_{kk'}(\mathbf{r}-\mathbf{r}'), \tag{2.44}$$

$$\{A_k(\mathbf{r}), A_{k'}(\mathbf{r}')\} = \{\Pi_k(\mathbf{r}), \Pi_{k'}(\mathbf{r}')\} = 0 \tag{2.45}$$

by the corresponding commutators multiplied by $(i\hbar)^{-1}$:

$$[\hat{A}_k(\mathbf{r}), \hat{\Pi}_{k'}(\mathbf{r}')] = i\hbar \delta^{\perp}_{kk'}(\mathbf{r}-\mathbf{r}'), \tag{2.46}$$

$$[\hat{A}_k(\mathbf{r}), \hat{A}_{k'}(\mathbf{r}')] = 0 = [\hat{\Pi}_k(\mathbf{r}), \hat{\Pi}_{k'}(\mathbf{r}')]. \tag{2.47}$$

Combining Eqs (2.41), (2.42) and (2.43) and substituting for $\mathbf{A}$ and Π the operators $\hat{\mathbf{A}}$ and $\hat{\Pi}$, respectively, yields the Hamiltonian $\hat{H}$ in the form

$$\hat{H} = \tfrac{1}{2}\int d^3r \left\{ \varepsilon_0^{-1}\hat{\Pi}^2(\mathbf{r}) + \mu_0^{-1}[\nabla\times\hat{\mathbf{A}}(\mathbf{r})]^2 \right\}. \tag{2.48}$$

According to Eqs (2.17), (2.32)–(2.34) (for $V=0$) and (2.43), the operators of the canonically conjugate fields are related to the operators of the magnetic and electric fields as $\hat{\mathbf{B}}=\nabla\times\hat{\mathbf{A}}$ and $\hat{\mathbf{E}}^{\perp}=\hat{\mathbf{E}}=-\hat{\Pi}/\varepsilon_0$, and the Hamiltonian (2.48) is nothing other than the energy of the electromagnetic field,

$$\hat{H} = \tfrac{1}{2}\int d^3r \left[\varepsilon_0 \hat{\mathbf{E}}^2(\mathbf{r}) + \mu_0^{-1}\hat{\mathbf{B}}^2(\mathbf{r})\right]. \tag{2.49}$$

From Eq. (2.47) it is not difficult to verify, on recalling the relation $\nabla\times\delta(\mathbf{r}) = \nabla\times\delta^{\perp}(\mathbf{r})$, that

$$[\hat{E}_k(\mathbf{r}), \hat{B}_{k'}(\mathbf{r}')] = -\epsilon_{kk'm}\frac{i\hbar}{\varepsilon_0}\frac{\partial \delta(\mathbf{r}-\mathbf{r}')}{\partial x_m}, \tag{2.50}$$

with ϵ_{ijk} being the (anti-symmetric) Levi–Civita tensor.[5]

---

5) Here, and in the following, we adopt the convention of summation over repeated vector-component indices.

2.2.2
Monochromatic-mode expansion

In the Heisenberg picture, the operators carry the full time dependence and evolve according to equations of motion that correspond, apart from possible ordering prescriptions, to the classical ones. In particular, the vector potential of the free electromagnetic field satisfies the operator-valued wave equation (2.27) for vanishing current. It can easily be proved that the canonical field equations of motion are

$$\dot{\hat{\mathbf{A}}}(\mathbf{r}) = \frac{1}{i\hbar}[\hat{\mathbf{A}}(\mathbf{r}), \hat{H}] = \frac{1}{\varepsilon_0} \hat{\mathbf{\Pi}}(\mathbf{r}), \tag{2.51}$$

$$\dot{\hat{\mathbf{\Pi}}}(\mathbf{r}) = \frac{1}{i\hbar}[\hat{\mathbf{\Pi}}(\mathbf{r}), \hat{H}] = \frac{1}{\mu_0} \Delta \hat{\mathbf{A}}(\mathbf{r}), \tag{2.52}$$

which are indeed equivalent to the wave equation

$$\Delta \hat{\mathbf{A}}(\mathbf{r}) - \frac{1}{c^2} \ddot{\hat{\mathbf{A}}}(\mathbf{r}) = 0. \tag{2.53}$$

2.2.2.1 Separation of variables

To solve the wave equation (2.53), we apply the familiar procedure of separation of variables and represent the vector potential in terms of monochromatic modes, that is, we write

$$\hat{\mathbf{A}}(\mathbf{r}, t) = \frac{1}{\sqrt{\varepsilon_0}} \sum_\lambda c_\lambda \mathbf{A}_\lambda(\mathbf{r}) \hat{q}_\lambda(t). \tag{2.54}$$

Inserting Eq. (2.54) into Eq. (2.53), we can easily see that the mode functions $\mathbf{A}_\lambda(\mathbf{r})$ must solve the Helmholtz equation

$$\Delta \mathbf{A}_\lambda(\mathbf{r}) + \frac{\omega_\lambda^2}{c^2} \mathbf{A}_\lambda(\mathbf{r}) = 0, \tag{2.55}$$

and the associated mode operators $\hat{q}_\lambda(t)$ evolve like harmonic oscillators, namely

$$\ddot{\hat{q}}_\lambda + \omega_\lambda^2 \hat{q}_\lambda = 0. \tag{2.56}$$

The constants c_λ in Eq. (2.54) are introduced for normalization purposes and will be specified later. Clearly, the mode functions must satisfy the condition $\nabla \mathbf{A}_\lambda = 0$, because of the Coulomb gauge.

Since the Laplace operator Δ in Eq. (2.55) is Hermitian, the orthogonality relation

$$\int d^3r \, \mathbf{A}_\lambda^*(\mathbf{r}) \mathbf{A}_{\lambda'}(\mathbf{r}) = 0 \quad (\omega_\lambda \neq \omega_{\lambda'}) \tag{2.57}$$

is valid, which may be proved by taking the product of the complex conjugate of Eq. (2.55) with $\mathbf{A}_{\lambda'}$, subtracting the corresponding equation, with λ and λ' exchanged, and integrating by parts. Obviously, orthogonality may also be derived for the functions $\mathbf{A}_\lambda$, $\mathbf{A}_{\lambda'}$ in place of $\mathbf{A}_\lambda^*$, $\mathbf{A}_{\lambda'}$ in Eq. (2.57). After orthogonalizing the mode functions $\mathbf{A}_\lambda$ which belong to the same frequency ω_λ, we may choose the mode functions in such a way that the ortho-normalization relation

$$|c_\lambda|^2 \int d^3r\, \mathbf{A}_\lambda^*(\mathbf{r}) \mathbf{A}_{\lambda'}(\mathbf{r}) = \delta_{\lambda\lambda'} \tag{2.58}$$

is valid. Further, the modes may be regarded as forming a complete set of functions in the space of transverse vector functions, so that the completeness relation may be given as

$$\sum_\lambda |c_\lambda|^2 \mathbf{A}_\lambda(\mathbf{r}) \otimes \mathbf{A}_\lambda^*(\mathbf{r}') = \delta^\perp(\mathbf{r}-\mathbf{r}'). \tag{2.59}$$

Clearly, the expansion of the vector potential in terms of monochromatic modes, as given in Eq. (2.54), is not attached to the Heisenberg picture, but can be performed in any picture, that is, we may write

$$\hat{\mathbf{A}}(\mathbf{r}) = \frac{1}{\sqrt{\varepsilon_0}} \sum_\lambda c_\lambda \mathbf{A}_\lambda(\mathbf{r}) \hat{q}_\lambda = \frac{1}{\sqrt{\varepsilon_0}} \sum_\lambda c_\lambda^* \mathbf{A}_\lambda^*(\mathbf{r}) \hat{q}_\lambda^\dagger, \tag{2.60}$$

without specifying the temporal evolution. Similarly, the formalism of mode expansion can be applied to the canonical momentum $\hat{\mathbf{\Pi}}$ to obtain[6]

$$\hat{\mathbf{\Pi}}(\mathbf{r}) = \sqrt{\varepsilon_0} \sum_\lambda c_\lambda^* \mathbf{A}_\lambda^*(\mathbf{r}) \hat{p}_\lambda = \sqrt{\varepsilon_0} \sum_\lambda c_\lambda \mathbf{A}_\lambda(\mathbf{r}) \hat{p}_\lambda^\dagger. \tag{2.61}$$

We have allowed for complex mode functions, because it is often convenient to use these rather than real ones. In that case, the canonically conjugate mode operators $\hat{q}_\lambda$ and $\hat{p}_\lambda$ are non-Hermitian operators in general, as is seen from Eqs (2.60) and (2.61). From Eqs (2.60) and (2.61), together with Eq. (2.58), we find that the $\hat{q}_\lambda^\dagger, \hat{p}_\lambda^\dagger$ and $\hat{q}_{\lambda'}, \hat{p}_{\lambda'}$ which belong to the same frequency are related to each other as

$$\hat{q}_\lambda^\dagger = \sum_{\lambda'} \left[\int d^3r\, c_\lambda \mathbf{A}_\lambda(\mathbf{r}) c_{\lambda'} \mathbf{A}_{\lambda'}(\mathbf{r})\right] \hat{q}_{\lambda'}, \tag{2.62}$$

$$\hat{p}_\lambda^\dagger = \sum_{\lambda'} \left[\int d^3r\, c_\lambda \mathbf{A}_\lambda(\mathbf{r}) c_{\lambda'} \mathbf{A}_{\lambda'}(\mathbf{r})\right]^* \hat{p}_{\lambda'}. \tag{2.63}$$

Note that in the λ' sums the terms with $\omega_{\lambda'} \neq \omega_\lambda$ are zero. In the case of real mode functions the $\hat{q}_\lambda$ and $\hat{p}_\lambda$ are obviously Hermitian operators

---

[6] Note that from Eqs (2.51) and (2.52) it follows that $\hat{\mathbf{\Pi}}$ obeys the same wave equation as the vector potential.

($\hat{q}_\lambda^\dagger = \hat{q}_\lambda$, $\hat{p}_\lambda^\dagger = \hat{p}_\lambda$). From the commutation relations (2.46) and (2.47) it follows that the $\hat{q}_\lambda$ and $\hat{p}_\lambda$ satisfy the (equal-time) commutation relations

$$[\hat{q}_\lambda, \hat{p}_{\lambda'}] = i\hbar \delta_{\lambda\lambda'}, \tag{2.64}$$

$$[\hat{q}_\lambda, \hat{q}_{\lambda'}] = 0 = [\hat{p}_\lambda, \hat{p}_{\lambda'}], \tag{2.65}$$

as may be proved by straightforward calculation, inverting the expansions (2.60) and (2.61). Expanding in Eq. (2.48) the canonically conjugate fields $\hat{\mathbf{A}}$ and $\hat{\mathbf{\Pi}}$ according to Eqs (2.60) and (2.61), respectively, after some lengthy but straightforward calculation we may express the Hamiltonian of the electromagnetic field in terms of the mode operators $\hat{q}_\lambda$ and $\hat{p}_\lambda$:

$$\hat{H} = \tfrac{1}{2} \sum_\lambda (\hat{p}_\lambda \hat{p}_\lambda^\dagger + \omega_\lambda^2 \hat{q}_\lambda \hat{q}_\lambda^\dagger). \tag{2.66}$$

As expected, $\hat{H}$ takes the familiar form of the Hamiltonian of an infinite set of uncoupled harmonic oscillators, whose excitation quanta of energy $\hbar \omega_\lambda$ are called photons (of type λ).

2.2.2.2 Photon annihilation and creation operators

Usually the photon annihilation and creation operators $\hat{a}_\lambda$ and $\hat{a}_\lambda^\dagger$ which are defined according to the relations[7]

$$\hat{q}_\lambda = \sqrt{\frac{\hbar}{2\omega_\lambda}} \left\{ \hat{a}_\lambda + \sum_{\lambda'} \left[\int d^3r\, c_\lambda \mathbf{A}_\lambda(\mathbf{r}) c_{\lambda'} \mathbf{A}_{\lambda'}(\mathbf{r}) \right]^* \hat{a}_{\lambda'}^\dagger \right\}, \tag{2.67}$$

$$\hat{p}_\lambda = i\sqrt{\frac{\hbar \omega_\lambda}{2}} \left\{ \hat{a}_\lambda^\dagger - \sum_{\lambda'} \left[\int d^3r\, c_\lambda \mathbf{A}_\lambda(\mathbf{r}) c_{\lambda'} \mathbf{A}_{\lambda'}(\mathbf{r}) \right] \hat{a}_{\lambda'} \right\} \tag{2.68}$$

are preferred to the (coordinate and momentum) operators $\hat{q}_\lambda$ and $\hat{p}_\lambda$. Expressing in Eqs (2.60) and (2.61) the operators $\hat{q}_\lambda$ and $\hat{p}_\lambda$ in terms of the operators $\hat{a}_\lambda$ and $\hat{a}_\lambda^\dagger$ yields the canonically conjugate field operators $\hat{\mathbf{A}}$ and $\hat{\mathbf{\Pi}}$ in the form

$$\hat{\mathbf{A}}(\mathbf{r}) = \sum_\lambda \mathbf{A}_\lambda(\mathbf{r}) \hat{a}_\lambda + \text{H.c.}, \tag{2.69}$$

$$\hat{\mathbf{\Pi}}(\mathbf{r}) = -\varepsilon_0 \sum_\lambda i\omega_\lambda \mathbf{A}_\lambda(\mathbf{r}) \hat{a}_\lambda + \text{H.c.}, \tag{2.70}$$

---

[7] Note that when real mode functions are used, then Eqs (2.67) and (2.68) reduce to the equations $\hat{q}_\lambda = \sqrt{\hbar/(2\omega_\lambda)}\,(\hat{a}_\lambda + \hat{a}_\lambda^\dagger)$ and $\hat{p}_\lambda = i\sqrt{\hbar \omega_\lambda/2}\,(\hat{a}_\lambda^\dagger - \hat{a}_\lambda)$. On the other hand, these equations can always be used to define canonically conjugate Hermitian operators $\hat{q}_\lambda'$ and $\hat{p}_\lambda'$, which are of course different from the operators $\hat{q}_\lambda$ and $\hat{p}_\lambda$ in Eqs (2.67) and (2.68) if complex mode functions are used.

where H.c. denotes the Hermitian conjugate. For notational convenience we have specified the c_λ to be

$$c_\lambda = \sqrt{\frac{2\omega_\lambda \varepsilon_0}{\hbar}}, \tag{2.71}$$

which implies that the ortho-normalization condition (2.58) reads as

$$\int d^3r\, \mathbf{A}_\lambda^*(\mathbf{r}) \mathbf{A}_{\lambda'}(\mathbf{r}) = \frac{\hbar}{2\omega_\lambda \varepsilon_0} \delta_{\lambda\lambda'}. \tag{2.72}$$

Inverting Eqs (2.69) and (2.70), we find that $\hat{a}_\lambda$ and $\hat{a}_\lambda^\dagger$ may be expressed in terms of the field operators $\hat{\mathbf{A}}$ and $\hat{\boldsymbol{\Pi}}$ as

$$\hat{a}_\lambda = \frac{1}{\hbar} \int d^3r\, \mathbf{A}_\lambda^*(\mathbf{r}) [\varepsilon_0 \omega_\lambda \hat{\mathbf{A}}(\mathbf{r}) + i\hat{\boldsymbol{\Pi}}(\mathbf{r})], \tag{2.73}$$

$$\hat{a}_\lambda^\dagger = \frac{1}{\hbar} \int d^3r\, \mathbf{A}_\lambda(\mathbf{r}) [\varepsilon_0 \omega_\lambda \hat{\mathbf{A}}(\mathbf{r}) - i\hat{\boldsymbol{\Pi}}(\mathbf{r})]. \tag{2.74}$$

As expected, Eqs (2.73) and (2.74) together with the commutation relations (2.46) and (2.47) imply bosonic commutation relations for the photonic annihilation and creation operators $\hat{a}_\lambda$ and $\hat{a}_\lambda^\dagger$:

$$[\hat{a}_\lambda, \hat{a}_{\lambda'}^\dagger] = \delta_{\lambda\lambda'}, \tag{2.75}$$

$$[\hat{a}_\lambda, \hat{a}_{\lambda'}] = 0 = [\hat{a}_\lambda^\dagger, \hat{a}_{\lambda'}^\dagger]. \tag{2.76}$$

Using Eqs (2.67) and (2.68) and expressing in (2.66) the canonically conjugate operators $\hat{q}_\lambda$ and $\hat{p}_\lambda$ in terms of the photon annihilation and creation operators $\hat{a}_\lambda$ and $\hat{a}_\lambda^\dagger$, we may represent the Hamiltonian of the electromagnetic field in the form

$$\hat{H} = \tfrac{1}{2} \sum_\lambda \hbar\omega_\lambda (\hat{a}_\lambda^\dagger \hat{a}_\lambda + \hat{a}_\lambda \hat{a}_\lambda^\dagger) = \sum_\lambda \hbar\omega_\lambda (\hat{n}_\lambda + \tfrac{1}{2}), \tag{2.77}$$

where

$$\hat{n}_\lambda = \hat{a}_\lambda^\dagger \hat{a}_\lambda \tag{2.78}$$

is the operator of the number of quanta of the λth oscillator, i.e., the operator of the number of photons in the λth field mode (for its eigenstates, see Section 3.1). Obviously, in the Heisenberg picture the equation of motion for $\hat{a}_\lambda$ reads

$$\dot{\hat{a}}_\lambda = \frac{1}{i\hbar}[\hat{a}_\lambda, \hat{H}] = -i\omega_\lambda \hat{a}_\lambda. \tag{2.79}$$

In Eq. (2.77), the term

$$E_v = \sum_\lambda \tfrac{1}{2}\hbar\omega_\lambda \tag{2.80}$$

is the energy of the electromagnetic vacuum. It is the sum of the ground-state energies of the harmonic oscillators associated with all the modes of the electromagnetic field.[8] Since there are infinitely many modes, E_V is infinite. In practice, this is of less importance, because only changes in the total energy are measurable. In the presence of macroscopic bodies, however, E_V depends on the geometry of the bodies and their arrangement and can be changed by changing the geometry. Such a (finite) change in the (infinite) vacuum energy can be thought of as being associated with a force acting on the bodies – the Casimir force[9] (Section 10.2.2). Provided that the (in general small) effect can be disregarded, the vacuum energy may be omitted.

2.2.2.3 The plane-wave expansion

For notational convenience, here and in the following a discrete mode structure is assumed, which requires that the electromagnetic field extends over a finite volume. In fact, the quantization volume is the whole universe, which implies a mode continuum. In order to apply the discrete-mode expansion formalism, in the first stage a large but finite rectangular box of volume $V = \mathcal{L}_1 \mathcal{L}_2 \mathcal{L}_3$ is commonly considered and periodic boundary conditions are assumed, namely

$$\mathbf{A}_\lambda(\mathbf{r}) = \mathbf{A}_\lambda(\mathbf{r} + \mathbf{L}), \qquad \mathbf{L} = (n_1 \mathcal{L}_1, n_2 \mathcal{L}_2, n_3 \mathcal{L}_3) \tag{2.81}$$

(n_i, integer). In the second stage, passing to the limit as $\mathcal{L}_i \to \infty$ ($i=1,2,3$), for the infinite quantization volume is allowed. The Helmholtz equation (2.55) together with the boundary conditions (2.81) is solved by traveling plane waves ($\lambda \to l, \sigma$)

$$c_{l,\sigma} \mathbf{A}_{l,\sigma}(\mathbf{r}) = V^{-1/2} \mathbf{e}_{l,\sigma} e^{i \mathbf{k}_l \mathbf{r}}, \tag{2.82}$$

where the (wave-number) vector $\mathbf{k}_l$ takes the discrete values

$$\mathbf{k}_l = 2\pi \left(\frac{l_1}{\mathcal{L}_1}, \frac{l_2}{\mathcal{L}_2}, \frac{l_3}{\mathcal{L}_3} \right) \tag{2.83}$$

($l = \{l_i\}$, l_i integer), and the dispersion relation

$$k_l^2 c^2 = \omega_l^2 \tag{2.84}$$

holds. Further, the Coulomb gauge $\nabla \mathbf{A}_\lambda = 0$ leads to the transversality condition

$$\mathbf{e}_{l,\sigma} \mathbf{k}_l = 0. \tag{2.85}$$

---
8) The state of the field in which all the oscillators are in the ground state, i. e., no photons are excited in any of the field modes, is called the vacuum state.
9) The Casimir force is an example of observable consequences of the interaction of matter with the electromagnetic vacuum (Chapter 10).

Thus for any given wave vector $\mathbf{k}_l$ two independent polarization vectors $\mathbf{e}_{l,\sigma}$ ($\sigma=1,2$) exist, which may be chosen orthogonal to each other,

$$\mathbf{e}_{l,\sigma}\mathbf{e}_{l,\sigma'} = \delta_{\sigma\sigma'}. \tag{2.86}$$

With regard to resonator-like systems, it is also usual to consider the electromagnetic field inside a finite cavity with perfectly reflecting walls. Physically, such boundary conditions cannot be realized rigorously, but only approximately in certain frequency ranges.[10]

Representing the electromagnetic field in terms of the traveling plane waves given in Eq. (2.82) [together with Eqs (2.71) and (2.84)], yields the vector potential (2.69) and the canonical momentum (2.70) in the form

$$\hat{\mathbf{A}}(\mathbf{r}) = \sum_{l,\sigma} \sqrt{\frac{\hbar}{2\varepsilon_0 c k_l V}} \, \mathbf{e}_{l,\sigma} e^{i\mathbf{k}_l \mathbf{r}} \hat{a}_{l,\sigma} + \text{H.c.}, \tag{2.87}$$

$$\hat{\mathbf{\Pi}}(\mathbf{r}) = -i \sum_{l,\sigma} \sqrt{\frac{\hbar \varepsilon_0 c k_l}{2V}} \, \mathbf{e}_{l,\sigma} e^{i\mathbf{k}_l \mathbf{r}} \hat{a}_{l,\sigma} + \text{H.c..} \tag{2.88}$$

In the limit as $V \to \infty$, Eqs (2.87) and (2.88) represent the fields in the whole space. According to Eq. (2.83), with increasing V the modes become more and more dense in the $\mathbf{k}$ domain. Defining new operators

$$\hat{a}_\sigma(\mathbf{k}) = \lim_{V \to \infty} \frac{\hat{a}_{l,\sigma}}{\sqrt{(\Delta k)^3}}, \tag{2.89}$$

where

$$\Delta k = \frac{2\pi}{V^{1/3}}, \tag{2.90}$$

we see that in the limit as $V \to \infty$ (i.e., $\Delta k \to 0$) the l sums in Eqs (2.87) and (2.88) approach integrals as follows:

$$\hat{\mathbf{A}}(\mathbf{r}) = \sum_\sigma \int d^3k \sqrt{\frac{\hbar}{2\varepsilon_0 c k (2\pi)^3}} \, \mathbf{e}_\sigma(\mathbf{k}) e^{i\mathbf{k}\mathbf{r}} \hat{a}_\sigma(\mathbf{k}) + \text{H.c.}, \tag{2.91}$$

and $\hat{\mathbf{\Pi}}(\mathbf{r})$ accordingly. The commutation relations for the operators $\hat{a}_\sigma(\mathbf{k})$ and $\hat{a}_\sigma^\dagger(\mathbf{k})$ may be found from the original commutation relations as given

---

[10] Nevertheless, the set of modes obtained from the boundary conditions that are realized by perfectly reflecting walls can be used to describe the electromagnetic field inside the cavity correctly. Note that the number of modes which contribute to the field can drastically increase with decreasing distance of the point of observation from the cavity walls, because of the wrong boundary conditions. Clearly, such a set of modes cannot be used to describe the field outside the cavity, in general.

in Eqs (2.75) and (2.76) together with Eq. (2.89). We derive

$$[\hat{a}_\sigma(\mathbf{k}), \hat{a}^\dagger_{\sigma'}(\mathbf{k}')] = \lim_{\Delta k \to 0} \frac{1}{(\Delta k)^3} [\hat{a}_{l,\sigma}, \hat{a}^\dagger_{l',\sigma'}]$$

$$= \delta_{\sigma\sigma'} \lim_{\Delta k \to 0} \frac{\delta_{ll'}}{(\Delta k)^3} = \delta_{\sigma\sigma'} \delta(\mathbf{k} - \mathbf{k}'). \tag{2.92}$$

Obviously, the operator

$$\hat{n}_\sigma(\mathbf{k}) = \hat{a}^\dagger_\sigma(\mathbf{k}) \hat{a}_\sigma(\mathbf{k}) \tag{2.93}$$

represents the operator of the photon-number density in the $\mathbf{k}$ domain for chosen polarization σ. Integration over all $\mathbf{k}$ yields the operator of the total number of photons of that polarization.

2.2.3
Nonmonochromatic modes

So far we have expressed the electromagnetic field in terms of monochromatic modes, i.e., solutions of the Helmholtz equation (2.55) and the associated harmonic-oscillator variables. The modes are characterized by the eigenfrequencies of the field (for the chosen boundary conditions). Frequently wave packets are observed, rather than monochromatic waves, and photonic operators which describe a wave packet as a whole may be more suited to the description of the electromagnetic field than those associated with the monochromatic waves forming the wave packet.

Applying a unitary transformation to the photon annihilation operators $\hat{a}_\lambda$ associated with the monochromatic modes, we may introduce new operators

$$\hat{a}'_\nu = \sum_\lambda U_{\nu\lambda} \hat{a}_\lambda \tag{2.94}$$

where the $U_{\nu\lambda}$ as the elements of a unitary matrix obey the relation

$$U^{-1}_{\nu\lambda} = U^*_{\lambda\nu} \tag{2.95}$$

Obviously, the operators $\hat{a}'_\nu$ and $\hat{a}'^\dagger_\nu$, respectively, are again annihilation and creation operators which satisfy bosonic commutation relations,

$$[\hat{a}'_\nu, \hat{a}'^\dagger_{\nu'}] = \delta_{\nu\nu'}, \tag{2.96}$$

$$[\hat{a}'_\nu, \hat{a}'_{\nu'}] = [\hat{a}'^\dagger_\nu, \hat{a}'^\dagger_{\nu'}] = 0, \tag{2.97}$$

as is easily seen from Eqs (2.75) and (2.76) in conjunction with Eq. (2.94) and its Hermitian conjugate. These operators can of course be regarded as being the annihilation and creation operators of photons, with

$$\hat{n}'_\nu = \hat{a}'^\dagger_\nu \hat{a}'_\nu \tag{2.98}$$

being the corresponding number operators. Clearly, the thus defined photons can exhibit quite different properties from those associated with the monochromatic modes.

By inverting the relation (2.94), i.e.,

$$\hat{a}_\lambda = \sum_\nu U^*_{\nu\lambda} \hat{a}'_\nu, \tag{2.99}$$

and expressing in Eqs (2.69) and (2.70) the operators $\hat{a}_\lambda, \hat{a}^\dagger_\lambda$ in terms of the new operators $\hat{a}'_\nu, \hat{a}'^\dagger_\nu$, we find [Titulaer and Glauber (1966)]

$$\hat{\mathbf{A}}(\mathbf{r}) = \sum_\nu \mathbf{A}'_\nu(\mathbf{r})\, \hat{a}'_\nu + \text{H.c.}, \tag{2.100}$$

and $\hat{\mathbf{\Pi}}(\mathbf{r})$ accordingly, where the nonmonochromatic mode functions $\mathbf{A}'_\nu(\mathbf{r})$ are defined by

$$\mathbf{A}'_\nu(\mathbf{r}) = \sum_\lambda U^*_{\nu\lambda} \mathbf{A}_\lambda(\mathbf{r}). \tag{2.101}$$

Recalling the ortho-normalization condition (2.72), we derive

$$\int d^3 r\, \mathbf{A}'^*_\nu(\mathbf{r}) \mathbf{A}'_{\nu'}(\mathbf{r}) = \frac{\hbar}{2\varepsilon_0} \sum_\lambda \omega_\lambda^{-1} U_{\nu\lambda} U^*_{\nu'\lambda}, \tag{2.102}$$

which reveals that the functions $\mathbf{A}'_\nu(\mathbf{r})$ are not orthogonal to each other in general. Introduction of the operators $\hat{a}'_\nu, \hat{a}'^\dagger_\nu$ in Eq. (2.77) yields the electromagnetic-field Hamiltonian in the form

$$\hat{H} = \tfrac{1}{2} \sum_{\nu,\nu'} \hbar \omega_{\nu\nu'} (\hat{a}'^\dagger_\nu \hat{a}'_{\nu'} + \hat{a}'_{\nu'} \hat{a}'^\dagger_\nu), \tag{2.103}$$

where

$$\omega_{\nu\nu'} = \sum_\lambda \omega_\lambda U_{\nu\lambda} U^*_{\nu'\lambda} = \omega^*_{\nu'\nu}. \tag{2.104}$$

As expected, the nonmonochromatic photons are coupled to each other and have no fixed energy in general.

In the Heisenberg picture, the photon annihilation operators associated with the monochromatic modes evolve as

$$\hat{a}_\lambda(t) = \hat{a}_\lambda e^{-i\omega_\lambda t}, \tag{2.105}$$

as can be seen from Eq. (2.79). Using Eq. (2.105), then from Eqs (2.94) and (2.100) we obtain

$$\hat{\mathbf{A}}(\mathbf{r},t) = \sum_\nu \mathbf{A}'_\nu(\mathbf{r},t)\, \hat{a}'_\nu + \text{H.c.}, \tag{2.106}$$

with

$$\mathbf{A}'_\nu(\mathbf{r},t) = \sum_\lambda U^*_{\nu\lambda} \mathbf{A}_\lambda(\mathbf{r}) e^{-i\omega_\lambda t} \tag{2.107}$$

in place of Eqs (2.100) and (2.101). The time-dependent nonmonochromatic modes $\mathbf{A}'_\nu(\mathbf{r},t)$ are also called spatio-temporal modes (of the vector potential).

Let the (excited) spatio-temporal modes $\mathbf{A}'_\nu(\mathbf{r},t)$ be nonoverlapping pulse-like wave packets,[11] each of them can be described by a mid-frequency Ω_ν and a slowly varying amplitude, i.e.,

$$\hat{\mathbf{A}}(\mathbf{r},t) = \sum_\nu \tilde{\mathbf{A}}_\nu(\mathbf{r},t)\,\hat{a}'_\nu e^{-i\Omega_\nu t} + \text{H.c.}, \tag{2.108}$$

where $(|\omega_\lambda - \Omega_\nu| \ll \Omega_\nu)$

$$\tilde{\mathbf{A}}_\nu(\mathbf{r},t) = \mathbf{A}'_\nu(\mathbf{r},t) e^{i\Omega_\nu t} = \sum_\lambda U^*_{\nu\lambda} \mathbf{A}_\lambda(\mathbf{r}) e^{-i(\omega_\lambda - \Omega_\nu)t}. \tag{2.109}$$

Obviously, nonmonochromatic modes of this type can be regarded, according to Eq. (2.102), as being approximately orthogonal to each other. In the same approximation, the operators $\hat{a}'_\nu$ and $\hat{a}'^\dagger_\nu$, which are associated with the νth wave packet, annihilate and create, respectively, a photon which carries an average energy of $\hbar\Omega_\nu$. Hence the operators

$$\hat{a}'_\nu(t) = \hat{a}'_\nu e^{-i\Omega_\nu t} \tag{2.110}$$

and their Hermitian conjugates in Eq. (2.108) can be regarded as being the operators in the Heisenberg picture, and the Hamiltonian is

$$\hat{H} \simeq \tfrac{1}{2}\sum_\nu \hbar\Omega_\nu(\hat{a}'^\dagger_\nu \hat{a}'_\nu + \hat{a}'_\nu \hat{a}'^\dagger_\nu). \tag{2.111}$$

2.3
Interaction of the electromagnetic field with charged particles

In Section 2.1 we have considered collections of charged particles interacting with the electromagnetic field, without specifying the different roles that they play in practice. Roughly speaking, there are two different kinds of matter in optics. The first kind comprises the charged particles that are subject to specific active (nonlinear) interactions with the electromagnetic field, whereas the second kind plays the role of a passive (linear) matter background. In contrast

---

[11] This is the case when, e.g., different wave packets have no common monochromatic modes or when the wave packets are spatially well separated.

to the first kind of matter, whose interaction with the electromagnetic field requires a dynamical description, the effect of the second kind is commonly described phenomenologically by introducing appropriate constitutive equations. Of course, these equations also result from a dynamical description of the interaction of the corresponding charged particles with the electromagnetic field. We therefore start with the dynamical description of all the charged particles involved in the interaction with the electromagnetic field.

2.3.1
Minimal coupling

For the quantization of the system of (microscopic) Maxwell's equations (2.1)–(2.4) [together with Eqs (2.14) and (2.15)] we return to the Lagrangian (2.35) in the Coulomb gauge which leads to the classical Hamiltonian

$$H = \sum_a \mathbf{p}_a \dot{\mathbf{r}}_a + \int d^3r \, \mathbf{\Pi}(\mathbf{r}) \dot{\mathbf{A}}(\mathbf{r}) - L, \tag{2.112}$$

where the canonical electromagnetic-field momentum $\mathbf{\Pi}$ is again given by Eq. (2.43), and

$$\mathbf{p}_a = \frac{\partial L}{\partial \dot{\mathbf{r}}_a} = m_a \dot{\mathbf{r}}_a + Q_a \mathbf{A}(\mathbf{r}_a) \tag{2.113}$$

is the canonical momentum of the ath charged particle. Combining Eqs (2.35) and (2.112)–(2.113) and going from the classical to the quantum-mechanical description, yields the Hamiltonian of the coupled radiation–matter system in the form

$$\hat{H} = \tfrac{1}{2} \int d^3r \, \{\varepsilon_0^{-1} \hat{\mathbf{\Pi}}^2(\mathbf{r}) + \mu_0^{-1} [\nabla \times \hat{\mathbf{A}}(\mathbf{r})]^2\}$$
$$+ \sum_a \frac{1}{2m_a} [\hat{\mathbf{p}}_a - Q_a \hat{\mathbf{A}}(\hat{\mathbf{r}}_a)]^2 + \hat{W}_{\text{Coul}}, \tag{2.114}$$

where the canonically conjugate variables are now Hilbert-space operators whose nonvanishing (equal-time) commutators are[12]

$$[\hat{r}_{ka}, \hat{p}_{k'a'}] = i\hbar \delta_{aa'} \delta_{kk'} \tag{2.115}$$

and [Eq. (2.46)]

$$[\hat{A}_k(\mathbf{r}), \hat{\Pi}_{k'}(\mathbf{r}')] = i\hbar \delta^\perp_{kk'}(\mathbf{r} - \mathbf{r}'). \tag{2.116}$$

According to Eqs (2.17), (2.32)–(2.34) and (2.43), the operators of the canonically conjugate fields are related to the operators of the magnetic and electric fields as $\hat{\mathbf{B}} = \nabla \times \hat{\mathbf{A}}$ and $\hat{\mathbf{E}}^\perp = -\hat{\mathbf{\Pi}}/\varepsilon_0$. Note that the commutation relation

---

[12] Note that $[\hat{\mathbf{p}}_a, \hat{\mathbf{A}}(\hat{\mathbf{r}}_a)] = [\hat{\mathbf{p}}_a, \hat{\mathbf{\Pi}}(\hat{\mathbf{r}}_a)] = 0$, because of the Coulomb gauge.

(2.50) does not change, even when $\hat{\mathbf{E}}$ is the full electric field composed of both a transverse and a longitudinal part.

The Hamiltonian (2.114) consists of three terms: the energy of the transverse electromagnetic field, the kinetic energy of the charged particles and their Coulomb energy.[13] The interaction of the particles with the transverse electromagnetic field, i.e., with that part of the field which is related to radiation, is introduced by replacing, in the kinetic energy, the mechanical particle momenta $m\hat{\mathbf{v}}_a$ with displaced canonical momenta according to the rule

$$m_a\hat{\mathbf{v}}_a = \hat{\mathbf{p}}_a - Q_a \hat{\mathbf{A}}(\hat{\mathbf{r}}_a) \tag{2.117}$$

[see Eq. (2.113)]. This kind of coupling is called minimal coupling. Obviously, the concept of mode expansion as developed in Section 2.2.2 can also be applied to the electromagnetic field interacting with charged particles.[14] In particular, photon annihilation and creation operators can be introduced, and $\hat{\mathbf{A}}$ and $\hat{\mathbf{\Pi}}$ can be represented as shown in Eqs (2.69) and (2.70), respectively. Accordingly, the energy of the transverse electromagnetic field can be given in the form (2.77).

It is often convenient to decompose the minimal-coupling Hamiltonian in the form

$$\hat{H} = \hat{H}_R + \hat{H}_C + \hat{H}_{int}, \tag{2.118}$$

where

$$\hat{H}_R = \tfrac{1}{2}\int d^3r \{\varepsilon_0^{-1}\hat{\mathbf{\Pi}}^2(\mathbf{r}) + \mu_0^{-1}[\nabla \times \hat{\mathbf{A}}(\mathbf{r})]^2\} = \sum_\lambda \hbar\omega_\lambda (\hat{n}_\lambda + \tfrac{1}{2}), \tag{2.119}$$

$$\hat{H}_C = \sum_a \frac{\hat{\mathbf{p}}_a^2}{2m_a} + \hat{W}_{Coul}, \tag{2.120}$$

$$\hat{H}_{int} = -\sum_a \frac{Q_a}{m_a}\hat{\mathbf{p}}_a\hat{\mathbf{A}}(\hat{\mathbf{r}}_a) + \sum_a \frac{Q_a^2}{2m_a}\hat{\mathbf{A}}^2(\hat{\mathbf{r}}_a). \tag{2.121}$$

Here $\hat{H}_R$ is the Hamiltonian of the transverse electromagnetic field, $\hat{H}_C$ is the Hamiltonian of the charged particles including their mutual interaction

---

[13] This decomposition of the Hamiltonian is independent of the chosen gauge and the choice of the canonically conjugate variables, unless an explicit time dependence is introduced.

[14] Note that mode expansion corresponds to a canonical transformation, which is of course not attached to the free electromagnetic field. Since the temporal evolution of the mode operators is now governed by the Hamiltonian of the coupled radiation–matter system, the spatio-temporal modes (2.107) introduced within the framework of nonmonochromatic mode expansion (Section 2.2.3) refer to freely propagating wave packets.

through the Coulomb coupling, and $\hat{H}_{\text{int}}$ describes the interaction between the two systems.

2.3.2
Multipolar coupling

The interaction between (localized) atomic systems (atoms, molecules, etc.) and the electromagnetic field, is typically treated in terms of the polarization and magnetization associated with the atomic charges. Moreover, in the phenomenological Maxwell theory these quantities play a fundamental role. Let us consider a collection of point-like particles which form an atomic system localized at some position r_A and define the polarization $\mathbf{P}_A$ such that

$$\nabla \mathbf{P}_A(\mathbf{r}) = -\rho(\mathbf{r}) + \rho_A(\mathbf{r}). \tag{2.122}$$

Here, the charge density $\rho_A = Q_A \delta(\mathbf{r} - \mathbf{r}_A)$ attributed to the total charge $Q_A = \sum_a Q_a$ of the atomic system can be regarded as being the charge density of the atomic system as a whole. Accordingly, the difference charge density $\rho - \rho_A$ represents the density of the "invisible" charges which give rise to a polarization of the system. We combine Eq. (2.122) with the Gaussian law (2.3) to obtain[15]

$$\nabla \mathbf{D}(\mathbf{r}) = \rho_A(\mathbf{r}), \tag{2.123}$$

where

$$\mathbf{D}(\mathbf{r}) = \varepsilon_0 \mathbf{E}(\mathbf{r}) + \mathbf{P}_A(\mathbf{r}) \tag{2.124}$$

is the displacement field, whose source is just the charge density $\rho_A(\mathbf{r})$ of the atomic system as a whole. Taking the derivative of Eq. (2.122) with respect to time and recalling the continuity equation, we find that

$$\nabla [\mathbf{j}(\mathbf{r}) - \mathbf{j}_A(\mathbf{r}) - \dot{\mathbf{P}}_A(\mathbf{r})] = 0, \tag{2.125}$$

where $\mathbf{j}_A = Q_A \dot{\mathbf{r}}_A \delta(\mathbf{r} - \mathbf{r}_A)$ is the current density associated with ρ_A. Note that $\dot{\rho}_A + \nabla \mathbf{j}_A = 0$. From Eq. (2.125) it follows that the vector field $\mathbf{j} - \mathbf{j}_A(\mathbf{r}) - \dot{\mathbf{P}}_A(\mathbf{r})$ can be given by the curl of another vector field $\mathbf{M}_A$ according to

$$\nabla \times \mathbf{M}_A(\mathbf{r}) = \mathbf{j}(\mathbf{r}) - \mathbf{j}_A(\mathbf{r}) - \dot{\mathbf{P}}_A(\mathbf{r}). \tag{2.126}$$

Obviously, $\mathbf{M}_A$ plays the role of the magnetization. Defining the magnetic field

$$\mathbf{H}(\mathbf{r}) = \mu_0^{-1} \mathbf{B}(\mathbf{r}) - \mathbf{M}_A(\mathbf{r}) \tag{2.127}$$

---

[15] When there are additional charged particles which are not included in the polarization, then the corresponding charge density must of course appear on the right-hand side in Eq. (2.123).

and using Eqs (2.124) and (2.126), we can rewrite the Maxwell equation (2.4) as

$$\nabla \times \mathbf{H}(\mathbf{r}) = \mathbf{j}_A(\mathbf{r}) + \dot{\mathbf{D}}(\mathbf{r}). \tag{2.128}$$

With regard to (2.126), the terms $\dot{\mathbf{P}}_A$ and $\nabla \times \mathbf{M}_A$ are also called polarization and magnetization currents respectively.

Let us restrict our attention to a neutral atomic system ($Q_A = 0$) whose position does not change ($\dot{\mathbf{r}}_A = 0$).[16] Substituting into Eqs (2.122) and (2.126) the expressions (2.14) and (2.15) for the charge and current densities, respectively, and using, e.g., the Fourier-integral representation of the δ function, it can be proved that

$$\mathbf{P}_A(\mathbf{r}) = \sum_a Q_a \int_0^1 ds \, (\mathbf{r}_a - \mathbf{r}_A) \delta[\mathbf{r} - \mathbf{r}_A - s(\mathbf{r}_a - \mathbf{r}_A)] \tag{2.129}$$

and

$$\mathbf{M}_A(\mathbf{r}) = \sum_a Q_a \int_0^1 ds \, s(\mathbf{r}_a - \mathbf{r}_A) \times \dot{\mathbf{r}}_a \delta[\mathbf{r} - \mathbf{r}_A - s(\mathbf{r}_a - \mathbf{r}_A)] \tag{2.130}$$

solve these equations.[17]

2.3.2.1 The multipolar-coupling Lagrangian

The total derivative with respect to time of a function of the generalized coordinates can of course be added to the Lagrangian L of a system to obtain a new Lagrangian L', which yields the same equations of motion and is therefore fully equivalent to the old Lagrangian. We return to the Lagrangian (2.35) and write it in the form

$$L = L_0 + L_{\text{int}}, \tag{2.131}$$

where

$$L_{\text{int}} = \int d^3r \, \mathbf{j}(\mathbf{r}) \mathbf{A}(\mathbf{r}). \tag{2.132}$$

Adding to L the derivative with respect to time of the function

$$F = -\int d^3r \, \mathbf{P}_A(\mathbf{r}) \mathbf{A}(\mathbf{r}) \tag{2.133}$$

yields the new Lagrangian

$$L' = L + \frac{dF}{dt} = L_0 + L'_{\text{int}} \tag{2.134}$$

[16] For nonvanishing charge density ρ_A that is not fixed, see Power and Thirunamachandran (1980).
[17] Note that Eqs (2.122) and (2.126), respectively, only determine the longitudinal part of $\mathbf{P}$ and the transverse part of $\mathbf{M}$.

with

$$L'_{int} = -\int d^3r \{\mathbf{P}_A(\mathbf{r})\dot{\mathbf{A}}(\mathbf{r}) - \mathbf{M}_A(\mathbf{r})[\nabla \times \mathbf{A}(\mathbf{r})]\}, \tag{2.135}$$

where we have used Eq. (2.126) and integrated by parts.[18] Thus with the use of Eqs (2.129) and (2.130)

$$\begin{aligned}L'_{int} = -\sum_a Q_a \int_0^1 ds \, \{&(\mathbf{r}_a - \mathbf{r}_A)\dot{\mathbf{A}}[\mathbf{r}_A + s(\mathbf{r}_a - \mathbf{r}_A)]\\&- [s(\mathbf{r}_a - \mathbf{r}_A) \times \dot{\mathbf{r}}_a][\nabla \times \mathbf{A}[\mathbf{r}_A + s(\mathbf{r}_a - \mathbf{r}_A)]]\}.\end{aligned} \tag{2.136}$$

It is worth noting that the transverse electromagnetic field appears in the new Lagrangian in terms of the field strengths $\mathbf{B} = \nabla \times \mathbf{A}$ and $\mathbf{E}^\perp = -\dot{\mathbf{A}}$. In particular, the charges interact with the transverse field via the coupling of the magnetization to the induction field and the coupling of the polarization to the transverse electric field,

$$\begin{aligned}L'_{int} &= \int d^3r \, [\mathbf{P}_A(\mathbf{r})\mathbf{E}^\perp(\mathbf{r}) + \mathbf{M}_A(\mathbf{r})\mathbf{B}(\mathbf{r})]\\&= \sum_a Q_a \int_0^1 ds \, \{(\mathbf{r}_a - \mathbf{r}_A)\mathbf{E}^\perp[\mathbf{r}_A + s(\mathbf{r}_a - \mathbf{r}_A)]\\&\quad + [s(\mathbf{r}_a - \mathbf{r}_A) \times \dot{\mathbf{r}}_a]\mathbf{B}[\mathbf{r}_A + s(\mathbf{r}_a - \mathbf{r}_A)]\}.\end{aligned} \tag{2.137}$$

This kind of coupling is called multipolar coupling. Taylor expansion of $\mathbf{E}^\perp[\mathbf{r}_A + s(\mathbf{r}_a - \mathbf{r}_A)]$ and $\mathbf{B}[\mathbf{r}_A + s(\mathbf{r}_a - \mathbf{r}_A)]$ around $\mathbf{r}_A$ expresses L'_{int} in terms of the familiar electric and magnetic multipole moments of the atomic charge and current distributions.

2.3.2.2 The multipolar-coupling Hamiltonian

It can easily be verified that the new canonical momenta are

$$\mathbf{\Pi}'(\mathbf{r}) = \frac{\delta L'}{\delta \dot{\mathbf{A}}(\mathbf{r})} = \varepsilon_0 \dot{\mathbf{A}}(\mathbf{r}) - \mathbf{P}_A^\perp(\mathbf{r}) = -\varepsilon_0 \mathbf{E}^\perp(\mathbf{r}) - \mathbf{P}_A^\perp(\mathbf{r}) \tag{2.138}$$

and

$$\begin{aligned}\mathbf{p}'_a &= \frac{\partial L'}{\partial \dot{\mathbf{r}}_a} = m_a \dot{\mathbf{r}}_a - Q_a \int_0^1 ds \, s(\mathbf{r}_a - \mathbf{r}_A) \times \{\nabla \times \mathbf{A}[\mathbf{r}_A + s(\mathbf{r}_a - \mathbf{r}_A)]\}\\&= m_a \dot{\mathbf{r}}_a - Q_a \int_0^1 ds \, s(\mathbf{r}_a - \mathbf{r}_A) \times \mathbf{B}[\mathbf{r}_A + s(\mathbf{r}_a - \mathbf{r}_A)],\end{aligned} \tag{2.139}$$

and the multipolar-coupling Hamiltonian thus is

$$H = \sum_a \mathbf{p}'_a \dot{\mathbf{r}}_a + \int d^3r \, \mathbf{\Pi}'(\mathbf{r})\dot{\mathbf{A}}(\mathbf{r}) - L'. \tag{2.140}$$

---

[18] Note that the relation $\mathbf{A}(\nabla \times \mathbf{M}_A) = \nabla(\mathbf{M}_A \times \mathbf{A}) + \mathbf{M}_A(\nabla \times \mathbf{A})$ is valid.

It again consists of the energy of the transverse electromagnetic field, the kinetic energy of the charged particles and their Coulomb energy, but with these energies being expressed in terms of the new variables.

Now quantization can be performed by regarding the canonically conjugate variables as operators and postulating the standard (equal-time) commutators such as

$$[\hat{r}_{ak}, \hat{p}'_{a'k'}] = i\hbar \delta_{aa'} \delta_{kk'} \tag{2.141}$$

and

$$[\hat{A}_k(\mathbf{r}), \hat{\Pi}'_{k'}(\mathbf{r}')] = i\hbar \delta^\perp_{kk'}(\mathbf{r} - \mathbf{r}'). \tag{2.142}$$

According to Eqs (2.138)–(2.140), the multipolar-coupling Hamiltonian takes the form

$$\hat{H} = \frac{1}{2} \int d^3r \left\{ \varepsilon_0^{-1} [\hat{\Pi}'(\mathbf{r}) + \hat{\mathbf{P}}^\perp_A(\mathbf{r})]^2 + \mu_0^{-1} [\nabla \times \hat{\mathbf{A}}(\mathbf{r})]^2 \right\}$$
$$+ \frac{1}{2} \sum_a \frac{1}{m_a} \left\{ \hat{\mathbf{p}}'_a + Q_a \int_0^1 ds\, s(\hat{\mathbf{r}}_a - \hat{\mathbf{r}}_A) \times [\nabla \times \hat{\mathbf{A}}[\mathbf{r}_A + s(\hat{\mathbf{r}}_a - \hat{\mathbf{r}}_A)]] \right\}^2 + \hat{W}_{\text{Coul}}. \tag{2.143}$$

Applying mode expansion to the vector potential $\hat{\mathbf{A}}$ and the canonical momentum field $\hat{\Pi}'$, all the formulas given in Section 2.2.2 for $\hat{\mathbf{A}}$ and $\hat{\Pi}$ are of course also valid with regard to $\hat{\mathbf{A}}$ and $\hat{\Pi}'$. In particular $\hat{\mathbf{A}}$ and $\hat{\Pi}'$ can be expressed in terms of photon annihilation and creation operators in exactly the same way as in Eqs (2.69) and (2.70).

The Hamiltonian (2.143) can be rewritten as a sum of three terms,

$$\hat{H} = \hat{H}_{R'} + \hat{H}_{C'} + \hat{H}_{\text{int}'}. \tag{2.144}$$

The first term in Eq. (2.144) is the Hamiltonian of the transverse electromagnetic field which consists of $\hat{\mathbf{B}}$ and $\varepsilon_0 \hat{\mathbf{E}}^\perp + \hat{\mathbf{P}}^\perp_A$ and can be given in the familiar form (2.77):

$$\hat{H}_{R'} = \frac{1}{2} \int d^3r \left\{ \varepsilon_0^{-1} \hat{\Pi}'^2(\mathbf{r}) + \mu_0^{-1} [\nabla \times \hat{\mathbf{A}}(\mathbf{r})]^2 \right\}$$
$$= \sum_\lambda \hbar \omega_\lambda (\hat{a}'^\dagger_\lambda \hat{a}'_\lambda + \tfrac{1}{2}). \tag{2.145}$$

The second term in Eq. (2.144) refers to the charged particles:

$$\hat{H}_{C'} = \sum_a \frac{\hat{\mathbf{p}}'^2_a}{2m_a} + \hat{W}_{\text{Coul}} + \frac{1}{2\varepsilon_0} \int d^3r\, \hat{\mathbf{P}}^{\perp 2}_A(\mathbf{r}). \tag{2.146}$$

From Eqs (2.122)–(2.124) it follows that

$$\varepsilon_0 \hat{\mathbf{E}}^\parallel(\mathbf{r}) = -\hat{\mathbf{P}}^\parallel_A(\mathbf{r}) \tag{2.147}$$

for a neutral atomic system, and hence $\hat{W}_{\text{Coul}}$ can be rewritten as [cf. Eq. (2.37)]

$$\hat{W}_{\text{Coul}} = \frac{1}{2\varepsilon_0} \int d^3r\, \hat{\mathbf{P}}_A^{\|\,2}(\mathbf{r}). \tag{2.148}$$

Combining Eqs (2.146) and (2.148) then yields the particle Hamiltonian in the form

$$\hat{H}_{C'} = \sum_a \frac{\hat{\mathbf{p}}_a'^2}{2m_a} + \frac{1}{2\varepsilon_0} \int d^3r\, \hat{\mathbf{P}}_A^2(\mathbf{r}). \tag{2.149}$$

The interaction between the electromagnetic field and the particles is governed by the final term in Eq. (2.144):

$$\hat{H}_{\text{int}'} = \frac{1}{\varepsilon_0} \sum_a Q_a \int_0^1 ds\, (\hat{\mathbf{r}}_a - \mathbf{r}_A)\hat{\mathbf{\Pi}}'[\mathbf{r}_A + s(\hat{\mathbf{r}}_a - \mathbf{r}_A)]$$

$$- \sum_a \frac{Q_a}{2m_a} \int_0^1 ds\, s\{[(\hat{\mathbf{r}}_a - \mathbf{r}_A)\times\hat{\mathbf{p}}_a'][\nabla\times\hat{\mathbf{A}}[\mathbf{r}_A + s(\hat{\mathbf{r}}_a - \mathbf{r}_A)]] + \text{H.c.}\}$$

$$+ \sum_a \frac{Q_a^2}{2m_a} \left\{ \int_0^1 ds\, s(\hat{\mathbf{r}}_a - \mathbf{r}_A)\times[\nabla\times\hat{\mathbf{A}}[\mathbf{r}_A + s(\hat{\mathbf{r}}_a - \mathbf{r}_A)]] \right\}^2. \tag{2.150}$$

The first term in Eq. (2.150) results from the interaction of the transverse displacement field $\varepsilon_0 \hat{\mathbf{E}}^\perp + \hat{\mathbf{P}}_A^\perp$ with the polarization $\hat{\mathbf{P}}_A$ of the atomic system. The second term results from the interaction between the magnetic induction field $\hat{\mathbf{B}}$ and the (paramagnetic) magnetization $\hat{\mathbf{M}}_A'$ of the atomic system, with $\hat{\mathbf{M}}_A'$ being obtained from $\hat{\mathbf{M}}_A$ by replacing the mechanical momenta $m_a\dot{\hat{\mathbf{r}}}_a$ of the particles with the canonical momenta $\hat{\mathbf{p}}_a'$. The final term is quadratic in the magnetic induction field $\hat{\mathbf{B}}$ and can be regarded as being the diamagnetic energy of the atomic system.

The above given results can easily be extended to an ensemble of (neutral) atomic systems at different (fixed) positions $\mathbf{r}_A$. The total polarization $\hat{\mathbf{P}}$ and the total magnetization $\hat{\mathbf{M}}$ are the sums of the polarizations $\hat{\mathbf{P}}_A$ and $\hat{\mathbf{M}}_A$ of the atomic systems:

$$\hat{\mathbf{P}}(\mathbf{r}) = \sum_A \hat{\mathbf{P}}_A(\mathbf{r}), \tag{2.151}$$

$$\hat{\mathbf{M}}(\mathbf{r}) = \sum_A \hat{\mathbf{M}}_A(\mathbf{r}), \tag{2.152}$$

and the multipolar-coupling Hamiltonian of the total system is

$$\hat{H} = \hat{H}_{R'} + \sum_A \hat{H}_A + \sum_{A,A'}{}' \hat{H}_{AA'} \tag{2.153}$$

$(A \neq A')$, where $\hat{H}_A = \hat{H}_{C'} + \hat{H}_{\text{int}'}$ with $\hat{H}_{C'}$ and $\hat{H}_{\text{int}'}$ being respectively given by Eqs (2.149) and (2.150), and $\hat{H}_{AA'}$ is the contact term

$$\hat{H}_{AA'} = \frac{1}{2\varepsilon_0} \int d^3r\, \hat{\mathbf{P}}_A(\mathbf{r})\hat{\mathbf{P}}_{A'}(\mathbf{r}), \tag{2.154}$$

2.3.2.3 The Power–Zienau unitary transformation

The transition from the minimal-coupling Hamiltonian to the multipolar-coupling Hamiltonian is a canonical transformation of the dynamical variables, corresponding to a unitary transformation. For a single atomic system, the unitary transformation operator reads[19]

$$\hat{U} = \exp\left[\frac{i}{\hbar} \int d^3r \, \hat{\mathbf{P}}_A(\mathbf{r}) \hat{\mathbf{A}}(\mathbf{r})\right]. \tag{2.155}$$

Obviously, this transformation, which is called the Power–Zienau transformation, does not change $\hat{\mathbf{r}}_a$ and $\hat{\mathbf{A}}$,

$$\hat{\mathbf{A}}'(\mathbf{r}) = \hat{U}\hat{\mathbf{A}}(\mathbf{r})\hat{U}^\dagger = \hat{\mathbf{A}}(\mathbf{r}), \tag{2.156}$$

$$\hat{\mathbf{r}}'_a = \hat{U}\hat{\mathbf{r}}_a\hat{U}^\dagger = \hat{\mathbf{r}}_a, \tag{2.157}$$

and it is not difficult to verify, on applying Eq. (C.10), that it transforms $\hat{\mathbf{\Pi}}$ and $\hat{\mathbf{p}}_a$ according to Eqs (2.138) and (2.139), respectively,

$$\hat{\mathbf{\Pi}}'(\mathbf{r}) = \hat{U}\hat{\mathbf{\Pi}}(\mathbf{r})\hat{U}^\dagger = \hat{\mathbf{\Pi}}(\mathbf{r}) - \hat{\mathbf{P}}_A^\perp(\mathbf{r}), \tag{2.158}$$

$$\hat{\mathbf{p}}'_a = \hat{U}\hat{\mathbf{p}}_a\hat{U}^\dagger = \hat{\mathbf{p}}_a - Q_a\hat{\mathbf{A}}(\hat{\mathbf{r}}_a)$$
$$- Q_a \int_0^1 ds\, s(\mathbf{r}_a - \mathbf{r}_A) \times \{\nabla \times \mathbf{A}[\mathbf{r}_A + s(\mathbf{r}_a - \mathbf{r}_A)]\}. \tag{2.159}$$

Hence expressing, in the minimal-coupling Hamiltonian (2.114), the original dynamical variables in terms of the transformed ones, yields the multipolar-coupling Hamiltonian (2.143). Note that $\hat{H}'_R = \hat{H}_{R'}$.

So far we have transformed the dynamical variables but have left the Hamiltonian unchanged. This type of Power–Zienau transformation should not be confused with another one in which both the dynamical variables and the Hamiltonian are transformed according to the prescriptions

$$\hat{\mathbf{\Pi}}''(\mathbf{r}) = \hat{U}^\dagger \hat{\mathbf{\Pi}}(\mathbf{r})\hat{U} = \hat{\mathbf{\Pi}}(\mathbf{r}) + \hat{\mathbf{P}}_A^\perp(\mathbf{r}), \tag{2.160}$$

$$\hat{\mathbf{p}}''_a = \hat{U}^\dagger \hat{\mathbf{p}}_a \hat{U} = \hat{\mathbf{p}}_a + Q_a\hat{\mathbf{A}}(\hat{\mathbf{r}}_a)$$
$$+ Q_a \int_0^1 ds\, s(\mathbf{r}_a - \mathbf{r}_A) \times \{\nabla \times \mathbf{A}[\mathbf{r}_A + s(\mathbf{r}_a - \mathbf{r}_A)]\}, \tag{2.161}$$

$$\hat{H}'' = \hat{U}^\dagger \hat{H} \hat{U}. \tag{2.162}$$

---

[19] The extension to many atomic systems is straightforward. The unitary operator is $\hat{U} = \prod_A \hat{U}_A$, with $\hat{U}_A$ from Eq. (2.155).

The new Hamiltonian $\hat{H}''$ expressed in terms of the new variables formally looks like the old minimal-coupling Hamiltonian $\hat{H}$ expressed in terms of the originally used variables. Expressing in $\hat{H}''$ the new variables in terms of the original ones, we arrive at a multipolar-coupling Hamiltonian

$$\hat{H}'' = \tfrac{1}{2}\int dr^3\left\{\varepsilon_0^{-1}[\hat{\mathbf{\Pi}}(\mathbf{r}) + \hat{\mathbf{P}}_A^\perp(\mathbf{r})]^2 + \mu_0^{-1}[\nabla\times\hat{\mathbf{A}}(\mathbf{r})]^2\right\}$$
$$+ \tfrac{1}{2}\sum_a \frac{1}{m_a}\left\{\hat{\mathbf{p}}_a + Q_a\int_0^1 ds\, s(\hat{\mathbf{r}}_a-\hat{\mathbf{r}}_A)\times[\nabla\times\hat{\mathbf{A}}[\mathbf{r}_A+s(\hat{\mathbf{r}}_a-\hat{\mathbf{r}}_A)]]\right\}^2 + \hat{W}_{\text{Coul}}.$$
(2.163)

The Hamiltonians (2.143) and (2.163) look the same at first glance but in fact they are different. Since the expectation value $\langle\hat{O}\rangle$ of a physical quantity associated with an operator $\hat{O}$ must not change, the use of the transformed Hamiltonian (2.163) requires that both the operator $\hat{O}$ and the density operator $\hat{\varrho}$ of the system must be transformed to

$$\hat{O}'' = \hat{U}^\dagger \hat{O}\hat{U} \tag{2.164}$$

and

$$\hat{\varrho}'' = \hat{U}^\dagger \hat{\varrho}\hat{U}, \tag{2.165}$$

so that

$$i\hbar\frac{d\langle\hat{O}\rangle}{dt} = \text{Tr}\{\hat{\varrho}[\hat{O},\hat{H}]\} = \text{Tr}\{\hat{\varrho}''[\hat{O}'',\hat{H}'']\}. \tag{2.166}$$

Further, the physical meaning of the canonical momenta in the Hamiltonians (2.143) and (2.163) are different. In particular, $\hat{\mathbf{\Pi}}'$ in Eq. (2.143) is related to the (transverse) displacement field $\varepsilon_0\hat{\mathbf{E}}^\perp + \hat{\mathbf{P}}_A^\perp$, whereas $\hat{\mathbf{\Pi}}$ in Eq. (2.163) – if $\hat{\varrho}$ be not transformed – is related to the (transverse) electric field $\hat{\mathbf{E}}^\perp$, and the interaction term that corresponds to the first term in Eq. (2.150) describes an interaction of the polarization $\hat{\mathbf{P}}_A$ of the atomic system with $\hat{\mathbf{E}}^\perp$ instead of $\hat{\mathbf{E}}^\perp + \varepsilon_0\hat{\mathbf{P}}_A^\perp$.

2.4
Interaction of the electromagnetic field with charged particles in the presence of dielectric media

In classical optics, dielectric matter is commonly described in terms of a phenomenologically introduced macroscopic dielectric susceptibility (or permittivity). This concept of macroscopic electrodynamics has the benefit of being universally valid, because it uses only general physical properties, without the need for involved *ab initio* calculations. Let $\mathbf{P}$ and ρ_P, respectively, be the (macroscopic) polarization and polarization charge density of a dielectric medium,

$$\nabla \mathbf{P}(\mathbf{r}) = -\rho_P(\mathbf{r}) \tag{2.167}$$

[Eq. (2.122) with $\mathbf{P}$ and ρ_P in place of $\mathbf{P}_A$ and $\rho - \rho_A$, respectively]. The displacement field

$$\mathbf{D}(\mathbf{r}) = \varepsilon_0 \mathbf{E}(\mathbf{r}) + \mathbf{P}(\mathbf{r}) \tag{2.168}$$

[Eq. (2.124) with $\mathbf{P}$ in place of $\mathbf{P}_A$] then satisfies the Maxwell equation

$$\nabla \mathbf{D}(\mathbf{r}) = \rho(\mathbf{r}) \tag{2.169}$$

[Eq. (2.123) with ρ in place of ρ_A], where $\rho(\mathbf{r})$ may be some additional charge density not included in ρ_P. Introduction of the displacement field (2.168) into the Maxwell equation (2.4) yields

$$\nabla \times \mathbf{H}(\mathbf{r}) = \mathbf{j}(\mathbf{r}) + \dot{\mathbf{D}}(\mathbf{r}) \tag{2.170}$$

[Eq. (2.128) with $\mathbf{j}$ in place of $\mathbf{j}_A$], with

$$\mathbf{H}(\mathbf{r}) = \frac{1}{\mu_0} \mathbf{B}(\mathbf{r}) \tag{2.171}$$

for nonmagnetic media. Obviously, the polarization current density is

$$\mathbf{j}_P(\mathbf{r}) = \dot{\mathbf{P}}(\mathbf{r}). \tag{2.172}$$

Let us consider an arbitrarily inhomogeneous medium whose polarization linearly and locally responds to the electric field and, for simplicity, restrict our attention to isotropic media. Causality and the dissipation-fluctuation theorem then require that

$$\mathbf{P}(\mathbf{r},t) = \varepsilon_0 \int_0^\infty d\tau\, \chi(\mathbf{r},\tau) \mathbf{E}(\mathbf{r}, t-\tau) + \mathbf{P}_N(\mathbf{r},t), \tag{2.173}$$

where $\chi(\mathbf{r},\tau)$ is the dielectric susceptibility (in the time domain) and $\mathbf{P}_N$ is the noise polarization associated with absorption. Substitution of this expression into Eq. (2.168) together with Fourier transformation converts this equation to[20]

$$\underline{\mathbf{D}}(\mathbf{r},\omega) = \varepsilon_0 \underline{\mathbf{E}}(\mathbf{r},\omega) + \underline{\mathbf{P}}(\mathbf{r},\omega) = \varepsilon_0 \varepsilon(\mathbf{r},\omega) \underline{\mathbf{E}}(\mathbf{r},\omega) + \underline{\mathbf{P}}_N(\mathbf{r},\omega), \tag{2.174}$$

[20] Here and in the following the Fourier transform $\underline{F}(\omega)$ of a real function $F(t)$ is defined according to the relation $F(t) = \int_0^\infty d\omega\, \underline{F}(\omega) e^{-i\omega t} + \text{c.c.}$.

and thus

$$\underline{P}(r, \omega) = \varepsilon_0[\varepsilon(r, \omega) - 1]\underline{E}(r, \omega) + \underline{P}_N(r, \omega), \qquad (2.175)$$

where the (relative) permittivity

$$\varepsilon(r, \omega) = 1 + \int_0^\infty d\tau\, e^{i\omega\tau} \chi(r, \tau) \qquad (2.176)$$

is a complex function of frequency in general, $\varepsilon(r, \omega) = \text{Re}\,\varepsilon(r, \omega) + i\text{Im}\,\varepsilon(r, \omega)$, the real and imaginary parts, which are respectively responsible for dispersion and absorption, being uniquely related to each other through the Kramers–Kronig relations

$$\text{Re}\,\varepsilon(r, \omega) - 1 = \frac{\mathcal{P}}{\pi} \int d\omega'\, \frac{\text{Im}\,\varepsilon(r, \omega')}{\omega' - \omega}, \qquad (2.177)$$

$$\text{Im}\,\varepsilon(r, \omega) = -\frac{\mathcal{P}}{\pi} \int d\omega'\, \frac{\text{Re}\,\varepsilon(r, \omega') - 1}{\omega' - \omega} \qquad (2.178)$$

($\mathcal{P}$, principal value). Further, $\varepsilon(r, \omega)$ as a function of complex ω satisfies the relation

$$\varepsilon(r, -\omega^*) = \varepsilon^*(r, \omega) \qquad (2.179)$$

and is holomorphic in the upper complex half-plane without zeros. In particular, it approaches unity in the high-frequency limit, i.e., $\varepsilon(r, \omega) \to 1$ if $|\omega| \to \infty$.

2.4.1
Nondispersing and nonabsorbing media

Let us suppose that the electromagnetic field under study extends over a frequency interval $(\Delta\omega)$ so that the ω integrals effectively run only over this interval, i.e.,

$$E(r, t) = \int_{(\Delta\omega)} d\omega\, E(r, \omega) e^{-i\omega t} + \text{c.c.} \qquad (2.180)$$

and $B(r, t)$ accordingly. Further, let us assume that the frequency interval is sufficiently small and sufficiently far from medium resonances, so that both dispersion and absorption may be disregarded. In this case, the permittivity becomes approximately real and independent of frequency,[21]

$$\varepsilon(r, \omega) \approx \varepsilon(r, \omega_0) \approx \varepsilon^*(r, \omega_0) \equiv \varepsilon(r) \qquad (2.181)$$

---

21) Note that a constant and real permittivity in the whole frequency domain can be realized only for unity permittivity, i.e, the vacuum. Otherwise, causality would be violated.

(ω_0, mid-frequency), and the noise polarization can be disregarded, $\underline{P}_N(\mathbf{r},\omega) \approx 0$. Integration of Eq. (2.174) over the frequency interval ($\Delta\omega$) then yields the approximately valid constitutive equation

$$\mathbf{D}(\mathbf{r}) = \varepsilon_0 \varepsilon(\mathbf{r}) \mathbf{E}(\mathbf{r}). \tag{2.182}$$

At this point the scalar potential V [Eq. (2.17)] and the vector potential $\mathbf{A}$ [Eq. (2.18)] can be introduced and the equations of motion for the potentials can be derived from the Maxwell equations (2.169) and (2.170) together with Eqs (2.171) and (2.182). It can be shown that these field equations and the Newtonian equations of motion for the particles can be derived from a Lagrangian as given by Eq. (2.22) except that the square of the electric field strength must be replaced according to

$$\mathbf{E}^2(\mathbf{r}) = [\dot{\mathbf{A}}(\mathbf{r}) + \nabla V(\mathbf{r})]^2 \mapsto \varepsilon(\mathbf{r})[\dot{\mathbf{A}}(\mathbf{r}) + \nabla V(\mathbf{r})]^2. \tag{2.183}$$

Using the generalized Coulomb gauge

$$\nabla[\varepsilon(\mathbf{r})\mathbf{A}(\mathbf{r})] = 0, \tag{2.184}$$

it is not difficult to verify that the equations for the vector potential and the scalar potential can be disentangled to obtain

$$\nabla \times \nabla \times \mathbf{A}(\mathbf{r}) + \frac{\varepsilon(\mathbf{r})}{c^2} \ddot{\mathbf{A}}(\mathbf{r}) = \mu_0 \mathbf{j}^\perp(\mathbf{r}) \tag{2.185}$$

and

$$-\varepsilon_0 \nabla[\varepsilon(\mathbf{r}) \nabla V(\mathbf{r})] = \rho(\mathbf{r}) \tag{2.186}$$

in place of Eqs (2.27) and (2.28), where the transverse current density is

$$\mathbf{j}^\perp(\mathbf{r}) = \mathbf{j}(\mathbf{r}) - \varepsilon_0 \varepsilon(\mathbf{r}) \nabla \dot{V}(\mathbf{r}) \tag{2.187}$$

in place of Eq. (2.29). From Eq. (2.186) it follows that the scalar potential V can be represented as a functional of the charge density ρ. Thus it can be expressed in terms of the coordinates $\mathbf{r}_a$ of the charged particles and no longer plays the role of an independent field variable. Note that $\mathbf{A}$ contains a longitudinal contribution in general, so that Eq. (2.33) and (2.34) are no longer valid.

The Lagrangian in the generalized Coulomb gauge (2.184) can then be taken from Eq. (2.35) together with Eq. (2.36) except that the term $|\dot{\mathbf{A}}|^2$ must be replaced according to

$$\dot{\mathbf{A}}^2(\mathbf{r}) \mapsto \varepsilon(\mathbf{r}) \dot{\mathbf{A}}^2(\mathbf{r}). \tag{2.188}$$

Hence we may write

$$L = \tfrac{1}{2} \int d^3r \left\{ \varepsilon_0 \varepsilon(\mathbf{r}) \dot{\mathbf{A}}^2(\mathbf{r}) - \mu_0^{-1}[\nabla \times \mathbf{A}(\mathbf{r})]^2 \right\}$$
$$+ \tfrac{1}{2} \sum_a m_a \dot{\mathbf{r}}_a^2 - W_{\text{Coul}} + \int d^3r\, \mathbf{j}(\mathbf{r}) \mathbf{A}(\mathbf{r}). \tag{2.189}$$

2.4 Dielectric background media

Note that W_{Coul} is again defined by Eq. (2.36), but with V now being the potential that solves Eq. (2.186):

$$W_{\text{Coul}} = \tfrac{1}{2}\int d^3r\,\rho(\mathbf{r})V(\mathbf{r}) = \tfrac{1}{2}\varepsilon_0\int d^3r\,\varepsilon(\mathbf{r})[\nabla V(\mathbf{r})]^2. \tag{2.190}$$

Let $\delta^{\perp(\varepsilon)}(\mathbf{r})$ be the generalized transverse δ function that projects onto the space of vector functions $\mathbf{g}(\mathbf{r})$ satisfying the condition $\nabla(\varepsilon\mathbf{g})=0$. Defining the functional derivative $\tilde{\delta}^{(\varepsilon)}F/\tilde{\delta}^{(\varepsilon)}\mathbf{g}(\mathbf{r})$ by replacing $\delta^{\perp}(\mathbf{r}-\mathbf{r}')$ with $\delta^{\perp(\varepsilon)}(\mathbf{r}-\mathbf{r}')$ in Eq. (2.40), it is seen that the relation between the canonical momentum and the vector potential,

$$\Pi(\mathbf{r}) = \frac{\tilde{\delta}^{(\varepsilon)}L}{\tilde{\delta}^{(\varepsilon)}\dot{\mathbf{A}}(\mathbf{r})} = \varepsilon_0\dot{\mathbf{A}}(\mathbf{r}), \tag{2.191}$$

looks like that in Eq. (2.43). Obviously, the relation (2.113) between the mechanical and canonical particle momenta does not change either. Now the Hamiltonian

$$H = \sum_a \mathbf{p}_a\dot{\mathbf{r}}_a + \int d^3r\,\varepsilon(\mathbf{r})\Pi(\mathbf{r})\dot{\mathbf{A}}(\mathbf{r}) - L \tag{2.192}$$

can be introduced which generalizes Eq. (2.112).

The transition from classical to quantum theory again consists of the replacement of the canonically conjugate c-number variables $\mathbf{r}_a$, $\mathbf{p}_a$, $\mathbf{A}(\mathbf{r})$ and $\Pi(\mathbf{r})$ by the Hilbert-space operators $\hat{\mathbf{r}}_a$, $\hat{\mathbf{p}}_a$, $\hat{\mathbf{A}}(\mathbf{r})$ and $\hat{\Pi}(\mathbf{r})$. The nonvanishing (equal-time) commutators are given by Eqs (2.115) and (2.116), but with $\delta^{\perp(\varepsilon)}(\mathbf{r}-\mathbf{r}')$ in place of $\delta^{\perp}(\mathbf{r}-\mathbf{r}')$ in Eq. (2.116), i.e.,[22]

$$\left[\hat{A}_k(\mathbf{r}), \hat{\Pi}_{k'}(\mathbf{r}')\right] = i\hbar\delta^{\perp(\varepsilon)}_{kk'}(\mathbf{r}-\mathbf{r}'), \tag{2.193}$$

and the Hamiltonian that corresponds to the minimal-coupling Hamiltonian (2.114) reads

$$\hat{H} = \tfrac{1}{2}\int d^3r\,\{\varepsilon(\mathbf{r})\varepsilon_0^{-1}\hat{\Pi}^2(\mathbf{r}) + \mu_0^{-1}[\nabla\times\hat{\mathbf{A}}(\mathbf{r})]^2\}$$
$$+ \sum_a \frac{1}{2m_a}[\hat{\mathbf{p}}_a - Q_a\hat{\mathbf{A}}(\hat{\mathbf{r}}_a)]^2 + \hat{W}_{\text{Coul}}. \tag{2.194}$$

The solutions of the time-independent wave equation

$$\nabla\times\nabla\times\mathbf{A}_\lambda(\mathbf{r}) - \frac{\omega_\lambda^2}{c^2}\varepsilon(\mathbf{r})\mathbf{A}_\lambda(\mathbf{r}) = 0 \tag{2.195}$$

---

22) When ε does not depend on $\mathbf{r}$, then $\varepsilon\delta^{\perp(\varepsilon)}(\mathbf{r})=\delta^{\perp}(\mathbf{r})$. Obviously, the commutation relation (2.193) does not lead to the commutation relation (2.46), whose violation indicates that a bandwidth-limited field is considered and not the complete one.

(generalized Helmholtz equation) can be used to introduce the monochromatic modes of the electromagnetic field in a spatially varying dielectric medium. In particular, they can be chosen to satisfy the generalized orthonormalization relation[23]

$$|c_\lambda|^2 \int d^3r\, \varepsilon(\mathbf{r}) \mathbf{A}_\lambda^*(\mathbf{r}) \mathbf{A}_{\lambda'}(\mathbf{r}) = \delta_{\lambda\lambda'}, \tag{2.196}$$

and the completeness relation reads

$$\sum_\lambda |c_\lambda|^2 \mathbf{A}_\lambda(\mathbf{r}) \otimes \mathbf{A}_\lambda^*(\mathbf{r}') = \boldsymbol{\delta}^{\perp(\varepsilon)}(\mathbf{r}-\mathbf{r}'). \tag{2.197}$$

Equations (2.196) and (2.197), respectively, replace Eqs (2.58) and (2.59) in Section 2.2.2. In particular, in bulk material ε does not depend on $\mathbf{r}$ and Eq. (2.195) reduces to Eq. (2.55) except that ε_0 must be replaced according to $\varepsilon_0 \mapsto \varepsilon\varepsilon_0$. Substitution of $\varepsilon\varepsilon_0$ for ε_0 in Eqs (2.87) and (2.88) thus yields the expansion of $\hat{\mathbf{A}}$ and $\hat{\boldsymbol{\Pi}}$ in terms of traveling plane waves, where the dispersion relation is $k_l^2 c^2 = \varepsilon\omega^2$ in place of Eq. (2.84). It should be pointed out that when ε does not depend on $\mathbf{r}$, then ε can be a (real) function of frequency.

Equations (2.69) and (2.70) can be used to express $\hat{\mathbf{A}}$ and $\hat{\boldsymbol{\Pi}}$ in terms of the monochromatic modes and the associated photon annihilation and creation operators, and the first contribution to the Hamiltonian (2.194) thus takes the form (2.77), as can be proved by straightforward calculation.[24] The Hamiltonian can then be decomposed, on symmetrizing $\hat{\mathbf{p}}_a \hat{\mathbf{A}}(\hat{\mathbf{r}}_a)$, in close analogy to Eqs (2.118)–(2.121). Finally, the Hamiltonian that corresponds to the multipolar-coupling Hamiltonian (2.143) can also be constructed.

2.4.2
Dispersing and absorbing media

As already mentioned, absorption is always associated with additional noise described by the noise polarization in the constitutive equation (2.173) [or (2.174)]. Hence, the corresponding dynamical variables of the medium must be included in the quantization scheme, which implies extension of the Hilbert space. For the sake of transparency let us first restrict our attention to the case when the dielectric medium is the only matter that is present.

2.4.2.1 The medium-assisted electromagnetic field

The medium-assisted electromagnetic field satisfies the Maxwell equations (2.1), (2.2), (2.169) and (2.170) for $\rho=\mathbf{j}=0$. Let us again restrict our attention to

---

[23] Note that with respect to the integration measure $\varepsilon(\mathbf{r})d^3r$ in the definition of scalar products, the differential operator $\varepsilon^{-1}(\mathbf{r})\nabla \times \nabla \times$ is Hermitian.

[24] Note that the photons defined in this way do not only refer to the transverse part of the electromagnetic field.

2.4 Dielectric background media

nonmagnetic matter, so that Eq. (2.171) holds. With regard to the constitutive equation (2.174), we convert the Maxwell equations by Fourier transformation to

$$\nabla \underline{\mathbf{B}}(\mathbf{r}, \omega) = 0, \tag{2.198}$$

$$\nabla \times \underline{\mathbf{E}}(\mathbf{r}, \omega) = i\omega \underline{\mathbf{B}}(\mathbf{r}, \omega), \tag{2.199}$$

$$\varepsilon_0 \nabla \varepsilon(\mathbf{r}, \omega) \underline{\mathbf{E}}(\mathbf{r}, \omega) = \underline{\rho}_N(\mathbf{r}, \omega), \tag{2.200}$$

$$\nabla \times \underline{\mathbf{B}}(\mathbf{r}, \omega) = \mu_0 \underline{\mathbf{j}}_N(\mathbf{r}, \omega) - i\frac{\omega}{c^2} \varepsilon(\mathbf{r}, \omega) \underline{\mathbf{E}}(\mathbf{r}, \omega), \tag{2.201}$$

where we have introduced the (Fourier transformed) noise charge density

$$\underline{\rho}_N(\mathbf{r}, \omega) = -\nabla \underline{\mathbf{P}}_N(\mathbf{r}, \omega) \tag{2.202}$$

and the noise current density

$$\underline{\mathbf{j}}_N(\mathbf{r}, \omega) = -i\omega \underline{\mathbf{P}}_N(\mathbf{r}, \omega), \tag{2.203}$$

which obey the continuity equation

$$\nabla \underline{\mathbf{j}}_N(\mathbf{r}, \omega) = i\omega \underline{\rho}_N(\mathbf{r}, \omega). \tag{2.204}$$

From the Maxwell equations (2.199) and (2.201) it follows that $\underline{\mathbf{E}}(\mathbf{r}, \omega)$ and $\underline{\mathbf{B}}(\mathbf{r}, \omega)$ can be represented in the form of

$$\underline{\mathbf{E}}(\mathbf{r}, \omega) = i\mu_0 \omega \int d^3r' \, G(\mathbf{r}, \mathbf{r}', \omega) \underline{\mathbf{j}}_N(\mathbf{r}', \omega) \tag{2.205}$$

and

$$\underline{\mathbf{B}}(\mathbf{r}, \omega) = (i\omega)^{-1} \nabla \times \underline{\mathbf{E}}(\mathbf{r}, \omega), \tag{2.206}$$

where the Green tensor $G(\mathbf{r}, \mathbf{r}', \omega)$ has to be determined from the equation

$$\nabla \times \nabla \times G(\mathbf{r}, \mathbf{r}', \omega) - \frac{\omega^2}{c^2} \varepsilon(\mathbf{r}, \omega) G(\mathbf{r}, \mathbf{r}', \omega) = \delta(\mathbf{r} - \mathbf{r}') \tag{2.207}$$

together with the boundary condition at infinity, $G(\mathbf{r}, \mathbf{r}', \omega) \to 0$ if $|\mathbf{r} - \mathbf{r}'| \to 0$ (for the properties of the Green tensor, see Appendix A). It is easily seen that the Maxwell equations (2.198) and (2.200) are satisfied identically.

The transition from classical to quantum theory and quantization of the medium-assisted electromagnetic field now consists in the replacement of the classical c-number fields by operator valued ones. Let $\hat{\mathbf{f}}(\mathbf{r}, \omega)$ [and $\hat{\mathbf{f}}^\dagger(\mathbf{r}, \omega)$] be the bosonic fields,

$$[\hat{f}_k(\mathbf{r}, \omega), \hat{f}_{k'}^\dagger(\mathbf{r}', \omega')] = \delta_{kk'} \delta(\mathbf{r} - \mathbf{r}') \delta(\omega - \omega'), \tag{2.208}$$

$$[\hat{f}_k(\mathbf{r},\omega), \hat{f}_{k'}(\mathbf{r}',\omega')] = 0, \tag{2.209}$$

which are attributed to the elementary excitations of the system composed of the electromagnetic field and the medium within the framework of linear electrodynamics. They can be regarded as playing the role of a set of (canonically conjugate) dynamical variables of the composed system, so that all quantities of the system can be expressed in terms of them. For this it is sufficient, to know the relation between $\underline{\hat{\mathbf{P}}}_N(\mathbf{r},\omega)$ und $\hat{\mathbf{f}}(\mathbf{r},\omega)$. The linear relation that ensures preservation of the fundamental equal-time commutation relations is

$$\underline{\hat{\mathbf{P}}}_N(\mathbf{r},\omega) = i\sqrt{\frac{\hbar\varepsilon_0}{\pi}} \operatorname{Im}\varepsilon(\mathbf{r},\omega)\, \hat{\mathbf{f}}(\mathbf{r},\omega). \tag{2.210}$$

Recalling Eq. (2.203) and using Eq. (2.210), we convert $\mathbf{E}(\mathbf{r},\omega)$ [Eq. (2.205)] and $\mathbf{B}(\mathbf{r},\omega)$ [Eq. (2.206)] into the quantum-mechanical operators

$$\underline{\hat{\mathbf{E}}}(\mathbf{r},\omega) = i\sqrt{\frac{\hbar}{\pi\varepsilon_0}}\frac{\omega^2}{c^2}\int d^3r'\, \sqrt{\operatorname{Im}\varepsilon(\mathbf{r}',\omega)}\, G(\mathbf{r},\mathbf{r}',\omega)\hat{\mathbf{f}}(\mathbf{r}',\omega) \tag{2.211}$$

and

$$\underline{\hat{\mathbf{B}}}(\mathbf{r},\omega) = (i\omega)^{-1}\nabla\times \underline{\hat{\mathbf{E}}}(\mathbf{r},\omega). \tag{2.212}$$

Integration over ω then yields the operators of the electric field and the magnetic induction field in the Schrödinger picture:

$$\hat{\mathbf{E}}(\mathbf{r}) = \int_0^\infty d\omega\, \underline{\hat{\mathbf{E}}}(\mathbf{r},\omega) + \text{H.c.}, \tag{2.213}$$

$$\hat{\mathbf{B}}(\mathbf{r}) = \int_0^\infty d\omega\, \underline{\hat{\mathbf{B}}}(\mathbf{r},\omega) + \text{H.c.}. \tag{2.214}$$

The operator of the displacement field reads

$$\hat{\mathbf{D}}(\mathbf{r}) = \int_0^\infty d\omega\, \underline{\hat{\mathbf{D}}}(\mathbf{r},\omega) + \text{H.c.}, \tag{2.215}$$

where, according to Eqs (2.174), (2.199), (2.201) and (2.203),

$$\underline{\hat{\mathbf{D}}}(\mathbf{r},\omega) = (\mu_0\omega^2)^{-1}\nabla\times\nabla\times\underline{\hat{\mathbf{E}}}(\mathbf{r},\omega). \tag{2.216}$$

Using very general properties of the permittivity and the Green tensor, it can be shown that $\hat{\mathbf{E}}$ and $\hat{\mathbf{B}}$ satisfy the correct (equal-time) commutation relation (2.50) (Appendix B). Obviously, the Hamiltonian of the composed system can be given in the form of

$$\hat{H} = \int d^3r \int_0^\infty d\omega\, \hbar\omega\, \hat{\mathbf{f}}^\dagger(\mathbf{r},\omega)\hat{\mathbf{f}}(\mathbf{r},\omega). \tag{2.217}$$

Further, the vector potential in the Coulomb gauge

$$\hat{\mathbf{A}}(\mathbf{r}) = \int_0^\infty d\omega\, \underline{\hat{\mathbf{A}}}(\mathbf{r},\omega) + \text{H.c.} \tag{2.218}$$

and the canonical momentum field

$$\hat{\mathbf{\Pi}}(\mathbf{r}) = -i\varepsilon_0 \int_0^\infty d\omega\, \omega \underline{\hat{\mathbf{A}}}(\mathbf{r},\omega) + \text{H.c.} \tag{2.219}$$

can be defined, where

$$\underline{\hat{\mathbf{A}}}(\mathbf{r},\omega) = \sqrt{\frac{\hbar}{\pi\varepsilon_0}} \frac{\omega}{c^2} \int d^3 r'\, \sqrt{\text{Im}\,\varepsilon(\mathbf{r}',\omega)}\, \mathbf{G}^\perp(\mathbf{r},\mathbf{r}',\omega) \hat{\mathbf{f}}(\mathbf{r}',\omega), \tag{2.220}$$

with $\mathbf{G}^\perp(\mathbf{r},\mathbf{r}')$ being the (from the left) one-sided transverse Green tensor.[25] It is not difficult to verify that $\hat{\mathbf{\Pi}} = -\varepsilon_0 \hat{\mathbf{E}}^\perp$ and $\nabla \times \hat{\mathbf{A}} = \hat{\mathbf{B}}$, and it can be shown that $\hat{\mathbf{A}}$ and $\hat{\mathbf{\Pi}}$ satisfy the well-known commutation relation (2.46) (Appendix B). Note that $\hat{\mathbf{\Pi}}$ and $-\varepsilon_0 \hat{\mathbf{E}}^\parallel$ are respectively the transverse part and the longitudinal part of a common vector field, and $\hat{\mathbf{E}}^\parallel$ can be attributed to a scalar potential $\hat{V}$,

$$-\nabla \hat{V}(\mathbf{r}) = \hat{\mathbf{E}}^\parallel(\mathbf{r}). \tag{2.221}$$

The expansion of the electromagnetic field in terms of the classical Green tensor and the dynamical-variable fields $\hat{\mathbf{f}}(\mathbf{r},\omega)$ and $\hat{\mathbf{f}}^\dagger(\mathbf{r},\omega)$ generalizes the mode expansion based on the (macroscopic) wave equation (2.195) for real permittivity.[26] It should be pointed out that Eqs (2.69) and (2.70) also apply to the transverse electromagnetic field in a medium, with the mode functions being determined from the (microscopic) wave equation (2.55). Clearly, the associated mode operators do not evolve freely, because of the interaction with the medium. Equations (2.218) and (2.219) [together with Eq. (2.220)] can be viewed as the solution of the interaction problem within the framework of linear electrodynamics, the original mode operators being expressed in terms of the operators of the elementary excitations of the composed system.[27]

2.4.2.2 The minimal-coupling Hamiltonian

When additional charged particles are present, then the interaction of the particles with the medium-assisted electromagnetic field can be described by the

---

25) Note that $\mathbf{G}^{\perp(\parallel)}(\mathbf{r},\mathbf{r}',\omega) = \int d^3 s\, \delta^{\perp(\parallel)}(\mathbf{r}-\mathbf{s})\mathbf{G}(\mathbf{s},\mathbf{r}',\omega).$
26) For an extension of the quantization scheme to magnetodielectric media characterized by both complex permittivities and complex permeabilities, see Ho, Buhmann, Knöll, Welsch, Scheel and Kästel (2003).
27) For an inclusion of nonlinear, absorbing media in the quantization scheme, see Scheel and Welsch (2006).

minimal-coupling Hamiltonian

$$\hat{H} = \int dr^3 \int_0^\infty d\omega\, \hbar\omega\, \hat{\mathbf{f}}^\dagger(\mathbf{r},\omega)\hat{\mathbf{f}}(\mathbf{r},\omega)$$
$$+ \sum_a \frac{1}{2m_a}[\hat{\mathbf{p}}_a - Q_a\hat{\mathbf{A}}(\hat{\mathbf{r}}_a)]^2 + \hat{W}_{\text{Coul}}, \quad (2.222)$$

where

$$\hat{W}_{\text{Coul}} = \hat{W}_{\text{Coul}}^C + \hat{W}_{\text{Coul}}^{CM}. \quad (2.223)$$

The first term in Eq. (2.222) is the energy of the electromagnetic field and the medium. The second term is the kinetic energy of the charged particles, with the vector potential $\hat{\mathbf{A}}(\mathbf{r})$ being expressed in terms of $\hat{\mathbf{f}}(\mathbf{r},\omega)$ and $\hat{\mathbf{f}}^\dagger(\mathbf{r},\omega)$ according to Eqs (2.218) and (2.220). The third term is the total Coulomb energy, which consists of the Coulomb energy of the charged particles, $\hat{W}_{\text{Coul}}^C$, and the Coulomb energy of interaction of the charged particles with the medium, $\hat{W}_{\text{Coul}}^{CM}$.

Let $\hat{\mathbf{E}}_M^\parallel$ be the contribution to $\hat{\mathbf{E}}^\parallel$ from the medium.[28] The interaction Coulomb energy can then be given by

$$\hat{W}_{\text{Coul}}^{CM} = \int d^3r\, \hat{\rho}(\mathbf{r})\hat{V}_M(\mathbf{r}), \quad (2.224)$$

where $\hat{\rho}$ is the charge density of the particles, and $\hat{V}_M$ is the Coulomb potential attributed to $\hat{\mathbf{E}}_M^\parallel$ such that, according to Eq. (2.221) ($\hat{\mathbf{E}}^\parallel \mapsto \hat{\mathbf{E}}_M^\parallel$, $\hat{V} \mapsto \hat{V}_M$), $\hat{\mathbf{E}}_M^\parallel = -\nabla\hat{V}_M$. In particular when the charged particles form a neutral atomic system, then $\hat{W}_{\text{Coul}}^{CM}$ can be expressed in terms of the polarization of the atomic system, $\hat{\rho} = -\nabla\hat{\mathbf{P}}_A$ [cf. Eq. (2.122)], as

$$\hat{W}_{\text{Coul}}^{CM} \to \hat{W}_{\text{Coul}}^{AM} = -\int d^3r\, \hat{\mathbf{P}}_A(\mathbf{r})\hat{\mathbf{E}}_M^\parallel(\mathbf{r}) = \frac{1}{\varepsilon_0}\int d^3r\, \hat{\mathbf{P}}_A(\mathbf{r})\hat{\mathbf{P}}_M^\parallel(\mathbf{r}), \quad (2.225)$$

where the relation $\hat{\mathbf{P}}_M^\parallel = -\varepsilon_0\hat{\mathbf{E}}_M^\parallel = \varepsilon_0\nabla\hat{V}_M$ has been used. In a straightforward calculation it can be shown (by means of the commutation relations in Appendix B) that the Hamiltonian (2.222) leads to both the operator-valued Maxwell equations (2.1), (2.2) and (2.169), (2.170) and the operator valued Newtonian equations of motion (2.16). Note that the scalar potential $\hat{V}_M(\mathbf{r})$ or, equivalently, $\hat{\mathbf{E}}_M^\parallel(\mathbf{r})$ must also be thought of as being expressed in terms of the dynamical variables $\hat{\mathbf{f}}(\mathbf{r},\omega)$ and $\hat{\mathbf{f}}^\dagger(\mathbf{r},\omega)$.

28) Note that $\hat{\mathbf{E}}_M^\parallel$ is defined according to Eqs (2.211) and (2.213) with the (from the left) one-sided longitudinal Green tensor $\mathbf{G}^\parallel$ in place of $\mathbf{G}$.

2.4.2.3 The multipolar-coupling Hamiltonian

In order to perform the transition from the minimal-coupling Hamiltonian to the multipolar-coupling Hamiltonian, we transform the variables by application of the unitary operator $\hat{U}$ defined by Eq. (2.155). According to Eqs (2.157) and (2.159), the relations

$$\hat{\mathbf{r}}'_a = \hat{U}\hat{\mathbf{r}}_a\hat{U}^\dagger = \hat{\mathbf{r}}_a \tag{2.226}$$

and

$$\hat{\mathbf{p}}'_a = \hat{U}\hat{\mathbf{p}}_a\hat{U}^\dagger = \hat{\mathbf{p}}_a - Q_a\hat{\mathbf{A}}(\hat{\mathbf{r}}_a)$$
$$- Q_a \int_0^1 ds\, s(\hat{\mathbf{r}}_a - \mathbf{r}_A) \times \{\nabla \times \mathbf{A}[\mathbf{r}_A + s(\hat{\mathbf{r}}_a - \mathbf{r}_A)]\} \tag{2.227}$$

are valid, and it is not difficult to prove that

$$\hat{\mathbf{f}}'(\mathbf{r},\omega) = \hat{U}\hat{\mathbf{f}}(\mathbf{r},\omega)\hat{U}^\dagger$$
$$= \hat{\mathbf{f}}(\mathbf{r},\omega) - \frac{i}{\hbar}\sqrt{\frac{\hbar}{\pi\varepsilon_0}\frac{\omega}{c^2}}\sqrt{\mathrm{Im}\,\varepsilon(\mathbf{r},\omega)} \int d^3r'\, \hat{\mathbf{P}}_A(\mathbf{r}') G^{\perp*}(\mathbf{r}',\mathbf{r},\omega). \tag{2.228}$$

Using Eqs (2.226)–(2.228) together with Eqs (2.218)–(2.220) and applying the relations (A.3) and (B.8), we express in Eq. (2.222) the old variables $\hat{\mathbf{r}}_a, \hat{\mathbf{p}}_a, \hat{\mathbf{f}}(\mathbf{r},\omega), \hat{\mathbf{f}}^\dagger(\mathbf{r},\omega)$ in terms of the new variables $\hat{\mathbf{r}}'_a = \hat{\mathbf{r}}_a, \hat{\mathbf{p}}'_a, \hat{\mathbf{f}}'(\mathbf{r},\omega), \hat{\mathbf{f}}'^\dagger(\mathbf{r},\omega)$. After some algebra we derive[29]

$$\hat{H} = \int d^3r \int_0^\infty d\omega\, \hbar\omega\, \hat{\mathbf{f}}'^\dagger(\mathbf{r},\omega)\hat{\mathbf{f}}'(\mathbf{r},\omega)$$
$$+ \sum_a \frac{1}{2m_a}\left\{\hat{\mathbf{p}}'_a + Q_a \int_0^1 ds\, s(\hat{\mathbf{r}}_a - \mathbf{r}_A) \times [\nabla \times \mathbf{A}'[\mathbf{r}_A + s(\hat{\mathbf{r}}_a - \mathbf{r}_A)]]\right\}^2$$
$$+ \frac{1}{\varepsilon_0}\int d^3r\, \hat{\mathbf{P}}_A(\mathbf{r})\hat{\mathbf{\Pi}}'(\mathbf{r}) + \frac{1}{2\varepsilon_0}\int d^3r\, \hat{\mathbf{P}}_A^{\perp 2}(\mathbf{r}) + \hat{W}'_{\mathrm{Coul}}. \tag{2.229}$$

Note that $\hat{\mathbf{A}}' = \hat{\mathbf{A}}$ [Eq. (2.156)],[30] $\hat{W}'_{\mathrm{Coul}} = \hat{W}_{\mathrm{Coul}}$ ($\hat{V}'_M = \hat{V}_M$, $\hat{\mathbf{P}}'_M = \hat{\mathbf{P}}_M$) and $\hat{\mathbf{\Pi}}' = \hat{\mathbf{\Pi}} - \hat{\mathbf{P}}_A^\perp$ [Eq. (2.158)]. In particular for a neutral atom, we may write

---

[29] Here the position $\mathbf{r}_A$ of the (neutral) atomic system is again assumed to be a given parameter. For an extension of the multipolar-coupling scheme to moving systems where $\mathbf{r}_A$ is also a dynamical variable ($\mathbf{r}_A \mapsto \hat{\mathbf{r}}_A$), which plays the role of the center-of-mass coordinate, see Buhmann, Knöll, Welsch and Ho (2004).

[30] The notation $\hat{\mathbf{A}}'$ is used to indicate that the vector potential must be thought of as being expressed in terms of $\hat{\mathbf{f}}'$ and $\hat{\mathbf{f}}'^\dagger$.

$$\frac{1}{2\varepsilon_0}\int d^3r\,\hat{\mathbf{P}}_A^{\prime\perp 2}(\mathbf{r}) + \hat{W}_{\text{Coul}}'$$
$$= \frac{1}{2\varepsilon_0}\int d^3r\,\hat{\mathbf{P}}_A^2(\mathbf{r}) + \frac{1}{\varepsilon_0}\int d^3r\,\hat{\mathbf{P}}_A(\mathbf{r})\hat{\mathbf{P}}_M^{\prime\parallel}(\mathbf{r}) \tag{2.230}$$

[see Eqs (2.223), (2.225), (2.148)].

The first term in Eq. (2.229) is the Hamiltonian of the medium and the (medium-assisted) electromagnetic field. The other terms can be regrouped to obtain the particle Hamiltonian $\hat{H}_{C'}$ defined by (2.149) and the interaction Hamiltonian

$$\hat{H}_{\text{int}'} = \hat{H}_{\text{int}'}^{(1)} + \hat{H}_{\text{int}'}^{(2)}, \tag{2.231}$$

where the term

$$\hat{H}_{\text{int}'}^{(1)} = \frac{1}{\varepsilon_0}\int d^3r\,\hat{\mathbf{P}}_A(\mathbf{r})\hat{\mathbf{P}}_M'(\mathbf{r}) \tag{2.232}$$

obviously corresponds to the sum over the contact terms in Eq. (2.153), and

$$\hat{H}_{\text{int}'}^{(2)} = \frac{1}{\varepsilon_0}\sum_a Q_a \int_0^1 ds\,(\hat{\mathbf{r}}_a - \mathbf{r}_A)\{\hat{\boldsymbol{\Pi}}'[\mathbf{r}_A + s(\hat{\mathbf{r}}_a - \mathbf{r}_A)] - \hat{\mathbf{P}}_M^{\prime\perp}[\mathbf{r}_A + s(\hat{\mathbf{r}}_a - \mathbf{r}_A)]\}$$

$$- \sum_a \frac{Q_a}{2m_a}\int_0^1 ds\,s\{[(\hat{\mathbf{r}}_a - \mathbf{r}_A)\times\hat{\mathbf{p}}_a'][\nabla\times\hat{\mathbf{A}}'[\mathbf{r}_A + s(\hat{\mathbf{r}}_a - \mathbf{r}_A)]] + \text{H.c.}\}$$

$$+ \sum_a \frac{Q_a^2}{2m_a}\left\{\int_0^1 ds\,s(\hat{\mathbf{r}}_a - \mathbf{r}_A)\times[\nabla\times\hat{\mathbf{A}}'[\mathbf{r}_A + s(\hat{\mathbf{r}}_a - \mathbf{r}_A)]]\right\}^2. \tag{2.233}$$

Equation (2.233) differs from Eq. (2.150) in the first term, which now describes the interaction of the polarization $\hat{\mathbf{P}}_A$ of the atomic system with the transformed transverse displacement field

$$\hat{\mathbf{D}}^{\prime\perp}(\mathbf{r}) = [\varepsilon_0\hat{\mathbf{E}}^\perp(\mathbf{r}) + \hat{\mathbf{P}}_M^\perp(\mathbf{r})]'$$
$$= -\hat{\boldsymbol{\Pi}}'(\mathbf{r}) + \hat{\mathbf{P}}_M^\perp(\mathbf{r}) = \varepsilon_0\hat{\mathbf{E}}^\perp(\mathbf{r}) + \hat{\mathbf{P}}_A^\perp(\mathbf{r}) + \hat{\mathbf{P}}_M^\perp(\mathbf{r}). \tag{2.234}$$

Note that all the primed quantities are defined according to the definitions of the unprimed quantities except that therein the old variables $\hat{\mathbf{f}}(\mathbf{r},\omega), \hat{\mathbf{f}}^\dagger(\mathbf{r},\omega)$ must be (formally) replaced by the new variables $\hat{\mathbf{f}}'(\mathbf{r},\omega), \hat{\mathbf{f}}'^\dagger(\mathbf{r},\omega)$.

2.5
Approximate interaction Hamiltonians

As we have seen in Sections 2.3 and 2.4, the terms which contain the interaction between the charged particles and the electromagnetic field look quite different in the minimal-coupling scheme and the multipolar-coupling scheme.

2.5.1
The electric-dipole approximation

In many cases of practical interest, when the interaction of the radiation field with the charged particles can be viewed as a (quasi-)resonant interaction with bound states of atomic systems, the quadratic contribution in the vector potential to the minimal-coupling term (2.121) may, as a good approximation, be ignored. An estimation shows [see, e.g., Schubert and Wilhelmi (1986)] that this approximation may be justified as long as the electric field is weak compared with the intra-atomic electric field ($E_{atom} \approx 10^{10}$ Vm^{-1}), to which the active charges are subjected, owing to their interaction with the atomic cores. Similarly, the nonlinear multipolar-coupling term in Eq. (2.150) may be disregarded [see, e.g., Loudon (1983)].

For bound atomic states the vector potential in the minimal-coupling energy (2.121) is commonly expanded in powers of $\hat{\mathbf{r}}_a - \mathbf{r}_A$.[31] In the electric-dipole approximation, only the zeroth-order term $\hat{\mathbf{A}}(\mathbf{r}_A)$ is retained and Eq. (2.121) simplifies to

$$\hat{H}_{int} = -\sum_a \frac{Q_a}{m_a} \hat{\mathbf{p}}_a \hat{\mathbf{A}}(\mathbf{r}_A) + \sum_a \frac{Q_a^2}{2m_a} \hat{\mathbf{A}}^2(\mathbf{r}_A). \tag{2.235}$$

In particular, when the quadratic term in the vector potential can be omitted (see also the remarks at the end of Section 2.5.2), then $\hat{H}_{int}$ further reduces to

$$\hat{H}_{int} = -\sum_a \frac{Q_a}{m_a} \hat{\mathbf{p}}_a \hat{\mathbf{A}}(\mathbf{r}_A) = \frac{i}{\hbar} [\hat{\mathbf{d}}, \hat{H}_C], \tag{2.236}$$

where $\hat{H}_C$ is given by Eq. (2.120) and

$$\hat{\mathbf{d}} = \sum_a Q_a (\hat{\mathbf{r}}_a - \mathbf{r}_A) \tag{2.237}$$

is the electric-dipole operator, which does not depend on $\mathbf{r}_A$ for a globally neutral atomic system. Recall that for bound atomic states the probabilities of finding the charges at places outside the atomic volume are effectively zero.

---

[31] Here the atomic position $\mathbf{r}_A$ is regarded as being a classical parameter.

Moreover, in the case of optical radiation the vector potential may be regarded as being slowly varying within the atomic volume. Clearly, if (owing to symmetry properties) certain dipole transitions are forbidden, gradients (evaluated at $\mathbf{r}_A$) of the vector potential must be taken into account, which are associated with magnetic-dipole transitions and higher-order (multipole) transitions.

In the multipolar-coupling scheme, the electric-dipole approximation is commonly understood as the approximation in which, in the multipolar-coupling energy (2.150), the magnetic-field terms, i.e., the terms containing $\nabla \times \hat{\mathbf{A}}[\mathbf{r}_A + s(\hat{\mathbf{r}}_a - \mathbf{r}_A)]$, are omitted and the canonical momentum field $\hat{\mathbf{\Pi}}'[\mathbf{r}_A + s(\hat{\mathbf{r}}_a - \mathbf{r}_A)]$ is taken at the atomic position $\mathbf{r}_A$,

$$\hat{H}_{\text{int}'} = \varepsilon_0^{-1} \hat{\mathbf{d}} \hat{\mathbf{\Pi}}'(\mathbf{r}_A) = -\hat{\mathbf{d}} \hat{\mathbf{E}}'^{\perp}(\mathbf{r}_A). \tag{2.238}$$

The generalization of Eqs (2.235), (2.236) and (2.238) to the case where the atomic system interacts with a medium-assisted electromagnetic field is straightforward. In particular, it is not difficult to see that the first term on the right-hand side in Eq. (2.233) and $\hat{H}_{\text{int}'}^{(1)}$ given by Eq. (2.232) can be combined to obtain

$$\hat{H}_{\text{int}'} = \varepsilon_0^{-1} \hat{\mathbf{d}} [\hat{\mathbf{\Pi}}'(\mathbf{r}_A) + \hat{\mathbf{P}}_M'^{\|}(\mathbf{r}_A)] = -\hat{\mathbf{d}} \hat{\mathbf{E}}'(\mathbf{r}_A), \tag{2.239}$$

where $\hat{\mathbf{E}}'(\mathbf{r}_A)$ is defined according to Eq. (2.213) together with Eq. (2.211) $[\hat{\mathbf{f}}'(\mathbf{r}, \omega) = \hat{U} \hat{\mathbf{f}}(\mathbf{r}, \omega) \hat{U}^\dagger$, cf. Eq. (2.228)].

If $\mathbf{r}_A$ is treated as a (quantum mechanical) dynamical variable, $(\mathbf{r}_A \mapsto \hat{\mathbf{r}}_A)$ the question may arise whether Eqs (2.235) and (2.238) [or (2.239)] (with $\hat{\mathbf{r}}_A$ in place of $\mathbf{r}_A$) are properly chosen interaction energies. Although both the minimal-coupling interaction energy (2.235) and the multipolar-coupling interaction energy (2.238) [or (2.239)] are referred to as interaction energies in the electric-dipole approximation, an essential difference between them becomes apparent. In contrast to Eq. (2.238) [or (2.239)], which leads to the correct electric part of the Lorentz force acting on an electric dipole, Eq. (2.235) fails.

Let $|m\rangle$ ($m=1,2,3,...$) and $\hbar\omega_m$, respectively, denote the eigenkets and eigenvalues of the respective atomic Hamiltonian, i.e., $\hat{H}_C|m\rangle = \hbar\omega_m|m\rangle$ in the minimal-coupling scheme and $\hat{H}_{C'}|m\rangle = \hbar\omega_m|m\rangle$ in the multipolar-coupling scheme. Introducing the atomic flip operators $\hat{A}_{mn} = |m\rangle\langle n|$, and using the orthogonality of the atomic states, we arrive at the commutation relation

$$[\hat{A}_{mn}, \hat{A}_{m'n'}] = \delta_{nm'} \hat{A}_{mn'} - \delta_{mn'} \hat{A}_{m'n}. \tag{2.240}$$

It is easily seen that $\hat{H}_{\text{int}}$ as given by Eq. (2.236) can be expressed in terms of the operators $\hat{A}_{mm}$ as

$$\hat{H}_{\text{int}} = -i\sum_{m,n}\omega_{mn}\hat{A}_{mn}\mathbf{d}_{mn}\hat{\mathbf{A}}(\mathbf{r}_A)$$
$$= -i\sum_{m,n}\sum_{\lambda}\omega_{mn}\mathbf{d}_{mn}\mathbf{A}_\lambda(\mathbf{r}_A)\hat{a}_\lambda\hat{A}_{mn} + \text{H.c.} \qquad (2.241)$$

($\omega_{mn} = \omega_m - \omega_n$, $\mathbf{d}_{mn} = \langle m|\hat{\mathbf{d}}|n\rangle$), where the second line is obtained from the first one by mode expansion of the vector potential according to Eq. (2.69). Introduction of the atomic flip operators and, according to Eq. (2.70), the mode expansion of the canonical momentum field converts Eq. (2.238) to

$$\hat{H}_{\text{int}'} = \varepsilon_0^{-1}\sum_{m,n}\hat{A}_{mn}\mathbf{d}_{mn}\hat{\boldsymbol{\Pi}}'(\mathbf{r}_A)$$
$$= -i\sum_{m,n}\sum_{\lambda}\omega_\lambda\mathbf{d}_{mn}\mathbf{A}_\lambda(\mathbf{r}_A)\hat{a}'_\lambda\hat{A}_{mn} + \text{H.c.} \qquad (2.242)$$

[$\hat{a}'_\lambda = \hat{U}\hat{a}_\lambda\hat{U}^\dagger$, cf. Eq. (2.156)]. Alternatively, in Eq. (2.239), expressing $\hat{\mathbf{E}}'(\mathbf{r}_A)$ in terms of the dynamical variables $\hat{\mathbf{f}}'(\mathbf{r},\omega)$ and $\hat{\mathbf{f}}'^\dagger(\mathbf{r},\omega)$ according to Eq. (2.213), together with Eq. (2.211), yields

$$\hat{H}_{\text{int}'} = -i\sqrt{\frac{\hbar}{\pi\varepsilon_0}}\sum_{m,n}\int_0^\infty d\omega\,\frac{\omega^2}{c^2}\int d^3r\,\mathbf{d}_{mn}\mathbf{G}(\mathbf{r}_A,\mathbf{r},\omega)\hat{\mathbf{f}}'(\mathbf{r},\omega)\hat{A}_{mn} + \text{H.c..}$$
$$(2.243)$$

Comparing Eqs (2.241) and (2.242), we see that the formal structure of $\hat{H}_{\text{int}'}$ is similar to that of $\hat{H}_{\text{int}}$ except that the atomic transition frequencies are replaced by the mode frequencies. However, it should be stressed that both the atomic flip operators (and the associated dipole matrix elements) and the photon annihilation and creation operators in the minimal-coupling scheme and the multipolar-coupling scheme are not the same. Since the atomic Hamiltonians in the two schemes are different [cf. Eqs (2.120) and (2.149)], the atomic flip operators (and the associated dipole matrix elements) refer to different atomic states in general. Accordingly, the photon number states [as the eigenstates of the photon number operators, cf. Section 3.1] are different in the two schemes, because the photonic operators in the two schemes are related to each other by the unitary transformation (2.155).

2.5.2
The rotating-wave approximation

In general, the light–matter interaction gives rise to a complicated (multi-level and multi-mode) set of coupled Heisenberg equations of motion for the atomic and photonic operators. However, in most practical cases the interaction processes are slow compared with the free (optical) oscillation of the radiation field. This allows (on the basis of the unperturbed solution) a classification of the various couplings with regard to efficient, resonant couplings and less efficient, off-resonant ones. Neglect of the latter is usually called the rotating-wave approximation.

Let us consider the multipolar-coupling Hamiltonian in electric-dipole approximation as given by Eq. (2.144) together with Eq. (2.242). If the radiation–matter interaction were absent, the operator products $\hat{A}_{mn}\hat{a}'_\lambda$ would vary as $\exp[i(\omega_{mn}-\omega_\lambda)t]$, which is approximately unity for resonant couplings satisfying the condition $\omega_\lambda \approx \omega_{mn}$. For a given mode of the radiation field the remaining operator products $\hat{A}_{m'n'}\hat{a}'_\lambda$ then vary with respect to these terms as $\exp[i(\omega_{m'n'}-\omega_{mn})t]$ $(\omega_{m'n'}\neq\omega_{mn})$. Let us suppose that the times $|\omega_{m'n'}-\omega_{mn}|^{-1}$ are sufficiently small compared with the characteristic time τ_{int} of the resonant radiation–matter interaction describing the dynamics of the system, $|\omega_{m'n'}-\omega_{mn}|^{-1}\ll\tau_{\text{int}}$, and let us further confine ourselves to a resolving-time scale $\delta\tau$ with $|\omega_{m'n'}-\omega_{mn}|^{-1}\ll\delta\tau\ll\tau_{\text{int}}$. On this time scale the exponentials $\exp[i(\omega_{m'n'}-\omega_{mn})t]$ are rapidly varying and average approximately to zero, so that we may neglect the terms proportional to $\hat{A}_{m'n'}\hat{a}'_\lambda$ in the Hamiltonian. Physically this means that resonant one-photon processes dominate the interaction between each relevant radiation field mode and the atomic system.

When the radiation field under study is near resonant with certain atomic transitions, then, in the rotating-wave approximation, the sum in Eq. (2.242) can be truncated by the requirement that $\omega_{mn}\approx\omega_\lambda$, which indicates that terms with $|\omega_{mn}-\omega_\lambda|\delta\tau\gg 1$ are disregarded. Since in this approximation the radiation–matter interaction cannot be treated with a time resolution better than $\delta\tau$, one may refer to this approximation as a kind of coarse-grained averaging. In the simplest case when the radiation field may be viewed as being near resonant with only a single atomic transition, e. g., $|1\rangle\leftrightarrow|2\rangle$ with $\omega_{21}\approx\omega_\lambda$ (see Fig. 2.3), the multi-level atomic system effectively reduces to a two-level system coupled to the radiation field and Eq. (2.242) approximates to[32]

$$\hat{H}_{\text{int}'} = -\mathbf{d}_{21}\hat{\mathbf{E}}'^{\perp(+)}(\mathbf{r}_A)\hat{A}_{12} + \text{H.c.}. \qquad (2.244)$$

Similarly, the minimal-coupling energy (2.241) approximately simplifies to

$$\hat{H}_{\text{int}} = -i\omega_{21}\mathbf{d}_{21}\hat{\mathbf{A}}^{(+)}(\mathbf{r}_A)\hat{A}_{12} + \text{H.c.}. \qquad (2.245)$$

[32] Here we have used the convention of decomposition of an operator $\hat{\mathbf{F}} = \hat{\mathbf{F}}^{(+)} + \hat{\mathbf{F}}^{(-)}$ into a positive-frequency part $\hat{\mathbf{F}}^{(+)}$ and a negative-frequency part $\hat{\mathbf{F}}^{(-)}$. When $\hat{\mathbf{F}}$ is a field operator whose mode decomposition is $\hat{\mathbf{F}} = \sum_\lambda \mathbf{F}_\lambda \hat{a}_\lambda + \text{H.c.}$, then $\hat{\mathbf{F}}^{(+)}$ is commonly identified with $\sum_\lambda \mathbf{F}_\lambda \hat{a}_\lambda$, and $\hat{\mathbf{F}}^{(-)} = (\hat{\mathbf{F}}^{(+)})^\dagger$.

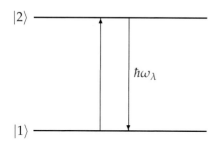

Fig. 2.3 Scheme of resonant one-photon absorption and emission. The atomic energy-level spacing $E_2 - E_1 = \hbar\omega_{21}$, and the energy of photons of mode λ satisfy the resonance condition $\omega_{21} \approx \omega_\lambda$.

Since the approximation implies that

$$i\omega_{21}\hat{\mathbf{A}}^{(+)}(\mathbf{r}) \simeq -\dot{\hat{\mathbf{A}}}^{(+)}(\mathbf{r}) \simeq \hat{\mathbf{E}}^{\perp(+)}(\mathbf{r}), \qquad (2.246)$$

we may rewrite Eq. (2.245) as

$$\hat{H}_{\text{int}} = -\mathbf{d}_{21}\hat{\mathbf{E}}^{\perp(+)}(\mathbf{r}_A)\hat{A}_{12} + \text{H.c.}, \qquad (2.247)$$

which reveals that in the rotating-wave approximation the minimal-coupling energy is effectively the same as the multipolar-coupling energy.[33] It should be pointed out that this is not necessarily true when off-resonant (virtual) transitions must be taken into account. This is typically the case when, with respect to the relevant radiation field modes, one-photon resonances are missing. To obtain effectively equivalent results in such a case, the $\hat{\mathbf{A}}^2$ term in the basic formula (2.235) for the minimal-coupling energy in electric-dipole approximation must not be omitted in general.[34]

33) Recall that when the atom interacts with a medium-assisted electromagnetic field and Eq. (2.244) [Eq. (2.247)] is used, then $\hat{\mathbf{E}}'^{\perp(\pm)}(\mathbf{r}_A)$ [$\hat{\mathbf{E}}^{\perp(\pm)}(\mathbf{r}_A)$] must be expressed in terms of $\hat{\mathbf{f}}'(\mathbf{r},\omega)$ [$\hat{\mathbf{f}}(\mathbf{r},\omega)$] and $\hat{\mathbf{f}}'^\dagger(\mathbf{r},\omega)$ [$\hat{\mathbf{f}}^\dagger(\mathbf{r},\omega)$] according to Eq. (2.213) together with Eq. (2.211).
34) Typical examples are the intermolecular energy transfer [see Ho, Knöll and Welsch (2002)] and the van der Waals potential [see Buhmann, Knöll, Welsch and Ho (2004)].

2.5.3
Effective Hamiltonians

If, for a given radiation field, one-photon resonances are missing, a large manifold of off-resonant terms in the m and n sums in, e. g., Eq. (2.242) must be taken into account, and a rotating-wave approximation in the simple form introduced above makes no sense. The efficiency of radiation–matter interaction in the domain where one-photon resonances are absent is usually small, except that strongly driven multi-photon resonances are allowed. In these cases the concept of so-called effective interaction Hamiltonians is widely used to simplify the problem by effectively reducing the number of coupled equations of motion to be solved. In particular, this concept is often applied in nonlinear optics.[35] Typical nonlinear optical processes are multi-photon absorption and emission, optical bistability, multi-wave mixing, such as the generation of higher-order harmonics of radiation field modes and related effects of frequency mixing, phase conjugation, etc.

To illustrate the concept, let us study the simplest case of near resonant two-photon absorption and emission (Fig. 2.4). For this purpose we consider the interaction of an atomic system with two parts of the radiation field, with the corresponding center frequencies Ω_1 and Ω_2 ($\Omega_2 \neq \Omega_1$) and the frequency of a certain atomic transition ω_{21} satisfying the two-photon resonance condition $\Omega_1 + \Omega_2 \approx \omega_{21}$. The frequencies Ω_1 and Ω_2 are assumed to be far from any atomic transition frequency ω_{lm}. According to Eq. (2.242), the relevant part of the interaction Hamiltonian may be written in the form[36]

$$\hat{H}_{\text{int}} = \hbar \sum_{N=1}^{2} \sum_{\lambda_N} \sum_{l,m} V_{lm\lambda_N} \hat{A}_{lm} \hat{a}_{\lambda_N} + \text{H.c.}, \tag{2.248}$$

where

$$V_{lm\lambda_N} = (i\hbar)^{-1} \mathbf{d}_{lm} \mathbf{A}_{\lambda_N}(\mathbf{r}_A) \omega_{\lambda_N}, \tag{2.249}$$

and the λ_1 and λ_2 sums run over the radiation field modes with $\omega_{\lambda_1} \approx \Omega_1$ and $\omega_{\lambda_2} \approx \Omega_2$, respectively. The Heisenberg equations of motion for the photonic operators $\hat{a}^{\dagger}_{\lambda_N}$ and the atomic operator $\hat{A}_{lm}$ are

$$\dot{\hat{a}}^{\dagger}_{\lambda_N} = (i\hbar)^{-1}[\hat{a}^{\dagger}_{\lambda_N}, \hat{H}_{\text{R}} + \hat{H}_{\text{int}}] = i\omega_{\lambda_N} \hat{a}^{\dagger}_{\lambda_N} + i \sum_{l,m} V_{lm\lambda_N} \hat{A}_{lm}, \tag{2.250}$$

$$\dot{\hat{A}}_{lm} = (i\hbar)^{-1}[\hat{A}_{lm}, \hat{H}_{\text{C}} + \hat{H}_{\text{int}}] = i\omega_{lm} \hat{A}_{lm} - i\hat{G}_{lm}, \tag{2.251}$$

---

[35] Nonlinear optics is considered, for example, in the books of Bloembergen (1965), Levenson and Kano (1988), Peřina (1991), Schubert and Wilhelmi (1986), Shen (1984), Shore (1990) and Stenholm (1984).

[36] Unless it should be explicitly distinguished between multipolar coupling and minimal coupling, the prime used to indicate multipolar-coupling quantities will be omitted for notational convenience.

2.5 Approximate interaction Hamiltonians

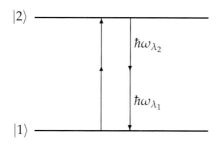

Fig. 2.4 Scheme of resonant two-photon absorption and emission. The atomic energy level spacing $E_2 - E_1 = \hbar\omega_{21}$ and the energies $\hbar\omega_{\lambda_1}$ and $\hbar\omega_{\lambda_2}$ of photons of modes λ_1 and λ_2, respectively, satisfy the two-photon resonance condition $\omega_{21} \approx \omega_{\lambda_1} + \omega_{\lambda_2}$.

where
$$\hat{G}_{lm} = \hbar^{-1}[\hat{A}_{lm}, \hat{H}_{int'}]$$
$$= \sum_{N=1}^{2}\sum_{\lambda_N}\sum_{k}\{[V_{mk\lambda_N}\hat{A}_{lk} - V_{kl\lambda_N}\hat{A}_{km}]\hat{a}_{\lambda_N} + [V^*_{km\lambda_N}\hat{A}_{lk} - V^*_{lk\lambda_N}\hat{A}_{km}]\hat{a}^\dagger_{\lambda_N}\}. \quad (2.252)$$

Formal solution of Eq. (2.251) yields
$$\hat{A}_{lm}(t) = e^{i\omega_{lm}(t-t')}\hat{A}_{lm}(t') - i\int_0^{t-t'} d\tau\, e^{i\omega_{lm}\tau}\hat{G}_{lm}(t-\tau). \quad (2.253)$$

We now substitute into Eq. (2.250) for $\hat{A}_{lm}$, the result of Eq. (2.253) together with Eq. (2.252) to obtain
$$\dot{\hat{a}}^\dagger_{\lambda_N} = i\omega_{\lambda_N}\hat{a}^\dagger_{\lambda_N} + i\sum_{l,m}V_{lm\lambda_N}e^{i\omega_{lm}(t-t')}\hat{A}_{lm}(t')$$
$$+ \sum_{M=1}^{2}\sum_{\lambda_N}\sum_{l,m,k}\int_0^{t-t'}d\tau\, e^{i\omega_{lm}\tau}V_{lm\lambda_N}\{[V_{mk\lambda_M}\hat{A}_{lk}(t-\tau)$$
$$- V_{kl\lambda_M}\hat{A}_{km}(t-\tau)]\hat{a}_{\lambda_M}(t-\tau) + [V^*_{km\lambda_M}\hat{A}_{lk}(t-\tau)$$
$$- V^*_{lk\lambda_M}\hat{A}_{km}(t-\tau)]\hat{a}^\dagger_{\lambda_M}(t-\tau)\}. \quad (2.254)$$

At this stage we apply the rotating-wave approximation. To pick out the relevant terms on the right-hand side of Eq. (2.254), we first remove the free motion from the atomic and photonic operators:

$$\hat{a}_{\lambda_N}(t) = e^{-i\omega_{\lambda_N} t}\hat{\tilde{a}}_{\lambda_N}(t), \quad \hat{A}_{lm}(t) = e^{i\omega_{lm}t}\hat{\tilde{A}}_{lm}(t). \tag{2.255}$$

Further, we restrict attention to a time scale during which, in comparison with the change of the system due to the two-photon light–matter interaction, the (free-motion) off-resonant exponentials are rapidly varying. Since on this time scale the rapidly-varying exponentials may be assumed to average approximately to zero, the terms associated with them may be disregarded. Using this argument, from careful inspection of Eq. (2.254), we see that for $N=1$ ($N=2$) the relevant two-photon coupling arises from the $\hat{A}_{lk}\hat{a}_{\lambda_M}$ and $\hat{A}_{km}\hat{a}_{\lambda_M}$ terms with $l=2, k=1$ and $m=1, k=2$ respectively, and $M=2$ ($M=1$). Thus we may approximately write ($n=1,2$)

$$\dot{\hat{a}}^\dagger_{\lambda_N} = i\omega_{\lambda_N}\hat{a}^\dagger_{\lambda_N} + \sum_{\lambda_M}\sum_m \int_0^{t-t'} d\tau \Big[V_{2m\lambda_N}V_{m1\lambda_M}e^{i(\omega_{1m}+\omega_{\lambda_M})\tau}$$
$$- V_{2m\lambda_M}V_{m1\lambda_N}e^{-i(\omega_{1m}+\omega_{\lambda_N})\tau}\Big]\hat{\tilde{A}}_{21}(t-\tau)\hat{\tilde{a}}_{\lambda_M}(t-\tau)e^{i(\omega_{21}-\omega_{\lambda_M})t}, \tag{2.256}$$

where $M=2$ if $N=1$ and vice versa; recall that $\omega_{21}\approx\omega_{\lambda_1}+\omega_{\lambda_2}$. Under the assumption made, in the τ integral in Eq. (2.256) we may now take the slowly-varying operators $\hat{\tilde{A}}_{21}(t-\tau)$ and $\hat{\tilde{a}}_{\lambda_M}(t-\tau)$ at $\tau=0$, and, after performing the τ integration, we may omit the rapidly-varying terms arising from the upper limit of integration. Using Eq. (2.255), we finally arrive at the following effective equations of motion describing the dynamics of resonant coupling of the two kinds of light modes to each other, via excitation of the atomic transition $|1\rangle\leftrightarrow|2\rangle$:

$$\dot{\hat{a}}^\dagger_{\lambda_1} = i\omega_{\lambda_1}\hat{a}^\dagger_{\lambda_1} + i\sum_{\lambda_2}\kappa_{21}^{\lambda_1\lambda_2}\hat{A}_{21}\hat{a}_{\lambda_2}, \tag{2.257}$$

$$\dot{\hat{a}}^\dagger_{\lambda_2} = i\omega_{\lambda_2}\hat{a}^\dagger_{\lambda_2} + i\sum_{\lambda_1}\kappa_{21}^{\lambda_2\lambda_1}\hat{A}_{21}\hat{a}_{\lambda_1}, \tag{2.258}$$

where the effective coupling parameter is

$$\kappa_{21}^{\lambda_1\lambda_2} = \kappa_{21}^{\lambda_2\lambda_1} = \sum_m \left(\frac{V_{2m\lambda_1}V_{m1\lambda_2}}{\omega_{1m}+\omega_{\lambda_2}} + \frac{V_{2m\lambda_2}V_{m1\lambda_1}}{\omega_{1m}+\omega_{\lambda_1}}\right). \tag{2.259}$$

It is easily seen that the equations of motion (2.257) and (2.258) can be obtained by starting from an effective interaction Hamiltonian as follows:[36]

$$\dot{\hat{a}}^\dagger_{\lambda_N} = (i\hbar)^{-1}[\hat{a}^\dagger_{\lambda_N}, \hat{H}_R + \hat{H}_{\text{int(eff)}}], \qquad (2.260)$$

where

$$\hat{H}_{\text{int(eff)}} = \hbar \sum_{\lambda_1,\lambda_2} \kappa_{21}^{\lambda_1\lambda_2} \hat{A}_{21} \hat{a}_{\lambda_1} \hat{a}_{\lambda_2} + \text{H.c.} \qquad (\omega_{21} \approx \omega_{\lambda_1} + \omega_{\lambda_2}). \qquad (2.261)$$

Clearly, the (effective) equations of motions for the photonic operators must be complemented by those for the atomic operators $\hat{A}_{21}(=\hat{A}^\dagger_{12})$, $\hat{A}_{11}$ and $\hat{A}_{22}$. Since the manipulations are quite similar to those leading to the photonic equations of motion, we omit the derivation here. A straightforward calculation shows that

$$\dot{\hat{A}}_{lm} = (i\hbar)^{-1}[\hat{A}_{lm}, \hat{H}_C + \hat{H}_{\text{int(eff)}}] \qquad (l = 1, 2, \ m = 1, 2). \qquad (2.262)$$

A generalization of the concept of effective interaction Hamiltonians to other (higher than second-order) resonant multi-photon interaction processes is straightforward. The iteration procedure must be repeated as long as the higher-order resonances sought appear explicitly in the equations of motion. Let us consider the three-photon resonant process shown schematically in Fig. 2.5. An example of this type of process is the optical parametric oscillator used for converting laser light with frequency $\omega_L (\approx \omega_{\lambda_3})$ into signal and idler light with frequencies $\omega_S (\approx \omega_{\lambda_1})$ and $\omega_I (\approx \omega_{\lambda_2})$ respectively. Compared with the process of two-photon absorption/emission, the state of the atom is not changed during this process. Performing the procedure outlined above yields an effective interaction Hamiltonian of the form

$$\hat{H}_{\text{int(eff)}} = \sum_n \hat{V}_n \hat{A}_{nn}, \qquad (2.263)$$

where

$$\hat{V}_n = \hbar \sum_{\lambda_1,\lambda_2,\lambda_3} \kappa_n^{\lambda_1\lambda_2\lambda_3} \hat{a}_{\lambda_1} \hat{a}_{\lambda_2} \hat{a}^\dagger_{\lambda_3} + \text{H.c..} \qquad (2.264)$$

From the total Hamiltonian $\hat{H} = \hat{H}_R + \hat{H}_C + \hat{H}_{\text{int(eff)}}$ we easily derive that $\dot{\hat{A}}_{nn} = 0$, which allows one to eliminate the atomic variables in the equation

---

[36] The use of the effective Hamiltonian drastically simplifies the problem of the two-photon resonance under study, since the multi-level atomic system is effectively reduced to a two-level system coupled to the radiation field through the effective interaction Hamiltonian, the multi-level structure of the atomic system being incorporated in the new, effective coupling parameter (2.259). It is, in general, small compared with the one-photon coupling, which is due to the quadratic dependence on the one-photon coupling parameter as well as the off-resonance denominators.

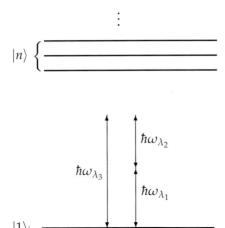

Fig. 2.5 Scheme of resonant three-photon interaction with the atom in the ground state. The photon energies $\hbar\omega_{\lambda_i}$ ($i = 1, 2, 3$) satisfy the resonance condition $\omega_{\lambda_3} \approx \omega_{\lambda_1} + \omega_{\lambda_2}$.

of motion for the photonic ones. In particular, when the atom is initially in its ground state it remains there for all time, and the temporal evolution of any photonic operator is then governed by the pure photonic Hamiltonian $\hat{H}_R + \hat{V}_1$. That is, the problem of parametric amplification is reduced to a purely photonic one.

2.6
Source-quantity representation of the electromagnetic field

As is well known, in the Heisenberg picture the equations of motion of the operators are closely related to the equations of motion of the corresponding classical quantities. The interaction of the electromagnetic field with charged particles gives rise to a coupled set of nonlinear equations of motion, which is, in general, hard to solve even in classical optics. In quantum optics the operator character, together with the corresponding commutation relations are responsible for typical quantum effects without classical counterparts and introduce additional difficulties. In this context, the question arises of how to transfer the standard rules of classical optics to quantum optics, for example, those concerning light detection and light propagation through various kinds of optical instruments such as resonator-like cavities, interferometers, beam splitters, etc. To deal with these and related problems, it may be useful to start from a radiation-field representation based on the formal solution of the (coupled radiation–matter) Heisenberg equations of motion (source-quantity

2.6 Source-quantity representation

representation). From Section 2.4 we know that such an approach enables us to include the effect of linear, causal background media in a consistent way in the theory. With respect to nonlinear atom–field interactions, source-quantity representations are of course not closed solutions, but they may be helpful in order to gain insight into the classical and quantum aspects of the problems under consideration.

Let us consider an ensemble of atomic systems and use the multipolar-coupling scheme to describe its interaction with the electromagnetic field in the electric-dipole approximation. We begin with the case of media being absent. The interaction energy is then simply the sum of the interaction energies (2.238), i.e.,

$$\hat{H}_{\text{int}} = \varepsilon_0^{-1} \int d^3r\, \hat{\tilde{P}}(\mathbf{r}) \hat{\Pi}(\mathbf{r}) = -\int d^3r\, \hat{\tilde{P}}(\mathbf{r}) \hat{E}^{\perp}(\mathbf{r}), \tag{2.265}$$

where[38]

$$\hat{\tilde{P}}(\mathbf{r}) = \sum_A \hat{\mathbf{d}}_A \delta(\mathbf{r} - \mathbf{r}_A) \tag{2.266}$$

($\hat{\mathbf{d}}_A$, electric-dipole operator of the Ath atomic system).[39] Since the concept of mode expansion applies, $\hat{\Pi}$ can be represented in the form of Eq. (2.70). It is not difficult to derive the Heisenberg equations of motion for the photonic operators attributed to the modes,

$$\dot{\hat{a}}_\lambda = \frac{1}{i\hbar}[\hat{a}_\lambda, \hat{H}] = -i\omega_\lambda \hat{a}_\lambda + \frac{1}{\hbar}\int d^3r'\, \omega_\lambda \mathbf{A}_\lambda^*(\mathbf{r}') \hat{\tilde{P}}(\mathbf{r}'). \tag{2.267}$$

The general, retarded solution of Eq. (2.267) is

$$\hat{a}_\lambda(t) = \hat{a}_{\lambda\,\text{free}}(t) + \hat{a}_{\lambda\,\text{s}}(t), \tag{2.268}$$

where

$$\hat{a}_{\lambda\,\text{free}}(t) = \hat{a}_{\lambda\,\text{free}}(t') e^{-i\omega_\lambda(t-t')}, \tag{2.269}$$

$$\hat{a}_{\lambda\,\text{s}}(t) = \frac{1}{\hbar}\int dt' \int d^3r'\, \Theta(t-t') e^{-i\omega_\lambda(t-t')} \omega_\lambda \mathbf{A}_\lambda^*(\mathbf{r}') \hat{\tilde{P}}(\mathbf{r}',t'), \tag{2.270}$$

$\Theta(t)$ being the unit step function. Note that $\hat{\tilde{P}}(\mathbf{r},t)$ also obeys a Heisenberg equation of motion, which is of course coupled to the photonic ones. The determination of the time evolution of the operator $\hat{\tilde{P}}$ therefore requires the solution of a system of coupled equations of motion. Let

$$\hat{\mathbf{F}}(\mathbf{r}) = \hat{\mathbf{F}}^{(+)}(\mathbf{r}) + \hat{\mathbf{F}}^{(-)}(\mathbf{r}), \quad \hat{\mathbf{F}}^{(+)}(\mathbf{r}) = \sum_\lambda \mathbf{F}_\lambda(\mathbf{r}) \hat{a}_\lambda, \quad \hat{\mathbf{F}}^{(-)} = \left(\hat{\mathbf{F}}^{(+)}\right)^\dagger \tag{2.271}$$

---

[38] Here the notation $\hat{\tilde{P}}(\mathbf{r})$ is used to indicate that $\hat{\tilde{P}}(\mathbf{r})$ is only the dipole contribution to $\hat{\mathbf{P}}(\mathbf{r})$.

[39] Recall that in the multipolar-coupling scheme $\hat{E}^\perp \equiv \hat{E}'^\perp$ is not necessarily the electric field strength.

be an appropriately chosen electromagnetic-field operator. Using Eqs (2.268) and (2.270), we may decompose $\hat{\mathbf{F}}^{(j)}$ ($j=\pm$) into a free-field and a source-field operator:

$$\hat{\mathbf{F}}^{(j)}(\mathbf{r},t) = \hat{\mathbf{F}}^{(j)}_{\text{free}}(\mathbf{r},t) + \hat{\mathbf{F}}^{(j)}_{s}(\mathbf{r},t), \tag{2.272}$$

where

$$\hat{\mathbf{F}}^{(+)}_{\text{free}}(\mathbf{r},t) = \sum_\lambda \mathbf{F}_\lambda(\mathbf{r})\, \hat{a}_{\lambda\,\text{free}}(t), \quad \hat{\mathbf{F}}^{(-)}_{\text{free}}(\mathbf{r},t) = [\hat{\mathbf{F}}^{(+)}_{\text{free}}(\mathbf{r},t)]^\dagger, \tag{2.273}$$

$$\hat{\mathbf{F}}^{(j)}_s(\mathbf{r},t) = \int dt' \int d^3r'\, \Theta(t-t') \mathbf{K}^{(j)}_{(F)}(\mathbf{r},t;\mathbf{r}',t')\hat{\vec{P}}(\mathbf{r}',t'). \tag{2.274}$$

Here the tensor-valued kernel function (propagation function) $\mathbf{K}^{(+)}_{(F)}(\mathbf{r},t;\mathbf{r}',t')$ is

$$\mathbf{K}^{(+)}_{(F)}(\mathbf{r},t;\mathbf{r}',t') = \frac{1}{\hbar}\sum_\lambda \omega_\lambda \mathbf{F}_\lambda(\mathbf{r}) \otimes \mathbf{A}^*_\lambda(\mathbf{r}')e^{-i\omega_\lambda(t-t')}, \tag{2.275}$$

and $\mathbf{K}^{(-)}_{(F)} = (\mathbf{K}^{(+)}_{(F)})^*$. Inserting Eq. (2.274) into (2.272) yields

$$\hat{\mathbf{F}}^{(j)}(\mathbf{r},t) = \int dt' \int d^3r'\, \Theta(t-t') \mathbf{K}^{(j)}_{(F)}(\mathbf{r},t;\mathbf{r}',t')\hat{\vec{P}}(\mathbf{r}',t') + \hat{\mathbf{F}}^{(j)}_{\text{free}}(\mathbf{r},t). \tag{2.276}$$

If we identify $\hat{\mathbf{F}}$ with the operator of the vector potential $\hat{\mathbf{A}}$, Eq. (2.276) ($\mathbf{F}_\lambda \mapsto \mathbf{A}_\lambda$) is the source-quantity representation of $\hat{\mathbf{A}}^{(j)}$. Analogously, if we are interested in the canonical momentum field $\hat{\mathbf{\Pi}}$, application of Eq. (2.276) ($\mathbf{F}_\lambda \mapsto -i\omega_\lambda \varepsilon_0 \mathbf{A}_\lambda$) yields the source-quantity representation of $\hat{\mathbf{\Pi}}^{(j)}$. It is not difficult to verify that the two kernel functions are related to each other as

$$-\varepsilon_0^{-1}\mathbf{K}^{(+)}_{(\Pi)}(\mathbf{r},t;\mathbf{r}',t') = -\frac{\partial}{\partial t}\mathbf{K}^{(+)}_{(A)}(\mathbf{r},t;\mathbf{r}',t')$$

$$= \frac{i}{\hbar}\sum_\lambda \omega_\lambda^2 \mathbf{A}_\lambda(\mathbf{r}) \otimes \mathbf{A}^*_\lambda(\mathbf{r}')\, e^{-i\omega_\lambda(t-t')}. \tag{2.277}$$

In practical calculations it is often useful to start from the field $\hat{\mathbf{F}}^{(j)}(\mathbf{r},t')$ at an appropriately chosen (finite) time $t=t'$ and to seek the field $\hat{\mathbf{F}}^{(j)}(\mathbf{r},t)$ at times $t \geq t'$ and/or $t \leq t'$. Applying Eq. (2.276) [together with Eqs (2.269) and (2.275)] yields

$$\hat{\mathbf{F}}^{(j)}(\mathbf{r},t) = \int_{t'}^{t} d\tau \int d^3r'\, \mathbf{K}^{(j)}_{(F)}(\mathbf{r},t;\mathbf{r}',\tau)\hat{\vec{P}}(\mathbf{r}',\tau) + \hat{\mathbf{F}}^{(j)}_{\text{free}}(\mathbf{r},t,t'), \tag{2.278}$$

where the [in comparison with Eq. (2.273)] modified free-field operators are now

$$\hat{\mathbf{F}}^{(+)}_{\text{free}}(\mathbf{r},t,t') = \sum_\lambda \mathbf{F}_\lambda(\mathbf{r})e^{-i\omega_\lambda(t-t')}\hat{a}_\lambda(t'),$$

$$\hat{\mathbf{F}}^{(-)}_{\text{free}}(\mathbf{r},t,t') = [\hat{\mathbf{F}}^{(+)}_{\text{free}}(\mathbf{r},t,t')]^\dagger. \tag{2.279}$$

Note that $\hat{\mathbf{F}}_{\text{free}}^{(+)}(\mathbf{r}, t) = \lim_{t' \to -\infty} \hat{\mathbf{F}}_{\text{free}}^{(+)}(\mathbf{r}, t, t')$. Taking the inverse of

$$\hat{\mathbf{A}}^{(+)}(\mathbf{r}, t') = \sum_\lambda \mathbf{A}_\lambda(\mathbf{r}) \hat{a}_\lambda(t'), \qquad (2.280)$$

we find [cf. Eq. (2.72)]

$$\hat{a}_\lambda(t') = \frac{2\varepsilon_0 \omega_\lambda}{\hbar} \int d^3 r'\, \mathbf{A}_\lambda^*(\mathbf{r}') \hat{\mathbf{A}}^{(+)}(\mathbf{r}', t'), \qquad (2.281)$$

and thus, on combining Eqs (2.279) and (2.281),

$$\hat{\mathbf{F}}_{\text{free}}^{(j)}(\mathbf{r}, t, t') = 2\varepsilon_0 \int d^3 r'\, \mathbf{K}_{(F)}^{(j)}(\mathbf{r}, t; \mathbf{r}', t') \hat{\mathbf{A}}^{(j)}(\mathbf{r}', t'). \qquad (2.282)$$

It should be pointed out that Eq. (2.278) [together with Eq. (2.282)] holds for both $t \geq t'$ and $t \leq t'$.

Equations (2.276) and (2.278) also remain valid in the case of medium-assisted electromagnetic fields, except that the propagation function cannot be obtained from Eq. (2.275) in general, but should be calculated on the basis of the formalism given in Section 2.4.2 in order to also allow for absorbing media.[40] To derive the source-quantity representation, we replace, according to Eq. (2.239), the interaction energy (2.265) with

$$\hat{H}_{\text{int}} = \varepsilon_0^{-1} \int d^3 r\, \hat{\vec{P}}(\mathbf{r})[\hat{\mathbf{\Pi}}(\mathbf{r}) + \hat{\mathbf{P}}_M^{\|}(\mathbf{r})] = -\int d^3 r\, \hat{\vec{P}}(\mathbf{r}) \hat{\mathbf{E}}(\mathbf{r}) \qquad (2.283)$$

and follow the line described above. Recalling Eq. (2.213) together with Eq. (2.211), instead of Eq. (2.267) we now have

$$\dot{\hat{\mathbf{f}}}(\mathbf{r}, \omega) = \frac{1}{i\hbar} [\hat{\mathbf{f}}(\mathbf{r}, \omega), \hat{H}]$$

$$= -i\omega \hat{\mathbf{f}}(\mathbf{r}, \omega) + \sqrt{\frac{\text{Im}\, \varepsilon(\mathbf{r}, \omega)}{\hbar \pi \varepsilon_0}} \frac{\omega^2}{c^2} \int d^3 r'\, \hat{\vec{P}}(\mathbf{r}') G^*(\mathbf{r}', \mathbf{r}, \omega), \qquad (2.284)$$

from which it follows that

$$\hat{\mathbf{f}}(\mathbf{r}, \omega, t) = \hat{\mathbf{f}}_{\text{free}}(\mathbf{r}, \omega, t) + \hat{\mathbf{f}}_s(\mathbf{r}, \omega, t), \qquad (2.285)$$

where

$$\hat{\mathbf{f}}_{\text{free}}(\mathbf{r}, \omega, t) = \hat{\mathbf{f}}_{\text{free}}(\mathbf{r}, \omega, t') e^{-i\omega(t-t')} \qquad (2.286)$$

---

40) When the medium can be approximately characterized by a (space-dependent) real permittivity, then the propagation function can be calculated by means of the generalized mode expansion described in Section 2.4.1. Note that in this case the integration measure $d^3 r'$ in Eqs (2.281) and (2.282) changes to $\varepsilon(\mathbf{r}') d^3 r'$.

and

$$\hat{f}_s(\mathbf{r},\omega,t) = \sqrt{\frac{\operatorname{Im}\varepsilon(\mathbf{r},\omega)}{\hbar\pi\varepsilon_0}} \frac{\omega^2}{c^2} \int dt'\, \Theta(t-t') \int d^3r'\, \hat{\mathbf{P}}(\mathbf{r}',t') \mathbf{G}^*(\mathbf{r}',\mathbf{r},\omega,t'). \tag{2.287}$$

Inserting $\hat{\mathbf{f}}(\mathbf{r},\omega,t)$ in Eq. (2.211), making use of Eq. (2.213), and applying the relation (A.3), we arrive at the source-quantity representation of $\hat{\mathbf{E}}(\mathbf{r},t)$ in the form of

$$\hat{\mathbf{E}}(\mathbf{r},t) = \hat{\mathbf{E}}_{\text{free}}(\mathbf{r},t) + \hat{\mathbf{E}}_s(\mathbf{r},t), \tag{2.288}$$

where

$$\hat{\mathbf{E}}_{\text{free}}(\mathbf{r},t) = \hat{\mathbf{E}}_{\text{free}}^{(+)}(\mathbf{r},t) + \hat{\mathbf{E}}_{\text{free}}^{(-)}(\mathbf{r},t), \tag{2.289}$$

$$\hat{\mathbf{E}}_{\text{free}}^{(+)}(\mathbf{r},t) = i\sqrt{\frac{\hbar}{\pi\varepsilon_0}} \int_0^\infty d\omega\, \frac{\omega^2}{c^2} \int d^3r'\, \sqrt{\operatorname{Im}\varepsilon(\mathbf{r}',\omega)}\, \mathbf{G}(\mathbf{r},\mathbf{r}',\omega) \hat{\mathbf{f}}_{\text{free}}(\mathbf{r}',\omega,t) \tag{2.290}$$

and

$$\hat{\mathbf{E}}_s(\mathbf{r},t) = \hat{\mathbf{E}}_s^{(+)}(\mathbf{r},t) + \hat{\mathbf{E}}_s^{(-)}(\mathbf{r},t), \tag{2.291}$$

$$\hat{\mathbf{E}}_s^{(+)}(\mathbf{r},t) = \frac{i}{\pi\varepsilon_0 c^2}$$
$$\times \int_0^\infty d\omega\, \omega^2 \int dt'\, \Theta(t-t') e^{-i\omega(t-t')} \int d^3r'\, \operatorname{Im}\mathbf{G}(\mathbf{r},\mathbf{r}',\omega) \hat{\mathbf{P}}(\mathbf{r}',t'). \tag{2.292}$$

Note that Eq. (2.291) can be rewritten as

$$\hat{\mathbf{E}}_s(\mathbf{r},t) = -\int dt' \int d^3r'\, \mathbf{D}_{\text{ret}}(\mathbf{r},t;\mathbf{r}',t') \hat{\mathbf{P}}(\mathbf{r}',t'), \tag{2.293}$$

where

$$\mathbf{D}_{\text{ret}}(\mathbf{r},t;\mathbf{r}',t') = \frac{i}{\pi\varepsilon_0 c^2} \Theta(t-t') \int d\omega\, \omega^2 \sin[\omega(t-t')] \mathbf{G}(\mathbf{r},\mathbf{r}',\omega) \tag{2.294}$$

is commonly called the retarded Green tensor.

To make contact with Eq. (2.274), we rewrite $\hat{\mathbf{E}}_s^{(+)}(\mathbf{r},t)$ as

$$\hat{\mathbf{E}}_s^{(+)}(\mathbf{r},t) = \int dt' \int d^3r'\, \Theta(t-t') \mathbf{K}_{(E)}^{(+)}(\mathbf{r},t;\mathbf{r}',t') \hat{\mathbf{P}}(\mathbf{r}',t') \tag{2.295}$$

and find that

$$K^{(+)}_{(E)}(\mathbf{r},t;\mathbf{r}',t') = \frac{i}{\pi\varepsilon_0 c^2}\int_0^\infty d\omega\, \omega^2 e^{-i\omega(t-t')}\mathrm{Im}\,G(\mathbf{r},\mathbf{r}',\omega). \tag{2.296}$$

Since in the absence of media, $\hat{\mathbf{E}}(\mathbf{r})$ simply reduces to $\hat{\mathbf{E}}^{\perp}(\mathbf{r})$, and the relation

$$K^{(+)}_{(\Pi)}(\mathbf{r},t;\mathbf{r}',t') = -\varepsilon_0 \lim_{\varepsilon\to 1} K^{(+)}_{(E)}(\mathbf{r},t;\mathbf{r}',t') \tag{2.297}$$

holds, which means that in Eq. (2.296) the Green tensor $G(\mathbf{r},\mathbf{r}',\omega)$ can be replaced by the well-known free-space Green tensor

$$G_0(\mathbf{r},\mathbf{r}',\omega) = \frac{c^2}{4\pi\omega^2}\left[\nabla\otimes\nabla + I\frac{\omega^2}{c^2}\right]\frac{e^{i\omega|\mathbf{r}-\mathbf{r}'|/c}}{|\mathbf{r}-\mathbf{r}'|}. \tag{2.298}$$

Clearly, Eqs (2.285)–(2.287) can also be used to find the source-quantity representation of quantities other than the electric field.

Source-quantity representations such as those given above may be regarded as basic equations describing the propagation of light through passive optical systems modeled by (more or less structured) dielectric bodies. The information about their action on the light propagation is contained in the space-time structure of the respective tensor-valued propagation function used in classical optics as well. In quantum optics these functions additionally determine the time-dependent field commutation relations and, in this way, the quantum statistical properties of the fields under consideration.

2.7
Time-dependent commutation relations

It should be stressed that, in classical optics, any light field may be attributed to sources, and hence in classical optics free-field terms such as the second term on the right-hand side of Eq. (2.276) may be omitted. In quantum optics the situation may be changed drastically. The field operators cannot be related to the source-field operators solely, but must also be related to the free-field operators. The latter are required for a correct description of the effects of quantum noise, at least of the vacuum. From a more general point of view, the free-field operators, which in general do not commute with the source-field operators, ensure the quantum-mechanical consistency of the theory.

Let us perform the calculations within the frame of microscopic electrodynamics by employing mode expansion. The equal-time commutation relations for field operators $\hat{\mathbf{F}}$ and $\hat{\mathbf{G}}$ of the type given in Eq. (2.271) can easily be constructed using the basic commutation relations (2.75) and (2.76) for the photonic operators $\hat{a}_\lambda$ and $\hat{a}^\dagger_\lambda$. The results are apparently the same as in the case of the corresponding free-field operators $\hat{\mathbf{F}}_{\text{free}}$ and $\hat{\mathbf{G}}_{\text{free}}$.[41] Moreover, the commutators of free-field operators at different times may also be constructed in this way, by including the time evolution [cf. Eq. (2.269)] in the calculation. We obtain

$$[\hat{F}_{k\,\text{free}}^{(\pm)}(\mathbf{r},t), \hat{G}_{k'\,\text{free}}^{(\pm)}(\mathbf{r}',t')] = 0, \tag{2.299}$$

$$[\hat{F}_{k\,\text{free}}^{(+)}(\mathbf{r},t), \hat{G}_{k'\,\text{free}}^{(-)}(\mathbf{r}',t')] = \sum_\lambda F_{k\lambda}(\mathbf{r}) G_{k'\lambda}^*(\mathbf{r}') e^{-i\omega_\lambda (t-t')}. \tag{2.300}$$

In particular, we see that $(\mathbf{F}_\lambda, \mathbf{G}_\lambda \mapsto \mathbf{\Pi}_\lambda = -i\omega_\lambda \varepsilon_0 \mathbf{A}_\lambda)$

$$[\hat{\Pi}_{k\,\text{free}}^{(+)}(\mathbf{r},t), \hat{\Pi}_{k'\,\text{free}}^{(-)}(\mathbf{r}',t')] = i\hbar\varepsilon_0 K_{(\Pi)kk'}^{(+)}(\mathbf{r},t;\mathbf{r}',t'), \tag{2.301}$$

where we have used Eq. (2.277). Since $\hat{\mathbf{A}}_{\text{free}}^{(\pm)} = \hat{\mathbf{\Pi}}_{\text{free}}^{(\pm)}/\varepsilon_0$, Eq. (2.301) also determines the commutation relations for the electric free field associated with $-\hat{\mathbf{A}}_{\text{free}}^{(\pm)}$. Let us, for example, consider the case of light propagating through free space. Using Eq. (2.88), we derive after some algebra

$$\varepsilon_0^{-2}[\hat{\Pi}_{k\,\text{free}}(\mathbf{r},t), \hat{\Pi}_{k'\,\text{free}}(\mathbf{r}',t')] = [\hat{E}_{k\,\text{free}}(\mathbf{r},t), \hat{E}_{k'\,\text{free}}(\mathbf{r}',t')]$$
$$= \frac{c\hbar}{4\varepsilon_0 \pi i}\left(\partial_k^{(r)}\partial_{k'}^{(r)} - \delta_{kk'}\Delta^{(r)}\right) \frac{\delta[|\mathbf{r}-\mathbf{r}'|-c(t-t')] - \delta[|\mathbf{r}-\mathbf{r}'|-c(t'-t)]}{|\mathbf{r}-\mathbf{r}'|}, \tag{2.302}$$

which reveals that the components of the electric free-field strength at different space-time points do not commute if the latter are on the light cone.

In contrast to equal-time commutation relations or free-field commutation relations, it is nontrivial to determine the commutators of field operators $\hat{\mathbf{F}}^{(\pm)}$ and $\hat{\mathbf{G}}^{(\pm)}$ or combinations of them at different times, because knowledge of the solution of the Heisenberg equations of motion of the coupled light–matter system is needed. However, source-quantity representations as given in Section 2.6 make it possible to express the commutation relations of radiation-field operators at different times in terms of commutators of free-field and source-quantity operators [Mollow (1973)]. Although the commutators of the source-quantity operators are in general not explicitly known, the general structure of the field commutators reveals important information on the effects of light propagation.

Let us consider the commutator of $\hat{F}_{k\,\text{free}}^{(j)}(\mathbf{r},t)$ and a source-quantity operator $\hat{Q}(t')$, which can be an arbitrarily chosen function of the canonically conjugate variables of the particles. Since the commutation rule

$$[\hat{a}_\lambda(t), \hat{Q}(t)] = 0 \tag{2.303}$$

41) For the derivation of equal-time commutation relations of the electromagnetic field in dispersing and absorbing media, see Appendix B.

holds, from Eqs (2.268)–(2.270) it follows that

$$[\hat{a}_{\lambda\,\text{free}}(t), \hat{Q}(t')]$$
$$= -\frac{1}{\hbar}\int d\tau \int d^3s\, \Theta(t'-\tau)e^{-i\omega_\lambda(t-\tau)}\omega_\lambda A^*_{l\lambda}(\mathbf{s})[\hat{P}_l(\mathbf{s},\tau), \hat{Q}(t')]. \quad (2.304)$$

We now multiply Eq. (2.304) by $F_{k\lambda}(\mathbf{r})$ and sum over λ. Recalling the definitions of the free-field operators $\hat{\mathbf{F}}^{(j)}_{\text{free}}$ [Eq. (2.273)] and the kernel $K^{(j)}_{(F)}$ [Eq. (2.275)], we arrive at the following representation of the desired commutator:

$$[\hat{F}^{(j)}_{k\,\text{free}}(\mathbf{r},t), \hat{Q}(t')]$$
$$= -\int d\tau \int d^3s\, \Theta(t'-\tau)K^{(j)}_{(F)kl}(\mathbf{r},t;\mathbf{s},\tau)[\hat{P}_l(\mathbf{s},\tau), \hat{Q}(t')]. \quad (2.305)$$

Equation (2.305) enables us to express commutators of free-field and source-field operators in terms of source-quantity commutators. Combining Eqs (2.305) and (2.274) yields

$$[\hat{F}^{(j)}_{k\,\text{free}}(\mathbf{r},t), \hat{G}^{(j')}_{k's}(\mathbf{r}',t')] = -\int d\tau \int d^3s \int d\tau' \int d^3s' \{\Theta(t'-\tau')\Theta(\tau'-\tau)$$
$$\times K^{(j)}_{(F)kl}(\mathbf{r},t;\mathbf{s},\tau)K^{(j')}_{(G)kl'}(\mathbf{r}',t';\mathbf{s}',\tau')[\hat{P}_l(\mathbf{s},\tau), \hat{P}_{l'}(\mathbf{s}',\tau')]\}. \quad (2.306)$$

Using the relation $\Theta(t-\tau)+\Theta(\tau-t)=1$, we may rewrite Eq. (2.306) as

$$[\hat{F}^{(j)}_{k\,\text{free}}(\mathbf{r},t), \hat{G}^{(j')}_{k's}(\mathbf{r}',t')] = \hat{C}^{(jj')}_{(FG)kk'}(\mathbf{r},t;\mathbf{r}',t') + \hat{D}^{(jj')}_{(FG)kk'}(\mathbf{r},t;\mathbf{r}',t'), \quad (2.307)$$

where $\hat{C}^{(jj')}_{(FG)kk'}$ and $\hat{D}^{(jj')}_{(FG)kk'}$ are respectively obtained from the integral on the right-hand side in Eq. (2.306) by introduction into it of $\Theta(t-\tau)$ and $\Theta(\tau-t)$. From an inspection of the three unit step functions in the expression for $\hat{D}^{(jj')}_{(FG)kk'}$ we readily verify that

$$\hat{D}^{(jj')}_{(FG)kk'}(\mathbf{r},t;\mathbf{r}',t') = 0 \quad \text{if} \quad t > t'. \quad (2.308)$$

We now turn to the problem of expressing radiation-field commutators in terms of free-field and source-quantity commutators. For this purpose, we decompose, according to Eq. (2.276), the operator product $\hat{F}^{(j)}_k(\mathbf{r},t)\hat{G}^{(j')}_{k'}(\mathbf{r}',t')$ as

$$\hat{F}^{(j)}_k(\mathbf{r},t)\hat{G}^{(j')}_{k'}(\mathbf{r}',t') = \hat{F}^{(j)}_{k\,\text{free}}(\mathbf{r},t)\hat{G}^{(j')}_{k'\,\text{free}}(\mathbf{r}',t') + \hat{F}^{(j)}_{ks}(\mathbf{r},t)\hat{G}^{(j')}_{k'\,\text{free}}(\mathbf{r}',t')$$
$$+ \hat{G}^{(j')}_{k's}(\mathbf{r}',t')\hat{F}^{(j)}_{k\,\text{free}}(\mathbf{r},t) + \hat{F}^{(j)}_{ks}(\mathbf{r},t)\hat{G}^{(j')}_{k's}(\mathbf{r}',t')$$
$$+ [\hat{F}^{(j)}_{k\,\text{free}}(\mathbf{r},t), \hat{G}^{(j')}_{k's}(\mathbf{r}',t')]. \quad (2.309)$$

Using Eq. (2.274) and Eqs (2.306), (2.307), we may prove that combining the last two terms on the right-hand side in Eq. (2.309) yields

$$\hat{F}^{(j)}_{ks}(\mathbf{r},t)\hat{G}^{(j')}_{k's}(\mathbf{r}',t') + [\hat{F}^{(j)}_{k\,\mathrm{free}}(\mathbf{r},t), \hat{G}^{(j')}_{k's}(\mathbf{r}',t')] = \hat{D}^{(j,j')}_{(F,G)kk'}(\mathbf{r},t;\mathbf{r}',t')$$
$$+ \int d\tau \int d^3s \int d\tau' \int d^3s'\, [\Theta(t-\tau)\Theta(t'-\tau')$$
$$\times K^{(j)}_{(F)kl}(\mathbf{r},t;\mathbf{s},\tau) K^{(j')}_{(G)k'l'}(\mathbf{r}',t';\mathbf{s}',\tau') \mathcal{T}_+ \hat{P}_l(\mathbf{s},\tau)\hat{P}_{l'}(\mathbf{s}',\tau')]. \quad (2.310)$$

The time-ordering symbols $\mathcal{T}_\pm$ are defined as follows. The symbol $\mathcal{T}_+$ time-orders the operators $\hat{A}_i(t_i)$ in an operator product $\hat{A}_1(t_1)\hat{A}_2(t_2)\cdots\hat{A}_n(t_n)$ with the latest time to the far left,

$$\mathcal{T}_+ \hat{A}_1(t_1)\hat{A}_2(t_2)\cdots\hat{A}_n(t_n)$$
$$= \hat{A}_{i_1}(t_{i_1})\hat{A}_{i_2}(t_{i_2})\cdots\hat{A}_{i_n}(t_{i_n}), \quad t_{i_1} > t_{i_2} > \cdots > t_{i_n}, \quad (2.311)$$

while the symbol $\mathcal{T}_-$ time-orders the operators $\hat{A}_i(t_i)$ with the latest time to the far right,

$$\mathcal{T}_- \hat{A}_1(t_1)\hat{A}_2(t_2)\cdots\hat{A}_n(t_n)$$
$$= \hat{A}_{i_1}(t_{i_1})\hat{A}_{i_2}(t_{i_2})\cdots\hat{A}_{i_n}(t_{i_n}), \quad t_{i_1} < t_{i_2} < \cdots < t_{i_n}. \quad (2.312)$$

From Eqs (2.309) and (2.310) we easily arrive at the commutation relation

$$[\hat{F}^{(j)}_k(\mathbf{r},t), \hat{G}^{(j')}_{k'}(\mathbf{r}',t')] = [\hat{F}^{(j)}_{k\,\mathrm{free}}(\mathbf{r},t), \hat{G}^{(j')}_{k'\,\mathrm{free}}(\mathbf{r}',t')]$$
$$+ \hat{D}^{(j,j')}_{(F,G)kk'}(\mathbf{r},t;\mathbf{r}',t') - \hat{D}^{(j',j)}_{(G,F)k'k}(\mathbf{r}',t';\mathbf{r},t). \quad (2.313)$$

Equation (2.313) reveals that the (time-dependent) commutators of fields that are attributed to sources, differ from the corresponding free-field commutators in the so called time-delayed contributions $\hat{D}^{(j,j')}_{(F,G)kk'}(\mathbf{r},t;\mathbf{r}',t')$ [and $\hat{D}^{(j',j)}_{(G,F)k'k}(\mathbf{r}',t';\mathbf{r},t)$] [Cresser (1984); Knöll, Vogel and Welsch (1987)]. It is worth noting that the terms which may contribute to the integral defining $\hat{D}^{(j,j')}_{(F,G)kk'}(\mathbf{r},t;\mathbf{r}',t')$ are time-ordered in such a way that $t<\tau<\tau'<t'$. This is just the time-ordering necessary for the propagation of light from the space point $\mathbf{r}$ (at time t) to the space point $\mathbf{r}'$ (at time t') via the space points of the sources, $\mathbf{s}$ (at times τ) and $\mathbf{s}'$ (at times τ'). Assuming that the experimental setup allows such a propagation, $\hat{D}^{(j,j')}_{(F,G)kk'}(\mathbf{r},t;\mathbf{r}',t')$ may be expected to be nonzero. Clearly, the farther away from the space points $\mathbf{r}$ and $\mathbf{r}'$ the atomic

sources are situated, the larger the time interval $|t' - t|$ becomes, for which the commutator is simply given by the free-field commutator. In this sense, far away from the sources the radiation field may be considered as being a free field. Recall that for equal times the free-field commutator is observed.

2.8
Correlation functions of field operators

In this section we turn to the problem of expressing correlation functions of field operators in terms of correlation functions of source-quantity operators and free-field operators. Since from the theory of photoelectric detection of light (Chapter 6) it is known that correlation functions of field operators subjected to normal and time orderings are observable, we demonstrate the method for the class of correlation functions given by[42]

$$G^{(m,n)}_{\{k_i k_j\}}(\{\mathbf{r}_i, t_i, \mathbf{r}_j, t_j\}) = \left\langle {}^\circ_\circ \prod_{i=1}^{m} \prod_{j=m+1}^{m+n} \hat{F}^{(-)}_{k_i}(\mathbf{r}_i, t_i) \hat{F}^{(+)}_{k_j}(\mathbf{r}_j, t_j) {}^\circ_\circ \right\rangle, \quad (2.314)$$

where $\langle \ldots \rangle = \mathrm{Tr}(\hat{\varrho} \ldots)$, $\hat{\varrho}$ being the density operator of the overall system consisting of radiation and matter, and ${}^\circ_\circ\ {}^\circ_\circ$ indicates the following operator orderings:

(i) normal ordering of the operators $\hat{F}^{(-)}_k$ and $\hat{F}^{(+)}_k$, with the operators $\hat{F}^{(-)}_k$ to the left of the operators $\hat{F}^{(+)}_k$;

(ii) $\mathcal{T}_+$ time ordering of the operators $\hat{F}^{(+)}_k$ and $\mathcal{T}_-$ time ordering of the operators $\hat{F}^{(-)}_k$.

We substitute into Eq. (2.314) for the field operators $\hat{F}^{(\pm)}_k$ the result of Eq. (2.272), so that the field operators $\hat{F}^{(\pm)}_k$ are decomposed into source-field operators $\hat{F}^{(\pm)}_{ks}$ and free-field operators $\hat{F}^{(\pm)}_{k\,\mathrm{free}}$. Applying the commutation relation (2.307), we rearrange the mixed operator products in such a way that the operators $\hat{F}^{(+)}_{k\,\mathrm{free}}$ are to the right of the operators $\hat{F}^{(+)}_{ks}$, and correspondingly the operators $\hat{F}^{(-)}_{k\,\mathrm{free}}$ are to the left of the operators $\hat{F}^{(-)}_{ks}$. To illustrate this procedure, let us consider the operator product $\hat{F}^{(+)}_{k_1}(\mathbf{r}_1, t_1)\hat{F}^{(+)}_{k_2}(\mathbf{r}_2, t_2)$. Applying

---

42) Here the abbreviated notation $G^{(m,n)}_{\{k_i k_j\}}(\{\mathbf{r}_i, t_i, \mathbf{r}_j, t_j\}) = G^{(m,n)}_{k_1 \ldots k_{m+n}}(\mathbf{r}_1, t_1, \ldots, \mathbf{r}_{m+n}, t_{m+n})$ is used.

Eqs (2.309) and (2.310) and recalling, Eq. (2.308), we readily derive

$$\begin{aligned}\mathcal{T}_+ \hat{F}_{k_1}^{(+)}(\mathbf{r}_1,t_1)\hat{F}_{k_2}^{(+)}(\mathbf{r}_2,t_2) &= \hat{F}_{k_1\,\text{free}}^{(+)}(\mathbf{r}_1,t_1)\hat{F}_{k_2\,\text{free}}^{(+)}(\mathbf{r}_2,t_2)\\ &+ \hat{F}_{k_1\,\text{s}}^{(+)}(\mathbf{r}_1,t_1)\hat{F}_{k_2\,\text{free}}^{(+)}(\mathbf{r}_2,t_2) + \hat{F}_{k_2\,\text{s}}^{(+)}(\mathbf{r}_2,t_2)\hat{F}_{k_1\,\text{free}}^{(+)}(\mathbf{r}_1,t_1)\\ &+ \int dt_1' \int d^3 r_1' \int dt_2' \int d^3 r_2' [\Theta(t_1-t_1')\,\Theta(t_2-t_2')\, K_{(F)k_1 k_1'}^{(+)}(\mathbf{r}_1,t_1;\mathbf{r}_1',t_1')\\ &\times K_{(F)k_2 k_2'}^{(+)}(\mathbf{r}_2,t_2;\mathbf{r}_2',t_2')\, \mathcal{T}_+\hat{P}_{k_1'}(\mathbf{r}_1',t_1')\hat{P}_{k_2'}(\mathbf{r}_2',t_2')]. \end{aligned} \quad (2.315)$$

Note that in the first term on the right-hand side of Eq. (2.315) the time-ordering symbol $\mathcal{T}_+$ may be left, because the free-field operators $\hat{F}_{k\,\text{free}}^{(+)}$ commute.

The $\mathcal{T}_+$ time ordering of the operator product $\hat{F}_{k_1}^{(+)}(\mathbf{r}_1,t_1)\hat{F}_{k_2}^{(+)}(\mathbf{r}_2,t_2)$ obviously rules out any time-delayed effect. Recalling Eq. (2.313) together with Eq. (2.299), the $\mathcal{T}_+$ time ordering may therefore be said to pick out the commuting parts of $\hat{F}_{k_1}^{(+)}(\mathbf{r}_1,t_1)$ and $\hat{F}_{k_2}^{(+)}(\mathbf{r}_2,t_2)$ in the product $\hat{F}_{k_1}^{(+)}(\mathbf{r}_1,t_1)\hat{F}_{k_2}^{(+)}(\mathbf{r}_2,t_2)$. Applying Eqs (2.272) and (2.274), we may write Eq. (2.315) and its Hermitian conjugate in the compact form

$$\mathcal{T}_\pm \hat{F}_{k_1}^{(\pm)}(\mathbf{r}_1,t_1)\hat{F}_{k_2}^{(\pm)}(\mathbf{r}_2,t_2) = \mathcal{O}_\pm \hat{F}_{k_1}^{(\pm)}(\mathbf{r}_1,t_1)\hat{F}_{k_2}^{(\pm)}(\mathbf{r}_2,t_2), \quad (2.316)$$

where $\mathcal{O}_+$ (respectively $\mathcal{O}_-$) introduces the following operator ordering in products of operators $\hat{F}_{k\,\text{s}}^{(+)}$ and $\hat{F}_{k\,\text{free}}^{(+)}$ (respectively $\hat{F}_{k\,\text{s}}^{(-)}$ and $\hat{F}_{k\,\text{free}}^{(-)}$), after decomposition of $\hat{F}_k^{(+)}$ (respectively $\hat{F}_k^{(-)}$) into source- and free-field parts:

(i) ordering of the operators $\hat{F}_{k\,\text{s}}^{(+)}$ and $\hat{F}_{k\,\text{free}}^{(+)}$ (respectively $\hat{F}_{k\,\text{s}}^{(-)}$ and $\hat{F}_{k\,\text{free}}^{(-)}$) with the operators $\hat{F}_{k\,\text{free}}^{(+)}$ (respectively $\hat{F}_{k\,\text{free}}^{(-)}$) to the right (respectively left) of the operators $\hat{F}_{k\,\text{s}}^{(+)}$ (respectively $\hat{F}_{k\,\text{s}}^{(-)}$);

(ii) substituting for the operators $\hat{F}_{k\,\text{s}}^{(+)}$ ($\hat{F}_{k\,\text{s}}^{(-)}$) the result of Eq. (2.274) and performing $\mathcal{T}_+$ ($\mathcal{T}_-$) time ordering of the source-quantity operators $\hat{P}_{k'}$ in the resulting source-quantity operator products before integrating with respect to the times t'.

Equation (2.316) may be extended to the case of higher-order operator products. It can be shown that [Knöll, Vogel and Welsch (1987)]

$$\mathcal{T}_\pm \prod_{i=1}^n \hat{F}_{k_i}^{(\pm)}(\mathbf{r}_i,t_i) = \mathcal{O}_\pm \prod_{i=1}^n \hat{F}_{k_i}^{(\pm)}(\mathbf{r}_i,t_i), \quad (2.317)$$

so combining Eqs (2.314) and (2.317) yields

$$G_{\{k_i k_j\}}^{(m,n)}(\{\mathbf{r}_i,t_i,\mathbf{r}_j,t_j\}) = \left\langle :\prod_{i=1}^m \prod_{j=m+1}^{m+n} \hat{F}_{k_i}^{(-)}(\mathbf{r}_i,t_i)\hat{F}_{k_j}^{(+)}(\mathbf{r}_j,t_j): \right\rangle, \quad (2.318)$$

where $\vdots\ \vdots$ indicates

(i) that the field operators are to be written in normal order (with the $\hat{F}_k^{(-)}$ to the left of the $\hat{F}_k^{(+)}$);

(ii) $\mathcal{O}_+$ (respectively $\mathcal{O}_-$) ordering of the operators $\hat{F}_k^{(+)}$ (respectively $\hat{F}_k^{(-)}$).

In practice, it is often desired to observe the properties of the light attributed to certain kinds of sources. This requires an observational scheme guaranteeing that at the observation points only this light is detected. In particular, when radiating sources are optically pumped and the light generated by them is studied, the range of observation must be outside the pump beam. Since, apart from the vacuum field, any real light field may be thought of as arising from sources, the remaining free field may be regarded as being the vacuum field. Hence the following conditions may be assumed to be satisfied:

$$\langle\cdots\hat{F}_{k\,\text{free}}^{(+)}\rangle = 0 = \langle\hat{F}_{k\,\text{free}}^{(-)}\cdots\rangle. \tag{2.319}$$

Recalling the definitions of the ordering symbols $\mathcal{O}_\pm$ in the $\vdots\ \vdots$ notation, we may omit the free-field operators in Eq. (2.318), and thus

$$G_{\{k_ik_j\}}^{(m,n)}(\{\mathbf{r}_i,t_i,\mathbf{r}_j,t_j\}) = \left\langle\ \vdots\ \prod_{i=1}^{m}\prod_{j=m+1}^{m+n}\hat{F}_{k_is}^{(-)}(\mathbf{r}_i,t_i)\hat{F}_{k_js}^{(+)}(\mathbf{r}_j,t_j)\ \vdots\ \right\rangle. \tag{2.320}$$

Equation (2.320) establishes that when the condition (2.319) is fulfilled, in the calculations of field correlation functions of the type defined by Eq. (2.314), the total field operators $\hat{F}_k^{(\pm)}$ may formally be replaced by the source-field operators $\hat{F}_{ks}^{(\pm)}$, the $\mathcal{T}_+$ and $\mathcal{T}_-$ time orderings originally concerning the operators $\hat{F}_k^{(+)}$ and $\hat{F}_k^{(-)}$, respectively, being transferred to the corresponding source-quantity operators $\hat{P}_k$.

As already mentioned, the effect of optical instruments is included in the actual structure of the propagation functions $K_{(F)}^{(\pm)}(\mathbf{r},t;\mathbf{r}',t')$ known from classical optics. However, there is an essential difference between classical and quantum optics; namely owing to the $\vdots\ \vdots$ ordering the (multi-time) convolution integrals arising from the source-field operators in Eq. (2.320) cannot generally be performed independently of each other, as is possible in classical optics. Hence in the case of quantum light fields the result of the integrations in Eq. (2.320) is in general expected to be different from that predicted from classical optics.

In many typical light-scattering problems, when the properties of the scattered light are sought and the dynamics of the sources is known, Eq. (2.320) directly applies to the study of the properties of scattered light (see also Chapter 11). For more complicated light–matter interaction processes, it is usually better to consider appropriately chosen field variables (the dynamics of which can often be treated on the basis of the concept of effective interaction Hamiltonians, see Section 2.5.3) rather than the atomic source quantities. Moreover, if there are additional optical instruments through which the light passes, the relevant light field may often be regarded as being known before it meets the optical instruments. The question then arises of how to relate the properties of the field observed behind the instruments (properties of the output field) to the properties of the field in front of the instruments (properties of the input field). In all these cases it is helpful to use Eq. (2.318) instead of Eq. (2.320) and to combine source-field parts and certain free-field parts to give new field variables (e. g., input/output fields; see also Chapter 9), which allow one to express correlation functions of the type given in Eq. (2.314) in terms of correlation functions of these field variables. This problem is nontrivial, since these field variables cannot be regarded as being a priori free-field variables with commuting positive- (and negative-) frequency parts. Clearly, the type of decomposition suitable depends on the physical situation.

References

Bloembergen, N. (1965) *Nonlinear Optics* (Benjamin, New York).

Buhmann, S.Y, L. Knöll, D.-G. Welsch and T.D. Ho (2004) *Phys. Rev. A* **70**, 052117.

Cresser, J.D. (1984) *Phys. Rev. A* **29**, 1984.

Cohen-Tannoudji, C., J. Dupont-Roc and G. Grynberg (1989) *Photons and Atoms* (Wiley, New York).

Glauber, R.J. and M. Lewenstein (1991) *Phys. Rev. A* **43**, 467.

Gruner, T. and D.-G. Welsch (1996) *Phys. Rev. A* **53**, 1818.

Haken, H. (1985) *Light*, Vol. 1 (North-Holland, Amsterdam).

Ho, T.D., L. Knöll and D.-G. Welsch (2002) *Phys. Rev. A* **65**, 043813.

Ho, T.D., S.Y. Buhmann, L. Knöll, D.-G. Welsch, S. Scheel and J.Kästel (2003) *Phys. Rev. A* **68**, 043816.

Huttner, B. and S.M. Barnett (1992) *Phys. Rev. A* **46**, 4306.

Knöll, L., W. Vogel and D.-G. Welsch (1987) *Phys. Rev. A* **36**, 3803.

Levenson, M.D. and S.S. Kano (1988) *Introduction to Nonlinear Laser Spectroscopy* (Academic Press, New York).

Loudon, R. (1983) *The Quantum Theory of Light* (Clarendon Press, Oxford).

Louisell, W.H. (1973) *Quantum Statistical Properties of Radiation* (Wiley, New York).

Meystre, P. and M. Sargent III (1990) *Elements of Quantum Optics* (Springer-Verlag, Berlin).

Milonni, P.W. (1994) *The Quantum Vacuum: An Introduction to Quantum Electrodynamics* (Academic Press, New York).

Mollow, B.R. (1973) *Phys. Rev. A* **8**, 2684.

Peřina, J. (1991) *Quantum Statistics of Linear and Nonlinear Optical Phenomena* (Reidel, Dordrecht).

Power, E.A. and T. Thirunamachandran (1980) *Proc. Roy. Soc. Lond. A* **372**, 265.

Scheel, S., L. Knöll and D.-G. Welsch (1998) *Phys. Rev. A* **58**, 700.

Scheel, S. and D.-G. Welsch (2006) *Phys. Rev. Letters* **96**, 073601.

Schubert, M. and B. Wilhelmi (1986) *Nonlinear Optics and Quantum Electronics* (Wiley, New York).

Shen, Y.R. (1984) *Principles of Nonlinear Optics* (Wiley, New York).

Shore, B.W. (1990) *Theory of Coherent Atomic Excitations* (Wiley, New York).

Stenholm, S. (1984) *Foundations of Laser Spectroscopy* (Wiley, New York).

Titulaer, U.M. and R.J. Glauber (1966) *Phys. Rev.* **145**, 1041.

3
Quantum states of bosonic systems

The radiation field is an example of an immaterial bosonic system. Examples of material bosonic systems which play an important role in quantum optics are the vibrations of trapped atoms and the internuclear vibrations of molecules inasmuch as they can be approximated by harmonic oscillators. Among the wide variety of possible quantum states of bosonic systems, there are some fundamental types that are of particular interest. First, such states may serve as a quantum-mechanical basis for representing the various observables of interest. Second, they may be regarded as typical examples for certain limiting cases of quantum noise, with special emphasis on nonclassical features. Although the definition of nonclassicality (Chapter 8) is a nontrivial task, we will use the term nonclassical in this chapter in a more generous way.

We will begin with the introduction of the number states (Section 3.1). Although they are well known, from standard quantum-mechanics textbooks, as energy eigenstates of harmonic oscillators, they will be seen to reveal the quite counter-intuitive feature of not showing the oscillatory behavior expected of classical harmonic oscillators. The classically expected oscillatory behavior is then shown to be realized by the coherent states (Section 3.2). The coherent states are Gaussian states to which the squeezed coherent states (often called quadrature-squeezed states) also belong (Section 3.3). Their quantum-noise properties are of great practical relevance for applications in measurement techniques below the standard quantum limit. Finally, the eigenstates of phase-rotated quadratures (Section 3.4) and phase states (Section 3.5) are introduced. Quadrature eigenstates play an important role in the context of homodyne detection (see Sections 6.5 and 7.1).

3.1
Number states

Let us consider a system of uncoupled harmonic oscillators whose Hamiltonian reads

$$\hat{H} = \sum_\lambda \hat{H}_\lambda, \qquad (3.1)$$

where

$$\hat{H}_\lambda = \hbar\omega_\lambda(\hat{n}_\lambda + \tfrac{1}{2}) \qquad (3.2)$$

is the single-oscillator Hamiltonian expressed in terms of the number operator

$$\hat{n}_\lambda = \hat{a}_\lambda^\dagger \hat{a}_\lambda \qquad (3.3)$$

of the λth oscillator, with $\hat{a}_\lambda$ and $\hat{a}_\lambda^\dagger$, respectively, being the annihilation and creation operators attributed to the oscillator which obey the bosonic commutation relations

$$[\hat{a}_\lambda, \hat{a}_{\lambda'}^\dagger] = \delta_{\lambda\lambda'} \qquad (3.4)$$

$$[\hat{a}_\lambda, \hat{a}_{\lambda'}] = 0 = [\hat{a}_\lambda^\dagger, \hat{a}_{\lambda'}^\dagger]. \qquad (3.5)$$

From Chapter 2 we know that mode expansion of the free radiation field just leads to Eqs (3.1)–(3.5) [cf. Eqs (2.75)–(2.78)]. Therefore, what we derive in the following in a very general context can be thought of as being the radiation field or other specific realizations of harmonic oscillators – also referred to as modes in the following – such as the (harmonic) motion of trapped atoms or (harmonic) molecular vibrations.

To proceed we will restrict ourselves first to a single oscillator, thereby omitting the index λ ($\hat{H}_\lambda \mapsto \hat{H}$), and we will then later consider the case of multimode systems. The eigenstates of the Hamiltonian (3.2) are of particular interest, since they naturally provide a complete and orthonormal set of basis states. To solve the eigenvalue problem for $\hat{H}$, it is sufficient to solve the eigenvalue problem for the number operator $\hat{n}$, since both operators commute, $[\hat{H}, \hat{n}] = 0$. The main steps for solving this standard problem of quantum mechanics may be summarized as follows. The eigenvalue equation reads

$$\hat{n}|\phi_n\rangle = n|\phi_n\rangle, \qquad (3.6)$$

with n being the eigenvalue and $|\phi_n\rangle$ the corresponding eigenvector. Since $\hat{n}$ is Hermitian, the eigenvalues n are real-valued and eigenvectors corresponding to different eigenvalues are orthogonal. Inserting $\hat{n}$ as given according to Eq. (3.3) in

$$\langle\phi_n|\hat{n}|\phi_n\rangle = n\langle\phi_n|\phi_n\rangle, \qquad (3.7)$$

we obtain

$$\langle\phi_n|\hat{a}^\dagger\hat{a}|\phi_n\rangle = n\langle\phi_n|\phi_n\rangle. \qquad (3.8)$$

Since $|\phi_n\rangle$ and $\hat{a}|\phi_n\rangle$ are Hilbert-space vectors, whose norms are non-negative, we immediately observe from Eq. (3.8) that the eigenvalue n must also be non-negative. Using the relation $[\hat{a}^l, \hat{n}] = l\hat{a}^l$, derived with the help of the bosonic commutator (3.4), we obtain

$$\hat{n}\hat{a}^l|\phi_n\rangle = (\hat{a}^l\hat{n} + [\hat{n}, \hat{a}^l])|\phi_n\rangle = (n-l)\hat{a}^l|\phi_n\rangle. \tag{3.9}$$

From this relation it is seen that, as long as $n - l \geq 0$, the state $\hat{a}^l|\phi_n\rangle$ is an eigenstate of $\hat{n}$ with eigenvalue $n-l$, which may be denoted by $|\phi_{n-l}\rangle = \hat{a}^l|\phi_n\rangle$. For a negative eigenvalue $n - l < 0$ it is required that $\hat{a}^l|\phi_n\rangle = 0$ in order to fulfill Eq. (3.9) under the constraint of non-negative eigenvalues. For $n=0$ and $l=1$ this requirement provides us with the relation

$$\hat{a}|\phi_0\rangle = 0, \tag{3.10}$$

which defines $|\phi_0\rangle$ as the ground state, having zero number of quanta, from which it follows that no further quantum can be annihilated. Analogously to Eq. (3.9), we may prove that

$$\hat{n}\hat{a}^{\dagger l}|\phi_n\rangle = (n+l)\hat{a}^{\dagger l}|\phi_n\rangle, \tag{3.11}$$

which states that $\hat{a}^{\dagger l}|\phi_n\rangle$ is an eigenstate of $\hat{n}$ with eigenvalue $n+l$, which we denote by $|\phi_{n+l}\rangle = \hat{a}^{\dagger l}|\phi_n\rangle$.

From Eqs (3.9) and (3.11) it is clear that $\hat{a}$ and $\hat{a}^\dagger$ decrease and increase the number of energy quanta by single quanta, respectively. Therefore, the operators $\hat{a}$ and $\hat{a}^\dagger$ are called annihilation and creation operators, respectively. Starting from the ground state $|\phi_0\rangle$ via Eq. (3.11) a ladder of eigenstates of the free Hamiltonian (3.2) is created by multiple application of the creation operator $\hat{a}^\dagger$. Since these states exhibit defined numbers of energy quanta they are usually called number states. Depending on the physical system under study they may correspond to eigenstates with a precise number of photons, phonons, or other elementary bosonic excitations.

Normalizing the states $|\phi_n\rangle$, we obtain the number states $|n\rangle = |\phi_n\rangle/\langle\phi_n|\phi_n\rangle$ as an orthonormal set of basis states. For this purpose we create the states $|n\rangle$ via Eq. (3.11) from the ground state $|0\rangle$ – also called the vacuum state – which we take to be normalized, $\langle 0|0\rangle = 1$,

$$|n\rangle = \mathcal{N}_n \hat{a}^{\dagger n}|0\rangle. \tag{3.12}$$

The value of $\mathcal{N}_n$ is determined by the normalization condition

$$\langle n|n\rangle = 1, \tag{3.13}$$

which by insertion of Eq. (3.12) into Eq. (3.13) reads

$$|\mathcal{N}_n|^2 \langle 0|\hat{a}^n \hat{a}^{\dagger n}|0\rangle = 1. \tag{3.14}$$

The simplest way of calculating the vacuum expectation value in Eq. (3.14) is to bring the operator product $\hat{a}^n \hat{a}^{\dagger n}$ into normal order. Applying Eq. (C.34) yields

$$\hat{a}^n \hat{a}^{\dagger n} = :\left(\hat{a} + \frac{\partial}{\partial \hat{a}^\dagger}\right)^n \hat{a}^{\dagger n}: = \sum_{l=0}^{n} \binom{n}{l} :\hat{a}^{n-l} \left(\frac{\partial}{\partial \hat{a}^\dagger}\right)^l \hat{a}^{\dagger n}:$$

$$= \sum_{l=0}^{n} \binom{n}{l} \frac{n!}{(n-l)!} \hat{a}^{\dagger n-l} \hat{a}^{n-l}, \tag{3.15}$$

so that we may rewrite Eq. (3.14) as

$$|\mathcal{N}_n|^2 \sum_{l=0}^{n} \binom{n}{l} \frac{n!}{(n-l)!} \langle 0|\hat{a}^{\dagger n-l} \hat{a}^{n-l}|0\rangle = 1. \tag{3.16}$$

Since quanta cannot be annihilated from the vacuum state, cf. Eq. (3.10), we conclude that only the term with $n-l=0$ contributes to the sum of Eq. (3.16), so that

$$\mathcal{N}_n = \frac{1}{\sqrt{n!}}, \tag{3.17}$$

where we have chosen $\mathcal{N}_n$ to be real-valued. Inserting Eq. (3.17) into Eq. (3.12), we obtain a rule for creating from the vacuum state all the number states $|n\rangle$, which are the eigenstates of the Hamiltonian (3.2),

$$|n\rangle = \frac{1}{\sqrt{n!}} \hat{a}^{\dagger n} |0\rangle. \tag{3.18}$$

This implies also the following relations for the (normalized) number states:

$$\hat{a}^\dagger |n\rangle = \frac{1}{\sqrt{n!}} \hat{a}^{\dagger n+1} |0\rangle = \sqrt{n+1} \, |n+1\rangle, \tag{3.19}$$

$$\hat{a} |n\rangle = \frac{1}{\sqrt{n!}} \hat{a} \hat{a}^{\dagger n} |0\rangle = \frac{1}{\sqrt{n!}} [\hat{a}, \hat{a}^{\dagger n}] |0\rangle$$

$$= \frac{n}{\sqrt{n!}} \hat{a}^{\dagger n-1} |0\rangle = \sqrt{n} \, |n-1\rangle, \tag{3.20}$$

where the relations (C.16) and (3.10) have been used for the derivation of Eq. (3.20).

As already mentioned, the number states $|n\rangle$ and $|m\rangle$ with $n \neq m$ are orthogonal, since they are eigenstates of the Hermitian number operator $\hat{n}$,

$$\langle m|n\rangle = \delta_{mn}. \tag{3.21}$$

This orthogonality may also be shown directly, by representing $|n\rangle$ and $|m\rangle$ in the form (3.18) and normally ordering the operators $\hat{a}$ and $\hat{a}^\dagger$. Moreover, the number states form a complete set of orthonormal vectors in the Hilbert space of the single-mode system,

$$\sum_{n=0}^{\infty} |n\rangle\langle n| = \hat{I}, \tag{3.22}$$

where $\hat{I}$ is the unity operator in this Hilbert space. It is now evident that, due to their orthonormality (3.21) and completeness (3.22), the number states are of potential use for many quantum-mechanical calculations.

3.1.1
Statistics of the number states

Let us consider the main properties of number states, focusing on the free radiation field expanded in terms of monochromatic modes. In this case the number states attributed to a mode of frequency ω may be called photon-number states since – for a mode of frequency ω – they are states with a precise number n of energy quanta $\hbar\omega$ which are commonly denoted as photons. The average energy of the mode in a number state is obtained by taking the expectation value $\langle n|\hat{H}|n\rangle$ with $\hat{H}$ according to Eq. (3.2). Since $\langle n|\hat{n}|n\rangle = n$, we readily obtain

$$\langle n|\hat{H}|n\rangle = \hbar\omega(n+\tfrac{1}{2}), \tag{3.23}$$

which shows that an energy $\hbar\omega$ is indeed associated with each photon. However, in the absence of photons, $n=0$, i.e., for the ground state, there is still the zero-point energy of the associated harmonic oscillator left: $\langle 0|\hat{H}|0\rangle = \hbar\omega/2$. This is due to the fact that quantities such as the coordinate and the momentum of the oscillator still reveal fluctuations in the ground state.

A rough measure of the fluctuation of a quantity $\hat{O}$ is the variance $\langle(\Delta\hat{O})^2\rangle$, where $\Delta\hat{O} = \hat{O} - \langle\hat{O}\rangle$. Since the number of quanta is well defined in a number state, the number variance for such a state of course vanishes,

$$\langle n|(\Delta\hat{n})^2|n\rangle = \langle n|\hat{n}^2|n\rangle - \langle n|\hat{n}|n\rangle^2 = 0, \tag{3.24}$$

as also does the energy variance.

Let us now consider the expectation value and the variance of a coordinate-like quantity such as the electric field of a single-mode radiation field,

$$\hat{\mathbf{E}}(\mathbf{r}) = i\omega[\hat{a}\,\mathbf{A}(\mathbf{r}) - \hat{a}^\dagger \mathbf{A}^*(\mathbf{r})] \tag{3.25}$$

[cf. Eq. (2.70)], where $\mathbf{A}(\mathbf{r})$ is the corresponding mode function. Using Eqs (3.19)–(3.21), we readily prove that when the mode is prepared in a photon-number state, then the mean value of the electric-field strength vanishes,

$$\langle n|\hat{\mathbf{E}}(\mathbf{r})|n\rangle = 0. \tag{3.26}$$

That is, a photon-number state is far from representing a (nonfluctuating) classical wave which would naturally reveal a nonvanishing electric field. In view of particle–wave dualism, this type of state is closely related to the particle nature of the radiation rather than to its wave nature. It will be seen later (Chapter 8) that it is a nontrivial problem to experimentally prepare radiation field modes in photon-number states. The expectation value of the intensity

$$\hat{I}(\mathbf{r}) = \hat{\mathbf{E}}^{(-)}(\mathbf{r})\hat{\mathbf{E}}^{(+)}(\mathbf{r}) \tag{3.27}$$

in a photon-number state is proportional to the number of photons in the field:

$$\langle n|\hat{I}(\mathbf{r})|n\rangle = \omega^2 |A(\mathbf{r})|^2 \langle n|\hat{a}^\dagger \hat{a}|n\rangle = \omega^2 |A(\mathbf{r})|^2 n. \tag{3.28}$$

Accordingly, the variance of the kth component of the electric-field strength is found to be

$$\langle n|[\Delta \hat{E}_k(\mathbf{r})]^2|n\rangle = \omega^2 |A_k(\mathbf{r})|^2 (2n+1), \tag{3.29}$$

from which the fluctuation of the electric field is seen to increase with the photon number n. The minimum noise is obtained for $n=0$, that is, in the case of the vacuum field we have

$$\langle 0|[\Delta \hat{E}_k(\mathbf{r})]^2|0\rangle = \omega^2 |A_k(\mathbf{r})|^2. \tag{3.30}$$

This clearly shows that even in the vacuum case, when no photons are present, there are quantum fluctuations of the field.

3.1.2
Multi-mode number states

We now turn to the more general case of a multi-mode system and remember that the Hamiltonian is additively composed of independent single-mode Hamiltonians [Eq. (3.1) together with Eq. (3.2)]. The additivity of the Hamiltonian leads to the fact that its eigenstates are simply products of the single-mode eigenstates, that is, $\hat{H}|\{n_\lambda\}\rangle = E_{\{n_\lambda\}}|\{n_\lambda\}\rangle$, where the eigenstates and energies are given by

$$|\{n_\lambda\}\rangle = \prod_\lambda |n_\lambda\rangle, \tag{3.31}$$

$$E_{\{n_\lambda\}} = \sum_\lambda \hbar \omega_\lambda (n_\lambda + \tfrac{1}{2}). \tag{3.32}$$

In the case of a radiation field, for example, one may consider an experiment where the total number of photons is measured regardless of their frequency or polarization, i.e., regardless of which mode they are in. The measured operator would then be the total-number operator, defined by

$$\hat{N} = \sum_\lambda \hat{n}_\lambda. \tag{3.33}$$

We can easily see that the total-number operator $\hat{N}$ commutes with the Hamiltonian (3.1) [together with Eq. (3.2)] and therefore has the same eigenvectors $|\{n_\lambda\}\rangle$,

$$\hat{N}|\{n_\lambda\}\rangle = N|\{n_\lambda\}\rangle, \tag{3.34}$$

with the total number of photons in the state $|\{n_\lambda\}\rangle$ being

$$N = \sum_\lambda n_\lambda. \tag{3.35}$$

However, since different combinations of numbers n_λ can lead to the same total number N, there is a degeneracy with respect to the operator $\hat{N}$. Taking into account Λ different modes, it can be readily seen that there are Λ^N possibilities of occupying the different single modes to obtain the total number N.

The noise properties of a multi-mode system in a number state are quite similar to those of a single-mode system. In particular, the variance of the total-photon number of a multi-mode radiation field vanishes, as does the mean value of the electric field:

$$\langle\{n_\lambda\}|(\Delta\hat{N})^2|\{n_\lambda\}\rangle = 0, \tag{3.36}$$
$$\langle\{n_\lambda\}|\hat{\mathbf{E}}(\mathbf{r})|\{n_\lambda\}\rangle = 0, \tag{3.37}$$

where $\hat{\mathbf{E}}(\mathbf{r})$ is now meant to be the multi-mode electric-field strength according to Eq. (2.70).

3.2
Coherent states

The quantum states that come closest to the classical ideal are the coherent states. As their name suggests, they indeed show a coherent amplitude, i.e., they reveal the classically expected oscillatory behavior of the harmonic-oscillator coordinates. In order to derive the coherent states and to provide the tools for the squeezed states (Section 3.3), we follow an approach based on unitary transformations. For the sake of clarity we again start from a single-mode system. Performing a unitary transformation $\hat{U}$ ($\hat{U}^\dagger = \hat{U}^{-1}$), the operator $\hat{a}$ transforms to $\hat{a}'$ as

$$\hat{a}' = \hat{U}\hat{a}\hat{U}^\dagger, \tag{3.38}$$

whereas the transformed number states $|n\rangle'$ read

$$|n\rangle' = \hat{U}|n\rangle. \tag{3.39}$$

Clearly, the transformed operators $\hat{a}'$ and $\hat{a}'^\dagger$ again obey the bosonic commutator relations (3.4) and (3.5). Defining the transformed number operator $\hat{n}' = \hat{a}'^\dagger \hat{a}'$ and taking into account Eqs (3.6), (3.19) and (3.20) we readily find the eigenvalue equation for the transformed number operator,

$$\hat{n}'|n\rangle' = n|n\rangle', \tag{3.40}$$

and the actions of the transformed creation and annihilation operators,

$$\hat{a}'^\dagger |n\rangle' = \sqrt{n+1}\,|n+1\rangle', \tag{3.41}$$
$$\hat{a}'|n\rangle' = \sqrt{n}\,|n-1\rangle'. \tag{3.42}$$

The transformed operators and states thus reveal the same algebraic relations as the original ones and therefore can also be used as a complete set of states to span the Hilbert space. In particular, from Eq. (3.42) we see that

$$\hat{a}'|0\rangle' = 0, \tag{3.43}$$

which defines a new ground state with respect to the transformed operators. Clearly, although their algebraic relations do not change, the physical properties of the transformed number states $|n\rangle'$ may drastically differ from the original number states $|n\rangle$. This, however, depends solely on the actual form of the applied unitary transformation $\hat{U}$.

The transformation leading to the coherent states is implemented by the displacement operator $\hat{D}(\alpha)$,

$$\hat{U} \equiv \hat{D}(\alpha) = \exp(\alpha \hat{a}^\dagger - \alpha^* \hat{a}), \tag{3.44}$$

with α being a complex c-number variable. Using Eq. (C.27), we may factorize the displacement operator to obtain its normally and anti-normally ordered forms, respectively,

$$\hat{D}(\alpha) = e^{\alpha \hat{a}^\dagger} e^{-\alpha^* \hat{a}} e^{-|\alpha|^2/2}, \tag{3.45}$$
$$\hat{D}(\alpha) = e^{-\alpha^* \hat{a}} e^{\alpha \hat{a}^\dagger} e^{|\alpha|^2/2}. \tag{3.46}$$

We may therefore write the transformed annihilation operator $\hat{a}'$ as

$$\hat{a}' = \hat{D}(\alpha) \hat{a} \hat{D}^\dagger(\alpha) = e^{\alpha \hat{a}^\dagger} e^{-\alpha^* \hat{a}} \hat{a}\, e^{\alpha^* \hat{a}} e^{-\alpha \hat{a}^\dagger} = e^{\alpha \hat{a}^\dagger} \hat{a}\, e^{-\alpha \hat{a}^\dagger}, \tag{3.47}$$

from which, together with the relation (C.9), it follows that

$$\hat{a}' = \hat{a} - \alpha. \tag{3.48}$$

This result now enables us to rewrite the definition of the transformed ground state (3.43) as

$$(\hat{a} - \alpha)\hat{D}(\alpha)|0\rangle = 0. \tag{3.49}$$

Denoting by $|\alpha\rangle = |0\rangle'$ the transformed ground state which depends parametrically on α,

$$|\alpha\rangle = \hat{D}(\alpha)|0\rangle, \tag{3.50}$$

we see from Eq. (3.49) that for each complex number α the state $|\alpha\rangle$ is a right-hand eigenstate of the non-Hermitian annihilation operator $\hat{a}$ with eigenvalue α,

$$\hat{a}|\alpha\rangle = \alpha|\alpha\rangle. \tag{3.51}$$

From Eq. (3.51) we further see that correspondingly $\langle\alpha|$ is a left-hand eigenstate of $\hat{a}^\dagger$,

$$\langle\alpha|\hat{a}^\dagger = \langle\alpha|\alpha^*. \tag{3.52}$$

The states $|\alpha\rangle$, being normalized to unity ($\langle\alpha|\alpha\rangle = 1$), are called coherent states or Glauber states [Schrödinger (1926); Klauder (1960); Glauber (1963a,b,c)]. The amplitude α determines a point in a complex phase space which corresponds to a coherent amplitude of the corresponding harmonic oscillation, i.e., $\langle\alpha|\hat{a}|\alpha\rangle = \alpha$. This phase-space amplitude or coherent excitation can be changed by use of the displacement operator, which can be seen by first considering the action of two subsequent displacements,

$$\hat{D}(\alpha)\hat{D}(\beta) = \hat{D}(\alpha + \beta)\exp[i\mathrm{Im}(\alpha\beta^*)]. \tag{3.53}$$

From this equation together with Eq. (3.50) we see that, apart from a phase factor, the action of a displacement operator $\hat{D}(\beta)$ on a coherent state $|\alpha\rangle$ creates a new coherent state with amplitude $\alpha + \beta$,

$$\hat{D}(\beta)|\alpha\rangle = e^{-i\mathrm{Im}(\alpha\beta^*)}|\alpha + \beta\rangle, \tag{3.54}$$

i.e., the operator $\hat{D}(\beta)$ displaces the phase-space amplitude of the coherent state by the amount β. In particular, this also shows that the ground state $|0\rangle$ can be regarded as being the coherent state of amplitude $\alpha = 0$, from which, by application of the displacement operator (3.44), all possible coherent states can be obtained, in agreement with Eq. (3.50).

The action of $\hat{a}^\dagger$ on $|\alpha\rangle$ and $\hat{a}$ on $\langle\alpha|$, respectively, can be derived as follows. Applying $\hat{a}^\dagger$ to $|\alpha\rangle = \hat{D}(\alpha)|0\rangle$ and using the normally ordered form of the displacement operator (3.45) yields

$$\begin{aligned}\hat{a}^\dagger|\alpha\rangle &= \hat{a}^\dagger \exp[(\hat{a}^\dagger - \tfrac{1}{2}\alpha^*)\alpha]\exp(-\alpha^*\hat{a})|0\rangle = \hat{a}^\dagger \exp[(\hat{a}^\dagger - \tfrac{1}{2}\alpha^*)\alpha]|0\rangle \\ &= \left(\frac{\partial}{\partial\alpha} + \frac{\alpha^*}{2}\right)\exp[(\hat{a}^\dagger - \tfrac{1}{2}\alpha^*)\alpha]|0\rangle,\end{aligned} \tag{3.55}$$

and hence

$$\hat{a}^\dagger |\alpha\rangle = \left(\frac{\partial}{\partial \alpha} + \frac{\alpha^*}{2}\right)|\alpha\rangle. \tag{3.56}$$

Accordingly, applying $\hat{a}$ to $\langle\alpha|$ and using the anti-normally ordered form (3.46) yields

$$\langle\alpha|\hat{a} = \langle\alpha|\left(\overleftarrow{\frac{\partial}{\partial \alpha^*}} + \frac{\alpha}{2}\right), \tag{3.57}$$

where the derivative is supposed to act to the left side.

Clearly, the coherent states $|\alpha\rangle$ can be expanded in terms of the number states $|n\rangle$ by use of their completeness relation (3.22),

$$|\alpha\rangle = \sum_{n=0}^{\infty} |n\rangle\langle n|\alpha\rangle. \tag{3.58}$$

The expansion coefficients $\langle n|\alpha\rangle$ can be calculated by means of Eqs (3.18) and (3.50), resulting in

$$\langle n|\alpha\rangle = \frac{\alpha^n}{\sqrt{n!}} e^{-|\alpha|^2/2}. \tag{3.59}$$

From Eqs (3.58) and (3.59) we immediately find that the number distribution of a coherent state is a Poissonian:

$$|\langle n|\alpha\rangle|^2 = \frac{(|\alpha|^2)^n}{n!} e^{-|\alpha|^2}, \tag{3.60}$$

with mean value and variance both being given by $|\alpha|^2$,

$$\langle\alpha|\hat{n}|\alpha\rangle = \langle\alpha|(\Delta\hat{n})^2|\alpha\rangle = |\alpha|^2. \tag{3.61}$$

Hence the number of quanta is a fluctuating quantity for a coherent state $|\alpha\rangle$.

We recall that the coherent states are eigenstates of a non-Hermitian operator. In comparison with the eigenstates of Hermitian operators, they therefore exhibit some unusual features. They are over-complete and nonorthogonal. Let us consider two coherent states $|\alpha\rangle$ and $|\beta\rangle$ (with $\alpha \neq \beta$) and calculate their overlap $\langle\beta|\alpha\rangle$. From Eqs (3.51) and (3.57) we obtain the relation

$$\langle\beta|\hat{a}|\alpha\rangle = \alpha\langle\beta|\alpha\rangle = \left(\frac{\partial}{\partial \beta^*} + \frac{\beta}{2}\right)\langle\beta|\alpha\rangle, \tag{3.62}$$

which represents a differential equation for $\langle\beta|\alpha\rangle$:

$$\frac{\partial}{\partial \beta^*}\langle\beta|\alpha\rangle = \left(\alpha - \frac{\beta}{2}\right)\langle\beta|\alpha\rangle. \tag{3.63}$$

Taking into account the boundary condition $\langle \alpha | \alpha \rangle = 1$, the solution is obtained as

$$\langle \beta | \alpha \rangle = \exp\left[-\tfrac{1}{2}|\alpha - \beta|^2 + \tfrac{1}{2}(\alpha \beta^* - \alpha^* \beta)\right], \tag{3.64}$$

and hence the squared modulus reads

$$|\langle \beta | \alpha \rangle|^2 = \exp(-|\alpha - \beta|^2). \tag{3.65}$$

Equation (3.65) clearly shows that for $\alpha \neq \beta$ the states $|\alpha\rangle$ and $|\beta\rangle$ are indeed not orthogonal to each other. However, if the values of α and β are sufficiently separated, so that $|\alpha - \beta| \gg 1$, they may be regarded as being approximately orthogonal.

To show that the coherent states resolve the identity, we recall the completeness relation for the number states, which enables us to write

$$\int d^2\alpha \, |\alpha\rangle\langle\alpha| = \sum_{n,m=0}^{\infty} \int d^2\alpha \, |n\rangle\langle n|\alpha\rangle\langle\alpha|m\rangle\langle m|, \tag{3.66}$$

where the integration is performed over the real and imaginary parts, $\alpha' \equiv \mathrm{Re}\,\alpha$ and $\alpha'' \equiv \mathrm{Im}\,\alpha$, respectively, of $\alpha = \alpha' + i\alpha''$,

$$d^2\alpha = d\alpha' d\alpha''. \tag{3.67}$$

We now use Eq. (3.59) and rewrite Eq. (3.66) as

$$\int d^2\alpha \, |\alpha\rangle\langle\alpha| = \sum_{n,m=0}^{\infty} \frac{|m\rangle\langle n|}{\sqrt{m!\, n!}} \int d^2\alpha \, \alpha^m \alpha^{*n} e^{-|\alpha|^2}. \tag{3.68}$$

The integral in Eq. (3.68) can be evaluated to be

$$\int d^2\alpha \, \alpha^m \alpha^{*n} e^{-|\alpha|^2} = \pi n! \delta_{mn}, \tag{3.69}$$

and we arrive at

$$\int d^2\alpha \, |\alpha\rangle\langle\alpha| = \pi \sum_{n=0}^{\infty} |n\rangle\langle n| = \pi \hat{I}, \tag{3.70}$$

from which we see that the identity can be resolved as

$$\frac{1}{\pi} \int d^2\alpha \, |\alpha\rangle\langle\alpha| = \hat{I}. \tag{3.71}$$

That is to say, any state $|\psi\rangle$ can be expanded in terms of the coherent states as follows:

$$|\psi\rangle = \frac{1}{\pi} \int d^2\alpha \, |\alpha\rangle\langle\alpha|\psi\rangle. \tag{3.72}$$

From Eq. (3.72) together with Eq. (3.65) we see that there is a nontrivial expansion of a coherent state in terms of coherent states, which indicates the over-completeness of the coherent states. To demonstrate the over-completeness more explicitly, let us consider a subset of coherent states whose modulus r of the complex amplitude, $\alpha = re^{i\varphi}$, is chosen to be constant. This represents the set of coherent states on a circle in phase space. Using Eq. (3.59) the number representation of these states is given as

$$|re^{i\varphi}\rangle = e^{-r^2/2} \sum_{n=0}^{\infty} \frac{r^n}{\sqrt{n!}} e^{in\varphi}|n\rangle. \tag{3.73}$$

Fourier transforming this equation with respect to the phase φ, we readily derive a representation of the number states $|n\rangle$ ($n=0,1,\ldots$) in terms of the coherent states on a circle,

$$|n\rangle = \frac{\sqrt{n!}}{2\pi r^n} e^{r^2/2} \int_0^{2\pi} d\varphi \, e^{-in\varphi}|re^{i\varphi}\rangle. \tag{3.74}$$

This result reveals that the complete number-state basis can be expressed in terms of the coherent states on any chosen circle in phase space. Equivalently, complete sets of coherent states can also be chosen on other contours, for example on a straight line [Adam, Földesi, and Janszky (1994)].

3.2.1
Statistics of the coherent states

To illustrate the main difference between coherent states and number states, let us focus on a radiation-field mode. We have already seen that in the case of the mode being in a coherent state $|\alpha\rangle$ the probability of finding n photons obeys a Poissonian distribution [Eq. (3.60)]. Hence both the mean number of photons and its variance are given by the squared modulus of the complex amplitude, $|\alpha|^2$. Next let us again consider a quantity of the type of the electric-field strength. From Eq. (3.25) together with Eqs (3.51) and (3.52) we derive for the mean value of the kth component of the electric-field strength

$$\langle \alpha | \hat{E}_k(\mathbf{r}) | \alpha \rangle = i\omega[A_k(\mathbf{r})\alpha - A_k^*(\mathbf{r})\alpha^*]. \tag{3.75}$$

That is, when the mode is prepared in a coherent state, then the mean electric-field strength looks like the electric-field strength of a coherent, classical mode with (complex) amplitude α. Notwithstanding this resemblance, there is a fundamental difference between a classical and a quantum mode in a coherent state, because of the vacuum noise inherent in the quantum system. Calculating the variance of the kth component of the electric-field strength, we easily derive

$$\langle \alpha | [\Delta \hat{E}_k(\mathbf{r})]^2 | \alpha \rangle = \omega^2 |A_k(\mathbf{r})|^2. \tag{3.76}$$

Comparing this with Eq. (3.30), we see that the noise of the electric field is indeed determined by the vacuum level, independent of the field amplitude. Rewriting Eq. (3.75) as

$$\langle \alpha | \hat{E}_k(\mathbf{r}) | \alpha \rangle = 2\omega |A_k(\mathbf{r})| |\alpha| \sin \varphi_{E_k}, \qquad (3.77)$$

where φ_{E_k} is the phase of the (kth component of the) mean value of the electric-field strength, and using Eq. (3.76), we can easily calculate the relative noise of the electric field, obtaining

$$\left\{ \frac{\langle \alpha | [\Delta \hat{E}_k(\mathbf{r})]^2 | \alpha \rangle}{[\langle \alpha | \hat{E}_k(\mathbf{r}) | \alpha \rangle]^2} \right\}^{\frac{1}{2}} = \frac{1}{2|\alpha| |\sin \varphi_{E_k}|}. \qquad (3.78)$$

Equation (3.78) reveals that (for $\sin \varphi_{E_k} \neq 0$) the relative noise decreases with increasing absolute value of α or, according to Eq. (3.61), with the square root of the mean photon number. The coherent states $|\alpha\rangle$ may therefore be regarded as being those quantum states that correspond most closely to classical, coherent waves. Without going into the detail of quantum coherence theory [see, e. g., Peřina (1985); Mandel and Wolf (1995)], we note that with a radiation field being prepared in a coherent state, normally ordered correlation functions factorize perfectly, that is, the coherence condition is satisfied up to any order:

$$\langle \alpha | (\hat{E}_k^{(-)})^m (\hat{E}_k^{(+)})^n | \alpha \rangle = (\langle \alpha | \hat{E}_k^{(-)} | \alpha \rangle)^m (\langle \alpha | \hat{E}_k^{(+)} | \alpha \rangle)^n. \qquad (3.79)$$

3.2.2
Multi-mode coherent states

The extension of the concept of coherent states to multi-mode systems is straightforward. Similar to the case of number states, the multi-mode coherent states $|\{\alpha_\lambda\}\rangle$ are simply obtained by taking the (direct) product of single-mode coherent states, that is,

$$|\{\alpha_\lambda\}\rangle = \prod_\lambda |\alpha_\lambda\rangle, \qquad (3.80)$$

and the identity operator in the multi-mode Hilbert space then reads

$$\prod_\lambda \left(\frac{1}{\pi} \int d^2\alpha_\lambda |\alpha_\lambda\rangle\langle\alpha_\lambda| \right) = \hat{I}. \qquad (3.81)$$

In view of a multi-mode radiation field, the mean value of the total number of photons in all modes, $\hat{N} = \sum_\lambda \hat{n}_\lambda$, and the corresponding photon-number variance are given by

$$\langle \{\alpha_\lambda\} | \hat{N} | \{\alpha_\lambda\} \rangle = \sum_\lambda |\alpha_\lambda|^2, \qquad (3.82)$$

$$\langle \{\alpha_\lambda\} | (\Delta \hat{N})^2 | \{\alpha_\lambda\} \rangle = \langle \{\alpha_\lambda\} | \hat{N} | \{\alpha_\lambda\} \rangle, \qquad (3.83)$$

and the mean value and the variance of the kth component of the electric-field strength are, respectively,

$$\langle\{\alpha_\lambda\}|\hat{E}_k(\mathbf{r})|\{\alpha_\lambda\}\rangle = \sum_\lambda i\omega_\lambda [A_{\lambda,k}(\mathbf{r})\alpha_\lambda - A^*_{\lambda,k}(\mathbf{r})\alpha^*_\lambda], \qquad (3.84)$$

$$\langle\{\alpha_\lambda\}|[\Delta\hat{E}_k(\mathbf{r})]^2|\{\alpha_\lambda\}\rangle = \sum_\lambda \omega_\lambda^2 |A_{\lambda,k}(\mathbf{r})|^2. \qquad (3.85)$$

To illustrate how a classical light pulse emerges from a radiation field in a coherent state $|\{\alpha_\lambda\}\rangle$, let us consider a multi-mode radiation field propagating in the positive x direction in free space. To take into account the temporal evolution of the (free) radiation field, we recall that in the Heisenberg picture the photon annihilation evolves as

$$\hat{a}_\lambda(t) = \hat{a}_\lambda e^{-i\omega_\lambda t}, \qquad (3.86)$$

where $\hat{a}_\lambda = \hat{a}_\lambda(0)$ is the annihilation operator at some initial time $t=0$. Assuming that the (nonevolving) initial state vector is a multi-mode coherent state as given in Eq. (3.80), and using the traveling-wave mode functions of frequency ω_l, polarization $\mathbf{e}_{l,\sigma}$ and quantization volume $\mathcal{AL}$,

$$\mathbf{A}_\lambda(\mathbf{r}) \mapsto \mathbf{A}_{l\sigma}(x) = \left(\frac{\hbar}{2\epsilon_0 \omega_l \mathcal{AL}}\right)^{\frac{1}{2}} \mathbf{e}_{l,\sigma} e^{i\omega_l x/c}, \qquad (3.87)$$

we derive from Eqs (3.86) and (3.87) the kth component of the electric-field operator as

$$\hat{E}_k(x,t) = i\sum_{l,\sigma} \sqrt{\frac{\hbar \omega_l}{2\epsilon_0 \mathcal{AL}}} (\mathbf{e}_{l,\sigma})_k \hat{a}_{l,\sigma} e^{-i\omega_l(t-x/c)} + \text{H.c.} \qquad (3.88)$$

[cf. Eqs (2.87) and (2.88)]. We now perform the limit of infinite propagation length, $\mathcal{L} \to \infty$, while the diameter or beam waist $\mathcal{A}$ is held constant. With increasing $\mathcal{L}$ the modes become more and more dense in the frequency domain, because $\omega_l = 2\pi c l/\mathcal{L}$. Defining the operators

$$\hat{a}_\sigma(\omega) = \lim_{\mathcal{L} \to \infty} \frac{\hat{a}_{l,\sigma}}{\sqrt{\Delta\omega}} \qquad (3.89)$$

with $\Delta\omega = 2\pi c/\mathcal{L}$ [cf. Eqs (2.89) and (2.90)], we see that in the limit $\mathcal{L} \to \infty$ the l sum in Eq. (3.88) can be written as an integral and the positive-frequency part of the electric-field operator becomes

$$\hat{E}_k^{(+)}(x,t) = i\sum_\sigma \int_0^\infty d\omega \sqrt{\frac{\hbar\omega}{4\pi\epsilon_0 c\mathcal{A}}} [\mathbf{e}_\sigma(\omega)]_k e^{-i\omega(t-x/c)} \hat{a}_\sigma(\omega). \qquad (3.90)$$

Correspondingly, the mean electric field in a coherent state reads

$$\langle \hat{E}_k(x,t) \rangle_{\text{coh}} = i \sum_\sigma \int_0^\infty d\omega \sqrt{\frac{\hbar\omega}{4\pi\epsilon_0 c \mathcal{A}}} [e_\sigma(\omega)]_k \alpha_\sigma(\omega) e^{-i\omega(t-x/c)} + \text{c.c.}$$

(3.91)

Equation (3.91) is capable of describing (the kth component of the electric-field strength of) a coherent light pulse. To give an example, let us consider the case where the polarization unit vectors $e_\sigma(\omega)$ are independent of frequency and suppose that the mode amplitudes are polarization independent,

$$\alpha_\sigma(\omega) \equiv \alpha(\omega) = |\alpha(\omega)| e^{i\varphi},$$

(3.92)

with the photon spectrum being of Gaussian form with center frequency $\bar{\omega}$ and spectral width $\Delta\omega \ll \bar{\omega}$,

$$|\alpha(\omega)|^2 = \frac{\langle \hat{N} \rangle_{\text{coh}}}{\sqrt{2\pi}\,\Delta\omega} \exp\left[-\frac{1}{2}\left(\frac{\omega - \bar{\omega}}{\Delta\omega}\right)^2\right],$$

(3.93)

where $\langle \hat{N} \rangle_{\text{coh}}$ is the total number of photons of the light pulse:

$$\langle \hat{N} \rangle_{\text{coh}} = \int_0^\infty d\omega\, |\alpha(\omega)|^2.$$

(3.94)

Combining Eqs (3.91)–(3.93) and taking into account that the spectral width of the electric-field strength is small compared with the center frequency ($\Delta\omega \ll \bar{\omega}$), we obtain for the mean electric field

$$\langle \hat{E}_k(x,t) \rangle_{\text{coh}} = 2\sqrt{\frac{\Delta\omega\,\hbar\bar{\omega}\,\langle \hat{N} \rangle_{\text{coh}}}{\epsilon_0 c \mathcal{A}\sqrt{2\pi}}} \exp\left\{-\left[\Delta\omega\left(t-\frac{x}{c}\right)\right]^2\right\} \sin\left[\bar{\omega}\left(t-\frac{x}{c}\right)-\varphi\right],$$

(3.95)

which represents an unpolarized, coherent Gaussian light pulse traveling in positive x direction.

3.2.3
Displaced number states

At this point it should be noted that the coherent states are a special class of states with respect to the transformation given in Eqs (3.39) and (3.44), since they are defined by the action of the displacement operator on the ground state $|0\rangle$. A broader class of states is obtained by considering the transformed states that emerge from the application of the displacement operator on arbitrary number states $|n\rangle$. Such states are denoted as displaced number states and they are defined via Eqs (3.39) and (3.44) as

$$|n, \alpha\rangle = \hat{D}(\alpha)|n\rangle.$$

(3.96)

From the general transformed eigenvalue equation (3.40) we can see that the displaced number states are eigenstates of the displaced number operator, i. e., $\hat{n}(\alpha)|n,\alpha\rangle = n|n,\alpha\rangle$, with

$$\hat{n}(\alpha) = \hat{D}(\alpha)\hat{n}\hat{D}^\dagger(\alpha) = (\hat{a}^\dagger - \alpha^*)(\hat{a} - \alpha), \qquad (3.97)$$

where we have made use of Eq. (3.48).

Clearly, for a fixed displacement α these states are orthonormal, just as the number states are:

$$\langle n,\alpha|m,\alpha\rangle = \langle n|\hat{D}^\dagger(\alpha)\hat{D}(\alpha)|m\rangle = \delta_{nm}, \qquad (3.98)$$

and for arbitrary α the identity operator can be resolved as

$$\sum_{n=0}^{\infty} |n,\alpha\rangle\langle n,\alpha| = \hat{I}. \qquad (3.99)$$

Moreover, for arbitrary n the identity can also be resolved by an integral over the displacement amplitude, in analogy with Eq. (3.71),

$$\frac{1}{\pi}\int d^2\alpha\, |n,\alpha\rangle\langle n,\alpha| = \hat{I}. \qquad (3.100)$$

Finally, their scalar product with number states can be shown to be expressible in terms of the Laguerre polynomials $L_n^{(k)}(x)$ as

$$\langle n|m,\alpha\rangle = (-\alpha^*)^{m-n}\sqrt{\frac{n!}{m!}}\, L_n^{(m-n)}(|\alpha|^2)\, e^{-|\alpha|^2/2} \qquad (m\geq n), \qquad (3.101)$$

and $\langle n|m,\alpha\rangle = (\langle m|n,-\alpha\rangle)^*$ for the coefficients with $m<n$.

3.3
Squeezed states

Another important class of quantum states are the squeezed states or more precisely the quadrature-squeezed states. For the purpose of deriving these states we return to the unitary transformation, Eqs (3.38) and (3.39), which was employed to derive the coherent states, and assume that the unitary operator $\hat{U}$ is now the squeeze operator,

$$\hat{U} \equiv \hat{S}(\zeta) = \exp\left[\tfrac{1}{2}(\zeta^*\hat{a}^2 - \zeta\hat{a}^{\dagger 2})\right], \qquad (3.102)$$

with ζ – the squeezing parameter – being a complex number. To obtain the relation between the transformed annihilation and creation operators $\hat{a}'$, $\hat{a}'^\dagger$ and the original ones, it is convenient to define the operators

$$\hat{G}(z) = \hat{S}^z(\zeta), \qquad (3.103)$$
$$\hat{a}(z) = \hat{G}(z)\hat{a}\hat{G}^\dagger(z), \qquad (3.104)$$

where z is a real number (the parameter ξ has been omitted for notational convenience). Comparing Eqs (3.38) and (3.104) we see that the original and the transformed annihilation operators are recovered from $\hat{a}(z)$ for $z=0$ and $z=1$, respectively,

$$\hat{a}(z)|_{z=0} = \hat{a}, \tag{3.105}$$
$$\hat{a}(z)|_{z=1} = \hat{a}'. \tag{3.106}$$

Using Eqs (3.102) and (3.103) the derivative of the operator (3.104) with respect to z can be obtained as

$$\frac{d\hat{a}(z)}{dz} = \frac{d\hat{G}(z)}{dz}\hat{a}\hat{G}^\dagger(z) + \hat{G}(z)\hat{a}\frac{d\hat{G}^\dagger(z)}{dz}$$
$$= [\hat{a}(z), \tfrac{1}{2}\{\xi\hat{a}^{\dagger 2}(z) - \xi^*\hat{a}^2(z)\}] = \tfrac{1}{2}\xi[\hat{a}(z), \hat{a}^{\dagger 2}(z)]. \tag{3.107}$$

Since Eq. (3.104) describes a unitary transformation, the operators $\hat{a}(z)$ and $\hat{a}^\dagger(z)$ again obey the bosonic commutator relation (3.4) and we obtain from Eq. (3.107)

$$\frac{d\hat{a}(z)}{dz} = \xi\hat{a}^\dagger(z), \quad \frac{d\hat{a}^\dagger(z)}{dz} = \xi^*\hat{a}(z). \tag{3.108}$$

The solution to the differential equations (3.108) reads

$$\hat{a}(z) = \hat{c}_1 e^{|\xi|z} + \hat{c}_2 e^{-|\xi|z}, \tag{3.109}$$

where the operators $\hat{c}_1$ and $\hat{c}_2$ are determined, according to Eqs (3.105) and (3.108), by the initial conditions

$$\hat{a}(z)|_{z=0} = \hat{c}_1 + \hat{c}_2 = \hat{a}, \tag{3.110}$$
$$\left.\frac{d\hat{a}(z)}{dz}\right|_{z=0} = |\xi|(\hat{c}_1 - \hat{c}_2) = \xi\hat{a}^\dagger. \tag{3.111}$$

Combining Eqs (3.109)–(3.111), after some algebra we finally arrive at ($\varphi_\xi = \arg(\xi)$)

$$\hat{a}(z) = \hat{a}\cosh(|\xi|z) + \hat{a}^\dagger e^{i\varphi_\xi}\sinh(|\xi|z), \tag{3.112}$$

from which we obtain, according to Eq. (3.106), the transformed operators $\hat{a}'$ and $\hat{a}'^\dagger$ in terms of the original ones:[1]

$$\hat{a}' = \mu\hat{a} + \nu\hat{a}^\dagger, \tag{3.113}$$
$$\hat{a}'^\dagger = \mu\hat{a}^\dagger + \nu^*\hat{a}, \tag{3.114}$$

1) Note that this is a SU(1,1) group transformation.

where the parameters μ and ν are defined by

$$\mu = \cosh|\xi|, \tag{3.115}$$
$$\nu = e^{i\varphi_\xi} \sinh|\xi|. \tag{3.116}$$

As mentioned before, since the operators $\hat{a}'$ and $\hat{a}'^\dagger$ are obtained from the operators $\hat{a}$ and $\hat{a}^\dagger$ by a unitary transformation, the (equal-time) commutator relation is preserved,

$$[\hat{a}', \hat{a}'^\dagger] = 1. \tag{3.117}$$

Inserting Eqs (3.113) and (3.114) into Eq. (3.117), it follows that the parameters μ and ν obey the following constraint:

$$\mu^2 - |\nu|^2 = 1, \tag{3.118}$$

which apparently is provided by their definitions (3.115) and (3.116).

As a unitary transformation $\hat{U}$ that can be used to define the squeezed coherent states, in a similar way to the coherent states (cf. Section 3.2), let us consider now the unitary transformation

$$\hat{U}(\xi, \beta) = \hat{S}(\xi)\hat{D}(\beta). \tag{3.119}$$

It first coherently displaces by an amplitude β and then squeezes with squeezing parameter ξ. Analogously to the definition of the coherent states [Eq. (3.50)], we may now define the states $|\xi, \beta\rangle$ by applying the transformation (3.119) onto the ground state $|0\rangle$,

$$|\xi, \beta\rangle = \hat{U}(\xi, \beta)|0\rangle = \hat{S}(\xi)\hat{D}(\beta)|0\rangle = \hat{S}(\xi)|\beta\rangle. \tag{3.120}$$

The states $|\xi, \beta\rangle$ (sometimes denoted by $|\mu, \nu; \beta\rangle$) are called squeezed coherent states [Stoler (1970, 1971); Yuen (1976); for reviews see Walls (1983); Loudon and Knight (1987)].

From the relations (3.120) we see that applying the displacement operator $\hat{D}(\beta)$ to the vacuum state $|0\rangle$ and then applying the squeeze operator $\hat{S}(\xi)$ to the resulting coherent state $|\beta\rangle$ yields the squeezed coherent state $|\xi, \beta\rangle$. However, we may arrive at the same result if we first apply the squeeze operator $\hat{S}(\xi)$ to the vacuum state $|0\rangle$ in order to generate the squeezed ground (or vacuum) state,

$$|\xi, 0\rangle = \hat{S}(\xi)|0\rangle \tag{3.121}$$

and then apply the transformed displacement operator $\hat{D}'(\beta)$ to this state:

$$|\xi, \beta\rangle = \hat{D}'(\beta)|\xi, 0\rangle = \hat{D}'(\beta)\hat{S}(\xi)|0\rangle, \tag{3.122}$$

where the transformed displacement operator, which has been used here, reads

$$\hat{D}'(\beta) = \hat{S}(\xi)\hat{D}(\beta)\hat{S}^\dagger(\xi) = \exp(\beta\hat{a}'^\dagger - \beta^*\hat{a}'). \tag{3.123}$$

By means of Eqs (3.113) and (3.114) it can be written in terms of the operators $\hat{a}$ and $\hat{a}^\dagger$ as

$$\hat{D}'(\beta) = \hat{D}(\beta') = \exp(\beta'\hat{a}^\dagger - \beta'^*\hat{a}), \tag{3.124}$$

where the transformed amplitude β' is given by

$$\beta' = \mu\beta - \nu\beta^*. \tag{3.125}$$

Hence the squeezed coherent states can also be obtained by first squeezing the ground state and then displacing it by a modified amplitude β':

$$|\xi,\beta\rangle = \hat{D}(\beta')\hat{S}(\xi)|0\rangle = \hat{D}(\beta')|\xi,0\rangle. \tag{3.126}$$

Taking into consideration that $0 = \hat{U}(\xi,\beta)\hat{a}|0\rangle = \hat{U}(\xi,\beta)\hat{a}\hat{U}^\dagger(\xi,\beta)|\xi,\beta\rangle = \hat{D}'(\beta)\hat{a}'\hat{D}'^\dagger(\beta)|\xi,\beta\rangle = (\hat{a}' - \beta)|\xi,\beta\rangle$ [cf. Eqs (3.47)–(3.49)], we see that the squeezed coherent states are the right-hand eigenstates of the transformed annihilation operator,

$$\hat{a}'|\xi,\beta\rangle = \beta|\xi,\beta\rangle, \tag{3.127}$$

which, in combination with Eq. (3.113), can be regarded as an alternative definition of these states [Yuen (1976)].[2]

Analogously to the coherent states, the squeezed coherent states are overcomplete and nonorthogonal. To prove that they resolve the identity with respect to the coherent amplitude, we use Eq. (3.120) and recall Eq. (3.71):

$$\frac{1}{\pi}\int d^2\beta\,|\xi,\beta\rangle\langle\xi,\beta| = \hat{S}(\xi)\left(\frac{1}{\pi}\int d^2\beta\,|\beta\rangle\langle\beta|\right)\hat{S}^\dagger(\xi)$$
$$= \hat{S}(\xi)\hat{I}\hat{S}^\dagger(\xi) = \hat{I}. \tag{3.128}$$

It is also easily seen that the nonorthogonality with respect to different coherent amplitudes but equal squeezing parameters is the same as for the coherent states:

$$\langle\xi,\alpha|\xi,\beta\rangle = \langle\alpha|\hat{S}^\dagger(\xi)\hat{S}(\xi)|\beta\rangle = \langle\alpha|\beta\rangle. \tag{3.129}$$

---

[2] In Yuen's approach to squeezed coherent states (also called two-photon coherent states) the parameter μ is chosen to be complex, μ and ν obeying the condition $|\mu|^2 - |\nu|^2 = 1$. Since here only the phase difference $\arg(\nu) - \arg(\mu)$ is relevant, without loss of generality, $\arg(\mu)$ may be chosen to be zero.

Without going into the details of calculation, we note that the squeeze operator $\hat{S}(\xi)$, Eq. (3.102), can be rewritten, on applying exponential-operator disentangling, as[3]

$$\hat{S}(\xi) = \exp\left(-\frac{\nu}{2\mu}\hat{a}^{\dagger 2}\right)\left(\frac{1}{\mu}\right)^{\hat{n}+\frac{1}{2}}\exp\left(\frac{\nu^*}{2\mu}\hat{a}^2\right), \qquad (3.130)$$

with μ and ν from Eqs (3.115) and (3.116). Hence, the squeezed ground state $|\xi,0\rangle$, Eq. (3.121), can be given in the form of

$$|\xi,0\rangle = \frac{1}{\sqrt{\mu}}\exp\left(-\frac{\nu}{2\mu}\hat{a}^{\dagger 2}\right)|0\rangle, \qquad (3.131)$$

and coherent displacement [according to Eqs (3.122) and (3.124)] then yields the squeezed coherent states in the form of

$$|\xi,\beta\rangle = \frac{1}{\sqrt{\mu}}\exp\left[-\frac{\nu}{2\mu}(\hat{a}^\dagger - \beta'^*)^2\right]|\beta'\rangle$$

$$= \frac{1}{\sqrt{\mu}}\exp\left[-\frac{1}{2}|\beta'|^2 + \beta'\hat{a}^\dagger - \frac{\nu}{2\mu}(\hat{a}^\dagger - \beta'^*)^2\right]|0\rangle, \qquad (3.132)$$

where β' is related to β according to Eq. (3.125). With the help of Eq. (3.132) it is not difficult to prove that the scalar products of the squeezed coherent states $|\xi,\beta\rangle$ with the coherent states $|\alpha\rangle$ and the number states $|n\rangle$, respectively, read[4]

$$\langle\alpha|\xi,\beta\rangle = \frac{1}{\sqrt{\mu}}\exp\left(-\frac{|\alpha|^2 + |\beta|^2}{2} + \frac{2\alpha^*\beta - \nu\alpha^{*2} + \nu^*\beta^2}{2\mu}\right), \qquad (3.133)$$

$$\langle n|\xi,\beta\rangle = \frac{[\nu/(2\mu)]^{\frac{n}{2}}}{\sqrt{\mu\, n!}}\exp\left[-\frac{1}{2}\left(|\beta|^2 - \frac{\nu^*}{\mu}\beta^2\right)\right]H_n\left(\frac{\beta}{\sqrt{2\mu\nu}}\right), \qquad (3.134)$$

with $H_n(x)$ being the Hermite polynomial.

3.3.1
Statistics of the squeezed states

To obtain the mean number of quanta, or in the case of a radiation-field mode, the mean photon number,

$$\langle\xi,\beta|\hat{n}|\xi,\beta\rangle = \langle\xi,\beta|\hat{a}^\dagger\hat{a}|\xi,\beta\rangle = \langle\beta|\hat{S}^\dagger(\xi)\hat{a}^\dagger\hat{a}\hat{S}(\xi)|\beta\rangle$$
$$= \langle\beta|\hat{S}^\dagger(\xi)\hat{a}^\dagger\hat{S}(\xi)\hat{S}^\dagger(\xi)\hat{a}\hat{S}(\xi)|\beta\rangle, \qquad (3.135)$$

---

3) Equation (3.130) can be proved correct, applying the differential-equation technique described previously in this section and showing that it leads exactly to Eqs (3.113) and (3.114).
4) Note that the relation $\sum_{n=0}^{\infty}\frac{1}{n!}H_n(x)t^n = \exp(-t^2 + 2tx)$ has been used to derive Eq. (3.134).

3.3 Squeezed states

we first note that $\hat{S}^\dagger(\xi) = \hat{S}(-\xi)$ and that $\mu(-\xi) = \mu(\xi)$ and $\nu(-\xi) = -\nu(\xi)$ [cf. Eqs (3.115) and (3.116)]. From these relations it follows that, on recalling Eq. (3.113),

$$\hat{S}^\dagger(\xi)\hat{a}\hat{S}(\xi) = \hat{S}(-\xi)\hat{a}\hat{S}^\dagger(-\xi) = \mu\hat{a} - \nu\hat{a}^\dagger, \tag{3.136}$$

and we obtain for Eq. (3.135)

$$\begin{aligned}\langle \xi, \beta | \hat{n} | \xi, \beta \rangle &= \langle \beta | (\mu\hat{a}^\dagger - \nu^*\hat{a})(\mu\hat{a} - \nu\hat{a}^\dagger) | \beta \rangle \\ &= \langle \beta | (\mu^2 \hat{a}^\dagger \hat{a} + |\nu|^2 \hat{a}\hat{a}^\dagger - \mu\nu \hat{a}^{\dagger 2} - \mu\nu^* \hat{a}^2) | \beta \rangle \\ &= |\beta'|^2 + |\nu|^2, \end{aligned} \tag{3.137}$$

where $\beta' = \mu\beta - \nu\beta^*$ [Eq. (3.125)]. Analogously, we find for the mean coherent amplitude

$$\langle \xi, \beta | \hat{a} | \xi, \beta \rangle = \langle \beta | \hat{S}^\dagger(\xi) \hat{a} \hat{S}(\xi) | \beta \rangle = \langle \beta | (\mu\hat{a} - \nu\hat{a}^\dagger) | \beta \rangle = \beta', \tag{3.138}$$

which means that, in the case of a single-mode radiation field, for example, the mean value of the electric-field strength takes the same form as when the mode is in the coherent state $|\beta'\rangle$ [cf. Eq.(3.75)], that is,

$$\langle \xi, \beta | \hat{E}_k(\mathbf{r}) | \xi, \beta \rangle = i\omega[A_k(\mathbf{r})\beta' - A_k^*(\mathbf{r})\beta'^*]. \tag{3.139}$$

For $\beta = 0$ (and hence $\beta' = 0$) we see from the last line in Eq. (3.137) that $|\nu|^2$ is the contribution to the mean number of quanta coming from the squeezed ground state $|\xi, 0\rangle$,

$$\langle \xi, 0 | \hat{n} | \xi, 0 \rangle = |\nu|^2. \tag{3.140}$$

The first term in the last line in Eq. (3.137) obviously constitutes the mean number of quanta coming from the coherent amplitude β' implemented when the squeezed ground state $|\xi, 0\rangle$ [Eq. (3.121)] in phase space is displaced by β' to generate the squeezed coherent state $|\xi, \beta\rangle$ [Eq. (3.122)], which is quite similar to the generation of the coherent state $|\beta'\rangle$ from the ordinary ground state $|0\rangle$. Hence $|\beta'|^2$ just corresponds to the mean number of quanta associated with the coherent part of the excitation of the system,

$$\langle \beta' | \hat{n} | \beta' \rangle = |\beta'|^2 = |\langle \xi, \beta | \hat{a} | \xi, \beta \rangle|^2. \tag{3.141}$$

Using Eqs (3.140) and (3.141), we may rewrite Eq. (3.137) as a sum of the mean number of quanta associated with the coherent part and the incoherent part of the excitation, respectively,

$$\langle \xi, \beta | \hat{n} | \xi, \beta \rangle = \langle \beta' | \hat{n} | \beta' \rangle + \langle \xi, 0 | \hat{n} | \xi, 0 \rangle. \tag{3.142}$$

3 Quantum states of bosonic systems

For the case of a radiation-field mode, the squeezed ground state $|\xi,0\rangle$ ($\xi \neq 0$) is often called the squeezed vacuum, in contrast to the ordinary vacuum $|0\rangle$, which is the state with zero photons. It should be pointed out that this does not hold for the squeezed vacuum $|\xi,0\rangle$: the mean number of photons in this state does not vanish ($|\nu|^2 \neq 0$).

To gain deeper insight into the statistics of squeezed coherent states, it may be useful to introduce the phase-rotated quadrature

$$\hat{x}(\varphi) = \hat{a}\, e^{i\varphi} + \hat{a}^\dagger e^{-i\varphi}, \tag{3.143}$$

which parametrically depends on φ. It can be readily proved that

$$[\hat{x}(\varphi), \hat{x}(\varphi')] = 2i \sin(\varphi - \varphi'). \tag{3.144}$$

Hence, two quadratures of orthogonal phases, i. e., $\varphi' = \varphi \pm \pi/2$, in a similar way to position and momentum, are canonically conjugate to each other, in the sense that

$$[\hat{x}(\varphi), \hat{x}(\varphi \pm \tfrac{1}{2}\pi)] = \mp 2i. \tag{3.145}$$

The phase-rotated quadrature can be used to represent various physical observables by supplementing a real-valued scaling factor and appropriate choice of the phase φ. For example, identifying $\hat{x}(\varphi)$ with a Cartesian component of the electric-field strength of a single-mode (free) radiation field, we would have

$$\hat{E}_k(\mathbf{r}) = i\omega[A_k(\mathbf{r})\hat{a} - A_k^*(\mathbf{r})\hat{a}^\dagger] = \omega|A_k(\mathbf{r})|\hat{x}(\varphi), \tag{3.146}$$

where the phase is given by

$$\varphi = \arg[A_k(\mathbf{r})] + \tfrac{1}{2}\pi. \tag{3.147}$$

From Eq. (3.138) we immediately obtain the expectation value of the quadrature as

$$\langle \xi,\beta|\hat{x}(\varphi)|\xi,\beta\rangle = \beta' e^{i\varphi} + \beta'^* e^{-i\varphi}, \tag{3.148}$$

which of course corresponds to Eq. (3.139).

Let us now consider the quadrature fluctuation by calculating the variance

$$\begin{aligned}\langle \xi,\beta|[\Delta\hat{x}(\varphi)]^2|\xi,\beta\rangle &= \langle \xi,\beta|\hat{x}^2(\varphi)|\xi,\beta\rangle - [\langle \xi,\beta|\hat{x}(\varphi)|\xi,\beta\rangle]^2 \\ &= [\langle \xi,\beta|(\hat{a}^\dagger\hat{a} + \hat{a}\hat{a}^\dagger)|\xi,\beta\rangle - 2|\langle \xi,\beta|\hat{a}|\xi,\beta\rangle|^2] \\ &\quad + 2\mathrm{Re}\big[\langle \xi,\beta|(\Delta\hat{a})^2|\xi,\beta\rangle\, e^{2i\varphi}\big]. \end{aligned} \tag{3.149}$$

Applying Eqs (3.136) and (3.138), we first calculate

$$\begin{aligned}\langle \xi, \beta|(\Delta \hat{a})^2|\xi, \beta\rangle &= \langle \beta|\hat{S}^\dagger(\xi)\hat{a}^2\hat{S}(\xi)|\beta\rangle - \beta'^2 = \langle \beta|(\mu\hat{a} - \nu\hat{a}^\dagger)^2|\beta\rangle - \beta'^2\\ &= \langle \beta|[\mu^2\hat{a}^2 + \nu^2\hat{a}^{\dagger 2} - \mu\nu(\hat{a}^\dagger\hat{a} + \hat{a}\hat{a}^\dagger)]|\beta\rangle - \beta'^2\\ &= -\mu\nu,\end{aligned} \quad (3.150)$$

so that, by insertion of Eq. (3.150) into (3.149), we obtain for the sought variance

$$\langle \xi, \beta|[\Delta\hat{x}(\varphi)]^2|\xi, \beta\rangle = |\mu e^{i\varphi} - \nu^* e^{-i\varphi}|^2. \quad (3.151)$$

Introducing the modulus $|\nu|$ and phase φ_ν ($= \varphi_\xi$) of ν, we may rewrite Eq. (3.151) as

$$\begin{aligned}\langle \xi, \beta|[\Delta\hat{x}(\varphi)]^2|\xi, \beta\rangle &= |\mu - |\nu|\exp[i(2\varphi + \varphi_\nu)]|^2\\ &= \left\{1 + 2|\nu|^2\left[1 - \sqrt{\frac{1+|\nu|^2}{|\nu|^2}}\cos(2\varphi + \varphi_\nu)\right]\right\},\end{aligned} \quad (3.152)$$

where, in order to express the variance solely in terms of ν, we have used the relation (3.118).

Equation (3.152) reveals that for a fixed value of ν (i.e., fixed ξ) the variance $\langle \xi, \beta|[\Delta\hat{x}(\varphi)]^2|\xi, \beta\rangle$ sensitively depends on the phase $2\varphi + \varphi_\nu$. Clearly, in the limiting case as ν goes to zero (or equivalently $\xi \to 0$), that is, when the squeezed coherent state $|\xi, \beta\rangle$ tends to the ordinary coherent state $|\beta\rangle$, this phase dependence vanishes and we obtain the ground-state quadrature fluctuation:

$$\lim_{\nu \to 0}\langle \xi, \beta|[\Delta\hat{x}(\varphi)]^2|\xi, \beta\rangle = \langle \beta|[\Delta\hat{x}(\varphi)]^2|\beta\rangle = \langle 0|[\Delta\hat{x}(\varphi)]^2|0\rangle = 1. \quad (3.153)$$

For nonvanishing squeezing parameter ($\nu \neq 0$), however, the fluctuation depends, for chosen φ_ν, crucially on φ, so that at certain values of φ the fluctuation may be larger or even smaller than the ground-state limit (3.153). From Eq. (3.152) it can be seen that the fluctuation is smaller than the ground-state limit, i.e., $\langle \xi, \beta|[\Delta\hat{x}(\varphi)]^2|\xi, \beta\rangle < 1$, for

$$\cos(2\varphi + \varphi_\nu) > \sqrt{\frac{|\nu|^2}{1+|\nu|^2}}, \quad (3.154)$$

and it is minimal for $\cos(2\varphi + \varphi_\nu) = 1$. That is, for the specific values of the phase φ given by

$$\varphi_{\min} = k\pi - \tfrac{1}{2}\varphi_\nu \quad (3.155)$$

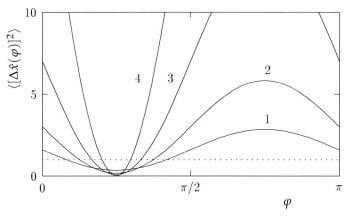

Fig. 3.1 The variance of the quadrature $\langle[\Delta\hat{x}(\varphi)]^2\rangle$ is shown for a squeezed state with $\varphi_v = -\pi/2$ and for $|v|^2 = 0.3$ (curve 1), 1 (2), 3 (3) and 10 (4); the ground-state noise level ($|v|^2 = 0$) is indicated by the dotted line. It can be seen that the noise is π-periodic with respect to the phase φ of the quadrature.

(k, integer), the quadrature fluctuation is reduced below the ground-state limit by the factor

$$\langle\xi,\beta|[\Delta\hat{x}(\varphi)]^2|\xi,\beta\rangle\big|_{\varphi=\varphi_{\min}} = e^{-2|\xi|}. \tag{3.156}$$

On the other hand, the fluctuation becomes larger than the ground-state limit, i.e., $\langle\xi,\beta|[\Delta\hat{x}(\varphi)]^2|\xi,\beta\rangle > 1$, for

$$\cos(2\varphi+\varphi_v) < \sqrt{\frac{|v|^2}{1+|v|^2}}. \tag{3.157}$$

Here the maximum fluctuation is observed at phases where $\cos(2\varphi+\varphi_v) = -1$, that is for values of φ given by

$$\varphi_{\max} = \tfrac{1}{2}(2k+1)\pi - \tfrac{1}{2}\varphi_v, \tag{3.158}$$

where the fluctuation is enhanced with respect to the ground state by the factor

$$\langle\xi,\beta|[\Delta\hat{x}(\varphi)]^2|\xi,\beta\rangle\big|_{\varphi=\varphi_{\max}} = e^{2|\xi|}. \tag{3.159}$$

We see that for certain phase values the quadrature noise can be "squeezed" below the ground-state (vacuum) level at the expense of increased noise for certain other phase values. From inspection of Eq. (3.152), this squeezing effect is seen to increase with $|v|$ ($=\cosh|\xi|$). The typical fluctuation behavior of a system in a squeezed coherent state is illustrated in Fig. 3.1. It clearly shows that, with decreasing noise for a given phase $\varphi_{\min}$, the noise for the phase $\varphi_{\max} = \varphi_{\min} + \pi/2$ can drastically increase. Moreover, the more strongly the noise is reduced below the ground-state (vacuum) level, the narrower the phase region around $\varphi_{\min}$, in which noise reduction is observed, becomes.

The behavior is closely related to Heisenberg's uncertainty principle for two observables $\hat{A}$ and $\hat{B}$,

$$\langle(\Delta\hat{A})^2\rangle\langle(\Delta\hat{B})^2\rangle \geq \tfrac{1}{4}|\langle[\hat{A},\hat{B}]\rangle|^2, \tag{3.160}$$

which, because of the commutator relation (3.144) for $\hat{A} = \hat{x}(\varphi)$ and $\hat{B} = \hat{x}(\varphi')$, reads

$$\langle[\Delta\hat{x}(\varphi)]^2\rangle\langle[\Delta\hat{x}(\varphi')]^2\rangle \geq \sin^2(\varphi - \varphi') \tag{3.161}$$

and holds for arbitrary quantum states. In the case when the two phases are $\varphi = \varphi_{\min}$ [Eq. (3.155)] and $\varphi' = \varphi_{\max}$ [Eq. (3.158)] – two phases that correspond to two orthogonal directions in phase space, so that the quadratures $\hat{x}(\varphi_{\min})$ and $\hat{x}(\varphi_{\max})$ are canonically conjugate to each other – then, in agreement with the commutator relation (3.145), the uncertainty relation (3.161) takes the form

$$\langle[\Delta\hat{x}(\varphi_{\min})]^2\rangle\langle[\Delta\hat{x}(\varphi_{\max})]^2\rangle \geq 1. \tag{3.162}$$

For squeezed coherent states, from Eqs (3.156) and (3.159) it follows that

$$\langle\xi,\beta|[\Delta\hat{x}(\varphi_{\min})]^2|\xi,\beta\rangle\langle\xi,\beta|[\Delta\hat{x}(\varphi_{\max})]^2|\xi,\beta\rangle = 1, \tag{3.163}$$

which shows that squeezed coherent states (as also coherent states) are minimum-uncertainty states. It should be pointed out that, in the more general case, where the squeezing phase φ_ν is not necessarily adjusted to the quadrature phases φ and $\varphi + \pi/2$, according to Eqs (3.155) and (3.158), from Eq. (3.151) the uncertainty product

$$\langle\xi,\beta|[\Delta\hat{x}(\varphi)]^2|\xi,\beta\rangle\langle\xi,\beta|[\Delta\hat{x}(\varphi+\tfrac{1}{2}\pi)]^2|\xi,\beta\rangle = [1 + 4\mu^2|\nu|^2\sin^2(2\varphi + \varphi_\nu)] \tag{3.164}$$

follows. Comparing this equation with Eq. (3.163), we find that the squeezed coherent states minimize the uncertainty product for $\hat{x}(\varphi)$ and $\hat{x}(\varphi + \pi/2)$ only when the phase of squeezing is related to φ by

$$2\varphi + \varphi_\nu = k\pi, \tag{3.165}$$

where k is an integer number [Schubert and Vogel (1978a)]. In this case $\langle\xi,\beta|[\Delta\hat{x}(\varphi)]^2|\xi,\beta\rangle$ and $\langle\xi,\beta|[\Delta\hat{x}(\varphi + \pi/2)]^2|\xi,\beta\rangle$ are just the extremal values [Eqs (3.156) and (3.159)]. The coherent states $|\alpha\rangle$, however, are minimum-uncertainty states, independent of the choice of the phase φ:

$$\langle\alpha|[\Delta\hat{x}(\varphi)]^2|\alpha\rangle\langle\alpha|[\Delta\hat{x}(\varphi+\tfrac{1}{2}\pi)]^2|\alpha\rangle = 1. \tag{3.166}$$

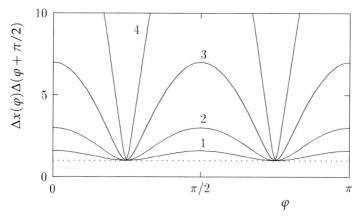

Fig. 3.2 The uncertainty product $\Delta x(\varphi)\Delta x(\varphi+\pi/2) \equiv \{\langle[\Delta\hat{x}(\varphi)]^2\rangle\langle[\Delta\hat{x}(\varphi+\pi/2)]^2\rangle\}^{\frac{1}{2}}$ as a function of φ for a squeezed state with $\varphi_\nu = -\pi/2$ and for $|\nu|^2 = 0.3$ (curve 1), 1 (2), 3 (3) and 10 (4). The case with $|\nu|^2 = 0$ (dotted line) corresponds to a coherent state that minimizes the uncertainty product for all values of φ.

The dependence on the phase φ of the uncertainty product of two canonically conjugated quadratures in the case of squeezed coherent states is illustrated in Fig. 3.2.

Recalling the definition of normally ordered operator products, we can easily prove the relation

$$\langle:[\Delta\hat{x}(\varphi)]^2:\rangle = \langle[\Delta\hat{x}(\varphi)]^2\rangle - \langle 0|[\Delta\hat{x}(\varphi)]^2|0\rangle, \tag{3.167}$$

which is valid for an arbitrary quantum state in which the system is prepared. Since in the case when the system is prepared in a squeezed coherent state $|\xi,\beta\rangle$ the quadrature variance $\langle\xi,\beta|[\Delta\hat{x}(\varphi)]^2|\xi,\beta\rangle$ becomes, for appropriately chosen values of φ, smaller than the vacuum limit $\langle 0|[\Delta\hat{x}(\varphi)]^2|0\rangle$, so that the normally ordered variance $\langle\xi,\beta|:[\Delta\hat{x}(\varphi)]^2:|\xi,\beta\rangle$ becomes negative. At this point it should be emphasized that the squeezed coherent states may be viewed as a typical but special class of states giving rise to squeezing. Quite general, a quantum state may be said to reveal squeezing if for certain values of the phase the normally ordered quadrature variance becomes negative:

$$\langle:[\Delta\hat{x}(\varphi)]^2:\rangle < 0. \tag{3.168}$$

3.3.2
Multi-mode squeezed states

The criterion for squeezing as given by Eq. (3.168) can also be used to study (complicated) multi-mode systems. Let

$$\hat{X} = \sum_\lambda c_\lambda \hat{x}_\lambda(\varphi_\lambda) \tag{3.169}$$

be the multi-mode quadrature operator. According to Eq. (3.168), squeezing is observed if

$$\langle :(\Delta \hat{X})^2: \rangle < 0. \tag{3.170}$$

In particular, a multi-mode radiation-field strength

$$\hat{F} = \sum_\lambda F_\lambda \hat{a}_\lambda + F_\lambda^* \hat{a}_\lambda^\dagger \tag{3.171}$$

[cf. Eq. (2.271)] can be regarded as a multi-mode quadrature, setting $c_\lambda = |F_\lambda|$ and $\varphi_\lambda = \arg F_\lambda$, and the squeezing criterion reads $\langle :(\Delta \hat{F})^2: \rangle < 0$, i.e, the normally ordered field variance must be negative.

A generalization of the single-mode squeeze operator defined by Eq. (3.102) to a squeeze operator acting on multi-mode systems is

$$\hat{S} = \exp\left[\sum_{\lambda,\lambda'} (\zeta^*_{\lambda\lambda'} \hat{a}_\lambda \hat{a}_{\lambda'} - \zeta_{\lambda\lambda'} \hat{a}_\lambda^\dagger \hat{a}_{\lambda'}^\dagger)\right]. \tag{3.172}$$

If the matrix $\zeta_{\lambda\lambda'}$ has only diagonal elements, $\zeta_{\lambda\lambda'} = \delta_{\lambda,\lambda'} \zeta_\lambda/2$, the multi-mode squeeze operator (3.172) reduces to a product of single-mode squeeze operators of the type (3.102), and its application to the ground state of the multi-mode system generates multi-mode squeezed vacuum states that are simply the (direct) products of single-mode squeezed states. Additional application of the multi-mode displacement operator then generates multi-mode squeezed coherent states, which are of course also direct-product states.

A system is said to genuinely feature multi-mode squeezing, if there are nonvanishing off-diagonal $\zeta_{\lambda\lambda'}$ that give rise to nonclassical correlations between the modes. To give a simple but illustrative example, let us consider the case of two modes ($\lambda = 1, 2$) being prepared in a two-mode squeezed vacuum state as a typical example of an entangled state (Section 8.5):

$$|\zeta, 0, 0\rangle = \hat{S}(\zeta)|0,0\rangle, \tag{3.173}$$

where $|0,0\rangle \equiv |0_1\rangle|0_2\rangle$ is the ordinary two-mode vacuum state and the two-mode squeeze operator $\hat{S}(\zeta)$ is a special case of the general operator (3.172) with $\zeta_{12} = \zeta$, $\zeta_{21} = \zeta^*$ and $\zeta_{11} = \zeta_{22} = 0$,

$$\hat{S}(\zeta) = \exp(\zeta^* \hat{a}_1 \hat{a}_2 - \zeta \hat{a}_1^\dagger \hat{a}_2^\dagger). \tag{3.174}$$

The explicit action of the two-mode squeeze operator on $\hat{a}_1$ and $\hat{a}_2$ can be found in a similar way to that described for the single-mode case. The result is

$$\hat{S}^\dagger(\zeta) \hat{a}_1 \hat{S}(\zeta) = \mu \hat{a}_1 - \nu \hat{a}_2^\dagger, \tag{3.175}$$
$$\hat{S}^\dagger(\zeta) \hat{a}_2 \hat{S}(\zeta) = \mu \hat{a}_2 - \nu \hat{a}_1^\dagger, \tag{3.176}$$

where μ and ν are related to ζ by Eqs (3.115) and (3.116), respectively, and in close analogy to Eq. (3.130), the two-mode squeeze operator can be disentangled to obtain

$$\hat{S} = \exp\left(-\frac{\nu}{\mu}\hat{a}_1^\dagger \hat{a}_2^\dagger\right)\left(\frac{1}{\mu}\right)^{\hat{n}_1+\hat{n}_2+1}\exp\left(\frac{\nu^*}{\mu}\hat{a}_1 \hat{a}_2\right). \tag{3.177}$$

Combining Eqs (3.173) and (3.177), we can easily see that the two-mode squeezed vacuum state can be represented as

$$|\zeta,0,0\rangle = \frac{1}{\mu}\exp\left(-\frac{\nu}{\mu}\hat{a}_1^\dagger \hat{a}_2^\dagger\right)|0,0\rangle = \frac{1}{\mu}\sum_{n=0}^{\infty}\left(-\frac{\nu}{\mu}\right)^n |n,n\rangle. \tag{3.178}$$

The generalization to two-mode squeezed coherent states with nonvanishing coherent amplitudes can be obtained in a straightforward way by additionally applying coherent displacement operators for both modes.

Let us consider a two-mode radiation-field strength

$$\hat{F} = F_1 \hat{a}_1 + F_1^* \hat{a}_1^\dagger + F_2 \hat{a}_2 + F_2^* \hat{a}_2^\dagger \tag{3.179}$$

$[F_\lambda = |F_\lambda|\exp(i\varphi_\lambda)]$. Taking the two-mode field to be in the squeezed vacuum state $|\zeta,0,0\rangle$ as given in Eq. (3.173) and using Eqs (3.175) and (3.176), we calculate the normally ordered variance of $\hat{F}$ to be

$$\langle \zeta,0,0|:(\Delta \hat{F})^2:|\zeta,0,0\rangle = 2(|F_1|^2+|F_2|^2)|\nu|^2$$
$$\times \left[1 - \frac{2|F_1 F_2|}{|F_1|^2+|F_2|^2}\sqrt{\frac{1+|\nu|^2}{|\nu|^2}}\cos(\varphi_1+\varphi_2+\varphi_\nu)\right]. \tag{3.180}$$

We see that the radiation may indeed be squeezed, because for appropriately chosen phases the value of $\langle:(\Delta \hat{F})^2:\rangle$ may become negative.

We identify $\hat{a}_1$ and $\hat{a}_2$ with the annihilation operators for modes of frequencies $\omega_1 = \omega_0 + \Delta\omega/2$ and $\omega_2 = \omega_0 - \Delta\omega/2$, respectively, and apply the model to the calculation of the normally ordered variance of the electric field $\hat{E}(x,t) = \hat{E}^{(+)}(x,t) + \hat{E}^{(-)}(x,t)$,

$$\hat{E}^{(+)}(x,t) = i\int_0^\infty d\omega \sqrt{\frac{\hbar\omega}{4\pi\epsilon_0 c A}}\, e^{-i\omega(t-x/c)}\hat{a}(\omega), \tag{3.181}$$

of a linearly polarized wave packet propagating in the positive x direction [see Eq. (3.90)]. The quantum state considered here is a squeezed vacuum of the form

$$|\psi\rangle_{sv} = \hat{S}|\psi\rangle_v, \tag{3.182}$$

where $|\psi\rangle_v$ denotes the ordinary vacuum state and the squeeze operator is given as

$$\hat{S} = \exp\left\{\int_0^\infty d\omega \left[\xi^*(\omega)\hat{a}(\omega_0+\omega)\hat{a}(\omega_0-\omega)\right.\right.$$
$$\left.\left. - \xi(\omega)\hat{a}^\dagger(\omega_0+\omega)\hat{a}^\dagger(\omega_0-\omega)\right]\right\}. \quad (3.183)$$

Such a squeeze operator correlates pairs of modes around the mid-frequency ω_0, quite similar to the previous example given in Eq. (3.174), however, integrated over all difference frequencies. Nevertheless, it can be easily seen that the following transformation holds in close analogy to Eqs (3.175) and (3.176)

$$\hat{S}^\dagger \hat{a}(\omega) \hat{S} = \mu(|\omega-\omega_0|)\hat{a}(\omega) - \nu(|\omega-\omega_0|)\hat{a}^\dagger(2\omega_0-\omega), \quad (3.184)$$

where $\mu(\omega)$ and $\nu(\omega)$ are defined via $\xi(\omega)$ as described in Eqs (3.115) and (3.116), respectively.

From this transformation can immediately be seen that

$$\langle \hat{E}^{(+)}(x,t)\rangle_{sv} = 0. \quad (3.185)$$

Further, by assuming a finite spectral width of squeezing in the sense that $\nu(\omega) \neq 0$ only for $0 \leq \omega \leq \Delta\omega$, where the spectral width is small compared with the mid-frequency, $\Delta\omega \ll \omega_0$, it can be shown that the field correlation functions read (to good approximation) as

$$\langle \hat{E}^{(-)}(x,t)\hat{E}^{(+)}(x',t')\rangle_{sv}$$
$$= \frac{\hbar\omega_0}{2\pi\epsilon_0 c \mathcal{A}} e^{i\omega_0(\tau-\tau')} \int_0^{\Delta\omega} d\omega |\nu(\omega)|^2 \cos[\omega(\tau-\tau')], \quad (3.186)$$

$$\langle \hat{E}^{(+)}(x,t)\hat{E}^{(+)}(x',t')\rangle_{sv}$$
$$= \frac{\hbar\omega_0}{2\pi\epsilon_0 c \mathcal{A}} e^{-i\omega_0(\tau+\tau')} \int_0^{\Delta\omega} d\omega \, \mu(\omega)\nu(\omega) \cos[\omega(\tau-\tau')], \quad (3.187)$$

where the notation $\tau = t - x/c$ has been used. Suppose that the times to be resolved are large compared with the inverse bandwidth of the squeezing spectrum, $\Delta\tau \gg (\Delta\omega)^{-1}$, and that the squeezing spectrum is sufficiently flat, i.e., $\nu(\omega) \simeq \bar{\nu}$ and $\mu(\omega) \simeq \bar{\mu}$. In this case the frequency integrals in Eqs (3.186) and (3.187) can be approximated as delta functions and we obtain:

$$\langle \hat{E}^{(-)}(x,t)\hat{E}^{(+)}(x',t')\rangle_{sv} = \frac{\hbar\omega_0}{2\epsilon_0 c \mathcal{A}} |\bar{\nu}|^2 \exp[i\omega_0(\tau-\tau')]\delta(\tau-\tau'), \quad (3.188)$$

$$\langle \hat{E}^{(+)}(x,t)\hat{E}^{(+)}(x',t')\rangle_{sv} = \frac{\hbar\omega_0}{2\epsilon_0 c \mathcal{A}} \bar{\mu}\bar{\nu} \exp[-i\omega_0(\tau+\tau')]\delta(\tau-\tau'). \quad (3.189)$$

A radiation field with mean value and correlation functions according to Eqs (3.185), (3.188) and (3.189) is usually called squeezed white noise [Gardiner (1991)]. Note that for an ordinary white-noise field, the relations $\langle \hat{E}^{(+)}\hat{E}^{(+)}\rangle = \langle \hat{E}^{(-)}\hat{E}^{(-)}\rangle = 0$ hold.

3.4
Quadrature eigenstates

So far we have studied the eigenstates of various kinds of operators, such as Hermitian number operators $\hat{n} = \hat{a}^\dagger \hat{a}$ (number states), non-Hermitian photon destruction operators $\hat{a}$ (coherent states), and linear combinations of photon destruction and creation operators $\mu\hat{a} + \nu\hat{a}^\dagger$ (squeezed coherent states). In this context, the question arises as to what are the eigenstates of the Hermitian phase-rotated quadrature operator $\hat{x}(\varphi)$ defined by Eq. (3.143).

Before going into detail and answering the question we first mention the limiting properties of the squeezed coherent states. For this purpose let us consider the commutator of the annihilation operator $\hat{a}' = \mu\hat{a} + \nu\hat{a}^\dagger$ and the quadrature operator,

$$[\hat{x}(\varphi), \hat{a}'] = [\hat{a}e^{i\varphi} + \hat{a}^\dagger e^{-i\varphi}, \mu\hat{a} + \nu\hat{a}^\dagger]$$
$$= |\nu|e^{-i\varphi}\left[e^{i(2\varphi+\varphi_\nu)} - \sqrt{\frac{1+|\nu|^2}{|\nu|^2}}\right]. \quad (3.190)$$

Choosing the phases as $2\varphi + \varphi_\nu = 2\pi k$ where k is an integer number, the commutator vanishes in the limit of infinite squeezing:

$$\lim_{|\nu|\to\infty} [\hat{x}(\varphi), \hat{a}'] = 0 \quad (2\varphi + \varphi_\nu = 2\pi k). \quad (3.191)$$

That is, in the considered limit the operators $\hat{x}(\varphi)$ and $\hat{a}'$ obviously have the same eigenstates. In the limit as $|\nu| \to \infty$ the eigenstates of $\hat{a}'$ represent ideally squeezed coherent states, so that (for appropriately chosen phase φ) the eigenstates of the quadrature operator can be viewed as ideally squeezed coherent states in that limit.

We now turn to the problem of deriving the explicit form of the (single-mode) quadrature eigenstates in terms of number states.[5] For this purpose let us first consider the case $\varphi = 0$, for which the form of $\hat{x}(\varphi)$ corresponds to the well known position operator. The corresponding eigenvalue equation expanded in terms of number states reads

$$\sqrt{n+1}\,\langle n+1|x\rangle + \sqrt{n}\,\langle n-1|x\rangle = x\langle n|x\rangle. \quad (3.192)$$

[5] Alternative derivations of the phase-rotated quadrature eigenstates in both number-state and coherent-state representations were given by Schubert and Vogel (1978b).

3.4 Quadrature eigenstates

The normalized solution of Eq. (3.192) can be written in terms of the Hermite polynomials as

$$\langle n|x\rangle = \psi_n(x) = (2^n n!\sqrt{2\pi})^{-\frac{1}{2}} H_n(x/\sqrt{2}) e^{-\frac{1}{4}x^2}. \tag{3.193}$$

For an arbitrary phase φ we employ the phase-rotation operator given by

$$\hat{U}(\varphi) = \exp(-i\varphi \hat{a}^\dagger \hat{a}), \tag{3.194}$$

with the help of which it may be easily proved that

$$\hat{x}(0) = \hat{U}^\dagger(\varphi)\hat{x}(\varphi)\hat{U}(\varphi). \tag{3.195}$$

The general eigenvalue problem can then be written as

$$\hat{x}(0)|x\rangle = \hat{U}^\dagger(\varphi)\hat{x}(\varphi)\hat{U}(\varphi)|x\rangle = x|x\rangle, \tag{3.196}$$

or, multiplying from the left-hand side by $\hat{U}(\varphi)$,

$$\hat{x}(\varphi)\hat{U}(\varphi)|x\rangle = x\hat{U}(\varphi)|x\rangle. \tag{3.197}$$

This shows that the eigenstates of $\hat{x}(\varphi)$ are simply given by

$$|x,\varphi\rangle = \hat{U}(\varphi)|x\rangle, \tag{3.198}$$

which in the number basis reads as, on recalling Eq. (3.193),

$$\langle n|x,\varphi\rangle = \psi_n(x) e^{-in\varphi}. \tag{3.199}$$

Clearly, the quadrature eigenstates $|x,\varphi\rangle$ can also be expressed in terms other than number states. In particular, in the case of a coherent-state representation we have

$$|x,\varphi\rangle = \frac{1}{\pi}\int d^2\alpha\, |\alpha\rangle\langle\alpha|x,\varphi\rangle. \tag{3.200}$$

The scalar product $\langle\alpha|x,\varphi\rangle$ can be obtained in different ways, one of which is by use of the number states and direct evaluation of the occurring sums (cf. footnote 4, p. 92). The result is

$$\langle\alpha|x,\varphi\rangle = (2\pi)^{-\frac{1}{4}} \exp\left[-\tfrac{1}{4}x^2 + x|\alpha|e^{-i(\varphi+\varphi_\alpha)}\right]$$
$$\times \exp\{-|\alpha|^2 \cos^2(\varphi+\varphi_\alpha) + \tfrac{1}{2}i|\alpha|^2 \sin[2(\varphi+\varphi_\alpha)]\}, \tag{3.201}$$

where $\varphi_\alpha = \arg(\alpha)$. Equation (3.201) reveals that the probability distribution for observing a value of the quadrature x when the system is prepared in a coherent state $|\alpha\rangle$ is a Gaussian,

$$|\langle\alpha|x,\varphi\rangle|^2 = \frac{1}{\sqrt{2\pi}} \exp\{-\tfrac{1}{2}[x - \langle\hat{x}(\varphi)\rangle]^2\}, \tag{3.202}$$

where the mean value reads

$$\langle \hat{x}(\varphi) \rangle = \langle \alpha | \hat{x}(\varphi) | \alpha \rangle = 2|\alpha| \cos(\varphi + \varphi_\alpha). \tag{3.203}$$

We finally note that, for given φ, the quadrature eigenstates are of course orthogonal and complete in the sense that

$$\langle x, \varphi | x', \varphi \rangle = \delta(x - x'), \tag{3.204}$$

and

$$\int dx \, |x, \varphi\rangle \langle x, \varphi| = \hat{I}, \tag{3.205}$$

which can be proved by using the explicit form of the quadrature eigenstates in the number-state or coherent-state representation as given above.

3.5
Phase states

As we know, the annihilation and creation operators $\hat{a}$ and $\hat{a}^\dagger$ correspond to the classical complex amplitudes α and α^*, respectively, according to the relations

$$\hat{a} \mapsto \alpha = |\alpha| e^{i\varphi}, \tag{3.206}$$

$$\hat{a}^\dagger \mapsto \alpha^* = |\alpha| e^{-i\varphi}. \tag{3.207}$$

Thus in classical physics it is straightforward to express the quantities of the system in terms of the amplitude and phase variables $|\alpha|$ and φ, respectively. Amplitude and phase seem to appear in this context as observable quantities and one may therefore ask for the quantum-mechanical operators which, in a sense, correspond to them. Attempts to introduce amplitude and phase variables in quantum mechanics are nearly as old as quantum mechanics itself. Since Dirac's introduction of amplitude and phase operators in 1927, a series of concepts have been developed. Here we concentrate on only a few of them and emphasize more the resulting phase states that are eigenstates of appropriately chosen phase operators.

In close analogy with the classical approach to the problem of defining amplitude and phase variables, Dirac introduced a phase operator $\hat{\phi}$ by factoring the annihilation and creation operators as follows:

$$\hat{a} = \hat{V} \sqrt{\hat{n}}, \quad \hat{a}^\dagger = \sqrt{\hat{n}} \, \hat{V}^\dagger, \tag{3.208}$$

where the operator $\hat{V}$ is regarded as being a unitary operator of the form

$$\hat{V} = e^{i\hat{\phi}} \tag{3.209}$$

[Dirac (1927)], with $\hat{\phi}$ being assumed to be the Hermitian phase operator,

$$\hat{\phi}^\dagger = \hat{\phi}. \tag{3.210}$$

However, difficulties arise here which are closely related to the fact that the operator $\hat{V}$ is actually not unitary [London (1926, 1927)] and therefore Eqs (3.208) and (3.209) do not define a Hermitian phase operator $\hat{\phi}$. It can be proved [Carruthers and Nieto (1968)] that $\langle 0|\hat{V}^\dagger\hat{V}|0\rangle = 0$, which contradicts the assumption of unitarity, $\hat{V}^\dagger\hat{V} = \hat{I}$. This result can be shown as follows. Applying Eq. (3.208) to a photon-number state $|n\rangle$ yields

$$\hat{V}|n\rangle = |n-1\rangle, \quad n = 1,2,\ldots. \tag{3.211}$$

In the case $n = 0$, due to the completeness of the number states, one may write

$$\hat{V}|0\rangle = \sum_{n=0}^{\infty} d_n |n\rangle. \tag{3.212}$$

From these equations one finds that

$$\hat{V}^\dagger|n\rangle = \sum_m |m\rangle\langle m|\hat{V}^\dagger|n\rangle = d_n^*|0\rangle + |n+1\rangle, \tag{3.213}$$

and hence for $n > 0$

$$\hat{V}^\dagger\hat{V}|n\rangle = \hat{V}^\dagger|n-1\rangle = d_{n-1}^*|0\rangle + |n\rangle. \tag{3.214}$$

Therefore, if $\hat{V}^\dagger\hat{V} = \hat{I}$ it follows that $d_n = 0$ for all n. This means that $\hat{V}|0\rangle = 0$ and therefore $\langle 0|\hat{V}^\dagger\hat{V}|0\rangle = 0$. Note that in contrast to $\hat{V}^\dagger\hat{V}$, $\hat{V}\hat{V}^\dagger$ is the identity operator: $\hat{V}\hat{V}^\dagger = \hat{I}$.

3.5.1
The eigenvalue problem of $\hat{V}$

Susskind and Glogower (1964) considered, according to Eq. (3.208), the exponential phase operator $\hat{V}$ but without assuming its unitarity:

$$\hat{V} = \widehat{e^{i\phi}}, \quad \hat{V}^\dagger = \left(\widehat{e^{i\phi}}\right)^\dagger \tag{3.215}$$

[see also Carruthers and Nieto (1968)]. To represent the operators $\hat{V}$ and $\hat{V}^\dagger$ in the number basis, we use the number representations of $\sqrt{\hat{n}}$, $\hat{a}$ and $\hat{a}^\dagger$, namely

$$\sqrt{\hat{n}} = \sum_{n=0}^{\infty} \sqrt{n}\, |n\rangle\langle n| = \sum_{n=0}^{\infty} \sqrt{n+1}\, |n+1\rangle\langle n+1|, \tag{3.216}$$

$$\hat{a} = \sum_{n=0}^{\infty} \sqrt{n+1}\, |n\rangle\langle n+1|, \tag{3.217}$$

$$\hat{a}^\dagger = \sum_{n=0}^{\infty} \sqrt{n+1}\, |n+1\rangle\langle n| \tag{3.218}$$

[cf. Eqs (3.19) and (3.20)]. Combining Eqs (3.217) and (3.216) and taking into account the completeness of the number states, we derive

$$\hat{a} = \sum_{n=0}^{\infty} \sum_{m=0}^{\infty} \sqrt{n+1} |n\rangle \langle n+1|m\rangle \langle m|$$

$$= \sum_{n=0}^{\infty} \sum_{m=1}^{\infty} \sqrt{m} |n\rangle \langle n+1|m\rangle \langle m| = \sum_{n=0}^{\infty} |n\rangle \langle n+1|\sqrt{\hat{n}}, \quad (3.219)$$

from which, together with Eq. (3.208), we may choose $\hat{V}$ as

$$\hat{V} = \sum_{n=0}^{\infty} |n\rangle \langle n+1|, \quad (3.220)$$

and the relation

$$\hat{V}\sqrt{\hat{n}} = \sqrt{\hat{n}+1}\,\hat{V} \quad (3.221)$$

holds.[6] Accordingly, we have

$$\hat{V}^{\dagger} = \sum_{n=0}^{\infty} |n+1\rangle \langle n|. \quad (3.222)$$

Applying $\hat{V}$ to a number state $|n\rangle$ gives, on using Eq. (3.220),

$$\hat{V}|n\rangle = |n-1\rangle. \quad (3.223)$$

In particular, the application of $\hat{V}$ to the ground state gives

$$\hat{V}|0\rangle = 0. \quad (3.224)$$

Combining Eqs (3.220) and (3.222), we derive

$$\hat{V}\hat{V}^{\dagger} = \hat{I}, \quad (3.225)$$
$$\hat{V}^{\dagger}\hat{V} = \hat{I} - |0\rangle \langle 0|. \quad (3.226)$$

Equations (3.225) and (3.226) imply that

$$[\hat{V}, \hat{V}^{\dagger}] = |0\rangle \langle 0|. \quad (3.227)$$

It should be noted that the nonunitarity and the noncommuting nature are only relevant for states $|\Psi\rangle$ having a significant overlap with the ground state (vacuum):

$$\langle \Psi|[\hat{V}, \hat{V}^{\dagger}]|\Psi\rangle = |\langle 0|\Psi\rangle|^2. \quad (3.228)$$

---

[6] Note that there is an ambiguity because of the undetermined term with $m=0$ in the second line of Eq. (3.219). From the first line of Eq. (3.219) an ambiguous definition of $\hat{V}$ can be given by supposing that $\hat{a} = (\hat{n}+1)^{1/2}\hat{V}$, which implies that $\hat{V}\hat{n}^{1/2} = (\hat{n}+1)^{1/2}\hat{V}$.

Let us now consider the eigenvalue problem for $\hat{V}$. Postulating

$$\hat{V}|\phi\rangle = e^{i\phi}|\phi\rangle \tag{3.229}$$

and expanding $|\phi\rangle$ in the number basis,

$$|\phi\rangle = \sum_{n=0}^{\infty} b_n |n\rangle, \tag{3.230}$$

we readily arrive, on using Eq. (3.220), at the recurrence relation

$$b_{n+1} = e^{i\phi} b_n, \tag{3.231}$$

which may be satisfied by choosing

$$b_n = b_0 e^{in\phi}. \tag{3.232}$$

The eigenstates of the operator $\hat{V}$ have therefore the form

$$|\phi\rangle = b_0 \sum_{n=0}^{\infty} e^{in\phi} |n\rangle. \tag{3.233}$$

As expected, application of $\hat{U}(\varphi)$, Eq. (3.194), onto $|\phi\rangle$ shifts ϕ to $\phi - \varphi$,

$$\hat{U}(\varphi)|\phi\rangle = |\phi - \varphi\rangle. \tag{3.234}$$

Furthermore, the states $|\phi\rangle$ obviously satisfy the periodicity condition

$$|\phi + 2\pi\rangle = |\phi\rangle, \tag{3.235}$$

and the identity can be resolved by these states:

$$|b_0 \sqrt{2\pi}|^{-2} \int_0^{2\pi} d\phi \, |\phi\rangle\langle\phi| = \sum_{n,m=0}^{\infty} |n\rangle\langle m| \frac{1}{2\pi} \int_0^{2\pi} d\phi \, \exp[i(n-m)\phi]$$

$$= \sum_{n,m=0}^{\infty} |n\rangle\langle m| \delta_{nm} = \hat{I}, \tag{3.236}$$

from which we may choose $b_0 = 1/\sqrt{2\pi}$.

Having in mind a classical picture of phase, the states

$$|\phi\rangle = \frac{1}{\sqrt{2\pi}} \sum_{n=0}^{\infty} e^{in\phi} |n\rangle, \tag{3.237}$$

may be regarded as quantum-mechanical phase states [cf. London (1926, 1927)]. The thus introduced phase is also called the canonical phase or London phase. In particular, Eq. (3.234) implies that a freely evolving phase state

$(\varphi \mapsto \omega t)$ would remain a phase state for all time. However, the states $|\phi\rangle$ are not orthogonal and, unfortunately, they cannot be normalized in a proper way. Indeed, we deduce from Eq. (3.237) that[7]

$$\langle \phi | \phi' \rangle = \frac{1}{2\pi} \sum_{n=0}^{\infty} \exp[-in(\phi - \phi')]$$

$$= \left\{ \frac{1}{4\pi} + \frac{1}{2}\delta(\phi - \phi') - \frac{i}{4\pi} \cot[\tfrac{1}{2}(\phi - \phi')] \right\} \qquad (3.238)$$

$(0 \leq |\phi - \phi'| < 2\pi)$. Clearly, the states $|\phi\rangle$ are not eigenstates of the operator $\hat{V}^\dagger$, which is significant for an analysis of states which substantially overlap with the ground state. Combining Eqs (3.222) and (3.237) we find that

$$\hat{V}^\dagger |\phi\rangle = e^{-i\phi} \left(|\phi\rangle - \frac{1}{\sqrt{2\pi}} |0\rangle \right). \qquad (3.239)$$

Nevertheless, the states $|\phi\rangle$ may be useful because they resolve the identity, Eq. (3.236). Hence any state $|\Psi\rangle$ can be expressed in terms of them:

$$|\Psi\rangle = \int_0^{2\pi} d\phi \, |\phi\rangle \langle \phi | \Psi \rangle. \qquad (3.240)$$

The nonorthogonality of the states $|\phi\rangle$ might be removed by using a finite-dimensional Hilbert space spanned by $r+1$ number states $\{|n\rangle\}$. In this truncated Hilbert space, a set of $r+1$ phase states $|\phi^{(r)}\rangle$ can be introduced as

$$|\phi_m^{(r)}\rangle = \frac{1}{\sqrt{r+1}} \sum_{n=0}^{r} e^{in\phi_m^{(r)}} |n\rangle, \qquad (3.241)$$

where

$$\phi_m^{(r)} = \phi_0^{(r)} + \frac{2m\pi}{r+1} \qquad (m = 0, 1, \ldots, r). \qquad (3.242)$$

[cf. Eq. (3.237)]. Here the phase $\phi_0^{(r)}$ is a reference phase whose value determines the choice of the 2π periodicity interval of the phase. For each finite r the states $|\phi_m^{(r)}\rangle$ are orthonormal and complete in the sense that

$$\langle \phi_m^{(r)} | \phi_{m'}^{(r)} \rangle = \delta_{mm'}, \qquad (3.243)$$

$$\sum_{m=0}^{r} |\phi_m^{(r)}\rangle \langle \phi_m^{(r)}| = \hat{I}. \qquad (3.244)$$

---

[7] Note that the relations $2\sum_{n=1}^{\infty} \sin(n\phi) = \cot(\phi/2)$ and $\sum_{n=1}^{\infty} \cos(n\phi) - 1/2 = \pi \sum_{n=-\infty}^{\infty} \delta(\phi - 2n\pi)$ are valid, the latter results from the identity $\sum_{n=-\infty}^{\infty} \exp(in\phi) = 2\pi \sum_{n=-\infty}^{\infty} \delta(\phi - 2n\pi)$.

Hence in the truncated Hilbert space a Hermitian phase operator can be defined as follows:

$$\hat{\phi}^{(r)} = \sum_{m=0}^{r} \phi_m^{(r)} |\phi_m^{(r)}\rangle \langle \phi_m^{(r)}|. \tag{3.245}$$

The limiting procedure $r \to \infty$ can then be performed at the end of all (c-number) calculations [Loudon (1973); Pegg and Barnett (1988, 1989); Barnett and Pegg (1992)].

3.5.2
Cosine and sine phase states

It is worth noting that the operators $\hat{V}$ and $\hat{V}^\dagger$ may be used to define Hermitian operator analogues of $\cos\phi$ and $\sin\phi$ as follows:

$$\hat{C} = \tfrac{1}{2}(\hat{V} + \hat{V}^\dagger), \tag{3.246}$$

$$\hat{S} = \tfrac{1}{2}i(\hat{V} - \hat{V}^\dagger). \tag{3.247}$$

Using Eqs (3.220) and (3.222), we see that

$$[\hat{V}, \hat{n}] = \hat{V}, \qquad [\hat{V}^\dagger, \hat{n}] = -\hat{V}^\dagger. \tag{3.248}$$

These commutation rules imply, on using Eqs (3.246) and (3.247), the following commutation rules for $\hat{C}$, $\hat{S}$ and $\hat{n}$:

$$[\hat{C}, \hat{n}] = i\hat{S}, \quad [\hat{S}, \hat{n}] = -i\hat{C}, \quad [\hat{C}, \hat{S}] = \tfrac{1}{2}i\hat{P}_0, \tag{3.249}$$

where $\hat{P}_0 = |0\rangle\langle 0|$. Hence according to Heisenberg's uncertainty principle (3.160), the following uncertainty relations can be deduced:

$$\Delta n \Delta C \geq \tfrac{1}{2}\langle \hat{S} \rangle, \quad \Delta n \Delta S \geq \tfrac{1}{2}\langle \hat{C} \rangle, \quad \Delta S \Delta C \geq \tfrac{1}{4}\langle \hat{P}_0 \rangle. \tag{3.250}$$

In particular, the third of these reveals that C and S can be accurately measured simultaneously only when the state, say $|\Psi\rangle$, has sufficiently small overlap with the ground state: $|\langle 0|\Psi\rangle|^2 \ll 1$. In other words, if the overlap cannot be disregarded, $\hat{C}$ and $\hat{S}$ are expected to give rise to two (Hermitian) phase operators $\hat{\phi}_C$ and $\hat{\phi}_S$ instead of the desired one-phase operator. Since $\hat{C}$ and $\hat{S}$ are well-defined Hermitian operators, their eigenvalues give possible results of measurements of C and S.

In order to solve the eigenvalue problem for $\hat{C}$,

$$\hat{C}|\cos\phi\rangle = C|\cos\phi\rangle, \tag{3.251}$$

we expand $|\cos\phi\rangle$ in the number basis,

$$|\cos\phi\rangle = \sum_{n=0}^{\infty} b_n |n\rangle. \tag{3.252}$$

Recalling Eqs (3.220) and (3.222), after some algebra we obtain the recurrence relations

$$2b_0 C = b_1, \quad 2b_{n+1} C = b_n + b_{n+2}. \tag{3.253}$$

The second of these is solved by

$$b_n = \alpha V^n + \beta V^{-n}, \quad C = \tfrac{1}{2}(V + V^{-1}) \tag{3.254}$$

for arbitrary values of α and β. To avoid divergence difficulties $|V|$ must be unity so that

$$V = e^{i\phi}, \quad C = \cos \phi. \tag{3.255}$$

To specify α and β, we note that b_0 can be chosen to be real so that $\alpha = \beta^*$ in Eq. (3.254). Making the substitution $b_0 \mapsto b_0 \sin \phi$ (b_0 real), it is seen from Eqs (3.254) and (3.255) that

$$b_n = b_0 \sin[(n+1)\phi]. \tag{3.256}$$

Hence Eqs (3.251) and (3.252) become

$$\hat{C} |\cos \phi\rangle = \cos \phi |\cos \phi\rangle, \tag{3.257}$$

$$|\cos \phi\rangle = b_0 \sum_{n=0}^{\infty} \sin[(n+1)\phi] |n\rangle. \tag{3.258}$$

Note that all independent solutions are contained in the interval $0 \le \phi < \pi$. By straightforward calculation, on recalling the formulae in footnote 7, p. 108, it can be shown that the $|\cos \phi\rangle$ form an orthonormal and complete set of basis vectors in the Hilbert space ($b_0 = \sqrt{2/\pi}$):

$$\langle \cos \phi | \cos \phi' \rangle = \delta(\phi - \phi'), \tag{3.259}$$

$$\int_0^{\pi} d\phi \, |\cos \phi\rangle \langle \cos \phi| = \hat{I}. \tag{3.260}$$

The solution of the eigenvalue problem for the sine operator $\hat{S}$ may be found in a very similar way [for details see Carruthers and Nieto (1968)]. The result may be written as

$$\hat{S} |\sin \phi\rangle = \sin \phi |\sin \phi\rangle, \tag{3.261}$$

$$|\sin \phi\rangle = \frac{1}{\sqrt{2\pi}} \sum_{n=0}^{\infty} \{\exp[i(n+1)\phi] - \exp[-i(n+1)(\phi - \pi)]\} |n\rangle, \tag{3.262}$$

$$\langle \sin \phi | \sin \phi' \rangle = \delta(\phi - \phi'), \tag{3.263}$$

$$\int_{-\pi/2}^{\pi/2} d\phi \, |\sin \phi\rangle \langle \sin \phi| = \hat{I}. \tag{3.264}$$

By power-series expansion, for each operator $\hat{C}$ and $\hat{S}$, Hermitian phase operators $\hat{\phi}_C$ $(=\hat{\phi}_C^\dagger)$ and $\hat{\phi}_S$ $(=\hat{\phi}_S^\dagger)$, respectively, can be defined as

$$\hat{\phi}_C \equiv \cos^{-1}\hat{C} = \tfrac{1}{2}\pi - \sum_{k=0}^{\infty} \frac{(-1)^k}{2k+1}\binom{-\tfrac{1}{2}}{k}\hat{C}^{2k+1}, \qquad (3.265)$$

$$\hat{\phi}_S \equiv \sin^{-1}\hat{S} = \sum_{k=0}^{\infty} \frac{(-1)^k}{2k+1}\binom{-\tfrac{1}{2}}{k}\hat{S}^{2k+1}. \qquad (3.266)$$

Note that $\hat{\phi}_C$ and $\hat{\phi}_S$ do not commute ($[\hat{\phi}_C,\hat{\phi}_S]\neq 0$). Hence two unitary operators $\hat{V}_C$ and $\hat{V}_S$ can be defined:

$$\hat{V}_C = e^{i\hat{\phi}_C} \qquad (\hat{V}_C^\dagger \hat{V}_C = \hat{V}_C \hat{V}_C^\dagger = \hat{I}), \qquad (3.267)$$

$$\hat{V}_S = e^{i\hat{\phi}_S} \qquad (\hat{V}_S^\dagger \hat{V}_S = \hat{V}_S \hat{V}_S^\dagger = \hat{I}), \qquad (3.268)$$

so that, combining Eqs (3.246) and (3.247) with the inverse of Eqs (3.265) and (3.266),

$$\hat{V} = \tfrac{1}{2}\left(e^{i\hat{\phi}_C} + e^{i\hat{\phi}_S} + e^{-i\hat{\phi}_C} - e^{-i\hat{\phi}_S}\right). \qquad (3.269)$$

This result together with Eq. (3.208) is the correct version of Dirac's postulate given in Eq. (3.209).

The cosine and sine operators of the phases of two modes can be used to define cosine and sine operators of the phase difference, by applying the addition theorems as follows:

$$\hat{C}_{12} = \hat{C}_1 \hat{C}_2 + \hat{S}_1 \hat{S}_2, \qquad (3.270)$$

$$\hat{S}_{12} = \hat{S}_1 \hat{C}_2 - \hat{S}_2 \hat{C}_1. \qquad (3.271)$$

From the commutation relations (3.249) it is easily shown that $\hat{C}_{12}$ and $\hat{S}_{12}$ commute with the total-number operator:

$$[\hat{C}_{12}, \hat{n}_1 + \hat{n}_2] = 0, \quad [\hat{S}_{12}, \hat{n}_1 + \hat{n}_2] = 0 \qquad (3.272)$$

(note that operators of different modes commute).

References

Adam, P., I. Földesi, and J. Janszky (1994) *Phys. Rev. A* **49**, 1281.

Barnett, S.M. and D.T. Pegg (1992) *J. Mod. Opt.* **39**, 1221.

Carruthers, P. and M.M. Nieto (1968) *Rev. Mod. Phys.* **40**, 411.

Dirac, P.A.M. (1927) *Proc. R. Soc. Lond. A* **114**, 243.

Gardiner, C.W. (1991) *Quantum Noise* (Springer Verlag, Berlin).

References

Glauber, R.J. (1963a) *Phys. Rev.* **130**, 2529.

Glauber, R.J. (1963b) *Phys. Rev.* **131**, 2766.

Glauber, R.J. (1963c) *Phys. Rev. Lett.* **10**, 84.

Klauder, J.R. (1960) *Ann. Phys.* **11**, 123.

London, F. (1926) *Z. Phys.* **37**, 915.

London, F. (1927) *Z. Phys.* **40**, 193.

Loudon, R. (1973) *The Quantum Theory of Light* (Oxford University Press, Oxford).

Loudon, R. and P.L. Knight (1987) *J. Mod. Opt.* **34**, 709.

Mandel, L. and E. Wolf (1995) *Optical Coherence and Quantum Optics* (Cambridge University Press, Cambridge).

Pegg, D.T. and S.M. Barnett (1988) *Europhys. Lett.* **6**, 483.

Pegg, D.T. and S.M. Barnett (1989) *Phys. Rev. A* **39**, 1665.

Peřina, J. (1985) *Coherence of Light* (Reidel, Dordrecht).

Schrödinger, E. (1926) *Naturwiss.* **14**, 664.

Schubert, M. and W. Vogel (1978a) *Phys. Lett. A* **68**, 321.

Schubert, M. and W. Vogel (1978b) *Wiss. Z. Univ. Jena, Math.-Naturwiss. Reihe* **27**, 179.

Stoler, D. (1970) *Phys. Rev. D* **1**, 3217.

Stoler, D. (1971) *Phys. Rev. D* **4**, 1925.

Susskind, L. and J. Glogower (1964) *Physics* **1**, 49.

Yuen, H.P. (1976) *Phys. Rev. A* **13**, 2226.

Walls, D.F. (1983) *Nature* **306**, 141.

4
Bosonic systems in phase space

In classical physics the state of a physical object and its dynamics can generally be illustrated by a time dependent probability density in phase space. The phase space is thereby spanned by the canonically conjugate variables as, for example, the position q and momentum p of a particle. The phase-space density $P(q,p,t)$ relates then to the probability dw at time t of observing the particle in the intervals dq and dp, centered around the values q and p, via the typical expression $dw = P(q,p,t)dqdp$.

Whereas this is perfectly appropriate in classical physics one encounters problems of interpretation in the quantum domain. Here Heisenberg's uncertainty relation $\Delta q \Delta p \geq \hbar/2$ prohibits one to consider the knowledge (i.e., observation) of both canonical variables at the same time with arbitrary precision. Proper probability densities in phase space, which are based on orthogonal projectors, may therefore be considered nonexistent. However, a complete description of the quantum mechanical state can still be obtained in phase space if one introduces phase-space functions in a wider sense. To do this, we may first generalize our description of a quantum mechanical system in order to include (classical) statistical uncertainties. This is readily obtained in terms of the statistical density operator, which includes both quantum and classical uncertainties in the inference of the properties of the considered physical object.

4.1
The statistical density operator

After the measurement of an observable with the observed outcome being O we encode our inferred knowledge of the physical state of the measured object in the form of the quantum state, which is chosen to be the eigenstate of the associated Hermitian operator $\hat{O}$ with eigenvalue O. Clearly this represents an idealized picture in that we assume a perfect detection of the observable. However, in general we may release these constraints to also allow statistical uncertainties in the measurement process itself. These uncertainties, being of classical nature, restrict the knowledge or information that can be gained in a measurement, since now a range of values of O may correspond to the measurement outcome.

To deal with such situations we have to incorporate classical statistics into our quantum mechanical description of the inferred state of a quantum mechanical object. Most naturally this is performed by turning to the statistical density operator which is a weighted sum of state projectors, with the weights having the properties of a probability distribution,

$$\hat{\varrho} = \sum_{\psi} P_\psi |\psi\rangle\langle\psi|. \tag{4.1}$$

Here P_ψ may be viewed as the probability of finding the system in the quantum state $|\psi\rangle$ and the sum goes over all possible or considered states. Clearly, the P_ψ being defined as probabilities have to satisfy the conditions

$$P_\psi \geq 0, \quad \sum_{\psi} P_\psi = 1. \tag{4.2}$$

In particular, a pure state $|\phi\rangle$ can be easily represented by a density operator by choosing the weights as $P_\psi = \delta_{\psi\phi}$. The density operator reduces then to the projector $\hat{\varrho} = |\phi\rangle\langle\phi|$.

The expectation value of a physical observable represented by the corresponding Hermitian operator $\hat{O}$ for the quantum state being described by a density operator reads as

$$\langle \hat{O} \rangle = \sum_{\psi} P_\psi \langle\psi|\hat{O}|\psi\rangle = \text{Tr}(\hat{\varrho}\hat{O}). \tag{4.3}$$

From Eq. (4.3) we can see that when the system is prepared in a statistical mixture of quantum states (4.1), obtaining the expectation value of an observable is in general a two-fold procedure. Firstly the quantum-mechanical expectation values of the observable $\hat{O}$ must be calculated for the states $|\psi\rangle$ and secondly these expectation values must be averaged according to the probabilities P_ψ in the usual (classical) way.

Using the density operator itself as the observable, $\hat{O} = \hat{\varrho}$, we obtain from Eq. (4.3) the special result

$$\langle \hat{\varrho} \rangle = \sum_{\psi,\psi'} P_\psi P_{\psi'} |\langle\psi'|\psi\rangle|^2, \tag{4.4}$$

where by using the relation $|\langle\psi|\psi'\rangle| \leq 1$ we may derive from (4.4) the inequality

$$\langle \hat{\varrho} \rangle = \text{Tr}\,\hat{\varrho}^2 \leq 1. \tag{4.5}$$

This inequality is in general regarded as a criterion for the statistical mixedness of a quantum state. When $\text{Tr}\,\hat{\varrho}^2 = 1$ we have a pure state, whereas with decreasing value of this trace the state becomes more and more statistically mixed. To show this in more detail, let us consider an orthogonal and complete set of states $|j\rangle$, i.e.,

$$\langle j|j'\rangle = \delta_{jj'}, \quad \sum_j |j\rangle\langle j| = \hat{I}. \tag{4.6}$$

By the use of this representation, the density operator may be written in the form

$$\hat{\varrho} = \sum_{j,j'} \varrho_{jj'} |j\rangle\langle j'|, \tag{4.7}$$

where $\varrho_{jj'} = \langle j|\hat{\varrho}|j'\rangle$ are the matrix elements of the density operator in the chosen representation, or in short, the density matrix. We can easily prove that Eq. (4.3) may be rewritten as

$$\langle \hat{O} \rangle = \sum_{j,j'} \varrho_{jj'} \langle j'|\hat{O}|j\rangle. \tag{4.8}$$

When the density operator $\hat{\varrho}$ corresponds to a pure quantum state without any statistical mixedness, $\hat{\varrho} = |\phi\rangle\langle\phi|$, the corresponding density matrix in the representation of states $|j\rangle$ reads

$$\varrho_{jj'} = \langle j|\phi\rangle\langle\phi|j'\rangle. \tag{4.9}$$

From Eq. (4.8) ($\hat{O} \mapsto \hat{\varrho}$) in this case we obtain for the statistical mixedness

$$\text{Tr}\,\hat{\varrho}^2 = \sum_{j,j'} |\langle j|\phi\rangle|^2 |\langle j'|\phi\rangle|^2 = 1, \tag{4.10}$$

due to the fact that the probability of finding the value j for the given state $|\phi\rangle$ is normalized to unity. In general one may observe that the moduli of the off-diagonal elements are smaller for a statistical mixture than those for a pure state, i.e., $|\rho_{jj'}| < \sqrt{|\rho_{jj}||\rho_{j'j'}|}$ for $j \neq j'$, which leads then to $\text{Tr}\,\hat{\varrho}^2 < 1$.

As is well known, in the Schrödinger picture the state vector $|\psi(t)\rangle$ obeys the Schrödinger equation

$$i\hbar \frac{d|\psi\rangle}{dt} = \hat{H}|\psi\rangle, \tag{4.11}$$

where $\hat{H}$ is the Hamiltonian of the system under consideration. By considering the pure-state case $\hat{\varrho} = |\psi\rangle\langle\psi|$ it is obvious that the density operator then obeys the following equation of motion:

$$i\hbar \frac{d\hat{\varrho}}{dt} = [\hat{H}, \hat{\varrho}]. \tag{4.12}$$

The use of density operators is typically useful when dealing with a system composed of interacting subsystems, where only one of the subsystems is of interest. Let us consider, for example, two interacting systems, such as a radiation field coupled to an atomic system, with Hamiltonian $\hat{H}=\hat{H}_1+\hat{H}_2+\hat{H}_{\text{int}}$. The Hilbert space of the total system is then the direct product of the Hilbert spaces of the two subsystems, that is, when $|j_1\rangle$ and $|j_2\rangle$ are forming complete sets of states of system 1 and 2, respectively, a complete set of states for the combined system is given by

$$|j_1,j_2\rangle = |j_1\rangle|j_2\rangle. \tag{4.13}$$

The expectation values of an observable $\hat{O}_k$ ($k=1,2$) associated with only one of the subsystems, say subsystem 1, is obtained using Eq. (4.8) as

$$\langle\hat{O}_1\rangle = \text{Tr}(\hat{\varrho}_1\hat{O}_1), \tag{4.14}$$

where the reduced density operator $\hat{\varrho}_1$, which describes subsystem 1 alone, is given by the trace with respect to subsystem 2,

$$\hat{\varrho}_1 = \text{Tr}_2\,\hat{\varrho} = \sum_{j_2}\langle j_2|\hat{\varrho}|j_2\rangle. \tag{4.15}$$

Note that, even when the overall system is in a pure quantum state, the subsystems, as represented by their reduced density operators, in general are not. If the overall system is closed, i.e., isolated from its environment, the corresponding density operator of the system will obey an equation of motion of the type (4.12). However, the dynamics of a system which is interacting with its environment, cannot in general be described by such a unitary time evolution.

4.2
Phase-space functions

Let $\hat{O}$ be an operator that may be a function of $\hat{a}$ and $\hat{a}^\dagger$, $\hat{O}=\hat{f}(\hat{a},\hat{a}^\dagger)$, and let us consider its expectation value when the system is described by the density operator $\hat{\varrho}$,

$$\langle\hat{O}\rangle = \langle\hat{f}(\hat{a},\hat{a}^\dagger)\rangle = \text{Tr}[\hat{\varrho}\,\hat{f}(\hat{a},\hat{a}^\dagger)]. \tag{4.16}$$

To perform the trace, a set of complete quantum states has to be chosen. To arrive at a phase-space description it is convenient to choose the coherent states $|\alpha\rangle$ as basis states. From Eq. (3.71) we obtain the coherent-state representation of the density operator as

$$\hat{\varrho} = \frac{1}{\pi^2}\int d^2\alpha\int d^2\beta\,\varrho(\alpha,\beta)|\alpha\rangle\langle\beta|, \tag{4.17}$$

where the density matrix in the coherent-state basis is given by

$$\varrho(\alpha, \beta) = \langle \alpha | \hat{\varrho} | \beta \rangle. \tag{4.18}$$

Inserting Eq. (4.17) into Eq. (4.16), we obtain the expectation value $\langle \hat{f}(\hat{a}, \hat{a}^\dagger) \rangle$ in the form of

$$\langle \hat{f}(\hat{a}, \hat{a}^\dagger) \rangle = \frac{1}{\pi^2} \int d^2\alpha \int d^2\beta \, \varrho(\alpha, \beta) \langle \beta | \hat{f}(\hat{a}, \hat{a}^\dagger) | \alpha \rangle. \tag{4.19}$$

By substituting $\hat{a}, \hat{a}^\dagger \mapsto \alpha, \alpha^*$ we may arrive at classical statistics, where the operator function which corresponds to the operator $\hat{O}$ will turn out to be a function in the phase-space spanned by the complex number α,

$$\hat{O} = \hat{f}(\hat{a}, \hat{a}^\dagger) \mapsto O = f(\alpha, \alpha^*) \equiv f(\alpha). \tag{4.20}$$

In this case the expectation value of the quantity O is obtained by the usual statistical averaging as

$$\langle O \rangle_{cl} = \int d^2\alpha \, P_{cl}(\alpha) f(\alpha), \tag{4.21}$$

where the phase-space function $P_{cl}(\alpha)$ is the classical probability density of observing the complex field amplitude α, which fully describes the (classical) state of the system. The question arises as to whether or not the quantum-mechanical expectation value (4.19) may be represented in a form similar to that of classical theory, Eq. (4.21). As we shall see below, Eq. (4.19) can indeed be rewritten in a form which formally looks like Eq. (4.21), provided that the operator under study, $\hat{O}$, is ordered in certain ways with respect to the operators $\hat{a}, \hat{a}^\dagger$. However, the phase-space functions found in this way cannot be viewed, in general, as being probability distribution functions.

4.2.1
Normal ordering: The P function

To arrive at one of these phase-space functions, let us assume that, by means of the commutation relation $[\hat{a}, \hat{a}^\dagger] = 1$, the operator $\hat{O} = \hat{f}(\hat{a}, \hat{a}^\dagger)$ is put into normal order. Normal order means in this context that in the resulting expression all the creation operators are positioned left of the annihilation operators. That is, if $\hat{f}^{(N)}(\hat{a}, \hat{a}^\dagger)$ is the resulting expression in normal order, we have the equivalence

$$\hat{f}(\hat{a}, \hat{a}^\dagger) \equiv \hat{f}^{(N)}(\hat{a}, \hat{a}^\dagger). \tag{4.22}$$

Furthermore, we may now define the associated c-number function $f^{(N)}(\alpha) \equiv f^{(N)}(\alpha, \alpha^*)$ by substituting in $\hat{f}^{(N)}(\hat{a}, \hat{a}^\dagger)$ for the operators $\hat{a}$ and $\hat{a}^\dagger$ the complex c numbers α and α^*, respectively. Obviously, $f^{(N)}(\alpha)$ is simply the diagonal matrix element of $\hat{f}(\hat{a}, \hat{a}^\dagger)$ with the coherent state $|\alpha\rangle$:

$$f^{(N)}(\alpha) = \langle\alpha|\hat{f}^{(N)}(\hat{a}, \hat{a}^\dagger)|\alpha\rangle = \langle\alpha|\hat{f}(\hat{a}, \hat{a}^\dagger)|\alpha\rangle. \tag{4.23}$$

Having this c-number function at hand, we now intend to express the operator $\hat{f}(\hat{a}, \hat{a}^\dagger)$ in terms of $f^{(N)}(\alpha)$ and other suitably chosen normally ordered operator functions. For this purpose, let us inspect the identity

$$f^{(N)}(\alpha) = \int d^2\beta\, \delta(\alpha - \beta) f^{(N)}(\beta), \tag{4.24}$$

where $\delta(\alpha)$ is the usual two-dimensional Dirac δ function for real and imaginary parts of the argument, i.e., ($\alpha = \alpha' + i\alpha''$),

$$\delta(\alpha) = \delta(\alpha')\delta(\alpha'') = \frac{1}{4\pi^2} \int dx \int dy\, \exp[i(\alpha' y + \alpha'' x)], \tag{4.25}$$

Substituting $\gamma = \pm(iy - x)/2$, we may rewrite the delta function (4.25) in a more convenient form, as an integral over the complex variable γ,

$$\delta(\alpha) = \frac{1}{\pi^2} \int d^2\gamma\, \exp(\alpha^*\gamma - \alpha\gamma^*) = \frac{1}{\pi^2} \int d^2\gamma\, \exp(\alpha\gamma^* - \alpha^*\gamma). \tag{4.26}$$

Inserting this expression for the delta function in Eq. (4.24), we obtain

$$f^{(N)}(\alpha) = \frac{1}{\pi^2} \int d^2\beta \int d^2\gamma\, f^{(N)}(\beta) \exp[(\alpha - \beta)^*\gamma - (\alpha - \beta)\gamma^*]. \tag{4.27}$$

Going from the associated c-number function $f^{(N)}(\alpha)$ back to the operator function $\hat{f}^{(N)}(\hat{a}, \hat{a}^\dagger)$, i.e., re-substituting $\alpha, \alpha^* \mapsto \hat{a}, \hat{a}^\dagger$, we see from Eq. (4.27) that the operator $\hat{f}^{(N)}(\hat{a}, \hat{a}^\dagger)$ may be represented as

$$\hat{f}^{(N)}(\hat{a}, \hat{a}^\dagger) = \frac{1}{\pi^2} \int d^2\beta \int d^2\gamma\, f^{(N)}(\beta) \exp\left[(\hat{a}^\dagger - \beta^*)\gamma\right] \exp\left[-(\hat{a} - \beta)\gamma^*\right]. \tag{4.28}$$

This substitution is allowed since, before replacing the c numbers by operators, we have factored the exponential function in order to obtain a normally ordered representation where the $\hat{a}^\dagger$ are located left of the $\hat{a}$.

Next, we introduce an operator-valued version of the Dirac δ function in straightforward generalization of Eq. (4.26),

$$\hat{\delta}(\hat{a} - \alpha) = \frac{1}{\pi^2} \int d^2\beta\, \exp[(\hat{a}^\dagger - \alpha^*)\beta - (\hat{a} - \alpha)\beta^*], \tag{4.29}$$

which may be also written as a Fourier transform of the displacement operator (3.44),

$$\hat{\delta}(\hat{a} - \alpha) = \frac{1}{\pi^2} \int d^2\beta\, \hat{D}(\beta) \exp(\alpha\beta^* - \alpha^*\beta). \tag{4.30}$$

Applying the normal-ordering prescription $\mathcal{N}$ onto the displacement operator we obtain[1]

$$\mathcal{N}\hat{D}(\alpha) \equiv :\hat{D}(\alpha): = e^{\hat{a}^\dagger \alpha} e^{-\hat{a}\alpha^*}. \tag{4.31}$$

Using the relations (4.30) and (4.31), we may then rewrite Eq. (4.28) as

$$\hat{f}(\hat{a},\hat{a}^\dagger) \equiv \hat{f}^{(N)}(\hat{a},\hat{a}^\dagger) = \int d^2\alpha\, f^{(N)}(\alpha) \mathcal{N}\hat{\delta}(\hat{a}-\alpha). \tag{4.32}$$

We now take the quantum-mechanical expectation value of both sides of Eq. (4.32) and obtain, on assuming that this operation and the integration can be interchanged, the sought result:

$$\langle \hat{f}(\hat{a},\hat{a}^\dagger)\rangle = \langle \hat{f}^{(N)}(\hat{a},\hat{a}^\dagger)\rangle = \int d^2\alpha\, P^{(N)}(\alpha)\, f^{(N)}(\alpha), \tag{4.33}$$

where the phase-space function $P^{(N)}(\alpha)$ is the expectation value of the operator delta function in normal order,

$$P^{(N)}(\alpha) = \langle :\hat{\delta}(\hat{a}-\alpha): \rangle. \tag{4.34}$$

Although Eqs (4.21) and (4.33) bear a great resemblance, there are essential differences between the two equations. Firstly, the c-number function $f^{(N)}(\alpha)$ in Eq. (4.33) is associated with the operator $\hat{f}(\hat{a},\hat{a}^\dagger)$ being transformed into its equivalent normally ordered form. That is, the complex numbers α and α^* are substituted for the operators in the operator function $\hat{f}^{(N)}(\hat{a},\hat{a}^\dagger)$ and not in the original form of the operator function $\hat{f}(\hat{a},\hat{a}^\dagger)$. Secondly, in general, the function $P^{(N)}(\alpha)$ cannot be regarded as a proper probability distribution function: $P^{(N)}(\alpha)$ can attain negative values that are not interpretable as probability densities and furthermore $P^{(N)}(\alpha)$ need not be a well-behaved function [for reviews, see Klauder and Sudarshan (1968); Peřina (1991)]. Notwithstanding these facts, in any case the function $P^{(N)}(\alpha)$ is normalized,

$$\int d^2\alpha\, P^{(N)}(\alpha) = 1, \tag{4.35}$$

which may readily be proved from Eq. (4.33) by choosing $\hat{f}(\hat{a},\hat{a}^\dagger) = \hat{I}$. The quantum-state representation based on the phase-space function $P^{(N)}(\alpha)$ is called the Glauber–Sudarshan representation [Glauber (1963); Sudarshan (1963)], $P^{(N)}(\alpha)$ being also called the P function, where a frequently used abbreviated notation is

$$P(\alpha) \equiv P^{(N)}(\alpha). \tag{4.36}$$

---

1) The process of normal ordering as described by $\mathcal{N}$ is not to be confused with a normally ordered, equivalent representation of an operator, such as represented by the relation $\hat{f}^{(N)}(\hat{a},\hat{a}^\dagger) = \hat{f}(\hat{a},\hat{a}^\dagger)$. $\mathcal{N}$ is not an equivalence operation, i.e., $\mathcal{N}\hat{f}(\hat{a},\hat{a}^\dagger) \neq \hat{f}(\hat{a},\hat{a}^\dagger)$.

The Glauber–Sudarshan representation is of special importance in the context of photodetection where the appearing expectation values contain normally ordered moments and correlations. Moreover, although the P function may be an ill-behaving function, this distribution proves to be very useful for formal derivations in connection with operator expectation values.

4.2.2
Anti-normal and symmetric ordering: The Q and the W function

The applicability of the concept of phase-space functions as outlined above, is of course not restricted to the case of normal order. For example, if the operator $\hat{O}$ can be put in anti-normal order by use of the commutator relation $[\hat{a}, \hat{a}^\dagger] = 1$, $\hat{O} = \hat{f}^{(A)}(\hat{a}, \hat{a}^\dagger)$, we may introduce the associated c-number function $f^{(A)}(\alpha)$ by substituting in $\hat{f}^{(A)}(\hat{a}, \hat{a}^\dagger)$ for the operators $\hat{a}$ and $\hat{a}^\dagger$ the c numbers α and α^*, respectively. Performing manipulations analogous to those leading to Eq. (4.28) now yields the corresponding expression in anti-normal order,

$$\hat{f}(\hat{a}, \hat{a}^\dagger) = \frac{1}{\pi^2} \int d^2\beta \int d^2\gamma\, f^{(A)}(\beta) \exp[-(\hat{a}-\beta)\gamma^*] \exp[(\hat{a}^\dagger - \beta^*)\gamma]. \quad (4.37)$$

Hence, instead of the normally ordered delta operator we now use the anti-normally ordered version,

$$\mathcal{A}\hat{\delta}(\hat{a} - \alpha) \equiv \ddagger\hat{\delta}(\hat{a} - \alpha)\ddagger = \frac{1}{\pi^2} \int d^2\beta\, \ddagger\hat{D}(\beta)\ddagger \exp(\alpha\beta^* - \alpha^*\beta), \quad (4.38)$$

so that the anti-normally ordered displacement operator reads

$$\ddagger\hat{D}(\alpha)\ddagger = e^{-\hat{a}\alpha^*} e^{\hat{a}^\dagger \alpha}. \quad (4.39)$$

We then obtain in analogy with Eq. (4.32)

$$\hat{f}(\hat{a}, \hat{a}^\dagger) = \int d^2\alpha\, f^{(A)}(\alpha) \mathcal{A}\hat{\delta}(\hat{a} - \alpha), \quad (4.40)$$

from which the relation for the expectation value is derived as

$$\langle \hat{f}(\hat{a}, \hat{a}^\dagger) \rangle = \langle \hat{f}^{(A)}(\hat{a}, \hat{a}^\dagger) \rangle = \int d^2\alpha\, P^{(A)}(\alpha) f^{(A)}(\alpha). \quad (4.41)$$

The phase-space function $P^{(A)}(\alpha)$, which is called the Husimi Q function, is the expectation value of the operator delta function in anti-normal order,

$$Q(\alpha) \equiv P^{(A)}(\alpha) = \langle \ddagger\hat{\delta}(\alpha - \hat{a})\ddagger \rangle. \quad (4.42)$$

On the other hand, taking the expectation value of the original operator delta function as defined by Eq. (4.29) obviously yields the phase space function suitable for averaging symmetrically ordered quantities,

$$\langle \hat{f}(\hat{a}, \hat{a}^\dagger) \rangle = \langle \hat{f}^{(S)}(\hat{a}, \hat{a}^\dagger) \rangle = \int d^2\alpha\, P^{(S)}(\alpha) f^{(S)}(\alpha), \quad (4.43)$$

where
$$W(\alpha) \equiv P^{(S)}(\alpha) = \langle \hat{\delta}(\alpha - \hat{a}) \rangle \tag{4.44}$$
is called the Wigner function, and $f^{(S)}(\alpha)$ is the c-number function associated with the operator $\hat{f}(\hat{a}, \hat{a}^\dagger)$ in symmetrical order. We will not give more details here, but instead, in the following we consider the more general case of so-called s ordering.

4.2.3
Parameterized phase-space functions

The phase-space functions considered above may be regarded as certain special cases of an operator $\hat{O}$ being put in a chosen order [Cahill and Glauber (1969); for a review, see also Peřina (1991)]. To generalize the concept of operator ordering, we may start with the displacement operator and define its s-ordered representation by

$$\hat{D}(\alpha;s) = \hat{D}(\alpha)e^{|\alpha|^2 s/2}, \tag{4.45}$$

which implies that

$$\hat{D}(\alpha;s) = \exp\left[\tfrac{1}{2}(s-s')|\alpha|^2\right]\hat{D}(\alpha;s'). \tag{4.46}$$

The case $s=0$ is then considered as symmetrical ordering, since we obtain the original displacement operator,

$$\hat{D}(\alpha;0) = \hat{D}(\alpha). \tag{4.47}$$

Moreover, comparing the expression (4.45) for the values $s=\pm 1$ with Eqs (4.31) and (4.39), we arrive at the following relations:

$$\hat{D}(\alpha;1) = e^{\alpha\hat{a}^\dagger}e^{-\alpha^*\hat{a}} = :\hat{D}(\alpha):, \tag{4.48}$$

$$\hat{D}(\alpha;-1) = e^{-\alpha^*\hat{a}}e^{\alpha\hat{a}^\dagger} = \ddagger\hat{D}(\alpha)\ddagger. \tag{4.49}$$

From Eqs (3.44)–(3.46) we see that choosing $s=0$, $s=1$ and $s=-1$ corresponds to putting the displacement operator in symmetrical, normal and anti-normal order, respectively. It should be pointed out that more general ordering procedures can be introduced, which unify s-ordering with other orderings such as standard and anti-standard ordering [Agarwal and Wolf (1968)].[2]

Let us now assume that the operator $\hat{O} = \hat{f}(\hat{a}, \hat{a}^\dagger)$ can be represented in any s-order $-1 \leq s \leq 1$ as[3]

$$\hat{f}(\hat{a}, \hat{a}^\dagger) = \int d^2\alpha\, f(\alpha;s)\hat{\delta}(\hat{a} - \alpha;s), \tag{4.50}$$

---

[2] For a detailed treatment of these unified ordering methods, see Agarwal and Wolf (1970).

[3] Note that the value of s is not necessarily restricted to the interval $-1,\ldots,1$.

where we have defined via Eq. (4.45) the general s-ordered operator delta function

$$\hat{\delta}(\hat{a} - \alpha; s) = \frac{1}{\pi^2} \int d^2\beta \, \hat{D}(\beta; s) \exp(\alpha\beta^* - \alpha^*\beta) \tag{4.51}$$

[for details about the existence of the representation (4.50), see Cahill and Glauber (1969)]. Obviously, the c-number function $f(\alpha; s)$ associated with the operator $\hat{f}(\hat{a}, \hat{a}^\dagger)$ in the chosen order, reduces in the special cases $s = 0, \pm 1$ to the familiar expressions $f(\alpha; 0) \equiv f^{(S)}(\alpha)$, $f(\alpha; 1) \equiv f^{(N)}(\alpha)$ and $f(\alpha; -1) \equiv f^{(A)}(\alpha)$. With the help of Eq. (4.50) the expectation value of an arbitrary operator $\hat{O} = \hat{f}(\hat{a}, \hat{a}^\dagger)$ may now be written as

$$\langle \hat{O} \rangle = \langle \hat{f}(\hat{a}, \hat{a}^\dagger) \rangle = \int d^2\alpha \, P(\alpha; s) f(\alpha; s), \tag{4.52}$$

where the s-parameterized phase-space function $P(\alpha; s)$ is defined as[4]

$$P(\alpha; s) = \langle \hat{\delta}(\hat{a} - \alpha; s) \rangle. \tag{4.53}$$

It is often useful to represent the s-ordered operator delta function in a somewhat different form. Expressing $\hat{D}(\alpha; s)$ in terms of $\hat{D}(\alpha; s')$ in Eq. (4.51) according to Eq. (4.46) and applying, with respect to $\hat{D}(\alpha; s')$, the inverse of Eq. (4.51), we may write

$$\hat{\delta}(\hat{a} - \alpha; s) = \frac{1}{\pi^2} \int d^2\beta \, \exp\left[\alpha\beta^* - \alpha^*\beta + \tfrac{1}{2}(s - s')|\beta|^2\right] \hat{D}(\beta; s')$$

$$= \frac{1}{\pi^2} \int d^2\beta \, \exp\left[\alpha\beta^* - \alpha^*\beta + \tfrac{1}{2}(s - s')|\beta|^2\right]$$

$$\times \int d^2\gamma \, \exp(\beta\gamma^* - \beta^*\gamma) \hat{\delta}(\hat{a} - \gamma; s'). \tag{4.54}$$

For $s \leq s'$ the integration over β can be performed[5] to obtain

$$\hat{\delta}(\hat{a} - \alpha; s) = \frac{2}{\pi(s' - s)} \int d^2\gamma \, \exp\left(-\frac{2|\alpha - \gamma|^2}{s' - s}\right) \hat{\delta}(\hat{a} - \gamma; s'). \tag{4.55}$$

For $s' = 1$ the operator $\hat{\delta}(\hat{a} - \gamma; 1)$ is the normally ordered form of the delta-function operator, and therefore the γ integration yields

$$\hat{\delta}(\hat{a} - \alpha; s) = \frac{2}{\pi(1 - s)} : \exp\left[-\frac{2(\hat{a}^\dagger - \alpha^*)(\hat{a} - \alpha)}{1 - s}\right] :, \tag{4.56}$$

which with the help of Eqs (3.47) and (3.48) may be rewritten in the form

$$\hat{\delta}(\hat{a} - \alpha; s) = \frac{2}{\pi(1 - s)} : \exp\left[-\frac{2\hat{n}(\alpha)}{1 - s}\right] :, \tag{4.57}$$

---

4) Recall that $W(\alpha) \equiv P^{(S)}(\alpha) \equiv P(\alpha; 0)$, $P(\alpha) \equiv P^{(N)}(\alpha) \equiv P(\alpha; 1)$ and $Q(\alpha) \equiv P^{(A)}(\alpha) \equiv P(\alpha; -1)$.
5) Note that the more general condition is Re $(s - s') \leq 0$.

with the displaced number operator being defined as

$$\hat{n}(\alpha) = \hat{D}(\alpha)\hat{n}\hat{D}^\dagger(\alpha). \tag{4.58}$$

Equation (4.57) can be further evaluated to obtain

$$\hat{\delta}(\hat{a} - \alpha; s) = \frac{2}{\pi(1-s)} : \exp\left[\frac{s+1}{s-1}\hat{n}(\alpha)\right]\exp[-\hat{n}(\alpha)]:$$

$$= \frac{2}{\pi(1-s)} \sum_{n=0}^{\infty} \left(\frac{s+1}{s-1}\right)^n : \frac{[\hat{n}(\alpha)]^n}{n!} \exp[-\hat{n}(\alpha)]:$$

$$= \frac{2}{\pi(1-s)} \sum_{n=0}^{\infty} \left(\frac{s+1}{s-1}\right)^n \hat{D}(\alpha) : \frac{\hat{n}^n}{n!} e^{-\hat{n}} : \hat{D}^\dagger(\alpha). \tag{4.59}$$

Since $|\langle n|\alpha\rangle|^2$ as given by Eq. (3.60) is the c-number function associated with $|n\rangle\langle n|$ in normal order, we find that

$$: \frac{\hat{n}^n}{n!} e^{-\hat{n}} : = |n\rangle\langle n|, \tag{4.60}$$

and Eq. (4.59) can be rewritten as

$$\hat{\delta}(\hat{a} - \alpha; s) = \frac{2}{\pi(1-s)} \sum_{n=0}^{\infty} \left(\frac{s+1}{s-1}\right)^n \hat{D}(\alpha)|n\rangle\langle n|\hat{D}^\dagger(\alpha), \tag{4.61}$$

equivalently

$$\hat{\delta}(\hat{a} - \alpha; s) = \frac{2}{\pi(1-s)} \hat{D}(\alpha)\left(\frac{s+1}{s-1}\right)^{\hat{n}} \hat{D}^\dagger(\alpha). \tag{4.62}$$

For $s=0$ the s-ordered operator delta function $\hat{\delta}(\hat{a}-\alpha;0)$ reduces to the ordinary (i. e., symmetrically ordered) operator delta function $\hat{\delta}(\hat{a}-\alpha)$ defined by Eq. (4.29), and from Eq. (4.62) it then follows that

$$\hat{\delta}(\hat{a} - \alpha) = 2\pi^{-1}\hat{D}(\alpha)(-1)^{\hat{n}}\hat{D}^\dagger(\alpha) = 2\pi^{-1}(-1)^{\hat{n}(\alpha)}. \tag{4.63}$$

That is, the operator delta function is given (apart from the factor $2/\pi$) by the displaced parity operator and the Wigner function is simply the expectation value of that operator:

$$W(\alpha) \equiv P(\alpha;0) = 2\pi^{-1}\langle\hat{D}(\alpha)(-1)^{\hat{n}}\hat{D}^\dagger(\alpha)\rangle = 2\pi^{-1}\langle(-1)^{\hat{n}(\alpha)}\rangle. \tag{4.64}$$

Equation (4.64) reveals that the Wigner function cannot be regarded as a probability distribution in general, because it may attain negative values,

$$-2\pi^{-1} \leq W(\alpha) \leq 2\pi^{-1}. \tag{4.65}$$

For $s=-1$ Eq. (4.61) reduces to[6]

$$\hat{\delta}(\hat{a}-\alpha;-1) = \pi^{-1}|\alpha\rangle\langle\alpha|. \tag{4.66}$$

Hence, the Q function is given (apart from the factor π^{-1}) by the c-number function associated with the density operator in normal order,

$$Q(\alpha) \equiv P(\alpha;-1) = \pi^{-1}\langle\alpha|\hat{\varrho}|\alpha\rangle, \tag{4.67}$$

from which it follows that

$$0 \leq Q(\alpha) \leq \pi^{-1}. \tag{4.68}$$

It is worth noting that, although the Q function does not allow an interpretation as a probability distribution of the complex amplitude α in the sense of classical theory, it has all the properties of a probability distribution. As can be seen from Eq. (4.61), for $s=1$ the corresponding operator delta function $\hat{\delta}(\hat{a}-\alpha;1)$ is not bounded, and therefore the P function $P(\alpha)\equiv P(\alpha;1)$ is not necessarily a well-behaved phase-space function.

It should be noted that Eq. (4.54) can be used to express the function $P(\alpha;s)$, Eq. (4.53), in terms of another function $P(\alpha;s')$:

$$P(\alpha;s) = \frac{1}{\pi^2}\int d^2\beta \, \exp\left[\alpha\beta^* - \alpha^*\beta + \tfrac{1}{2}(s-s')|\beta|^2\right] \int d^2\gamma \, \exp(\beta\gamma^* - \beta^*\gamma) P(\gamma;s'). \tag{4.69}$$

For $s \leq s'$ (or $\operatorname{Re} s \leq \operatorname{Re} s'$) the integration over β can again be performed to obtain

$$P(\alpha;s) = \frac{2}{\pi(s'-s)} \int d^2\gamma \, P(\gamma;s') \exp\left(-\frac{2|\alpha-\gamma|^2}{s'-s}\right). \tag{4.70}$$

In the opposite case when $s' < s$ (or $\operatorname{Re} s' < \operatorname{Re} s$), the integration over γ should be done first in Eq. (4.69) in order to avoid having to deal with singular expressions.

4.3
Operator expansion in phase space

Equation (4.50) can be viewed as an expansion of an operator $\hat{f}(\hat{a},\hat{a}^\dagger)$ in terms of the generalized projectors $\hat{\delta}(\hat{a}-\alpha;s)$. Whereas for $s=0,\pm 1$ it is clear, in principle, how to obtain the associated c-number functions $f(\alpha;s)$, for arbitrary values of s we still require a recipe.

---

[6] Note that for $s=-1$, from comparison of Eq. (4.57) and (4.66), it follows that $|\alpha\rangle\langle\alpha| = \,:\exp[-\hat{n}(\alpha)]:$.

4.3.1
Orthogonalization relations

For the purpose of deriving such a prescription we may consider the following relation obtained by taking the trace of Eq. (4.50) multiplied by $\hat{\delta}(\hat{a}-\beta;-s)$,

$$\text{Tr}[\hat{f}(\hat{a},\hat{a}^\dagger)\hat{\delta}(\hat{a}-\beta;-s)] = \int d^2\alpha\, f(\alpha;s)\, \text{Tr}[\hat{\delta}(\hat{a}-\alpha;s)\hat{\delta}(\hat{a}-\beta;-s)]. \quad (4.71)$$

To obtain the trace on the right-hand side of Eq. (4.71) we first calculate the trace (via Fourier transformation) contained therein over the displacement operators. Using Eqs (3.53) and (4.45) we may write

$$\text{Tr}[\hat{D}(\alpha;s)\hat{D}(\beta;-s)] = \exp\left[\tfrac{1}{2}s(|\alpha|^2 - |\beta|^2)\right]\text{Tr}[\hat{D}(\alpha)\hat{D}(\beta)]$$
$$= \exp\left[\tfrac{1}{2}s(|\alpha|^2 - |\beta|^2) + i\,\text{Im}(\alpha\beta^*)\right]\text{Tr}[\hat{D}(\alpha+\beta)]. \quad (4.72)$$

We calculate the trace of the displacement operator using the coherent-state basis and applying Eqs (3.54) and (4.26):

$$\text{Tr}[\hat{D}(\alpha)] = \frac{1}{\pi}\int d^2\beta\, \langle\beta|\hat{D}(\alpha)|\beta\rangle$$
$$= e^{-|\alpha|^2/2}\frac{1}{\pi}\int d^2\beta\, \exp(\alpha\beta^* - \alpha^*\beta) = \pi\delta(\alpha). \quad (4.73)$$

Hence, Eq. (4.72) takes the form of

$$\text{Tr}[\hat{D}(\alpha;s)\hat{D}(\beta;-s)] = \pi\delta(\alpha+\beta), \quad (4.74)$$

and combining Eqs (4.51) and (4.74) yields

$$\text{Tr}[\hat{\delta}(\hat{a}-\alpha;s)\hat{\delta}(\hat{a}-\beta;-s)] = \pi^{-1}\delta(\alpha-\beta). \quad (4.75)$$

Note that Eq. (4.73) implies that

$$\text{Tr}[\hat{\delta}(\hat{a}-\alpha;s)] = \pi^{-1}. \quad (4.76)$$

Inserting the orthogonalization relation (4.75) into Eq. (4.71), we see that $f(\alpha;s)$ may be represented as

$$f(\alpha;s) = \pi\,\text{Tr}[\hat{f}(\hat{a},\hat{a}^\dagger)\hat{\delta}(\hat{a}-\alpha;-s)]. \quad (4.77)$$

Equation (4.77) may be viewed as the sought prescription for calculating the c-number function $f(\alpha;s)$ associated with the operator $\hat{f}(\hat{a},\hat{a}^\dagger)$ in s order from $\hat{f}(\hat{a},\hat{a}^\dagger)$ via the s-ordered delta operator. Substitution of the expression (4.77) into Eq. (4.50) yields the operator expansion in the phase space

$$\hat{f}(\hat{a},\hat{a}^\dagger) = \pi\int d^2\alpha\, \text{Tr}[\hat{f}(\hat{a},\hat{a}^\dagger)\hat{\delta}(\hat{a}-\alpha;-s)]\hat{\delta}(\hat{a}-\alpha;s). \quad (4.78)$$

Equivalently, we may expand $\hat{f}(\hat{a}, \hat{a}^\dagger)$ in terms of the s-ordered displacement operator $\hat{D}(\alpha; s)$. Recalling Eq. (4.51), it is not difficult to see that Eq. (4.78) can be rewritten as

$$\hat{f}(\hat{a}, \hat{a}^\dagger) = \frac{1}{\pi} \int d^2\alpha \ \mathrm{Tr}[\hat{f}(\hat{a}, \hat{a}^\dagger)\hat{D}(-\alpha; -s)]\hat{D}(\alpha; s). \tag{4.79}$$

4.3.2
The density operator in phase space

If we now identify in Eq. (4.78) [together with Eq. (4.77)] the operator $\hat{f}(\hat{a}, \hat{a}^\dagger)$ with the density operator $\hat{\varrho}$, we obtain the following representation of the density operator:

$$\hat{\varrho} = \int d^2\alpha \ \varrho(\alpha; s)\hat{\delta}(\hat{a} - \alpha; s), \tag{4.80}$$

where the c-number function associated with the density operator in s order reads

$$\varrho(\alpha; s) = \pi \, \mathrm{Tr}[\hat{\varrho}\hat{\delta}(\hat{a} - \alpha; -s)] = \pi\langle\hat{\delta}(\hat{a} - \alpha; -s)\rangle. \tag{4.81}$$

Comparing Eq. (4.81) with (4.53), we see that the phase-space function $P(\alpha; s)$ is (apart from the factor π^{-1}) identical to $\varrho(\alpha; -s)$,

$$P(\alpha; s) = \pi^{-1}\varrho(\alpha; -s). \tag{4.82}$$

Therefore, the density operator itself can be represented via Eq. (4.80) as $(s \to -s)$

$$\hat{\varrho} = \pi \int d^2\alpha \ P(\alpha; s)\hat{\delta}(\hat{a} - \alpha; -s). \tag{4.83}$$

Note that from Eq. (4.83) the phase-space distribution $P(\alpha; s)$ may be seen explicitly to be normalized to unity,

$$\int d^2\alpha \ P(\alpha; s) = 1, \tag{4.84}$$

because of $\mathrm{Tr}(\hat{\varrho}) = 1$ and Eq. (4.76).

In particular, from Eqs (4.52) and (4.82) we see that in calculating the expectation value of an operator $\hat{O} = \hat{f}(\hat{a}, \hat{a}^\dagger)$ by "averaging" the c-number function $f(\alpha; 1) \equiv f^{(N)}(\alpha, \alpha^*)$, which is associated with the operator $\hat{f}(\hat{a}, \hat{a}^\dagger)$ put in normal order, the required phase-space function $P(\alpha; 1) \equiv P(\alpha)$ is determined by the c-number function $\varrho(\alpha; -1)$ associated with the density operator $\hat{\varrho}$ put in anti-normal order (Glauber–Sudarshan representation), and vice versa. Only for symmetrical order $(s = 0)$ are the c-number functions $f(\alpha; 0)$ and $\varrho(\alpha; 0)$

associated with the operators $\hat{f}(\hat{a}, \hat{a}^\dagger)$ and $\hat{\varrho}$, respectively, both put into symmetrical order (Wigner representation).

Since expectation values of normally ordered operators are of particular importance in the context of quantities observable in optical photodetection experiments (Chapter 6), the Glauber–Sudarshan representation

$$\hat{\varrho} = \pi \int d^2\alpha \, P(\alpha; 1) \hat{\delta}(\hat{a} - \alpha; -1) \tag{4.85}$$

is often used in quantum optics. Substitution of the expression (4.66) into Eq. (4.85) yields $[P(\alpha) \equiv P(\alpha; 1)]$

$$\hat{\varrho} = \int d^2\alpha \, P(\alpha) |\alpha\rangle\langle\alpha|, \tag{4.86}$$

which is conceptually different from the straightforward representation of the density operator in terms of coherent states, as given in Eq. (4.17). The expectation value of an operator $\hat{O} = \hat{f}(\hat{a}, \hat{a}^\dagger)$ may then be written with the help of Eq. (4.86) as

$$\langle \hat{f}(\hat{a}, \hat{a}^\dagger) \rangle = \int d^2\alpha \, P(\alpha) \, \text{Tr}[|\alpha\rangle\langle\alpha|\hat{f}(\hat{a}, \hat{a}^\dagger)]$$

$$= \int d^2\alpha \, P(\alpha) \langle\alpha|\hat{f}(\hat{a}, \hat{a}^\dagger)|\alpha\rangle, \tag{4.87}$$

from which we also immediately recognize via Eq. (4.52) the result (4.23):

$$f(\alpha; 1) = \langle\alpha|\hat{f}(\hat{a}, \hat{a}^\dagger)|\alpha\rangle. \tag{4.88}$$

As already mentioned, $P(\alpha)$ can be highly singular. In particular, in the case of nonclassical states (Chapter 8), such as for example squeezed states, the calculation of $P(\alpha)$ may also lead to expressions that are not well behaved and are hard to interpret. However, using the phase-space representation defined by $\hat{\delta}(\hat{a} - \alpha; 1)$,

$$\hat{\varrho} = \pi \int d^2\alpha \, P(\alpha; -1) \hat{\delta}(\hat{a} - \alpha; 1), \tag{4.89}$$

leads, according to Eq. (4.67), to the well-behaved Q function, $Q(\alpha) \equiv P(\alpha, -1) = \pi^{-1} \langle\alpha|\hat{\varrho}|\alpha\rangle$, which is suitable for the calculation of expectation values of anti-normally ordered operators.

We therefore observe a trade-off for the representation of the density operator in terms of phase-space functions that correspond to normal and anti-normal order. Either the phase-space function is well-behaved (Husimi Q function) and the associated delta operator may be problematic, or the phase-space function may be ill-behaved (Glauber–Sudarshan P function), whereas the delta operator is a simple projector on a coherent state. We note, however,

4 Bosonic systems in phase space

that the ill-behaved and possibly singular expression do not formally represent serious obstacles for most derivations. Only concrete evaluations may turn out to be rather cumbersome.

The formal analogy between phase-space functions and classical statistics encourages the introduction of a formalism similar to that used in the usual probability theory. In particular, it is useful to introduce characteristic (generating) functions, by defining

$$\Phi(\alpha;s) = \langle \hat{D}(\alpha;s) \rangle. \tag{4.90}$$

According to Eqs (4.46) and (4.90), $\Phi(\alpha;s)$ and $\Phi(\alpha;s')$ are related to each other by

$$\Phi(\alpha;s) = \exp\left[\tfrac{1}{2}(s-s')|\alpha|^2\right]\Phi(\alpha;s'). \tag{4.91}$$

Equations (4.51), (4.53) and (4.90) reveal that, as usual, the characteristic function is the Fourier transform of the phase-space function, that is,

$$\Phi(\alpha;s) = \int d^2\beta\, P(\beta;s) \exp(\alpha\beta^* - \alpha^*\beta), \tag{4.92}$$

and vice versa,

$$P(\alpha;s) = \frac{1}{\pi^2} \int d^2\beta\, \Phi(\beta;s) \exp(\alpha\beta^* - \alpha^*\beta). \tag{4.93}$$

From the operator expansion (4.79) it can be seen that the characteristic function $\Phi(\alpha;s)$ also carries the full information about the quantum state. As in ordinary probability theory, $\Phi(\alpha;s)$ allows one to generate the various s-ordered moments $\langle \hat{a}^{\dagger k}\hat{a}^l \rangle_s$

$$\langle \hat{a}^{\dagger k}\hat{a}^l \rangle_s = \int d^2\alpha\, \alpha^{*k}\alpha^l P(\alpha;s) = \frac{\partial^k}{\partial \beta^k} \frac{\partial^l}{\partial(-\beta^*)^l} \Phi(\beta;s)\Big|_{\beta=0}. \tag{4.94}$$

Here the s-ordered product $\{\hat{a}^{\dagger k}\hat{a}^l\}_s$ is defined by

$$\{\hat{a}^{\dagger k}\hat{a}^l\}_s = \frac{\partial^k}{\partial \alpha^k} \frac{\partial^l}{\partial(-\alpha^*)^l} \hat{D}(\alpha;s)\Big|_{\alpha=0}. \tag{4.95}$$

In Eq. (4.95) expressing $\hat{D}(\alpha;s)$ in terms of $\hat{D}(\alpha;s')$ according to Eq. (4.46) and differentiating, we may relate the s-ordered operator product to an s'-ordered operator product. After some algebra we derive

$$\{\hat{a}^{\dagger m}\hat{a}^n\}_s = \sum_{k=0}^{\min(m,n)} k! \binom{m}{k}\binom{n}{k} \left(\frac{s'-s}{2}\right)^k \{\hat{a}^{\dagger m-k}\hat{a}^{n-k}\}_{s'}. \tag{4.96}$$

The generalization of the concept of phase-space representation to multi-mode systems is straightforward. An extension of this concept to other than $\hat{a}$ and $\hat{a}^\dagger$ as basic operators is outlined in Section 5.2.2, in which the problem of formulating equations of motion (of Fokker–Planck type) for phase-space functions is studied. Moreover, we note that problems arising from singular behavior of the P function in the study of nonclassical states may be avoided by using generalized P representations [Drummond and Gardiner (1980); Gardiner (1983, 1991)],

$$\hat{\varrho} = \int_\mathcal{D} \mathrm{d}\mu(\alpha,\beta)\,\hat{\Lambda}(\alpha,\beta) P(\alpha,\beta), \tag{4.97}$$

where the operator $\hat{\Lambda}(\alpha,\beta)$ is given by

$$\hat{\Lambda}(\alpha,\beta) = \frac{|\alpha\rangle\langle\beta^*|}{\langle\beta^*|\alpha\rangle}, \tag{4.98}$$

and $\mathrm{d}\mu(\alpha,\beta)$ is the integration measure defining different classes of possible representations, with $\mathcal{D}$ being the domain of integration. In particular, it can be shown that the representation with measure

$$\mathrm{d}\mu(\alpha,\beta) = \mathrm{d}^2\alpha\,\mathrm{d}^2\beta \tag{4.99}$$

and integration over the whole complex plane always exists for a physical density operator and that $P(\alpha,\beta)$ can always be chosen positive, in which case it is called the positive P representation:

$$P(\alpha,\beta) = \frac{1}{4\pi^2}\exp\!\left(-\tfrac{1}{4}|\alpha-\beta^*|^2\right)\langle\tfrac{1}{2}(\alpha+\beta^*)|\hat{\varrho}|\tfrac{1}{2}(\alpha+\beta^*)\rangle. \tag{4.100}$$

4.3.3
Some elementary examples

To illustrate the theory, let us consider the Glauber–Sudarshan representation for some elementary quantum states, as introduced in Chapter 3. In the case of a coherent state $|\alpha_0\rangle$ we may immediately formulate the density operator as

$$\hat{\varrho} = |\alpha_0\rangle\langle\alpha_0| = \int \mathrm{d}^2\alpha\,\delta(\alpha-\alpha_0)\,|\alpha\rangle\langle\alpha|. \tag{4.101}$$

Comparing this equation with Eq. (4.86), we can obviously see that the Glauber–Sudarshan P function is

$$P(\alpha) = \delta(\alpha-\alpha_0). \tag{4.102}$$

From the point of view of classical statistics this function appears to have no fluctuations. In quantum theory such an interpretation is of course wrong.

From Sec. 3.2 we know that a system in a coherent state is indeed noisy, so that the appearance of a delta function in Eq. (4.102) should not mislead one to associate with it a deterministic behavior of the quantum system. In fact, this is only an effect of the chosen operator order, since other distributions with $s<1$ reveal a nonvanishing variance around the value α_0. We may furnish this, for example, by calculating the variance of the excitation number $\hat{n} = \hat{a}^\dagger \hat{a}$. To apply the Glauber–Sudarshan representation it is necessary to put the operator $(\Delta \hat{n})^2$ in normal order,

$$(\Delta \hat{n})^2 = (\hat{a}^\dagger \hat{a})^2 - \langle \hat{a}^\dagger \hat{a} \rangle^2 = \hat{a}^{\dagger 2} \hat{a}^2 + \hat{a}^\dagger \hat{a} - \langle \hat{a}^\dagger \hat{a} \rangle^2, \qquad (4.103)$$

The evaluation of the expectation value of Eq. (4.103) is then performed as follows:

$$\langle (\Delta \hat{n})^2 \rangle = \int d^2\alpha \, P(\alpha)(|\alpha|^4 + |\alpha|^2) - \left[\int d^2\alpha \, P(\alpha) |\alpha|^2 \right]^2, \qquad (4.104)$$

which with $P(\alpha) = \delta(\alpha - \alpha_0)$ [Eq. (4.102)] just leads to the familiar result $\langle (\Delta \hat{n})^2 \rangle = \langle \hat{n} \rangle = |\alpha_0|^2$. Clearly, the normally ordered variance $\langle : (\Delta \hat{n})^2 : \rangle$, and generally any normally ordered moment of a mean-value deviation, vanishes in the case of the δ peaked distribution function.

Next, let us consider a thermal state. In the case of radiation a thermal state serves as an example of so-called chaotic light. As is well known, if a harmonic oscillator of frequency ω is in thermal equilibrium with a heat bath of temperature T, the density operator $\hat{\varrho}$ may be written in the form

$$\hat{\varrho} = \frac{\exp[-\hbar \omega \hat{n} / (k_B T)]}{\text{Tr}\{\exp[-\hbar \omega \hat{n} / (k_B T)]\}} = \frac{1}{n_{\text{th}} + 1} \left(\frac{n_{\text{th}} + 1}{n_{\text{th}}} \right)^{-\hat{n}}, \qquad (4.105)$$

where the mean number of thermal photons, $n_{\text{th}} = \langle \hat{n} \rangle$, is given by the familiar formula

$$n_{\text{th}} = \left[\exp\left(\frac{\hbar \omega}{k_B T} \right) - 1 \right]^{-1}. \qquad (4.106)$$

Note that for a thermal state the mean coherent amplitude vanishes: $\langle \hat{a} \rangle = 0$. To calculate $P(\alpha)$, we recall that $P(\alpha)$ is determined by the c-number function $\varrho(\alpha; -1)$ associated with the density operator $\hat{\varrho}$ put in anti-normal order [Eq. (4.82) with $s=1$]:

$$P(\alpha) = \pi^{-1} \varrho(\alpha; -1). \qquad (4.107)$$

To put $\hat{\varrho}$ in anti-normal order, we note that, after a straightforward but somewhat lengthy calculation [using, e.g., Eq. (4.96)], the anti-normally ordered form of an exponential operator $\exp(-z \hat{a}^\dagger \hat{a})$ may be written as

$$e^{-\hat{a}^\dagger \hat{a} z} = e^z \sum_{k=0}^{\infty} \frac{(1-e^z)^k}{k!} \hat{a}^k \hat{a}^{\dagger k}. \qquad (4.108)$$

From Eqs (4.105) and (4.108) the anti-normally ordered form of $\hat{\varrho}$ is then found to be

$$\hat{\varrho} = \frac{1}{n_{\text{th}}} \sum_{k=0}^{\infty} \frac{(-1)^k}{k!} \frac{\hat{a}^k \hat{a}^{\dagger k}}{n_{\text{th}}^k}, \qquad (4.109)$$

so that we obtain for the associated c-number function

$$\varrho(\alpha; -1) = \frac{1}{n_{\text{th}}} \sum_{k=0}^{\infty} \frac{1}{k!} \left(-\frac{|\alpha|^2}{n_{\text{th}}}\right)^k = \frac{1}{n_{\text{th}}} \exp\left(-\frac{|\alpha|^2}{n_{\text{th}}}\right). \qquad (4.110)$$

Hence in this case the P function is a well-behaved Gaussian:

$$P(\alpha) = \frac{1}{\pi n_{\text{th}}} \exp\left(-\frac{|\alpha|^2}{n_{\text{th}}}\right). \qquad (4.111)$$

In the case of radiation one may also think of the more general situation where the thermal light of mean photon number n_{th} is superimposed on coherent light of (complex) amplitude α_0, so that $\langle \hat{n} \rangle = n_{\text{th}} + |\alpha_0|^2$ and $\langle \hat{a} \rangle = \alpha_0 \neq 0$. The superposition of the thermal state by a coherent one may be represented by a displacement of the distribution in phase space by the coherent amplitude α_0. The displaced P function is then obtained as

$$P(\alpha) = \frac{1}{\pi n_{\text{th}}} \exp\left(-\frac{|\alpha - \alpha_0|^2}{n_{\text{th}}}\right), \qquad (4.112)$$

which obviously corresponds to the density operator

$$\hat{\varrho} = \hat{D}(\alpha_0) \frac{1}{n_{\text{th}} + 1} \left(\frac{n_{\text{th}} + 1}{n_{\text{th}}}\right)^{-\hat{a}^\dagger \hat{a}} \hat{D}^\dagger(\alpha_0)$$

$$= \frac{1}{n_{\text{th}} + 1} \left(\frac{n_{\text{th}} + 1}{n_{\text{th}}}\right)^{-(\hat{a}^\dagger - \alpha_0^*)(\hat{a} - \alpha_0)}. \qquad (4.113)$$

Note that, for coherent light, $n_{\text{th}} = 0$, Eq. (4.112) reduces to the P function in Eq. (4.102), whereas for chaotic light, $\alpha_0 = 0$, it reduces to Eq. (4.111). The superposition of chaotic light with coherent light may be viewed as a simple model for characterizing the properties of single-mode laser light.

The above P functions exhibit all the properties of ordinary probability distribution functions. The only difference from classical statistics is that they apply to the calculation of normally ordered expectation values only. Since negative values of normally ordered variances, which are related to negative values of $P(\alpha)$, indicate nonclassical states (Chapter 8), quantum states such as coherent states or thermal states with a well-behaved P function may therefore be said to have a classical analog.

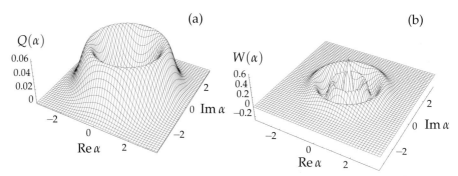

Fig. 4.1 Phase-space functions for the number state $|n=4\rangle$. Part (a) shows the Q function and part (b) the Wigner function, which reveals additional oscillatory and partially negative contributions.

As an example of a typical quantum state having no classical analog let us consider a number state

$$\hat{\varrho} = |n\rangle\langle n|. \tag{4.114}$$

To calculate the P function, we apply Eq. (4.53) directly and use Eq. (4.51) to obtain

$$P(\alpha) = \frac{1}{\pi^2} \int d^2\beta \, \exp(\alpha\beta^* - \alpha^*\beta) \langle n|\hat{D}(\beta;1)|n\rangle, \tag{4.115}$$

where $\langle n|\hat{D}(\beta;1)|n\rangle$ can be calculated as

$$\langle n|\hat{D}(\beta;1)|n\rangle = \langle n|e^{\beta\hat{a}^\dagger}e^{-\beta^*\hat{a}}|n\rangle = \sum_{k=0}^{n} \binom{n}{k} \frac{(-1)^k}{k!} |\beta|^{2k}. \tag{4.116}$$

Inserting this result into Eq. (4.115), reproducing the factors $|\beta|^{2k}$ by derivatives with respect to α and α^* and performing the remaining integration, we obtain the P function as

$$P(\alpha) = \sum_{k=0}^{n} \binom{n}{k} \frac{1}{k!} \frac{\partial^k}{\partial \alpha^k} \frac{\partial^k}{\partial \alpha^{*k}} \delta(\alpha). \tag{4.117}$$

As expected, $P(\alpha)$ is highly singular and bears no resemblance to a proper probability distribution function. Nevertheless, it may be used to calculate normally ordered expectation values.

The Wigner function $W(\alpha)$ can be obtained from Eq. (4.115) by replacing therein $\hat{D}(\beta;1)$ with $\hat{D}(\beta) = \hat{D}(\beta;0)$ [cf. Eqs (4.51) and (4.53)]. Recalling Eq. (4.45), we may therefore write

$$W(\alpha) = \frac{1}{\pi^2} \int d^2\beta \, \exp(\alpha\beta^* - \alpha^*\beta - \tfrac{1}{2}|\beta|^2) \langle n|\hat{D}(\beta;1)|n\rangle. \tag{4.118}$$

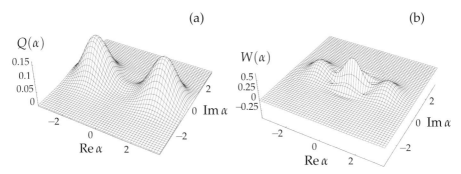

Fig. 4.2 Phase-space functions for a superposition of two coherent states $|\psi\rangle = N(|\alpha\rangle + |-\alpha\rangle)$ with $\alpha = 2$. Compared with the Q function (a) the Wigner function (b) shows negative values between the peaks of the two coherent states, which are signatures of their mutual quantum interference.

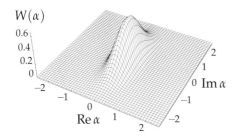

Fig. 4.3 Wigner function for a squeezed ground state, i.e., $|\beta, \tilde{\zeta}\rangle$ where $\beta = 0$, and the squeezing parameter is $\tilde{\zeta} = 0.5$.

Substituting the expansion as given by Eq. (4.116) for $\langle n|\hat{D}(\beta;1)|n\rangle$, we can calculate the β-integral to obtain

$$W(\alpha) = 2\pi^{-1}(-1)^n e^{-2|\alpha|^2} L_n(4|\alpha|^2) \tag{4.119}$$

($L_n(x)$ is the Laguerre polynomial). Since $L_n(4|\alpha|^2)$ can take positive and negative values, the Wigner function – although well behaved – takes positive and negative values as well. As already mentioned, the Q function, $Q(\alpha) \equiv P(\alpha; -1)$, is well behaved and positive in any case. In the above example, with the system being in a number state, we easily find that

$$Q(\alpha) = \frac{1}{\pi}|\langle\alpha|n\rangle|^2 = \frac{1}{\pi}\frac{|\alpha|^{2n}}{n!}e^{-|\alpha|^2}. \tag{4.120}$$

Examples of the Q function for a number state and a superposition of two coherent states are shown in Figs 4.1(a) and 4.2(a), respectively. Their Wigner function counterparts are shown in Fig. 4.1(b) for the number state and in Fig. 4.2(b) for the coherent-state superposition. It is clearly observable that, in general, the Wigner function reveals sharper structures as compared with the Q function. Moreover, negative values occur in the Wigner function, being a signature of quantum interference effects. The Wigner function of a squeezed ground state is shown in Fig. 4.3.

References

Agarwal, G.S. and E. Wolf (1968) *Phys. Rev. Lett.* **21**, 180.

Agarwal, G.S. and E. Wolf (1970) *Phys. Rev. D* **2**, 2161; 2187; 2206.

Cahill, K.E. and R. Glauber (1969) *Phys. Rev.* **177**, 1857; 1882.

Drummond, P.D. and C.W. Gardiner (1980) *Phys. Rev. A* **13**, 2353.

Gardiner, C.W. (1983) *Handbook of Stochastic Methods* (Springer-Verlag, Berlin).

Gardiner, C.W. (1991) *Quantum Noise* (Springer-Verlag, Berlin).

Glauber, R.J. (1963) *Phys. Rev.* **131**, 2766.

Klauder, J.R. and E.C.G. Sudarshan (1968) *Fundamentals of Quantum Optics* (Benjamin, New York).

Peřina, J. (1991) *Quantum Statistics of Linear and Nonlinear Optical Phenomena* (Reidel, Dordrecht).

E.C.G. Sudarshan (1963) *Phys. Rev. Lett.* **10**, 277.

5
Quantum theory of damping

In the study of physical problems it is often convenient to subdivide the "universe" into two parts, namely the system of particular interest and the environment, which is viewed as a large collection of systems with a very large number of degrees of freedom giving rise to (quasi-)continua of states. In many cases of practical relevance, when the (weak) interaction between the chosen system and the environment is of the type where each member of the collection of environmental systems is weakly (infinitesimally) disturbed, and a finite disturbance of the system is the result of the large number of these systems, the effect of the environment on the system can be described within the framework of damping theory (also called relaxation theory). In this case the system is also called a dynamic system, and the environment is called a dissipative system, a (heat) bath or reservoir.

Let us consider, for example, an excited single-mode optical field in a leaky cavity (Chapter 9). For sufficiently weak contact of the cavity mode (system) with the continuum of modes (reservoir) outside the cavity, the cavity mode (with photon annihilation operator $\hat{a}$) undergoes an exponential decay:

$$\langle \hat{a}(t) \rangle = \langle \hat{a}(t') \rangle \exp\left[-\left(i\omega + \tfrac{1}{2}\Gamma\right)(t-t')\right] \qquad (t-t' \geq 0). \tag{5.1}$$

This result is of course also valid in classical optics, where $\hat{a}$ reduces to the complex mode amplitude. Clearly, Eq. (5.1) cannot be valid in the sense of an operator equation. It is readily seen that if the operator $\hat{a}(t)$ had the damped solution $\hat{a}(t) = \hat{a}(t')\exp[-(i\omega+\Gamma/2)(t-t')]$, the fundamental rules of quantum mechanics would be violated, because with increasing time $t-t'$ the commutator $[\hat{a}(t), \hat{a}^\dagger(t)] \sim \exp[-\Gamma(t-t')]$ would approach zero, and hence Heisenberg's uncertainty principle would be violated. The reason for this unsatisfactory result lies of course in the fact that the fluctuations of the field outside the cavity, which feed noise into the cavity field, are ignored. Indeed, a quantum-mechanically consistent equation of motion is the quantum Langevin equation[1]

$$\dot{\hat{a}} = -\left(i\omega + \tfrac{1}{2}\Gamma\right)\hat{a} + \hat{f}(t) \tag{5.2}$$

---

1) Note that when thermal noise must be taken into account, then the Langevin equation is already required in classical optics.

Quantum Optics, Third, revised and extended edition. Werner Vogel and Dirk-Gunnar Welsch
Copyright © 2006 WILEY-VCH Verlag GmbH & Co. KGaA, Weinheim
ISBN: 3-527-40507-0

[cf. Eq. (9.80)], where the operator-valued Langevin noise source $\hat{f}(t)$ is just the quantum noise generator needed in order to preserve the correct commutation relation $[\hat{a}(t), \hat{a}^\dagger(t)] = 1$, which can be proved by direct integration of Eq. (5.2), which yields

$$\hat{a}(t) = \hat{a}(t') \exp\left[-\left(i\omega + \tfrac{1}{2}\Gamma\right)(t-t')\right]$$
$$+ \int_{t'}^{t} d\tau \, \hat{f}(\tau) \exp\left[-\left(i\omega + \tfrac{1}{2}\Gamma\right)(t-\tau)\right] \qquad (t-t' \geq 0). \qquad (5.3)$$

Assuming that at some (initial) time t' the commutation relation $[\hat{a}(t'), \hat{a}^\dagger(t')] = 1$ holds, we easily find from Eq. (5.3) that for any time t with $t \geq t'$ the commutator $[\hat{a}(t), \hat{a}^\dagger(t)]$ reads

$$[\hat{a}(t), \hat{a}^\dagger(t)] = \exp[-\Gamma(t-t')]$$
$$+ \exp\left[-\left(i\omega + \tfrac{1}{2}\Gamma\right)(t-t')\right] \int_{t'}^{t} d\tau \, [\hat{a}(t'), \hat{f}^\dagger(\tau)] \exp\left[-\left(-i\omega + \tfrac{1}{2}\Gamma\right)(t-\tau)\right]$$
$$+ \exp\left[-\left(-i\omega + \tfrac{1}{2}\Gamma\right)(t-t')\right] \int_{t'}^{t} d\tau \, [\hat{f}(\tau), \hat{a}^\dagger(t')] \exp\left[-\left(i\omega + \tfrac{1}{2}\Gamma\right)(t-\tau)\right]$$
$$+ e^{-\Gamma t} \int_{t'}^{t} d\tau \int_{t'}^{t} d\tau' \, [\hat{f}(\tau), \hat{f}^\dagger(\tau')] \exp\left[i\omega(\tau-\tau') + \tfrac{1}{2}\Gamma(\tau+\tau')\right]. \qquad (5.4)$$

Let us further assume the time-dependent commutation relations

$$[\hat{f}(t_1), \hat{f}^\dagger(t_2)] \exp[i\omega(t_1 - t_2)] = \Gamma \delta(t_1 - t_2), \qquad (5.5)$$
$$[\hat{a}(t_1), \hat{f}^\dagger(t_2)] = 0 \qquad (t_2 > t_1) \qquad (5.6)$$

[cf. Eqs (9.117) and (9.126)]. Combining Eqs (5.4)–(5.6) yields the correct commutation relation at time t:

$$[\hat{a}(t), \hat{a}^\dagger(t)] = 1. \qquad (5.7)$$

It should be pointed out that in Eq. (5.2) the term $\hat{f}(t)$ can reasonably be interpreted as noise when its expectation value vanishes, $\langle \hat{f}(t) \rangle = 0$. In the case of a bosonic dissipative system and if $\hat{f}(t)$ is linear in the mode operators, this is observed for the vacuum state or a thermal state. The case where the dissipative system is in a coherent state is closely related to that where a classical driving force is applied to the dynamic system. Clearly, there are various intermediate situations, in which we may be interested. It is worth noting that the existence and form of the damping term in Eq. (5.2) has nothing to do with the state of the reservoir. Thus damping will occur even when $\hat{f}(t)$ is in a coherent state. In this context we recall that the consistency of Eq. (5.2) with the principles of quantum mechanics only requires the validity of the commutation relations (5.5) and (5.6).

The results briefly discussed above correspond to the (appropriately specified) results of Markovian damping theory, which will be developed in this chapter [Weisskopf and Wigner (1932); Zwanzig (1960); Fajn and Chanin (1969); Louisell (1973); Gardiner (1991)]. The concept of damping theory not only applies to radiation-field quantities, but it is also used to describe the damping behavior of atomic systems (e. g., due to spontaneous emission). Within the framework of Markovian damping theory, we derive quantum Langevin equations and expressions for mean values depending on a single time argument (Section 5.1). Master equations and related equations, such as Fokker–Planck equations, are considered in Section 5.2. The methods are applied to damped harmonic oscillators (Section 5.3) and to damped two-level atoms (Section 5.4). Moreover, the quantum regression theorem is derived (Section 5.5) in order to handle correlation functions, which usually depend on more than one time argument.

5.1
Quantum Langevin equations and one-time averages

There are various, equivalent approaches to the Markovian damping theory. For example, one can start from the Heisenberg equations of motions or the Schrödinger equation for the composed system. Here we prefer to work in the Heisenberg picture to derive quantum Langevin equations, by introducing the Born and Markov approximations in the formal solution for the operators of the dynamic system.

5.1.1
Hamiltonian

Let us consider some dynamic system – referred to as "the system" in the following – described by a Hamiltonian $\hat{H}_{\text{sys}}$ and assume that there is a reservoir described by a Hamiltonian $\hat{H}_{\text{res}}$, the coupling of the reservoir to the system being defined by $\hat{H}_{\text{int}}$. The total Hamiltonian $\hat{H}$ is then

$$\hat{H} = \hat{H}_{\text{sys}} + \hat{H}_{\text{res}} + \hat{H}_{\text{int}}. \tag{5.8}$$

We further assume that $\hat{H}_{\text{int}}$ may be written as a sum of products,

$$\hat{H}_{\text{int}} = \hbar \sum_i \hat{A}_i \hat{R}_i, \tag{5.9}$$

where $\hat{A}_i$ and $\hat{R}_i$ are operator functions of the system and reservoir operators, respectively. If $|\nu\rangle$ are an orthonormal and complete set of reservoir Hilbert-space vectors that are also eigenkets of $\hat{H}_{\text{res}}$, we may write

$$\hat{H}_{\text{res}} = \hbar \sum_{\nu} \omega_{\nu} \hat{r}_{\nu\nu}, \tag{5.10}$$

$$\hat{R}_i = \sum_{\nu,\nu'} V^{(i)}_{\nu\nu'} \hat{r}_{\nu\nu'}, \tag{5.11}$$

where

$$\hat{r}_{\nu\nu'} = |\nu\rangle\langle\nu'|, \tag{5.12}$$

and $V^{(i)}_{\nu\nu'}$ are the corresponding coupling-matrix elements. Since the reservoir is thought of as having a very large number of degrees of freedom giving rise to (quasi-)continua of energy eigenstates, the index ν contains (quasi-)continuous parts, so that parts of the above ν sums are effectively integrals.

To exemplify the notation used, let us consider the case of a harmonic oscillator (system) which is coupled to a reservoir, which consists of a large number of other harmonic oscillators. If we make the identifications

$$\hat{A}_1 = \hat{a}^\dagger, \qquad \hat{A}_2 = \hat{A}_1^\dagger, \tag{5.13}$$

$$\hat{R}_1 = \sum_{\alpha}\sum_{n_\alpha} V_{n_\alpha n_\alpha + 1} \hat{r}_{n_\alpha n_\alpha + 1}, \qquad \hat{R}_2 = \hat{R}_1^\dagger, \tag{5.14}$$

where

$$V_{n_\alpha n_\alpha + 1} = \kappa_\alpha \sqrt{n_\alpha + 1}, \tag{5.15}$$

with α labeling the oscillators, and introduce the boson annihilation and creation operators $\hat{b}_\alpha$ and $\hat{b}_\alpha^\dagger$, respectively, via

$$\hat{b}_\alpha = \sum_{n_\alpha} \sqrt{n_\alpha + 1}\, \hat{r}_{n_\alpha n_\alpha + 1}, \tag{5.16}$$

then Eqs (5.9) and (5.10) take the more conventional forms ($\hat{n}_\alpha = \hat{b}_\alpha^\dagger \hat{b}_\alpha$)

$$\hat{H}_{\text{int}} = \hbar \sum_{\alpha} \kappa_\alpha \hat{a}^\dagger \hat{b}_\alpha + \text{H.c.}, \tag{5.17}$$

$$\hat{H}_{\text{res}} = \sum_{\alpha} \hbar \omega_\alpha \hat{n}_\alpha. \tag{5.18}$$

Equation (5.17) can be viewed, for example, as a model of an interaction energy used to describe a possible (energy) damping mechanism for a radiation mode in a cavity. This is because of the interaction of the radiation mode with the absorbing cavity walls, which in the case of sufficiently low excitation, might be described by a model of appropriately chosen ensembles of harmonic oscillators. Light in the cavity mode could be scattered in the wall excitations, and vice versa, thermal and vacuum fluctuations in the walls could be scattered into the cavity mode. The net effect of the first kind of process may be regarded as being the annihilation of a photon at frequency ω and the simultaneous creation of a wall boson at frequency ω_α. Accordingly, the second kind of process describes the annihilation of a wall boson at ω_α and the creation of a cavity photon at ω. Equation (5.17) can also be regarded as a model of an interaction energy to describe the damping of a radiation mode in a leaky cavity (Chapter 9). In this case one could visualize the net effect of the scattering process from the cavity mode as the annihilation of a cavity photon and the simultaneous creation of an output photon. Scattering into the cavity mode then consists of the annihilation of an input photon and the creation of a cavity photon. Clearly, many other damping mechanisms that can be described on the basis of Eq. (5.17) could be envisaged. In this equation, the coupling coefficients κ_α characterize the strength of the system–reservoir coupling, which of course depends on the actual interaction mechanism. Note that interaction processes of the types $\hat{b}_\alpha \hat{a}$ and $\hat{a}^\dagger \hat{b}_\alpha^\dagger$ are ignored (rotating-wave approximation), because they usually give rise to only small off-resonant effects.

5.1.2
Heisenberg equations of motion

After this excursion, let us return to the more general model based on Eqs (5.8)–(5.11), in which the system, the reservoir and the interactions of both subsystems are not specified. If $\hat{C}$ is an arbitrary system operator, its equation of motion in the Heisenberg picture is

$$\dot{\hat{C}} = -\frac{i}{\hbar}[\hat{C}, \hat{H}_{\text{sys}}] - i\sum_i [\hat{C}, \hat{A}_i]\hat{R}_i. \tag{5.19}$$

In the following it will be convenient to separate from the system operators the slowly varying amplitude operators, by assuming that the system operators $\hat{A}_i$ may be chosen in such a way that

$$\hat{A}_i(t) = \tilde{\hat{A}}_i(t)e^{-i\omega_i t}, \tag{5.20}$$

where positive, negative and zero values of the ω_i are possible. Accordingly, we write[2]

$$\hat{C}(t) = \tilde{\hat{C}}(t)e^{-i\omega_C t}. \tag{5.21}$$

---

2) If the operator $\hat{C}$ consists of more than one rapidly oscillating component, the following calculations should be thought of as being performed for each component separately. At the end of the calculations the result for the overall operator may then be found by appropriate superposition.

Using Eqs (5.20) and (5.21), we may rewrite Eq. (5.19) as

$$\frac{d\hat{C}}{dt} = i\omega_C\hat{C} - \frac{i}{\hbar}[\hat{C}, \hat{H}_{\text{sys}}] - i\sum_i [\hat{C}, \hat{A}_i] e^{-i\omega_i t}\hat{R}_i. \tag{5.22}$$

The equations of motion for the reservoir operators $\hat{r}_{vv'}$ are

$$\dot{\hat{r}}_{vv'} = i\omega_{vv'}\hat{r}_{vv'} - i\sum_j e^{-i\omega_j t}\hat{A}_j[\hat{r}_{vv'}, \hat{R}_j], \tag{5.23}$$

where $\omega_{vv'} = \omega_v - \omega_{v'}$.

We formally solve Eq. (5.23) to obtain

$$\hat{r}_{vv'}(t) = \exp[i\omega_{vv'}(t-t')]\hat{r}_{vv'}(t')$$
$$- i\sum_j \int_{t'}^{t} d\tau\, e^{-i\omega_j \tau}\hat{A}_j(\tau) \exp[i\omega_{vv'}(t-\tau)][\hat{r}_{vv'}(\tau), \hat{R}_j(\tau)]. \tag{5.24}$$

Combining Eqs (5.24) and (5.11) yields

$$\hat{R}_i(t) = \hat{F}_i(t)$$
$$- i\sum_j \int_{t'}^{t} d\tau\, e^{-i\omega_j \tau}\hat{A}_j(\tau) \sum_{vv'} V_{vv'}^{(i)} \exp[i\omega_{vv'}(t-\tau)][\hat{r}_{vv'}(\tau), \hat{R}_j(\tau)], \tag{5.25}$$

where the operators

$$\hat{F}_i(t) = \exp\left[\frac{i}{\hbar}\hat{H}_{\text{res}}(t-t')\right]\hat{R}_i(t') \exp\left[-\frac{i}{\hbar}\hat{H}_{\text{res}}(t-t')\right]$$
$$= \sum_{vv'} V_{vv'}^{(i)} \exp[i\omega_{vv'}(t-t')]\hat{r}_{vv'}(t') \tag{5.26}$$

are free-reservoir operators, whose time dependence is governed by $\hat{H}_{\text{res}}$. In Eqs (5.24) and (5.25), t' is an appropriately chosen initial time. We now insert the result (5.25) into Eq. (5.22) to obtain[3]

$$\frac{d\hat{C}}{dt} = i\omega_C\hat{C} - \frac{i}{\hbar}[\hat{C}, \hat{H}_{\text{sys}}]$$
$$- \sum_{i,j}[\hat{C}, \hat{A}_i]e^{-i\omega_i t}\int_{t'}^{t} d\tau\, e^{-i\omega_j \tau}\hat{A}_j(\tau) \sum_{v,v'} V_{vv'}^{(i)} \exp[i\omega_{vv'}(t-\tau)][\hat{r}_{vv'}(\tau), \hat{R}_j(\tau)]$$
$$- i\sum_i [\hat{C}, \hat{A}_i]e^{-i\omega_i t}\hat{F}_i(t). \tag{5.27}$$

Equation (5.27) is exact so far. In the particular case where the reservoir can be approximated by a collection of harmonic oscillators and Eq. (5.17) is valid,

---
3) For notational convenience, we omit the time argument of system operators when it is t.

the ν, ν' sum in the τ integral in Eq. (5.27) simply gives rise to a c number. In this case, Eq. (5.27) is an equation of motion for system operators only, because the time dependence of the free-reservoir operators $\hat{F}_i(t)$ is known. The definition of $\hat{F}_i(t)$ in terms of the $\hat{r}_{\nu\nu'}(t')$ ensures that these operators may be specified in the sense of initial conditions.

5.1.3
Born and Markov approximations

If under more general conditions the (ν, ν')-sum term in the τ integral in the second line in Eq. (5.27) gives rise to certain kinds of reservoir operators, this integral can only be approximately decoupled from the reservoir operators. Let us suppose that the (free) reservoir is in equilibrium, that is to say, reservoir correlation functions are stationary:

$$\langle \hat{F}_i(t)\hat{F}_j(t-\tau)\rangle = \langle \hat{F}_i(\tau)\hat{F}_j(0)\rangle. \tag{5.28}$$

Under the assumption that the interaction between the system and the reservoir is weak, and because of the large number of degrees of freedom of the reservoir, any "elementary" reservoir variable $\hat{r}_{\nu\nu'}$ can only be weakly perturbed by the system. Therefore, we only take into account effects up to second order in the interaction energy between the system and the reservoir (Born approximation). In this approximation, we then replace, on recalling Eqs (5.25) and (5.26), the remaining (ν, ν')-sum term by its stationary expectation value, $[\hat{F}_i(t), \hat{F}_j(t-\tau)] \mapsto \langle [\hat{F}_i(\tau), \hat{F}_j(0)]\rangle$. Changing the variable $(t-\tau \mapsto \tau)$, we obtain

$$\begin{aligned}\frac{d\hat{\tilde{C}}}{dt} &= i\omega_C \hat{\tilde{C}} - \frac{i}{\hbar}[\hat{\tilde{C}}, \hat{H}_{\text{sys}}] \\ &\quad - \sum_{i,j}[\hat{\tilde{C}}, \hat{\tilde{A}}_i]e^{-i(\omega_i+\omega_j)t}\int_0^{t-t'} d\tau\, e^{i\omega_j\tau}\langle[\hat{F}_i(\tau), \hat{F}_j(0)]\rangle \hat{\tilde{A}}_j(t-\tau) \\ &\quad - \sum_i[\hat{\tilde{C}}, \hat{\tilde{A}}_i]e^{-i\omega_i t}\hat{F}_i(t).\end{aligned} \tag{5.29}$$

Note that when the reservoir is a sample of harmonic oscillators, Eq. (5.29) is of course exact.

Since any physical correlation decays for a sufficiently large time delay, we have

$$\lim_{\tau\to\infty}\langle \hat{F}_i(\tau)\hat{F}_j(0)\rangle = \lim_{\tau\to\infty}\langle \hat{F}_i(\tau)\rangle\langle \hat{F}_j(0)\rangle, \tag{5.30}$$

hence

$$\lim_{\tau\to\infty}\langle[\hat{F}_i(\tau), \hat{F}_j(0)]\rangle = 0. \tag{5.31}$$

In fact $\langle[\hat{F}_i(\tau), \hat{F}_j(0)]\rangle$ may be expected to be nonzero only during a finite time interval of length τ_c – the correlation time. As long as we require that $t - t' \gg \tau_c$, we may therefore extend the upper limit of the τ integral in Eq. (5.29) to infinity, with little error. Moreover, we require that τ_c be small on a time scale in which the system is changed owing to the coupling to the reservoir (and to intra-system couplings included in $\hat{H}_{\text{sys}}$). Equivalently, the bandwidth of the reservoir spectrum involved in the coupling to a system quantity is required to be large compared with the bandwidth of the system-quantity spectrum [Lax (1966)]. In this case, in the τ integral in Eq. (5.29) the (slowly varying) system operators at time $t - \tau$, $\hat{A}_j(t - \tau)$, may be replaced by the corresponding operators at time t, $\hat{A}_j(t)$. In this approximation, which is usually called the Markov approximation, Eq. (5.29) simplifies to

$$\frac{d\hat{C}}{dt} = i\omega_C \hat{C} - \frac{i}{\hbar}[\hat{C}, \hat{H}_{\text{sys}}]$$
$$- \sum_{i,j}[\hat{C}, \hat{A}_i]\hat{A}_j \exp[-i(\omega_i + \omega_j)t] \int_0^\infty d\tau\, e^{i\omega_j \tau} \langle[\hat{F}_i(\tau), \hat{F}_j(0)]\rangle$$
$$- i\sum_i [\hat{C}, \hat{A}_i] e^{-i\omega_i t} \hat{F}_i(t). \tag{5.32}$$

Note that, in the Markov approximation, the right-hand side of Eq. (5.32) no longer contains time integrals with system variables for times earlier than the present. Thus the temporal change of system quantities at an arbitrarily chosen time is now determined by system quantities at the same time. In other words, the fluctuations of the system are smoothed out on a time scale τ_d during which the system is damped (decay time); so that the system will lose its past memory on this time scale. Clearly, in this approximation the time derivative of any (slowly) varying system operator in an equation of motion of the type (5.32) must be thought of as the (averaged) time rate of change on a time scale Δt, with

$$\tau_c \ll \Delta t \ll \tau_d. \tag{5.33}$$

In this sense, one also refers to the Markov approximation as a coarse-grained averaging.

5.1.4
Quantum Langevin equations

Recall that the introduction of slowly varying system variables implies some kind of rotating-wave approximation. Indeed, from inspection of Eq. (5.32) we see that in the (i, j) sum only those terms are important for which $(\omega_i + \omega_j)\tau_c \ll 1$, because in the Markov approximation the physically allowed resolving time must be reasonably large compared with the reservoir correlation time. To pick out these terms, we introduce a function $\Delta(\omega)$ as

$$\Delta(\omega) = \begin{cases} 1 & \text{if } \omega\tau_c \ll 1, \\ 0 & \text{otherwise,} \end{cases} \tag{5.34}$$

so that Eq. (5.32) effectively becomes

$$\frac{d\hat{\tilde{C}}}{dt} = i\omega_C \hat{\tilde{C}} - \frac{i}{\hbar}[\hat{\tilde{C}}, \hat{H}_{\text{sys}}] \\ - \sum_{i,j} \Gamma_{ij}[\hat{\tilde{C}}, \hat{\tilde{A}}_i]\hat{\tilde{A}}_j \Delta(\omega_i + \omega_j) - i\sum_i [\hat{\tilde{C}}, \hat{\tilde{A}}_i]e^{-i\omega_i t}\hat{F}_i(t), \tag{5.35}$$

where[4]

$$\Gamma_{ij} = \int_0^\infty d\tau\, e^{-i\omega_i \tau} \langle [\hat{F}_i(\tau), \hat{F}_j(0)] \rangle. \tag{5.36}$$

Note that the $e^{-i\omega_i t}\hat{F}_i(t)$ are effectively slowly varying reservoir operators. Going back to the complete Heisenberg operators $\hat{C}$, $\hat{A}_i$ and $\hat{A}_j$ [cf. Eqs (5.20) and (5.21)], we obtain

$$\frac{d\hat{C}}{dt} = -\frac{i}{\hbar}[\hat{C}, \hat{H}_{\text{sys}}] - \sum_{i,j} \Gamma_{ij}[\hat{C}, \hat{A}_i]\hat{A}_j \Delta(\omega_i + \omega_j) - i\sum_i [\hat{C}, \hat{A}_i]\hat{F}_i(t). \tag{5.37}$$

It should be pointed out that Eq. (5.37) is written in a form where both the $\hat{F}_i(t)$ and $\hat{A}_j$ are on the right of the commutators $[\hat{C}, \hat{A}_i]$. From the derivation of Eq. (5.37) it is clear that there is no need for such an order. The only essential point is that for a given i the order of $[\hat{C}, \hat{A}_i]$ and $\hat{F}_i(t)$ must coincide with the order of $[\hat{C}, \hat{A}_i]$ and $\hat{A}_j$.

The equation of motion (5.37) [or (5.35)] may be viewed as a quantum Langevin equation [Lax (1966)]. The terms proportional to $\operatorname{Re}\Gamma_{ij}$ are responsible for damping, with $\operatorname{Re}\Gamma_{ij}$ being the damping rates, whereas the terms proportional to $\operatorname{Im}\Gamma_{ij}$ typically shift the unperturbed system frequencies. The terms depending on the $\hat{F}_i(t)$ may be interpreted as noise sources, provided that the $\hat{F}_i(t)$ are proper noise generators, $\langle \hat{F}_i(t) \rangle = 0$.

It is worth noting that, if the reservoir commutator terms appearing in the derivation of the quantum Langevin equation are c numbers, the damping terms arise (in the Markov approximation) without any particular specification of the state of the reservoir and without the factorization assumption in Eq. (5.37). Clearly, the interpretation of the $\hat{F}_i(t)$ as proper noise generators requires that the reservoir is in a state that is incoherent in the sense that $\langle \hat{F}_i(t) \rangle = 0$ and that expectation values of products of system operators and the

---

[4] In our approximation we may let $e^{i\omega_j \tau} \simeq e^{-i\omega_i \tau}$.

$\hat{F}_i(t)$ are factorized at some appropriately chosen initial time $t=t'$. When the reservoir is in a coherent state – a right-hand eigenstate of $\hat{F}_i(t)$ – we have a situation that closely corresponds to classical driving fields being applied to the system.

In general, quantum Langevin equations of the form (5.37) are hard to handle. Apart from more or less complicated (nonlinear) intra-system couplings (included in $\hat{H}_{\text{sys}}$), there are mixed operator products $[\hat{C}, \hat{A}_i]\hat{F}_i(t)$. The commutators $[\hat{C}, \hat{A}_i]$ do not reduce, in general, to c numbers but may represent more or less complicated system operators. Provided that the $\hat{F}_i(t)$ are proper noise generators, equations of motion of the type (5.37) may be interpreted as quantum Langevin equations with multiplicative noise.[5] Clearly, the expectation values of $[\hat{C}, \hat{A}_i]\hat{F}_i(t)$ cannot be taken, in general, to be zero. In order to derive more tractable equations of motion from Eq. (5.37), the averaging of which directly leads to averaged system quantities, we have to try to express the expectation values of $[\hat{C}, \hat{A}_i]\hat{F}_i(t)$ solely in terms of expectation values of system operators. In the general case of more or less unspecified system and reservoir variables it is reasonable to perform the calculations under the same assumptions and approximations leading, in general, to Eq. (5.37).

For this purpose, in Eq. (5.35) we represent the (slowly varying) commutator $[\hat{\tilde{C}}, \hat{\tilde{A}}_i]$ in the form

$$[\hat{\tilde{C}}(t), \hat{\tilde{A}}_i(t)] = [\hat{\tilde{C}}(t_-), \hat{\tilde{A}}_i(t_-)] + \int_{t_-}^{t} d\tau \, \frac{d}{d\tau}[\hat{\tilde{C}}(\tau), \hat{\tilde{A}}_i(\tau)], \tag{5.38}$$

where $t - t_- > 0$, and we require that

$$\tau_c \ll t - t_- \ll \tau_d. \tag{5.39}$$

In Eq. (5.38) we substitute for $d[\hat{\tilde{C}}(\tau), \hat{\tilde{A}}_i(\tau)]/d\tau$ the result (5.35) with $[\hat{\tilde{C}}, \hat{\tilde{A}}_i]$ instead of $\hat{\tilde{C}}$. Introducing the change of variable $t - \tau \mapsto \tau$ and omitting terms proportional to $t - t_-$, we derive[6]

$$-i \sum_i [\hat{\tilde{C}}, \hat{\tilde{A}}_i] e^{-i\omega_i t} \hat{F}_i(t)$$

$$\simeq -\sum_{i,j} e^{-i(\omega_i + \omega_j)t} \int_0^{t-t_-} d\tau \, [[\hat{\tilde{C}}(t-\tau), \hat{\tilde{A}}_i(t-\tau)], \hat{\tilde{A}}_j(t-\tau)] e^{i\omega_j \tau} \hat{F}_j(t-\tau) \hat{F}_i(t)$$

$$-i \sum_i [\hat{\tilde{C}}(t_-), \hat{\tilde{A}}_i(t_-)] e^{-i\omega_i t} \hat{F}_i(t). \tag{5.40}$$

We evaluate the τ integral in Eq. (5.40) within the same approximation scheme used to derive Eq. (5.35). Replacing the reservoir operator products by their

---

5) Products of system and noise operators appear.
6) On the time scale of the validity of Eq. (5.35) $t - t_-$ is negligibly small.

(equilibrium) averages $\langle \hat{F}_j(0)\hat{F}_i(\tau)\rangle$, extending the upper limit of the τ integral to infinity, replacing the (slowly varying) system operators at time $t-\tau$ by their values at time t (Markov approximation), and recalling the definition (5.34) of the function $\Delta(\omega)$, we may write

$$-i\sum_i [\hat{C},\hat{A}_i]e^{-i\omega_i t}\hat{F}_i(t)$$

$$\simeq -\sum_{i,j}[[\hat{C},\hat{A}_i],\hat{A}_j]\Delta(\omega_i+\omega_j)\int_0^\infty d\tau\, e^{-i\omega_i\tau}\langle \hat{F}_j(0)\hat{F}_i(\tau)\rangle$$

$$-i\sum_i [\hat{C}(t_-),\hat{A}_i(t_-)]e^{-i\omega_i t}\hat{F}_i(t). \tag{5.41}$$

We now combine Eqs (5.35), (5.36) and (5.41) to finally obtain

$$\frac{d\hat{C}}{dt} = i\omega_C \hat{C} - \frac{i}{\hbar}[\hat{C},\hat{H}_{\mathrm{sys}}]$$

$$-\sum_{i,j}\Delta(\omega_i+\omega_j)\{\Gamma_{ij}^+[\hat{C},\hat{A}_i]\hat{A}_j - \Gamma_{ji}^-\hat{A}_j[\hat{C},\hat{A}_i]\}$$

$$-i\sum_i [\hat{C}(t_-),\hat{A}_i(t_-)]e^{-i\omega_i t}\hat{F}_i(t), \tag{5.42}$$

where

$$\Gamma_{ij}^+ = \int_0^\infty d\tau\, e^{-i\omega_i\tau}\langle \hat{F}_i(\tau)\hat{F}_j(0)\rangle, \tag{5.43}$$

$$\Gamma_{ji}^- = \int_0^\infty d\tau\, e^{-i\omega_i\tau}\langle \hat{F}_j(0)\hat{F}_i(\tau)\rangle. \tag{5.44}$$

Note that $\Gamma_{ij} = \Gamma_{ij}^+ - \Gamma_{ji}^-$ [see Eqs (5.36), (5.43) and (5.44)].

The equation of motion (5.42) may again be interpreted as a quantum Langevin equation, which (within the approximation scheme used) is equivalent to Eq. (5.35). The terms proportional to $\mathrm{Re}\,\Gamma_{ij}^+$ and $\mathrm{Re}\,\Gamma_{ji}^-$ are damping terms. It should be noted that the operator Langevin force

$$\hat{F}_C(t) = -i\sum_i [\hat{C}(t_-),\hat{A}_i(t_-)]e^{-i\omega_i t}\hat{F}_i(t) \tag{5.45}$$

satisfies

$$\langle \hat{F}_C(t)\rangle = 0, \tag{5.46}$$

because of the assumption that $\langle \hat{F}_i(t)\rangle = 0$. Clearly, Eq. (5.46) must hold in the Markov approximation. If Eq. (5.35) is solved in terms of the past system values and those of the noise generators $\hat{F}_i(t)$, the system variables do not depend on the values of the noise generators in the future. In Eq. (5.45) the system and noise variables are taken at the times t_- and t respectively. Since $t - t_- \gg \tau_c$, and the noise variables are only correlated over a time interval τ_c, the system and noise variables cannot be correlated over the time interval $t - t_-$, and hence

$$\langle \hat{\tilde{F}}_C(t) \rangle = -i \sum_i \langle [\hat{\tilde{C}}(t_-), \hat{\tilde{A}}_i(t_-)] \hat{F}_i(t) \rangle e^{-i\omega_i t}$$

$$\simeq -i \sum_i \langle [\hat{\tilde{C}}(t_-), \hat{\tilde{A}}_i(t_-)] \rangle \langle \hat{F}_i(t) \rangle e^{-i\omega_i t} = 0. \tag{5.47}$$

In this way, the noise effect of the reservoir on the expectation value of a system operator $\hat{C}$ is, in contrast to Eq. (5.35), completely contained in the damping terms.

Going back to the complete Heisenberg operators, Eq. (5.42) may be written as

$$\frac{d\hat{C}}{dt} = -\frac{i}{\hbar}[\hat{C}, \hat{H}_{\text{sys}}]$$
$$- \sum_{i,j} \Delta(\omega_i + \omega_j)\{\Gamma_{ij}^+ [\hat{C}, \hat{A}_i]\hat{A}_j - \Gamma_{ji}^- \hat{A}_j[\hat{C}, \hat{A}_i]\} + \hat{F}_C(t), \tag{5.48}$$

where

$$\hat{F}_C(t) = \hat{\tilde{F}}_C(t) e^{-i\omega_C t} \tag{5.49}$$

and $\langle \hat{F}_C(t) \rangle = 0$. On the basis of the quantum Langevin equation in the form (5.48) together with Eq. (5.46), we can easily derive the equations of motion for (one-time) expectation values of system operators:

$$\frac{d\langle \hat{C} \rangle}{dt} = -\frac{i}{\hbar} \langle [\hat{C}, \hat{H}_{\text{sys}}] \rangle$$
$$- \sum_{i,j} \Delta(\omega_i + \omega_j)\{\Gamma_{ij}^+ \langle [\hat{C}, \hat{A}_i]\hat{A}_j \rangle - \Gamma_{ji}^- \langle \hat{A}_j[\hat{C}, \hat{A}_i] \rangle\}. \tag{5.50}$$

It should be pointed out that the Langevin equation (5.37) or (5.48) effectively represents a system of coupled equations. Clearly, the combinations of system operators, such as $\hat{C}' \equiv [\hat{C}, \hat{A}_i]\hat{A}_j$ and $\hat{C}'' \equiv \hat{A}_j[\hat{C}, \hat{A}_i]$ [cf. Eq. (5.48)], also satisfy Langevin equations of the type considered, so that, in general, a more or less complicated hierarchy of operator equations of motion is created. Accordingly, the mean-value equation of motion (5.50) also represents a system of coupled equations of motion.

5.2
Master equations and related equations

So far we have dealt with the Heisenberg picture. The equation of motion for the expectation value of any system operator, Eq. (5.50), is of course indepen-

5.2.1
Master equations

Without loss of generality we may assume that at some (initial) time $t=t'$ the Heisenberg picture agrees with the Schrödinger picture. In the Heisenberg picture the expectation value of any operator $\hat{M}$ at time t is

$$\langle \hat{M}(t) \rangle = \text{Tr}[\hat{\varrho}(t')\hat{M}(t)], \tag{5.51}$$

where $\hat{\varrho}(t')$ is the (initial) density operator. Recalling that

$$\hat{M}(t) = \hat{U}^\dagger(t,t')\hat{M}(t')\hat{U}(t,t'), \tag{5.52}$$

where $\hat{U}(t,t')$ is the time evolution operator, and using the cyclic properties of the trace, we may rewrite Eq. (5.51) as

$$\langle \hat{M}(t) \rangle = \text{Tr}[\hat{\varrho}(t)\hat{M}(t')], \tag{5.53}$$

where

$$\hat{\varrho}(t) = \hat{U}(t,t')\hat{\varrho}(t')\hat{U}^\dagger(t,t'). \tag{5.54}$$

Equation (5.53) is just the prescription for calculating expectation values in the Schrödinger picture, in which the full time dependence is included in the density operator $\hat{\varrho}(t)$, whereas the operators, apart from an external, explicit time dependence, are time independent $[\hat{M}(t)=\hat{M}(t')]$.

Let us identify the operator $\hat{M}$ by a system operator $\hat{C}$. In this case we may first take the trace over the reservoir in Eq. (5.53) and then take the remaining trace over the system:

$$\langle \hat{C} \rangle = \text{Tr}_{\text{sys}}[\hat{\sigma}(t)\hat{C}(t')], \tag{5.55}$$

where

$$\hat{\sigma}(t) = \text{Tr}_{\text{res}}\, \hat{\varrho}(t) \tag{5.56}$$

is the (reduced) density operator for the system. Hence we may write

$$\frac{d\langle \hat{C} \rangle}{dt} = \text{Tr}_{\text{sys}}\left[\frac{d\hat{\sigma}(t)}{dt} \hat{C}(t')\right]. \tag{5.57}$$

Substituting the result (5.50) into Eq. (5.57) for $d\langle\hat{C}\rangle/dt$, recalling Eq. (5.55), and using the cyclic properties of the trace, we can easily derive

$$\text{Tr}_{\text{sys}}\left[\frac{d\sigma(t)}{dt}\hat{C}(t')\right] = \text{Tr}_{\text{sys}}\left\{\left[-\frac{i}{\hbar}[\hat{H}_{\text{sys}}(t'),\hat{\sigma}(t)]\right.\right.$$
$$-\sum_{i,j}\Delta(\omega_i+\omega_j)\{\Gamma_{ij}^+[\hat{A}_i(t')\hat{A}_j(t')\hat{\sigma}(t)-\hat{A}_j(t')\hat{\sigma}(t)\hat{A}_i(t')]$$
$$\left.\left.-\Gamma_{ji}^-[\hat{A}_i(t')\hat{\sigma}(t)\hat{A}_j(t')-\hat{\sigma}(t)\hat{A}_j(t')\hat{A}_i(t')]\}\right]\hat{C}(t')\right\}. \quad (5.58)$$

Since Eq. (5.58) holds for any system operator $\hat{C}$, we deduce that

$$\frac{d\hat{\sigma}}{dt} = -\frac{i}{\hbar}[\hat{H}_{\text{sys}},\hat{\sigma}] - \sum_{i,j}\Delta(\omega_i+\omega_j)\{\Gamma_{ij}^+[\hat{A}_i,\hat{A}_j\hat{\sigma}]-\Gamma_{ji}^-[\hat{A}_i,\hat{\sigma}\hat{A}_j]\}. \quad (5.59)$$

In Eq. (5.59) the notation $\hat{A}_i$ is used for the (now time-independent) system operators $\hat{A}_i(t')$ in the Schrödinger picture. The equation of motion (5.59) for the system density operator is called the master equation.

In practical calculations it may be advantageous to use this equation in an appropriately chosen representation in order to get c-number equations instead of the operator equation given above. A standard way is to write down Eq. (5.59) in the basis of the eigenkets of the system Hamiltonian, $|n\rangle$. The result is a set of coupled first-order differential equations for the density-matrix elements $\sigma_{nm}=\langle n|\hat{\sigma}|m\rangle$.

5.2.2
Fokker–Planck equations

There is another way of formulating c-number equations on the basis of Eq. (5.59), namely by using the concept of phase-space functions [Gordon (1967); Lax (1968); Lax and Yuen (1968); Louisell and Marburger (1968)], introduced in Chapter 4 for the case of a pair of boson basic operators, $\hat{a}$ and $\hat{a}^\dagger$. The extension to other basic operators is straightforward [see, e.g., Louisell (1973)]. For this purpose, let us consider a basis set of noncommuting (system) operators $\hat{a}_1, \hat{a}_2, \ldots, \hat{a}_n$, so that any system operator $\hat{C}$ may be viewed as an operator function of the $\hat{a}_\nu$: $\hat{C} = \hat{C}(\hat{a}_1,\ldots,\hat{a}_n) \equiv \hat{C}(\hat{a}_\nu)$. Let us further suppose that, by means of the commutation (or anti-commutation) relations for the $\hat{a}_\nu$, the operator $\hat{C}$ is put, with regard to the $\hat{a}_\nu$, in a given order: $\hat{C}=\hat{C}^{(O)}(\hat{a}_1,\ldots,\hat{a}_n)\equiv\hat{C}^{(O)}(\hat{a}_\nu)$. Substituting into $\hat{C}=\hat{C}^{(O)}(\hat{a}_\nu)$ for the operators $\hat{a}_\nu$ the c numbers α_ν, we obtain the associated c-number function $C^{(O)}(\alpha_\nu)$. The α_ν are real or complex, depending on whether or not the corresponding $\hat{a}_\nu$ are Hermitian operators. We now use the identity

$$C^{(O)}(\gamma_\nu) = \int d\alpha_1 \cdots \int d\alpha_n\, \delta(\gamma_\nu-\alpha_\nu)C^{(O)}(\alpha_\nu), \quad (5.60)$$

with the n-dimensional δ function $\delta(\alpha_\nu)$ defined by

$$\delta(\alpha_\nu) = \int \frac{\mathrm{d}\beta_1}{N_1} \cdots \int \frac{\mathrm{d}\beta_n}{N_n} \exp(i\beta_1\alpha_1 + \cdots + i\beta_n\alpha_n), \tag{5.61}$$

where $\mathrm{d}\alpha_\nu/N_\nu = \mathrm{d}\alpha_\nu/2\pi$ if α_ν is real, and $(\mathrm{d}\alpha_\nu/N_\nu)(\mathrm{d}\alpha_{\nu'}/N_{\nu'}) = \mathrm{d}^2\alpha_\nu/\pi^2$, with $\mathrm{d}^2\alpha_\nu = \mathrm{d}\mathrm{Re}\alpha_\nu\,\mathrm{d}\mathrm{Im}\alpha_\nu$, and $\exp(i\alpha_\nu\beta_\nu + i\alpha_{\nu'}\beta_{\nu'}) = \exp(i\alpha_\nu\beta_\nu + i\alpha_\nu^*\beta_\nu^*)$ if $\alpha_{\nu'} = \alpha_\nu^*$ ($\nu' \neq \nu$), that is, if $\hat{a}_{\nu'} = \hat{a}_\nu^\dagger$. Going from $C^{(\mathcal{O})}(\gamma_\nu)$ back to $\hat{C}^{(\mathcal{O})}(\hat{a}_\nu)$, we may represent the operator $\hat{C}$ (in the chosen order) as

$$\hat{C} = \int \mathrm{d}\alpha_1 \cdots \int \mathrm{d}\alpha_n\, C^{(\mathcal{O})}(\alpha_\nu)\, \mathcal{O}\hat{\delta}(\hat{a}_\nu - \alpha_\nu), \tag{5.62}$$

where the symbol $\mathcal{O}$ indicates the chosen order and

$$\hat{\delta}(\hat{a}_\nu - \alpha_\nu) = \int \frac{\mathrm{d}\beta_1}{N_1} \cdots \int \frac{\mathrm{d}\beta_n}{N_n} \exp[i(\hat{a}_1 - \alpha_1)\beta_1 + \cdots + i(\hat{a}_n - \alpha_n)\beta_n] \tag{5.63}$$

is the n-dimensional operator δ function. Hence the expectation value of $\hat{C}$ may be represented in the form

$$\langle\hat{C}\rangle = \int \mathrm{d}\alpha_1 \cdots \int \mathrm{d}\alpha_n\, P^{(\mathcal{O})}(\alpha_\nu) C^{(\mathcal{O})}(\alpha_\nu), \tag{5.64}$$

where the phase-space function $P^{(\mathcal{O})}(\alpha_\nu)$ is defined by

$$P^{(\mathcal{O})}(\alpha_\nu) = \langle \mathcal{O}\hat{\delta}(\hat{a}_\nu - \alpha_\nu) \rangle \tag{5.65}$$

[cf. Eq. (4.53)]. Note that combining Eqs (5.63) and (5.65) equivalently yields the following representation of $P^{(\mathcal{O})}(\alpha_\nu)$:

$$P^{(\mathcal{O})}(\alpha_\nu) = \int \frac{\mathrm{d}\beta_1}{N_1} \cdots \int \frac{\mathrm{d}\beta_n}{N_n} \exp(-i\alpha_1\beta_1 - \cdots - i\alpha_n\beta_n)\Phi^{(\mathcal{O})}(\beta_\nu), \tag{5.66}$$

where

$$\Phi^{(\mathcal{O})}(\beta_\nu) = \langle \mathcal{O}\hat{D}(\beta_\nu) \rangle, \tag{5.67}$$

with the operator $\hat{D}(\beta_\nu)$ being defined as

$$\hat{D}(\beta_\nu) = \exp(i\beta_1\hat{a}_1 + \ldots + i\beta_n\hat{a}_n). \tag{5.68}$$

Here $\Phi^{(\mathcal{O})}(\beta_\nu)$ may be viewed as a characteristic (generating function) [cf. Eq. (4.90)]. In particular, the expectation values of products of operators $\hat{a}_\nu$ in the chosen order (that is, $\mathcal{O}$-ordered moments) are simply obtained from $\Phi^{(\mathcal{O})}(\beta_\nu)$ by differentiation, e.g.,

$$\frac{\partial^{l_{\nu_1}+\ldots+l_{\nu_n}}}{\partial(i\beta_{\nu_1})^{l_{\nu_1}}\cdots\partial(i\beta_{\nu_n})^{l_{\nu_n}}}\Phi^{(O)}(\beta_{\nu_j})\bigg|_{\beta_{\nu_j}=0}=\left\langle \mathcal{O}\prod_j(\hat{a}_{\nu_j})^{l_{\nu_j}}\right\rangle \tag{5.69}$$

[cf. Eq. (4.94)].

Now let us turn to the problem of deriving an equation of motion for the phase-space function $P^{(O)}(\alpha_\nu,t)$, where the argument t explicitly indicates its time dependence. From Eq. (5.65) we can easily see that, in the Schrödinger picture, the time derivative of $P^{(O)}(\alpha_\nu,t)$ may be expressed in terms of the time derivative of the (system) density operator $\hat{\sigma}(t)$ as

$$\frac{\partial}{\partial t}P^{(O)}(\alpha_\nu,t)=\mathrm{Tr}_{\mathrm{sys}}\left[\frac{d\hat{\sigma}(t)}{dt}\mathcal{O}\hat{\delta}(\hat{a}_\nu-\alpha_\nu)\right]. \tag{5.70}$$

Substituting the result of Eq. (5.59) into Eq. (5.70) for $d\hat{\sigma}/dt$ and using the cyclic properties of the trace, we derive

$$\frac{\partial}{\partial t}P^{(O)}(\alpha_\nu,t)=\mathrm{Tr}_{\mathrm{sys}}\left[\hat{\sigma}(t)\hat{K}\left(\frac{\partial}{\partial\alpha_\nu}\right)\delta(\alpha_\nu)\right], \tag{5.71}$$

where

$$\hat{K}\left(\frac{\partial}{\partial\alpha_\nu}\right)=-\frac{i}{\hbar}\left[\mathcal{O}\hat{D}\left(i\frac{\partial}{\partial\alpha_\nu}\right),\hat{H}_{\mathrm{sys}}\right]$$
$$-\sum_{i,j}\Delta(\omega_i+\omega_j)\left\{\Gamma_{ij}^+\left[\mathcal{O}\hat{D}\left(i\frac{\partial}{\partial\alpha_\nu}\right),\hat{A}_i\right]\hat{A}_j-\Gamma_{ji}^-\hat{A}_j\left[\mathcal{O}\hat{D}\left(i\frac{\partial}{\partial\alpha_\nu}\right),\hat{A}_i\right]\right\},$$
$$\tag{5.72}$$

with the operator $\hat{D}$ from Eq. (5.68). In Eqs (5.71) and (5.72) the relation

$$\hat{\delta}(\hat{a}_\nu-\alpha_\nu)=\exp\left(-\hat{a}_1\frac{\partial}{\partial\alpha_1}-\ldots-\hat{a}_n\frac{\partial}{\partial\alpha_n}\right)\delta(\alpha_\nu) \tag{5.73}$$

has been used, which may be proved correct by expanding the exponential operator $\exp(i\beta_1\hat{a}_1+\ldots+i\beta_n\hat{a}_n)$ in Eq. (5.63) in a power series and using the relation $(-i\beta_\nu)^l\exp(-i\beta_\nu\alpha_\nu)=\partial^l\exp(-i\beta_\nu\alpha_\nu)/\partial(\alpha_\nu)^l$. Since in Eq. (5.72) the operators $\hat{H}_{\mathrm{sys}}$, $\hat{A}_i$ and $\hat{A}_j$ are operator functions of the $\hat{a}_\nu$, the operator $\hat{K}$ is also an operator function of the $\hat{a}_\nu$: $\hat{K}(\partial/\partial\alpha_\nu)\equiv\hat{K}(\partial/\partial\alpha_\nu,\hat{a}_\nu)$. Putting the $\hat{a}_\nu$ in $\hat{K}$ into the chosen order, $\hat{K}(\partial/\partial\alpha_\nu,\hat{a}_\nu)=\hat{K}^{(O)}(\partial/\partial\alpha_\nu,\hat{a}_\nu)$, and recalling Eq. (5.62), we may rewrite Eq. (5.71) as

$$\frac{\partial}{\partial t}P^{(O)}(\alpha_\nu,t)$$
$$=\mathrm{Tr}_{\mathrm{sys}}\left[\hat{\sigma}(t)\int d\alpha_1'\cdots\int d\alpha_n' K^{(O)}\left(\frac{\partial}{\partial\alpha_\nu},\alpha_\nu'\right)\mathcal{O}\hat{\delta}(\hat{a}_\nu-\alpha_\nu')\delta(\alpha_\nu)\right], \tag{5.74}$$

where $K^{(O)}(\partial/\partial\alpha_\nu, \alpha'_\nu)$ is the c-number function associated with $\hat{K}$ in the chosen order. Defining the function $L^{(O)}(\partial/\partial\alpha_\nu, \alpha'_\nu)$ by

$$K^{(O)}\left(\frac{\partial}{\partial\alpha_\nu}, \alpha'_\nu\right) = L^{(O)}\left(\frac{\partial}{\partial\alpha_\nu}, \alpha'_\nu\right) \exp\left(-\alpha'_1\frac{\partial}{\partial\alpha_1} - \cdots - \alpha'_n\frac{\partial}{\partial\alpha_n}\right), \quad (5.75)$$

recalling Eq. (5.73) ($\hat{a}_\nu \mapsto \alpha_{\nu'}$), and using Eq. (5.65), we finally obtain from Eq. (5.74) the desired equation of motion for the function $P^{(O)}(\alpha_\nu, t)$:

$$\frac{\partial}{\partial t} P^{(O)}(\alpha_\nu, t) = L^{(O)}\left(\frac{\partial}{\partial\alpha_\nu}, \alpha_\nu\right) P^{(O)}(\alpha_\nu, t). \quad (5.76)$$

From the given derivation it is clear that this partial differential equation, which is equivalent to the infinite set of ordinary differential equations for the density-matrix elements $\sigma_{nm}(t)$, contains, in general, all orders of derivatives with respect to the α_ν. In many cases of practical interest derivatives higher than a certain order may be ignored. Frequently the designation Fokker–Planck equation is used when only derivatives up to second order occur.

5.3
Damped harmonic oscillator

Let us apply the theory developed above to the important and illustrative case of a harmonic oscillator undergoing energy relaxation. For this example we shall formulate the Langevin, master and Fokker–Planck equations. Moreover, the effects of an additional dephasing process, not related to energy relaxation, will be studied.

5.3.1
Langevin equations

Assuming that the dominant energy relaxation mechanism arises from one-quantum transitions in the harmonic oscillator (of frequency ω, described by the annihilation and creation operators $\hat{a}$ and $\hat{a}^\dagger$ respectively), we have

$$\hat{H}_{int} = \hbar\hat{V}\hat{a}^\dagger + \text{H.c.}, \quad (5.77)$$

where

$$\hat{V} = \sum_{\nu,\nu'} V_{\nu\nu'}\hat{r}_{\nu\nu'}. \quad (5.78)$$

Comparing Eq. (5.77) with Eq. (5.9), we may make the identifications

$$\hat{A}_1 = \hat{a}^\dagger, \quad \hat{A}_2 = \hat{A}_1^\dagger = \hat{a}, \quad (5.79)$$

$$\hat{R}_1 = \hat{V}, \quad \hat{R}_2 = \hat{R}_1^\dagger = \hat{V}^\dagger \quad (5.80)$$

[cf. Eqs (5.13)–(5.16)]. Applying Eq. (5.37) [together with Eq. (5.36)] to an arbitrary system operator $\hat{C}=\hat{C}(\hat{a},\hat{a}^\dagger)$ yields[7]

$$\dot{\hat{C}} = -\frac{i}{\hbar}[\hat{C},\hat{H}_{\text{sys}}]$$
$$-\tfrac{1}{2}(\Gamma_{12}-\Gamma_{12}^*)([\hat{C},\hat{a}^\dagger]\hat{a}+\hat{a}^\dagger[\hat{C},\hat{a}]) - \tfrac{1}{2}(\Gamma_{12}+\Gamma_{12}^*)([\hat{C},\hat{a}^\dagger]\hat{a}-\hat{a}^\dagger[\hat{C},\hat{a}])$$
$$- i[\hat{C},\hat{a}^\dagger]\hat{F}(t) - i\hat{F}^\dagger(t)[\hat{C},\hat{a}], \qquad (5.81)$$

where

$$\Gamma_{12} = \int_0^\infty d\tau\, e^{i\omega\tau}\langle[\hat{F}(\tau),\hat{F}^\dagger(0)]\rangle \qquad (5.82)$$

and [cf. Eq. (5.26)]

$$\hat{F}(t) = \exp\!\left[\frac{i}{\hbar}\hat{H}_{\text{res}}(t-t')\right]\hat{V}(t')\exp\!\left[-\frac{i}{\hbar}\hat{H}_{\text{res}}(t-t')\right]$$
$$= \sum_{\nu,\nu'} V_{\nu\nu'}\,\hat{r}_{\nu\nu'}(t')\exp[i\omega_{\nu'\nu}(t-t')]. \qquad (5.83)$$

Combining Eqs (5.82) and (5.83) yields

$$\Gamma_{12} = \sum_{\nu,\nu'}|V_{\nu\nu'}|^2[(\sigma_{\text{res}})_{\nu\nu}-(\sigma_{\text{res}})_{\nu'\nu'}]\zeta(\omega-\omega_{\nu'\nu}), \qquad (5.84)$$

where the ζ function is defined by

$$\zeta(x) = \int_0^\infty dy\, e^{ixy} = \pi\delta(x) + i\frac{\mathcal{P}}{x}. \qquad (5.85)$$

In Eq. (5.84) the relation [recall Eq. (5.12)]

$$\langle\hat{r}_{\nu\nu'}(t')\rangle = \text{Tr}(\hat{\sigma}_{\text{res}}|\nu\rangle\langle\nu'|) = (\sigma_{\text{res}})_{\nu'\nu} = \delta_{\nu\nu'}(\sigma_{\text{res}})_{\nu\nu} \qquad (5.86)$$

has been used, where $\hat{\sigma}_{\text{res}}$ is the (diagonal) density operator of the reservoir in equilibrium.

Defining

$$\Gamma = \Gamma_{12} + \Gamma_{12}^* = w_\downarrow - w_\uparrow, \qquad (5.87)$$

where

$$w_\downarrow = 2\pi\sum_{\nu,\nu'}|V_{\nu\nu'}|^2(\sigma_{\text{res}})_{\nu\nu}\delta(\omega-\omega_{\nu'\nu}), \qquad (5.88)$$

$$w_\uparrow = 2\pi\sum_{\nu,\nu'}|V_{\nu\nu'}|^2(\sigma_{\text{res}})_{\nu'\nu'}\delta(\omega-\omega_{\nu'\nu}), \qquad (5.89)$$

$$\Delta = \frac{1}{2i}(\Gamma_{12}-\Gamma_{12}^*) = \mathcal{P}\sum_{\nu,\nu'}|V_{\nu\nu'}|^2[(\sigma_{\text{res}})_{\nu\nu}-(\sigma_{\text{res}})_{\nu'\nu'}](\omega-\omega_{\nu'\nu})^{-1}, \qquad (5.90)$$

---

[7] Note that in the term arising from Eq. (5.37) for $i=2$, $j=1$ the order of both $[\hat{C},\hat{A}_i]$, $\hat{F}_i$ and $[\hat{C},\hat{A}_i]$, $\hat{A}_j$ is changed; cf. the comment below Eq. (5.37).

we may rewrite Eq. (5.81) as

$$\dot{\hat{C}} = -\frac{i}{\hbar}[\hat{C}, \hat{H}_{\text{sys}} + \hbar\Delta\hat{a}^\dagger\hat{a}]$$
$$-\tfrac{1}{2}\Gamma([\hat{C},\hat{a}^\dagger]\hat{a} - \hat{a}^\dagger[\hat{C},\hat{a}]) - i[\hat{C},\hat{a}^\dagger]\hat{F}(t) - i\hat{F}^\dagger(t)[\hat{C},\hat{a}]. \quad (5.91)$$

In the particular case where $\hat{C}=\hat{a}$ this result reduces to

$$\dot{\hat{a}} = -\frac{i}{\hbar}[\hat{a}, \hat{H}_{\text{sys}} + \hbar\Delta\hat{a}^\dagger\hat{a}] - \tfrac{1}{2}\Gamma\hat{a} - i\hat{F}(t) \quad (5.92)$$

[cf. Eq. (5.2)].

We can see that the real part of Γ_{12} gives rise to damping. As we shall see later, $nw_\downarrow$ and $nw_\uparrow$ are just the familiar probabilities per unit time for the oscillator transitions $|n\rangle \to |n-1\rangle$ and $|n-1\rangle \to |n\rangle$, respectively. The imaginary part of Γ_{12} is seen to lead to a shift in the frequency of the (unperturbed) oscillator. Since $\hat{H}_{\text{sys}}$ contains the (unperturbed) oscillator energy $\hbar\omega_0\hat{a}^\dagger\hat{a}$, we may simply let $\hbar\omega_0\hat{a}^\dagger\hat{a} + \hbar\Delta\hat{a}^\dagger\hat{a} = \hbar\omega\hat{a}^\dagger\hat{a}$ with $\omega=\omega_0+\Delta$. That is, the frequency shift may be thought of as being included in the (renormalized) system Hamiltonian: $\hat{H}_{\text{sys}} + \hbar\Delta\hat{a}^\dagger\hat{a} \mapsto \hat{H}_{\text{sys}}$.

When the reservoir may be viewed as a large collection of harmonic oscillators and Eqs (5.15)–(5.16) apply, from Eqs (5.83) and (5.84) we arrive, after some calculation,[8] at

$$\hat{F}(t) = \sum_\alpha \kappa_\alpha \hat{b}_\alpha(t') \exp[-i\omega_\alpha(t-t')], \quad (5.93)$$

$$\Gamma_{12} = \sum_\alpha |\kappa_\alpha|^2 \zeta(\omega - \omega_\alpha), \quad (5.94)$$

and hence

$$\Gamma = \Gamma_{12} + \Gamma_{12}^* = w_\downarrow - w_\uparrow = 2\pi\sum_\alpha |\kappa_\alpha|^2 \delta(\omega - \omega_\alpha), \quad (5.95)$$

$$\Delta = \frac{1}{2i}(\Gamma_{12} - \Gamma_{12}^*) = P\sum_\alpha |\kappa_\alpha|^2(\omega - \omega_\alpha)^{-1}, \quad (5.96)$$

where

$$w_\downarrow = 2\pi\sum_\alpha |\kappa_\alpha|^2 (\bar{n}_\alpha + 1)\delta(\omega - \omega_\alpha), \quad (5.97)$$

$$w_\uparrow = 2\pi\sum_\alpha |\kappa_\alpha|^2 \bar{n}_\alpha \delta(\omega - \omega_\alpha), \quad (5.98)$$

$$\bar{n}_\alpha = \langle \hat{b}_\alpha^\dagger(t')\hat{b}_\alpha(t')\rangle = \text{Tr}(\hat{\sigma}_{\text{res}}\hat{b}_\alpha^\dagger\hat{b}_\alpha). \quad (5.99)$$

[8] We apply the relations $\sum_{vv'}|V_{vv'}|^2(\sigma_{\text{res}})_{vv} = \sum_\alpha \sum_{n_\alpha} |\kappa_\alpha|^2 (n_\alpha+1)\,\text{Tr}(\hat{\sigma}_{\text{res}}|n_\alpha\rangle\langle n_\alpha|) = \sum_\alpha |\kappa_\alpha|^2 (\bar{n}_\alpha+1)$ and, correspondingly, $\sum_{vv'}|V_{vv'}|^2(\sigma_{\text{res}})_{v'v'} = \sum_\alpha |\kappa_\alpha|^2 \bar{n}_\alpha$.

Equation (5.92) together with Eqs (5.93) and (5.95) and $\hat{H}_{\text{sys}} + \hbar\Delta \hat{a}^\dagger \hat{a} \mapsto \hat{H}_{\text{sys}}$ just corresponds to the Langevin equation for a damped cavity mode:

$$\dot{\hat{a}} = -\frac{i}{\hbar}[\hat{a}, \hat{H}_{\text{sys}}] - \tfrac{1}{2}\Gamma \hat{a} - i\hat{F}(t) \qquad (5.100)$$

[cf. Eq. (9.80)]. This correspondence also includes the commutation relations. Using Eq. (5.93) and recalling the commutation rules for boson operators at equal times, we may write

$$[\hat{F}(t_1), \hat{F}^\dagger(t_2)]\exp[i\omega(t_1-t_2)] = \sum_\alpha |\kappa_\alpha|^2 \exp[-i(\omega_\alpha - \omega)(t_1 - t_2)]. \qquad (5.101)$$

Since the reservoir is assumed to be a large collection of harmonic oscillators, we may assume that their frequencies are closely spaced, so that the α sum may be changed to an integral:

$$\sum_\alpha \ldots \mapsto \int_0^\infty d\omega'\, \varrho(\omega') \ldots, \qquad (5.102)$$

with $\varrho(\omega')d\omega'$ being the number of oscillators between ω' and $\omega'+d\omega'$. Hence Eq. (5.101) becomes ($\omega' - \omega = \Omega$, $d\omega' = d\Omega$)

$$[\hat{F}(t_1), \hat{F}^\dagger(t_2)]e^{i\omega(t_1-t_2)} = \int_{-\omega}^\infty d\Omega\, \varrho(\omega + \Omega)|\kappa(\omega + \Omega)|^2 e^{-i\Omega(t_1-t_2)}. \qquad (5.103)$$

In the Markov approximation the bandwidth of $f(\Omega) \equiv \varrho(\omega+\Omega)|\kappa(\omega+\Omega)|^2$ must be large compared with Γ, and $\exp[-i\Omega(t_1 - t_2)]$ is, in comparison with $f(\Omega)$, a rapidly varying function of Ω. We may therefore remove $f(\Omega)|_{\Omega=0} = \varrho(\omega)|\kappa(\omega)|^2$ from the integral and extend the lower limit to $-\infty$. Equation (5.103) therefore becomes, on using Eq. (5.95) together with Eq. (5.102),

$$[\hat{F}(t_1), \hat{F}^\dagger(t_2)]e^{i\omega(t_1-t_2)} = \Gamma\delta(t_1 - t_2), \qquad (5.104)$$

[cf. Eq. (5.5)]. The damping rate Γ is now

$$\Gamma = 2\pi\rho(\omega)|\kappa(\omega)|^2. \qquad (5.105)$$

Note that $[\hat{F}(t_1), \hat{F}(t_2)] = [\hat{F}^\dagger(t_1), \hat{F}^\dagger(t_2)] = 0$. To find the time-dependent commutation relations for system and reservoir variables, we note that if Eq. (5.92) [or, more generally, Eq. (5.91)] is solved in order to express system operators in terms of their past values and the reservoir operators in the past, it is clear that system operators at time t_1 do not depend on reservoir operators at time t_2 if $t_2 > t_1$. In other words, system variables may be regarded as independent of reservoir variables in the future. Hence, representing commutators of the type $[\hat{C}(t_1), \hat{F}(t_2)]$ or $[\hat{C}(t_1), \hat{F}^\dagger(t_2)]$ in terms of commutators of reservoir variables as given in Eq. (5.104), we find that they vanish, provided that $t_2 > t_1$ and $t_2 - t_1 \gg \tau_c$ (time scale Γ^{-1}):

$$[\hat{C}(t_1), \hat{F}(t_2)] = [\hat{C}(t_1), \hat{F}^\dagger(t_2)] = 0 \qquad (t_2 > t_1) \tag{5.106}$$

[cf. Eq. (9.126)]. We recall that Eqs (5.104) and (5.106) are based on a fluctuation operator of the type (5.93). In cases where the more general expression (5.83) must be used, Eqs (5.104) and (5.106) may be regarded as being valid in the sense of (reservoir) averages.

Let us return to Eq. (5.91) and suppose that $\hat{F}(t)$ is a proper noise generator, $\langle \hat{F} \rangle = 0$. Under the assumptions made when deriving Eq. (5.48), we may represent Eq. (5.91) in a form corresponding to Eq. (5.48). Straightforward calculations yield ($\hat{H}_{sys} + \hbar \Delta \hat{a}^\dagger \hat{a} \mapsto \hat{H}_{sys}$)

$$\begin{aligned}\dot{\hat{C}} &= -\frac{i}{\hbar}[\hat{C}, \hat{H}_{sys}] + \tfrac{1}{2}w_\downarrow (\hat{a}^\dagger[\hat{C}, \hat{a}] - [\hat{C}, \hat{a}^\dagger]\hat{a}) \\&\quad + \tfrac{1}{2}w_\uparrow (\hat{a}[\hat{C}, \hat{a}^\dagger] - [\hat{C}, \hat{a}]\hat{a}^\dagger) + \hat{F}_C(t) \\&= -\frac{i}{\hbar}[\hat{C}, \hat{H}_{sys}] + \tfrac{1}{2}w_\downarrow (2\hat{a}^\dagger \hat{C} \hat{a} - \hat{C}\hat{a}^\dagger \hat{a} - \hat{a}^\dagger \hat{a} \hat{C}) \\&\quad + \tfrac{1}{2}w_\uparrow (2\hat{a}\hat{C}\hat{a}^\dagger - \hat{C}\hat{a}\hat{a}^\dagger - \hat{a}\hat{a}^\dagger \hat{C}) + \hat{F}_C(t), \end{aligned}\tag{5.107}$$

where[9]

$$\langle \hat{F}_C(t) \rangle = 0; \tag{5.108}$$

$w_\downarrow$ and $w_\uparrow$ are defined by Eqs (5.97) and (5.98), respectively.

5.3.2
Master equations

According to the procedure outlined for proceeding from Eq. (5.48) to Eq. (5.59), from Eq. (5.107) the equation of motion for the reduced density operator $\hat{\sigma}$ may now be derived as

$$\begin{aligned}\frac{d\hat{\sigma}}{dt} &= \frac{i}{\hbar}[\hat{\sigma}, \hat{H}_{sys}] + \tfrac{1}{2}w_\downarrow ([\hat{a}, \hat{\sigma}\hat{a}^\dagger] - [\hat{a}^\dagger, \hat{a}\hat{\sigma}]) \\&\quad + \tfrac{1}{2}w_\uparrow ([\hat{a}^\dagger, \hat{\sigma}\hat{a}] - [\hat{a}, \hat{a}^\dagger \hat{\sigma}]) \\&= \frac{i}{\hbar}[\hat{\sigma}, \hat{H}_{sys}] + \tfrac{1}{2}w_\downarrow (2\hat{a}\hat{\sigma}\hat{a}^\dagger - \hat{a}^\dagger \hat{a}\hat{\sigma} - \hat{\sigma}\hat{a}^\dagger \hat{a}) \\&\quad + \tfrac{1}{2}w_\uparrow (2\hat{a}^\dagger \hat{\sigma}\hat{a} - \hat{a}\hat{a}^\dagger \hat{\sigma} - \hat{\sigma}\hat{a}\hat{a}^\dagger). \end{aligned}\tag{5.109}$$

---
[9] Here the noise operator $\hat{F}_C(t)$ is understood as being constructed for this specific case according to the prescription given in Section 5.1.4 in Eqs (5.45) and (5.49).

Writing Eq. (5.109) in the basis defined by the eigenkets $|n\rangle$ of the number operator $\hat{a}^\dagger\hat{a}$, after some algebra, the following system of equations of motion for the density-matrix elements $\sigma_{nm} = \langle n|\hat{\sigma}|m\rangle$ is obtained:

$$\dot{\sigma}_{nn} = \frac{i}{\hbar} \langle n|[\hat{\sigma}, \hat{H}_{\text{sys}}]|n\rangle + w_\downarrow[(n+1)\sigma_{n+1\,n+1} - n\,\sigma_{nn}]$$
$$+ w_\uparrow[n\sigma_{n-1\,n-1} - (n+1)\sigma_{nn}], \tag{5.110}$$

$$\dot{\sigma}_{nm} = \frac{i}{\hbar} \langle n|[\hat{\sigma}, \hat{H}_{\text{sys}}]|m\rangle - \gamma_{nm}\sigma_{nm}$$
$$+ w_\downarrow[(n+1)(m+1)]^{1/2}\sigma_{n+1\,m+1} + w_\uparrow(nm)^{1/2}\sigma_{n-1\,m-1} \tag{5.111}$$

$(n \neq m)$, where

$$\gamma_{nm} = \tfrac{1}{2}[(n+m)w_\downarrow + (n+m+2)w_\uparrow]. \tag{5.112}$$

When the system is an unperturbed harmonic oscillator, then the number states are the energy eigenstates, and representation of the master equation in the number basis allows an interpretation of the various relaxation terms.[10] Obviously, $w_\downarrow(n+1)$ and $w_\uparrow n$ are the (energy relaxation) rates for the $|n+1\rangle \to |n\rangle$ and $|n-1\rangle \to |n\rangle$ transitions, respectively. The rate γ_{nm} ($n \neq m$) describes the dephasing (decay of the off-diagonal element σ_{nm}), and the rates $w_\downarrow\sqrt{(n+1)(m+1)}$ and $w_\uparrow\sqrt{nm}$ describe reservoir-induced couplings between the off-diagonal elements, $\sigma_{n+1\,m+1} \leftrightarrow \sigma_{nm}$ and $\sigma_{n-1\,m-1} \leftrightarrow \sigma_{nm}$, respectively.

5.3.3
Fokker–Planck equations

Let us consider the s-parameterized phase-space functions introduced in Chapter 4, which apply to the calculation of averages of operator functions in s order. Instead of following the general procedure outlined in Section 5.2.2, we apply the formalism specific to s-parameterized phase-space functions.

Let us first consider the characteristic function $\Phi(\alpha) \equiv \Phi(\alpha; s=0)$ of the Wigner function $W(\alpha)$. Recalling Eqs (4.47) and (4.90), we can write the time derivative of $\Phi(\alpha)$ (in the Schrödinger picture) as

$$\frac{\partial \Phi(\alpha)}{\partial t} = \text{Tr}\left[\hat{D}(\alpha)\frac{d\hat{\sigma}}{dt}\right]. \tag{5.113}$$

---

10) In that case the infinite system of balance equations (5.110) can be solved by means of the generating function $G(x,t) = \sum_n (x+1)^{n+1}\sigma_{nn}(t)$ [Montroll and Shuler (1957)], which is easily seen to obey the simple equation

$$\frac{\partial G}{\partial t} + x\frac{\partial G}{\partial x}[w_\downarrow - (x+1)w_\uparrow] = (w_\downarrow - w_\uparrow)\,G.$$

Substituting the expression (5.109) into Eq. (5.113) for $d\hat{\varrho}/dt$ and using the cyclic properties of the trace, we see that expectation values of products of $\hat{D}$ and $\hat{a}$ and $\hat{a}^\dagger$ are to be considered. With the help of Eqs (4.46) and (4.48) it is not difficult to prove that [cf. Eqs (3.56) and (3.57)]

$$\hat{a}^\dagger \hat{D}(\alpha) = \left(\frac{\alpha^*}{2} + \frac{\partial}{\partial \alpha}\right) \hat{D}(\alpha), \quad \hat{D}(\alpha)\hat{a} = [\hat{a}^\dagger \hat{D}(-\alpha)]^\dagger. \tag{5.114}$$

Similarly, from Eqs (4.46) and (4.49) it follows that

$$\hat{a}\hat{D}(\alpha) = \left(\frac{\alpha}{2} - \frac{\partial}{\partial \alpha^*}\right) \hat{D}(\alpha), \quad \hat{D}(\alpha)\hat{a}^\dagger = [\hat{a}\hat{D}(-\alpha)]^\dagger. \tag{5.115}$$

Combining Eqs (5.113) and (5.109), on assuming that $\hat{H}_{\text{sys}} = \hbar \omega \hat{a}^\dagger \hat{a}$, and applying the rules given in Eqs (5.114) and (5.115), after some straightforward calculation, we derive the following evolution equation for the characteristic function of the Wigner function:

$$\frac{\partial \Phi(\alpha)}{\partial t} = -\left[\left(\tfrac{1}{2}\Gamma + i\omega\right)\alpha^* \frac{\partial}{\partial \alpha^*} + \left(\tfrac{1}{2}\Gamma - i\omega\right)\alpha \frac{\partial}{\partial \alpha}\right] \Phi(\alpha)$$
$$- \tfrac{1}{2}\Gamma \left(1 + \frac{2w_\uparrow}{\Gamma}\right) |\alpha|^2 \Phi(\alpha). \tag{5.116}$$

Note that here the relation $\Gamma = w_\downarrow - w_\uparrow$ has been used [cf. Eq. (5.87)].

Equation (5.116) can easily be extended to the case of arbitrary s order. For this purpose we note that, according to Eq. (4.91), the relation

$$\frac{\partial \Phi(\alpha;s)}{\partial t} = \frac{\partial}{\partial t}[e^{\frac{1}{2}s|\alpha|^2}\Phi(\alpha)] = e^{\frac{1}{2}s|\alpha|^2}\frac{\partial \Phi(\alpha)}{\partial t} \tag{5.117}$$

holds. Hence, the evolution equation for $\Phi(\alpha;s)$ can be obtained from Eq. (5.116) by multiplying the expression on the right-hand side of that equation by $e^{\frac{1}{2}s|\alpha|^2}$. Taking into account that

$$e^{\frac{1}{2}s|\alpha|^2} \alpha \frac{\partial \Phi(\alpha)}{\partial \alpha} = \alpha \frac{\partial \Phi(\alpha;s)}{\partial \alpha} - \tfrac{1}{2}s|\alpha|^2 \Phi(\alpha;s), \tag{5.118}$$

we can easily see that $\Phi(\alpha;s)$ obeys the partial differential equation

$$\frac{\partial \Phi(\alpha;s)}{\partial t} = -\left[\left(\tfrac{1}{2}\Gamma + i\omega\right)\alpha^* \frac{\partial}{\partial \alpha^*} + \left(\tfrac{1}{2}\Gamma - i\omega\right)\alpha \frac{\partial}{\partial \alpha}\right] \Phi(\alpha;s)$$
$$- \tfrac{1}{2}\Gamma \left(1 - s + \frac{2w_\uparrow}{\Gamma}\right) |\alpha|^2 \Phi(\alpha;s). \tag{5.119}$$

Since the characteristic function $\Phi(\alpha;s)$ and the phase-space function $P(\alpha;s)$ are related to each other by the Fourier transformation (4.93), the evolution equation for $P(\alpha;s)$ can be obtained by Fourier transformation of Eq. (5.119). Using the relations

$$\frac{1}{\pi^2}\int d^2\beta\, \exp(\alpha\beta^* - \alpha^*\beta)|\beta|^2\Phi(\beta;s) = -\frac{\partial^2 P(\alpha;s)}{\partial\alpha\partial\alpha^*} \tag{5.120}$$

and

$$\frac{1}{\pi^2}\int d^2\beta\, \exp(\alpha\beta^* - \alpha^*\beta)\beta^*\frac{\partial\Phi(\beta;s)}{\partial\beta^*}$$
$$= \frac{1}{\pi^2}\frac{\partial}{\partial\alpha}\int d^2\beta\, \exp(\alpha\beta^* - \alpha^*\beta)\frac{\partial\Phi(\beta;s)}{\partial\beta^*} = -\frac{\partial}{\partial\alpha}[\alpha P(\alpha;s)], \tag{5.121}$$

we can easily see that the sought Fokker–Planck equation can be given in the form of[11]

$$\frac{\partial P(\alpha;s)}{\partial t} = \left[(\tfrac{1}{2}\Gamma + i\omega)\frac{\partial}{\partial\alpha}\alpha + (\tfrac{1}{2}\Gamma - i\omega)\frac{\partial}{\partial\alpha^*}\alpha^*\right]P(\alpha;s)$$
$$+ \tfrac{1}{2}\Gamma\left(1 - s + \frac{2w_\uparrow}{\Gamma}\right)\frac{\partial^2 P(\alpha;s)}{\partial\alpha\partial\alpha^*}. \tag{5.122}$$

Note that the free motion can easily be removed by the ansatz

$$P(\alpha,t;s) = P(\tilde{\alpha}e^{-i\omega t}, t; s) \equiv \tilde{P}(\tilde{\alpha}, t; s). \tag{5.123}$$

It is not difficult to prove that the equation of motion of $\tilde{P}(\tilde{\alpha};s)$ reads

$$\frac{\partial \tilde{P}(\tilde{\alpha};s)}{\partial t} = \tfrac{1}{2}\Gamma\left(\frac{\partial}{\partial\tilde{\alpha}}\tilde{\alpha} + \frac{\partial}{\partial\tilde{\alpha}^*}\tilde{\alpha}^*\right)\tilde{P}(\tilde{\alpha};s) + \tfrac{1}{2}\Gamma\left(1 - s + \frac{2w_\uparrow}{\Gamma}\right)\frac{\partial^2 \tilde{P}(\alpha;s)}{\partial\tilde{\alpha}\partial\tilde{\alpha}^*}. \tag{5.124}$$

5.3.4
Radiationless dephasing

The formulae derived in Sections 5.3.1–5.3.3 are based on an interaction Hamiltonian of the type (5.77). This model is commonly used to describe the effect of (lowest-order) energy relaxation, the phase relaxation (dephasing) being attributed to the energy relaxation. To see this, we recall that from inspection of the (balance) equations (5.110) for the diagonal density-matrix elements σ_{nn} the energy relaxation rates w_{nl} for the transitions $|n\rangle \to |l\rangle$ ($l = n-1, n+1$) are found as

$$w_{nl} = n w_\downarrow \delta_{l\,n-1} + (n+1)w_\uparrow \delta_{l\,n+1}. \tag{5.125}$$

On the other hand, the phase relaxation rates γ_{nm} responsible for the decay of the (phase-sensitive) off-diagonal density-matrix elements σ_{nm} ($n \neq m$) are

11) For methods of solution, see, e. g., Louisell (1973).

determined solely by the energy relaxation rates; that is,

$$\gamma_{nm} = \tfrac{1}{2}\sum_l (w_{nl} + w_{ml}) \quad (n \ne m), \tag{5.126}$$

as can readily be seen by substituting the result (5.125) into Eq. (5.126) for w_{nl} (and w_{ml}) and comparing the result of this with Eq. (5.111). We note that the relations (5.126) between the energy relaxation rates and the associated phase relaxation rates is very general and also applicable to other than harmonic-oscillator systems. In particular, they typically describe the effect of radiative decay.

In many cases of radiationless relaxations, however, there are decay channels giving rise to pure dephasings. Let us suppose that the interaction energy between the system oscillator and the reservoir contains an additional term bilinear in the system operator [Diestler (1976); Paerschke, Süsse and Welsch (1980)]:

$$\hat{H}'_{\text{int}} = \hbar \hat{V}' \hat{a}^\dagger \hat{a} \tag{5.127}$$

[with $(\hat{V}')^\dagger = \hat{V}'$].[12] According to Eq. (5.9), we make the identifications

$$\hat{A}_3 = \hat{a}^\dagger \hat{a}, \tag{5.128}$$

$$\hat{R}_3 = \hat{V}', \tag{5.129}$$

and define, according to Eq. (5.26), the noise generator

$$\hat{F}'(t) = \exp\left[\frac{i}{\hbar}\hat{H}_{\text{res}}(t-t')\right] \hat{V}'(t') \exp\left[-\frac{i}{\hbar}\hat{H}_{\text{res}}(t-t')\right] \tag{5.130}$$

[with $\langle \hat{F}'(t) \rangle = 0$]. Applying Eq. (5.48) then yields

$$\begin{aligned}-\Delta(\omega_3+\omega_3)&(\Gamma^+_{33}[\hat{C},\hat{A}_3]\hat{A}_3 - \Gamma^-_{33}\hat{A}_3[\hat{C},\hat{A}_3])\\ &= -i\Delta'[\hat{C},\hat{a}^\dagger \hat{a}\hat{a}^\dagger \hat{a}] - \gamma'[[\hat{C},\hat{a}^\dagger \hat{a}],\hat{a}^\dagger \hat{a}],\end{aligned} \tag{5.131}$$

where

$$\gamma' = \tfrac{1}{2}\int_0^\infty d\tau\, \langle \hat{F}'(\tau)\hat{F}'(0)\rangle + \text{c.c.}, \tag{5.132}$$

$$i\Delta' = \tfrac{1}{2}\int_0^\infty d\tau\, \langle \hat{F}'(\tau)\hat{F}'(0)\rangle - \text{c.c.}. \tag{5.133}$$

The second term on the right-hand side of Eq. (5.131) gives rise to an additional damping, whereas the first describes the associated frequency shift.

---

[12] Note that a term $\hbar \hat{V}'' \hat{a}\hat{a}^\dagger$ can be thought of as being included in $\hat{H}'_{\text{int}}$ and $\hat{H}_{\text{res}}$, because $\hbar \hat{V}'' \hat{a}\hat{a}^\dagger$ can be rewritten as $\hbar \hat{V}'' \hat{a}\hat{a}^\dagger = \hbar \hat{V}'' + \hbar \hat{V}'' \hat{a}^\dagger \hat{a}$.

Note that in contrast to the case of energy relaxation [with the interaction energy being given by Eq. (5.77)], the frequency-shift term in Eq. (5.131) cannot be included in the (unperturbed) Hamiltonian of the system oscillator, because of its quadratic dependence on the number operator. Ignoring this small effect, we find the complete Langevin equation for any system operator $\hat{C}$ by adding the damping term in Eq. (5.131) to the damping terms on the right-hand side of Eq. (5.107) (cf. footnote 9, p. 155):

$$\dot{\hat{C}} = -\frac{i}{\hbar}[\hat{C}, \hat{H}_{\text{sys}}] - \gamma'[[\hat{C}, \hat{a}^\dagger \hat{a}], \hat{a}^\dagger \hat{a}]$$
$$+ \tfrac{1}{2}w_\downarrow \left(\hat{a}^\dagger[\hat{C}, \hat{a}] - [\hat{C}, \hat{a}^\dagger]\hat{a}\right) + \tfrac{1}{2}w_\uparrow \left(\hat{a}[\hat{C}, \hat{a}^\dagger] - [\hat{C}, \hat{a}]\hat{a}^\dagger\right) + \hat{F}_C(t). \quad (5.134)$$

If $\hat{C}$ is a function of the number operator, we simply have $[\hat{C}, \hat{a}^\dagger \hat{a}] = 0$, and hence in the equation of motion the damping terms only arise from energy relaxation. The simplest example is the equation of motion for the number operator itself ($\hat{C} = \hat{n} = \hat{a}^\dagger \hat{a}$):

$$\dot{\hat{n}} = -\frac{i}{\hbar}[\hat{n}, \hat{H}_{\text{sys}}] - \Gamma \hat{n} + w_\uparrow + \hat{F}_n(t) \quad (5.135)$$

(note that $\Gamma = w_\downarrow - w_\uparrow$). The dephasing term (proportional to γ') in Eq. (5.134) becomes relevant if the chosen system operator $\hat{C}$ is phase-dependent. For example, when $\hat{C} = \hat{a}$, we have

$$\dot{\hat{a}} = -\frac{i}{\hbar}[\hat{a}, \hat{H}_{\text{sys}}] - (\tfrac{1}{2}\Gamma + \gamma')\hat{a} + \hat{F}_a(t). \quad (5.136)$$

In an analogous way to the derivation of Eq. (5.109), from Eq. (5.134) we derive the corresponding master equation:

$$\frac{d\hat{\sigma}}{dt} = -\frac{i}{\hbar}[\hat{\sigma}, \hat{H}_{\text{sys}}] - \gamma'[[\hat{\sigma}, \hat{a}^\dagger \hat{a}], \hat{a}^\dagger \hat{a}]$$
$$+ \tfrac{1}{2}w_\downarrow ([\hat{a}, \hat{\sigma}\hat{a}^\dagger] - [\hat{a}^\dagger, \hat{a}\hat{\sigma}]) + \tfrac{1}{2}w_\uparrow ([\hat{a}^\dagger, \hat{\sigma}\hat{a}] - [\hat{a}, \hat{a}^\dagger \hat{\sigma}]). \quad (5.137)$$

Representing Eq. (5.137) in the form given in Eqs (5.110) and (5.111), we can easily see that in the equations of motion for the (phase-sensitive) off-diagonal density-matrix elements σ_{nm} ($n \neq m$) the dephasing rates are now

$$\gamma_{nm} = \tfrac{1}{2}[(n+m)w_\downarrow + (n+m+2)w_\uparrow] + \gamma'(n-m)^2. \quad (5.138)$$

The second term on the right-hand side of Eq. (5.137) can be incorporated into the Fokker–Planck equation (5.124), by applying the scheme outlined in Section 5.3.3. The result is

$$\frac{\partial P(\alpha;s)}{\partial t} = \left[(\tfrac{1}{2}\Gamma + i\omega)\frac{\partial}{\partial \alpha}\alpha + (\tfrac{1}{2}\Gamma - i\omega)\frac{\partial}{\partial \alpha^*}\alpha^*\right]P(\alpha;s)$$
$$+ \tfrac{1}{2}\Gamma\left(1 - s + \frac{2w_\uparrow}{\Gamma}\right)\frac{\partial^2 P(\alpha;s)}{\partial \alpha \partial \alpha^*} - \gamma'\left(\alpha\frac{\partial}{\partial \alpha} - \alpha^*\frac{\partial}{\partial \alpha^*}\right)^2 P(\alpha;s).$$
$$(5.139)$$

An alternative way is to put the second term on the right-hand side of Eq. (5.137) into anti-normal order:[13]

$$[[\hat{\sigma}, \hat{a}^\dagger \hat{a}], \hat{a}^\dagger \hat{a}] = \hat{a} \frac{\partial}{\partial \hat{a}} \left(\hat{a} \frac{\partial \hat{\sigma}^{(A)}}{\partial \hat{a}} \right) + \left[\frac{\partial}{\partial \hat{a}^\dagger} \left(\frac{\partial \hat{\sigma}^{(A)}}{\partial \hat{a}^\dagger} \hat{a}^\dagger \right) \right] \hat{a}^\dagger - 2\hat{a} \frac{\partial^2 \hat{\sigma}^{(A)}}{\partial \hat{a} \partial \hat{a}^\dagger} \hat{a}^\dagger \qquad (5.140)$$

($\hat{\sigma}^{(A)}$, density operator in anti-normal order). Thus, the associated c-number function is just the last term in Eq. (5.139) in the case when $s=1$, because of $P(\alpha;1) = \pi^{-1}\sigma(\alpha;-1)$ [Eq. (4.82)]. Now Eq. (4.70) can be used to show that this result is also valid for other values of s.

5.4
Damped two-level system

In the study of the resonant interaction between atomic systems and light it is often sufficient to model the atomic systems – according to the number of resonant transitions that are effectively involved in the respective interaction process – by few-level systems. The simplest example is the two-level model widely used in quantum optics. As we will see, it also serves as a very illustrative example in the study of damped atomic systems. Standard equations which describe a driven two-level system that is subject to relaxation are the optical Bloch equations. We shall formulate them in Section 5.4.2 and also outline the extension of the theory to multi-level atomic systems.

5.4.1
Basic equations

Denoting the two quantum states of the (unperturbed) atomic system by $|1\rangle$ (ground state) and $|2\rangle$ (excited state), we may define the system operators $\hat{A}_{mn} = |m\rangle\langle n|$ ($\{m,n\}=1,2$). A complete set of system operators is $\hat{A}_{11}$ (or $\hat{A}_{22}$), $\hat{A}_{12}$ and $\hat{A}_{21}$, by virtue of the completeness relation $\hat{A}_{11} + \hat{A}_{22} = \hat{I}$. Note that for operator products the relation $\hat{A}_{mn}\hat{A}_{kl} = \delta_{nk}\hat{A}_{ml}$ holds. The (unperturbed) system Hamiltonian may be written as

$$\hat{H}_{\text{sys}} = \hbar\omega_1 \hat{A}_{11} + \hbar\omega_2 \hat{A}_{22}. \qquad (5.141)$$

Let us first study the case where the atomic system undergoes energy relaxation, so that, by close analogy with Eq. (5.77), the coupling of the system to the reservoir may be assumed to be based on the interaction energy

$$\hat{H}_{\text{int}} = \hbar(\hat{V}_{21}\hat{A}_{21} + \hat{V}_{12}\hat{A}_{12}), \qquad (5.142)$$

[13] Replace $\hat{\sigma}$ by $\hat{\sigma}^{(A)}$ and apply Eqs (C.16) and (C.17).

where $\hat{V}_{21}$ may be thought of as being given in the form (5.78) (with $V_{\nu\nu'} \mapsto V_{21,\nu\nu'}$). According to the notation in Eq. (5.9), we now make the identifications[14]

$$\hat{A}_1 = \hat{A}_{21}, \quad \hat{A}_2 = \hat{A}_{12} = \hat{A}_1^\dagger, \tag{5.143}$$

$$\hat{R}_1 = \hat{V}_{21}, \quad \hat{R}_2 = \hat{V}_{12} = \hat{R}_1^\dagger, \tag{5.144}$$

and define, according to Eq. (5.26),

$$\hat{F}_{mn}(t) = \exp\left[\frac{i}{\hbar}\hat{H}_{\text{res}}(t-t')\right] \hat{V}_{mn}(t') \exp\left[-\frac{i}{\hbar}\hat{H}_{\text{res}}(t-t')\right]. \tag{5.145}$$

The further calculations are very similar to those leading to Eqs (5.81)–(5.91). Applying the result (5.37) together with Eq. (5.36) to an arbitrary operator function $\hat{C} = \hat{C}(\hat{A}_{mn})$, after some algebra we obtain the following Langevin equation:

$$\begin{aligned}\dot{\hat{C}} = &-\frac{i}{\hbar}[\hat{C}, \hat{H}_{\text{sys}}] - \tfrac{1}{2}(\Gamma_{12} - \Gamma_{12}^*)([\hat{C}, \hat{A}_{21}]\hat{A}_{12} + \hat{A}_{21}[\hat{C}, \hat{A}_{12}]) \\ &- \tfrac{1}{2}(\Gamma_{12} + \Gamma_{12}^*)([\hat{C}, \hat{A}_{21}]\hat{A}_{12} - \hat{A}_{21}[\hat{C}, \hat{A}_{12}]) \\ &- i[\hat{C}, \hat{A}_{21}]\hat{F}_{21}(t) - i\hat{F}_{12}(t)[\hat{C}, \hat{A}_{12}],\end{aligned} \tag{5.146}$$

where

$$\Gamma_{12} = \int_0^\infty d\tau\, e^{i\omega_{21}\tau} \langle [\hat{F}_{21}(\tau), \hat{F}_{12}(0)] \rangle \tag{5.147}$$

and $\omega_{21} = \omega_2 - \omega_1$. Defining

$$\Gamma = \Gamma_{12} + \Gamma_{12}^*, \tag{5.148}$$

$$i\Delta = \tfrac{1}{2}(\Gamma_{12} - \Gamma_{12}^*) \tag{5.149}$$

[cf. Eqs (5.87)–(5.90)] and using the identity $[\hat{C}, \hat{A}_{21}]\hat{A}_{12} + \hat{A}_{21}[\hat{C}, \hat{A}_{12}] = [\hat{C}, \hat{A}_{22}]$, we may rewrite Eq. (5.146) as

$$\begin{aligned}\dot{\hat{C}} = &-\frac{i}{\hbar}[\hat{C}, \hat{H}_{\text{sys}} + \hbar\Delta\hat{A}_{22}] - \tfrac{1}{2}\Gamma([\hat{C}, \hat{A}_{21}]\hat{A}_{12} - \hat{A}_{21}[\hat{C}, \hat{A}_{12}]) \\ &- i[\hat{C}, \hat{A}_{21}]\hat{F}_{21}(t) - i\hat{F}_{12}(t)[\hat{C}, \hat{A}_{12}] \\ = &-\frac{i}{\hbar}[\hat{C}, \hat{H}_{\text{sys}} + \hbar\Delta\hat{A}_{22}] - \tfrac{1}{2}\Gamma(\hat{C}\hat{A}_{22} + \hat{A}_{22}\hat{C} - 2\hat{A}_{21}\hat{C}\hat{A}_{12}) \\ &- i[\hat{C}, \hat{A}_{21}]\hat{F}_{21}(t) - i\hat{F}_{12}(t)[\hat{C}, \hat{A}_{12}],\end{aligned} \tag{5.150}$$

[14] Note that the frequencies ω_i introduced in Eq. (5.20) through $\hat{A}_i(t) = \tilde{\hat{A}}_i(t)e^{-i\omega_i t}$ must be distinguished from the frequencies defined, according to Eq. (5.141), by the energy eigenvalues of the two-level system. The flip operators $\hat{A}_{mn}$ evolve as $\hat{A}_{mn}(t) = \tilde{\hat{A}}_{mn}(t)e^{i\omega_{mn}t}$, where $\omega_{mn} = \omega_m - \omega_n$, with $\omega_{m(n)}$ according to Eq. (5.141).

where the damping rate Γ may be represented in the form (5.87), together with Eqs (5.88) and (5.89) ($V_{\nu\nu'} \mapsto V_{21,\nu\nu'}$) or, in the particular case where the reservoir is described by a sample of harmonic oscillators, Eqs (5.97) and (5.98) ($\kappa_\alpha \mapsto \kappa_{21,\alpha}$). Furthermore, from inspection of Eqs (5.141) and (5.150) we see that the effect of a frequency shift may again be thought of as being included in the system Hamiltonian ($\hat{H}_{sys} + \hbar\Delta\hat{A}_{22} \mapsto \hat{H}_{sys}$).

If all the premises on which Eq. (5.48) is based are satisfied, we may turn from the Langevin equation (5.150) to a Langevin equation of the type (5.48). By close analogy with Eq. (5.107), we obtain ($\hat{H}_{sys} + \hbar\Delta\hat{A}_{22} \mapsto \hat{H}_{sys}$) the result (cf. footnote 9, p. 155)

$$\begin{aligned}\dot{\hat{C}} &= -\frac{i}{\hbar}[\hat{C}, \hat{H}_{sys}] + \tfrac{1}{2}w_{21}(\hat{A}_{21}[\hat{C}, \hat{A}_{12}] - [\hat{C}, \hat{A}_{21}]\hat{A}_{12}) \\ &\quad + \tfrac{1}{2}w_{12}(\hat{A}_{12}[\hat{C}, \hat{A}_{21}] - [\hat{C}, \hat{A}_{12}]\hat{A}_{21}) + \hat{F}_C(t) \\ &= -\frac{i}{\hbar}[\hat{C}, \hat{H}_{sys}] + \tfrac{1}{2}w_{21}(2\hat{A}_{21}\hat{C}\hat{A}_{12} - \hat{C}\hat{A}_{22} - \hat{A}_{22}\hat{C}) \\ &\quad + \tfrac{1}{2}w_{12}(2\hat{A}_{12}\hat{C}\hat{A}_{21} - \hat{C}\hat{A}_{11} - \hat{A}_{11}\hat{C}) + \hat{F}_C(t)\end{aligned} \tag{5.151}$$

with

$$\langle \hat{F}_C(t) \rangle = 0, \tag{5.152}$$

and $w_{21} = w_\downarrow$, $w_{12} = w_\uparrow$. Clearly, the energy relaxation rates w_{21} and w_{12} are just the transition probabilities per unit time for the transitions $|2\rangle \to |1\rangle$ and $|1\rangle \to |2\rangle$, respectively [cf. Eqs (5.88) and (5.89) or Eqs (5.97) and (5.98)].

To find the master equation for the reduced (two-level) density operator $\hat{\sigma}$ we again apply the procedure leading from Eq. (5.48) to Eq. (5.59). In this way, we derive from Eq. (5.151)

$$\frac{d\hat{\sigma}}{dt} = -\frac{i}{\hbar}[\hat{H}_{sys}, \hat{\sigma}] + \tfrac{1}{2}w_{21}([\hat{A}_{12}, \hat{\sigma}\hat{A}_{21}] - [\hat{A}_{21}, \hat{A}_{12}\hat{\sigma}]) + \tfrac{1}{2}w_{12}([\hat{A}_{21}, \hat{\sigma}\hat{A}_{12}] - [\hat{A}_{12}, \hat{A}_{21}\hat{\sigma}]). \tag{5.153}$$

Let us now allow for an additional dephasing channel. We assume that the interaction energy between the two-level system and the reservoir has non-vanishing diagonal matrix elements with respect to the quantum states of the system:[15]

$$\hat{H}'_{int} = \hat{V}_{22}\hat{A}_{22} \tag{5.154}$$

[cf. Eq. (5.127)]. Making the identifications

$$\hat{A}_3 = \hat{A}_{22}, \tag{5.155}$$
$$\hat{R}_3 = \hat{V}_{22}, \tag{5.156}$$

---

15) Note that a term in the form of $\hbar\hat{V}_{11}\hat{A}_{11}$ can be thought of as being included in $\hat{H}'_{int}$ and $\hat{H}_{res}$, because $\hbar\hat{V}_{11}\hat{A}_{11}$ can be rewritten as $\hbar\hat{V}_{11}\hat{A}_{11} = \hbar\hat{V}_{11} + (\hbar\hat{V}_{22} - \hat{V}_{11})\hat{A}_{22}$.

and taking Eq. (5.145) into account, we apply Eq. (5.48) to obtain

$$-\Delta(\omega_3 + \omega_3)\left(\Gamma_{33}^+[\hat{C}, \hat{A}_3]\hat{A}_3 - \Gamma_{33}^-\hat{A}_3[\hat{C}, \hat{A}_3]\right)$$
$$= -i\Delta'[\hat{C}, \hat{A}_{22}] - \gamma'[[\hat{C}, \hat{A}_{22}], \hat{A}_{22}], \tag{5.157}$$

where

$$\gamma' = \tfrac{1}{2} \int_0^\infty d\tau \, \langle \hat{V}_{22}(\tau)\hat{V}_{22}(0) \rangle + \text{c.c.}, \tag{5.158}$$

$$i\Delta' = \tfrac{1}{2} \int_0^\infty d\tau \, \langle \hat{V}_{22}(\tau)\hat{V}_{22}(0) \rangle - \text{c.c.}. \tag{5.159}$$

Including the frequency shift [first term on the right-hand side of Eq. (5.157)] into the system Hamiltonian and adding the damping term [second term on the right-hand side of Eq. (5.157)] to the damping terms in Eq. (5.151), the Langevin equation is (cf. footnote 9, p. 155)

$$\dot{\hat{C}} = -\frac{i}{\hbar}[\hat{C}, \hat{H}_{\text{sys}}] - \gamma'[[\hat{C}, \hat{A}_{22}], \hat{A}_{22}] + \tfrac{1}{2}w_{21}\left(\hat{A}_{21}[\hat{C}, \hat{A}_{12}] - [\hat{C}, \hat{A}_{21}]\hat{A}_{12}\right)$$
$$+ \tfrac{1}{2}w_{12}\left(\hat{A}_{12}[\hat{C}, \hat{A}_{21}] - [\hat{C}, \hat{A}_{12}]\hat{A}_{21}\right) + \hat{F}_C(t). \tag{5.160}$$

The corresponding master equation is then easily found to be

$$\frac{d\hat{\sigma}}{dt} = -\frac{i}{\hbar}[\hat{H}_{\text{sys}}, \hat{\sigma}] - \gamma'[[\hat{\sigma}, \hat{A}_{22}], \hat{A}_{22}] + \tfrac{1}{2}w_{21}\left([\hat{A}_{12}, \hat{\sigma}\hat{A}_{21}] - [\hat{A}_{21}, \hat{A}_{12}\hat{\sigma}]\right)$$
$$+ \tfrac{1}{2}w_{12}\left([\hat{A}_{21}, \hat{\sigma}\hat{A}_{12}] - [\hat{A}_{12}, \hat{A}_{21}\hat{\sigma}]\right). \tag{5.161}$$

5.4.2
Optical Bloch equations

If the radiation field to which a two-level system is coupled effectively acts as a reservoir, Eq. (5.151) [or Eq. (5.153)] applies directly. If, for example, a two-level system is (resonantly) driven by a monochromatic, (laser-like) coherent radiation field, the application of Eq. (5.151) (to describe the effects of coherent excitation) makes no sense, because its conditions of validity are obviously violated. Let us assume that the coupling of the radiation field to the two-level system is described by an interaction energy of the form (2.247) and that the spectral-mode density is sufficiently flat. In this case the two-level system is linearly coupled to a large sample of harmonic oscillators and the Langevin equation (5.150) may be more appropriate to the problem than is Eq. (5.151).

Let us suppose that the two-level system is coupled to both the radiation field and an (unspecified) radiationless reservoir with proper noise generators. We may write the interaction energy as

$$\hat{H}_{\text{int}} = \hat{H}_{\text{int}}^{(r)} + \hat{H}_{\text{int}}^{(nr)}, \tag{5.162}$$

$$\hat{H}_{\text{int}}^{(r)} = \hbar(\hat{V}_{21}^{(r)}\hat{A}_{21} + \text{H.c.}), \tag{5.163}$$

$$\hat{H}_{\text{int}}^{(nr)} = \hbar(\hat{V}_{21}^{(nr)}\hat{A}_{21} + \text{H.c.}) + \hbar(\hat{V}_{11}^{(nr)}\hat{A}_{11} + \hat{V}_{22}^{(nr)}\hat{A}_{22}). \tag{5.164}$$

Here (r) and (nr) indicate the radiative and nonradiative contributions, respectively. In particular, using Eq. (2.247), we may represent $\hat{V}_{21}^{(r)}$ as

$$\hat{V}_{21}^{(r)} = -\hbar^{-1}\mathbf{d}_{21}\hat{\mathbf{E}}^{(+)} \tag{5.165}$$

[recall that $\hat{\mathbf{E}}^{(+)} = \hat{\mathbf{E}}^{\perp(+)}$ in the absence of media]. Since the radiation field and the radiationless reservoir represent two systems which are independent of each other, they affect the Langevin equations additively. Treating the effect of the radiation field in the approximation of the Langevin equation (5.150) and that of the radiationless reservoir in the approximation of the Langevin equation (5.160), we easily see that the (combined) Langevin equation for any system operator $\hat{C}$ may be written in the form

$$\dot{\hat{C}} = -\frac{i}{\hbar}[\hat{C}, \hat{H}_{\text{sys}}] + \tfrac{1}{2}(\Gamma^{(r)} + w_{21}^{(nr)})(\hat{A}_{21}[\hat{C}, \hat{A}_{12}] - [\hat{C}, \hat{A}_{21}]\hat{A}_{12})$$
$$+ \tfrac{1}{2}w_{12}^{(nr)}(\hat{A}_{12}[\hat{C}, \hat{A}_{21}] - [\hat{C}, \hat{A}_{12}]\hat{A}_{21}) - \gamma'[[\hat{C}, \hat{A}_{22}], \hat{A}_{22}]$$
$$- i[\hat{C}, \hat{A}_{21}]\hat{F}_{21}^{(r)}(t) - i\hat{F}_{12}^{(r)}(t)[\hat{C}, \hat{A}_{12}] + \hat{F}_C^{(nr)}(t) \tag{5.166}$$

together with

$$\langle \hat{F}_C^{(nr)}(t)\rangle = 0 \tag{5.167}$$

and [cf. Eqs (5.145) and (5.165)]

$$\hat{F}_{21}^{(r)}(t) = -\hbar^{-1}\mathbf{d}_{21}\hat{\mathbf{E}}_{\text{free}}^{(+)}(t), \quad \hat{F}_{12}^{(r)}(t) = [\hat{F}_{21}^{(r)}(t)]^\dagger, \tag{5.168}$$

where $\hat{\mathbf{E}}_{\text{free}}^{(+)}(t)$ evolves freely. The radiative decay rate $\Gamma^{(r)}$, which is given by Eqs (5.148) and (5.147) with $\hat{F}_{21} = \hat{F}_{21}^{(r)}$ (and $\hat{F}_{12} = \hat{F}_{12}^{(r)}$) from Eq. (5.168), is the rate of spontaneous emission (Section 10.1). Note that the dephasing rate γ' can only arise from the radiationless reservoir.

From Eqs (5.166) and (5.167) the equation of motion for the expectation value of any system variable $\langle \hat{C}\rangle$ is then

$$\frac{d\langle\hat{C}\rangle}{dt} = -\frac{i}{\hbar}\langle[\hat{C}, \hat{H}_{\text{sys}}]\rangle$$
$$+ \tfrac{1}{2}(\Gamma^{(r)} + w_{21}^{(nr)})(\langle\hat{A}_{21}[\hat{C}, \hat{A}_{12}]\rangle - \langle[\hat{C}, \hat{A}_{21}]\hat{A}_{12}\rangle)$$
$$+ \tfrac{1}{2}w_{12}^{(nr)}(\langle\hat{A}_{12}[\hat{C}, \hat{A}_{21}]\rangle - \langle[\hat{C}, \hat{A}_{12}]\hat{A}_{21}\rangle) - \gamma'\langle[[\hat{C}, \hat{A}_{22}], \hat{A}_{22}]\rangle$$
$$- i\langle[\hat{C}, \hat{A}_{21}]\hat{F}_{21}^{(r)}(t)\rangle - i\langle\hat{F}_{12}^{(r)}(t)[\hat{C}, \hat{A}_{12}]\rangle, \tag{5.169}$$

which corresponds to the following master equation:[16]

$$\frac{d\hat{\sigma}}{dt} = -\frac{i}{\hbar}[\hat{H}_{\text{sys}}, \hat{\sigma}] + \tfrac{1}{2}(\Gamma^{(r)} + w_{21}^{(\text{nr})})([\hat{A}_{12}, \hat{\sigma}\hat{A}_{21}] - [\hat{A}_{21}, \hat{A}_{12}\hat{\sigma}])$$
$$+ \tfrac{1}{2}w_{12}^{(\text{nr})}([\hat{A}_{21}, \hat{\sigma}\hat{A}_{12}] - [\hat{A}_{12}, \hat{A}_{21}\hat{\sigma}]) - \gamma'[[\hat{\sigma}, \hat{A}_{22}], \hat{A}_{22}]$$
$$- i[\hat{A}_{21}, \hat{\sigma}\hat{F}_{21}^{(r)}(t)] - i[\hat{A}_{12}, \hat{F}_{12}^{(r)}(t)\hat{\sigma}]. \qquad (5.170)$$

Further manipulation of Eq. (5.169) or Eq. (5.170) requires knowledge of the radiation-field state. In particular, if the radiation field is in the vacuum state, $\langle \cdots \hat{F}_{21}^{(r)} \rangle = \langle \hat{F}_{12}^{(r)} \cdots \rangle = 0$, the last two terms of Eq. (5.169) disappear. Averaging Eq. (5.170) over the radiation field, the last two terms do not contribute and the result corresponds to the master equation (5.161) with $w_{21} = \Gamma^{(r)} + w_{21}^{(\text{nr})}$ and $w_{12} = w_{12}^{(\text{nr})}$.

Let us assume that the free (external) radiation field can be given in the form of

$$\hat{\mathbf{E}}_{\text{free}}^{(+)}(t) = \sum_\lambda i\omega_\lambda \mathbf{A}_\lambda(\mathbf{r})\hat{a}_\lambda e^{i\omega_\lambda(t-t')}, \qquad (5.171)$$

[cf. Eq. (2.70)], and its density operator (at a chosen initial time t') is

$$\hat{\sigma}^{(r)} = \int d^2\alpha\, P(\alpha)|\alpha\rangle\langle\alpha|, \qquad (5.172)$$

with $P(\alpha)$ being the multi-mode P function and $|\alpha\rangle$ the multi-mode coherent states. Here we have introduced the abbreviated notations

$$\alpha \equiv \{\alpha_\lambda\}, \quad d^2\alpha \equiv \prod_\lambda d^2\alpha_\lambda. \qquad (5.173)$$

Using Eq. (5.172), in Eq. (5.169) or Eq. (5.170) we may substitute, on recalling Eqs (5.168) and (5.171), the corresponding c numbers for the operators $\hat{F}_{21}^{(r)}(t)$ and $\hat{F}_{12}^{(r)}(t)$

$$F_{21}^{(r)}(t) = -\hbar^{-1}\mathbf{d}_{21} \sum_\lambda i\omega_\lambda \mathbf{A}_\lambda(\mathbf{r})\alpha_\lambda e^{-i\omega_\lambda(t-t')} \qquad (5.174)$$

and $F_{12}^{(r)}(t) = [F_{21}^{(r)}(t)]^*$ respectively, because $\hat{F}_{21}^{(r)}(t)|\alpha\rangle = F_{21}^{(r)}(t)|\alpha\rangle$. In this way we may represent the expectation value of any system operator, $\langle \hat{C}(t)\rangle$, as

$$\langle \hat{C}(t)\rangle = \int d^2\alpha\, P(\alpha)\langle \hat{C}(t;\alpha)\rangle, \qquad (5.175)$$

---

16) Note that $\hat{\sigma}$ still contains operators of the radiation field, and the average over the field must be performed to obtain the reduced density operator $\text{Tr}_{\text{rad}}\,\hat{\sigma}(t)$ of the two-level system.

where $\langle \hat{C}(t;\alpha) \rangle$ obeys the equation of motion (5.169) with the c-number functions $F_{21}^{(r)}(t)$ and $F_{12}^{(r)}(t)$ instead of the operator functions $\hat{F}_{21}^{(r)}(t)$ and $\hat{F}_{12}^{(r)}(t)$, respectively. With regard to Eq. (5.170), we may represent the reduced density operator of the two-level system as $[\mathrm{Tr}_{\mathrm{rad}}\,\hat{\sigma}(t) \mapsto \hat{\sigma}(t)]$

$$\hat{\sigma}(t) = \int d^2\alpha\, P(\alpha)\hat{\sigma}(t;\alpha). \tag{5.176}$$

Obviously, $\hat{\sigma}(t;\alpha)$ obeys the master equation (5.170) with the c-number functions $F_{21}^{(r)}(t)$ and $F_{12}^{(r)}(t)$ instead of the operator functions $\hat{F}_{21}^{(r)}(t)$ and $\hat{F}_{12}^{(r)}(t)$, respectively, and the expectation value $\langle \hat{C}(t;\alpha) \rangle$ can be calculated by means of $\hat{\sigma}(t;\alpha)$ in the usual way.

Let us now consider the master equation in the basis of the eigenkets $|m\rangle$ ($m=1,2$) of $\hat{H}_{\mathrm{sys}}$ [Eq. (5.141)]. From Eq. (5.170) we find the following system of coupled equations of motion for the density-matrix elements $\sigma_{mn}(t;\alpha)$:[17]

$$\dot{\sigma}_{22}(t;\alpha) = -(\Gamma^{(r)} + w_{21}^{(nr)})\sigma_{22}(t;\alpha) + w_{12}^{(nr)}\sigma_{11}(t;\alpha)$$
$$- iF_{21}^{(r)}(t)\sigma_{12}(t;\alpha) + iF_{12}^{(r)}(t)\sigma_{21}(t;\alpha), \tag{5.177}$$

$$\dot{\sigma}_{11}(t;\alpha) = -w_{12}^{(nr)}\sigma_{11}(t;\alpha) + (\Gamma^{(r)} + w_{21}^{(nr)})\sigma_{22}(t;\alpha)$$
$$+ iF_{21}^{(r)}(t)\sigma_{12}(t;\alpha) - iF_{12}^{(r)}(t)\sigma_{21}(t;\alpha), \tag{5.178}$$

$$\dot{\sigma}_{21}(t;\alpha) = -\left[i\omega_{21} + \tfrac{1}{2}(\Gamma^{(r)} + w_{21}^{(nr)} + w_{12}^{(nr)}) + \gamma'\right]\sigma_{21}(t;\alpha)$$
$$+ iF_{21}^{(r)}(t)[\sigma_{22}(t;\alpha) - \sigma_{11}(t;\alpha)], \tag{5.179}$$

$$\dot{\sigma}_{12}(t;\alpha) = \dot{\sigma}_{21}^*(t;\alpha). \tag{5.180}$$

After solving Eqs (5.177)–(5.180), the reduced density matrix elements $\sigma_{mn}(t)$ can be obtained, according to Eq. (5.176), by averaging the $\sigma_{mn}(t;\alpha)$ with $P(\alpha)$:[18]

$$\sigma_{mn}(t) = \int d^2\alpha\, P(\alpha)\sigma_{mn}(t;\alpha). \tag{5.181}$$

In the semi-classical limit when the (external) radiation field can be treated approximately as classical, the distribution $P(\alpha)$ may be viewed as an ordinary probability distribution for the (complex) field amplitudes α.

---

[17] Taking into account that $\langle \hat{A}_{mn} \rangle = \sigma_{nm}$, the density-matrix equations of motion can also be obtained from Eq. (5.169).

[18] Note that the solution of Eqs (5.177)–(5.180) is not trivial, because $F_{mn}^{(r)}(t)$ represents, in general, a set of (arbitrary) functions of time.

The equations of motion (5.177)–(5.180) are usually called the optical Bloch equations. They are often used in the form

$$\dot{n}(t;\alpha) = -(\Gamma^{(r)} + w_{21}^{(nr)} + w_{12}^{(nr)})[n(t;\alpha) - n_0]$$
$$- 2iF_{21}^{(r)}(t)\sigma_{12}(t;\alpha) + 2iF_{12}^{(r)}(t)\sigma_{21}(t;\alpha), \quad (5.182)$$

$$\dot{\sigma}_{21}(t;\alpha) = -[i\omega_{21} + \tfrac{1}{2}(\Gamma^{(r)} + w_{21}^{(nr)} + w_{12}^{(nr)}) + \gamma']\sigma_{21}(t;\alpha)$$
$$+ iF_{21}^{(r)}(t)n(t;\alpha), \quad (5.183)$$

$$\dot{\sigma}_{12}(t;\alpha) = \dot{\sigma}_{21}^*(t;\alpha). \quad (5.184)$$

Here the population inversion $n = \sigma_{22} - \sigma_{11}$ and the relation $1 = \sigma_{11} + \sigma_{22}$ has been used, so that

$$\sigma_{11} = \tfrac{1}{2}(1-n), \quad \sigma_{22} = \tfrac{1}{2}(1+n), \quad (5.185)$$

and

$$n_0 = -\frac{\Gamma^{(r)} + w_{21}^{(nr)} - w_{12}^{(nr)}}{\Gamma^{(r)} + w_{21}^{(nr)} + w_{12}^{(nr)}} = -1 + \frac{2w_{12}^{(nr)}}{\Gamma^{(r)} + w_{21}^{(nr)} + w_{12}^{(nr)}} \quad (5.186)$$

is the equilibrium inversion in the absence of the (external) radiation field.

The generalization to the case where more than two quantum states of the atomic system are involved in a (resonant) light–matter interaction process is straightforward. In particular, the overall rate of depopulation of the mth atomic level, $(\Gamma_{mm})_{\mathrm{dep}}$, is simply the sum over the depopulation rates (energy relaxation rates) for the allowed decay channels:

$$(\Gamma_{mm})_{\mathrm{dep}} = \sum_n{}' w_{mn}, \quad (5.187)$$

where w_{mn} is the probability per unit time for the transition $|m\rangle \to |n\rangle$; the notation $\sum'$ means that $n \neq m$. Accordingly, the overall filling rate of the mth state, $(\Gamma_{mm})_{\mathrm{fill}}$, is

$$(\Gamma_{mm})_{\mathrm{fill}} = \sum_n{}' w_{nm}. \quad (5.188)$$

In the case of a two-level system, from Eqs (5.177)–(5.180) these rates are seen to be

$$(\Gamma_{11})_{\mathrm{dep}} = w_{12}^{(nr)}, \quad (\Gamma_{22})_{\mathrm{dep}} = \Gamma^{(r)} + w_{21}^{(nr)}, \quad (5.189)$$

$$(\Gamma_{11})_{\mathrm{fill}} = \Gamma^{(r)} + w_{21}^{(nr)}, \quad (\Gamma_{22})_{\mathrm{fill}} = w_{12}^{(nr)}. \quad (5.190)$$

As already mentioned, the effect of radiative damping described by the rate $\Gamma^{(r)}$ corresponds to spontaneous emission (Section 10.1). Clearly, if the effect of the (external) radiation field on the two-level system were treated in a relaxational approximation [in the sense of Eq. (5.160)], the radiative effects of induced emission and absorption would appear and would also contribute to the depopulation and filling rates.

Generally, the damping (dephasing) rates $\Gamma_{mn} = \Gamma_{nm}$ of the off-diagonal density-matrix elements σ_{mn} ($m \neq n$) may be represented as

$$\Gamma_{mn} = \tfrac{1}{2} {\sum_k}' (w_{mk} + w_{nk}) + \gamma'_{mn}. \tag{5.191}$$

It is worth noting that each allowed (energy relaxation) transition from the states $|m\rangle$ and $|n\rangle$ to the states $|k\rangle$ ($|m\rangle \to |k\rangle$, $k \neq m$ and $|n\rangle \to |k\rangle$, $k \neq n$) gives rise to a contribution of half the transition rate to the overall dephasing rate Γ_{mn}. Deviations from the dephasing rate thus determined may result from pure phase relaxations (rates γ'_{mn}), which are typically observed in the case of a system embedded in a dense medium. In the case of the considered two-level system, from Eqs (5.177)–(5.180) we obtain

$$\Gamma_{12} = \tfrac{1}{2}(\Gamma^{(r)} + w_{21}^{(nr)} + w_{12}^{(nr)}) + \gamma'. \tag{5.192}$$

5.5
Quantum regression theorem

The description of quantum-statistical processes requires not only knowledge of the temporal evolution of the averages of the system variables, but also knowledge of their various multi-time correlations. To determine these, we can again start from the quantum Langevin equations. The Markov approximation used in the derivation of the Langevin equations will enable us to effectively reduce the problem of calculating (multi-time) correlation functions of system variables to that of calculating one-time averages [Lax (1967)].

Let us first consider a two-time correlation function of two system operators $\hat{C}_2$ and $\hat{C}_1$, $\langle \hat{C}_2(t_2) \hat{C}_1(t_1) \rangle$. Using Eq. (5.48), we may write

$$\begin{aligned}\frac{d\hat{C}_2(t_2)}{dt_2} &= -\frac{i}{\hbar}[\hat{C}_2(t_2), \hat{H}_{sys}(t_2)] \\ &\quad - \sum_{i,j} \Delta(\omega_i + \omega_j)\{\Gamma_{ij}^+[\hat{C}_2(t_2), \hat{A}_i(t_2)]\hat{A}_j(t_2) \\ &\quad - \Gamma_{ji}^- \hat{A}_j(t_2)[\hat{C}_2(t_2), \hat{A}_i(t_2)]\} + \hat{F}_{C_2}(t_2),\end{aligned} \tag{5.193}$$

with $\langle \hat{F}_{C_2}(t_2) \rangle = 0$. At this point we recall that, in the Markov approximation, the system variables at time t_1 and noise variables at time t_2 may be assumed to be uncorrelated with each other provided that $t_2 > t_1$ ($t_2 - t_1 \gg \tau_c$); in particular,

$$\langle \hat{F}_{C_2}(t_2) \hat{C}_1(t_1) \rangle = 0 \quad \text{if} \quad t_2 > t_1 \tag{5.194}$$

[cf. the derivation of Eq. (5.48)]. Multiplying Eq. (5.193) by $\hat{C}_1(t_1)$ and using Eq. (5.194), we can readily see that the correlation function $\langle \hat{C}_2(t_2)\hat{C}_1(t_1) \rangle$ obeys the equation of motion

$$\frac{\partial}{\partial t_2} \langle \hat{C}_2(t_2)\hat{C}_1(t_1) \rangle = -\frac{i}{\hbar} \langle [\hat{C}_2(t_2), \hat{H}_{\text{sys}}(t_2)]\hat{C}_1(t_1) \rangle$$
$$- \sum_{i,j} \Delta(\omega_i + \omega_j)\{\Gamma_{ij}^+ \langle [\hat{C}_2(t_2), \hat{A}_i(t_2)]\hat{A}_j(t_2)\hat{C}_1(t_1) \rangle$$
$$- \Gamma_{ji}^- \langle \hat{A}_j(t_2)[\hat{C}_2(t_2), \hat{A}_i(t_2)]\hat{C}_1(t_1) \rangle \} \qquad (5.195)$$

$(t_2 > t_1)$. If $t_1 > t_2$, the equation of motion $\partial \langle \hat{C}_2(t_2)\hat{C}_1(t_1) \rangle / \partial t_1 = \ldots$ is found analogously. Note that Eq. (5.195) effectively represents a set of coupled equations for two-time correlation functions of system variables, because the correlation functions $\langle [\hat{C}_2(t_2), \hat{H}_{\text{sys}}(t_2)]\hat{C}_1(t_1) \rangle$, $\langle [\hat{C}_2(t_2), \hat{A}_i(t_2)]\hat{A}_j(t_2)\hat{C}_1(t_1) \rangle$ and $\langle \hat{A}_j(t_2)[\hat{C}_2(t_2), \hat{A}_i(t_2)]\hat{C}_1(t_1) \rangle$ on the right-hand side also obey equations of motion of the same type, and so on.

Although the mean value $\langle \hat{C}_2(t_2) \rangle$ and, with regard to t_2, the correlation function $\langle \hat{C}_2(t_2)\hat{C}_1(t_1) \rangle$ obey equations of motion of the same type, the relevant initial conditions are in general different. Introducing the system operator

$$\hat{C}(t_1) = \hat{C}_2(t_2)\hat{C}_1(t_1)|_{t_2=t_1} \equiv \lim_{t_2 \to t_1^+} \hat{C}_2(t_2)\hat{C}_1(t_1), \qquad (5.196)$$

the initial condition for $\langle \hat{C}_2(t_2)\hat{C}_1(t_1) \rangle$, namely

$$\langle \hat{C}_2(t_2)\hat{C}_1(t_1) \rangle|_{t_2=t_1} = \langle \hat{C}(t_1) \rangle, \qquad (5.197)$$

must be determined by solving the equation of motion (5.50) for $\langle \hat{C}(t_1) \rangle$ with appropriately chosen initial value at time $t_1 = t'$:

$$\frac{d\langle \hat{C} \rangle}{dt_1} = -\frac{i}{\hbar} \langle [\hat{C}, \hat{H}_{\text{sys}}] \rangle$$
$$- \sum_{i,j} \Delta(\omega_i + \omega_j)(\Gamma_{ij}^+ \langle [\hat{C}, \hat{A}_i]\hat{A}_j \rangle - \Gamma_{ji}^- \langle \hat{A}_j[\hat{C}, \hat{A}_i] \rangle). \qquad (5.198)$$

To formulate the above results in the language of master equations, we proceed in close analogy with Section 5.2.1. Going from the Heisenberg to the Schrödinger picture, we may write

$$\langle \hat{C}_2(t_2)\hat{C}_1(t_1) \rangle = \text{Tr}_{\text{sys}}[\hat{\sigma}(t')\hat{C}_2(t_2)\hat{C}_1(t_1)] = \text{Tr}_{\text{sys}}[\hat{\sigma}^{(2)}(t_2)\hat{C}_2(t')], \qquad (5.199)$$

where

$$\hat{\sigma}^{(2)}(t_2) = \text{Tr}_{\text{res}}\, \hat{\varrho}^{(2)}(t_2), \qquad (5.200)$$
$$\hat{\varrho}^{(2)}(t_2) = \hat{U}(t_2, t_1)\hat{C}_1(t')\hat{\varrho}(t_1)\hat{U}^\dagger(t_2, t_1). \qquad (5.201)$$

Here $\hat{U}(t_2, t_1)$ is the time-evolution operator and $\hat{\varrho}(t_1)$ is the ordinary Schrödinger density operator at time t_1 [cf. Eq. (5.54)]. Note that

$$\hat{\sigma}^{(2)}(t_2)\big|_{t_2=t_1} = \hat{C}_1(t')\hat{\sigma}(t_1). \tag{5.202}$$

Substituting into

$$\frac{\partial}{\partial t_2}\langle \hat{C}_2(t_2)\hat{C}_1(t_1)\rangle = \mathrm{Tr}_{\mathrm{sys}}\left[\frac{d\hat{\sigma}^{(2)}(t_2)}{dt_2}\hat{C}_2(t')\right] \tag{5.203}$$

for $\partial\langle \hat{C}_2(t_2)\hat{C}_1(t_1)\rangle/\partial t_2$ the result (5.195) and using the cyclic properties of the trace, we can readily prove that $\hat{\sigma}^{(2)}(t_2)$ $(t_2>t_1)$ obeys the master equation (5.59):

$$\frac{d\hat{\sigma}^{(2)}}{dt_2} = -\frac{i}{\hbar}[\hat{H}_{\mathrm{sys}}, \hat{\sigma}^{(2)}]$$
$$-\sum_{i,j}\Delta(\omega_i+\omega_j)(\Gamma^+_{ij}[\hat{A}_i, \hat{A}_j\hat{\sigma}^{(2)}] - \Gamma^-_{ji}[\hat{A}_i, \hat{\sigma}^{(2)}\hat{A}_j]), \tag{5.204}$$

but with the initial condition

$$\hat{\sigma}^{(2)}(t_2)\big|_{t_2=t_1} = \hat{C}_1\hat{\sigma}(t_1). \tag{5.205}$$

Note that $\hat{\sigma}(t_1)$ is the ordinary system-density operator which obeys the above master equation with appropriately chosen initial values at the chosen initial time t' $(t'<t_1)$.

The extension of the above results to time-ordered correlation functions of higher order,

$$G^{(n)}(t_n, t_{n-1}\ldots, t_1) = \langle \hat{C}_n(t_n)\hat{C}_{n-1}(t_{n-1})\cdots\hat{C}_1(t_1)\rangle \tag{5.206}$$

$(t_n > t_{n-1} > \cdots > t_1 > t')$, is straightforward. Defining

$$\hat{\sigma}^{(n)}(t_n) = \mathrm{Tr}_{\mathrm{res}}\,\hat{\varrho}^{(n)}(t_n), \tag{5.207}$$
$$\hat{\varrho}^{(n)}(t_n) = \hat{U}(t_n, t_{n-1})\hat{C}_{n-1}\hat{\varrho}^{(n-1)}(t_{n-1})\hat{U}^\dagger(t_n, t_{n-1}), \tag{5.208}$$

one can prove that

$$G^{(n)}(t_n, t_{n-1}\ldots, t_1) = \mathrm{Tr}_{\mathrm{sys}}[\hat{\sigma}^{(n)}(t_n)\hat{C}_n]. \tag{5.209}$$

The determination of $\hat{\sigma}^{(n)}(t_n)$ requires the calculation of $\hat{\sigma}^{(n-1)}(t_{n-1})$, the determination of $\hat{\sigma}^{(n-1)}(t_{n-1})$ requires the calculation of $\hat{\sigma}^{(n-2)}(t_{n-2})$, and so on, where $\hat{\sigma}^{(k)}(t_k)$, $k=1,2,\ldots,n$, $(t_k>t_{k-1})$ obeys the master equation (5.59):

$$\frac{d\hat{\sigma}^{(k)}}{dt_k} = -\frac{i}{\hbar}[\hat{H}_{\text{sys}}, \hat{\sigma}^{(k)}] \\ - \sum_{i,j} \Delta(\omega_i + \omega_j)(\Gamma_{ij}^+[\hat{A}_i, \hat{A}_j\hat{\sigma}^{(k)}] - \Gamma_{ji}^-[\hat{A}_i, \hat{\sigma}^{(k)}\hat{A}_j]), \tag{5.210}$$

with the initial condition

$$\hat{\sigma}^{(k)}(t_k)\big|_{t_k=t_{k-1}} = \hat{C}_{k-1}\hat{\sigma}^{(k-1)}(t_{k-1}). \tag{5.211}$$

Recall that $\hat{\sigma}^{(1)}(t_1) \equiv \hat{\sigma}(t_1)$ is the ordinary (physical) density operator of the system, and the kth order correlation function $G^{(k)}$ is, according to Eq. (5.209), just determined by $\hat{\sigma}^{(k)}$. Although the equation of motion for $\hat{\sigma}^{(k)}$ agrees with that for the density operator $\hat{\sigma}$, $\hat{\sigma}^{(k)}$ does not, in general, have the typical properties of a density operator, such as $\text{Tr}_{\text{sys}}\hat{\sigma} = 1$. In practice, one has to solve the master equation under the most general initial conditions and specify the solution step by step according to the initial conditions (5.211), by starting from an appropriately chosen (physical) system-density operator at the initial time t'. The content of Eqs (5.209)–(5.211) is usually called the quantum regression theorem.

References

Diestler, D.J. (1976) *Chem. Phys. Lett.* **39**, 37.

Fajn, V.M. and J.I. Chanin (1969) *Quantenelektronik* (Teubner, Leipzig).

Gardiner, C.W. (1991) *Quantum Noise* (Springer-Verlag, Berlin).

Gordon, J.P. (1967) *Phys. Rev.* **161**, 367.

Lax, M. (1966) *Phys. Rev.* **145**, 110.

Lax, M. (1967) *Phys. Rev.* **157**, 213.

Lax, M. (1968) *Phys. Rev.* **172**, 350.

Lax, M. and H. Yuen (1968) *Phys. Rev.* **172**, 362.

Louisell, W.H. (1973) *Quantum Statistical Properties of Radiation* (Wiley, New York).

Louisell, W.H. and J. H. Marburger (1968) *Phys. Rev.* **186**, 174.

Montroll, E.W. and K.E. Shuler (1957) *J. Chem. Phys.* **26**, 454.

Paerschke, H., K.-E. Süsse and D.-G. Welsch (1980) *Phys. Stat. Sol. (b)* **98**, 253.

Weisskopf, V.G. and E. Wigner (1932) *Z. Physik* **63**, 54.

Zwanzig, R. (1960) *J. Chem. Phys.* **33**, 1338.

6
Photoelectric detection of light

To study the properties of light experimentally, detectors operating on the basis of the (internal) photoelectric effect are usually illuminated to produce an electric output signal, such as the number of emitted photoelectrons or the photocurrent. The question arises as to what properties of light are measurable in this way and how the measured properties of the photoelectric output signal of such a detector can be related to the properties of the light in the input channel.

The photons monitored are of course annihilated. If the light to be detected is required to be available (with the true quantum statistics of certain observables) after the detection, a so-called quantum nondemolition (QND) measurement is required, which does not modify the measured quantity. As is well known, during a measurement the quantum state is changed and thus the statistics of some observables are also changed. In a quantum nondemolition measurement the measurement-assisted perturbation of the system does not affect the observable which is desired to be determined, but is confined to other quantities [Landau and Peierls (1931); Braginsky, Vorontsov and Khalili (1977); Thorne, Drever, Caves, Zimmermann and Sandberg (1978); Thorne, Caves, Sandberg, Zimmermann and Drever (1979); Caves, Thorne, Drever, Sandberg and Zimmermann (1980); Caves (1983); see also the example in Section 12.6.2].

6.1
Photoelectric counting

Let us consider, as a simple model of a photodetector, a large sample of N atomic systems (light-absorbing centers) that are capable of absorbing light through the photoemission of electrons in a certain time interval t, $t + \Delta t$, see Fig. 6.1. Clearly, the statistical distribution of the observed photoelectric counts is expected to be closely related to the quantum-statistical properties of the incident light. In this way, photoelectric counting experiments have been successfully used to study various properties of light.

In the following we suppose that the total number N of atomic systems is large compared with the mean number $\bar{n}$ of emitted electrons: $\bar{n} \ll N$. Thus

Quantum Optics, Third, revised and extended edition. Werner Vogel and Dirk-Gunnar Welsch
Copyright © 2006 WILEY-VCH Verlag GmbH & Co. KGaA, Weinheim
ISBN: 3-527-40507-0

6 Photoelectric detection of light

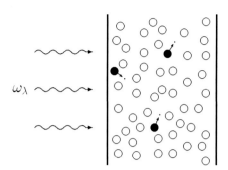

Fig. 6.1 Model of a photodetector. A sample of atomic systems (empty circles) is illuminated by light (frequency ω_λ). During a certain time interval a few of these centers (filled circles) eject photoelectrons (dots) due to the absorption of light quanta.

we may assume that each atomic system which is involved in the process of photoemission of electrons emits not more than one electron. Under these assumptions the main features of the theory may be developed by applying Dirac's perturbation theory to the basic process of light absorption and combining the corresponding results with methods of classical statistics with respect to the ensemble of photoelectrons generated by the absorption processes [Mandel (1958, 1959, 1963); Kelley and Kleiner (1964); Glauber (1965, 1966, 1972); Lax and Zwanziger (1973)].

6.1.1
Quantum-mechanical transition probabilities

To calculate the quantum-mechanical transition probabilities of photoemission of electrons in a chosen time interval $t, t + \Delta t$, we start from the Hamiltonian

$$\hat{H} = \hat{H}_0 + \hat{H}_{\text{int}}, \tag{6.1}$$

where $\hat{H}_0$ is assumed to be composed of the Hamiltonian of the radiation field together with the sources to which the radiation field is attributed (including the corresponding coupling term) and the Hamiltonian of the atomic systems of the photodetector. If the wave functions of different detector atoms may be assumed not to overlap, the coupling of the radiation field to the detector atoms may be written as

$$\hat{H}_{\text{int}} = \sum_{i=1}^{N} \hat{H}_{\text{int}}^{(i)}, \tag{6.2}$$

where $\hat{H}_{\text{int}}^{(i)}$ describes the interaction of the ith atomic system with the radiation field.

In the interaction picture the temporal evolution of the overall system may be calculated from the density-operator equation of motion

$$i\hbar \frac{d\hat{\sigma}(\tau)}{d\tau} = [\hat{H}_{int}(\tau), \hat{\sigma}(\tau)], \tag{6.3}$$

where the interaction Hamiltonian in this picture reads

$$\hat{H}_{int}(\tau) = \exp\left[\frac{i}{\hbar}\hat{H}_0(\tau - t)\right] \hat{H}_{int} \exp\left[-\frac{i}{\hbar}\hat{H}_0(\tau - t)\right], \tag{6.4}$$

with t denoting the initial time. By means of the ansatz

$$\hat{\sigma}(\tau) = \hat{U}(\tau, t)\hat{\sigma}(t)\hat{U}^\dagger(\tau, t), \tag{6.5}$$

from Eq. (6.3) the time-evolution operator $\hat{U}(\tau, t)$ is seen to satisfy the equation of motion

$$i\hbar \frac{d\hat{U}(\tau, t)}{d\tau} = \hat{H}_{int}(\tau)\hat{U}(\tau, t) \tag{6.6}$$

together with the initial condition

$$\hat{U}(t, t) = 1, \tag{6.7}$$

and the formal solution of Eq. (6.6) reads

$$\hat{U}(t + \Delta t, t) = \mathcal{T}_+ \exp\left[-\frac{i}{\hbar}\int_t^{t+\Delta t} d\tau\, \hat{H}_{int}(\tau)\right]. \tag{6.8}$$

In the following we assume that the photodetector system is in its ground state $|G\rangle$ at the (initial) time t and that the overall density operator is initially factorized as

$$\hat{\sigma}(t) = |G\rangle\langle G|\hat{\varrho}, \tag{6.9}$$

where $\hat{\varrho}$ is the density operator (at time t) of the radiation field with the sources included. The probability of a photoelectric transition from the ground state of the photodetector system, $|G\rangle$, at time t to the continuum of final states $|F\rangle$ at time $t+\Delta t$ may then be written as

$$P_{\{F\}}(t, \Delta t) = \sum_{\{F\}} \text{Tr}\{\hat{\varrho}\langle G|\hat{U}^\dagger(t + \Delta t, t)|F\rangle\langle F|\hat{U}(t + \Delta t, t)|G\rangle\}. \tag{6.10}$$

We now ask for the probability of the photoemission of m electrons from a sub-ensemble of m atomic systems (with one electron each) of the overall N-atom system. The state vectors $|G\rangle$ and $|F\rangle$, which are products of state vectors of the individual atoms, may be given in the form

$$|G\rangle = \prod_{l=1}^{m} |g_{i_l}\rangle, \quad |F\rangle = \prod_{l=1}^{m} |f_{i_l}\rangle. \tag{6.11}$$

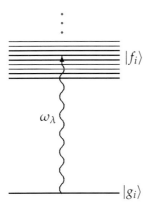

Fig. 6.2 Level scheme of a light-absorbing atomic center. The relevant frequencies ω_λ of the radiation field are in resonance with the transition from the atomic ground state $|g_i\rangle$ to the manifold of (final) excited states $|f_i\rangle$.

Here the state vectors of the ith atomic system are denoted by $|g_i\rangle$ and $|f_i\rangle$, where the $|f_i\rangle$ are assumed to represent a manifold of (quasi-continuum) states, cf. Fig. 6.2. The set of indices $\{i_l\}$ stands for the sub-ensemble considered and $|F\rangle\langle F|$ projects onto the Hilbert space of that ensemble. By calculating the probability in lowest-order perturbation theory, we may approximate the time-evolution operator (6.8) by the mth-order expansion term:

$$\hat{U}(t+\Delta t, t) = \frac{1}{m!} T_+ \left[-\frac{i}{\hbar} \int_t^{t+\Delta t} d\tau \sum_{l=1}^m \hat{H}_{int}^{(i_l)}(\tau) \right]^m. \tag{6.12}$$

Combining Eqs (6.10)–(6.12) then yields $[p_{\{F\}}(t, \Delta t) \mapsto p_{i_1,\ldots,i_m}(t, \Delta t)]$

$$p_{i_1,\ldots,i_m}(t, \Delta t) = \sum_{f_{i_1}, f_{i_2}, \ldots, f_{i_m}} \mathrm{Tr}\Bigg\{ \hat{\varrho} \Bigg[T_- \prod_{l=1}^m \frac{i}{\hbar} \int_t^{t+\Delta t} d\tau_l \, \langle g_{i_l} | \hat{H}_{int}^{(i_l)}(\tau_l) | f_{i_l} \rangle$$

$$\times \Bigg[T_+ \prod_{l=1}^m \left(-\frac{i}{\hbar}\right) \int_t^{t+\Delta t} d\tau_l \, \langle f_{i_l} | \hat{H}_{int}^{(i_l)}(\tau_l) | g_{i_l} \rangle \Bigg] \Bigg\}. \tag{6.13}$$

Treating the coupling of the radiation to the ith atomic system (at position $\mathbf{r}_i$) in the electric-dipole approximation and the rotating-wave approximation, from Eq. (2.247) we find that[1]

$$\langle f_i | \hat{H}_{int}^{(i)}(\tau) | g_i \rangle = -\exp\left[i\omega_{f_i g_i}(\tau - t) \right] \mathbf{d}_{f_i g_i} \hat{\mathbf{E}}^{(+)}(\mathbf{r}_i, \tau). \tag{6.14}$$

---

[1] Note that $\hat{\mathbf{E}}^{(\pm)} = (\hat{\mathbf{E}}^{(\pm)})^\perp$. For notational convenience we omit the superscript $\perp$.

In this manner, we derive

$$\sum_{f_i}\langle g_i|\hat{H}_{\mathrm{int}}^{(i)}(\tau)|f_i\rangle\langle f_i|\hat{H}_{\mathrm{int}}^{(i)}(\tau')|g_i\rangle = \hat{\mathbf{E}}^{(-)}(\mathbf{r}_i,\tau)\mathbf{S}^{(i)}(\tau-\tau')\hat{\mathbf{E}}^{(+)}(\mathbf{r}_i,\tau'), \quad (6.15)$$

where

$$\mathbf{S}^{(i)}(\tau-\tau') = \sum_{f_i} \mathbf{d}_{g_i f_i} \otimes \mathbf{d}_{f_i g_i} \exp[-i\omega_{f_i g_i}(\tau-\tau')]. \quad (6.16)$$

Since the sum in Eq. (6.16) involves summation over all possible orientations of the electric transition-dipole moments, $\mathbf{S}^{(i)}$ may be assumed to reduce to

$$\mathbf{S}^{(i)} = I S^{(i)}(\tau - \tau'). \quad (6.17)$$

Inserting Eq. (6.15) together with Eq. (6.17) into Eq. (6.13), we may write

$$p_{i_1,\ldots,i_m}(t,\Delta t)$$
$$= \left\langle {}^{\circ}_{\circ}\prod_{l=1}^{m}\int_{t}^{t+\Delta t}d\tau_l \int_{t}^{t+\Delta t}d\tau'_l\, S^{(i_l)}(\tau_l-\tau'_l)\hat{\mathbf{E}}^{(-)}(\mathbf{r}_{i_l},\tau_l)\hat{\mathbf{E}}^{(+)}(\mathbf{r}_{i_l},\tau'_l) {}^{\circ}_{\circ}\right\rangle, \quad (6.18)$$

where $\langle\cdots\rangle \equiv \mathrm{Tr}(\hat{\varrho}\ldots)$. Recall that the ${}^{\circ}_{\circ}\ {}^{\circ}_{\circ}$ notation indicates that field operators are to be written in normal order (creation operators to the left of annihilation operators) and time order (time arguments increasing to the right in products of creation operators and to the left in products of annihilation operators). Note that the operator ordering must of course be performed before evaluating the time integrals. Obviously, $p_{i_1,\ldots,i_m}(t,\Delta t)$ may be regarded as the joint probability for photoemission of m electrons from m atomic systems (situated at positions $\mathbf{r}_{i_l}$, $l=1,2,\ldots,m$) in the time interval t, $t+\Delta t$, each atomic subsystem contributing a single electron.

Now let us suppose that the atoms belong to different photodetectors operating during appropriately chosen time intervals t_{i_l}, $t_{i_l}+\Delta t_{i_l}$ ($l=1,2,\ldots,m$) and we are interested in the joint probability $p_{i_1,\ldots,i_m}(t_{i_1},\Delta t_{i_1},\ldots,t_{i_m},\Delta t_{i_m})$ of photoemission of an electron from the i_1th atom in the time interval t_{i_1}, $t_{i_1}+\Delta t_{i_1}$, an electron from the i_2th atom in the time interval t_{i_2}, $t_{i_2}+\Delta t_{i_2}$, and so forth, so that altogether m electrons are emitted. To treat this case, it is convenient to introduce in the interaction Hamiltonian (6.2) switching-operation functions explicitly indicating the time intervals of interaction of the various atomic systems with the light to be detected:

$$\hat{H}_{\mathrm{int}} \mapsto \hat{H}_{\mathrm{int}}(\tau) = \sum_{i=1}^{N} \hat{H}_{\mathrm{int}}^{(i)}(\tau), \quad (6.19)$$

where

$$\hat{H}_{\mathrm{int}}^{(i)}(\tau) = \Theta(\tau - t_i)\Theta(t_i + \Delta t_i - \tau)\hat{H}_{\mathrm{int}}^{(i)}. \quad (6.20)$$

Further calculations can then be performed analogously to that leading to Eq. (6.18). The result is

$$p_{i_1,\ldots,i_m}(t_{i_1}, \Delta t_{i_1}, \ldots, t_{i_m}, \Delta t_{i_m})$$
$$= \left\langle {}^\circ_\circ \prod_{l=1}^{m} \int_{t_{i_l}}^{t_{i_l}+\Delta t_{i_l}} d\tau_l \int_{t_{i_l}}^{t_{i_l}+\Delta t_{i_l}} d\tau_l' \, S^{(i_l)}(\tau_l - \tau_l') \hat{E}^{(-)}(\mathbf{r}_{i_l}, \tau_l) \hat{E}^{(+)}(\mathbf{r}_{i_l}, \tau_l') {}^\circ_\circ \right\rangle, \quad (6.21)$$

which is the natural generalization of Eq. (6.18). In the particular case where the time intervals are the same ($t_{i_l} = t$, $\Delta t_{i_l} = \Delta t$, $l = 1, 2, \ldots, m$), Eq. (6.21) reduces to Eq. (6.18).

In many cases of practical interest one deals with so-called broad-band photodetectors, which means that

$$S^{(i)}(\tau - \tau') \approx S^{(i)} \delta(\tau - \tau'). \quad (6.22)$$

To clarify this point, let us consider the probability of photoemission of a single electron by a chosen atomic system. Specifying Eq. (6.21) yields[2]

$$p_i(t, \Delta t) = \int_t^{t+\Delta t} d\tau \int_t^{t+\Delta t} d\tau' \, S^{(i)}(\tau - \tau') \langle \hat{E}^{(-)}(\mathbf{r}_i, \tau) \hat{E}^{(+)}(\mathbf{r}_i, \tau') \rangle. \quad (6.23)$$

Introducing the Fourier transformation

$$S^{(i)}(\tau - \tau') = \frac{1}{2\pi} \int d\omega \, e^{-i\omega(\tau - \tau')} \underline{S}^{(i)}(\omega), \quad (6.24)$$

we may represent Eq. (6.23) as

$$p_i(t, \Delta t) = \frac{1}{2\pi} \int d\omega \, \underline{S}^{(i)}(\omega) \underline{K}(\mathbf{r}_i, \omega, t, \Delta t), \quad (6.25)$$

where

$$\underline{K}(\mathbf{r}_i, \omega, t, \Delta t) = \int d\tau \int d\tau' \left[e^{-i\omega(\tau-\tau')} \right.$$
$$\left. \times \Theta(\tau - t)\Theta(t + \Delta t - \tau)\Theta(\tau' - t)\Theta(t + \Delta t - \tau') \langle \hat{E}^{(-)}(\mathbf{r}_i, \tau) \hat{E}^{(+)}(\mathbf{r}_i, \tau') \rangle \right]. \quad (6.26)$$

Let us now assume that in the relevant frequency range of $\underline{K}(\mathbf{r}_i, \omega, t, \Delta t)$ the spectral response function $\underline{S}^{(i)}(\omega)$ is slowly varying. In other words, we assume that the bandwidth $\Delta \omega_S$ of $\underline{S}^{(i)}(\omega)$ is large compared with the bandwidth $\Delta \omega_K$ of $\underline{K}(\mathbf{r}, \omega, t, \Delta t)$:

$$\Delta \omega_S \gg \Delta \omega_K, \quad (6.27)$$

[2] The ${}^\circ_\circ\,{}^\circ_\circ$ ordering prescription is superfluous here since only one positive (negative) frequency operator occurs.

where $\Delta\omega_K$ may be estimated as

$$\Delta\omega_K \approx \max\{\Delta\omega_R, (\Delta t)^{-1}\}, \tag{6.28}$$

with $\Delta\omega_R$ being the bandwidth of the radiation under study. Clearly, if the condition (6.27) is satisfied (broad-band photodetector), in Eq. (6.25) the spectral response function $\underline{S}^{(i)}(\omega)$ may be regarded as approximately independent of ω [$\underline{S}^{(i)}(\omega) \simeq \underline{S}^{(i)}(\omega_0) \equiv \underline{S}^{(i)}$, ω_0 being an appropriately chosen mid-band frequency], so that it can be taken outside the integral. Now, it is easily seen that this approximation just corresponds to substituting the result (6.22) into Eq. (6.23) for $S^{(i)}(\tau - \tau')$:

$$p_i(t, \Delta t) = \underline{S}^{(i)} \int_t^{t+\Delta t} d\tau \, \langle \hat{\mathbf{E}}^{(-)}(\mathbf{r}_i, \tau) \hat{\mathbf{E}}^{(+)}(\mathbf{r}_i, \tau) \rangle. \tag{6.29}$$

6.1.2
Photoelectric counting probabilities

We now turn to the problem of determining the probability $P_m(t, \Delta t)$ of emission of m photoelectrons in the time interval $t, t + \Delta t$ within the framework of classical probability theory.[3] Applying Bernoulli's scheme, we introduce (independent) random variables n_i ($i = 1, 2, \ldots, N$), each of which has two realizations $n_i = 0, 1$, where $n_i = 1$ if in the time interval $t, t + \Delta t$ the ith atomic system contributes an electron to the electrons ejected by the overall system, and $n_i = 0$ otherwise. The random variable for the total number of photoelectrons emitted during the time interval $t, t + \Delta t$ may then be introduced as

$$n = \sum_{i=1}^{N} n_i. \tag{6.30}$$

The probability $P_m(t, \Delta t)$ may be derived from the characteristic (generating) function[4]

$$y(x, t, \Delta t) = \overline{(1+x)^n} = \sum_{m=0}^{\infty} P_m(t, \Delta t)(1+x)^m, \tag{6.31}$$

from which the relation

$$P_m(t, \Delta t) = \frac{1}{m!} \frac{\partial^m}{\partial x^m} y(x, t, \Delta t) \bigg|_{x=-1} \tag{6.32}$$

---

[3] For methods of statistics see, e. g., van Kampen (1981) or Gardiner (1983).
[4] Here and in the following, over-bars indicate classical statistical averaging.

is seen to be valid. To calculate the characteristic function $y(x, t, \Delta t)$ we expand it in a Taylor series:

$$y(x, t, \Delta t) = \sum_{m=0}^{\infty} \frac{1}{m!} F_m(t, \Delta t) x^m, \qquad (6.33)$$

where

$$F_m(t, \Delta t) = \left. \frac{\partial^m}{\partial x^m} y(x, t, \Delta t) \right|_{x=0}. \qquad (6.34)$$

Combining Eqs (6.34) and (6.31) obviously yields

$$F_m(t, \Delta t) = \overline{\prod_{j=1}^{m} [n - (j-1)]}; \qquad (6.35)$$

that is, the F_m are the factorial moments of the random number n of photoelectrons. Recalling Eq. (6.30) and taking into account that $n_i = n_i^2$, we may rewrite Eq. (6.35) as ($m < N$)

$$F_m(t, \Delta t) = {\sum_{i_1,\ldots,i_m}}' \overline{n_{i_1} \cdots n_{i_m}} = {\sum_{i_1,\ldots,i_m}}' p_{i_1,\ldots,i_m}(t, \Delta t), \qquad (6.36)$$

where the notation $\sum'$ implies excluding from the summation terms with equal indices i_j.

The $p_{i_1,\ldots,i_m}(t, \Delta t)$ may be identified with the (quantum-mechanical) joint probabilities as given in Eq. (6.18). We thus may write, with little error,[5]

$$F_m(t, \Delta t) = \langle \, \substack{\circ \\ \circ} \, [\hat{\Gamma}(t, \Delta t)]^m \, \substack{\circ \\ \circ} \, \rangle, \qquad (6.37)$$

where

$$\hat{\Gamma}(t, \Delta t) = \sum_{i=1}^{N} \int_t^{t+\Delta t} d\tau \int_t^{t+\Delta t} d\tau' \, S^{(i)}(\tau - \tau') \hat{E}^{(-)}(\mathbf{r}_i, \tau) \hat{E}^{(+)}(\mathbf{r}_i, \tau'), \qquad (6.38)$$

which for a broad-band photodetector reduces to

$$\hat{\Gamma}(t, \Delta t) = \sum_{i=1}^{N} S^{(i)} \int_t^{t+\Delta t} d\tau \, \hat{E}^{(-)}(\mathbf{r}_i, \tau) \hat{E}^{(+)}(\mathbf{r}_i, \tau) \qquad (6.39)$$

[see Eq. (6.22)]. When the photosensitive centers are localized in a sufficiently small range of space whose linear dimensions are small compared with those of the slowly varying amplitude of light, the detector is said to be point-like.

---

5) Note that, with regard to Eq. (6.36) (primed sum $\sum'$), the error made in Eq. (6.37) together with Eq. (6.38) is of the order of magnitude m/N and vanishes for $N \to \infty$.

In this case, the expression on the right-hand side of Eq. (6.39) becomes proportional to the time-integrated intensity of the radiation field at the position of detection **r**:

$$\hat{\Gamma}(t, \Delta t) = \tilde{\zeta}\hat{I}(\mathbf{r}, t, \Delta t), \tag{6.40}$$

$$\hat{I}(\mathbf{r}, t, \Delta t) = \int_{t}^{t+\Delta t} d\tau\, \hat{I}(\mathbf{r}, \tau), \tag{6.41}$$

$$\hat{I}(\mathbf{r}, t) = \hat{\mathbf{E}}^{(-)}(\mathbf{r}, t)\hat{\mathbf{E}}^{(+)}(\mathbf{r}, \tau). \tag{6.42}$$

The proportional factor $\tilde{\zeta}$, which for equal detector atoms ($S \equiv S^{(i)} = S^{(j)}$) is given by $\tilde{\zeta} = NS$, is also called the detection efficiency.

Substituting into Eq. (6.33) for $F_m(t, \Delta t)$ the expression given in Eq. (6.37), we may represent the characteristic function $y(x, t, \Delta t)$ in the form ($N \to \infty$)

$$y(x, t, \Delta t) = \left\langle : \sum_{m=0}^{\infty} \frac{1}{m!}[\hat{\Gamma}(t, \Delta t)x]^m : \right\rangle = \langle : \exp[\hat{\Gamma}(t, \Delta t)x] : \rangle. \tag{6.43}$$

Recalling Eq. (6.32), from Eq. (6.43) we can easily derive the sought photoelectric counting probabilities

$$P_m(t, \Delta t) = \frac{1}{m!}\langle : [\hat{\Gamma}(t, \Delta t)]^m \exp[-\hat{\Gamma}(t, \Delta t)] : \rangle. \tag{6.44}$$

It is worth noting that $P_m(t, \Delta t)$ resembles a Poissonian distribution, which is not surprising because of the assumption that $\bar{n}/N \to 0$. In the case of a classical nonfluctuating light field[6] or, in the case of a quantum field in a coherent state, the statistics $P_m(t, \Delta t)$ is indeed Poissonian. However, in other cases the resemblance of $P_m(t, \Delta t)$ in Eq. (6.44) to a Poissonian is only formal.

For some calculations it may be convenient to use the exponential characteristic function

$$u(z, t, \Delta t) = \overline{e^{nz}} = \sum_{m=0}^{\infty} P_m(t, \Delta t)e^{mz}, \tag{6.45}$$

so that $P_m(t, \Delta t)$ is simply the Fourier transform of the characteristic function:

$$P_m(t, \Delta t) = \frac{1}{2\pi}\int_{0}^{2\pi} dx\, u(ix, t, \Delta t)e^{-imx}. \tag{6.46}$$

We further see that the (ordinary) moments of the number of emitted electrons may be derived from the characteristic function $u(z, t, \Delta t)$ as

$$\overline{n^m} = \frac{\partial^m}{\partial z^m} u(z, t, \Delta t)\Big|_{z=0}. \tag{6.47}$$

---

6) In the case of a classical light field the (quantum-mechanical) ordering prescription : : becomes meaningless and the symbol $\langle \cdots \rangle$ simply introduces classical statistical averaging.

To calculate $u(z,t,\Delta t)$ we note that, comparing the definitions of $u(z,t,\Delta t)$, Eq. (6.45), and of $y(x,t,\Delta t)$, Eq. (6.31), yields the relation

$$u(z,t,\Delta t) = y(e^z - 1, t, \Delta t). \tag{6.48}$$

Hence we may represent the characteristic function $u(z,t,\Delta t)$, by using Eq. (6.43), in the form

$$u(z,t,\Delta t) = \langle {}^{\circ}_{\circ} \exp[\hat{\Gamma}(t,\Delta t)(e^z - 1)] {}^{\circ}_{\circ} \rangle. \tag{6.49}$$

We briefly extend the above results to the case where M photodetectors are involved in a detection experiment. For this purpose, let us assume that the first detector operates during the time interval $t_1, t_1 + \Delta t_1$, the second during the time interval $t_2, t_2 + \Delta t_2$, and so forth, and let us ask for the joint probability $P_{\{m_l\}}(\{t_l, \Delta t_l\}) \equiv P_{m_1,\ldots,m_M}(t_1, \Delta t_1, \ldots, t_M, \Delta t_M)$ for m_1 photoelectrons ejected by the first detector during the time interval $t_1, t_1 + \Delta t_1$, m_2 photoelectrons by the second detector during the time interval $t_2, t_2 + \Delta t_2$, and so forth. To determine this probability, we start from the characteristic function

$$y(\{t_l, \Delta t_l\}) = \sum_{\{m_l\}} P_{\{m_l\}}(\{t_l, \Delta t_l\}) \prod_{l=1}^{M}(1 + x_l)^{m_l}, \tag{6.50}$$

which is obviously the natural generalization of Eq. (6.31). Further calculations may now be performed analogously to those for a single detector, using Eq. (6.21). We therefore omit the details here and simply present the result, which, in the generalization of Eqs (6.37) and (6.44), is

$$F_{\{m_l\}}(\{t_l, \Delta t_l\}) = \langle {}^{\circ}_{\circ} \prod_{l=1}^{M} [\hat{\Gamma}^{(l)}(t_l, \Delta t_l)]^{m_l} {}^{\circ}_{\circ} \rangle, \tag{6.51}$$

$$P_{\{m_l\}}(\{t_l, \Delta t_l\}) = \langle {}^{\circ}_{\circ} \prod_{l=1}^{M} \frac{1}{m_l!} [\hat{\Gamma}^{(l)}(t_l, \Delta t_l)]^{m_l} \exp[-\hat{\Gamma}^{(l)}(t_l, \Delta t_l)] {}^{\circ}_{\circ} \rangle, \tag{6.52}$$

$\hat{\Gamma}^{(l)}(t_l, t_l + \Delta t_l)$ being defined according to Eq. (6.38)–(6.40). Clearly, if the photodetectors used are identical (equal efficiencies) the superscript (l) essentially characterizes the positions of the detectors involved in the detection scheme. It should be pointed out that the derivation of Eq. (6.52) outlined here implies that $P_{\{m_l\}}(\{t_l, \Delta t_l\})$ may also be applied to the study of correlations of counts recorded by a single photodetector operating during a sequence of nonoverlapping time intervals $t_l, t_l + \Delta t_l$ ($l = 1, 2, \ldots, M$), $t_i + \Delta t_i < t_{i+1}$.

From inspection of Eq. (6.52) [together with Eqs (6.38)–(6.40)] we see that the photoelectric counting probability distribution $P_{\{m_l\}}(\{t_l, \Delta t_l\})$ is determined by all orders of normally and time-ordered electric-field correlation functions $G^{(r,s)}_{\{k_ik_j\}}(\{\mathbf{r}_i, t_i, \mathbf{r}_j, t_j\})$ of the type defined by Eq. (2.314) for $r = s$ ($\hat{\mathbf{F}}^{(\pm)} \mapsto \hat{\mathbf{E}}^{(\pm)}$).

The appearance of normally and time-ordered correlation functions implies that Eq. (6.52) can be rewritten in the source-quantity representation by replacing the $\genfrac{}{}{0pt}{}{\circ}{\circ}\genfrac{}{}{0pt}{}{\circ}{\circ}$ ordering with the $\genfrac{}{}{0pt}{}{\bullet}{\bullet}\genfrac{}{}{0pt}{}{\bullet}{\bullet}$ ordering (see Section 2.8). In particular, when at the position of the photodetector, the free field may be assumed to be in the vacuum state, so that the conditions

$$\langle \cdots \hat{E}^{(+)}_{k\,\mathrm{free}} \rangle = \langle \hat{E}^{(-)}_{k\,\mathrm{free}} \cdots \rangle = 0 \tag{6.53}$$

hold [see Eq. (2.319)], the $\genfrac{}{}{0pt}{}{\bullet}{\bullet}\genfrac{}{}{0pt}{}{\bullet}{\bullet}$ ordering allows one to omit all the free field terms [cf. Eq. (2.320)]. In this way, $P_{\{m_l\}}(\{t_l, \Delta t_l\})$ may be expressed in terms of time-ordered (atomic) source-quantity correlation functions according to Eq. (2.320). Note that after power-series expansion of the operator exponentials $\exp[-\hat{\Gamma}^{(l)}(t_l, \Delta t_l)]$ and expressing, according to Eq. (2.274), the source-field operators in terms of the atomic operators, in the resulting source-quantity correlation functions, the atomic operators must be time ordered ($\mathcal{T}_\pm$).

6.1.3
Counting moments and correlations

Let us consider some statistical properties of the photoelectric counts typically recorded by a point-like broad-band photodetector. From Eq. (6.44) [or Eqs (6.47) and (6.49)] together with Eqs (6.40)–(6.42) it is not difficult to see that the mean number of counts recorded during the time interval $t, t+\Delta t$ is proportional to the time-integrated intensity of the radiation field:

$$\overline{n(t, \Delta t)} = \sum_{m=0}^{\infty} m\, P_m(t, \Delta t) = \xi \langle \hat{I}(\mathbf{r}, t, \Delta t) \rangle. \tag{6.54}$$

In particular, in the short-time domain we have

$$\overline{n(t, \Delta t)} = \xi \Delta t \langle \hat{I}(\mathbf{r}, t) \rangle. \tag{6.55}$$

This equation enables us to define the photocounting rate

$$R = \overline{n(t, \Delta t)}(\Delta t)^{-1} = \xi \langle \hat{I}(\mathbf{r}, t) \rangle, \tag{6.56}$$

which is independent of Δt and proportional to the mean intensity of light.

Next, let us consider the variance of counts

$$\overline{[\Delta n(t, \Delta t)]^2} = \overline{n^2(t, \Delta t)} - \left[\overline{n(t, \Delta t)}\right]^2. \tag{6.57}$$

Applying again Eq. (6.44) [or Eqs (6.47) and (6.49)] together with Eqs (6.40)–(6.42), we can easily derive

$$\overline{n^2(t, \Delta t)} = \sum_{m=0}^{\infty} m^2 P_m(t, \Delta t) = \overline{n(t, \Delta t)} + \xi^2 \langle \genfrac{}{}{0pt}{}{\circ}{\circ} \hat{I}^2(\mathbf{r}, t, \Delta t) \genfrac{}{}{0pt}{}{\circ}{\circ} \rangle, \tag{6.58}$$

which reads in detail as

$$\overline{n^2(t, \Delta t)} = \overline{n(t, \Delta t)} + \xi^2 \int_t^{t+\Delta t} d\tau \int_t^{t+\Delta t} d\tau' \, G^{(2)}(\mathbf{r}, \tau, \mathbf{r}, \tau'), \tag{6.59}$$

with

$$G^{(2)}(\mathbf{r}, \tau, \mathbf{r}, \tau')$$
$$= \langle [\mathcal{T}_- \hat{E}_k^{(-)}(\mathbf{r}, \tau) \hat{E}_{k'}^{(-)}(\mathbf{r}, \tau')] [\mathcal{T}_+ \hat{E}_{k'}^{(+)}(\mathbf{r}, \tau') \hat{E}_k^{(+)}(\mathbf{r}, \tau)] \rangle \tag{6.60}$$

being the normally and time-ordered (second-order) intensity correlation function of the light under study. Substituting into Eq. (6.57) for $\overline{n^2}$ and $\overline{n}^2$ the results of Eqs (6.58) and (6.54), respectively, we derive

$$\overline{[\Delta n(t, \Delta t)]^2} = \overline{n(t, \Delta t)} + \xi^2 \langle \, {}^\circ_\circ [\Delta \hat{I}(\mathbf{r}, t, \Delta t)]^2 \, {}^\circ_\circ \rangle. \tag{6.61}$$

Note that in the short-time domain

$$\Delta \hat{I}(\mathbf{r}, t, \Delta t) = \Delta t \, \Delta \hat{I}(\mathbf{r}, t), \tag{6.62}$$

so that Eq. (6.61) reduces to

$$\overline{[\Delta n(t, \Delta t)]^2} = \overline{n(t, \Delta t)} + \xi^2 (\Delta t)^2 \langle : [\Delta \hat{I}(\mathbf{r}, t)]^2 : \rangle. \tag{6.63}$$

From inspection of Eq. (6.61) [or Eq. (6.63)] two kinds of noise are seen to contribute to the variance of the number of ejected photoelectrons. The (shot-noise) term $\overline{n}$ obviously results from the photon-number fluctuation according to a (classical) Poissonian distribution. The term proportional to $\langle \, {}^\circ_\circ (\Delta \hat{I})^2 \, {}^\circ_\circ \rangle$ reflects the noise of the detected light. As mentioned above, in the case when the light may be regarded as being a classical nonfluctuating one, in Eq. (6.61) [or Eq. (6.63)] the second term vanishes and hence the variance is equal to that of classical particles with Poissonian statistics: $\overline{(\Delta n)^2} = \overline{n}$. In classical optics the normally and time-ordered variance of the integrated light intensity simply reduces to the ordinary variance, $\langle \, {}^\circ_\circ (\Delta \hat{I})^2 \, {}^\circ_\circ \rangle \to \langle (\Delta I)^2 \rangle_{cl}$, which of course cannot be negative. Thus for a classical noisy radiation field $\langle (\Delta I)^2 \rangle_{cl} > 0$, which implies a counting statistics of super-Poissonian type: $\overline{(\Delta n)^2} > \overline{n}$. In quantum optics the situation may change drastically because $\langle \, {}^\circ_\circ (\Delta \hat{I})^2 \, {}^\circ_\circ \rangle$ may attain also negative values. Hence so-called nonclassical light with $\langle \, {}^\circ_\circ (\Delta \hat{I})^2 \, {}^\circ_\circ \rangle < 0$ gives rise to sub-Poissonian counting statistics: $\overline{(\Delta n)^2} < \overline{n}$ (for more details see Chapter 8).

Let us now proceed with the calculation of the two-time correlation of counts registered during the time intervals $t, t + \Delta t$ and $t + \tau, t + \tau + \Delta t$. If the

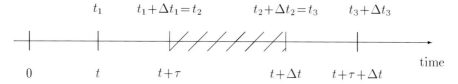

Fig. 6.3 Subdivision of two overlapping time intervals $t, t+\Delta t$ and $t+\tau, t+\tau+\Delta t$ (below time axis) into three nonoverlapping intervals (above time axis).

two intervals do not overlap ($\tau > \Delta t$), we may simply calculate the sought correlation by means of the probability $P_{m_1,m_2}(t, \Delta t, t+\tau, \Delta t)$ as given in Eq. (6.52) (for $M=2$):

$$\overline{n(t,\Delta t)n(t+\tau,\Delta t)} = \sum_{m_1,m_2=0}^{\infty} m_1 m_2 \, P_{m_1,m_2}(t, \Delta t, t+\tau, \Delta t)$$

$$= \zeta^2 \langle \, {}_\circ^\circ \hat{I}(\mathbf{r},t,\Delta t)\hat{I}(\mathbf{r},t+\tau,\Delta t) \, {}_\circ^\circ \rangle, \qquad \tau > \Delta t. \qquad (6.64)$$

In order to take account of the self-correlation in the case of overlapping time intervals ($\tau < \Delta t$), we subdivide the whole time interval $t, t+\tau+\Delta t$ into three nonoverlapping time intervals $t_1, t_1+\Delta t_1, t_2, t_2+\Delta t_2, t_3, t_3+\Delta t_3$, where

$$t_1 = t, \qquad t_2 = t+\tau, \qquad t_3 = t+\Delta t,$$
$$\Delta t_1 = \tau, \qquad \Delta t_2 = \Delta t - \tau, \qquad \Delta t_3 = \tau, \qquad (6.65)$$

see Fig. 6.3. The probability $P_{m_1,m_2,m_3}(t_1, \Delta t_1, t_2, \Delta t_2, t_3, \Delta t_3)$ for the emission of m_1 photoelectrons during the time interval $t_1, t_1+\Delta t_1$, m_2 photoelectrons during the time interval $t_2, t_2+\Delta t_2$, and m_3 photoelectrons during the time interval $t_3, t_3+\Delta t_3$ may then be taken from Eq. (6.52) ($M=3$). Since the numbers of photoelectrons ejected during the time intervals $t=t_1, t+\Delta t=t_1+\Delta t_1+\Delta t_2$ and $t+\tau=t_2, t+\tau+\Delta t=t_2+\Delta t_2+\Delta t_3$ are m_1+m_2 and m_2+m_3 respectively, the sought correlation of counts is just the average of $(m_1+m_2)(m_2+m_3)$,

$$\overline{n(t,\Delta t)n(t+\tau,\Delta t)}$$
$$= \sum_{m_1,m_2,m_3=0}^{\infty} (m_1+m_2)(m_2+m_3) \, P_{m_1,m_2,m_3}(t_1,\Delta t_1,t_2,\Delta t_2,t_3,\Delta t_3), \quad (6.66)$$

from which it follows that ($\tau < \Delta t$)

$$\overline{n(t,\Delta t)n(t+\tau,\Delta t)} = \overline{n^2(t_2,\Delta t_2)} + \overline{n(t_1,\Delta t_1)n(t_3,\Delta t_3)}$$
$$+ \overline{n(t_1,\Delta t_1)n(t_2,\Delta t_2)} + \overline{n(t_2,\Delta t_2)n(t_3,\Delta t_3)}. \quad (6.67)$$

The mean-square number and the two-time correlations of counts on the right-hand side of Eq. (6.67) may now be calculated by applying the relations (6.58) and (6.64) respectively. Recalling the relations (6.65), after straightforward calculation we arrive at

$$\overline{n(t,\Delta t)n(t+\tau,\Delta t)} = \Theta(\Delta t - \tau)\overline{n(t+\tau,\Delta t-\tau)} \\ + \xi^2 \langle {\textstyle{\circ\atop\circ}} \hat{I}(\mathbf{r},t,\Delta t)\hat{I}(\mathbf{r},t+\tau,\Delta t) {\textstyle{\circ\atop\circ}} \rangle. \tag{6.68}$$

Note that when τ goes to zero, then Eq. (6.68) simply reduces to Eq. (6.58). In the short-time limit Eq. (6.68) reduces to

$$\overline{n(t,\Delta t)n(t+\tau,\Delta t)} = \xi\Theta(\Delta t-\tau)(\Delta t - \tau)\langle \hat{I}(\mathbf{r},t)\rangle \\ + \xi^2(\Delta t)^2 \langle {\textstyle{\circ\atop\circ}} \hat{I}(\mathbf{r},t)\hat{I}(\mathbf{r},t+\tau) {\textstyle{\circ\atop\circ}} \rangle. \tag{6.69}$$

Until now we have considered moments and correlations of photoelectric counts up to second order. It is evident that higher-order moments and correlations may be derived analogously. In particular, in the short-time limit the Mth order correlation of counts registered by different photodetectors is given by

$$\overline{\prod_{l=1}^{M} n(t_l,\Delta t_l)} = G^{(M)}(\{\mathbf{r}_l,\tau_l\}) \prod_{l=1}^{M} \xi_l \Delta t_l, \tag{6.70}$$

where

$$G^{(M)}(\{\mathbf{r}_l,t_l\}) = G^{(m,n)}(\{\mathbf{r}_i,t_i,\mathbf{r}_j,t_j\})|_{m=n=M} \\ = \left\langle {\textstyle{\circ\atop\circ}} \prod_{l=1}^{M} \hat{\mathbf{E}}^{(-)}(\mathbf{r}_l,t_l)\hat{\mathbf{E}}^{(+)}(\mathbf{r}_l,t_l) {\textstyle{\circ\atop\circ}} \right\rangle \tag{6.71}$$

is the normally and time-ordered Mth-order intensity correlation function.

In practical measurements the quantities to be observed are often photocurrents. The question therefore arises of how to relate the photoelectrons to the photocurrent. In the case of amplification this is a rather difficult problem, because of the complex mechanism of multiplying the photoelectrons (primarily generated through absorption of light) that build up the amplified photocurrent.[8] Within the framework of a deterministic description of the multiplication process [see, e. g., Carmichael (1987); Huttner and Ben-Aryeh (1988)], in the simplest case of constant gain factor g the total number of electrons

---

[8] Photomultiplier-like devices typically operate by multiplying the photoelectrons that are primarily generated. Since this multiplication process, in general, introduces additional noise, the sensitivity of the receiver is determined by the noise of both the light and the gain process. Consequently, statistical concepts for dealing with the multiplication process need to be applied [see, e. g., McIntyre (1966); Personick (1971); Conradi (1972); Teich, Matsuo and Saleh (1986a,b); Kühn and Welsch (1991)].

available (after multiplication) in a time interval $t, t + \Delta t$ is gn, with n being the number of primarily generated electrons. The photocurrent may then be introduced as

$$i(t) = gen(t, \Delta t)(\Delta t)^{-1} \tag{6.72}$$

(e, electron charge). The photocurrent moments and correlations thus read [9]

$$\overline{\prod_{l=1}^{M} i(t_l)} = \left(\frac{ge}{\Delta t}\right)^M \overline{\prod_{l=1}^{M} n(t_l, \Delta t)}. \tag{6.73}$$

Note that the photocurrent time resolution Δt is usually required to be sufficiently small so that the short-time results apply on the right-hand side of Eq. (6.73). Otherwise, the photocurrent only yields information about field properties averaged over the time-resolution interval.

6.2
Photoelectric counts and photons

We have seen that the statistics of photoelectrons recorded by a point-like, broad-band photodetector may be expressed in terms of normally and time-ordered intensity correlation functions of the radiation field at the points of observation. Let us now turn to the problem of relating the counting statistics to the photon-number statistics, expressing the time-integrated correlation functions of the incident radiation in terms of moments and correlations of photon (number) operators [Mandel (1966); Peřina, Saleh and Teich (1983); Fleischhauer and Welsch (1991)].

6.2.1
Detection scheme

For this purpose, we consider radiation traveling along the positive x direction into point-like broad-band photodetectors of used detection areas $\mathcal{A}_l$, the direction of polarization being perpendicular to the x axis (see Fig. 6.4 for a single detector). Application of Eqs (6.40)–(6.42) yields

$$\hat{\Gamma}^{(l)}(t_l, \Delta t_l) = \xi_l \hat{I}(x_l, t_l, \Delta t_l), \tag{6.74}$$

$$\hat{I}(x_l, t_l, \Delta t_l) = \int_{t_l}^{t_l + \Delta t_l} d\tau \, \hat{I}(x_l, \tau), \tag{6.75}$$

$$\hat{I}(x_l, \tau) = \hat{E}^{(-)}(x_l, \tau) \hat{E}^{(+)}(x_l, \tau), \tag{6.76}$$

---

9) When different photodetectors are used and Eq. (6.73) applies, then the gain factors may be different for different detectors.

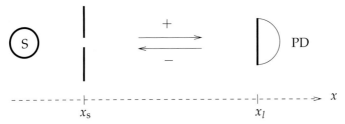

Fig. 6.4 Light produced by a source S travels along the positive x direction into a photodetector PD.

where x_l is the position of the lth detection area. To calculate intensity correlation functions of the type relevant for photodetection,

$$G^{(M)}(\{x_l, t_l\}) = \left\langle {}^{\circ}_{\circ} \prod_{l=1}^{M} \hat{E}^{(-)}(x_l, t_l) \hat{E}^{(+)}(x_l, t_l) {}^{\circ}_{\circ} \right\rangle \tag{6.77}$$

[see Eq. (6.71)], we assume that the light source is on the left-hand side of the photodetectors. That is, $x_s < x_l$, where x_s is the position of the right-hand boundary of the light source. At any point x with $x_s \leq x \leq x_l$ the operator of the (total) electric-field strength obviously consists of two parts:[10]

$$\hat{E}^{(\pm)}(x, t) = \hat{E}_+^{(\pm)}(x, t) + \hat{E}_-^{(\pm)}(x, t), \tag{6.78}$$

$$\hat{E}_\pm^{(+)}(x, t) = i \int_0^\infty d\omega \sqrt{\frac{\hbar\omega}{4\pi\varepsilon_0 c \mathcal{A}}} e^{\pm i\omega x/c} \hat{a}_\pm(\omega, t), \quad \hat{E}_\pm^{(-)} = \left(\hat{E}_\pm^{(+)}\right)^\dagger, \tag{6.79}$$

where the operators $\hat{a}_\pm(\omega)$ are introduced for the field modes propagating in the positive (+) and negative (−) x direction. With regard to the position of the light source, $\hat{E}_+^{(\pm)}$ and $\hat{E}_-^{(\pm)}$ are respectively the outgoing and incoming fields.

According to the detection scheme considered, we may assume that, at the points of observation, the free field represents the vacuum field, so that at these points the condition (2.319) is satisfied. Thus applying Eq. (2.320) enables us to replace the ${}^{\circ}_{\circ}\,{}^{\circ}_{\circ}$ ordering by the $\colon\colon$ ordering in Eq. (6.77). Since for $x_s \leq x \leq x_l$ only fields that propagate in the positive x direction can be attributed to sources, we may complement the source-field parts by the free-field parts to obtain the full-field parts $\hat{E}_+^{(\pm)}$ and replace the $\colon\colon$ ordering by the ${}^{\circ}_{\circ}\,{}^{\circ}_{\circ}$ ordering. Assuming that back-actions of the photodetectors on the sources may be ignored, from the arguments given in Section 2.7 it follows that the $\hat{E}_+^{(\pm)}$ may be regarded as being effectively free fields, because the

---

10) Equations (6.78) and (6.79) are easily obtained by specifying the plane-wave expansion given in Section 2.2.2.3 [see also Eq. (3.90)].

time-dependent commutation relations are the same relations as in the free-field case.[11] Thus, the $\genfrac{}{}{0pt}{}{\circ}{\circ}\genfrac{}{}{0pt}{}{\circ}{\circ}$ ordering simply reduces to normal ordering and we may rewrite Eq. (6.77) as

$$G^{(M)}(\{x_l, t_l\}) = \left\langle : \prod_{l=1}^{M} \hat{E}_+^{(-)}(x_l, t_l) \hat{E}_+^{(+)}(x_l, t_l) : \right\rangle. \tag{6.80}$$

Hence, replacing Eqs (6.74)–(6.76) according to

$$\hat{\Gamma}^{(l)}(t_l, \Delta t_l) \mapsto \xi_l \hat{I}_+(x_l, t_l, \Delta t_l), \tag{6.81}$$

$$\hat{I}_+(x_l, t_l, \Delta t_l) = \int_{t_l}^{t_l + \Delta t_l} d\tau\, \hat{I}_+(x_l, \tau), \tag{6.82}$$

$$\hat{I}_+(x_l, \tau) = \hat{E}_+^{(-)}(x_l, \tau) \hat{E}_+^{(+)}(x_l, \tau), \tag{6.83}$$

we may omit the time-ordering prescriptions, and Eq. (6.52) can be given in the form

$$P_{\{m_l\}}(\{t_l, \Delta t_l\}) = \left\langle : \prod_{l=1}^{M} \frac{1}{m_l!} [\xi_l \hat{I}_+(x_l, t_l, \Delta t_l)]^{m_l} \exp[-\xi_l \hat{I}_+(x_l, t_l, \Delta t_l)] : \right\rangle. \tag{6.84}$$

The field relevant for photodetection is the field impinging on the lth detector in the measurement time interval $t_l, t_l + \Delta t_l$.

6.2.2
Mode expansion

Defining the periodic field

$$\hat{E}_D^{(\pm)}(x_l, \tau) = \hat{E}_D^{(\pm)}(x_l, \tau + \Delta t_l) \tag{6.85}$$

in such a way that in the time interval $t_l, t_l + \Delta t_l$ the field $\hat{E}_D^{(\pm)}(x_l, \tau)$ is equal to the actual incoming field $\hat{E}_+^{(\pm)}(x_l, \tau)$,

$$\hat{E}_D^{(\pm)}(x_l, \tau) = \hat{E}_+^{(\pm)}(x_l, \tau) \qquad (t_l < \tau < t_l + \Delta t_l), \tag{6.86}$$

we may represent $\hat{E}_D^{(\pm)}(x_l, \tau)$ by Fourier decomposition as

$$\hat{E}_D^{(+)}(x_l, \tau) = \frac{i}{\sqrt{\Delta t_l}} \sum_\mu \sqrt{\frac{\hbar \omega_\mu}{2\varepsilon_0 c \mathcal{A}}}\, \hat{a}_\mu \exp\left[-i\omega_\mu\left(\tau - \frac{x_l}{c}\right)\right], \tag{6.87}$$

$$\hat{E}_D^{(-)} = (\hat{E}_D^{(+)})^\dagger,$$

---

[11] Examples of time-dependent commutation relations for incoming and outgoing fields of the type considered here are treated in more detail in Chapter 9.

$$\hat{a}_\mu = -\frac{i}{\sqrt{\Delta t_l}} \sqrt{\frac{2\varepsilon_0 c \mathcal{A}}{\hbar \omega_\mu}} \int_{t_l}^{t_l+\Delta t_l} d\tau \, \hat{E}_+^{(+)}(x_l, \tau) \exp\left[i\omega_\mu \left(\tau - \frac{x_l}{c}\right)\right], \quad (6.88)$$

where $\omega_\mu = 2\pi\mu/\Delta t_l$, ($\mu = 0, 1, 2, \ldots$). Apart from the values at the boundary times $\tau = t_l$ and $\tau = t_l + \Delta t_l$, the field $\hat{E}_D^{(\pm)}(x_l, \tau)$ as given in Eqs (6.87) and (6.88) does indeed agree with $\hat{E}_+^{(\pm)}(x_l, \tau)$ in the time interval t_l, $t_l + \Delta t_l$. Note that the two boundary values are meaningless, because we are only interested in quantities integrated over the time interval t_l, $t_l + \Delta t_l$.

It is appropriate to regard the operators $\hat{a}_\mu^\dagger$ and $\hat{a}_\mu$ as photon creation and annihilation operators respectively, so that the operators $\hat{n}_\mu = \hat{a}_\mu^\dagger \hat{a}_\mu$ represent photon-number operators. To show this, let us calculate the commutator $[\hat{a}_\mu, \hat{a}_{\mu'}^\dagger]$. Recalling that $\hat{E}_+^{(\pm)}$ may be regarded as being an effectively free field and using Eq. (6.79), we find that[12]

$$[\hat{E}_+^{(+)}(x_l, \tau), \hat{E}_+^{(-)}(x_l, \tau')] e^{i\omega_0(\tau-\tau')}$$
$$= [\hat{E}_{+\,\text{free}}^{(+)}(x_l, \tau), \hat{E}_{+\,\text{free}}^{(-)}(x_l, \tau')] e^{i\omega_0(\tau-\tau')}$$
$$= \frac{\hbar}{4\pi\varepsilon_0 c \mathcal{A}} \int_0^\infty d\omega \, \omega \, e^{-i(\omega-\omega_0)(\tau-\tau')} \simeq \frac{\hbar \omega_0}{2\pi\varepsilon_0 c \mathcal{A}} \delta(\tau - \tau'). \quad (6.89)$$

Here we have assumed that the bandwidth of the detected (optical) radiation is small compared with some center frequency ω_0, and we have confined ourselves to resolving times large compared with ω_0^{-1}. Combination of Eqs (6.88) and (6.89) then gives the well-known boson commutation rule

$$[\hat{a}_\mu, \hat{a}_{\mu'}^\dagger] = \delta_{\mu\mu'}. \quad (6.90)$$

We combine Eqs (6.82), (6.86) and (6.87) to obtain[13]

$$\xi_l \hat{I}_+(x_l, t_l, \Delta t_l) = \frac{\xi_l}{2\varepsilon_0 c \mathcal{A}} \sum_\mu \hbar\omega_\mu \hat{n}_\mu \simeq \eta_l \sum_\mu \hat{n}_\mu, \quad (6.91)$$

with $\eta_l = \xi_l \hbar\omega_0/(2\varepsilon_0 c \mathcal{A})$ being the detection efficiency (quantum efficiency) with regard to the number of photons to be detected. It is worth noting that the photon picture introduced here allows an interpretation closely related to the conventional procedure of field quantization in a (finite) volume $\mathcal{V}$. This can easily be seen by performing in the time integral in Eq. (6.82) [together with Eqs (6.86) and (6.87)] the change of variables $\tau = t_l + (x_l - x)/c$. Because of the condition $t_l \leq \tau \leq t_l + \Delta t_l$, the range of variation of x is $x_l - \mathcal{L} \leq x \leq x_l$,

---

[12] Note that $\hat{E}_{+\,\text{free}}^{(\pm)}(x,t)$ is given by Eq. (6.79), with $\hat{a}_+(\omega,t)$ being replaced according to $\hat{a}_+(\omega,t) \mapsto \hat{a}_+(\omega)e^{-i\omega t}$, and $[\hat{a}_+(\omega), \hat{a}_+^\dagger(\omega')] = \delta(\omega-\omega')$.

[13] Note that $2\varepsilon_0 c \hat{I}_+(x_l, t)$ may be regarded as the operator of the energy flux density falling on a photodetector with entrance plane at x_l.

where $\mathcal{L} = c\Delta t$. Hence, the radiation field interacting with a photodetector at position x_l during the time interval t_l, $t_l + \Delta t_l$ is just the field inside the volume $\mathcal{V} = \mathcal{A}\mathcal{L}$ with $\mathcal{L} = c\Delta t_l$, provided that the detection area coincides with $\mathcal{A}$, i.e., $\mathcal{A}_l = \mathcal{A}$. In this case $\mathcal{V}$ may indeed be regarded as the natural quantization volume for defining photons of the field relevant for photodetection.

The above used mode decomposition corresponds to a monochromatic mode expansion as considered in Section 2.2.2. As shown in Section 2.2.3, unitary transformations can be applied to the photon creation and annihilation operators associated with the monochromatic waves to obtain new operators that are associated with wave packets (nonmonochromatic modes). Obviously, Eq. (6.91) also applies to nonoverlapping wave packets, $\hat{n}_\mu$ being the photon-number operator assigned to the μth wave packet.

It might often be convenient to represent the operator of the number of photons falling on a photodetector at position x_l during the time interval t_l, $t_l + \Delta t_l$ in a form closely related to a continuous mode expansion as given in Eq. (6.79). For this purpose, let us consider the so-called detection operator

$$\hat{a}(x_l, t) = \left(\frac{1}{2\pi}\right)^{\frac{1}{2}} \int_0^\infty d\omega\, \hat{a}_+(\omega, t) e^{i\omega x_l/c}. \tag{6.92}$$

Combining Eqs (6.82) and (6.79), taking into account that the bandwidth of the light has been assumed to be small compared with the center frequency, and using Eq. (6.92), we may (approximately) rewrite Eq. (6.82) to obtain

$$\xi_l \hat{I}_+(x_l, t_l, \Delta t_l) = \eta_l \int_{t_l}^{t_l + \Delta t_l} d\tau\, \hat{a}^\dagger(x_l, \tau) \hat{a}(x_l, \tau). \tag{6.93}$$

Comparing Eqs (6.93) and (6.91), we find that the operator of the total number of photons falling on the photodetector during the time interval t_l, $t_l + \Delta t_l$ may be represented in the form

$$\sum_\mu \hat{n}_\mu = \int_{t_l}^{t_l + \Delta t_l} d\tau\, \hat{a}^\dagger(x_l, \tau) \hat{a}(x_l, \tau). \tag{6.94}$$

The operator $\hat{a}$ ($\hat{a}^\dagger$) may therefore be regarded as being the (photon per unit time)$^{1/2}$ units annihilation (creation) operator of the radiation field incident on the photodetector (with entrance plane at x_l). Accordingly, the operator $\hat{a}^\dagger \hat{a}$ may be viewed as the corresponding number operator in photon per unit time units. Note that the time dependence of $\hat{a}_+(\omega, t)$ [in Eq. (6.92)] must be determined from the solution of the full light-source interaction problem.

6.2.3
Photon-number statistics

Let us consider a single photodetector with entrance plane at x_l. According to Eqs (6.84) and (6.91), we can represent the probabilities for the photoelectrons ejected during the time interval t_l, $t_l + \Delta t_l$ in the form

$$P_m = \langle \hat{P}_m(\{\hat{a}_\mu, \hat{a}_\mu^\dagger\}) \rangle, \qquad (6.95)$$

where[14]

$$\hat{P}_m = \hat{P}_m(\{\hat{a}_\mu, \hat{a}_\mu^\dagger\}) = :\frac{1}{m!}(\eta \hat{n})^m e^{-\eta \hat{n}}:, \qquad (6.96)$$

$$\hat{n} = \sum_\mu \hat{n}_\mu \qquad (6.97)$$

($\eta \equiv \eta_l$). Here and in the following we omit the arguments t_l and Δt_l, which indicate the chosen measurement interval. Let us view $\hat{P}_m$ as a function of the detection efficiency η. It is not difficult to prove that expansion of the exponential in Eq. (6.96) gives

$$\hat{P}_m(\{\hat{a}_\mu, \hat{a}_\mu^\dagger\}) = \sum_{n=0}^{\infty} P_{m|n}(\eta) \, \hat{P}_n(\{\hat{a}_\mu, \hat{a}_\mu^\dagger\})\big|_{\eta=1}, \qquad (6.98)$$

where

$$P_{m|n}(\eta) = \begin{cases} \binom{n}{m} \eta^m (1-\eta)^{n-m} & \text{if } n \geq m, \\ 0 & \text{if } n < m. \end{cases} \qquad (6.99)$$

In order to calculate $\hat{P}_n$ for $\eta=1$, we note that the associated c-number function for $s=1$ (normal order) reads

$$P_n(\{\alpha_\mu, \alpha_\mu^*\}; 1)\big|_{\eta=1} = \frac{1}{n!}\left(\sum_\mu |\alpha_\mu|^2\right)^n \exp\left(-\sum_\mu |\alpha_\mu|^2\right)$$

$$= \sum_{\{n_\mu\}} \prod_\mu \frac{|\alpha_\mu|^{2n_\mu}}{n_\mu!} e^{-|\alpha_\mu|^2} = \sum_{\{n_\mu\}} \prod_\mu |\langle \alpha_\mu | n_\mu \rangle|^2, \quad \sum_\mu n_\mu = n. \qquad (6.100)$$

Recalling that $|\langle \alpha_\mu | n_\mu \rangle|^2$ is the c-number function associated with the number-state projector $|n_\mu\rangle\langle n_\mu|$ in normal order [cf. Eq. (4.60)], we see that $\hat{P}_n$ for $\eta=1$ is a multi-mode number-state projector:

$$\hat{P}_n(\{\hat{a}_\mu, \hat{a}_\mu^\dagger\})\big|_{\eta=1} = \sum_{\{n_\mu\}} |\{n_\mu\}\rangle\langle\{n_\mu\}|, \quad \sum_\mu n_\mu = n. \qquad (6.101)$$

---

[14] Note that the set of $\hat{P}_m$ is an example of a positive operator valued measure (POVM), because each Hermitian operator $\hat{P}_m$ is a non-negative operator, and $\sum_m \hat{P}_m = \hat{I}$. The main difference between POVMs and von Neumann's projection valued measures, is that the elements of POVMs are not necessarily orthogonal projectors, so that the corresponding probability distributions cannot be infinitely sharply peaked in general [for details see, e.g., Helstrom (1976)].

Thus for $\eta<1$, $\hat{P}_m$ in Eq. (6.98) is a statistical mixture of orthogonal projectors that project onto multi-mode number states whose total photon numbers n satisfy the condition that $n\geq m$.

Combining Eqs (6.95), (6.98) and (6.101), we may write

$$P_m = \sum_{n=0}^{\infty} P_{m|n}(\eta) p_n, \qquad (6.102)$$

where

$$p_n = \sum_{\{n_\mu\}} \langle\{n_\mu\}|\hat{\varrho}|\{n_\mu\}\rangle, \qquad \sum_\mu n_\mu = n, \qquad (6.103)$$

is the probability that the incident light contains n photons. In Eq. (6.102), $P_{m|n}$ is the probability of detecting m photoelectrons conditioned on n incident photons. Clearly, in order to observe m photoelectrons the number n of incident photons must not be smaller than m. The joint probability of n incident photons being available and m photoelectrons being recorded is then $P_{m|n} p_n$. Summing over n yields the marginal probability P_m of recording m photoelectrons. When $\eta\to 1$ then $P_{m|n}\to\delta_{mn}$, and $P_m\to p_m$, i.e., when the detection efficiency is equal to unity (perfect detection), then the observed probability distribution is the (total) photon-number probability distribution of the incident radiation in the chosen detection interval.[15]

Equation (6.102) [together with Eq. (6.99)] is a Bernoulli transformation, which yields the counting probabilities in terms of the photon-number probabilities. In practice the photon-number probabilities are desired to be determined. The problem can formally be solved by the inverse Bernoulli transformation

$$p_m = \sum_{n=0}^{\infty} P_{m|n}(\eta^{-1}) P_n, \qquad (6.104)$$

the derivation of which is similar to that of the Bernoulli transformation except that one has to start from $\hat{P}_m$ for $\eta=1$. From Eq. (6.99) it is easily seen that the coefficients $P_{m|n}(\eta^{-1})$ in Eq. (6.104) are not bounded for $\eta<0.5$ and hence the reconstruction of p_m from P_m leads, in practice, to an error explosion.

Equation (6.95) together with Eq. (6.96) can be used to relate the moments of counts to the (relevant) moments of the (total) photon number in a straight-

---

15) When $\eta<1$ then some of the incident photons are lost and P_m can be regarded as being the photon-number probability distribution after losing the (nondetected) photons.

forward manner. In particular, in close analogy with Eqs (6.55) and (6.63) we derive

$$\bar{n} = \eta \langle \hat{n} \rangle, \tag{6.105}$$

$$\overline{(\Delta n)^2} = \eta \langle \hat{n} \rangle + \eta^2 \langle :(\Delta \hat{n})^2: \rangle = \eta(1-\eta)\langle \hat{n} \rangle + \eta^2 \langle (\Delta \hat{n})^2 \rangle. \tag{6.106}$$

This result reveals that the variance of the number of photoelectrons is not solely determined by the variance of the number of incident photons but also by the mean photon number. Only in the limit as $\eta \to 1$ (perfect detection) are the two variances directly related to each other. In the context of the study of nonclassical light, it is often useful to introduce Fano factors and to express the result of Eq. (6.106) in terms of them. Defining the electronic Fano factor F_{el} by

$$F_{el} = \frac{\overline{(\Delta n)^2}}{\bar{n}} \tag{6.107}$$

and accordingly the photonic Fano factor F_{ph} as

$$F_{ph} = \frac{\langle (\Delta \hat{n})^2 \rangle}{\langle \hat{n} \rangle}, \tag{6.108}$$

we can easily see that Eq. (6.106) implies that

$$F_{el} - 1 = \eta(F_{ph} - 1). \tag{6.109}$$

Therefore, when the observed statistics of electrons is sub-Poissonian ($F_{el} < 1$), one may conclude that the photon statistics is also ($F_{ph} < 1$).

It is not difficult to extend Eq. (6.95) to the case where more than one detector is used. Recalling Eq. (6.52), the joint probability $P_{\{m_l\}}$ for m_1 photoelectrons ejected by the first detector during the time interval $t_1, t_1 + \Delta t_1$, m_2 photoelectrons by the second detector during the time interval $t_2, t_2 + \Delta t_2$, and so forth, is

$$P_{\{m_l\}} = \langle \hat{P}_{\{m_l\}} \rangle, \tag{6.110}$$

where

$$\hat{P}_{\{m_l\}} = \prod_{l=1}^{M} \hat{P}_{m_l}(\{\hat{a}_{\mu_l}, \hat{a}^{\dagger}_{\mu_l}\}). \tag{6.111}$$

Here $\hat{P}_{m_l}$ is defined according to Eq. (6.96) where $\hat{a}_{\mu_l}$ ($\hat{a}^{\dagger}_{\mu_l}$) is the photon annihilation (creation) operator associated with the μ_lth mode relevant for the lth detector. Accordingly,

$$\hat{n}_l = \sum_{\mu_l} \hat{a}^{\dagger}_{l\mu_l} \hat{a}_{l\mu_l} \tag{6.112}$$

is the operator of the number of photons falling on the lth detector during its operation-time interval t_l, $t_l+\Delta t_l$.

6.3
Nonperturbative corrections

The approach to the quantum theory of photodetection developed in the preceding sections is based on a perturbative treatment of the interaction between the light to be detected and the sample of detector atoms. It may therefore lead to unphysical consequences if the conditions of validity of perturbation theory are not taken into account. To illustrate the inadequacies that might appear, let us again consider a light field traveling along the positive x axis into a (broad-band) photodetector of (used) detection area $\mathcal{A}_l$ at position x_l and suppose that the direction of polarization is fixed. From inspection of Eq. (6.105) it can be seen that the value of η_l must not exceed unity. Otherwise the number of photoelectrons would exceed the number of incident photons. However, since η_l is proportional to the number of detector atoms, there is no upper bound. This inadequacy is of course a result of illicitly extending the range of validity of perturbation theory. Clearly, in the perturbative approach developed, the attenuation of radiation due to the detection process is completely disregarded. Hence the result (6.105) may be regarded as being valid for sufficiently small values of η_l ($\eta_l \ll 1$), so that only a small fraction of incident photons can be absorbed by the photodetector. On the other hand, when (low-intensity) quantum light fields are detected, one is interested in high detection efficiency. The question therefore arises of how to extend the perturbative results in order to include nonperturbative corrections. This problem has been studied in particular for the case of a single-mode cavity field [Mollow (1968); Scully and Lamb, Jr. (1969); Selloni, Schwendimann, Quattropani and Baltes (1978); Srinivas and Davies (1981); Ueda (1990)]. The common situation with photodetection is that the light emitted from a source propagates in free space and falls onto the photodetector, which is spatially well separated from the source. To describe the propagation of light from the source to the detector followed by light attenuation during the detection process, a multi-mode theory is required [Chmara (1987); Fleischhauer and Welsch (1991)].

We recall that the formulae for the photoelectric counting distributions have been derived under the assumption that the number of detector atoms substantially exceeds the (mean) number of photons to be detected. In this case the probability of a detector atom being excited during the time interval of detection is very small so that saturation effects are meaningless. The interaction of each detector atom with the radiation field may therefore be treated within the framework of lowest-order perturbation theory with little error. However, the radiation field actually interacting with a given detector atom is not the unperturbed field as given by the outgoing field from the sources incident on the entrance plane of the detector, but it is a modified version of it, because of the interaction of the field with the remaining detector atoms. Clearly, with increasing η_l this modification becomes more and more pronounced. The intensity of light can no longer be regarded as (approximately) constant within the photosensitive detection slab and the treatment of the photodetector as a point-like device needs more careful consideration.

Going back to Eq. (6.39) and performing the summation over the detector atoms in the sense of an integration ($S \equiv S^{(i)} = S^{(j)}$), $\xi_l \hat{I}_+(x_l, t_l, \Delta t_l)$ in Eq. (6.81) must be replaced according to

$$\xi_l \hat{I}_+(x_l, t_l, \Delta t_l) \mapsto S_l \int_{t_l}^{t_l+\Delta t_l} d\tau \int_{x_l}^{x_l+D_l} dx\, \sigma_l(x) \hat{I}_+(x, \tau), \qquad (6.113)$$

where $\hat{I}_+(x, \tau)$ is defined according to Eq. (6.83), $\sigma_l(x)$ is the atomic number density (per unit length) and D_l is the thickness of the photoelectrically sensitive slab. Note that

$$\xi_l = S_l N_l = S_l \int_{x_l}^{x_l+D_l} dx\, \sigma_l(x), \qquad (6.114)$$

where N_l is the number of atoms within the volume $\mathcal{A}_l D_l$ of the photosensitive detection slab. Application of Eq. (6.54) together with the replacement (6.113) yields the mean number of counts recorded during the time interval $t_l, t_l + \Delta t_l$:

$$\bar{n}(t_l, \Delta t_l) = S_l \int_{t_l}^{t_l+\Delta t_l} d\tau \int_{x_l}^{x_l+D_l} dx\, \sigma_l(x) \langle \hat{I}_+(x, \tau) \rangle. \qquad (6.115)$$

We now take into account that the light field propagating from the entrance plane at x_l to a plane at x ($x_l < x < x_l + D_l$) loses intensity owing to its interaction with the (absorbing) detector atoms. According to the Lambert–Beer law we may write

$$\langle \hat{I}_+(x, t) \rangle = \langle \hat{I}_+(x_l, t) \rangle e^{-\kappa_l(x-x_l)}, \qquad (6.116)$$

so that

$$\bar{n}(t_l, \Delta t_l) = \xi'_l \langle \hat{I}_+(x_l, t_l, \Delta t_l) \rangle, \qquad (6.117)$$

where $\hat{I}_+(x_l, t_l, \Delta t_l)$ again refers to the entrance plane of the detector, but the new, renormalized efficiency reads

$$\xi'_l = S_l \int_{x_l}^{x_l+D_l} dx\, e^{-\kappa_l(x-x_l)} \sigma_l(x). \qquad (6.118)$$

Assuming that the detector atoms are uniformly distributed [$\sigma_l(x) = N_l/D_l$], from Eq. (6.118) we evaluate

$$\xi'_l = \frac{\xi_l}{\kappa_l D_l}\left(1 - e^{-\kappa_l D_l}\right). \tag{6.119}$$

Hence the detection efficiency which relates the number of counts to the number of incident photons [see Eq. (6.91)] must also be replaced by a renormalized one:

$$\eta_l \mapsto \eta'_l = \frac{\hbar\omega_0}{2\varepsilon_0 c A} \frac{\xi_l}{\kappa_l D_l}\left(1 - e^{-\kappa_l D_l}\right). \tag{6.120}$$

Let us assume that the losses described by the extinction coefficient κ_l only result from the absorption of light by the detector atoms. In this case κ_l is proportional to ξ_l. Since the mean number $\bar{n}$ of counts must tend to the fraction $(A_l/A)\langle\hat{n}\rangle$ of incident photons as $\kappa_l D_l$ goes to infinity, from Eq. (6.105) together with Eq. (6.120) we conclude that

$$\kappa_l D_l = \xi_l \frac{\hbar\omega_0}{2\varepsilon_0 c A_l}. \tag{6.121}$$

In this way, we may rewrite Eqs (6.119) and (6.120) as

$$\xi'_l = \frac{2\varepsilon_0 c A_l}{\hbar\omega_0}\left[1 - \exp\left(-\xi_l \frac{\hbar\omega_0}{2\varepsilon_0 c A_l}\right)\right], \tag{6.122}$$

$$\eta'_l = \frac{A_l}{A}\left[1 - \exp\left(-\eta \frac{A}{A_l}\right)\right]. \tag{6.123}$$

Clearly, for small values of η_l ($\eta_l \ll 1$) the well-known result from perturbation theory emerges ($\xi'_l \approx \xi_l$, $\eta'_l \approx \eta_l$). Moreover, from Eqs (6.105) and (6.123) it is easily found that the number of counts cannot exceed the number of incident photons, even when the value of η becomes large.

Although the above arguments are somewhat intuitive and mainly concern the mean number of counts, they suggest that (under the assumptions made) the nonperturbative extension of the results given in the preceding sections simply consists in interpreting the detection efficiencies (ξ_l and η_l) as renormalized ones, which, in particular, implies that the forms of the counting distribution functions need not be changed. We omit the general proof here and refer the reader to the literature [Fleischhauer and Welsch (1991)].

6.4 Spectral detection

In photodetection experiments the light under study frequently passes through a more or less complicated setup of optical instruments before it is registered by photodetectors. Spectral properties of light may be observed by appropriately combining photodetectors with frequency-sensitive devices, such as spectral filters of Fabry–Perot type. Another typical example is the superimposition of two or more light fields by means of beam splitters and the detection of the combined fields. Interferometric detection schemes of this type have been successfully used in quantum-state measurement (Chapter 7).

The field relevant for photodetection is that at the entrance plane of the detector. If the light produced by some kind of source passes through an optical instrument before it is detected, the field impinging on the detector is the field transformed by the action of the instrument. As long as absorption may be disregarded, a Fabry–Perot spectral apparatus or a beam splitter may be regarded as a four-port device, whose action may be treated in quantum optics within the concept of field quantization in a dielectric with space-dependent refractive index, as developed in Section 2.4.1. Moreover, both a Fabry–Perot device and a beam splitter may be viewed as a type of multi-slab dielectric configuration, the significant difference between the two devices being the frequency sensitivity.

6.4.1
Radiation-field modes

For clarity, and to avoid rather lengthy derivations and formulae, we confine ourselves to the simplest model of a nonabsorbing four-port device, namely a dielectric plate of real refractive index n and thickness d [Knöll, Vogel and Welsch (1986, 1990); Ley and Loudon (1987)]. Let us consider the scheme in Fig. 6.5. Assuming fixed directions of propagation and polarization of the radiation under study, we may omit vector indices and the electric-field strength can be given by

$$\hat{E}(x) = \hat{E}^{(+)}(x) + \hat{E}^{(-)}(x), \tag{6.124}$$

$$\hat{E}^{(+)}(x) = i \int dk\, c|k| A(k,x)\hat{a}(k), \qquad \hat{E}^{(-)} = \left(\hat{E}^{(+)}\right)^\dagger. \tag{6.125}$$

The Helmholtz equation (2.195) for determining the mode functions $A(k,x)$ simplifies to

$$\frac{\partial^2}{\partial x^2} A(k,x) + n^2(x) k^2 A(k,x) = 0, \qquad k^2 = \frac{\omega^2}{c^2}, \tag{6.126}$$

where, according to our model of a dielectric plate,

$$n(x) = \sqrt{\varepsilon(x)} = \begin{cases} n & \text{if } -\tfrac{1}{2}d \leq x \leq \tfrac{1}{2}d, \\ 1 & \text{if } |x| > \tfrac{1}{2}d. \end{cases} \tag{6.127}$$

Accounting for the boundary conditions at the surfaces of discontinuity $x = \pm d/2$ and for the normalization condition (2.196) [together with Eq. (2.71)],

6.4 Spectral detection

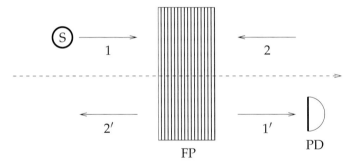

Fig. 6.5 Scheme of a spectral measurement. The light emitted by the source (S) passes through a frequency-sensitive (multi-slab dielectric) device of Fabry–Perot type (FP) in a photodetector (PD).

a straightforward calculation yields the following expressions for the mode functions

$$A(k,x) = \sqrt{\frac{\hbar}{4\pi\varepsilon_0 \omega \mathcal{A}}} \begin{cases} e^{ikx} + \underline{R}(\omega)e^{-ikx} & \text{if } x \leq -\tfrac{1}{2}d, \\ \underline{T}(\omega)e^{ikx} & \text{if } x \geq \tfrac{1}{2}d \end{cases} \quad (6.128)$$

for $k > 0$, and

$$A(k,x) = \sqrt{\frac{\hbar}{4\pi\varepsilon_0 \omega \mathcal{A}}} \begin{cases} \underline{T}(\omega)e^{ikx} & \text{if } x \leq -\tfrac{1}{2}d, \\ e^{ikx} + \underline{R}(\omega)e^{-ikx} & \text{if } x \geq \tfrac{1}{2}d \end{cases} \quad (6.129)$$

for $k < 0$, with $\mathcal{A}$ being an appropriately chosen normalization area. In Eqs (6.128) and (6.129), $\underline{T}(\omega)$ and $\underline{R}(\omega)$ are the spectral transmission and reflection response functions of the dielectric plate, respectively, which, in agreement with the Airy formulae [see, e.g., Born and Wolf (1959)], are

$$\underline{T}(\omega) \exp\left[-i(d_{\text{opt}} - d)\frac{\omega}{c}\right] = \frac{1 - r^2}{1 - r^2 \exp(2i\omega d_{\text{opt}}/c)}, \quad (6.130)$$

$$\underline{R}(\omega) \exp\left(id\frac{\omega}{c}\right) = -r + r \exp\left[i(d_{\text{opt}} + d)\frac{\omega}{c}\right]\underline{T}(\omega), \quad (6.131)$$

$$r^2 = \left(\frac{n-1}{n+1}\right)^2, \quad (6.132)$$

where $d_{\text{opt}} = nd$ is the optical path through the dielectric plate. It can easily be proved that $\underline{T}(\omega)$ and $\underline{R}(\omega)$ satisfy the conditions

$$|\underline{T}(\omega)|^2 + |\underline{R}(\omega)|^2 = 1, \quad (6.133)$$

$$\underline{T}(\omega)\underline{R}^*(\omega) + \underline{T}^*(\omega)\underline{R}(\omega) = 0. \quad (6.134)$$

Although Eqs (6.128) and (6.129) are derived for the simple case of a dielectric plate, their forms may be regarded as being valid also for more complicated multi-slab four-port devices. Clearly, the formulae for the spectral transmission and reflection response functions may become more complicated than those given in Eqs (6.130) and (6.131).

With regard to the considered dielectric plate, the (complex) poles Ω_m of the spectral transmission response function $\underline{T}(\omega)$ determined from the equation

$$1 - r^2 \exp\left(2i\Omega_m \frac{d_{opt}}{c}\right) = 0 \tag{6.135}$$

read

$$\Omega_m = \omega_m - \tfrac{1}{2}i\Gamma, \tag{6.136}$$

where (m, integer)

$$\omega_m = m\pi \frac{c}{d_{opt}}, \tag{6.137}$$

$$\Gamma = -\frac{c}{d_{opt}} \ln r^2. \tag{6.138}$$

In particular, for very high reflectance ($r^2 \approx 1$) the plate obviously acts as a spectral filter. In this case Eq. (6.138) may be approximated as

$$\Gamma = \frac{c}{d_{opt}}(1 - r^2), \tag{6.139}$$

so that the passband width Γ becomes small compared with the setting frequencies ω_m and the distance $\Delta\omega = \pi c/d_{opt}$ between neighboring setting frequencies. From Eqs (6.130), (6.136) and (6.139) the behavior of $\underline{T}(\omega)$ for frequencies in the vicinity of a chosen setting frequency ω_m is then found to be

$$\underline{T}(\omega) \exp\left[-i(d_{opt} - d)\frac{\omega}{c}\right] = \frac{\tfrac{1}{2}\Gamma}{\tfrac{1}{2}\Gamma - i(\omega - \omega_m)}. \tag{6.140}$$

That is, the spectral transmission response function becomes very effective in discriminating against values of ω different from the setting frequency ω_m.

6.4.2
Input-output relations

From inspection of Eqs (6.128) and (6.129) we see that two kinds of mode functions are to be considered, describing incoming waves from the left and right, each of which is partly reflected and transmitted.[16] Using Eqs (6.128) and

16) In Fig. 6.5 it is assumed that the light under study is produced by a source on the left of the dielectric plate. In practical measurements, one of course tries to avoid directing the reflected part of the light into the source, for example by inclined incidence of light and appropriately arranged diaphragms, so that the reflected part of the light cannot strike the source.

(6.129), we now decompose the field outside the plate in incoming and outgoing fields:

$$\hat{E}(x) = \sum_{\nu=1}^{2} [\hat{E}_\nu^{(+)}(x) + \hat{E}_\nu'^{(+)}(x)] + \text{H.c.,} \qquad (6.141)$$

where ($|x| \geq d/2$)

$$\hat{E}_\nu^{(+)}(x) = i \int_0^\infty d\omega \sqrt{\frac{\hbar\omega}{4\pi\varepsilon_0 c \mathcal{A}}} e^{ik_\nu x} \hat{a}_\nu(\omega), \qquad (6.142)$$

$$\hat{E}_\nu'^{(+)}(x) = i \int_0^\infty d\omega \sqrt{\frac{\hbar\omega}{4\pi\varepsilon_0 c \mathcal{A}}} e^{ik_\nu x} \hat{a}_\nu'(\omega), \qquad (6.143)$$

and $k_1 = \omega/c$, $k_2 = -\omega/c$. Here the photonic operators of the outgoing field, $\hat{a}_\nu'(\omega)$, are related to the photonic operators of the incoming field, $\hat{a}_\nu(\omega) = c^{-1/2}\hat{a}(k)$, according to the input-output relations

$$\hat{a}_1'(\omega) = \underline{T}(\omega)\hat{a}_1(\omega) + \underline{R}(\omega)\hat{a}_2(\omega), \qquad (6.144)$$

$$\hat{a}_2'(\omega) = \underline{R}(\omega)\hat{a}_1(\omega) + \underline{T}(\omega)\hat{a}_2(\omega). \qquad (6.145)$$

Note that, from Eqs (6.133) and (6.134), it follows that Eqs (6.144) and (6.145) represent a unitary transformation of the input operators $\hat{a}_\nu$ into the output operators $\hat{a}_\nu'$. Thus the bosonic commutation relations are preserved:

$$[\hat{a}_\nu(\omega), \hat{a}_{\nu'}^\dagger(\omega')] = [\hat{a}_\nu'(\omega), \hat{a}_{\nu'}'^\dagger(\omega')] = \delta_{\nu\nu'}\delta(\omega-\omega'), \qquad (6.146)$$

$$[\hat{a}_\nu(\omega), \hat{a}_{\nu'}(\omega')] = [\hat{a}_\nu'(\omega), \hat{a}_{\nu'}'(\omega')] = 0. \qquad (6.147)$$

Applying the input-output relations (6.144) and (6.145) and combining Eqs (6.142) and (6.143), we can express the outgoing fields in terms of the incoming fields. The outgoing fields then read

$$\hat{E}_1'^{(+)}(x) = \int dt'\, T(t') \hat{E}_1^{(+)}(x+ct') + \int dt'\, R(t') \hat{E}_2^{(+)}(-x-ct'), \qquad (6.148)$$

$$\hat{E}_2'^{(+)}(x) = \int dt'\, R(t') \hat{E}_1^{(+)}(-x+ct') + \int dt'\, T(t') \hat{E}_2^{(+)}(x-ct'), \qquad (6.149)$$

where the transmission response function $T(t)$ and the reflection response function $R(t)$ are given by

$$T(t) = \frac{1}{2\pi} \int_0^\infty d\omega\, \underline{T}(\omega) e^{-i\omega t} \qquad (6.150)$$

and

$$R(t) = \frac{1}{2\pi} \int_0^\infty d\omega\, \underline{R}(\omega) e^{-i\omega t}. \qquad (6.151)$$

Equations (6.148) and (6.149) are operator equations which are not related to any specific quantum-mechanical picture of temporal evolution. In particular, in the Heisenberg picture, where the temporal evolution of the operators formally looks like the classical one, we may replace the $\hat{E}_\nu^{(+)}(x)$ according to

$$\hat{E}_1^{(+)}(x) \mapsto \hat{E}_1^{(+)}(x,t) = \hat{\mathcal{E}}_1^{(+)}(t - x/c), \qquad (6.152)$$

$$\hat{E}_2^{(+)}(x) \mapsto \hat{E}_2^{(+)}(x,t) = \hat{\mathcal{E}}_2^{(+)}(t + x/c), \qquad (6.153)$$

and $\hat{E}_\nu'^{(+)}(x)$ accordingly. Thus Eqs (6.148) and (6.149) take the form

$$\hat{\mathcal{E}}_1'^{(+)}(t) = \int dt'\, T(t - t')\, \hat{\mathcal{E}}_1^{(+)}(t') + \int dt'\, R(t - t')\, \hat{\mathcal{E}}_2^{(+)}(t'), \qquad (6.154)$$

$$\hat{\mathcal{E}}_2'^{(+)}(t) = \int dt'\, R(t - t')\hat{\mathcal{E}}_1^{(+)}(t') + \int dt'\, T(t - t')\hat{\mathcal{E}}_2^{(+)}(t'). \qquad (6.155)$$

In Fig. 6.5 it is assumed that the field relevant for photodetection is that in the channel 1′, so that Eqs (6.81)–(6.84) with

$$\hat{E}_+^{(\pm)}(x_l, t_l) = \hat{E}_1'^{(\pm)}(x_l, t_l) = \hat{\mathcal{E}}_1'^{(\pm)}(t_l - x_l/c) \qquad (6.156)$$

apply. Since in classical optics the vacuum in the input channel 2 in Fig. 6.5 is irrelevant, the spectrally filtered field may be expressed as a convolution of the incoming unfiltered field with the transmission response function of the spectral apparatus [Eq. (6.154) with c numbers instead of operators and $\mathcal{E}_2^{(+)}(t')=0$]. In quantum optics the situation is changed because now both incoming fields are needed to ensure that the commutation relations are not violated, and thus the full operator equation (6.154) must be taken into account.

6.4.3
Spectral correlation functions

Detection of light behind a spectral apparatus yields the physically accessible information on its spectral properties.[17] Thus the introduced spectra are often called physical spectra [Eberly and Wódkiewicz (1977)]. In contrast to these, intrinsic spectral properties may be defined mathematically by a Fourier analysis of the light [Metha and Wolf (1967)].

Spectral properties are commonly expressed in terms of spectrally resolved correlation functions. According to Eqs (6.80)–(6.136), in photocounting measurements of spectrally filtered light, normally ordered intensity correlation

---

[17] Alternatively, the unfiltered light is detected and the photocurrent is electronically filtered.

functions of the type[18]

$$G^{(M)}(\{x_l, t_l\}) = \left\langle : \prod_{l=1}^{M} \hat{\mathcal{E}}_1'^{(-)}(\tau_l) \hat{\mathcal{E}}_1'^{(+)}(\tau_l) : \right\rangle \qquad (\tau_l = t_l - x_l/c) \qquad (6.157)$$

can be detected, with $\hat{\mathcal{E}}_1'^{(+)}$ [and $\hat{\mathcal{E}}_1'^{(-)} = (\hat{\mathcal{E}}_1'^{(+)})^\dagger$] being given by Eq. (6.154). Since we may assume (see Fig. 6.5) that the source only contributes to the $\hat{\mathcal{E}}_1^{(\pm)}$, we may omit the $\hat{\mathcal{E}}_2^{(\pm)}$ on following the arguments given in Section 6.2.1. Introducing the normally ordered correlation functions of the incoming unfiltered light,

$$\Gamma^{(m,n)}(\{t_i, t_j\}) = \left\langle : \prod_{i=1}^{m} \prod_{j=m+1}^{n} \hat{\mathcal{E}}_1^{(-)}(t_i) \hat{\mathcal{E}}_1^{(+)}(t_j) : \right\rangle, \qquad (6.158)$$

we may rewrite Eq. (6.157), on using Eq. (6.154), as

$$G^{(M)}(\{x_l, t_l\}) = \int dt_1' \int dt_{M+1}' \, T_1^*(\tau_1 - t_1') T_1(\tau_1 - t_{M+1}') \cdots$$
$$\cdots \int dt_M' \int dt_{2M}' \, T_M^*(\tau_M - t_M') T_M(\tau_M - t_{2M}') \, \Gamma^{(M,M)}(t_1', \ldots, t_{2M}'). \qquad (6.159)$$

Here we have allowed for different spectral filters in order to study the correlation behavior of different frequency components. This can be realized in an appropriately extended scheme that uses spectral filters in combination with beam splitters. The simplest example of such an experimental setup is the following. The outgoing radiation field from the light source is subdivided into two parts by means of a beam splitter. These parts may then be used as input fields for two spectral filters that differ in setting frequency. The correlations between the corresponding output fields may finally be detected in a two-photodetector photocounting experiment. From Eq. (6.159) we see that the normally ordered intensity correlation functions $G^{(M)}$ observed behind the filters may be expressed in terms of convolutions of normally ordered field correlation functions $\Gamma^{(m,n)}$ of the incoming unfiltered light with the transmission response functions of the filters.

In some cases it may be useful to decompose the incoming unfiltered light into the free-field and the source-field,

$$\hat{\mathcal{E}}_1^{(\pm)}(t) = \hat{\mathcal{E}}_{s1}^{(\pm)}(t) + \hat{\mathcal{E}}_{\text{free}\,1}^{(\pm)}(t). \qquad (6.160)$$

[18] Here it is assumed that there are no back-actions of the equipment on the light source. Otherwise, time orderings must be considered, i.e., the normal ordering must be replaced by the $\substack{\circ\circ\\\circ\circ}$ ordering according to Eq. (6.77).

When the free field represents the vacuum, then Eq. (2.320) applies and Eq. (6.158) can be rewritten as

$$\Gamma^{(m,n)}(\{t_i, t_j\}) = \left\langle : \prod_{i=1}^{m} \prod_{j=m+1}^{m+n} \hat{\mathcal{E}}_{s1}^{(-)}(t_i) \hat{\mathcal{E}}_{s1}^{(+)}(t_j) : \right\rangle, \tag{6.161}$$

where now, according to the $::$ ordering prescription, normal ordering and time ordering (with regard to the atomic source-quantity operators) must be considered.

As already mentioned, spectral properties of light are frequently introduced by means of Fourier decomposition of the field under study, which allows one to define radiation-field correlation functions in the frequency domain. Let us consider the normally ordered correlation function $\Gamma^{(m,n)}$ as given in Eq. (6.158). Introducing the Fourier transform

$$\underline{\Gamma}^{(m,n)}(\{\omega_i, \omega_j\}) = \int \frac{dt_1}{2\pi} e^{-i\omega_1 t_1} \cdots \int \frac{dt_{m+n}}{2\pi} e^{i\omega_{m+n} t_{m+n}} \Gamma^{(m,n)}(t_1, \ldots, t_{m+n}), \tag{6.162}$$

we may relate, on recalling Eq. (6.150), the measured correlation functions (6.159) to the Fourier transforms of the (normally ordered) correlation functions of the incoming unfiltered light as follows:

$$G^{(M)}(\{t_l, x_l\}) = \int d\omega_1 \int d\omega_{M+1} e^{i\tau_1(\omega_1 - \omega_{M+1})} \underline{T}_1^*(\omega_1) \underline{T}_1(\omega_{M+1}) \cdots$$
$$\cdots \int d\omega_M \int d\omega_{2M} e^{i\tau_M(\omega_M - \omega_{2M})} \underline{T}_M^*(\omega_M) \underline{T}_M(\omega_{2M}) \underline{\Gamma}^{(M,M)}(\omega_1, \ldots, \omega_{2M}). \tag{6.163}$$

In the simplest case of measuring the intensity of the filtered light,

$$I = G^{(1)}(t, x) = \langle \mathcal{E}_1^{\prime(-)}(t - x/c) \mathcal{E}_1^{\prime(+)}(t - x/c) \rangle, \tag{6.164}$$

application of Eq. (6.163) yields

$$I = \int d\omega \int d\omega' \, e^{i(\omega - \omega')(t - x/c)} \underline{T}^*(\omega) \underline{T}(\omega') \underline{\Gamma}^{(1,1)}(\omega, \omega'). \tag{6.165}$$

In particular, in the steady-state regime we may write

$$\Gamma^{(1,1)}(t, t') = \Gamma^{(1,1)}(t - t', 0), \tag{6.166}$$

and thus from Eq. (6.162) it follows that

$$\underline{\Gamma}^{(1,1)}(\omega, \omega') = \delta(\omega - \omega') S_1(\omega), \tag{6.167}$$

where $S_1(\omega)$ is the Wiener–Khintchine spectrum (power spectrum) of the light under study. Substitution of this expression into Eq. (6.165) then yields

$$I = \int d\omega\, |\underline{T}(\omega)|^2 S_1(\omega). \tag{6.168}$$

If $|T^2(\omega)|$ becomes sufficiently effective in discriminating against values of ω different from the chosen setting frequency ω_m [cf. Eq. (6.140)], the power spectrum at $\omega = \omega_m$ is detected:

$$I \simeq |\underline{T}(\omega_m)|^2 S_1(\omega_m). \tag{6.169}$$

6.5
Homodyne detection

From Section 6.1 we know that photodetectors respond to the intensity of the incident light. Thus correlation functions of the type $G^{(m,n)}$ with $m = n$ can be observed. In order to measure phase-sensitive properties of light, interferometric methods are required. In the four-port basic scheme (Fig. 6.6), a signal field is combined through a beam splitter with a reference field and the superimposed fields impinge on the photodetectors. Then, however, the correlation functions which are detected, contain contributions of correlation functions $G^{(m,n)}$ of the signal field also with $m \neq n$.

In homodyne detection, a highly stable reference field is used which has the same mid-frequency as the signal field. The reference field, also called local oscillator, is usually prepared in a coherent state of large photon number [see, e. g., Yuen and Shapiro (1978); Shapiro and Yuen (1979); Mandel (1982); Schumaker (1984); Yurke (1985); Walker and Caroll (1984); Carmichael (1987); Walker (1987)]. The observed counting statistics, which vary with the difference in phase between the signal field and the local oscillator, reflect the quantum statistics of the signal field and can be used – under certain circumstances – to obtain the quantum state of the signal field (Chapter 7).

6.5.1
Fields combining through a nonabsorbing beam splitter

The scheme in Fig. 6.6 corresponds, for inclined incidence, to the scheme in Fig. 6.5 supplemented by a second light source. Describing the beam splitter by the simple model of a nonabsorbing dielectric plate, the basic formulas derived in Section 6.4 for a spectral filter also apply to a beam splitter, except that there is no need for specifying the dependence on frequency of the spectral transmittance and reflectance $\underline{T}$ and $\underline{R}$, respectively. The dielectric-plate model leads to input-output relations of the type given in Eqs (6.144) and (6.145), which are valid for a symmetric four-port device where the phases $\phi_{\underline{T}}$

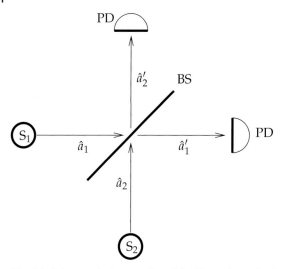

Fig. 6.6 Scheme of a beam splitter BS with two incoming light fields produced by the sources S_1 and S_2. The outgoing fields can be detected by the photodetectors PD.

and ϕ_R of $\underline{T}$ and $\underline{R}$, respectively, satisfy, according to Eq. (6.134), the condition $\phi_T - \phi_R = \pm \pi/2$. In the more general case of an asymmetric, linear four-port device, the transmittance and reflectance from the one side of the device may be expected to be different from those from the other side.

6.5.1.1 Input-output relations

For notational reasons it is convenient to combine the photonic operators $\hat{a}_\nu$ ($\hat{a}'_\nu$) to a two-component vector $\hat{\mathbf{a}} = (\hat{a}_\nu, \hat{a}'_\nu)^T$ so that the equations can be given in a compact form. The extension of the input-output relations (6.144) and (6.145) to an asymmetric device can then be written in the form of

$$\hat{\mathbf{a}}'(\omega) = \mathbf{U}(\omega)\hat{\mathbf{a}}(\omega), \tag{6.170}$$

where the elements of the 2×2 matrix are given as follows:

$$U_{11}(\omega) = \underline{T}(\omega), \quad U_{12}(\omega) = \underline{R}'(\omega), \tag{6.171}$$
$$U_{21}(\omega) = \underline{R}(\omega), \quad U_{22}(\omega) = \underline{T}'(\omega). \tag{6.172}$$

Here $\underline{T}$ and $\underline{R}$ are the transmittance and reflectance of the beam splitter from one side, and $\underline{T}'$ and $\underline{R}'$ are those from the other side. From inspection of Eq. (6.170) we easily see that the bosonic commutation relations are preserved if the transformation matrix $\mathbf{U}$ is a U(2) group matrix:

$$\mathbf{U}^+(\omega) = \mathbf{U}^{-1}(\omega), \tag{6.173}$$

which implies that the following conditions hold for the $U_{\nu\nu'}$:

$$|U_{11}(\omega)|^2 + |U_{12}(\omega)|^2 = |\underline{T}(\omega)|^2 + |\underline{R}'(\omega)|^2 = 1, \tag{6.174}$$

$$|U_{21}(\omega)|^2 + |U_{22}(\omega)|^2 = |\underline{R}(\omega)|^2 + |\underline{T}'(\omega)|^2 = 1, \tag{6.175}$$

$$U_{11}(\omega)U_{21}^*(\omega) + U_{12}(\omega)U_{22}^*(\omega) = \underline{T}(\omega)\underline{R}^*(\omega) + \underline{R}'(\omega)\underline{T}'^*(\omega) = 0, \tag{6.176}$$

from which it follows that

$$|\underline{T}'(\omega)| = |\underline{T}(\omega)|, \quad |\underline{R}'(\omega)| = |\underline{R}(\omega)|, \tag{6.177}$$

$$\phi_T(\omega) - \phi_R(\omega) + \phi_{T'}(\omega) - \phi_{R'}(\omega) = \pm\pi. \tag{6.178}$$

The input-output relations in Eq. (6.170) enables one to calculate arbitrary correlation functions of the outgoing fields from the correlation functions of the incoming fields.

With the redefinitions

$$\varphi(\omega) = \tfrac{1}{2}[\phi_T(\omega) + \phi_{T'}(\omega)], \tag{6.179}$$

$$\varphi_T(\omega) = \tfrac{1}{2}[\phi_T(\omega) - \phi_{T'}(\omega)], \tag{6.180}$$

$$\varphi_R(\omega) = \tfrac{1}{2}[2\phi_{R'}(\omega) - \phi_T(\omega) - \phi_{T'}(\omega)] = \tfrac{1}{2}[\phi_{R'}(\omega) - \phi_R(\omega) \pm \pi], \tag{6.181}$$

and

$$\underline{T}(\omega) = e^{i\varphi_T(\omega)} \cos\vartheta(\omega), \tag{6.182}$$

$$\underline{R}(\omega) = e^{i\varphi_R(\omega)} \sin\vartheta(\omega) \tag{6.183}$$

($0 \leq \vartheta \leq \pi/2$) the matrix $\boldsymbol{U}$ may be rewritten as

$$\boldsymbol{U}(\omega) = e^{i\varphi(\omega)} \begin{pmatrix} \underline{T}(\omega) & \underline{R}(\omega) \\ -\underline{R}^*(\omega) & \underline{T}^*(\omega) \end{pmatrix}. \tag{6.184}$$

It is not difficult to prove that it can be decomposed as follows:

$$\boldsymbol{U}(\omega) = e^{i\varphi(\omega)} \boldsymbol{U}_z^{(\varphi_+)}(\omega) \boldsymbol{U}_y^{(2\vartheta)}(\omega) \boldsymbol{U}_z^{(\varphi_-)}(\omega), \tag{6.185}$$

where

$$\boldsymbol{U}_z^{(\varphi_\pm)}(\omega) = \begin{pmatrix} e^{i\varphi_\pm(\omega)/2} & 0 \\ 0 & e^{-i\varphi_\pm(\omega)/2} \end{pmatrix}, \tag{6.186}$$

$$\boldsymbol{U}_y^{(2\vartheta)}(\omega) = \begin{pmatrix} \cos\vartheta(\omega) & \sin\vartheta(\omega) \\ -\sin\vartheta(\omega) & \cos\vartheta(\omega) \end{pmatrix}, \tag{6.187}$$

and $\varphi_\pm = \varphi_T \pm \varphi_R$. Obviously, the phase factor $e^{i\varphi}$ separated from the transformation matrix U can be thought of as being included in the definition of the operators $\hat{a}_\nu$ and can therefore be omitted. Thus the transformation realized by a beam splitter can always be regarded as an SU(2) group transformation.

6.5.1.2 Quantum-state transformation

To obtain the quantum state of the outgoing fields, the question arises as to which quantum-state transformation corresponds to the operator input–output relations in Eq. (6.170). To answer it, we note that a U(2) or SU(2) group transformation implies a unitary operator transformation according to

$$\hat{\mathbf{a}}'(\omega) = \hat{U}^\dagger \hat{\mathbf{a}}(\omega) \hat{U} = \mathbf{U}(\omega) \hat{\mathbf{a}}(\omega), \tag{6.188}$$

where the unitary operator $\hat{U}$ can be given by

$$\hat{U} = \exp\left\{-i \int_0^\infty d\omega\, [\hat{\mathbf{a}}^\dagger(\omega)]^T \mathbf{\Phi}(\omega) \hat{\mathbf{a}}(\omega)\right\}. \tag{6.189}$$

(the symbol T introduces transposition). Here the 2×2 Hermitian matrix $\mathbf{\Phi}$ is related to the 2×2 unitary matrix $\mathbf{U}$ as[19]

$$e^{-i\mathbf{\Phi}(\omega)} = \mathbf{U}(\omega). \tag{6.190}$$

Let $\hat{\varrho}$ be the density operator of the quantum state of the incoming fields. The effect of the four-port device can then equivalently be described by leaving the photonic operators $\hat{a}_\nu$ unchanged ($\hat{a}'_\nu = \hat{a}_\nu$) but transforming the input-state density operator $\hat{\varrho}$ to obtain the output-state density operator $\hat{\varrho}'$ as

$$\hat{\varrho}' = \hat{U} \hat{\varrho} \hat{U}^\dagger. \tag{6.191}$$

Since $\hat{\varrho}$ can be regarded as being an operator functional of the photonic operators $\hat{a}_\nu(\omega)$ and $\hat{a}^\dagger_\nu(\omega)$, $\hat{\varrho} = \hat{\varrho}[\hat{\mathbf{a}}(\omega), \hat{\mathbf{a}}^\dagger(\omega)]$, from Eq. (6.191) together with Eq. (6.188) it follows that the transformed density operator can be given by

$$\hat{\varrho}'[\hat{\mathbf{a}}(\omega), \hat{\mathbf{a}}^\dagger(\omega)] = \hat{\varrho}[\hat{U}\hat{\mathbf{a}}(\omega)\hat{U}^\dagger, \hat{U}\hat{\mathbf{a}}^\dagger(\omega)\hat{U}^\dagger]$$
$$= \hat{\varrho}[\mathbf{U}^+(\omega)\hat{\mathbf{a}}(\omega), \mathbf{U}^T(\omega)\hat{\mathbf{a}}^\dagger(\omega)]. \tag{6.192}$$

It is often useful and illustrative to describe quantum states in terms of phase-space functions, such as the familiar s-parameterized phase-space functions (Chapter 4). Equation (6.192) implies that an s-parameterized phase-space functional $P[\boldsymbol{\alpha}(\omega), \boldsymbol{\alpha}^*(\omega); s]$ is transformed into

$$P'[\boldsymbol{\alpha}(\omega), \boldsymbol{\alpha}^*(\omega); s] = P[\mathbf{U}^+(\omega)\boldsymbol{\alpha}(\omega), \mathbf{U}^T(\omega)\boldsymbol{\alpha}^*(\omega); s]. \tag{6.193}$$

---
[19] Eqs (6.188)–(6.190) can be proved in a similar way to the SU(1,1) transformation formulas in Section 3.3.

In Eq. (6.193) we have used the fact that application of the unitary transformation under consideration implies preservation of operator ordering, i.e., the annihilation and creation operators are not mixed by the quantum-state transformation.

For practical purposes a description of the incoming and outgoing radiation in terms of discrete modes is frequently preferred. We divide the frequency axis into sufficiently small intervals of mid-frequencies ω_m and widths $\Delta\omega_m$. Now we define the discrete photonic input operators

$$\hat{a}_m = \sqrt{\Delta\omega_m}\, \hat{a}(\omega_m), \tag{6.194}$$

and the discrete photonic output operators $\hat{a}'_m$ accordingly, and assign to each pair of operators $\hat{a}_m$ and $\hat{a}'_m$ the input-output relations (6.170) with the 2×2 matrix $\mathbf{U}_m = \mathbf{U}(\omega_m)$,

$$\hat{\mathbf{a}}'_m = \mathbf{U}_m \hat{\mathbf{a}}_m. \tag{6.195}$$

According to Eq. (6.189) the unitary operator $\hat{U}$ then reads as

$$\hat{U} = \prod_m \hat{U}_m, \tag{6.196}$$

where

$$\hat{U}_m = \exp[-i(\hat{\mathbf{a}}_m^\dagger)^T \mathbf{\Phi}_m \hat{\mathbf{a}}_m] \tag{6.197}$$

with $\mathbf{\Phi}_m = \mathbf{\Phi}(\omega_m)$. Obviously, the discrete-mode concept also applies to nonmonochromatic modes that extend over frequency intervals in which the transmittance and reflectance can be regarded as being (approximately) constant.

The decomposition of each transformation matrix $\mathbf{U}_m$ according to Eqs (6.185)–(6.187) corresponds to an equivalent decomposition of the associated transformation operator $\hat{U}_m$. In particular, for a single mode in each input channel ($m=1$) the U(2) transformation operator can be factorized as

$$\hat{U} = \exp(i\varphi \hat{N})\exp(i\varphi_+\hat{L}_z)\exp(2i\vartheta\hat{L}_y)\exp(i\varphi_-\hat{L}_z), \tag{6.198}$$

where $\hat{N}=\hat{n}_1+\hat{n}_2$, and[20]

$$\hat{L}_y = \tfrac{1}{2i}(\hat{a}_1^\dagger \hat{a}_2 - \hat{a}_2^\dagger \hat{a}_1), \tag{6.199}$$

$$\hat{L}_z = \tfrac{1}{2}(\hat{a}_1^\dagger \hat{a}_1 - \hat{a}_2^\dagger \hat{a}_2). \tag{6.200}$$

---

[20] Note that $\hat{L}_y$, $\hat{L}_z$ and $\hat{L}_x=(\hat{a}_1^\dagger \hat{a}_2+\hat{a}_2^\dagger \hat{a}_1)/2$ are the generators of the SU(2) group, which satisfy the angular-momentum commutation relations ($\hbar=1$). The operator $\hat{L}_0=\hat{N}/2$ completes the set of generators of the U(2) group.

It can be further disentangled to obtain [Wódkiewicz and Eberly (1985)]

$$\hat{U} = \left(e^{i\varphi}\mathcal{T}\right)^{\hat{n}_1} \exp\left(-e^{i\varphi}\mathcal{R}^*\hat{a}_2^\dagger\hat{a}_1\right) \exp\left(e^{-i\varphi}\mathcal{R}\hat{a}_1^\dagger\hat{a}_2\right) \left(e^{-i\varphi}\mathcal{T}\right)^{-\hat{n}_2}. \quad (6.201)$$

Let us consider the simple case where the incoming radiation is prepared in a two-mode coherent state $|\Psi\rangle = |\boldsymbol{\alpha}\rangle$.[21] The state vector of the field in the two output channels, $|\Psi\rangle'$, may then be written as

$$|\Psi\rangle' = \hat{U}|\boldsymbol{\alpha}\rangle \equiv \hat{U}|\alpha_1,\alpha_2\rangle. \quad (6.202)$$

Using Eqs (3.50) and (3.44) we have

$$|\Psi\rangle' = \hat{U}\hat{D}(\boldsymbol{\alpha})\hat{U}^\dagger|0,0\rangle, \quad (6.203)$$

where

$$\hat{D}(\boldsymbol{\alpha}) = \exp(\boldsymbol{\alpha}^T\hat{\mathbf{a}}^\dagger - \text{H.c.}) \quad (6.204)$$

(note that $\hat{U}|0,0\rangle = |0,0\rangle$). It is not difficult to prove that

$$\hat{U}\hat{D}(\boldsymbol{\alpha})\hat{U}^\dagger = \exp[\boldsymbol{\alpha}^T(\hat{U}\hat{\mathbf{a}}\hat{U}^\dagger)^\dagger - \text{H.c.}] = \exp(\boldsymbol{\alpha}'^T\hat{\mathbf{a}}^\dagger - \text{H.c.}), \quad (6.205)$$

where

$$\boldsymbol{\alpha}' = \mathbf{U}\boldsymbol{\alpha}, \quad (6.206)$$

which implies that a coherent state $|\Psi\rangle = |\boldsymbol{\alpha}\rangle$ is transformed into a coherent state $|\Psi\rangle' = |\boldsymbol{\alpha}'\rangle$, where the coherent amplitudes are transformed as the operators in Eq. (6.170).

6.5.2
Fields combining through an absorbing beam splitter

Input-output relations of the type given in Eq. (6.170) can only be valid for nonabsorbing four-port devices. Since for absorbing devices an inequality of the type of $|T|^2 + |R|^2 < 1$ is valid, the characteristic transformation matrix $\mathbf{U}$ defined by Eq. (6.172) cannot be unitary. Application of Eq. (6.170) would not preserve the bosonic commutation relations and thus Eqs (6.146) and (6.147) would be violated. To preserve the commutation relations, Eq. (6.170) must be supplemented with a term that relates the operators $\hat{a}'_\nu(\omega)$ of the outgoing radiation fields to bosonic device operators $\hat{g}_\nu(\omega)$:

$$\hat{\mathbf{a}}'(\omega) = \mathbf{T}(\omega)\hat{\mathbf{a}}(\omega) + \mathbf{A}(\omega)\hat{\mathbf{g}}(\omega). \quad (6.207)$$

[21] For explicit formulas for the transformation of number states, see Yurke, McCall and Klauder (1986); Campos, Saleh and Teich (1989).

Here and in the following the symbol T (in place of U) is used for the 2×2 characteristic transformation matrix [which contains the transmittance and reflectance according to Eq. (6.172)] and A is the 2×2 characteristic absorption matrix. The two matrices obey the equation

$$T(\omega)T^+(\omega) + A(\omega)A^+(\omega) = I \qquad (6.208)$$

(I, identity matrix), which replaces Eq. (6.173). Obviously it ensures the required preservation of the commutation relations.

As shown in Section 6.4, input-output relations of the type given in Eq. (6.170) can be derived by applying mode decomposition on the basis of the (macroscopic) Helmholtz equation (2.195). They are valid for narrow-bandwidth light far from medium resonances. Input-output relations of the type given in Eq. (6.207) apply to optical fields at arbitrary frequencies and bandwidths. They can be derived by applying the more general source-quantity representation introduced in Section 2.6, p. 63 and following [for details, see Gruner and Welsch (1996); Khanbekyan, Knöll and Welsch (2003)]. In this way the matrices T and A can be expressed in terms of the space- and frequency-profile of the complex permittivity of the device. When the imaginary part of the permittivity is disregarded, then the characteristic absorption matrix vanishes and the input-output relations (6.170) are recognized.

The input-output relations (6.207) can be used to calculate various moments and correlations of the outgoing fields from the moments and correlations of the incoming fields and the device excitations. Let us ask for the quantum-state transformation that corresponds to the operator input-output relations (6.207). Due to the absorption we will definitely not be able to construct any unitary transformation which acts on the electromagnetic field operators alone. But we may look for one in the larger Hilbert space which comprises both the electromagnetic field and the device. Defining the four-component vector $\hat{\mathbf{b}}(\omega)$ with $\hat{b}_1(\omega) = \hat{a}_1(\omega)$, $\hat{b}_2(\omega) = \hat{a}_2(\omega)$, $\hat{b}_3(\omega) = \hat{g}_1(\omega)$, $\hat{b}_4(\omega) = \hat{g}_2(\omega)$, we may extend the input-output relations (6.207) to the form

$$\hat{\mathbf{b}}'(\omega) = U(\omega)\hat{\mathbf{b}}(\omega), \qquad (6.209)$$

where $U(\omega)$ is a unitary 4×4-matrix, hence $U(\omega)U^\dagger(\omega) = I$. After separation of some phases from the matrices T and A and their inclusion in the input operators $\hat{a}_\nu(\omega)$ and $\hat{g}_\nu(\omega)$, the matrix U can be regarded (for each ω) as an element of the group SU(4) and it can be expressed in terms of the matrices T and A as [Knöll, Scheel, Schmidt, Welsch and Chizhov (1999)]

$$U(\omega) = \begin{pmatrix} T(\omega) & A(\omega) \\ -S(\omega)C^{-1}(\omega)T(\omega) & C(\omega)S^{-1}(\omega)A(\omega) \end{pmatrix}, \qquad (6.210)$$

where

$$C(\omega) = \sqrt{T(\omega)T^+(\omega)} \qquad (6.211)$$

and

$$S(\omega) = \sqrt{A(\omega)A^+(\omega)}. \tag{6.212}$$

Obviously, Eq. (6.209) [together with Eq. (6.210)] yields the input-output relations (6.207) and it is not difficult to prove that U is a unitary matrix.

The matrix transformation in Eq. (6.209) can then be realized as a unitary operator transformation according to Eqs (6.188)–(6.190) [$\hat{a} \mapsto \hat{b}$ and U from Eq. (6.210)]. Instead the operators may be left unchanged but the quantum state (of the overall system) is transformed according to Eqs (6.191)–(6.193). Projecting $\hat{\varrho}'$ onto the Hilbert space of the radiation field then yields the density operator of the outgoing fields

$$\begin{aligned}\hat{\varrho}'_F[\hat{a}(\omega), \hat{a}^\dagger(\omega)] &= \mathrm{Tr}_D\{\hat{\varrho}'[\hat{b}(\omega), \hat{b}^\dagger(\omega)]\} \\ &= \mathrm{Tr}_D\{\hat{\varrho}[U^+(\omega)\hat{b}(\omega), U^T(\omega)\hat{b}^\dagger(\omega)]\},\end{aligned} \tag{6.213}$$

where Tr_D means the trace with respect to the device. Accordingly, an s-parameterized phase-space functional of the outgoing radiation fields is obtained from the corresponding phase-space functional of the overall system by the functional integral

$$\begin{aligned}P'_F[\alpha(\omega), \alpha^*(\omega); s] &= \int \mathcal{D}\gamma \, P'[\beta(\omega), \beta^*(\omega); s] \\ &= \int \mathcal{D}\gamma \, P[U^+(\omega)\beta(\omega), U^T(\omega)\beta^*(\omega); s],\end{aligned} \tag{6.214}$$

where $\beta_1(\omega) = \alpha_1(\omega)$, $\beta_2(\omega) = \alpha_2(\omega)$, $\beta_3(\omega) = \gamma_1(\omega)$, $\beta_4(\omega) = \gamma_2(\omega)$, with the $\alpha_\nu(\omega)$ and $\gamma_\nu(\omega)$ being the phase-space variables of the radiation and the device, respectively, and the functional integration (notation $\mathcal{D}\gamma$) is taken over the continua of the complex phase-space variables $\gamma_1(\omega)$ and $\gamma_2(\omega)$ of the device.

Let us again consider the simple case of coherent-state transformation,[22] restricting our attention to (quasi-)monochromatic fields, so that it is sufficient to consider only a single frequency component. Suppose the incoming fields and the device are prepared in coherent states, i.e., $|\Psi\rangle = |\beta\rangle \equiv |\alpha_1, \alpha_2, \gamma_1, \gamma_2\rangle$. Then the transformed state $|\Psi\rangle' = \hat{U}|\beta\rangle$ is, according to Eqs (6.202)–(6.206), again a coherent state, $|\Psi\rangle' = |\beta'\rangle$, where $\beta' = U\beta$ with U from Eq. (6.210). Applying Eq. (6.213), we can easily see that the outgoing fields are prepared in coherent states:

$$\hat{\varrho}'_F = |\alpha'\rangle\langle\alpha'|, \qquad \alpha' = T\alpha + A\gamma. \tag{6.215}$$

[22] For explicit formulas for the transformation of number states, see Knöll, Scheel, Schmidt, Welsch and Chizhov (1999).

Thus the coherent amplitudes α'_ν of the outgoing fields are not only determined by the characteristic transformation matrix but also by the characteristic absorption matrix, provided that the device is excited ($\gamma_\nu \neq 0$).

6.5.3
Unbalanced four-port homodyning

As already mentioned, homodyne detection renders it possible to measure phase-sensitive properties of light which are not accessible by direct photodetection. To illustrate this, let us consider the four-port scheme sketched in Fig. 6.6, p. 206.

6.5.3.1 Basic relations

We assume that a single-mode signal field ($\hat{a}_1 = \hat{a}$) prepared in a quantum state $\hat{\varrho}$ is combined, through an (almost) lossless beam splitter, with a (mode-matched) local-oscillator field ($\hat{a}_2 = \hat{a}_\mathrm{L}$) prepared in a coherent state $|\alpha_\mathrm{L}\rangle$, $\alpha_\mathrm{L} = |\alpha_\mathrm{L}|e^{i\varphi_\mathrm{L}}$, so that the two-mode quantum state reads

$$\hat{\sigma} = \hat{\varrho}|\alpha_\mathrm{L}\rangle\langle\alpha_\mathrm{L}|. \tag{6.216}$$

The superimposed light ($\hat{a}'_1$) is recorded by a photodetector, the resulting number of emitted (electronically processed) photoelectrons being the unbalanced homodyne detection output. The modes may be thought of as being pulse-like and the detection-time interval is assumed to cover the full mode extension.

The mean number of photoelectric counts and the variance of counts can be calculated by applying the results of Section 6.2.3. Using Eq. (6.105) and applying the input-output relations (6.195) [with $\boldsymbol{U}_m$ according to Eq. (6.184) for $\varphi = 0$], the mean number of counts is given by

$$\bar{n} = \eta\langle\hat{n}'_1\rangle, \tag{6.217}$$

where

$$\begin{aligned}\hat{n}'_1 = \hat{a}'^\dagger_1\hat{a}'_1 &= (\mathcal{T}\hat{a} + \mathcal{R}\hat{a}_\mathrm{L})^\dagger(\mathcal{T}\hat{a} + \mathcal{R}\hat{a}_\mathrm{L}) \\ &= |\mathcal{T}|^2\hat{n} + |\mathcal{R}|^2\hat{n}_\mathrm{L} + (\mathcal{T}^*\mathcal{R}\hat{a}^\dagger\hat{a}_\mathrm{L} + \mathrm{H.c.}).\end{aligned} \tag{6.218}$$

Substituting the result (6.218) into Eq. (6.217) for $\hat{n}'_1$ and recalling Eq. (6.216), we obtain

$$\bar{n} = \eta[|\mathcal{T}|^2\langle\hat{n}\rangle + |\mathcal{R}|^2|\alpha_\mathrm{L}|^2 + |\mathcal{T}||\mathcal{R}||\alpha_\mathrm{L}|\langle\hat{x}(\varphi)\rangle], \tag{6.219}$$

where the phase φ of the phase-rotated quadrature operator of the signal mode,

$$\hat{x}(\varphi) = \hat{a}e^{i\varphi} + \hat{a}^\dagger e^{-i\varphi}, \tag{6.220}$$

is given by

$$\varphi = \varphi_T - \varphi_R - \varphi_L. \tag{6.221}$$

In Eq. (6.219) the first term in the square brackets is the contribution from the signal-field photon number reduced by $|\mathcal{T}|^2$ at the beam splitter and the second term is the contribution from the local-oscillator photon number reduced by $|\mathcal{R}|^2$ at the beam splitter. The third term obviously results from the interference between the local-oscillator field and the signal field and is closely related to the (mean) electric-field strength of the signal field (the rapidly varying phase in the field operator being replaced by the phase parameter φ). In particular, when the signal and the local oscillator come from the same source, then the phase parameter φ can be controlled easily, so that the dependence on φ of $\hat{x}(\varphi)$ can be controlled by shifting the phase difference between the signal and the local oscillator in the input ports of the beam splitter.

Next let us consider the variance of counts which, according to Eq. (6.106), reads

$$\overline{(\Delta n)^2} = \eta \langle \hat{n}'_1 \rangle + \eta^2 \langle :(\Delta \hat{n}'_1)^2:\rangle. \tag{6.222}$$

Using Eq. (6.218), we may represent $\Delta \hat{n}'_1$ as

$$\Delta \hat{n}'_1 = |\mathcal{T}|^2 \Delta \hat{n} + |\mathcal{R}|^2 \Delta \hat{n}_L + [\mathcal{T}^* \mathcal{R}(\hat{a}^\dagger \hat{a}_L - \langle \hat{a}^\dagger \hat{a}_L\rangle) + \text{H.c.}]. \tag{6.223}$$

Straightforward calculation then leads to the result

$$\overline{(\Delta n)^2} = \bar{n} + \eta^2 \{|\mathcal{T}|^2 |\mathcal{R}|^2 |\alpha_L|^2 \langle :[\Delta \hat{x}(\varphi)]^2:\rangle$$
$$+ 2|\mathcal{T}|^3 |\mathcal{R}||\alpha_L| \langle :\Delta \hat{n} \Delta \hat{x}(\varphi):\rangle + |\mathcal{T}|^4 \langle :(\Delta \hat{n})^2:\rangle \}. \tag{6.224}$$

From Eq. (6.224) the shot-noise level is seen to be determined by the mean number of counts $\bar{n}$. The effect of the signal-field fluctuation is described by the terms in the curly brackets. Whereas the first term results from the fluctuation of the phase-rotated quadrature, which is closely related to the noise of the electric-field strength, the third term results from the photon-number fluctuation. The second term represents the correlation of photon-number and phase-rotated quadrature. Depending on the (quantum) noise properties of the signal field, the noise level of the number of counts can be increased above the shot-noise level as well as reduced below it. In particular, noise reduction is typically a nonclassical effect (Chapter 8).

For example, when the strength of the local-oscillator field greatly exceeds that of the signal field, the mean number of counts (and hence the shot-noise level) is determined by the local-oscillator field,

$$\bar{n} \simeq \eta |\mathcal{R}|^2 |\alpha_L|^2, \tag{6.225}$$

and the effect of the signal-field noise on the variance of counts comes from the term which is related to the normally ordered variance of the phase-rotated quadrature [first term in the curly brackets in Eq. (6.224)]:

$$\overline{(\Delta n)^2} \simeq \bar{n} + \eta^2 |\mathcal{T}|^2 |\mathcal{R}|^2 |\alpha_L|^2 \langle :[\Delta \hat{x}(\varphi)]^2: \rangle. \tag{6.226}$$

If

$$\langle :[\Delta \hat{x}(\varphi)]^2: \rangle < 0, \tag{6.227}$$

then the noise level of the number of photoelectric counts created by the superimposed light is below the shot-noise level. The condition (6.227) indicates that the phase-rotated quadrature fluctuations are squeezed below the vacuum noise level (cf. Section 3.3). We see that homodyne detection is a useful method for the experimental study of squeezed light.

6.5.3.2 Displaced photon-number statistics

Let $\hat{F}$ be an operator that is an arbitrary function of $\hat{a}'_1$ and $\hat{a}'_1{}^\dagger$ in normal order. From the first row in Eq. (6.218) it then follows that

$$\hat{F} = :f(\hat{a}'_1, \hat{a}'_1{}^\dagger): = :f[(\mathcal{T}\hat{a} + \mathcal{R}\hat{a}_L), (\mathcal{T}\hat{a} + \mathcal{R}\hat{a}_L)^\dagger]:. \tag{6.228}$$

Recalling Eq. (6.216), the expectation value of $\hat{F}$ reads

$$\langle \hat{F} \rangle = \mathrm{Tr}\{\hat{\varrho} :f[\mathcal{T}(\hat{a}-\alpha), \mathcal{T}^*(\hat{a}-\alpha)^\dagger]:\}, \tag{6.229}$$

where

$$\alpha = -\frac{\mathcal{R}}{\mathcal{T}} \alpha_L. \tag{6.230}$$

It is not difficult to prove that, on using Eqs (3.47) and (3.48),

$$\mathcal{T}(\hat{a}-\alpha) = |\mathcal{T}| e^{-i\varphi_T \hat{n}(\alpha)} \hat{a}(\alpha) e^{i\varphi_T \hat{n}(\alpha)}, \tag{6.231}$$

where

$$\hat{a}(\alpha) = \hat{a} - \alpha = \hat{D}(\alpha)\hat{a}\hat{D}^\dagger(\alpha), \tag{6.232}$$

and $\hat{n}(\alpha) = \hat{a}^\dagger(\alpha)\hat{a}(\alpha)$ is the displaced photon-number operator of the signal mode. Combining Eqs (6.229) and (6.231), we derive

$$\langle \hat{F} \rangle = \mathrm{Tr}\{\hat{\varrho}' :f(|\mathcal{T}|\hat{a}, |\mathcal{T}|\hat{a}^\dagger):\}, \tag{6.233}$$

where

$$\hat{\varrho}' = \hat{D}^\dagger(\alpha) e^{i\varphi_T \hat{n}(\alpha)} \hat{\varrho} e^{-i\varphi_T \hat{n}(\alpha)} \hat{D}(\alpha). \tag{6.234}$$

When a signal and a strong local oscillator, $|\alpha_L| \to \infty$, are mixed by a beam splitter with high transmittance, $|T| \to 1$, and low reflectance, $|\mathcal{R}| \to 0$, such that the product $|\mathcal{R}\alpha_L|$ is finite, then

$$\langle \hat{F} \rangle \to \mathrm{Tr}\{\hat{\varrho}' : f(\hat{a}, \hat{a}^\dagger):\}; \tag{6.235}$$

that is, the quantum state of the outgoing signal is just given by the density operator $\hat{\varrho}'$ in Eq. (6.234). In particular, for $\varphi_T = 0$ a (coherent) displacement of the quantum state of the signal is realized. Otherwise, only the diagonal matrix element of $\hat{\varrho}'_1$ in the number basis are given by the diagonal matrix elements of $\hat{\varrho}_1$ in the displaced number basis, because

$$\langle m|\hat{\varrho}'|n\rangle \to \langle m, \alpha|\hat{\varrho}|n, \alpha\rangle\, e^{i\varphi_T(m-n)}, \tag{6.236}$$

with the

$$|n, \alpha\rangle = \hat{D}(\alpha)|n\rangle \tag{6.237}$$

being the displaced photon-number states, which are the eigenstates of the displaced photon-number operator (Section 3.2.3). The result reveals that homodyne detection can be used to measure the displaced photon-number statistics of the signal field [Wallentowitz and Vogel (1996); Banaszek and Wódkiewicz (1996)]. Identifying the operator $\hat{F}$ with the generalized projector $\hat{P}_m$ [Eq. (6.96)], whose expectation value is the probability of recording m counts, $P_m = \langle \hat{P}_m \rangle$ [Eq. (6.95)], application of Eqs (6.228), (6.229) and (6.231) yields

$$\hat{P}_m = :\frac{1}{m!}[\eta_T \hat{n}(\alpha)]^m e^{-\eta_T \hat{n}(\alpha)}:, \tag{6.238}$$

where

$$\eta_T = \eta|T|^2 \tag{6.239}$$

is the (overall) quantum efficiency with which the displaced photon-number statistics can be measured. Hence $P_m \to p_m(\alpha) = \langle m, \alpha|\hat{\varrho}|m, \alpha\rangle$ if $\eta_T \to 1$. In practice $|T|^2$ is always smaller than one, so that η_T is smaller than η, which itself is smaller than one, in general. Note that precise measurement of the displaced photon-number statistics in unbalanced homodyning requires highly efficient photodetectors which can discriminate between m and $m+1$ photons.

In order to obtain the displaced photon-number statistics $p_m(\alpha)$ for all (relevant) displacements α, a succession of ensemble measurements must be performed, α being controlled by the local-oscillator complex amplitude α_L [cf. Eq. (6.230)]. Whereas for a chosen value of α the quantity $p_m(\alpha)$ is an ordinary probability for the displaced photon number m of the signal mode, for a chosen value of m it can be regarded (apart from the factor π) as a probability distribution for α,

$$p_m(\alpha) = \pi\langle \hat{P}(\alpha)\rangle, \tag{6.240}$$

which corresponds to the POVM[23]

$$\hat{P}_{\hat\sigma}(\alpha) = \pi^{-1}\hat{D}(\alpha)\hat{\sigma}\hat{D}^\dagger(\alpha) \tag{6.241}$$

with

$$\hat{\sigma} = |m\rangle\langle m|. \tag{6.242}$$

Obviously, it generalizes the Q function ($m=0$) to $m>0$ and thus contains (for each value of m) all knowable information on the signal-mode quantum state. Measurement of the displaced photon-number statistics as a function of α is therefore expected to yield more data than the minimum necessary for reconstructing it (see Section 7.3.2).

6.5.4
Balanced four-port homodyning

The calculation of the photon number in the second output channel of the beam splitter in Fig. 6.6, p. 206, which can be performed analogously to the calculation of the photon number in the first output channel, Eq. (6.218), yields

$$\hat{n}'_2 = \hat{a}'^\dagger_2 \hat{a}'_2 = (-\mathcal{R}^*\hat{a} + \mathcal{T}^*\hat{a}_L)^\dagger(-\mathcal{R}^*\hat{a} + \mathcal{T}^*\hat{a}_L)$$
$$= |\mathcal{R}|^2\hat{n} + |\mathcal{T}|^2\hat{n}_L - (\mathcal{T}^*\mathcal{R}\hat{a}^\dagger\hat{a}_L + \text{H.c.}). \tag{6.243}$$

When, in a balanced scheme, a 50%:50% beam splitter is used, then the relation $|\mathcal{T}|=|\mathcal{R}|=1/\sqrt{2}$ is valid, and hence the difference photon number $\langle\hat{n}'_1\rangle - \langle\hat{n}'_2\rangle$ is proportional to the phase-rotated quadrature of the signal mode,

$$\langle\hat{n}'_1\rangle - \langle\hat{n}'_2\rangle = |\alpha_L|\langle\hat{x}(\varphi)\rangle, \tag{6.244}$$

as is easily seen by combining Eqs (6.218) and (6.243). This result suggests that it is advantageous to use a balanced scheme and to measure the difference events or the corresponding difference of the photocurrents of the detectors in the two output channels in order to eliminate the intensities of the two input fields from the measured output. The method is also called balanced homodyning and can be used advantageously in order to suppress perturbing effects due to classical excess noise of the local oscillator. In particular, when the local oscillator is much stronger than the signal field, then even small intensity

---

[23] Note that for any density operator $\hat{\sigma}$ the (continuous) set of operators $\hat{P}_{\hat\sigma}(\alpha)$ represents a POVM, and it can be shown that $\langle\hat{P}_{\hat\sigma}(\alpha)\rangle$, which is also called the propensity, can be given by a convolution of the Wigner function of the signal with the Wigner function of a reference system prepared in a quantum state $\hat{\sigma}$ [see, e.g., Walker (1987)]. The reference system which acts as a filter is also called the quantum ruler.

fluctuations of the local oscillator may significantly disturb the signal-mode quadrature components which are required to be observed. For example, if the quadrature-component variance is intended to be derived from the variance of events measured by a single detector, the classical noise of the local oscillator and the quantum noise of the signal would contribute to the measured data in the same manner, so that the two effects are hardly distinguishable. Since in the two output channels identical classical-noise effects are observed, in the balanced scheme they eventually cancel in the measured signal, due to the subtraction procedure.

From the arguments given so far it is only established that for chosen phase parameters the mean value of the measured difference events or photocurrents is proportional to the expectation value of a quadrature component of the signal mode. From a more careful (quantum-mechanical) analysis it can be shown that in perfect balanced homodyning the quadrature-component statistics of the signal mode are indeed measured, provided that the local oscillator is sufficiently strong compared with the mean number of signal photons [Carmichael (1987); Braunstein (1990); Vogel and Grabow (1993)]. To derive the statistics of difference events, we start from the joint-event probability P_{m_1,m_2} of measuring m_1 and m_2 events in the two output channels in Fig. 6.6. Applying Eqs (6.110) and (6.111) together with Eq. (6.96), we have

$$P_{m_1,m_2} = \langle \hat{P}_{m_1,m_2} \rangle, \tag{6.245}$$

where the POVM operators

$$\hat{P}_{m_1,m_2} = \, : \prod_{l=1}^{2} \frac{(\eta_l \hat{n}'_l)^{m_l}}{m_l!} e^{-\eta_l \hat{n}'_l} : , \tag{6.246}$$

with $\hat{n}'_1$ and $\hat{n}'_2$ given by Eqs (6.218) and (6.243), respectively (η_l, quantum efficiency for detecting a photon in the lth output channel). The probability for the difference events

$$\Delta m = m_1 - m_2 \tag{6.247}$$

is obtained from the joint probability as

$$P_{\Delta m} = \langle \hat{P}_{\Delta m} \rangle, \tag{6.248}$$

where

$$\hat{P}_{\Delta m} = \sum_{m_2} \hat{P}_{m_2+\Delta m, m_2}. \tag{6.249}$$

Introducing the c-number functions $P_{m_1,m_2}(\alpha_1, \alpha_2; 1)$ and $P_{\Delta m}(\alpha_1, \alpha_2; 1)$ that are associated with $\hat{P}_{m_1,m_2}$ and $\hat{P}_{\Delta m}$, respectively, in normal order ($s=1$), we may write

$$P_{m_1,m_2}(\alpha_1,\alpha_2;1) = \prod_{l=1}^{2} \frac{(\eta_l|\alpha'_l|^2)^{m_l}}{m_l!} e^{-\eta_l|\alpha'_l|^2}, \quad (6.250)$$

$$P_{\Delta m}(\alpha_1,\alpha_2;1) = \sum_{m_2} P_{m_2+\Delta m, m_2}(\alpha_1,\alpha_2;1), \quad (6.251)$$

where

$$|\alpha'_1|^2 = |\mathcal{T}\alpha_1 + \mathcal{R}\alpha_2|^2, \quad (6.252)$$
$$|\alpha'_2|^2 = |-\mathcal{R}^*\alpha_1 + \mathcal{T}^*\alpha_2|^2. \quad (6.253)$$

Substituting the result (6.250) into Eq. (6.251) for $P_{m_2+\Delta m, m_2}$ and performing the m_2 summation, we derive

$$P_{\Delta m}(\alpha_1,\alpha_2;1) = \left(\frac{\eta_1|\alpha'_1|^2}{\eta_2|\alpha'_2|^2}\right)^{\frac{\Delta m}{2}} I_{\Delta m}\left(2\sqrt{\eta_1|\alpha'_1|^2\eta_2|\alpha'_2|^2}\right) e^{-\eta_1|\alpha'_1|^2} e^{-\eta_2|\alpha'_2|^2}, \quad (6.254)$$

where $I_n(z)$ is the modified Bessel function [note that $I_{-n}(z)=I_n(z)$]. Since the local-oscillator mode is prepared, according to Eq. (6.216), in a coherent state, we may set $\alpha_2=\alpha_L$, so that $P_{\Delta m}(\alpha_1,\alpha_2;1)$ effectively becomes a function solely of $\alpha\equiv\alpha_1$:[24]

$$P_{\Delta m}(\alpha_1,\alpha_2;1) \mapsto P_{\Delta m}(\alpha,\alpha_L;1) \equiv P_{\Delta m}(\alpha;1). \quad (6.255)$$

Now we consider $P_{\Delta m}(\alpha;1)$ in the limit of a strong local oscillator when $|\alpha_L|^2$ is large compared with the average number of photons in the signal field. Let us assume that the detectors are identical,

$$\eta_1 = \eta_2 = \eta, \quad (6.256)$$

and consider a 50%:50% beam splitter,

$$|\mathcal{T}| = |\mathcal{R}| = \sqrt{\tfrac{1}{2}}, \quad (6.257)$$

which implies that Eqs (6.252) and (6.253) reduce to

$$|\alpha'_1|^2 = \tfrac{1}{2}|\alpha + e^{-i(\varphi_T-\varphi_R)}\alpha_L|^2, \quad (6.258)$$
$$|\alpha'_2|^2 = \tfrac{1}{2}|\alpha - e^{-i(\varphi_T-\varphi_R)}\alpha_L|^2. \quad (6.259)$$

---

24) Note that $P_{\Delta m}(\alpha;1)$ is the c-number function associated with the operator $\hat{P}_{\Delta m}$ averaged with respect to the local oscillator.

Using the asymptotic expression of the modified Bessel function,[25]

$$I_n(z) \simeq \frac{1}{\sqrt{2\pi z}} \exp\left(z - \frac{n^2}{2z}\right) \tag{6.260}$$

($|z| \to \infty$, $n \to \infty$, $\sqrt{|n^2/z|}$ finite) and the approximations

$$\left(\frac{|\alpha_1'|^2}{|\alpha_2'|^2}\right)^{\frac{\Delta m}{2}} = \exp\left(\Delta m \ln\left|\frac{\alpha + \lambda \alpha_L}{\alpha - \lambda \alpha_L}\right|\right) \simeq \exp\left[2\Delta m \frac{\text{Re}\{\lambda \alpha_L \alpha^*\}}{|\alpha_L|^2}\right], \tag{6.261}$$

$$\exp(2\eta|\alpha_1'||\alpha_2'| - \eta|\alpha_1'|^2 - \eta|\alpha_2'|^2) = \exp\left[-\tfrac{1}{2}\eta(|\alpha + \lambda \alpha_L| - |\alpha - \lambda \alpha_L|)^2\right]$$

$$\simeq \exp\left\{-\frac{2\eta}{|\alpha_L|^2}[\text{Re}(\lambda \alpha_L \alpha^*)]^2\right\} \tag{6.262}$$

($\lambda = e^{-i(\varphi_T - \varphi_R)}$), Eq. (6.254) reduces to

$$P_{\Delta m}(\alpha;1) = \frac{1}{\sqrt{2\pi\eta|\alpha_L|^2}} \exp\left\{-\frac{[\Delta m - \eta(|\alpha_L|e^{i\varphi}\alpha + \text{c.c.})]^2}{2\eta|\alpha_L|^2}\right\}, \tag{6.263}$$

with φ being defined by Eq. (6.221). Introducing the phase-rotated quadrature operator $\hat{x}(\varphi)$ [Eq. (6.220)], we may rewrite Eq. (6.263) as

$$P_{\Delta m}(\alpha;1) = \frac{1}{\sqrt{2\pi\eta|\alpha_L|^2}} \exp\{-\tfrac{1}{2}\eta[\Delta m/(\eta|\alpha_L|) - \langle\alpha|\hat{x}(\varphi)|\alpha\rangle]^2\}. \tag{6.264}$$

Comparing Eq. (6.264) with Eq. (3.202), we see that for perfect detection ($\eta=1$) the function $P_{\Delta m}(\alpha;1)$ is (apart from the factor $|\alpha_L|^{-1}$) the c-number function associated with a phase-rotated quadrature projector of the signal mode, thus

$$\hat{P}_{\Delta m} = \frac{1}{|\alpha_L|}|x,\varphi\rangle\langle x,\varphi| \leftrightarrow P_{\Delta m} = \frac{1}{|\alpha_L|}p(x,\varphi), \tag{6.265}$$

with x being related to the number of difference events Δm according to

$$x = \frac{\Delta m}{|\alpha_L|}. \tag{6.266}$$

Hence the measured difference-event probability distribution is a (scaled) phase-rotated quadrature distribution. Note that when the local oscillator is sufficiently strong, then $\Delta m/|\alpha_L|$ is effectively continuous. In other words,

---

25) Equation (6.260) can be obtained from a saddle-point analysis of the integral representation
$I_n(z) = (z/2)^n/[\Gamma(1/2)\Gamma(n+1/2)]\int_{-1}^{1} dt\,(1-t^2)^{n-1/2}e^{zt}$, $n > 1/2$
[Freyberger, Vogel and Schleich (1993)].

single-photon resolution is not needed in order to measure the (continuous) quadrature-component statistics with high accuracy. In this case, highly efficient linear response photodiodes can be used, which do not discriminate between single photons but which nearly reach 100% quantum efficiency.

In the case of nonperfect detection ($\eta < 1$) the function $P_{\Delta m}(\alpha; 1)$ is (apart from the factor $\eta |\alpha_L|^{-1}$) a convolution of the c-number functions associated with the phase-rotated quadrature projectors (of chosen φ) of the signal mode with a Gaussian [Vogel and Grabow (1993)]. That is, $\hat{P}_{\Delta m}$ is a statistical mixture of phase-rotated quadrature projectors:

$$\hat{P}_{\Delta m} = \frac{1}{\eta |\alpha_L|} \int dx'\, q(x-x'; \eta)\, |x', \varphi\rangle \langle x', \varphi|, \qquad (6.267)$$

where

$$q(x; \eta) = \sqrt{\frac{\eta}{2\pi(1-\eta)}}\, \exp\left[-\frac{\eta x^2}{2(1-\eta)}\right] \qquad (6.268)$$

and

$$x = \frac{\Delta m}{\eta |\alpha_L|}. \qquad (6.269)$$

Accordingly, the measured difference-event probability distribution is a convolution of the phase-rotated quadrature distribution $p(x', \varphi)$ with the Gaussian $q(x-x'; \eta)$:

$$P_{\Delta m} = \frac{1}{\eta |\alpha_L|}\, p\!\left(x = \frac{\Delta m}{\eta |\alpha_L|}, \varphi; \eta\right), \qquad (6.270)$$

$$p(x, \varphi; \eta) = \int dx'\, q(x-x'; \eta)\, p(x', \varphi). \qquad (6.271)$$

Note that when $\eta \to 1$ then $q(x-x'; \eta) \to \delta(x-x')$ and the convolution exactly reduces to $p(x, \varphi)$. The Gaussian [with variance $\sigma^2 = (1-\eta)/\eta$] obviously reflects the noise associated with nonperfect detection.[26] The distribution $p(x, \varphi; \eta)$ corresponds to the phase-rotated quadrature $\hat{X}(\varphi)$ of a superposition of the phase-rotated quadrature of the signal, $\hat{x}(\varphi)$, and that of an additional (Gaussian) noise source, $\hat{x}_N(\varphi)$,

$$\hat{X}(\varphi) = \sqrt{\eta}\, \hat{x}(\varphi) + \sqrt{1-\eta}\, \hat{x}_N(\varphi), \qquad (6.272)$$

so that $p(X, \varphi) = p(x/\sqrt{\eta}, \varphi; \eta)/\sqrt{\eta}$. In particular, Eq. (6.272) reveals that the effect of nonperfect detection can be modeled by assuming a (virtual) beam splitter placed in front of a perfect detector, since Eq. (6.272) exactly corresponds to a beam-splitter transformation [Yurke and Stoler (1987)]. In this

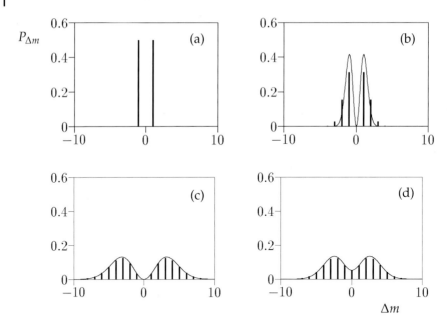

Fig. 6.7 Difference statistics $P_{\Delta m}$ for a signal mode prepared in a one-photon number state for various mean photon numbers of the local oscillator: $|\alpha_L|^2 = 0$ (a), 0.5 (b) and 5 (c,d); $\eta = 1$ in (a)–(c) and $\eta = 0.75$ in (d). The solid curves represent the (scaled) phase-rotated quadrature distribution $p(x, \varphi; \eta)$, Eq. (6.271). [After Vogel and Grabow (1993).]

case the fields are only partly detected, together with a fraction of vacuum noise introduced through the "unused" input ports of the beam splitters.

To illustrate the method let us consider the case where the signal field is prepared in a one-photon number state. In order to calculate the probabilities $P_{\Delta m}$ for arbitrary strength of the local oscillator, we note that when the P-function of the signal mode is known, then "averaging" $P_{\Delta m}(\alpha; 1)$ with the P-function immediately yields $P_{\Delta m}$:

$$P_{\Delta m} = \int d^2\alpha \, P(\alpha) P_{\Delta m}(\alpha; 1). \tag{6.273}$$

Calculating $P_{\Delta m}(\alpha; 1)$ according to Eqs (6.254) and (6.255), substituting for the $P(\alpha)$ the result (4.117) with $n = 1$ and integrating by parts, we derive

$$P_{\Delta m} = \left\{ 1 - \tfrac{1}{2}(\eta_1 + \eta_2) + \left[\frac{\Delta m}{|\alpha_L|} - \tfrac{1}{2}(\eta_1 - \eta_2)|\alpha_L| \right]^2 \right\} \left(\frac{\eta_1}{\eta_2} \right)^{\frac{\Delta m}{2}}$$
$$\times I_{\Delta m}\left(\sqrt{\eta_1 \eta_2}\,|\alpha_L|^2\right) \exp\left[-\tfrac{1}{2}|\alpha_L|^2(\eta_1 + \eta_2)\right], \tag{6.274}$$

26) Note that for $\eta = 1/2$ the Gaussian is the (phase-independent) phase-rotated quadrature distribution of the vacuum ($\sigma^2 = 1/2$).

which in the limit of a strong local oscillator and equal detector efficiencies reduces, in agreement with Eqs (6.270) and (6.271), to

$$P_{\Delta m} = \frac{1}{\sqrt{2\pi\eta|\alpha_L|^2}} \left[1 - \eta + \frac{(\Delta m)^2}{|\alpha_L|^2}\right] \exp\left[-\frac{(\Delta m)^2}{2\eta|\alpha_L|^2}\right]. \tag{6.275}$$

In Fig. 6.7 the difference-event probability $P_{\Delta m}$ as given in Eq. (6.274) for a signal field prepared in a one-photon number state $|n=1\rangle$, is shown for various values of the mean photon number of the local oscillator, $|\alpha_L|^2$, and the result is compared with the phase-rotated quadrature distribution as given by Eq. (6.271). We see that in the limit as $|\alpha_L| \to 0$ the difference-count statistics simply reflect the fact that the probability of detecting the input photon in one of the output channels is $1/2$, so that $P_{\Delta m} = 1/2$ for $\Delta m = \pm 1$ and $P_{\Delta m} = 0$ otherwise. It is worth noting that (for a signal prepared in a one-photon state) only very few local-oscillator photons are needed to find the difference-event probabilities close to the phase-rotated quadrature probabilities.

6.5.5
Balanced eight-port homodyning

From Eq. (6.265) together with Eq. (6.221) we readily find that $p(x, \varphi + \Delta\varphi)$ can be observed in a separate measurement, in which a phase shifter is inserted, e.g., in the local-oscillator beam such that $\varphi_L \to \varphi_L - \Delta\varphi$. In this way, by varying $\Delta\varphi$ from measurement to measurement, the phase-rotated quadrature distribution can be determined as a function of the phase parameter φ. In particular, to determine the distributions of two $\pi/2$ shifted quadrature components ("position" and "momentum") two separate measurements with phases φ and $\varphi \pm \pi/2$ are needed. At this point the question arises of what is measured when the two measurements are combined in an eight-port detection scheme as shown in Fig. 6.8. Straightforward application of the input-output relations (6.195) [with U_m according to Eq. (6.184) for $\varphi = 0$] yields, on assuming identical beam splitters,

$$\hat{a}'_1 = |\mathcal{T}|^2 \hat{a}_1 + \mathcal{RT}^* \hat{a}_2 - |\mathcal{R}|^2 \hat{a}_3 - \mathcal{R}^* \mathcal{T} \hat{a}_4, \tag{6.276}$$

$$\hat{a}'_2 = -i\mathcal{R}^* \mathcal{T}^* \hat{a}_1 + i\mathcal{T}^{*2} \hat{a}_2 - \mathcal{R}^* \mathcal{T}^* \hat{a}_3 + \mathcal{R}^{*2} \hat{a}_4, \tag{6.277}$$

$$\hat{a}'_3 = -i|\mathcal{R}|^2 \hat{a}_1 + i\mathcal{RT}^* \hat{a}_2 + |\mathcal{T}|^2 \hat{a}_3 - \mathcal{R}^* \mathcal{T} \hat{a}_4, \tag{6.278}$$

$$\hat{a}'_4 = \mathcal{RT} \hat{a}_1 + \mathcal{R}^2 \hat{a}_2 + \mathcal{RT} \hat{a}_3 + \mathcal{T}^2 \hat{a}_4, \tag{6.279}$$

from which the output photon-number operators $\hat{n}'_l = \hat{a}'^\dagger_l \hat{a}'_l$ ($l = 1, \ldots, 4$) can easily be obtained. The operators $\hat{P}_{\{m_l\}}$ of the POVM for the measured joint-event probabilities $P_{\{m_l\}}$ are then given by

$$\hat{P}_{\{m_l\}} = : \prod_{l=1}^{4} \frac{(\eta_l \hat{n}'_l)^{m_l}}{m_l!} e^{-\eta_l \hat{n}'_l} : \tag{6.280}$$

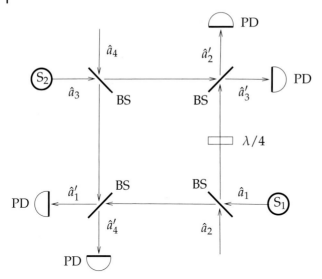

Fig. 6.8 Eight-port detection scheme of four beam splitters BS and a $\lambda/4$ phase shifter. Light from the sources S_1 and S_2 are fed into two input ports, whereas two input ports are unused. The outgoing fields can be detected by the photodetectors PD.

[cf. Eq. (6.246)] and the associated c-number functions in normal order read, on using Eqs (6.276)–(6.279) and recalling that the input channels 2 and 4 are unused,[27]

$$P_{\{m_l\}}(\alpha_1, \alpha_3; 1) = \,:\prod_{l=1}^{4} \frac{(\eta_l |\alpha'_l|^2)^{m_l}}{m_l!} e^{-\eta_l |\alpha'_l|^2} : \qquad (6.281)$$

[cf. Eq. (6.250)], where

$$|\alpha'_1|^2 = ||\mathcal{T}|^2 \alpha_1 - |\mathcal{R}|^2 \alpha_3|^2, \qquad (6.282)$$

$$|\alpha'_2|^2 = |\mathcal{R}|^2 |\mathcal{T}|^2 |\alpha_1 - i\alpha_3|^2, \qquad (6.283)$$

$$|\alpha'_3|^2 = ||\mathcal{R}|^2 \alpha_1 + i|\mathcal{T}|^2 \alpha_2|^2, \qquad (6.284)$$

$$|\alpha'_4|^2 = |\mathcal{R}|^2 |\mathcal{T}|^2 |\alpha_1 + \alpha_3|^2. \qquad (6.285)$$

Introducing the POVM for the joint probability of the difference counts $\Delta m_1 = m_1 - m_4$ and $\Delta m_2 = m_3 - m_2$,

$$\hat{P}_{\Delta m_1, \Delta m_2} = \sum_{m_4, m_6} \hat{P}_{m_4 + \Delta m_1, m_4, m_6 + \Delta m_2, m_6} \qquad (6.286)$$

[27] Strictly speaking, $P_{\{m_l\}}(\alpha_1, \alpha_3; 1)$ is the c-number function associated with $\hat{P}_{\{m_l\}}$ averaged with respect to the unused input channels ($\alpha_2 = \alpha_4 = 0$).

[cf. Eq. (6.249)], whose associated c-number function in normal order reads

$$P_{\Delta m_1, \Delta m_2}(\alpha_1, \alpha_3; 1) = K_{\Delta m_1}(\alpha_1, \alpha_3) K_{\Delta m_2}(\alpha_1, \alpha_3), \quad (6.287)$$

where

$$K_{\Delta m_1}(\alpha_1, \alpha_3) = \left(\frac{\eta_1 |\alpha'_1|^2}{\eta_4 |\alpha'_4|^2}\right)^{\frac{\Delta m_1}{2}} I_{\Delta m_1}\left(2\sqrt{\eta_1 |\alpha'_1|^2 \eta_4 |\alpha'_4|^2}\right) e^{-\eta_1 |\alpha'_1|^2} e^{-\eta_4 |\alpha'_4|^2}, \quad (6.288)$$

and $K_{\Delta m_2}(\alpha_1, \alpha_3)$ accordingly [cf. Eq. (6.254)]. If we again assume that 50%:50% beam splitters are used, so that

$$|\alpha'_1|^2 = \tfrac{1}{4}|\alpha_1 - \alpha_3|^2, \quad |\alpha'_4|^2 = \tfrac{1}{4}|\alpha_1 + \alpha_3|^2, \quad (6.289)$$

$$|\alpha'_3|^2 = \tfrac{1}{4}|\alpha_1 + i\alpha_3|^2, \quad |\alpha'_2|^2 = \tfrac{1}{4}|\alpha_1 - i\alpha_3|^2, \quad (6.290)$$

and that the detectors have equal efficiencies, the functions $K_{\Delta m_1}(\alpha_1, \alpha_3)$ and $K_{\Delta m_2}(\alpha_1, \alpha_3)$ are

$$K_{\Delta m_1}(\alpha_1, \alpha_3) = \left|\frac{\alpha_1 - \alpha_3}{\alpha_1 + \alpha_3}\right|^{\Delta m_1} I_{\Delta m_1}(\tfrac{1}{2}\eta|\alpha_1^2 - \alpha_3^2|) \exp\left[-\tfrac{1}{2}\eta(|\alpha_1|^2 + |\alpha_3|^2)\right], \quad (6.291)$$

and

$$K_{\Delta m_2}(\alpha_1, \alpha_3) = \left|\frac{\alpha_1 + i\alpha_3}{\alpha_1 - i\alpha_3}\right|^{\Delta m_2} I_{\Delta m_2}(\tfrac{1}{2}\eta|\alpha_1^2 + \alpha_3^2|) \exp\left[-\tfrac{1}{2}\eta(|\alpha_1|^2 + |\alpha_3|^2)\right]. \quad (6.292)$$

We now regard the fields in the input channels 1 and 3 as being the signal field and a strong local oscillator prepared in a coherent state $|\alpha_L\rangle$, respectively, so that $(\alpha = \alpha_1)$

$$P_{\Delta m_1, \Delta m_2}(\alpha_1, \alpha_3; 1) \mapsto P_{\Delta m_1, \Delta m_2}(\alpha, \alpha_L; 1) \equiv P_{\Delta m_1, \Delta m_2}(\alpha; 1). \quad (6.293)$$

[cf. Eq. (6.255)]. Using the asymptotic expression (6.260) for the modified Bessel function and performing similar approximations as in the derivation of Eq. (6.263), we deduce that

$$P_{\Delta m_1, \Delta m_2}(\alpha; 1) = \frac{1}{\pi \eta |\alpha_L|^2} \exp\left\{-\frac{[2\Delta m_1 + \eta(\alpha_L^* \alpha + \text{c.c.})]^2}{4\eta |\alpha_L|^2}\right\}$$
$$\times \exp\left\{-\frac{[2\Delta m_2 + i\eta(\alpha_L^* \alpha - \text{c.c.})]^2}{4\eta |\alpha_L|^2}\right\}, \quad (6.294)$$

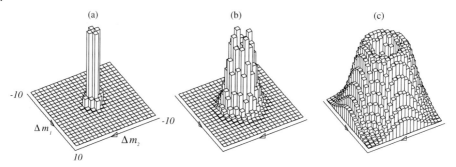

Fig. 6.9 Joint-difference statistics $P_{\Delta m_1, \Delta m_2}$ for a signal mode prepared in a one-photon number state for $\eta = 1$ and various mean photon numbers of the local oscillator: $|\alpha_L|^2 = 0.25$ (a), 4 (b) and 25 (c). [After Freyberger, Vogel and Schleich (1993).]

which can be rewritten as

$$P_{\Delta m_1, \Delta m_2}(\alpha; 1) = \frac{1}{\pi \eta |\alpha_L|^2} \exp\left[-\eta \left| \frac{i\Delta m_2 - \Delta m_1}{\eta \alpha_L^*} - \alpha \right|^2 \right]. \quad (6.295)$$

We compare Eq. (6.295) with Eq. (3.65) and see that for perfect detection ($\eta = 1$) the function $P_{\Delta m_1, \Delta m_2}(\alpha; 1)$ is (apart from the factor $|\alpha_L|^{-2}$) the Q function of the signal mode, thus

$$\hat{P}_{\Delta m_1, \Delta m_2} = \frac{1}{|\alpha_L|^2} |\alpha\rangle\langle\alpha| \leftrightarrow P_{\Delta m_1, \Delta m_2} = \frac{1}{|\alpha_L|^2} Q(\alpha), \quad (6.296)$$

with α being given by

$$\alpha = \frac{i\Delta m_2 - \Delta m_1}{\alpha_L^*}. \quad (6.297)$$

That is, for a strong local oscillator and perfect detection the measured joint probability distribution of the difference events, $P_{\Delta m_1, \Delta m_2}$, is the (scaled) Q function of the signal field [Freyberger and Schleich (1992)].[28] In Fig. 6.9 the dependence on the strength of the local oscillator of the (unnormalized) probabilities is shown for the case of a signal field prepared in a one-photon number state. It is not difficult to prove that for nonperfect detection ($\eta < 1$) the measured probability distribution $P_{\Delta m_1, \Delta m_2}$ is a (scaled) s-parameterized phase-space function $P(\alpha; s)$ with $s < -1$ [Leonhardt and Paul (1993)]:

$$P_{\Delta m_1, \Delta m_2} = \frac{1}{\eta^2 |\alpha_L|^2} P\left(\alpha = \frac{i\Delta m_2 - \Delta m_1}{\eta \alpha_L^*}; 1 - \frac{2}{\eta}\right). \quad (6.298)$$

28) It can be shown that six-port homodyne detection is the minimum in order to determine the Q function [Zucchetti, Vogel and Welsch (1996)].

Note that for $s < -1$ the function $P(\alpha; s)$ is a convolution of the Q function $Q(\alpha) = P(\alpha; -1)$ with a Gaussian [cf. Eq. (4.70)], which for $\eta = 1/2$ is just the vacuum Q function.

To relate the results of measurements in an eight-port scheme to the results of two independent four-port-scheme measurements which yield the probability distributions of two $\pi/2$ shifted quadrature components, we note that Eq. (6.294) can be rewritten as

$$P_{\Delta m_1, \Delta m_2}(\alpha; 1) = \prod_{k=1}^{2} P^{(k)}_{\Delta m_k}(\alpha; 1), \tag{6.299}$$

where $P^{(k)}_{\Delta m_k}(\alpha; 1)$ is given by

$$P^{(k)}_{\Delta m_k}(\alpha; 1) = \frac{1}{\sqrt{2\pi \eta_k |\alpha_L|^2}} \exp\{-\tfrac{1}{2}\eta_k[\Delta m_k/(\eta_k |\alpha_L|) - \langle \alpha | \hat{x}(\varphi_k) | \alpha \rangle]^2\}, \tag{6.300}$$

where $\varphi_1 = \pi - \varphi_L$, $\varphi_2 = \varphi_1 + \pi/2$ and $\eta_1 = \eta_2 = \eta/2$. A comparison with Eq. (6.264) reveals that the $P^{(k)}_{\Delta m_k}(\alpha; 1)$ and the associated operators $\hat{P}^{(k)}_{\Delta m_k}$ just define the POVMs for measuring two $\pi/2$ shifted phase-rotated quadrature distributions in balanced four-port homodyning with the detection efficiency η reduced to $\eta/2$, which reflects the effect of the photon vacuum in the two unused input ports in the eight-port detection scheme. We thus obtain, on recalling Eqs (6.267) and (6.268),

$$\hat{P}_{\Delta m_1, \Delta m_2} = \frac{1}{(\eta/2)^2 |\alpha_L|^2} \prod_{k=1}^{2} \int dx'_k \, q(x_k - x'_k; \tfrac{1}{2}\eta) |x'_k, \varphi_k\rangle \langle x'_k, \varphi_k|; \tag{6.301}$$

that is,

$$P_{\Delta m_1, \Delta m_2} = \frac{1}{(\eta/2)^2 |\alpha_L|^2} p(x_1, x_2, \varphi_1, \varphi_2; \tfrac{1}{2}\eta). \tag{6.302}$$

We see that in an eight-port scheme the joint probability distribution function of the signal-field quadratures are observed to be convoluted with the product of two Gaussians, which for perfect detection simply result from the photon vacuum in the unused input ports.[29] One could say that the vacuum input is needed to "smooth out" the two $\pi/2$ shifted quadratures to obtain simultaneously measurable quantities, whose joint probability distribution is just given

---

29) Note that if the modes in these ports were prepared in some states $\hat{\sigma}$ ($\hat{\sigma} \neq |0\rangle\langle 0|$), then the measured probability distribution would correspond to the POVM $\pi^{-1}\hat{D}(\alpha)\hat{\sigma}\hat{D}^\dagger(\alpha)$ (cf. the last paragraph in Section 6.5.3).

by the Q function,
$$Q(\alpha) = 4p(x_1, x_2, \varphi_1, \varphi_2; \tfrac{1}{2}), \tag{6.303}$$
where $\alpha = e^{i\varphi_L}(ix_2 - x_1)/2$.

The eight-port scheme in Fig. 6.8 has also been used for phase measurements [Noh, Fougères and Mandel (1991, 1992, 1993)]. Based on this measurement scheme phase operators have been operationally defined and their properties have been observed. In the limit when one of the input fields is a strong local oscillator, one can measure the phase distribution of the other input field (signal) in the form of the radially integrated Q function [Noh, Fougères and Mandel (1993)].

6.5.6
Homodyne correlation measurement

Let $\hat{\mathcal{E}}^{(\pm)}(t)$ be the (positive and negative frequency parts of the) operator of the electric-field strength of the signal field [cf. Eq. (6.152)]. We may consider a time-resolved intensity measurement that is performed by unbalanced homodyning with a strong local oscillator, cf. Section 6.5.3. In this case we have

$$\overline{n(t, \Delta t)} \simeq \xi \Delta t |\mathcal{R}|^2 |\tilde{\mathcal{E}}_L|^2 \tag{6.304}$$

and

$$\overline{[\Delta n(t, \Delta t)]^2} \simeq \bar{n} + \xi^2 (\Delta t)^2 |\mathcal{T}|^2 |\mathcal{R}|^2 |\tilde{\mathcal{E}}_L|^2 \langle : [\Delta \hat{\mathcal{E}}(t - x/c, \varphi)]^2 : \rangle \tag{6.305}$$

in place of Eqs (6.225) and (6.226), respectively, where $\tilde{\mathcal{E}}_L$ is the amplitude of the electric field of the local oscillator and

$$\hat{\mathcal{E}}(t, \varphi) = \hat{\mathcal{E}}^{(+)}(t) e^{i(\omega_0 t + \varphi)} + \text{H.c.} = \hat{\mathcal{E}}^{(+)}(t, \varphi) + \text{H.c.}. \tag{6.306}$$

Here $\hat{\mathcal{E}}^{(\pm)}(t) e^{\pm i\omega_0 t}$ are just the slowly varying amplitude operators of the electric field of the signal (ω_0, mid-frequency). In the operators $\hat{\mathcal{E}}^{(\pm)}(t, \varphi)$ the fast time dependence is replaced with the dependence on the local oscillator phase φ. From Eqs (6.304) and (6.305) it follows that the relative deviation of the observed noise from the shot-noise level due to the fluctuation of the signal field is

$$\frac{\overline{[\Delta n(t, \Delta t)]^2} - \overline{n(t, \Delta t)}}{\overline{n(t, \Delta t)}} \simeq \xi \Delta t |\mathcal{T}|^2 \langle : [\Delta \hat{\mathcal{E}}(t - x/c, \varphi)]^2 : \rangle. \tag{6.307}$$

We see that the magnitude of the effect is limited by the detection efficiency ξ. For many applications this kind of limitation may be irrelevant. For example, when the light field under study is strong and highly collimated so that efficient detection is possible.[30]

[30] Note that the overall detection efficiency ξ may be less than the efficiency ξ_{PD} of the photodetector used, $\xi \leq \xi_{PD}$, because of various losses between light source and detector.

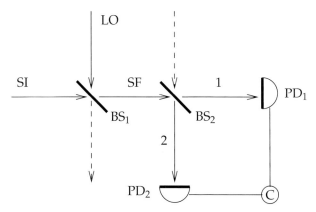

Fig. 6.10 Scheme of a homodyne correlation measurement. The signal field (SI) is combined through a beam splitter (BS_1) with the local oscillator (LO). The superimposed field (SF) is split, on using a second beam splitter (BS_2), into two fields detected in channels 1 and 2 by means of two detectors (PD_1, PD_2) together with a correlator (C) to measure the intensity correlation. The dashed arrows indicate vacuum input.

It is of course of interest also to study the fluctuation behavior of the electric field of light that does not satisfy such conditions. A typical example is the extremely weak resonance fluorescence from a single atom, where the low collection efficiency of the fluorescence renders it almost impossible to determine the fluctuations of the electric field from measurement of the sub-Poissonian statistics [Mandel (1982)]. In these cases an alternative way of measuring the noise of the signal field is a homodyne measurement of the intensity correlations of the superimposed light by using a weak local oscillator [Vogel (1991, 1995); Carmichael, Castro-Beltran, Foster, and Orozco (2000)], that is, the local-oscillator field is of the same order of magnitude as the signal field.[31] In this type of observation scheme, one of the detectors in Fig. 6.6, p. 206, is replaced by a correlation apparatus with two detectors, as shown in Fig. 6.10.

The observed second-order correlation of counts can be found by straightforwardly applying the results given in Sections 6.1.3, 6.2.1 and 6.5.1:

$$\frac{\overline{n(t_1, \Delta t_1) n(t_2, \Delta t_2)}}{\varsigma_1 \varsigma_2 \Delta t_1 \Delta t_2} = |\mathcal{T}_2|^2 |\mathcal{R}_2|^2 \Gamma'^{(2,2)}(t_1 - x_1/c, t_2 - x_2/c_2), \quad (6.308)$$

[31] Homodyne correlation measurements have also been studied for a strong local oscillator [Ou, Hong and Mandel (1987)]. In this case the classical noise of the local oscillator becomes a serious limitation of the method. For a weak local oscillator the corresponding noise effects can be suppressed [Vogel (1995)].

where

$$\Gamma'^{(2,2)}(t_1,t_2) = \langle \hat{\mathcal{E}}_1'^{(-)}(t_1)\hat{\mathcal{E}}_1'^{(-)}(t_2)\hat{\mathcal{E}}_1'^{(+)}(t_2)\hat{\mathcal{E}}_1'^{(+)}(t_1)\rangle$$
$$= \langle :\hat{I}_1'(t_1)\hat{I}_1'(t_2): \rangle \qquad (6.309)$$

is the normally ordered intensity correlation of the superimposed field $\mathcal{E}_1'$ falling on the beam splitter BS_2 in Fig. 6.10. Expressing it in terms of the fields $\mathcal{E} = \mathcal{E}_1$ and $\mathcal{E}_L = \mathcal{E}_2$ fed into the input ports of the beam splitter BS_1, yields

$$\hat{I}_1'(t) = |\mathcal{T}_1|^2 \hat{I} + |\mathcal{R}_1|^2 \hat{I}_L + [\mathcal{T}_1^* \mathcal{R}_1 \hat{\mathcal{E}}^{(-)}(t)\hat{\mathcal{E}}_L^{(+)}(t) + \text{H.c.}] \qquad (6.310)$$

[cf. Eq. (6.218)]. Now let us consider the steady-state regime. For this purpose, we write

$$\Gamma'^{(2,2)}(t',t) = \Gamma'^{(2,2)}(t+\tau,t), \qquad t' = t+\tau, \qquad (6.311)$$

and take the limit as $t \to \infty$ to introduce the function $\Gamma'^{(2,2)}(\tau)$, which depends only on the time difference τ:

$$\Gamma'^{(2,2)}(\tau) \equiv \lim_{t\to\infty} \Gamma'^{(2,2)}(t+\tau,t). \qquad (6.312)$$

In particular, assuming that the correlation decays for sufficiently large time difference τ, we have[32]

$$\Gamma'^{(2,2)}(\infty) = \langle \hat{I}' \rangle^2. \qquad (6.313)$$

To obtain a quantitative measure of the relative change of correlation, we now compare the zero-time value $\Gamma'^{(2,2)}(0)$ with the long-time value $\Gamma'^{(2,2)}(\infty)$:

$$\frac{\Gamma'^{(2,2)}(0) - \Gamma'^{(2,2)}(\infty)}{\Gamma'^{(2,2)}(\infty)} = \frac{\langle :(\Delta\hat{I}')^2: \rangle}{\langle \hat{I}' \rangle^2}. \qquad (6.314)$$

Note that the relative change in the intensity correlation as introduced in Eq. (6.314) does not depend on the detection efficiency.

In the special case where the local oscillator is strong, Eq. (6.314) reduces to

$$\frac{\Gamma'^{(2,2)}(0) - \Gamma'^{(2,2)}(\infty)}{\Gamma'^{(2,2)}(\infty)} = \frac{|\mathcal{T}_1|^2}{|\mathcal{R}_1|^2} \frac{\langle :[\Delta\hat{\mathcal{E}}(\varphi)]^2: \rangle}{|\tilde{\mathcal{E}}_L|^2}. \qquad (6.315)$$

Clearly, the relative change in the intensity correlation is small when the signal field is weak compared with the (strong) local oscillator. Large effects can be

---

32) Since steady-state one-time expectation values do not depend on time, the time argument $t = \infty$ is omitted.

observed when the strength of the local oscillator becomes comparable to that of the signal. In this case we find that

$$\Gamma'^{(2,2)}(\infty) = \left[|\mathcal{T}_1|^2 \langle \hat{I} \rangle + |\mathcal{R}_1|^2 |\tilde{\mathcal{E}}_L|^2 + |\mathcal{T}_1||\mathcal{R}_1||\tilde{\mathcal{E}}_L|\langle \hat{\mathcal{E}}(\varphi) \rangle\right]^2, \quad (6.316)$$

$$\Gamma'^{(2,2)}(0) - \Gamma'^{(2,2)}(\infty) = |\mathcal{T}_1|^2 |\mathcal{R}_1|^2 |\tilde{\mathcal{E}}_L|^2 \langle :[\Delta\hat{\mathcal{E}}(\varphi)]^2: \rangle \\
+ 2|\mathcal{T}_1|^3|\mathcal{R}_1||\tilde{\mathcal{E}}_L|\langle :\Delta\hat{I}\Delta\hat{\mathcal{E}}(\varphi): \rangle + |\mathcal{T}_1|^4 \langle :(\Delta\hat{I})^2: \rangle. \quad (6.317)$$

From Eq. (6.317), we can see that in such a measurement various kinds of terms contribute to the observed intensity correlation. The first term results from the signal-field noise, whereas the third term describes the effect of the intensity noise of the signal field. The second term results from the correlation of the intensity and field-strength noise of the signal. The three terms, which contribute significantly to the observed intensity correlation when the local oscillator is weak, can be separated from each other. From Eq. (6.306) together with Eq. (6.221) it is seen that they differ in periodicity with regard to the phase φ of the local oscillator. For a detailed discussion of the separation of the three contributions to the measured signal, see Vogel (1995).

6.5.7
Normally ordered moments

The homodyne correlation measurement scheme can be advanced, as illustrated in Fig. 6.11, to detect higher-order normally ordered signal-field correlations and moments with unequal powers of annihilation and creation operators [Shchukin and Vogel (2005)].[33] It is evident that the scheme in Fig. 6.11 can easily be further extended, by replacing each of the photodetectors with another beam splitter and two detectors in its output channels and so forth. Needless to say that the extension is limited by the fact that the output signals are reduced, which necessarily requires a longer measurement time to ensure the desired signal-to-noise ratio. To overcome this problem, one can measure the homodyne correlation data with a strong local oscillator and analyze the data in a balanced manner [Shchukin and Vogel (2006)].

Comparing the setup in Fig. 6.11 with that in Fig. 6.10, it is obvious that the detector pairs $PD_{1,2}$ are equivalent to each other in the two schemes. The detector pair $PD_{3,4}$ acts in a quite similar way, except that the transmission and reflection coefficients of the entrance beam splitter exchange their roles. Hence we may use the same notation as in Section 6.5.6. We begin with the

---

[33] For sampling normally ordered moments from the phase-rotated quadrature statistics measured in balanced four-port homodyning, see Section 7.5.

6 Photoelectric detection of light

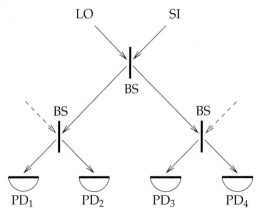

Fig. 6.11 Scheme for the determination of normally ordered correlation functions by homodyne correlation measurements. The notations correspond to those in Fig. 6.10.

analysis of the mean number of counts of one detector, say PD_1:

$$\frac{n(t_1, \Delta t_1)}{\varsigma_1 \Delta t_1} = |\mathcal{T}_2|^2 \langle \hat{I}'_1(t_1 - x_1/c) \rangle \equiv |\mathcal{T}_2|^2 \Gamma^{(1,1)}(t_1 - x_1/c, \varphi), \tag{6.318}$$

where $\Gamma^{(1,1)}(t, \varphi)$ corresponds to the expectation value of the intensity operator given by Eq. (6.310):

$$\Gamma^{(1,1)}(t, \varphi) = |\mathcal{T}_1|^2 I(t) + |\mathcal{R}_1|^2 I_L(t) + [\mathcal{T}_1^* \mathcal{R}_1 \langle \hat{\mathcal{E}}^{(-)}(t, \varphi) \rangle \tilde{\mathcal{E}}_L + \text{H.c.}]. \tag{6.319}$$

For the following considerations we have explicitly indicated the dependence of the measured intensity on the phase φ of the local oscillator, which is again assumed to be in a coherent state with (real) amplitude $\tilde{\mathcal{E}}_L$. The measured set of data can be analyzed by Fourier transform to obtain

$$\Gamma_m^{1,1}(t, \phi) = \frac{1}{2\pi} \int_0^{2\pi} d\varphi \, e^{im(\phi - \varphi)} \Gamma_m^{1,1}(t, \varphi) \tag{6.320}$$

($m = 0, \pm 1$). Using Eq. (6.319) together with (6.306), we can easily calculate the Fourier components $\Gamma_m^{1,1}(t, \phi)$. The phase-independent component is given by

$$\Gamma_0^{1,1}(t, \phi) = |\mathcal{T}_1|^2 I(t) + |\mathcal{R}_1|^2 I_L(t), \tag{6.321}$$

and the first-order components read as

$$\Gamma_{-1}^{1,1}(t, \phi) = \mathcal{T}_1^* \mathcal{R}_1 \tilde{\mathcal{E}}_L \langle \hat{\mathcal{E}}^{(-)}(t, \phi) \rangle, \quad \Gamma_{+1}^{1,1}(t, \phi) = \mathcal{T}_1 \mathcal{R}_1^* \tilde{\mathcal{E}}_L \langle \hat{\mathcal{E}}^{(+)}(t, \phi) \rangle. \tag{6.322}$$

Hence the field strength moments of first order can be inferred from the measured data as follows:

$$\langle \hat{\mathcal{E}}^{(-)}(t, \phi) \rangle = \frac{\Gamma_{-1}^{1,1}(t, \phi)}{\mathcal{T}_1^* \mathcal{R}_1 \tilde{\mathcal{E}}_L}, \quad \langle \hat{\mathcal{E}}^{(+)}(t, \phi) \rangle = \frac{\Gamma_{+1}^{1,1}(t, \phi)}{\mathcal{T}_1 \mathcal{R}_1^* \tilde{\mathcal{E}}_L}. \tag{6.323}$$

That is to say, the first-order moments of the positive and negative frequency parts of the signal field, which are complex conjugate to each other, are obtained, independently from each other, from different Fourier coefficients.

In the next step one may consider the detection of equal-time coincidences recorded with two detectors, say PD_1 and PD_2, leading to $\Gamma^{(2,2)}(t,\varphi) \equiv \Gamma^{(2,2)}(t,t)$, with $\Gamma^{(2,2)}(t,t)$ according to Eq. (6.309). Proceeding in this way, one may eventually detect the equal-time coincidences of n photodetectors, $\Gamma^{(n,n)}(t,\varphi)$, in dependence on the phase φ of the local oscillator. Performing a Fourier analysis of the measured data,

$$\underline{\Gamma}_m^{(n,n)}(t,\phi) = \frac{1}{2\pi} \int_0^{2\pi} d\varphi\, e^{im(\phi-\varphi)} \Gamma^{(n,n)}(t,\varphi) \tag{6.324}$$

($m \leq n$), one may directly obtain the nth-order moment

$$\langle [\hat{\mathcal{E}}^{(\pm)}(t,\phi)]^n \rangle \sim \underline{\Gamma}_{\pm n}^{(n,n)}(t,\phi), \tag{6.325}$$

which cannot be observed by direct photodetection. The Fourier components $\underline{\Gamma}_{\pm m}^{(n,n)}(t,\phi)$ ($m < n$) contain linear combinations of different moments, in general. By a stepwise increase in the number of detectors, however, only one new type of moment is added in each step. In this manner it becomes possible to derive all the moments of the order $m+n$,

$$G^{(m,n)} = \langle [\hat{\mathcal{E}}^{(-)}(t,\phi)]^m [\hat{\mathcal{E}}^{(+)}(t,\phi)]^n \rangle, \tag{6.326}$$

by homodyne correlation measurement with a weak local oscillator [for details, see Shchukin and Vogel (2005)]. Allowing for a strong local oscillator, the method can be generalized to measure normally ordered space-time dependent correlation functions [Shchukin and Vogel (2006)].

References

Banaszek, K. and K. Wódkiewicz (1996) *Phys. Rev. Lett.* **76**, 4344.

Born, M. and E. Wolf (1959) *Principles of Optics* (Pergamon, London).

Braginsky, V.B., Y.I. Vorontsov and F.Y. Khalili (1977) *Sov. Phys. - JETP* **46**, 705.

Braunstein, S.L. (1990) *Phys. Rev. A* **42**, 474.

Campos, R.A., B.E.A. Saleh and M.C. Teich (1989) *Phys. Rev. A* **40**, 1371.

Carmichael, H.J. (1987) *J. Opt. Soc. Am. B***4**, 1588.

Carmichael, H.J., H.M. Castro-Beltran, G.T. Foster, and L.A. Orozco (2000) *Phys. Rev. Lett.* **85**, 1855.

Caves, C.M. (1983) In *Quantum Optics, Experimental Gravitation and Measurement Theory*, eds P. Meystre and M.O. Scully (Plenum, New York), p. 567.

Caves, C.M., K.S. Thorne, R.W.P. Drever, V.D. Sandberg and M. Zimmermann (1980) *Rev. Mod. Phys.* **52**, 341.

Chmara, W. (1987) *J. Mod. Opt.* **34**, 455.

Conradi, J. (1972) *IEEE Trans. Electron. Devices* **19**, 713.

Eberly, J.H. and K. Wódkiewicz (1977) *J. Opt. Soc. Am.* **67**, 1252.

Fleischhauer, M. and D.-G. Welsch (1991) *Phys. Rev. A* **44**, 747.

Freyberger, M. and W.P. Schleich (1992) *Phys. Rev. A* **47**, R30.

Freyberger, M., K. Vogel and W.P. Schleich (1993) *Phys. Lett. A* **176**, 41.

Gardiner, C.W. (1983) *Handbook of Stochastic Methods* (Springer-Verlag, Berlin).

Gardiner, C.W. and C.M. Savage (1984) *Opt. Commun.* **50**, 173.

Glauber, R.J. (1965) In *Quantum Optics and Electronics*, eds C. DeWitt, A. Blandin and C. Cohen-Tannoudji (Gordon and Breach, New York), p. 144.

Glauber, R.J. (1966) *Phys. Lett.* **21**, 650.

Glauber, R. (1972) In *Laser Handbook*, Vol. 1, eds F.T. Arecchi and E.O. Schulz-Dubois (North-Holland, Amsterdam), p. 1.

Gruner, T. and D.-G. Welsch (1996) *Phys. Rev. A* **54**, 1661.

Helstrom, C.W. (1976) *Quantum Detection and Estimation Theory* (Academic Press, New York).

Huttner, B. and Y. Ben-Aryeh (1988) *Opt. Commun.* **69**, 93.

Kelley, P.L. and W.H. Kleiner (1964) *Phys. Rev.* **136**, A316.

Khanbekyan, M., L. Knöll and D.-G. Welsch (2003) *Phys. Rev. A* **67**, 063812.

Knöll, L., S. Scheel, E. Schmidt, D.-G. Welsch and A.V. Chizhov (1999) *Phys. Rev. A* **59**, 4716.

Knöll, L., W. Vogel and D.-G. Welsch (1986) *J. Opt. Soc. Am. B* **3**, 1315.

Knöll, L., W. Vogel and D.-G. Welsch (1990) *Phys. Rev. A* **42**, 503.

Kühn, H. and D.-G. Welsch (1991) *Phys. Rev. Lett.* **67**, 580.

Landau, L. and R. Peierls (1931) *Z. Phys.* **69**, 56.

Lax, M. and M. Zwanziger (1973) *Phys. Rev. A* **7**, 750.

Leonhardt, U. and H. Paul (1993) *Phys. Rev. A* **48**, 4598.

Ley, M. and R. Loudon (1987) *J. Mod. Opt.* **34**, 227.

Mandel, L. (1958) *Proc. Phys. Soc. Lond.* **72**, 1037.

Mandel, L. (1959) *Proc. Phys. Soc. Lond.* **74**, 233.

Mandel, L. (1963) In *Progress in Optics*, Vol. 2, ed. E. Wolf (North-Holland, Amsterdam), p. 181.

Mandel, L. (1966) *Phys. Rev.* **144**, 1071.

Mandel, L. (1982) *Phys. Rev. Lett.* **49**, 136.

McIntyre, R.J. (1966) *IEEE Trans. Electron. Devices* **13**, 164.

Metha, C.L. and E. Wolf (1967) *Phys. Rev.* **157**, 1188.

Mollow, B.R. (1968) *Phys. Rev.* **136**, 1896.

Noh, J.W., A. Fougères and L. Mandel (1991) *Phys. Rev. Lett.* **67**, 1426.

Noh, J.W., A. Fougères and L. Mandel (1992) *Phys. Rev. A* **45**, 424.

Noh, J.W., A. Fougères and L. Mandel (1993) *Phys. Rev. Lett.* **71**, 2579.

Ou, Z.Y., C.K. Hong and L. Mandel (1987) *Phys. Rev. A* **36**, 192.

Peřina, J., B.E.A. Saleh and M.C. Teich (1983) *Opt. Commun.* **48**, 212.

Personick, S.D. (1971) *Bell. Syst. Tech. J.* **50**, 167.

Schumaker, B.L. (1984) *Opt. Lett.* **9**, 189.

Scully, M.O. and W.E. Lamb, Jr. (1969) *Phys. Rev.* **179**, 368.

Selloni, A., P. Schwendimann, A. Quattropani and H.P. Baltes (1978) *J. Phys. A* **11**, 1427.

Shapiro, J.H. and H.P. Yuen (1979) *IEEE Trans. Inform. Theor.* **25**, 179.

Shchukin, E. and W. Vogel (2005) *Phys. Rev. A* **72**, 043808.

Shchukin, E. and W. Vogel (2006) arXiv:quant-ph/0602124.

Srinivas, M.D. and E.B. Davies (1981) *Optica Acta* **28**, 981.

Teich, M.C., K. Matsuo and B.E.A. Saleh (1986a) *IEEE J. Quant. Electron.* **22**, 1184.

Teich, M.C., K. Matsuo and B.E.A. Saleh (1986b) *IEEE Trans. Electron. Devices* **33**, 1511.

Thorne, K.S., C.M. Caves, V.D. Sandberg, M. Zimmermann and R.W.P. Drever (1979) In *Sources of Gravitational Radiation*, ed. L. Smarr (Cambridge University, Cambridge), p. 49.

Thorne, K.S., R.W.P. Drever, C.M. Caves, M. Zimmermann and V.D. Sandberg (1978) *Phys. Rev. Lett.* **40**, 667.

Ueda, M. (1990) *Phys. Rev. A* **41**, 3875.

Unruh, W.G. (1978) *Phys. Rev. D* **18**, 1764.

van Kampen, N.G. (1981) *Stochastic Processes in Physics and Chemistry* (Elsevier, Amsterdam).

Vogel, W. (1991) *Phys. Rev. Lett.* **67**, 2450.

Vogel, W. (1995) *Phys. Rev. A* **51**, 4160.

Vogel, W. and J. Grabow (1993) *Phys. Rev. A* **47**, 4227.

Walker, N.G. and J.E. Caroll (1984) *Electron. Lett* **20**, 981.

Walker, N.G. (1987) *J. Mod. Opt.* **34**, 15.

Wallentowitz, S. and W. Vogel (1996) *Phys. Rev. A* **53**, 4528.

Wódkiewicz, K. and J.H. Eberly (1985) *J. Opt. Soc. Am. B* **2**, 458.

Yuen, H.P., and J.H. Shapiro (1978) *IEEE Trans. Inf. Theory* **24**, 657.

Yurke, B. (1985) *Phys. Rev. A* **32**, 300; 311.

Yurke, B., S.L. McCall and J.R. Klauder (1986) *Phys. Rev. A* **33**, 4033.

Yurke, B. and D. Stoler (1987) *Phys. Rev. A* **36**, 1955.

Zucchetti, A., W. Vogel and D.–G. Welsch (1996) *Phys. Rev. A* **54**, 856.

7
Quantum-state reconstruction

The state of a quantum object is commonly described by a normalized Hilbert-space vector $|\Psi\rangle$ or, more generally, by a density operator $\hat{\varrho}$ which is a Hermitian and non-negative valued Hilbert-space operator of trace one. The Hilbert space of the object is usually spanned up by an orthonormalized set of basic vectors $|A\rangle$ representing the eigenvectors (eigenstates) of Hermitian operators $\hat{A}$ associated with a complete set of simultaneously measurable observables (physical quantities) of the object.[1] The eigenvalues A of these operators are the values of the observables which can be registered in a measurement. Here we must distinguish between an individual (single) and an ensemble measurement (i.e., in principle, an infinitely large number of repeated measurements on identically prepared objects). Performing a single measurement on the object, a totally unpredictable value A is observed, in general, and the state of the object has collapsed to the state $|A\rangle$ according to the von Neumann's projection definition of a measurement [von Neumann (1932)]. If the same measurement is repeated immediately after the first measurement (on the same object), the result is now very predictable – the same value A as in the first measurement, is observed. Obviously, owing to the first measurement, the object has been prepared in the state $|A\rangle$. Repeating the measurement many times on an identically prepared object, the relative rate at which the result A is observed approaches the diagonal density-matrix element $\langle A|\hat{\varrho}|A\rangle$ as the number of measurements tends to infinity. Measuring $\langle A|\hat{\varrho}|A\rangle$ for all values of A, the statistics of $\hat{A}$ (and of any function of $\hat{A}$) are known. To completely describe the quantum state, i.e., to determine all the quantum-statistical properties of the object, knowledge of all density-matrix elements $\langle A|\hat{\varrho}|A'\rangle$ is needed. In particular, the off-diagonal elements essentially determine the statistics of such sets of observables $\hat{B}$ that are not compatible with $\hat{A}$ ($[\hat{A},\hat{B}] \neq 0$) and cannot be measured simultaneously with $\hat{A}$. Obviously, the statistics of $\hat{B}$ can also be obtained directly – similar to the statistics of $\hat{A}$ – from an ensemble measurement yielding the diagonal density-matrix elements $\langle B|\hat{\varrho}|B\rangle$ in the basis of the eigenvectors $|B\rangle$ of $\hat{B}$. Now one can

---

1) For notational convenience we write $\hat{A}$, without further specifying the quantities belonging to the set.

Quantum Optics, Third, revised and extended edition. Werner Vogel and Dirk-Gunnar Welsch
Copyright © 2006 WILEY-VCH Verlag GmbH & Co. KGaA, Weinheim
ISBN: 3-527-40507-0

proceed to consider other sets of observables which are not compatible with $\hat{A}$ and $\hat{B}$. When a set of observables $\hat{A}_i$, also called a *quorum*, is found, such that the density operator can be represented in the form of [2]

$$\hat{\varrho} = \sum_i \hat{C}_i \langle \hat{A}_i \rangle, \tag{7.1}$$

then, in principle, all knowable information on the quantum state of the system can be obtained [Fano (1957)]. In particular, when in a chosen basis $|\Lambda\rangle$ the matrix elements $\langle \Lambda | \hat{C}_i | \Lambda' \rangle$ are bounded for all values of i, then the density-matrix elements $\langle \Lambda | \hat{\varrho} | \Lambda' \rangle$ can be sampled directly from the measured $\langle \hat{A}_i \rangle$ according to the relation

$$\langle \Lambda | \hat{\varrho} | \Lambda' \rangle = \sum_i \langle \Lambda | \hat{C}_i | \Lambda' \rangle \langle \hat{A}_i \rangle. \tag{7.2}$$

Roughly speaking, there have been two routes used to collect measurable data for reconstructing the quantum state of an object under consideration. In the first, which closely follows the method given above, a succession of (ensemble) measurements is made such that a set of noncommutative object observables is measured which carries the complete information about the quantum state. A typical example is the reconstruction of the quantum state from the [in balanced four-port homodyning (Section 6.5.4) measurable] probability distributions $p(x, \varphi) = \langle x, \varphi | \hat{\varrho} | x, \varphi \rangle$ of the phase-rotated quadratures $\hat{x}(\varphi)$ [Eq. (3.143)]. Here the projectors $|x, \varphi\rangle\langle x, \varphi|$ for all phases φ within a π interval play the role of the $\hat{A}_i$ in Eqs (7.1) (Section 7.1.1).

In the second, the object is coupled to a reference system (whose quantum state is well known) such that the measurement of observables of the composite system corresponds to "simultaneous" measurement of noncommutative observables of the object. In this case the number of observables (of the composite system) which must be measured in a succession of (ensemble) measurements can be reduced drastically, but at the expense of the image sharpness of the object. As a result of the additional noise introduced by the reference system, only fuzzy measurements on the object can be performed, which just makes a "simultaneous" measurement of incompatible object observables feasible. A typical example is the [in balanced eight-port homodyning (Section 6.5.5) measurable] Q function, which is given by the diagonal density-matrix elements in the coherent-state basis, $Q(\alpha) = \pi^{-1} \langle \alpha | \hat{\varrho} | \alpha \rangle$ [Eq. (4.67)]. Here, the POVM operators $\pi^{-1} |\alpha\rangle\langle\alpha|$ play the role of the $\hat{A}_i$ in Eqs (7.1) [Eq. (4.83) for $s=-1$]. The Q function can already be obtained from one ensemble measurement of the complex amplitude α in balanced eight-port homodyning, which

---

2) Here we have assumed a discrete set of observables. For continuous sets of observables, the sum in Eq. (7.1) must be replaced by an integral.

corresponds to a fuzzy measurement of the "joint" probability distribution of two canonically conjugated observables $\hat{x}(\varphi)$ and $\hat{x}(\varphi+\pi/2)$ of the object.

Since the sets of quantities measured via the one or the other route (or an appropriate combination of both) carry the complete information on the quantum state of the object, they can be regarded, in a sense, as representations of the quantum state, which can be more or less close to familiar quantum-state representations, such as the density matrix in the number-state basis or a phase-space function. In any case, the question arises of how to reconstruct, from the measured data, specific quantum-state representations (or specific quantum-statistical properties of the object for which a direct measurement scheme is not available). Again, there have been two typical methods of solving the problem. In the first, equations that relate the measured quantities to the desired quantities are derived and solved either analytically or numerically in order to obtain the desired quantities in terms of the measured quantities. In practice, the measured data are often incomplete, i.e., not all quantities needed for a precise reconstruction are measured,[3] and moreover, the measured data are inaccurate. Obviously, any experiment can only run for a finite time, which prevents one, in principle, from performing an infinite number of repeated measurements in order to obtain precise expectation values. These inadequacies give rise to systematic and statistical errors of the reconstructed quantities, which can be quantified in terms of confidence intervals.

In the second approach, statistical methods are used from the very beginning in order to obtain the best *a posteriori* estimation of the desired quantities on the basis of the available (i.e., incomplete and/or inaccurate) data measured. However, the price to pay may be high. Whereas in the first concept linear equations are typically to be handled and estimates of the desired quantities (including statistical errors) can often be directly sampled from the measured data, application of purely statistical methods, such as the principle of maximum entropy or Bayesian inference, require the treatment of nonlinear equations and a reconstruction in real time is, in general, impossible. Here we concentrate on the first concept, restricting our attention to single-mode systems [for a review on quantum-state reconstruction, see Welsch, Vogel and Opatrný (1999)].

7.1
Optical homodyne tomography

From Section 6.5.4 we know that, in a succession of (phase-shifted) balanced four-port homodyne measurements, the phase-rotated quadrature distribu-

---

[3] To compensate for incomplete data, some *a priori* knowledge of the quantum state is needed, in general.

tion $p(x, \varphi)$ of a signal-field mode can be obtained for various values of the phase parameter φ, provided that 100% detection efficiency is realized. As we will see, the quantum state is then known when $p(x, \varphi)$ is known for all values of φ within a π interval [Vogel and Risken (1989)]. That is to say, all quantum-statistical properties can be obtained from the quadrature-component distributions measured in a π interval.

7.1.1
Quantum state and phase-rotated quadratures

In order to relate the phase-rotated quadrature distributions

$$p(x, \varphi) = \langle x, \varphi | \hat{\varrho} | x, \varphi \rangle = \mathrm{Tr}\left(\hat{\varrho} | x, \varphi \rangle \langle x, \varphi |\right) \tag{7.3}$$

to the quantum state, we note that the projector

$$|x, \varphi\rangle \langle x, \varphi| = \hat{\delta}[\hat{x}(\varphi) - x] = \frac{1}{2\pi} \int dy\, e^{-iyx} e^{iy\hat{x}(\varphi)} \tag{7.4}$$

can be expressed in terms of the displacement operator $\hat{D}$ [Eq. (3.44)] as follows:

$$|x, \varphi\rangle \langle x, \varphi| = \frac{1}{2\pi} \int dy\, e^{-iyx} \hat{D}(iye^{-i\varphi}). \tag{7.5}$$

We now introduce the characteristic (generating) function $\Psi(y, \varphi)$ of the probability distribution $p(x, \varphi)$ via the Fourier transform

$$p(x, \varphi) = \frac{1}{2\pi} \int dy\, e^{-iyx} \Psi(y, \varphi), \tag{7.6}$$

$$\Psi(y, \varphi) = \int dx\, e^{iyx} p(x, \varphi). \tag{7.7}$$

Recalling the definition (3.143) of $\hat{x}(\varphi)$, it is not difficult to prove the following symmetry relations

$$p(-x, \varphi \pm \pi) = p(x, \varphi), \tag{7.8}$$
$$\Psi(-y, \varphi \pm \pi) = \Psi(y, \varphi). \tag{7.9}$$

Substituting the expression (7.4) into Eq. (7.3) and comparing with Eq. (7.6), we can easily see that $\Psi(y, \varphi)$ is the expectation value of a displacement operator:

$$\Psi(y, \varphi) = \mathrm{Tr}\left\{\hat{\varrho} \exp[iy\hat{x}(\varphi)]\right\} = \mathrm{Tr}\left[\hat{\varrho} \hat{D}(iye^{-i\varphi})\right]. \tag{7.10}$$

On the other hand, the characteristic function $\Phi(\alpha; s)$ of an s-parameterized phase-space function $P(\alpha; s)$ is given by

$$\Phi(\alpha; s) = \mathrm{Tr}\left[\hat{\varrho} \hat{D}(\alpha; s)\right] = e^{\frac{1}{2}s|\alpha|^2} \mathrm{Tr}\left[\hat{\varrho} \hat{D}(\alpha)\right] \tag{7.11}$$

[Eq. (4.91) together with Eqs (4.90) and (4.47)]. Comparing Eqs (7.10) and (7.11), we find that

$$\Psi(y,\varphi) = e^{-\frac{1}{2}sy^2}\Phi(iye^{-i\varphi};s). \tag{7.12}$$

As expected, $\Psi(y,\varphi)$ can be directly determined from $\Phi(\alpha;s)$. To determine $\Phi(\alpha;s)$ from $\Psi(y,\varphi)$, we note that the relation $\alpha = |\alpha|e^{i\varphi_\alpha} = iye^{-i\varphi}$, with real y, implies that

$$|\alpha|e^{in\pi} = y, \qquad \varphi_\alpha + \varphi = \tfrac{1}{2}(2n+1)\pi. \tag{7.13}$$

From Eqs (7.12) and (7.13) together with the symmetry relation (7.9), it then follows that for chosen values of α and s, $\Phi(\alpha;s)$ can be obtained from $\Psi(y,\varphi)$ according to

$$\Phi(\alpha;s) = e^{\frac{1}{2}s|\alpha|^2}\Psi(|\alpha|, \tfrac{1}{2}\pi - \varphi_\alpha). \tag{7.14}$$

We can see that for a fixed value of the quadrature phase φ the function $\Phi(\alpha;s)$ can be determined from $\Psi(y,\varphi)$ only along a line in the complex α plane which is at an angle of $\varphi_\alpha = \pi/2 - \varphi$ to the real axis. That is, complete information on the quantum state, as given by the knowledge of $\Phi(\alpha;s)$ for all values of φ_α in a 2π interval can only be provided by knowing $\Psi(y,\varphi)$ for all values of φ in a π interval. This just expresses the fact that the quantum state contains potential information on measurements of all possible observables, whereas $\Psi(y,\varphi)$ for chosen φ contains that information only for a specific choice of the observable $\hat{x}(\varphi)$.

The correspondence between the phase-space function $P(\alpha;s)$ and the phase-rotated quadrature distributions $p(x,\varphi)$ is analogous. Whereas $p(x,\varphi)$ can be determined from $P(\alpha;s)$, the determination of $P(\alpha;s)$ requires knowledge of $p(x,\varphi)$ for all values of φ within a π interval. The first statement is again not surprising, because $p(x,\varphi)$ is the expectation value of $|x,\varphi\rangle\langle x,\varphi|$ and any expectation value can be calculated by means of a phase-space function $P(\alpha;s)$. Combining Eqs (7.6) and (7.12) yields

$$p(x,\varphi) = \frac{1}{2\pi}\int dy\, \exp(-iyx - \tfrac{1}{2}sy^2)\Phi(iye^{-i\varphi};s). \tag{7.15}$$

Using Eq. (4.92) in Eq. (7.15) to express $\Phi(\alpha;s)$ in terms of $P(\alpha;s)$, we obtain

$$p(x,\varphi) = \frac{1}{2\pi}\int dy\, \exp(-iyx - \tfrac{1}{2}sy^2)\int d^2\alpha\, \exp[iy\langle\alpha|\hat{x}(\varphi)|\alpha\rangle]P(\alpha;s), \tag{7.16}$$

where the relation

$$\langle\alpha|\hat{x}(\varphi)|\alpha\rangle = \alpha e^{i\varphi} + \alpha^* e^{-i\varphi} \tag{7.17}$$

[cf. Eq. (3.203)] has been used. When we assume that $s \geq 0$ (or $\operatorname{Re} s \geq 0$), then in Eq. (7.16) the integration over y can be performed to obtain

$$p(x,\varphi) = \int d^2\alpha \, p[x - \langle \alpha | \hat{x}(\varphi) | \alpha \rangle; s] P(\alpha; s), \tag{7.18}$$

where the function $p(x;s)$ is defined by

$$p(x;s) = \frac{1}{2\pi} \int dy \, \exp(-iyx - \tfrac{1}{2}sy^2) = \frac{1}{\sqrt{2\pi s}} \exp\left(-\frac{x^2}{2s}\right), \tag{7.19}$$

which in the limit as $s \to 0$ reduces to a δ function:

$$\lim_{s \to 0} p(x;s) = \delta(x). \tag{7.20}$$

In Eq. (7.18) the Gaussian $p[x - \langle \alpha | \hat{x}(\varphi) | \alpha \rangle; s]$ is obviously the associated c-number function of the projector $|x, \varphi\rangle\langle x, \varphi|$ put in s order ($s \geq 0$).[4] In particular, when $s = 1$ (Glauber–Sudarshan representation) the function $p[x - \langle \alpha | \hat{x}(\varphi) | \alpha \rangle; 1]$ is just the phase-rotated quadrature distribution of a coherent state $|\alpha\rangle$ [cf. Eq. (3.202)].

Let us now turn to the problem of determining $P(\alpha; s)$ from $p(x, \varphi)$. Substituting in Eq. (4.93) for $\Phi(\beta; s)$ the result of Eq. (7.14) and using Eq. (7.7), we derive, on changing the integration variables and applying the symmetry relation (7.8),

$$P(\alpha;s) = \frac{1}{\pi^2} \int_\pi d\varphi \int dy \, |y| e^{\frac{1}{2}sy^2} \exp[-iy\langle \alpha | \hat{x}(\varphi) | \alpha \rangle] \int dx \, e^{iyx} p(x,\varphi). \tag{7.21}$$

From inspection of Eq. (7.21) we see that when $s < 0$ (or $\operatorname{Re} s < 0$), then the y integration can be performed separately to obtain

$$P(\alpha;s) = \int_\pi d\varphi \int dx \, \tilde{p}[x - \langle \alpha | \hat{x}(\varphi) | \alpha \rangle; s] p(x,\varphi), \tag{7.22}$$

where we have introduced the function

$$\tilde{p}(x;s) = \frac{2}{\pi^2} \int_0^\infty dr \, r e^{\frac{1}{2}sr^2} \cos(rx) = \frac{2}{\pi^2|s|} F\left(1, \frac{1}{2}; -\frac{x^2}{2|s|}\right) \tag{7.23}$$

[$F(a, b; x)$, confluent hyper-geometric function]. Note that the integral of this function vanishes:

$$\int dx \, \tilde{p}(x;s) = 0. \tag{7.24}$$

---

4) Note that if one uses Eq. (7.18) together with (7.19) for values of s with $s<0$, then one has to deal with a highly singular function $p(x;s)$, which reflects the fact that the associated c-number function of the projector $|x, \varphi\rangle\langle x, \varphi|$ put in s order with $s<0$ is not well behaved. In practice this difficulty can be overcome by preserving the original order of integration.

7.1 Optical homodyne tomography

For $s > 0$ the application of Eq. (7.22) together with Eq. (7.23) would require dealing with a highly singular function $\tilde{p}(x;s)$. We recall that, for positive values of s, depending on the analytic form of $p(x,\varphi)$, the distribution $P(\alpha;s)$ may become highly singular as well, whereas for negative values of s it is well behaved in any case.

Applying Eq. (4.79) to the density operator $\hat{\varrho}$ and and making use of Eqs (7.11) and (7.14), we arrive, on changing the integration variables and applying the symmetry relation (7.9), at the following expansion of $\hat{\varrho}$

$$\hat{\varrho} = \frac{1}{\pi}\int_\pi d\varphi \int dy \, |y| e^{\frac{1}{2}sr^2} \Psi(-y,\varphi) \hat{D}(iye^{-i\varphi}; -s)$$

$$= \frac{1}{\pi}\int_\pi d\varphi \int dy \, |y| \Psi(-y,\varphi) \exp[iy\hat{x}(\varphi)], \tag{7.25}$$

which can be represented in the equivalent form as

$$\hat{\varrho} = \frac{2}{\pi}\int_\pi d\varphi \int_0^\infty dy \, |y| \left\{ \langle \cos[y\hat{x}(\varphi)] \rangle \cos[y\hat{x}(\varphi)] + \langle \sin[y\hat{x}(\varphi)] \rangle \sin[y\hat{x}(\varphi)] \right\}. \tag{7.26}$$

A comparison with Eq. (7.1) shows that the operators $\cos[y\hat{x}(\varphi)]$ and $\sin[y\hat{x}(\varphi)]$, for all positive values of y and all values of φ within a π interval, play the role of the operators $\hat{A}_i$ in Eq. (7.1). Expressing in Eq. (7.25) $\Psi(-y,\varphi)$ in terms of $p(x,\varphi)$ according to Eq. (7.7), we may (formally) relate $\hat{\varrho}$ to $p(x,\varphi)$ as

$$\hat{\varrho} = \int_\pi d\varphi \int dx \, \hat{K}(x,\varphi) p(x,\varphi), \tag{7.27}$$

where the kernel operator is given by

$$\hat{K}(x,\varphi) = \frac{1}{\pi}\int dy \, |y| \exp\{iy[\hat{x}(\varphi) - x]\} \tag{7.28}$$

[note that $\hat{K}(-x, \varphi \pm \pi) = \hat{K}(x,\varphi)$]. A comparison with Eq. (7.1) shows that now the projectors $|x,\varphi\rangle\langle x,\varphi|$ for all values of x and all values of φ within a π interval play the role of the operators $\hat{A}_i$ in Eq. (7.1).

The extension of Eqs (7.25) and (7.27) to multi-mode fields is straightforward. The single-mode phase-rotated quadrature distributions are replaced by multi-mode joint probability distributions and the kernel operators are replaced by the corresponding multi-mode kernel operators, which are the direct products of the single-mode kernel operators. In particular, for a two-mode system Eq. (7.25) extends to

$$\hat{\varrho} = \frac{1}{\pi^2}\int_\pi d\varphi_1 \int_\pi d\varphi_2 \int dy_1 \int dy_2 \, \{|y_1||y_2| \\ \times \Psi(-y_1,-y_2,\varphi_1,\varphi_2) \exp[iy_1\hat{x}_1(\varphi_1) + iy_2\hat{x}_2(\varphi_2)]\}, \tag{7.29}$$

and Eqs (7.27) and (7.28) extend to

$$\hat{\varrho} = \int_\pi d\varphi_1 \int_\pi d\varphi_2 \int dx_1 \int dx_2 \, \hat{K}(x_1, x_2, \varphi_1, \varphi_2) p(x_1, x_2, \varphi_1, \varphi_2), \quad (7.30)$$

$$\hat{K}(x_1, x_2, \varphi_1, \varphi_2)$$
$$= \frac{1}{\pi^2} \int dy_1 \int dy_2 \, |y_1||y_2| \exp\{iy_1[\hat{x}_1(\varphi_1) - x_1] + iy_2[\hat{x}_2(\varphi_2) - x_2]\}. \quad (7.31)$$

Instead, the probability distributions $p(x, \vartheta, \varphi_1, \varphi_2)$ of the weighted sums

$$\hat{x}(\vartheta, \varphi_1, \varphi_2) = \hat{x}_1(\varphi_1) \cos \vartheta + \hat{x}_2(\varphi_2) \sin \vartheta \quad (7.32)$$

can be considered, with ϑ being within a $\pi/2$ interval. The probability distributions of the sums are related to the joint probability distributions as

$$p(x, \vartheta, \varphi_1, \varphi_2) = \int dx_1 \int dx_2 \, p(x_1, x_2, \varphi_1, \varphi_2) \delta(x - x_1 \cos \vartheta - x_2 \sin \vartheta), \quad (7.33)$$

from which it follows that their characteristic functions (Fourier transforms) $\Psi(y_1, y_2, \varphi_1, \varphi_2)$ and $\Psi(y, \vartheta, \varphi_1, \varphi_2)$ are related to each other as

$$\Psi(y_1 = y \cos \vartheta, y_2 = y \sin \vartheta, \varphi_1, \varphi_2) = \Psi(y, \vartheta, \varphi_1, \varphi_2). \quad (7.34)$$

7.1.2
Wigner function

Application of Eq. (7.18) to the case where $s=0$ yields the phase-rotated quadrature distributions expressed in terms of the Wigner function. With the help of Eq. (7.20) it is not difficult to prove, on changing the integration variables, that

$$p(x, \varphi) = \frac{1}{2} \int dy \, W[\tfrac{1}{2}x \cos \varphi + y \sin \varphi + i(-\tfrac{1}{2}x \sin \varphi + y \cos \varphi)]. \quad (7.35)$$

Equation (7.35) reveals that the (scaled) phase-rotated quadrature distributions can be regarded as marginals of the Wigner function. An integral relation of the form given in Eq. (7.35) is also called Radon transformation. Inverse Radon transformation then yields the Wigner function in terms of the phase-rotated quadrature distributions for all phases within a π interval.

Application of Eq. (7.21) [together with Eq. (7.17)] to the case where $s = 0$ yields the inverse Radon transformation in the form of

$$W(\alpha) = \frac{1}{\pi^2} \int_\pi d\varphi \int dy \int dx \, |y| \exp[iy(x - 2\alpha_R \cos \varphi + 2\alpha_I \sin \varphi)] p(x, \varphi). \quad (7.36)$$

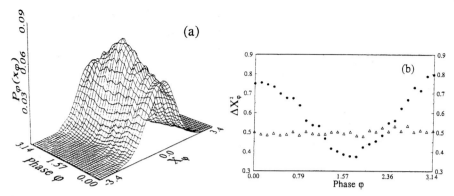

Fig. 7.1 (a) In balanced four-port homodyning measured phase-rotated quadrature distributions $[X_\varphi = x/\sqrt{2},\ P_\varphi(X_\varphi) = \sqrt{2}\,p(x=\sqrt{2}X_\varphi)]$ of a squeezed state. (b) Measured variances $\Delta X_\varphi^2 = \langle [\Delta \hat{x}(\varphi)]^2 \rangle / 2$ of the phase-rotated quadratures: circles, squeezed state; triangle, vacuum state. In the experiment 4000 repeated measurements of the photoelectron difference number at 27 values of the relative phase φ were made. [After Smithey, Beck, Raymer and Faridani (1993).]

Performing the y integral first would lead to an integral kernel which is not well-behaved. In the numerical calculation, regularization techniques can of course be used in order to overcome this difficulty. In the filtered back projection algorithm, the y integral is truncated such that

$$W(\alpha) \simeq W_C = \int_\pi d\varphi \int dx\, K_C(x - 2\alpha_R \cos\varphi + 2\alpha_I \sin\varphi) p(x,\varphi), \quad (7.37)$$

where

$$K_C(x) = \frac{1}{\pi^2} \int_{-y_C}^{y_C} dy\, |y| e^{iyx} \quad (7.38)$$

($y_C > 0$). In the first experimental demonstration of the method [Smithey, Beck, Raymer and Faridani (1993)], a pulsed signal field was superimposed by a pulsed local-oscillator field much stronger than the signal field, and the measurements and reconstructions were performed for a squeezed signal field (Figs 7.1 and 7.2) and for a vacuum signal field, the field mode detected being selected by the spatio-temporal mode of the local-oscillator field.

As mentioned in Section 6.5.4, the measured phase-rotated quadrature distributions do not correspond, in general, to the true signal mode, but they must be regarded as the distributions of a superposition of the signal and an additional noise source [Eq. (6.272)], because of nonperfect detection. Substituting in Eq. (7.35) for $p(x,\varphi)$ the measured distributions $p(x,\varphi;\eta)$ with $\eta < 1$ [Eq. (6.271)] and performing the inverse Radon transform on them, the Wigner function of a noise-assisted signal field is effectively reconstructed. Equivalently, the reconstructed Wigner function can be regarded

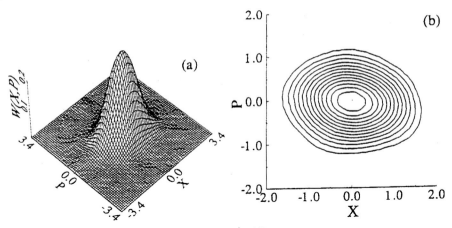

Fig. 7.2 Wigner distributions $W(X,P) = 2^{-1}W[\alpha = 2^{-1/2}(X+iP)]$ reconstructed from the measured rotated-phase quadrature distributions for a squeezed state, viewed in (a) 3D and as (b) contour plots, with equal numbers of constant-height contours. The reconstruction is performed by using inverse Radon transformation. [After Smithey, Beck, Raymer and Faridani (1993).]

as an s-parameterized phase-space function of the true signal field, however with $s<0$.

From Eq. (6.271) together with Eq. (6.268) it follows that the characteristic function $\Psi(y,\varphi;\eta)$ of $p(x,\varphi;\eta)$ typically measured when $\eta<1$ [Eq. (7.7) with $p(x,\varphi;\eta)$ in place of $p(x,\varphi)$] is related to the characteristic function $\Psi(y,\varphi)$ of $p(x,\varphi)$ as

$$\Psi(y,\varphi;\eta) = \exp\left[\tfrac{1}{2}(1-\eta^{-1})y^2\right]\Psi(y,\varphi). \tag{7.39}$$

Combining Eqs (7.12) and (7.39), we see that $\Psi(y,\varphi;\eta)$ and $\Phi(\beta;s)$ are related to each other as

$$\Psi(y,\varphi;\eta) = \exp\left[-\tfrac{1}{2}(s-1+\eta^{-1})y^2\right]\Phi(iye^{-i\varphi};s). \tag{7.40}$$

Hence, in the exponentials in Eqs (7.16) and (7.21) for s performing the substitution $s-1+\eta^{-1}$ yields the relations between $P(\alpha;s)$ and $p(x,\varphi;\eta)$. From Eq. (7.21) it then follows that, on recalling Eq. (7.17),

$$P(\alpha;s) = \frac{1}{\pi^2}\int_\pi d\varphi \int dy \int dx \, \{\exp[\tfrac{1}{2}(s-1+\eta^{-1})y^2] \\ \times |y|\exp[iy(x-\alpha_R\cos\varphi+\alpha_I\sin\varphi)]p(x,\varphi;\eta)\}. \tag{7.41}$$

Obviously, when $s=1-\eta^{-1}$, then Eq. (7.41) takes the form of Eq. (7.36),

$$P(\alpha;s=1-\eta^{-1})$$
$$=\frac{1}{\pi^2}\int_\pi d\varphi \int dy \int dx\,|y|\exp[iy(x-2\alpha_R\cos\varphi+2\alpha_I\sin\varphi)]p(x,\varphi;\eta). \tag{7.42}$$

Accordingly, Eq. (7.35) takes the form

$$p(x,\varphi;\eta)=$$
$$\tfrac{1}{2}\int dy\,P\bigl[\tfrac{1}{2}x\cos\varphi+y\sin\varphi+i(-\tfrac{1}{2}x\sin\varphi+y\cos\varphi);s=1-\eta^{-1}\bigr]. \tag{7.43}$$

Thus, in Eq. (7.35) replacing $p(x,\varphi)$ with $p(x,\varphi;\eta)$, the inverse Radon transformation yields the signal-mode phase-space function $P(\alpha;s=1-\eta^{-1})$ in place of the Wigner function.

7.2
Density matrix in phase-rotated quadrature basis

Let us reconstruct the density-matrix elements in a phase-rotated quadrature basis of chosen phase φ, $\langle x,\varphi|\hat\varrho|x',\varphi\rangle$, from the phase-rotated quadrature distributions $p(\tilde x,\tilde\varphi)$ for all phases $\tilde\varphi$ in a π interval. The most straightforward way is to extend the expression (7.5) for $|x\rangle\langle x|$ to $|x'\rangle\langle x|$. Recalling the commutation relation (3.145), we may write, according to the position representation of the momentum operator in quantum mechanics,

$$\frac{\partial}{\partial x}|x,\varphi\rangle = -\frac{i}{2}\hat x(\varphi-\tfrac{1}{2}\pi)|x,\varphi\rangle, \tag{7.44}$$

from which it follows that

$$|x,\varphi\rangle = \exp\left[-\frac{i}{2}(x-x')\hat x(\varphi-\tfrac{1}{2}\pi)\right]|x',\varphi\rangle$$
$$= \hat D\bigl[\tfrac{1}{2}(x-x')e^{-i\varphi}\bigr]|x',\varphi\rangle. \tag{7.45}$$

Hence we may write

$$|x',\varphi\rangle\langle x,\varphi| = \hat D\bigl[\tfrac{1}{2}(x'-x)e^{-i\varphi}\bigr]|x,\varphi\rangle\langle x,\varphi|. \tag{7.46}$$

Combining Eqs (7.5) and (7.46) and applying the relation (3.53), it is not difficult to prove that

$$|x',\varphi\rangle\langle x,\varphi| = \frac{1}{2\pi}\int dy\,e^{-iy(x+x')/2}\hat D\bigl[(\tfrac{1}{2}(x'-x)+iy)e^{-i\varphi}\bigr], \tag{7.47}$$

so that the density matrix in a phase-rotated quadrature basis can be given by ($x = x_1 - x_2$, $x' = x_1 + x_2$)

$$\langle x_1 - x_2, \varphi | \hat{\varrho} | x_1 + x_2, \varphi \rangle = \frac{1}{2\pi} \int dy \, e^{-ix_1 y} \Psi(y, x_2, \varphi), \tag{7.48}$$

where the characteristic function $\Psi(y, x_2, \varphi)$ reads as

$$\Psi(y, x_2, \varphi) = \text{Tr}\{\hat{\varrho}\hat{D}[(x_2 + iy)e^{-i\varphi}]\}. \tag{7.49}$$

Recalling Eq. (7.10), it can be related to the characteristic function of a phase-rotated quadrature distribution,

$$\Psi(y, x_2, \varphi) = \Psi(\tilde{y}, \tilde{\varphi}), \tag{7.50}$$

with

$$\tilde{y} = \sqrt{y^2 + x_2^2}, \qquad \tilde{\varphi} = \varphi - \arg(y - ix_2). \tag{7.51}$$

Thus, combining Eqs (7.48) and (7.50) and making use of Eq. (7.7), we obtain the density-matrix elements from the phase-rotated quadrature distributions by means of a two-fold Fourier transformation [Kühn, Welsch and Vogel (1994)]:

$$\langle x_1 - x_2, \varphi | \hat{\varrho} | x_1 + x_2, \varphi \rangle = \frac{1}{2\pi} \int dy \, e^{-ix_1 y} \int dx \, e^{i\tilde{y}x} p(x, \tilde{\varphi}). \tag{7.52}$$

Note that Eq. (7.52) can be used to obtain the density matrix in different representations, by varying the phase φ of the quadrature component defining the basis. Furthermore, Eq. (7.52) can also be extended, in principle, to imperfect detection, expressing $\Psi(\tilde{y}, \tilde{\varphi})$ in terms of $\Psi(\tilde{y}, \tilde{\varphi}; \eta)$ according to Eq. (7.39). As an illustration of the method, in Fig. 7.3 the reconstructed density matrices in (a) the "position" basis, $\varphi = 0$, and (b) the "momentum" basis, $\varphi = \pi/2$, of a squeezed vacuum state are shown.[5] The phases $\varphi = 0$ and $\varphi = \pi/2$ coincide with the phases of minimal and maximal field noise respectively.

Using Eq. (7.12), we rewrite Eq. (7.50) as

$$\Psi(y, x_2, \varphi) = e^{-\frac{1}{2}s\tilde{y}^2} \Phi(i\tilde{y}e^{-i\tilde{\varphi}}; s) = e^{-\frac{1}{2}s\tilde{y}^2} \Phi[(x_2 + iy)e^{-i\varphi}; s]. \tag{7.53}$$

Substituting this expression into Eq. (7.48) and using Eq. (4.92), we may express the density matrix in a phase-rotated quadrature basis in terms of an

---

5) The homodyne data were recorded by T. Coudreau, A.Z. Khoury and E. Giacobino, using the experimental setup reported by Lambrecht, Coudreau, Steinberg and Giacobino (1996). In the experiment the quadratures were measured at 48 phases and at each phase 7812 measurements were performed. The reconstruction is based on the numerical algorithm by Zucchetti, Vogel, Tasche and Welsch (1996).

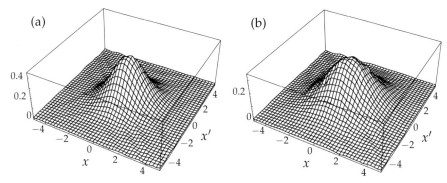

Fig. 7.3 From measured phase-rotated quadrature distributions, reconstructed (real) density matrix $\langle x, \varphi | \hat{\varrho} | x', \varphi \rangle$ of a squeezed state in (a) the "position" basis ($\varphi = 0$) and (b) the "momentum" basis ($\varphi = \pi/2$).

s-parameterized phase-space function:

$$\langle x_1 - x_2, \varphi | \hat{\varrho} | x_1 + x_2, \varphi \rangle$$
$$= \frac{1}{2\pi} \int dy \, e^{-ix_1 y - \frac{1}{2} s \tilde{y}^2} \int d^2\alpha \, \exp\left[(x_2 + iy) e^{-i\varphi} \alpha^* - \text{c.c.}\right] P(\alpha; s), \tag{7.54}$$

which, on changing integration variables, can be rewritten as

$$\langle x_1 - x_2, \varphi | \hat{\varrho} | x_1 + x_2, \varphi \rangle = \frac{1}{2\pi} \int dz \, \{ e^{-ix_1 z - \frac{1}{2} s \tilde{z}^2}$$
$$\times \int dx \int dy \, e^{2i(xz - x_2 y)} P[x \cos\varphi + y \sin\varphi + i(-x \sin\varphi + y \cos\varphi); s] \}, \tag{7.55}$$

with $\tilde{y} = \tilde{y}(y, x_2)$ and $\tilde{z} = \tilde{z}(z, x_2)$ being defined according to Eq. (7.51). If $s > 0$ (or $\text{Re}\, s > 0$), then in Eq. (7.55) the integration over z can be performed separately to reduce the three-fold integral transformation to a two-fold one. In the limit as $s \to 0$ a δ function appears and thus the integration over x can also be performed. In this way we find that the density-matrix elements can be obtained from the Wigner function by a single Fourier transformation as follows [Wigner (1932)]:

$$\langle x_1 - x_2, \varphi | \hat{\varrho} | x_1 + x_2, \varphi \rangle$$
$$= \frac{1}{2} \int dy \, e^{-2ix_2 y} W\left[\tfrac{1}{2} x_1 \cos\varphi + y \sin\varphi + i(-\tfrac{1}{2} x_1 \sin\varphi + y \cos\varphi)\right], \tag{7.56}$$

which, when $x_2 = 0$, just reduces to Eq. (7.35).

7.3
Density matrix in the number basis

If a phase-space function $P(\alpha;s)$ is known, then the density-matrix elements in the number basis, $\langle m|\hat{\varrho}|n\rangle$, can be calculated, in principle, by applying Eq. (4.85) and calculating the c-number function $\pi\langle m|\hat{\delta}(\hat{a}-\alpha;-s)|n\rangle$ associated with the operator $|n\rangle\langle m|$ in s order. From Section 6.5.5 we know that $P(\alpha;s)$ can be measured directly for $s \leq -1$. Thus, the procedure outlined could be used to infer $\langle m|\hat{\varrho}|n\rangle$ from the measured $P(\alpha;s)$. Since the resulting reconstruction formula is not suited for statistical sampling [according to Eq. (7.2)] and the inaccuracies of the measured $P(\alpha;s)$ can give rise to an error explosion in the reconstructed density-matrix elements, the method has been of less experimental relevance. Instead of using integration techniques, one could try to apply equivalent differentiation techniques. In particular, combining Eq. (4.67) with Eqs (3.22) and (3.59), we easily derive

$$Q(\alpha) = \pi^{-1} e^{-|\alpha|^2} \sum_{m,n} \frac{1}{\sqrt{m!n!}} \alpha^{*m} \alpha^n \langle m|\hat{\varrho}|n\rangle, \tag{7.57}$$

from which it follows that

$$\langle m|\hat{\varrho}|n\rangle = \frac{\pi}{\sqrt{m!n!}} \frac{\partial^{m+n}}{\partial \alpha^{*m} \partial \alpha^n} e^{|\alpha|^2} Q(\alpha)\Big|_{\alpha=\alpha^*=0}. \tag{7.58}$$

Clearly, the inaccuracies in the measured Q function prevent one from carrying out the derivatives with sufficient precision, in general. Therefore, much effort has been made to obtain the density matrix in the number basis from measurable data with sufficient precision and as directly as possible.

7.3.1
Sampling from quadrature components

It turns out that the density-matrix elements in the number basis can be inferred from the measured phase-rotated quadrature distributions in a very direct way. In the number basis Eq. (7.27) reads as

$$\varrho_{mn} = \langle m|\hat{\varrho}|n\rangle = \int_\pi d\varphi \int dx\, K_{mn}(x,\varphi) p(x,\varphi), \tag{7.59}$$

where the kernel function (also called the sampling function) can be written as

$$K_{mn}(x,\varphi) = \frac{1}{\pi} \int dy\, |y| \langle m| \exp\{iy[\hat{x}(\varphi) - x]\}|n\rangle. \tag{7.60}$$

Recalling the relation (3.195) [together with Eq. (3.194)], we find that $K_{mn}(x,\varphi)$ takes the form of

$$K_{mn}(x,\varphi) = e^{-i(m-n)\varphi} f_{mn}(x), \tag{7.61}$$

where

$$f_{mn}(x) = \frac{1}{\pi} \int dy \, |y| \langle m| \exp[iy(\hat{x} - x)]|n\rangle \tag{7.62}$$

$[\hat{x} \equiv \hat{x}(0)]$. Recalling the relation $\hat{K}(x, \varphi + \pi) = \hat{K}(-x, \varphi)$, we see that

$$f_{mn}(-x) = (-1)^{m-n} f_{mn}(x). \tag{7.63}$$

In order to find an expression of $f_{mn}(x)$, which is more suitable for practical applications than that in Eq. (7.62), let us express the phase-rotated quadrature distributions in terms of the density-matrix elements in the number basis,

$$p(x, \varphi) = \sum_{k,l} \langle x, \varphi|k\rangle \langle l|x, \varphi\rangle \varrho_{kl}. \tag{7.64}$$

Recalling Eqs (3.193) and (3.199), we may rewrite Eq. (7.64) as

$$p(x, \varphi) = \sum_{k,l} g_{kl}(x) e^{i(k-l)\varphi} \varrho_{kl}, \tag{7.65}$$

where

$$g_{kl}(x) = \psi_k(x) \psi_l(x) \tag{7.66}$$

$[\psi_k(x) = \langle k|x\rangle]$. We now multiply Eq. (7.65) by $f_{mn}(x) e^{-i(m-n)\varphi}$ and integrate with respect to φ and x. Using the symmetry relation (7.63), we can easily derive

$$\int_\pi d\varphi \int dx \, p(x, \varphi) f_{mn}(x) e^{-i(m-n)\varphi}$$

$$= \frac{1}{2} \int_{2\pi} d\varphi \int dx \, p(x, \varphi) f_{mn}(x) e^{-i(m-n)\varphi}$$

$$= \sum_{k,l} \left[\pi \int dx \, g_{kl}(x) f_{mn}(x) \right] \varrho_{kl}, \quad k - l = m - n. \tag{7.67}$$

Comparing with Eq. (7.59) [together with Eq. (7.61)], we find that $f_{mn}(x)$ obeys the integral equation

$$\pi \int dx \, g_{kl}(x) f_{mn}(x) = \delta_{km} \delta_{ln}, \quad k - l = m - n. \tag{7.68}$$

Thus, the sought function $f_{mn}(x)$ is orthonormal to the product of wave functions $\psi_k(x)\psi_l(x)$. Obviously, $\psi_k(x)$ is the ordinary (i.e., regular) solution of the harmonic-oscillator Schrödinger equation for the kth energy level,

$$\left[\frac{d^2}{dx^2} + \left(k + \frac{1}{2} - \frac{1}{4}x^2\right) \right] \phi_k(x) = 0 \tag{7.69}$$

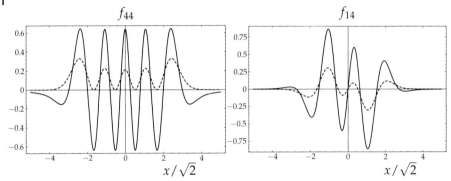

Fig. 7.4 Examples of the kernel function $\pi f_{mn}(x)$ for sampling the density-matrix elements ϱ_{mn} in the number basis from the phase-rotated quadrature distributions according to Eq. (7.59) together with Eq. (7.61) for $m=n=4$ (left figure) and $m=1$, $n=4$ (right figure). For comparison, the product of the wave functions $\psi_m(x)\psi_n(x)/\sqrt{2}$ is shown (dashed lines). [After Leonhardt (1997).]

$[\phi_k(x) = \psi_k(x)]$. It is worth noting that the solution of the integral equation (7.68) can be given in the form of the derivative of a product of regular and irregular wave functions [Leonhardt, Munroe, Kiss, Richter and Raymer (1996); Richter (1996a)],

$$f_{mn}(x) = \frac{d}{dx}[\psi_m(x)\chi_n(x)], \qquad (7.70)$$

where $\chi_n(x)$ is an irregular solution of the wave equation (7.69) $[\phi_n(x) = \chi_n(x)]$ which must be chosen such that

$$\psi_n(x)\frac{d\chi_n(x)}{dx} - \frac{d\psi_n(x)}{dx}\chi_n(x) = \frac{2}{\pi} \qquad (7.71)$$

(for a proof, see Appendix D). It should be pointed out that the ambiguities of the kernel function leave enough room to choose the most convenient form [for details, see Leonhardt (1997)]. Examples are shown in Fig. 7.4.

Equation (7.59) reveals that, according to Eq. (7.2), the density-matrix elements in the number basis can be sampled directly from the measured phase-rotated quadrature statistics. Each density-matrix element ϱ_{mn} can be regarded as a statistical average of the bounded kernel function $K_{mn}(x,\varphi)$.[6] In an experiment each outcome x of $\hat{x}(\varphi)$, with $\varphi \in [0,\pi)$, contributes individually to ϱ_{mn}, so that ϱ_{mn} is gradually building up during the data collection. That is to say, ϱ_{mn} can be sampled from a sufficiently large set of homodyne data in real time, and the mean value obtained from different experiments can be expected to be normal-Gaussian distributed around the true value, because of the central-limit theorem. Moreover, the sampling method can also

---

[6] For a numerical implementation of Eq. (7.70), see Leonhardt, Munroe, Kiss, Richter and Raymer (1996).

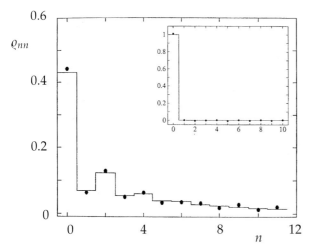

Fig. 7.5 From the phase-rotated quadrature distributions according to Eq. (7.59) reconstructed photon-number distribution of a squeezed vacuum and the vacuum state (inset). Solid points refer to experimental data, histograms to theory. [After Schiller, Breitenbach, Pereira, Müller and Mlynek (1996).]

be used to estimate the statistical error. Experimentally, the method was first successfully applied to the determination of the density matrix of squeezed light generated by a continuous-wave optical parametric amplifier [Schiller, Breitenbach, Pereira, Müller and Mlynek (1996)]. The reconstructed diagonal density-matrix elements are shown in Fig. 7.5.

Since in a realistic experiment the quantum efficiency η is always less than unity, insertion into Eq. (7.59) of the measured phase-rotated quadrature distributions $p(x,\varphi;\eta)$ [in place of the exact distributions $p(x,\varphi)$] yields the density matrix of a noise-assisted state in general (cf. the last paragraph in Section 7.1.2). If $\eta > 0.5$ it is possible to compensate for detection losses by introducing a modified sampling function which depends on η such that performing the sampling algorithm on the real measured (i.e., the smeared) phase-rotated quadrature distributions $p(x,\varphi;\eta)$ yield the correct quantum state [D'Ariano, Leonhardt and Paul (1995)]. Another approach to the problem of loss compensation is that the sampling function is left unchanged and, at the first stage, the density matrix of the quantum state which corresponds to the measured distributions $p(x,\varphi;\eta)$ is reconstructed. After that, at the second stage, the true density matrix is calculated from the reconstructed one using an inverse Bernoulli transformation [Kiss, Herzog and Leonhardt (1995)].

Since the density matrix in any basis contains the full information about the quantum state of the system under consideration, all quantum-statistical properties can be inferred from it. Let $\hat{F}$ be an operator whose expectation value

$$\langle \hat{F} \rangle = \sum_{mn} F_{nm} \varrho_{mn}, \quad (7.72)$$

is to be determined. One may be tempted to calculate it from the reconstructed density matrix. However, an experimentally determined density matrix always suffers from various inaccuracies which can propagate (and increase) in the calculation process. Therefore, it may be advantageous to determine directly the quantities of interest from the measured data, without reconstructing the whole quantum state. In particular, in Eq. (7.72) substituting the integral representation (7.59) for ϱ_{mn}, one can try to obtain an integral representation

$$\langle \hat{F} \rangle = \int_\pi d\varphi \int dx \, K_F(x,\varphi) p(x,\varphi) \quad (7.73)$$

suited to the direct sampling of $\langle \hat{F} \rangle$, from the quadrature component distributions $p(x,\varphi)$. It is worth noting that the kernel function $K_F(x,\varphi)$ is not defined uniquely by the integral relation (7.73), but only up to a function $\Theta(x,\varphi)$ that satisfies the integral equation

$$\int_\pi d\varphi \int dx \, \Theta(x,\varphi) p(x,\varphi) = 0. \quad (7.74)$$

Hence, if the integral kernel $K_F(x,\varphi)$ which is obtained from Eq. (7.72), together with Eq. (7.59), is unbounded for $|x| \to \infty$ such that the x integral in Eq. (7.73) does not exist for any normalizable state, it cannot be concluded that $\langle \hat{F} \rangle$ cannot be sampled from $p(x,\varphi)$, since a different, bounded kernel may exist. Obviously, the ambiguity mentioned is also true for the kernel function $K_{mn}(x,\varphi)$ in Eq. (7.59). In particular, applying normally-ordered moment expansion (Section 7.5), $K_{mn}(x,\varphi)$ can be represented in the equivalent form of Eq. (7.95), which is not necessarily suitable for statistical sampling.

7.3.2
Reconstruction from displaced number states

From Section 6.5.3 we know that in unbalanced homodyning the photon-number distribution of the transmitted signal mode is, under certain conditions, the displaced photon-number distribution of the signal mode, $p_m(\alpha)$, the displacement parameter $\alpha = |\alpha|e^{i\varphi}$ being controlled by the local-oscillator complex amplitude. Expanding the density operator in the number basis, $p_m(\alpha)$ can be related to the density matrix of the signal mode as

$$p_m(\alpha) = \langle m,\alpha|\hat{\varrho}|m,\alpha\rangle = \sum_{k,n} \langle m,\alpha|k\rangle \langle n|m,\alpha\rangle \varrho_{kn}, \quad (7.75)$$

where the expansion coefficients $\langle n|m,\alpha\rangle$ can be taken from Eq. (3.101). Equation (7.75) can always be inverted in order to obtain ϱ_{mn} in terms of $p_k(\alpha)$. Combining Eqs (4.62) and (4.80) [together with Eq. (4.81)], we may write

$$\varrho_{mn} = \sum_{k=0}^{\infty} \int d^2\alpha \, K_{mn}^k(\alpha) p_k(\alpha), \tag{7.76}$$

where

$$K_{mn}^k(\alpha) = \frac{2}{(1-s)} \left(\frac{s+1}{s-1}\right)^k \langle m|\hat{\delta}(\hat{a}-\alpha;-s)|n\rangle. \tag{7.77}$$

It can be shown that $K_{mn}^k(\alpha)$ is bounded for $-1 < s \leq 0$, which offers the possibility of direct sampling of ϱ_{mn} from the displaced number probability distribution $p_m(\alpha)$ [Mancini, Man'ko and Tombesi (1997)].

Since $p_m(\alpha)$ as a function of α for chosen m already determines the quantum state, it is clear that when m is allowed to be varying, then – in contrast to Eq. (7.76) – $p_m(\alpha)$ need not be known for all complex values of α in order to reconstruct the density-matrix elements ϱ_{kn} from $p_m(\alpha)$. In particular, it is sufficient to know $p_m(\alpha)$ for all values of m and all phases $\varphi = \arg(\alpha)$, $|\alpha|$ being fixed [Leibfried, Meekhof, King, Monroe, Itano and Wineland (1996), Opatrný and Welsch (1997)]. For chosen $|\alpha|$ we can regard $p_m(\alpha)$ as a function of φ and introduce the Fourier coefficients

$$p_m^k(|\alpha|) = \frac{1}{2\pi} \int_0^{2\pi} d\varphi \, e^{ik\varphi} p_m(\alpha) \tag{7.78}$$

($k = 0, 1, 2, \ldots$), which are related to the density-matrix elements whose row and column indices differ by k. Substituting Eqs (7.75) into Eq. (7.78) and using Eq. (3.101), we can easily derive

$$p_m^k(|\alpha|) = \sum_{n=0}^{\infty} G_{mn}^k(|\alpha|) \varrho_{n+k\,n}, \tag{7.79}$$

where

$$G_{mn}^k(|\alpha|) = e^{-|\alpha|^2} \frac{m!}{\sqrt{n!(n+k)!}} L_m^{n-m}(|\alpha|^2) L_m^{k+n-m}(|\alpha|^2)|\alpha|^{2n+k-2m}. \tag{7.80}$$

Inverting Eq. (7.79) for each value of k yields the sought density-matrix elements. Unfortunately, no analytical solution has been found. However, Eq. (7.79) can be inverted numerically, setting $\varrho_{mn} = 0$ for $m, n > n_{\max}$ and using, e. g., least-squares inversion. The method was first used to reconstruct the density matrix of the center-of-mass motion of a trapped ion [Leibfried, Meekhof, King, Monroe, Itano and Wineland (1996)]. Details on the measurement techniques and an example of a reconstructed density matrix are given in Section 13.5.2 (Fig. 13.11, p. 477).

7.4
Local reconstruction of phase-space functions

The displaced photon-number statistics $p_m(\alpha)$ which are measurable in unbalanced homodyning (Section 6.5.3) can be used for a point-wise reconstruction of s-parameterized phase-space functions $P(\alpha;s)$ [Wallentowitz and Vogel (1996); Banaszek and Wódkiewicz (1996)]. From Eq. (4.53) together with (4.62) it is easily seen that

$$P(\alpha;s) = \frac{2}{\pi(1-s)} \sum_{m=0}^{\infty} \left(\frac{s+1}{s-1}\right)^m p_m(\alpha). \tag{7.81}$$

Hence, in principle, all the phase-space functions $P(\alpha;s)$, with $s<1$, can be obtained from $p_m(\alpha)$ for each phase-space point α in a very direct way, without integral transformations. However, for unknown statistics $p_m(\alpha)$ it is advantageous, in practice, to require nonexploding coefficients, $|(s+1)/(s-1)|\leq 1$, which are obtained for $s\leq 0$ only. In particular when $s=-1$, then Eq. (7.81) reduces to the well-known result that $Q(\alpha) = \pi^{-1} p_0(\alpha)$, with $p_0(\alpha) = \langle \alpha|\hat{\varrho}|\alpha\rangle$. Further, choosing $s=0$ in Eq. (7.81), we arrive at the Wigner function,

$$W(\alpha) = \frac{2}{\pi} \sum_{m=0}^{\infty} (-1)^m p_m(\alpha). \tag{7.82}$$

Equation (7.82) reflects the fact that the Wigner function is proportional to the expectation value of the displaced parity operator [cf. Eq. (4.64)].

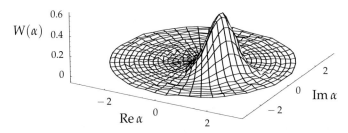

Fig. 7.6 The reconstructed Wigner function $W(\alpha)$ for a weak coherent state with approximately one photon. [After Banaszek, Radzewicz, Wódkiewicz and Krasiński (1999).]

Experimentally, the method was first applied to the reconstruction of the Wigner function of the center-of-mass motion of a trapped ion [Leibfried, Meekhof, King, Monroe, Itano and Wineland (1996)], see Section 13.5.2 (Fig. 13.10). For light, the method was demonstrated experimentally as well [Banaszek, Radzewicz, Wódkiewicz and Krasiński (1999)], an example is given in Fig. 7.6. The use of the method for light is hampered by the problem that it is difficult to discriminate adjacent photon numbers in photodetection needed for determining the Wigner function according to Eq. (7.82). To determine the Wigner function by this method for more general light fields a cascaded homodyne detection scheme has been proposed [Kis, Kiss, Janszky, Adam, Wallentowitz and Vogel (1999)], which combines the unbalanced scheme with balanced homodyning, allowing the determination of the desired photon-number statistics of the displaced field.

7.5
Normally ordered moments

It is often sufficient to know some moments of the creation and annihilation operators rather than the overall quantum state. By means of Eq. (4.96) s-ordered moments can be expressed in terms of normally ordered moments and vice versa, so that we may restrict our attention to normally ordered moments. We apply Eqs (3.19) and (3.20) and express the normally ordered moments in terms of the density-matrix elements in the number basis:

$$\langle \hat{a}^{\dagger m} \hat{a}^n \rangle = \sum_{k=0}^{\infty} \langle k | \hat{\varrho} \hat{a}^{\dagger m} \hat{a}^n | k \rangle$$

$$= \sum_{k=n}^{\infty} \sqrt{k(k-1)\cdots(k-n+1)} \sqrt{(k+1)(k+2)\cdots(k+m-n)}\, \varrho_{k\,k+m-n}$$

$$= \sum_{k=n}^{\infty} \sqrt{\frac{(k+m-n)!}{(k-n)!}}\, \varrho_{k\,k+m-n}. \tag{7.83}$$

In Eq. (7.83) substituting for $\varrho_{k\,k+m-n}$ the expression in Eq. (7.59), we expect that $\langle \hat{a}^{\dagger m} \hat{a}^n \rangle$ can be expressed in terms of $p(x,\varphi)$ as

$$\langle \hat{a}^{\dagger m} \hat{a}^n \rangle = \int_\pi d\varphi \int dx\, M_{mn}(x,\varphi) p(x,\varphi). \tag{7.84}$$

Indeed, it can be shown that a kernel function $M_{mn}(x,\varphi)$ which solves Eq. (7.84) together with Eq. (7.83) reads [Richter (1996b)]

$$M_{mn}(x,\varphi) = M_{mn}(x) e^{i(m-n)\varphi}, \tag{7.85}$$

where

$$M_{mn}(x) = \frac{m!n!}{\pi\sqrt{2^{m+n}(m+n)!}} H_{m+n}(x/\sqrt{2}). \tag{7.86}$$

Using Eq. (7.64) together with Eq. (3.199), we may write

$$\int_\pi d\varphi \int dx\, e^{i(m-n)\varphi} H_{m+n}(x/\sqrt{2}) p(x,\varphi)$$

$$= \sum_{k,l} (\pi 2^{k+l} k! l!)^{-1/2} \varrho_{kl} I_{m+n\,kl} \frac{1}{2} \int_{2\pi} d\varphi\, e^{i(m-n-l+k)\varphi}$$

$$= \sum_{k,l} (\pi 2^{k+l} k! l!)^{-1/2} \varrho_{kl} I_{m+n\,kl} \pi \delta_{m-n\,l-k}, \tag{7.87}$$

where

$$I_{klm} = \int dx\, e^{-x^2} H_k(x) H_l(x) H_m(x)$$

$$= \frac{\sqrt{\pi}\, 2^n k! l! m!}{(n-k)!(n-l)!(n-m)!} \delta_{k+l+m\,2n}, \tag{7.88}$$

with n being a (non-negative) integer. Starting from the first line in Eq. (7.87) we have used the symmetry relation (7.8) and the property of the Hermite polynomials that $H_n(-x) = (-1)^n H_n(x)$. Substitution of the expression (7.88) into Eq. (7.87) and comparison of the result with Eq. (7.83) then shows the validity of Eq. (7.84) together with Eqs (7.85) and (7.86). Equation (7.84) offers the possibility of direct sampling of normally ordered moments from the phase-rotated quadrature distributions.[7]

The extension of the method to the reconstruction of normally-ordered moments of multi-mode fields from the corresponding joint phase-rotated quadrature distributions is straightforward. Instead, they can also be inferred from combined distributions, following the procedure outlined in the last paragraph of Section 7.1.1. The method was first used to experimentally demonstrate the determination of the ultrafast two-time photon number correlation of a nanosecond optical pulse [McAlister and Raymer (1997)]. In the experiment, two femtosecond local-oscillator pulses were used and phase-rotated sum quadratures $\hat{x}(\vartheta, \varphi_1, \varphi_2)$ [Eq. (7.32)] were measured. From these the normalized second-order coherence function

$$g^{(2)}(t_1, t_2) = \frac{\langle \hat{a}_1^\dagger \hat{a}_2^\dagger \hat{a}_1 \hat{a}_2 \rangle}{\langle \hat{a}_1^\dagger \hat{a}_1 \rangle \langle \hat{a}_2^\dagger \hat{a}_2 \rangle} \tag{7.89}$$

was computed (Fig. 7.7).[8]

---

[7] It should be noted that knowledge of $p(x,\varphi)$ at all phases within a π interval is not necessary to reconstruct a chosen $\langle \hat{a}^{\dagger m} \hat{a}^n \rangle$ and therefore the φ integral in Eq. (7.84) can be replaced by a sum. It was shown that $\langle \hat{a}^{\dagger m} \hat{a}^n \rangle$ can already be obtained from $p(x,\varphi)$ at $N = m+n+1$ different phases φ_k.

[8] Here, $\hat{a}_1$ and $\hat{a}_2$ are the photon destruction operators of the non-monochromatic modes defined by the local-oscillator pulses centered at times t_1 and t_2.

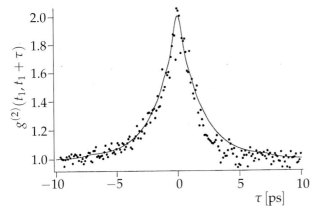

Fig. 7.7 The second-order coherence, Eq. (7.89), experimentally determined via balanced four-port homodyne detection (dots) and from the measured optical spectrum (solid line). The value of t_1 is set to occur near the maximum of the signal pulse. [After McAlister and Raymer (1997).]

Let $\hat{F}$ be an operator whose expectation value $\langle \hat{F} \rangle$ possesses the normally-ordered moment expansion

$$\langle \hat{F} \rangle = \sum_{mn} c_{mn} \langle \hat{a}^{\dagger m} \hat{a}^n \rangle. \tag{7.90}$$

In order to relate $\langle \hat{F} \rangle$ to $p(x, \varphi)$, according to the sampling formula (7.73), the kernel function $K_F(x, \varphi)$ may be calculated by substituting for $\langle \hat{a}^{\dagger m} \hat{a}^n \rangle$ in Eq. (7.90) the integral representation (7.84) [together with Eqs (7.85) and (7.86)], i.e.,

$$K_F(x, \varphi) = \sum_{m,n} c_{mn} M_{mn}(x) e^{i(m-n)\varphi}. \tag{7.91}$$

Recall that $K_F(x, \varphi)$ is only determined up to a function $\Theta(x, \varphi)$ that satisfies the integral equation (7.74).

We identify the operator $\hat{F}$ with the flip operator

$$\hat{A}_{nm} = |n\rangle\langle m| \tag{7.92}$$

in the number basis. It can easily be proved correct, by using Eq. (3.59) and calculating the c-number function $A_{nm}(\alpha; 1) = \langle \alpha | n \rangle \langle m | \alpha \rangle$ of $\hat{A}_{nm}$ in normal order, that $\hat{A}_{nm}$ can be written as[9]

$$\hat{A}_{nm} = \frac{1}{\sqrt{n!m!}} : \hat{a}^{\dagger n} \hat{a}^m e^{-\hat{n}} : . \tag{7.93}$$

---

9) Note that for $m=n$ Eq. (7.93) reduces to Eq. (4.60).

Equation (7.93) implies that the density-matrix elements in the number basis can be given in the form of the series (7.90),

$$\varrho_{mn} = \langle \hat{A}_{nm} \rangle = \sum_l \frac{(-1)^l}{l!\sqrt{n!m!}} \langle \hat{a}^{\dagger n+l} \hat{a}^{m+l} \rangle, \qquad (7.94)$$

which reveals that, when the whole manifold of moments $\langle \hat{a}^{\dagger m} \hat{a}^n \rangle$ is known, then the quantum state is also known in principle.[10] Application of Eq. (7.91) just yields the sampling formula (7.59), where the kernel function is given [up to a function $\Theta(x, \varphi)$] by

$$K_{mn}(x, \varphi) = e^{-i(m-n)\varphi} \sum_l \frac{(-1)^l}{l!\sqrt{n!m!}} M_{n+l\,m+l}(x). \qquad (7.95)$$

7.6
Canonical phase statistics

Direct sampling of quantities from the phase-rotated quadrature statistics may be used advantageously when no other method for direct detection of the quantities is available. A typical example is the canonical phase. Whereas the photon number can be measured by direct photodetection, in principle, there has been no apparatus for direct detection of the canonical phase. Therefore, the question arises as to whether or not the phase statistics can be sampled from the phase-rotated quadrature statistics.

Let Ψ_k be the exponential phase moments, i.e., the Fourier components of the phase distribution $p(\phi)$,

$$p(\phi) = \frac{1}{2\pi} \sum_{k=-\infty}^{\infty} e^{-ik\phi} \Psi_k, \qquad (7.96)$$

$$\Psi_k = \int_{2\pi} d\phi \, e^{ik\phi} p(\phi), \qquad (7.97)$$

where $p(\phi) = \langle \phi | \hat{\varrho} | \phi \rangle$, with $|\phi\rangle$ being given by Eq. (3.237). It is not difficult to prove that the substitution of this expression into Eq. (7.97) yields

$$\Psi_k = \sum_n \varrho_{n+k\,n} = \langle \hat{V}^k \rangle \qquad (7.98)$$

if $k \geq 0$, and $\Psi_{-k} = \Psi_k^*$ if $k < 0$, with the operator $\hat{V}$ being defined in Eq. (3.220). The sampling function for $\varrho_{n+k\,n}$ can be taken from Eq. (7.95) for $m = n+k$. By

---

10) Since the moments are not necessarily bounded, the expansion of the density-matrix elements according to Eq. (7.94) does not necessarily converge. The problem of non-convergence may be overcome by analytic continuation of appropriately chosen generating functions [for details, see Herzog (1996)].

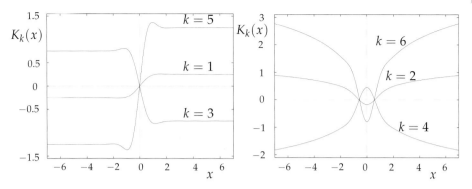

Fig. 7.8 Examples of the kernel function $f_k(x) = 2K_k(x/\sqrt{2})$ for sampling the exponential moments of the canonical phase from the phase-rotated quadrature distributions according to Eq. (7.99). [After Dakna, Opatrný and Welsch (1998).]

summing the result over n, we then obtain the sampling function for the kth exponential phase moment. Thus, we arrive at the result that [Dakna, Opatrný and Welsch (1998)]

$$\Psi_k = \int_\pi d\varphi \int dx\, e^{-ik\varphi} f_k(x) p(x, \varphi), \qquad (7.99)$$

where

$$f_k(x) = \sum_{n,l} \frac{(-1)^l}{l!\sqrt{n!(n+k)!}} M_{n+l\,n+l+k}(x) + \Theta_k(x). \qquad (7.100)$$

According to Eq. (7.74), we have introduced a function $\Theta_k(x)$ in order to take into account the ambiguity of the kernel function. It should be chosen so that $f_k(x)$ takes the simplest possible form best suited for sampling Ψ_k from $p(x, \varphi)$. In particular, $\Theta_k(x)$ can be an arbitrary polynomial of degree $k' = k - 2n$ (n, integer).

To obtain insight into the structure of the kernel function, we note that Eq. (7.99) applies to quantum and classical systems in a unified way and bridges the gap between quantum and classical phase. It is not difficult to prove that Eq. (7.35) can be rewritten as

$$p(x, \varphi) = \int d^2\alpha\, W(\alpha) \delta[x - 2|\alpha|\cos(\varphi + \phi_\alpha)]. \qquad (7.101)$$

Substitution of this expression into Eq. (7.99) yields ($d^2\alpha = r\,dr\,d\phi$, $\phi_\alpha \mapsto \phi$)

$$\Psi_k = \int_\pi d\varphi \int_0^\infty r\,dr \int_{2\pi} d\phi\, e^{-ik(\varphi-\phi)} f_k(2r\cos\varphi) W(re^{i\phi}). \qquad (7.102)$$

Classically, the phase distribution is simply given by the radially integrated Wigner function (i.e., the radially integrated classical phase-space probability

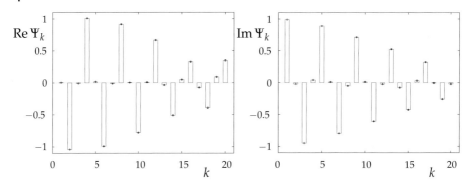

Fig. 7.9 From the phase-rotated quadrature distributions according to Eq. (7.99) reconstructed exponential phase moments of a phase-squeezed state. [After Dakna, Breitenbach, Mlynek, Opatrný, Schiller and Welsch (1998).]

distribution function),

$$p(\phi) = \int_0^\infty r\,dr\, W(re^{i\phi}), \qquad (7.103)$$

so that from Eq. (7.97) it follows that

$$\Psi_k = \int_{2\pi} d\phi \int_0^\infty r\,dr\, e^{ik\phi} W(re^{i\phi}). \qquad (7.104)$$

We compare Eq. (7.104) with (7.102) and find that the classical kernel function $f_k(x, \varphi)$ observed for $|x| \to \infty$ satisfies the integral equation

$$\int_\pi d\varphi\, e^{-ik\varphi} f_k(2r\cos\varphi) = 1 \qquad (7.105)$$

for all r. As can be verified by direct substitution, a solution of Eq. (7.105) is ($k > 0$)

$$f_k(x) = \begin{cases} \frac{1}{2}(-1)^{(k-1)/2} k\,\mathrm{sign}\,x & \text{if } k \text{ odd,} \\ \pi^{-1}(-1)^{(k+2)/2} k\ln(x/\sqrt{2}) & \text{if } k \text{ even.} \end{cases} \qquad (7.106)$$

Already from the classical kernel (7.106) it can be seen that $\sum_{k=-\infty}^{+\infty} e^{ik(\varphi-\phi)} f_k(x)$ does not exist, and hence the canonical phase distribution $p(\phi)$ itself cannot be sampled from $p(x, \varphi)$ without knowledge of the state.

Choosing $\Theta_k(x)$ in Eq. (7.100) such that $f_k(x)$ approaches the classical limit in the form given by Eq. (7.105) ensures that Eq. (7.99) is suitable for statistical sampling of the exponential phase moments from the phase-rotated quadrature distributions. Examples of $f_k(x)$ are shown in Fig. 7.8. The figure reveals

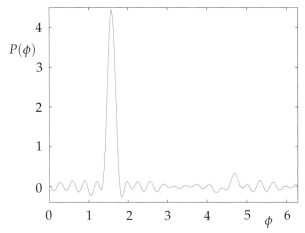

Fig. 7.10 The canonical phase distribution calculated from 20 reconstructed exponential phase moments as shown in Fig. 7.9. [After Dakna, Breitenbach, Mlynek, Opatrný, Schiller and Welsch (1998).]

that $f_k(x)$ rapidly approaches the classical limit [Eq. (7.106)] and differs from it only in a small interval around the origin, which defines the quantum regime. Obviously, the extension of the interval is just of the order of magnitude of the vacuum fluctuation. Recall that the quantum regime is realized when the state under study has a substantial overlap with the vacuum state.

Figure 7.9 shows the application of the method to the experimental determination of the exponential phase moments of phase-squeezed light [Dakna, Breitenbach, Mlynek, Opatrný, Schiller and Welsch (1998)]. The canonical phase distribution can then be obtained by straightforward summation according to Eq. (7.96), as shown in Fig. 7.10. It should be pointed out that sampling of the first exponential phase moment and the photon-number variance is already sufficient to verify fundamental phase-number uncertainties. The method also applies to the determination of the cosine- and sine-phase statistics associated with the Hermitian operators $\hat{C}$ [Eq. (3.246)] and $\hat{S}$ [Eq. (3.247)]. In particular, the mean values of the cosine phase and the sine phase, respectively, are simply given by the first exponential phase moment according to the relations

$$\langle \hat{C} \rangle = \tfrac{1}{2}(\Psi_1 + \Psi_1^*) \tag{7.107}$$

and

$$\langle \hat{S} \rangle = \tfrac{1}{2i}(\Psi_1 - \Psi_1^*). \tag{7.108}$$

Generally, the moments $\langle \hat{C}^n \rangle$ and $\langle \hat{S}^n \rangle$ ($n \geq 2$) can be obtained from the exponential phase moments and vacuum-assisted density-matrix elements.[11] Finally, the method can also be extended to multi-mode fields. In particular, sampling of the two-mode exponential moments

$$\Psi_{12}^{kl} = \langle \hat{V}_1^k \hat{V}_2^l \rangle \tag{7.109}$$

may be suitable for determining the difference-phase statistics of the modes.

References

Banaszek, K. and K. Wódkiewicz (1996) *Phys. Rev. Lett.* **76**, 4344.

Banaszek, K., C. Radzewicz, K. Wódkiewicz and J.S. Krasiński (1999) *Phys. Rev. A* **60**, 674.

Dakna, M., T. Opatrný and D.-G. Welsch (1998) *Opt. Commun.* **148**, 355.

Dakna, M., G. Breitenbach, J. Mlynek, T. Opatrný, S. Schiller and D.-G. Welsch (1998) *Opt. Commun.* **152**, 289.

D'Ariano, G.M., U. Leonhardt and H. Paul (1995) *Phys. Rev. A* **52**, R1801.

Fano, U. (1957) *Rev. Mod. Phys.* **29**, 74.

Herzog, U. (1996) *Phys. Rev. A* **53**, 2889.

Kis, Z., T. Kiss, J. Janszky, P. Adam, S. Wallentowitz and W. Vogel (1999) *Phys. Rev. A* **59**, R39.

Kiss, T., U. Herzog and U. Leonhardt (1995) *Phys. Rev. A* **52**, 2433.

Kühn, H., D.-G. Welsch and W. Vogel (1994) *J. Mod. Opt.* **41**, 1607.

Lambrecht, A., T. Coudreau, A.M. Steinberg and E. Giacobino (1996) *Europhys. Lett.* **36**, 93.

Leibfried, D., D.M. Meekhof, B.E. King, C. Monroe, W.M. Itano and D.J. Wineland (1996) *Phys. Rev. Lett.* **77**, 4281 (1996).

Leonhardt, U., M. Munroe, T. Kiss, Th. Richter and M.G. Raymer (1996) *Opt. Commun.* **127**, 144.

Leonhardt, U. (1997) *Measuring the Quantum State of Light* (Cambridge University Press).

Mancini, S., V.I. Man'ko and P. Tombesi (1997) *J. Mod. Opt.* **44**, 2281.

McAlister, D.F., and M.G. Raymer (1997) *Phys. Rev. A* **55**, R1609.

von Neumann, J. (1932) *Mathematische Grundlagen der Quantenmechanik* (Springer-Verlag, Berlin).

Opatrný, T., and D.-G. Welsch (1997) *Phys. Rev. A* **55**, 1462.

Richter, Th. (1996a) *Phys. Lett. A* **211**, 327.

Richter, Th. (1996b) *Phys. Rev. A* **53**, 1197.

Schiller, S., G. Breitenbach, S.F. Pereira, T. Müller and J. Mlynek (1996) *Phys. Rev. Lett.* **77**, 2933.

Smithey, D.T., M. Beck, M.G. Raymer and A. Faridani (1993) *Phys. Rev. Lett.* **70**, 1244.

Vogel, K. and H. Risken (1989) *Phys. Rev. A* **40**, 2847.

Wallentowitz, S. and W. Vogel (1996) *Phys. Rev. A* **53**, 4528.

Welsch, D.-G., W. Vogel and T. Opatrný (1999) in *Progress in Optics*, Vol. XXXIX, ed. E. Wolf (Elsevier, Amsterdam), p. 63.

Wigner, E.P. (1932) *Phys. Rev.* **40**, 749.

Zucchetti, A., W. Vogel, M. Tasche and D.-G. Welsch (1996) Phys. Rev. A **54**, 1678.

---

11) The kernel functions for sampling the exponential phase moments and the vacuum-assisted density-matrix elements can of course be combined to kernel functions that are directly related to $\langle \hat{C}^n \rangle$ and $\langle \hat{S}^n \rangle$.

8
Nonclassicality and entanglement of bosonic systems

From the point of view of classical optics it is theoretically possible that a radiation field is free from any kind of noise. In practice, however, the production of radiation always involves the generation of some noise. Moreover, from Chapter 3, we know that the quantum nature of radiation is unavoidably connected with noise. Hence any radiation field may be said to be noisy. As long as the methods of classical statistics are sufficient to model the properties of a given radiation field with little error, the field may be said to have a classical counterpart with regard to these properties. Under certain circumstances the quantum features can dominate the properties of the field so that a description by methods of classical statistics fails. This implies that, with regard to certain properties, a counterpart of the quantum state of the field in classical physics does not exist. For this reason such states are called nonclassical.

For bosonic systems other than radiation, the situation is quite similar. In Section 8.1 we provide some background for understanding the nonclassical effects considered in Section 8.2. Since the early experiments showing evidence of nonclassical states were performed with light, we focus on nonclassical light therein. As we shall see, the effects can be characterized by inequalities which contradict classical statistics. Typical examples are the inequalities to characterize photon anti-bunching, sub-Poissonian photon statistics and squeezing. Extending this route, in Sections 8.3 and 8.4 we present a unified concept of measurement-based nonclassicality criteria for bosonic systems, either by means of observable characteristic functions (Section 8.3) or measurable moments (Section 8.4) – quantities whose detection is studied in Chapters 6 and 7. Entanglement of bipartite bosonic systems is considered in Section 8.5, which provides a unified concept of measurement-based entanglement criteria with respect to the negativity of the partial transposition of the system density operator.

8.1
Quantum states with classical counterparts

From the theory of the photoelectric detection of light (Chapter 6) we know that in standard photodetection experiments normally[1] and time-ordered correlation functions of the type

$$G^{(l,l)}(\{\mathbf{r}_i, t_i\}) = \left\langle : \prod_{i=1}^{l} \hat{I}(\mathbf{r}_i, t_i) : \right\rangle, \tag{8.1}$$

where

$$\hat{I}(\mathbf{r}, t) = \hat{\mathbf{E}}^{(-)}(\mathbf{r}, t)\hat{\mathbf{E}}^{(+)}(\mathbf{r}, t), \tag{8.2}$$

can be observed. Furthermore, normally and time-ordered correlation functions of the more general type

$$G^{(m,n)}_{k_1 \ldots k_{m+n}}(\mathbf{r}_1, t_1, \ldots, \mathbf{r}_{m+n}, t_{m+n})$$

$$= \left\langle \left[\mathcal{T}_- \prod_{i=1}^{m} \hat{E}^{(-)}_{k_i}(\mathbf{r}_i, t_i) \right] \left[\mathcal{T}_+ \prod_{j=m+1}^{m+n} \hat{E}^{(+)}_{k_j}(\mathbf{r}_j, t_j) \right] \right\rangle \quad (m \neq n) \tag{8.3}$$

can be recorded by homodyne correlation techniques (Section 6.5.7) or sampled from the rotated-phase quadrature distributions measured in balanced four-port homodyning (Section 7.5).

The corresponding classical correlation functions are obtained by replacing the field operators with stochastic c-number variables ($\hat{E}^{(\pm)}_k \mapsto E^{(\pm)}_k$), for example,

$$G^{(l,l)}(\{\mathbf{r}_i, t_i\})_{\text{cl}} = \left\langle \prod_{i=1}^{l} I(\mathbf{r}_i, t_i) \right\rangle_{\text{cl}}, \tag{8.4}$$

$\langle \cdots \rangle_{\text{cl}}$ indicates classical-statistical averaging. Clearly, in classical physics the ordering prescriptions become superfluous. To perform the classical averaging in Eq. (8.4), the lth order joint probability distribution function $p_{\text{cl}}[\{I(\mathbf{r}_i), t_i\}]$ for the intensity values $I(\mathbf{r}_i)$ at times t_i, $i = 1, 2, \ldots, l$, is required:[2]

$$G^{(l,l)}_{\text{cl}}(\{\mathbf{r}_i, t_i\}) = \int dI(\mathbf{r}_1)\, I(\mathbf{r}_1) \cdots \int dI(\mathbf{r}_l)\, I(\mathbf{r}_l)\, p_{\text{cl}}[\{I(\mathbf{r}_i), t_i\}]. \tag{8.5}$$

---

1) Recall that the normal-ordering prescription results from the fact that photodetectors usually operate on the basis of light absorption. The observation of anti-normally ordered correlation functions would require emission detectors [Mandel (1966)].
2) For methods of statistics see, e. g., van Kampen (1981) or Gardiner (1983).

Let us consider the normally ordered lth moment of the intensity at a chosen space point $\mathbf{r}$,

$$G^{(l,l)}(\mathbf{r},t) = \langle :\hat{I}^l(\mathbf{r},t): \rangle, \tag{8.6}$$

which in classical theory can be calculated by means of the marginal probability distribution function $p_{\text{cl}}[I(\mathbf{r}),t]$ as

$$G_{\text{cl}}^{(l,l)}(\mathbf{r},t) = \int dI(\mathbf{r})\, p_{\text{cl}}[I(\mathbf{r}),t] I^l(\mathbf{r}). \tag{8.7}$$

Performing a mode expansion (Section 2.2)[3]

$$\mathbf{E}^{(+)}(\mathbf{r}) = \sum_\lambda i\omega_\lambda \mathbf{A}_\lambda(\mathbf{r})\alpha_\lambda, \quad \mathbf{E}^{(-)}(\mathbf{r}) = [\mathbf{E}^{(+)}(\mathbf{r})]^*, \tag{8.8}$$

we may regard the sample of complex mode amplitudes, $\{\alpha_\lambda\}$, as a sample of stochastic variables with probability distribution $p_{\text{cl}}(\{\alpha_\lambda\},t)$. Hence Eq. (8.7) may be rewritten as

$$G_{\text{cl}}^{(l,l)}(\mathbf{r},t) = \int d^2\{\alpha_\lambda\}\, p_{\text{cl}}(\{\alpha_\lambda\},t) I^l(\mathbf{r};\{\alpha_\lambda\}), \tag{8.9}$$

where $d^2\{\alpha_\lambda\} = \prod_\lambda d^2\alpha_\lambda$, and the notation

$$I(\mathbf{r};\{\alpha_\lambda\}) = \mathbf{E}^{(-)}(\mathbf{r};\{\alpha_\lambda\}) \mathbf{E}^{(+)}(\mathbf{r};\{\alpha_\lambda\}) \equiv \mathbf{E}^{(-)}(\mathbf{r})\mathbf{E}^{(+)}(\mathbf{r}) \tag{8.10}$$

indicates that $\mathbf{E}^{(\pm)}$ in the form (8.8) depends on a set of random variables α_λ. The probability distribution of the mode amplitudes, $p_{\text{cl}}(\{\alpha_\lambda\},t)$, is of course well behaved and positive semi-definite.

We now turn to the quantum mechanical moments as defined in Eq. (8.6). To calculate them, it is advantageous to apply the concept of phase-space functions (Chapter 4). In particular, in the Glauber–Sudarshan representation (P representation) the density operator $\hat{\varrho}$ is [cf. Eq. (4.86)]

$$\hat{\varrho} = \int d^2\{\alpha_\lambda\} P(\{\alpha_\lambda\},t)|\{\alpha_\lambda\}\rangle\langle\{\alpha_\lambda\}|, \tag{8.11}$$

so that

$$G^{(l,l)}(\mathbf{r},t) = \int d^2\{\alpha_\lambda\} P(\{\alpha_\lambda\},t) I^l(\mathbf{r};\{\alpha_\lambda\}). \tag{8.12}$$

Here $I(\mathbf{r};\{\alpha_\lambda\})$ is again given by Eq. (8.10) with $\mathbf{E}^{(\pm)}(\mathbf{r};\{\alpha_\lambda\})$ from Eq. (8.8), where $\mathbf{E}^{(+)}(\mathbf{r};\{\alpha_\lambda\})$ solves the eigenvalue equation

$$\hat{\mathbf{E}}^{(+)}(\mathbf{r})|\{\alpha_\lambda\}\rangle = \mathbf{E}^{(+)}(\mathbf{r};\{\alpha_\lambda\})|\{\alpha_\lambda\}\rangle. \tag{8.13}$$

3) Recall that $\mathbf{E} = \mathbf{E}^\perp$ in free space.

Comparing the classical result (8.9) with the quantum-mechanical one (8.12), we find that the latter can be obtained within the framework of classical noise theory, provided that the chosen $p_{cl}(\{\alpha_\lambda\}, t)$ satisfies the condition that

$$p_{cl}(\{\alpha_\lambda\}, t) = P(\{\alpha_\lambda\}, t). \tag{8.14}$$

This requires that the P function must exhibit all the properties of a classical probability measure. Otherwise, there is no way to model the observed field statistics by means of a classical probability distribution function $p_{cl}(\{\alpha_\lambda\}, t)$. For this reason, a radiation field with a distribution $P(\{\alpha_\lambda\})$ which is not well behaved in the sense of a classical probability distribution may be called, with regard to the normally ordered moments considered, a nonclassical field [Titulaer and Glauber (1965); Mandel (1986)].

It should be emphasized that, even when the condition (8.14) is satisfied, the radiation field considered remains a quantum field and can differ essentially from the classical counterpart described by the probability distribution $p_{cl}(\{\alpha_\lambda\}, t)$. In particular, the vacuum noise as a pure quantum effect is always present. The close analogy between the quantum mechanical and classical descriptions has only been established with regard to normally ordered moments. Were one to consider other than normally ordered expectation values and try to calculate them using the classical model with the $p_{cl}(\{\alpha_\lambda\}, t)$ given above, Eq. (8.14), the result would be wrong in general. As is well known, different orderings give rise to different phase-space functions, and hence to different classical models. However, since measurements with photodetectors usually yield normally ordered quantities, the point of view taken above is of practical relevance.

Clearly, the concept of a classical counterpart also applies to bosonic systems other than radiation. Matter systems of harmonic-oscillator type, such as atoms moving in harmonic potentials, are typical examples. The expectation value of any system operator expressed in terms of the respective annihilation and creation operators and given in its normally ordered form, i.e., $\hat{O}(\{\hat{a}_\lambda\}, \{\hat{a}_\lambda^\dagger\}) \equiv \; :\hat{O}(\{\hat{a}_\lambda\}, \{\hat{a}_\lambda^\dagger\}):$, can be obtained, in principle, by using the P representation (8.11):

$$\langle \hat{O}(\{\hat{a}_\lambda\}, \{\hat{a}_\lambda^\dagger\})\rangle = \int d^2\{\alpha_\lambda\} \, P(\{\alpha_\lambda\}) O(\{\alpha_\lambda\}, \{\alpha_\lambda^*\}). \tag{8.15}$$

If, according to Eq. (8.14), the P function can be viewed as being a probability measure, the quantum state in which the system is prepared has a classical counterpart. It should be pointed out that system observables are normally defined in a form that is not normally ordered ($\hat{O} \neq \; :\hat{O}:$). Replacing them with their normally ordered forms ($\hat{O} \mapsto \; :\hat{O}:$) corresponds to discarding the noise effects in the ground state. In any case, the uncertainty product for the ground state represents the minimum noise level needed to obey Heisenberg's uncertainty principle. Thus the noise in the ground state is throughout a quantum effect and has no classical counterpart.

To account for this situation, one may regard a state as being nonclassical if it shows one of the following two features [Vogel (2000)]: First, the P function is not a probability density and thus Eq. (8.14) is violated. Second, a state is also nonclassical when its noise level is close to the ground-state noise, that is, when for the interpretation of measured data, the recorded noise level does not significantly exceed the minimum noise required according to Heisenberg's uncertainty relations. This definition relates the normally ordering prescription, which effectively eliminates the ground-state noise in representative observables, to an additional assumption concerning the measurable noise effects. Note that instead of the second condition another one can be formulated: A quantum state is nonclassical if the photon number is small [Mandel (1986)]. The question may arise if there is some difference between these conditions. Of course for a sufficiently small photon number we expect that the noise of the quantum state cannot be much larger than the ground-state level. However, the inverse is not true. Even the noise of a quantum state with a large photon number may be at the ground-state noise level. This behavior is well known for the coherent states. Thus the question may arise of whether or not the coherent states show observable nonclassical signatures. Usually the coherent states are considered to be those states which are closest to the classical ones. In particular, it has been shown that the coherent states are the only pure quantum states whose P function can be interpreted as a probability density [Hillery (1985)]. Thus they have a classical counterpart with respect to the first condition.

On the other hand, when one wishes to measure, e. g., the quadrature noise of a coherent state, it turns out to be at the ultimate quantum limit set by Heisenberg's uncertainty relation. Thus the coherent state is of course nonclassical in the sense of the second condition, even if the photon number is large. It is noteworthy in this context that, with respect to so-called weak measurements[4], another signature of nonclassicality can be considered. Instead of the normally ordered distributions, in such measurements the real part of the standard-ordered phase-space distribution is appropriate to predict the observable values. This distribution may attain negative values for both coherent states [Johansen (2004a)] and thermal states [Johansen and Luis (2004)]. In the latter case the occurrence of significant negative values of the phase-space distribution is limited to mean thermal photon numbers smaller than one. For the coherent states, however, the negative values survive even for large photon numbers. This seems to support the second condition for nonclassicality discussed above.

---

4) For the underlying measurement principle we refer the reader to Johansen (2004).

For many applications the first condition is of more practical relevance than the second one, hence we will concentrate on the properties of the P function. Since all pure quantum states – except the coherent ones – are nonclassical in this sense, it is of great importance to formulate the theory to include statistical mixtures of quantum states. In addition, since the P function may be highly singular, it is necessary to relate the nonclassicality conditions to observable quantities.

8.2
Nonclassical light

The development of laser techniques has allowed one to perform precise experiments demonstrating the nonclassical character of light. In particular, resonance fluorescence of a single atom created for the first time the possibility of observing photon antibunching and sub-Poissonian photon statistics. Later on, quadrature squeezing was demonstrated for the first time by four-wave mixing. Various methods for generating nonclassical light have already been established. Hence it would exceed the scope of this section to consider all of them. In the following we will consider some of the early experiments.

8.2.1
Photon anti-bunching

As is well known, the second-order (normally and time-ordered) intensity correlation function

$$G^{(2,2)}(\mathbf{r}, t + \tau, \mathbf{r}, t) = \langle {}^\circ_\circ \hat{I}(\mathbf{r}, t + \tau) \hat{I}(\mathbf{r}, t) {}^\circ_\circ \rangle$$
$$= \langle \hat{E}^{(-)}_{k_1}(\mathbf{r}, t) \hat{E}^{(-)}_{k_2}(\mathbf{r}, t + \tau) \hat{E}^{(+)}_{k_2}(\mathbf{r}, t + \tau) \hat{E}^{(+)}_{k_1}(\mathbf{r}, t) \rangle$$
(8.16)

can be determined from a photocounting-correlation measurement (Section 6.1). In practice, the measurement can be performed in an experiment of Hanbury Brown–Twiss type as shown schematically in Fig. 8.1. The radiation field under study is decomposed by a beam splitter into two parts of equal mean intensity, and the coincidences of the light in the two output channels are detected.

From the point of view of classical optics, Eq. (8.16) would take the form

$$G^{(2,2)}_{\mathrm{cl}}(t + \tau, t) = \langle I(t + \tau) I(t) \rangle_{\mathrm{cl}} \tag{8.17}$$

($I = \mathbf{E}^{(-)} \mathbf{E}^{(+)}$). Here and in the following, we drop the spatial argument $\mathbf{r}$ for the sake of notational convenience. Introducing the joint probability distribution function $p_{\mathrm{cl}}(I_1, t_1, I_2, t_2)$ for the two intensities I_1 and I_2 at times t_1 and t_2

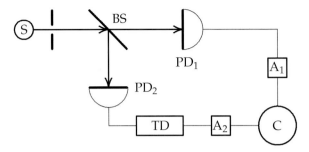

Fig. 8.1 Scheme of an experiment of Hanbury Brown–Twiss type for measuring the intensity correlation (S, light source; BS, beam splitter; PD$_1$, PD$_2$, photodetectors; A$_1$, A$_2$, amplifiers; TD, delay-time control; C, correlator).

respectively, we may write

$$G_{cl}^{(2,2)}(t+\tau,t) = \int dI \int dI'\, II'\, p_{cl}(I, t+\tau, I', t). \tag{8.18}$$

Applying the Schwarz inequality yields

$$\int dI \int dI'\, II'\, p_{cl}(I, t+\tau, I', t)$$
$$\leq \left[\int dI\, I^2 p_{cl}(I, t+\tau)\right]^{\frac{1}{2}} \left[\int dI'\, I'^2 p_{cl}(I', t)\right]^{\frac{1}{2}}, \tag{8.19}$$

where $p_{cl}(I, t)$ is the marginal probability distribution function. Let us now consider the intensity correlation function under steady-state conditions:

$$G^{(2)}(\tau) = \lim_{t\to\infty} G^{(2,2)}(t+\tau, t). \tag{8.20}$$

In the steady-state regime we obviously have

$$p_{cl}(I) = \lim_{t\to\infty} p_{cl}(I, t) = \lim_{t\to\infty} p_{cl}(I, t+\tau), \tag{8.21}$$

and the inequality (8.19) may be written as

$$G_{cl}^{(2)}(\tau) \leq G_{cl}^{(2)}(0). \tag{8.22}$$

This result reveals that in classical optics the second-order intensity correlation function of a steady-state radiation field has a nonpositive initial slope. In other words, the probability of observing equal-time coincidences is greater than that of observing time-delayed coincidences. We see that when a stationary radiation field satisfies

$$G^{(2)}(\tau) \leq G^{(2)}(0), \tag{8.23}$$

a classical counterpart which also satisfies this condition can be found. The radiation-field property introduced by the inequality (8.23) is called photon bunching. It is obvious that the observation of more equal-time coincidences than time-delayed coincidences implies that the photons to be detected[5] show a tendency to arrive in bunches.

It is now straightforward to define a nonclassical property by the requirement that

$$G^{(2)}(\tau) > G^{(2)}(0). \tag{8.24}$$

Since this inequality is violated by any classical light field, it is obvious that a light field is far from being classical when the steady-state intensity correlation function (as a function of the delay time τ) has a positive initial slope. The joint probability of observing two counts in the two detection channels simultaneously, is smaller than the probability in the case of a finite time delay. This effect is called photon anti-bunching, because one could say that, within the photon concept of light, the photons have a tendency to arrive separated from each other. The experimental demonstration of photon anti-bunching may therefore be regarded as being a very direct proof of the photon nature of light. It is straightforward to prove that for an (effectively free) field prepared in a coherent state, the equation

$$G^{(2)}(\tau)_{\text{coh}} = G^{(2)}(0)_{\text{coh}} \tag{8.25}$$

holds. Hence a coherent field just marks the boundary between (classical) bunching and (nonclassical) anti-bunching.

The resonance fluorescence from a single two-level atom (Chapter 11) is a typical example which demonstrates photon anti-bunching [Carmichael and Walls (1976); Kimble and Mandel (1976)]. It is quite simple to understand the appearance of photon anti-bunching in resonance fluorescence. Let us assume that a two-level atom that is (continuously) irradiated by a laser beam, whose frequency is tuned to the atomic transition frequency is, at a certain time, in its ground state. Owing to the interaction with the laser field, the atom undergoes a transition from the ground state to the excited one. The excited atom can emit a fluorescence photon, which may be observed in a direction different from that of the exciting laser beam. Since the emission of a photon is connected with a transition of the atom into the ground state, two photons cannot be emitted simultaneously, which implies that for zero delay the intensity correlation function (8.16) is expected to tend to zero:

$$\lim_{\tau \to 0} G^{(2,2)}(t+\tau, t) = \langle : \hat{I}^2(t) : \rangle = 0. \tag{8.26}$$

---

5) For the relation between the counting and the photon-number distributions, see Section 6.2.

Clearly, in the further course of time the atom can again be excited and return to the ground state through the emission of a photon. Consequently, the joint probability of emitting a photon at time t and a photon at time $t+\tau$ is expected to increase with increasing delay τ. In other words, the intensity correlation function (8.16) is expected to have a positive initial slope,[6] which is just the criterion for photon anti-bunching, Eq. (8.24).

Indeed, photon anti-bunching was first measured in resonance fluorescence [Kimble, Dagenais and Mandel (1977)]. In the experimental scheme (Fig. 8.1) for the light source, an atomic beam was used whose density was small enough to ensure that, at most, one atom was involved in resonance fluorescence. In particular, when the mean number of atoms in the interaction volume is less than unity, the fluorescence from a single atom is, apart from background scattering, the dominant contribution to the observed signal. Later, the effect of photon anti-bunching in resonance fluorescence was observed using a single ion in a Paul trap [Diedrich and Walther (1987); Schubert, Siemers, Blatt, Neuhauser and Toschek (1992)].

8.2.2
Sub-Poissonian light

Let us consider the mean number of counts and the variance of counts as given by Eqs (6.54) and (6.61), respectively. The classical version of Eq. (6.61) is

$$\overline{[\Delta n(t,\Delta t)]^2} = \overline{n(t,\Delta t)} + \xi^2 \langle [\Delta I(t,\Delta t)]^2 \rangle_{\text{cl}}. \tag{8.27}$$

Here $\langle [\Delta I(t,\Delta t)]^2 \rangle_{\text{cl}}$ is an ordinary variance, which cannot be negative,

$$\langle [\Delta I(t,\Delta t)]^2 \rangle_{\text{cl}} \geq 0, \tag{8.28}$$

and hence the following classical inequality is deduced:

$$\overline{[\Delta n(t,\Delta t)]^2} \geq \overline{n(t,\Delta t)}. \tag{8.29}$$

The equality sign in Eq. (8.29) corresponds to the lowest noise level in classical photocounting, the so-called shot-noise level. It is attained in the limit of a nonfluctuating classical field being detected. Indeed, in this case, the photocounting distribution (6.44) reduces to a Poissonian:

$$P_m(t,\Delta t) = \frac{1}{m!} [\Gamma(t,\Delta t)]^m e^{-\Gamma(t,\Delta t)}, \tag{8.30}$$

$$\Gamma(t,\Delta t) = \xi \int_t^{t+\Delta t} d\tau\, I(\tau). \tag{8.31}$$

---

6) For details, see Section 11.2.2.

For a quantized radiation field the second term in Eq. (6.61) is not necessarily non-negative, because of the operator ordering. Thus in quantum optics the classically established inequalities (8.28) and (8.29) may be violated, and the variance of the counts can become smaller than the mean value of counts:

$$\langle {}^\circ_\circ [\Delta \hat{I}(t, \Delta t)]^2 {}^\circ_\circ \rangle < 0, \tag{8.32}$$

$$\overline{[\Delta n(t, \Delta t)]^2} < \overline{n(t, \Delta t)}. \tag{8.33}$$

In other words, if the condition (8.33) is satisfied, the photocounting distribution is narrower than a Poissonian one, which implies in particular that the noise of the counts is reduced below the shot-noise level. From the above it is clear that light which gives rise to sub-Poissonian counting statistics – also called sub-Poissonian light – is nonclassical. In this context we recall that, as shown in Section 6.2.3, sub-Poissonian counting statistics are associated with sub-Poissonian photon-number statistics of the light to be detected.

For a free field in a coherent state we have

$$\langle :[\Delta \hat{I}(t, \Delta t)]^2: \rangle_{\text{coh}} = 0, \tag{8.34}$$

so that Eq. (6.61) reduces to

$$\overline{[\Delta n(t, \Delta t)]^2} = \overline{n(t, \Delta t)}. \tag{8.35}$$

We see that a field prepared in a coherent state just gives rise to a Poissonian counting statistics, which implies that the noise in photocounting is given by the shot noise. In this sense, a radiation field in a coherent state corresponds to a noise-free classical field.

In a short-time measurement, Eq. (6.63) is valid and the condition for sub-Poissonian light is

$$\langle :[\Delta \hat{I}(t)]^2: \rangle < 0. \tag{8.36}$$

Applying the relation

$$\langle :[\Delta \hat{I}(t)]^2: \rangle = \langle :\hat{I}^2(t): \rangle - \langle \hat{I}(t) \rangle^2, \tag{8.37}$$

the condition (8.36) can be rewritten as

$$\langle :\hat{I}^2(t): \rangle < \langle \hat{I}(t) \rangle^2, \tag{8.38}$$

which for a steady-state radiation field agrees with the limit $\tau \to \infty$ of the condition (8.24) for photon anti-bunching. In this sense the nonclassical properties of photon anti-bunching and sub-Poissonian statistics appear to be closely related to each other. We would like to note that the condition (8.36) may be given, on using the P representation, in the form

$$\int d^2\{\alpha_\lambda\} P(\{\alpha_\lambda\}, t)[\Delta I(\{\alpha_\lambda\})]^2 < 0 \tag{8.39}$$

(cf. Section 8.1). Since $[\Delta I(\{\alpha_\lambda\})]^2$ is always non-negative, this inequality can only be satisfied when the P function attains negative values. As expected, in the case of sub-Poissonian light $P(\{\alpha_\lambda\})$ cannot be positive semi-definite, and a classical-statistical description would fail.

The sub-Poissonian effect was first demonstrated in resonance fluorescence [Short and Mandel (1983)], by using the same light source as in the antibunching experiment outlined in Section 8.2.1. For a finite detection-time interval Δt the condition for sub-Poissonian light, Eq. (8.32), reads

$$\int_t^{t+\Delta t} d\tau \int_t^{t+\Delta t} d\tau' \left[\langle \,_\circ^\circ \hat{I}(\tau)\hat{I}(\tau') \,_\circ^\circ \rangle - \langle \hat{I}(\tau) \rangle \langle \hat{I}(\tau') \rangle \right] < 0. \tag{8.40}$$

Since the experiment was performed under steady-state conditions, the intensity correlation function $G^{(2,2)}(\tau,\tau') = \langle \,_\circ^\circ \hat{I}(\tau)\hat{I}(\tau') \,_\circ^\circ \rangle$ depends only on the time difference: $G^{(2,2)}(\tau,\tau') = G^{(2)}(\tau - \tau')$. Introducing the sum and difference times, we may perform the integration over $\tau + \tau'$ to simplify the condition (8.40) as follows:

$$\int_0^{\Delta t} d\tau \, [G^{(2)}(\tau) - \langle \hat{I} \rangle^2](\Delta t - \tau) < 0. \tag{8.41}$$

For sufficiently small Δt this condition is fulfilled, in the case of resonance fluorescence from a two-level atom (Section 11.2.2), due to the fact that $G^{(2)}(0) = 0$. For an arbitrary measurement time Δt the temporal evolution of $G^{(2)}(\tau)$ during the measurement ($0 \leq \tau \leq \Delta t$) becomes important. If the inequality

$$G^{(2)}(\tau) < \langle \hat{I} \rangle^2 \equiv G^{(2)}(\infty) \quad (0 \leq \tau < \infty) \tag{8.42}$$

is satisfied, the light is sub-Poissonian for any measurement time. In resonance fluorescence from a single two-level atom, this condition is fulfilled for a weak driving field (Section 11.2.2). In more general cases of strong driving fields the intensity correlation function may be oscillatory and there is no one-to-one correspondence between the effects of photon anti-bunching and sub-Poissonian statistics, in general.

The first experimental demonstration of sub-Poissonian light was followed by a number of further experiments. For example, in a Franck–Hertz experiment [Teich and Saleh (1985)] a regular electron beam was created by making use of Coulomb repulsion. This beam excites a sample of mercury atoms which subsequently produces Franck–Hertz light, whose photon statistics partly reveal the sub-Poissonian nature of the exciting electron beam. Based on spontaneous parametric down-conversion it became possible to realize a localized single photon state [Hong and Mandel (1986)]. The idler photon of the spontaneously created photon pair was used to trigger an optical shutter in the signal channel, such that the signal field incident on a photodetector in the selected time interval is, with a high probability, prepared in a ngle photon state. A laser was shown to produce sub-Poissonian light when the pump fluctuations are suppressed below the shot-noise level [Machida, Yamamoto and Itaya (1987); Machida and Yamamoto (1989)].

8.2.3
Squeezed light

As shown in Section 3.3, the noise of an appropriately chosen field-strength quantity

$$\hat{\mathbf{F}}(\mathbf{r}) = \sum_\lambda [\mathbf{F}(\mathbf{r})\hat{a}_\lambda + \mathbf{F}^*(\mathbf{r})\hat{a}_\lambda^\dagger], \tag{8.43}$$

such as for example the electric field strength, of a radiation field prepared in a squeezed coherent state can be reduced, depending upon the phase,[7] below the vacuum level, i. e.,

$$\langle [\Delta\hat{\mathbf{F}}(\mathbf{r},t)]^2 \rangle < \langle [\Delta\hat{\mathbf{F}}(\mathbf{r},t)]^2 \rangle_{\text{vac}}, \tag{8.44}$$

which can be rewritten as

$$\langle :[\Delta\hat{\mathbf{F}}(\mathbf{r},t)]^2: \rangle < 0 \tag{8.45}$$

[cf. Eq. (3.170)]. We recall that, as a consequence of the noise reduction below the vacuum level for a certain phase value, noise enhancement appears when the phase is shifted by a value of $\pm\pi/2$. In particular for a plane-wave field propagating in the positive x direction the noise reduction (and the $\pi/2$-shifted noise enhancement) is π periodical in $\omega t - kx$.

The inequality (8.45) can be used as a definition of squeezed light, and is independent of the particular choice of the state of the field. Clearly, squeezing is a nonclassical effect, as can be seen by going over to a classical description:

$$\langle :[\Delta\hat{\mathbf{F}}(\mathbf{r},t)]^2: \rangle \;\longmapsto\; \langle [\Delta\mathbf{F}(\mathbf{r},t)]^2 \rangle_{\text{cl}} > 0. \tag{8.46}$$

The normally ordered field variance reduces to an ordinary variance, which cannot be negative. Note that the boundary between classical and nonclassical behavior is again given by a radiation field prepared in a coherent state. It is easily proved that, in this case,

$$\langle :[\Delta\hat{\mathbf{F}}(\mathbf{r},t)]^2: \rangle_{\text{coh}} = 0. \tag{8.47}$$

In terms of the P representation, the squeezing condition (8.45) becomes

$$\int d^2\{\alpha_\lambda\}\, P(\{\alpha_\lambda\},t)[\Delta\mathbf{F}(\mathbf{r};\{\alpha_\lambda\})]^2 < 0, \tag{8.48}$$

---

[7] Here, by "phase" an ordinary c-number phase parameter is understood; cf., e. g., Eq. (3.152).

which shows that (because $[\Delta \mathbf{F}(\mathbf{r}; \{\alpha_\lambda\})]^2 \geq 0$) for squeezed light, a classical-statistical model with a positive semi-definite probability distribution function would fail.

The definition of squeezing as given by the inequality (8.45) is based on the normally ordered second-order moment of the field quantity $\hat{\mathbf{F}}$ (second-order squeezing). This inequality can be extended to normally ordered higher-order moments in order to introduce higher-order squeezing. From arguments analogous to those given above it is obvious that a radiation field satisfying a condition of the form

$$\langle :[\Delta \hat{\mathbf{F}}(\mathbf{r},t)]^{2n}: \rangle < 0 \qquad (8.49)$$

($n > 0$, integer) is nonclassical.

Next let us consider the expectation value of the operator exponential $\exp[\Delta \hat{\mathbf{F}}(\mathbf{r})z]$. We put it in normal order, on using Eq. (C.26), to obtain

$$\langle \exp[\Delta \hat{\mathbf{F}}(\mathbf{r},t)z] \rangle = \langle :\exp[\Delta \hat{\mathbf{F}}(\mathbf{r},t)z]: \rangle \exp\{\tfrac{1}{2}z^2 \langle [\Delta \hat{\mathbf{F}}(\mathbf{r},t)]^2 \rangle_{\text{vac}}\}. \qquad (8.50)$$

By power-series expansion and comparison of the expansion coefficients of z^{2n} ($n=1,2,3,\ldots$), we deduce that

$$\langle [\Delta \hat{\mathbf{F}}(\mathbf{r})]^{2n} \rangle - \frac{(2n)!}{2^n n!} \langle [\Delta \hat{\mathbf{F}}(\mathbf{r})]^2 \rangle_{\text{vac}}^n$$
$$= \frac{(2n)!}{2^n (n-1)!} \langle :[\Delta \hat{\mathbf{F}}(\mathbf{r})]^2: \rangle \langle [\Delta \hat{\mathbf{F}}(\mathbf{r})]^2 \rangle_{\text{vac}}^{n-1} + \ldots \qquad (8.51)$$

Taking the inequality (8.49) into account, we see from Eq. (8.51) that

$$\langle [\Delta \hat{\mathbf{F}}(\mathbf{r})]^{2n} \rangle - \frac{(2n)!}{2^n n!} \langle [\Delta \hat{\mathbf{F}}(\mathbf{r})]^2 \rangle_{\text{vac}}^n < 0 \qquad (8.52)$$

is a condition which can be satisfied by nonclassical light. Recalling that the field-strength probability distribution of the vacuum field is a Gaussian with zero mean, we may rewrite the inequality (8.52) to obtain

$$\langle [\Delta \hat{\mathbf{F}}(\mathbf{r},t)]^{2n} \rangle < \langle [\Delta \hat{\mathbf{F}}(\mathbf{r},t)]^{2n} \rangle_{\text{vac}}, \qquad (8.53)$$

which is a generalization of the (second-order) squeezing condition (8.44) in the sense of a condition for higher-order squeezing ($n \geq 2$) [Hong and Mandel (1985)].

It should be noted that, for a field with a Gaussian field-strength distribution, second-order squeezing is always accompanied by higher-order squeezing. In general, the latter depends sensitively on the shape of the field-strength distribution rather than simply on its width as is the case for second-order squeezing. Since for higher-order squeezing, higher-order moments of the field strength become important, even small changes in the wings of the field-strength distribution may significantly influence the effect of higher-order squeezing [Vogel (1990)].

When we are looking for mechanisms which can be used to generate squeezed light, it may be helpful to recall the structure of the unitary squeeze operator introduced in Section 3.3:

$$\hat{S}(\xi) = \exp\left[\tfrac{1}{2}(\xi^*\hat{a}^2 - \xi\hat{a}^{\dagger 2})\right] \tag{8.54}$$

in the single-mode case [cf. Eq. (3.102)] and

$$\hat{S}(\xi_{12}) = \exp(\xi^*\hat{a}_1\hat{a}_2 - \xi\hat{a}_1^\dagger\hat{a}_2^\dagger) \tag{8.55}$$

in the two-mode case [cf. Eq. (3.174)]. Applying $\hat{S}(\xi)$ to a (single-mode) coherent state yields a single-mode squeezed (coherent) state. Accordingly, a two-mode squeezed state is obtained by applying $\hat{S}(\xi_{12})$ to a (two-mode) coherent state. One may therefore expect that squeezed light can be produced in an optical process governed by an effective interaction Hamiltonian of the type

$$H_{\text{int(eff)}} = \hbar(\lambda\hat{a}^{\dagger 2} + \lambda^*\hat{a}^2) \tag{8.56}$$

(degenerate case) and

$$H_{\text{int(eff)}} = \hbar(\lambda_{12}\hat{a}_1^\dagger\hat{a}_2^\dagger + \lambda_{12}^*\hat{a}_1\hat{a}_2) \tag{8.57}$$

(nondegenerate case). The unitary time evolution operator would correspond to a squeeze operator and hence a field initially prepared in the vacuum state or a coherent state should be squeezed in the further course of time.

Clearly, an optical process governed by an interaction Hamiltonian of the type (8.56) or (8.57) cannot actually be found. The simultaneous generation of two c-number-coupled photons can only be regarded as an approximation to a realistic process, in which the creation of the two photons is unavoidably connected with the destruction of other photons or atomic excitations. In this sense, the structure of the squeeze operator can only give a general orientation. Moreover, squeezing is not necessarily connected with the special squeezed states introduced by a squeeze operator of the type given above.

In Section 2.5.3 we have introduced effective interaction Hamiltonians for nonlinear processes that do not (explicitly) depend on atomic variables. For example, let us consider the process of four-wave mixing, which, in analogy with Eq. (2.264), is governed by the effective Hamiltonian

$$\hat{H}_{\text{int(eff)}} = \hbar\kappa^{(4)}\hat{a}_1\hat{a}_2\hat{a}_3^\dagger\hat{a}_4^\dagger + \text{H.c.}. \tag{8.58}$$

The elementary process is seen to be the creation (destruction) of two photons in the modes 3 and 4 and the simultaneous destruction (creation) of two photons in the modes 1 and 2. If the modes 1 and 2 represent strong coherent

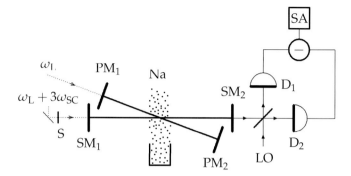

Fig. 8.2 Experimental scheme for squeezed-light generation by four-wave mixing. A ring dye laser pumps a Na atomic beam at the pump cavity (mirrors PM_1, PM_2) resonance frequency. The pumped Na atoms generate four-wave mixing gain in the squeezing cavity (mirrors SM_1, SM_2). The squeezed cavity noise is detected by a balanced homodyne detector (detectors D_1, D_2) and observed on a spectrum analyzer SA. The classical four-wave-mixing gain is measured by opening the shutter S and injecting, into the squeezing cavity, a beam that is frequency-shifted from the pump by $3\omega_{SC}$ (ω_{SC} is the mode-spacing frequency of the cavity). [After Slusher, Hollberg, Yurke, Mertz and Valley (1985).]

pump fields, then one may expect the replacement of the operators $\hat{a}_1$ and $\hat{a}_2$ with c numbers α_1 and α_2 to be a suitable approximation:

$$\hat{H}_{\text{int(eff)}} \simeq \hbar \kappa^{(4)} \alpha_1 \alpha_2 \hat{a}_3^\dagger \hat{a}_4^\dagger + \text{H.c.}. \tag{8.59}$$

As early as 1979, the process of (degenerate, i.e., $\hat{a}_3 = \hat{a}_4 = \hat{a}$) four-wave mixing was proposed to be a hopeful candidate for producing squeezed light [Yuen and Shapiro (1979)].

For the first time, squeezed light was experimentally realized by nondegenerate four-wave mixing, the nonlinear medium being sodium atoms in an optical cavity [Slusher, Hollberg, Yurke, Mertz and Valley (1985)]. The scheme of the experimental apparatus is shown in Fig. 8.2. The nondegenerate process in an optical cavity was used in order to enhance the four-wave mixing gain and to restrict the frequency spectrum to a low-noise region. A c.w. single-mode ring dye laser was used to pump a beam of sodium atoms at a frequency tuned above the hyperfine components in the D_2 resonance of Na at 589 nm. The pump was focused in the Na beam and enhanced in a confocal build-up cavity (formed by mirrors PM_1 and PM_2 in Fig. 8.2). The squeezed light was generated in a confocal cavity (formed by mirrors SM_1 and SM_2 in Fig. 8.2) by a linear combination of the conjugate pairs of photons generated by the four-wave-mixing process. Using a single-ended cavity avoids additional vacuum noise entering the cavity, and highly correlated photon pairs can be produced at frequencies which are symmetrically shifted with respect to the pump frequency ω_L:

$$\omega_{3,4} = \omega_L \pm n\omega_{SC}, \tag{8.60}$$

where n is an integer and ω_{SC} is the mode-spacing frequency of the cavity.

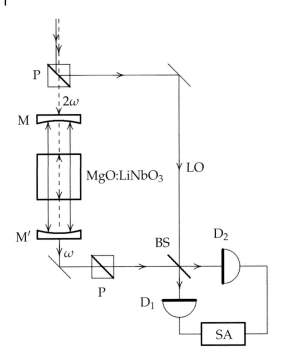

Fig. 8.3 Scheme of the apparatus for squeezed-light generation by degenerate parametric down-conversion (M, M', mirrors of the OPO cavity; P, polarizer; BS, beam splitter; LO, local oscillator; D_1, D_2, photodiodes; SA, spectrum analyzer). The lithium niobate crystal has dual-band anti-reflection coatings to minimize loss at ω and 2ω. The pump at 2ω is obtained by frequency-doubling from ω to 2ω with a crystal of $Ba_2NaNb_5O_{15}$ inside the cavity of a frequency-stabilized neodymium-doped yttrium aluminium garnet laser, whose emission also acts as the local oscillator in homodyne detection. [After Wu, Kimble, Hall and Wu (1986).]

In the experiment the reflectivities were 0.995 for SM_1 and 0.98 for SM_2, and the measured photon pairs were shifted by $\pm 3\omega_{SC}$, since at these frequencies the spontaneous emission noise in the tails of the homogeneous line shape of the sodium atoms is sufficiently weak to observe squeezing. To measure the cavity four-wave-mixing power gain in the classical limit, a portion of the pump beam was shifted by $3\omega_{SC}$ and coupled into the squeezing cavity by the opening of the shutter S in Fig. 8.2. Radiation from the SM_2 output mirror was detected by balanced homodyne detection (Section 6.5.4), with an unshifted portion of the pump beam used as the local oscillator LO in Fig. 8.2. A 7% reduction of the noise level for homodyne detection below the vacuum (shot-noise) level was measured, which corresponds, under the experimental conditions, to nearly 20% squeezing [for improved results see Slusher, Yurke, Grangier, La Porta, Walls and Reid (1987)].

Substantial squeezing was produced by means of an optical parametric oscillator (OPO) [Wu, Kimble, Hall and Wu (1986); Wu, Xiao and Kimble (1987)]. In the experiments, a scheme as shown in Fig. 8.3 was used with a triply resonant cavity and the pump mode (frequency $\omega_l=2\omega$) and the nearly degenerate signal and idler fields (frequencies $\omega_s \approx \omega_i \approx \omega$) being simultaneously resonant. Squeezing was generated below the oscillation threshold of the OPO. A Nd:YAG ring laser with a nonlinear crystal inside the cavity was used to generate both the pump (frequency 2ω) and the local oscillator (frequency ω), which were orthogonally polarized with respect to each other. The two beams were separated from each other by a polarizer, and the pump beam was directed onto the OPO cavity. One mirror of this cavity was fractionally transparent for the frequency 2ω, and the other mirror, used as the output port for the produced squeezed light, was fractionally transparent for the frequency ω. After superimposing the local oscillator, the squeezing was measured in a balanced homodyne detection scheme. Correcting for the efficiencies, the field noise was found to be squeezed to $80-90\%$ of the vacuum level.

Squeezed light, with a reduction of the noise in one quadrature below the vacuum noise level, has been widely considered to be of interest for applications such as the improvement of the sensitivity of measurements. An improvement in the performance of an interferometer by use of squeezed light was first demonstrated experimentally by a Mach–Zehnder interferometer for the detection of the phase modulation in the arms of the interferometer [Xiao, Wu and Kimble (1987)]. In this experiment an improvement in the signal-to-noise ratio of 3 dB relative to the shot-noise limit was achieved. In another experiment, squeezed light was used for improving the sensitivity in polarization measurements, with a polarization interferometer analogous to a Mach–Zehnder interferometer [Grangier, Slusher, Yurke and La Porta (1987)]. Squeezed light was also used to improve the sensitivity in intensity measurements [Xiao, Wu and Kimble (1988)] and in spectroscopy [Polzik, Carri and Kimble (1992)].

8.3
Nonclassical characteristic functions

The nonclassical effects considered so far are typically related to second-order moments and correlation functions of special observables. For a complete characterization of the nonclassical properties of a system, however, the quantum state as a whole should be considered. To do so, we will use the characteristic function of the Glauber–Sudarshan P function, which can be expressed in terms of the characteristic functions of the observable quadrature distributions. For the sake of transparency we will first restrict our attention to single-mode systems. The extension to multi-mode systems is straightforward but rather involved. As we know, entanglement is a nonclassical effect observed in multi-mode systems. We will deal with two-mode entanglement in Section 8.5 and present observable criteria for it.

8.3.1
The Bochner theorem

We are interested in a reformulation of the nonclassicality condition

$$P(\alpha) \neq p_{\text{cl}}(\alpha), \tag{8.61}$$

that is the failure of the P function to show the properties of a probability density. Since $P(\alpha)$ may become highly singular, see, e.g., the situation for the number state in Eq. (4.117), another form of the condition is required which can be used in experiments. Such a form can be found [Vogel (2000); Richter and Vogel (2002)] on the basis of an old theorem formulated by Bochner (1933), which provides the following conditions for a continuous function to be the Fourier transform of a probability density, i.e., a (classical) characteristic function: A continuous function $\Phi(\alpha)$ which obeys the conditions $\Phi(0) = 1$ and $\Phi(\alpha) = \Phi^*(-\alpha)$ is the characteristic function of a probability density iff $\Phi(\alpha)$ is non-negative. That is, for arbitrary complex numbers α_i and ζ_i the condition

$$\sum_{i,j=1}^{n} \Phi(\alpha_i - \alpha_j)\zeta_j^*\zeta_i \geq 0 \tag{8.62}$$

must be fulfilled for any integer n.

When we identify $\Phi(\alpha)$ with the Fourier transform of the P function [cf. Eq. (4.92)],[8]

$$\Phi(\alpha) = \int d^2\beta \, P(\beta) \exp(\alpha\beta^* - \alpha^*\beta), \tag{8.63}$$

a violation of the requirements according to the Bochner theorem is equivalent to the failure of the P function to be a probability density. Hence we may conclude that a quantum state is nonclassical iff a violation of the Bochner condition (8.62) can be found for appropriately chosen values of the "vector" components α_i and ζ_i for some value of n:

$$\sum_{i,j=1}^{n} \Phi(\alpha_i - \alpha_j)\zeta_j^*\zeta_i < 0. \tag{8.64}$$

The nonclassicality condition (8.64) is still not useful for practical applications, and it requires some further effort to formulate nonclassicality criteria

[8] For convenience, throughout this chapter we use the notation $\Phi(\alpha) \equiv \Phi(\alpha; s = 1)$.

in a form applicable to measured data. First, the characteristic function $\Phi(\alpha)$ is not directly accessible. Second, it is not very useful for handling the high-dimensional problem, which is covered in the Bochner theorem, in the form given by the inequality (8.64). The first problem can be resolved by relating the characteristic function $\Phi(\alpha)$ to the characteristic function $\Psi(y,\varphi)$ of the observable quadrature distribution $p(x,\varphi)$, as defined by Eq. (7.7), via

$$\Psi(y,\varphi) = e^{-\frac{1}{2}y^2}\Phi(iye^{-i\varphi}) \tag{8.65}$$

[cf. Eq. (7.12) for $s=1$]. To resolve the second problem, the condition (8.64) can be used to derive conditions that are expressed in terms of $\Phi(\alpha)$, or equivalently, in terms of $\Psi(y,\varphi)$. Alternatively, nonclassicality conditions in terms of measurable moments can be derived.

8.3.2
First-order nonclassicality

Let us first analyze the inequality (8.64) for the simplest case $n=2$, leading to the condition for first-order nonclassicality.[9] Determining the extreme values of the function in the inequality (8.64) (for $n=2$) with respect to the occurring phase difference and the ratio $|\zeta_1|/|\zeta_2|$, we derive the condition

$$|\Phi(\alpha)| > 1, \tag{8.66}$$

which must be fulfilled in order to obtain nonclassicality of first order. Making use of Eq. (8.65), we may express the condition in terms of the characteristic function of the quadrature distribution as

$$|\Psi(y,\varphi)| > \Psi_{\text{gr}}(y,\varphi), \tag{8.67}$$

where we have made use of the fact that the characteristic function $\Psi_{\text{gr}}(y,\varphi)$ of the ground-state quadrature distribution reads

$$\Psi_{\text{gr}}(y,\varphi) = e^{-\frac{1}{2}y^2}. \tag{8.68}$$

Hence a quantum state is nonclassical (of first order) if the absolute value of its characteristic function of the observable quadrature distribution exceeds the characteristic function in the ground or vacuum state [Vogel (2000)]. From the derivation it is obvious that the condition (8.67) is only a sufficient one. It is worth noting that this very simple condition already covers a number of conditions specific to different states, such as for example number states, quadrature squeezed states and even or odd coherent states. In Fig. 8.4(a) the effect of first-order nonclassicality is illustrated for a Fock state and an even coherent state.

---

9) Note that for $n=1$ the inequality (8.64) cannot be fulfilled since $\Phi(0)=1$ and $|\zeta_1|^2 \geq 0$. Hence we identify the number $k=n-1$ with the order of nonclassicality.

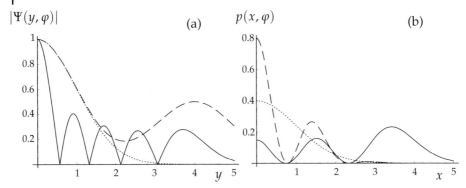

Fig. 8.4 The absolute value of the characteristic function $|\Psi(y,\varphi)|$ (a) and the quadrature distribution $p(x,\varphi)$ (b) are shown for different quantum states: the number state $|n=4\rangle$ (full line), the even coherent state $|\alpha\rangle_+ \sim (|\alpha\rangle + |-\alpha\rangle)$ for $\alpha=2$ and $\varphi=\pi/2$ (dashed line), and the ground state $|0\rangle$ (dotted line).

Let us consider the effects of first-order nonclassicality from the viewpoint of the quadrature distribution $p(x,\varphi)$. For the examples mentioned above the absolute value of the characteristic function, $|\Psi(y,\varphi)|$, clearly decays more slowly than the ground-state value. Fourier transforming back to the quadrature distribution, one expects structures in $p(x,\varphi)$ that are narrower than those typical of the ground state distribution. This is illustrated in Fig. 8.4(b). The first-order-nonclassicality condition may be regarded, in a sense, as being a generalization of the quadrature-squeezing condition. Whereas the latter describes the narrowing of the whole quadrature distribution, the former already applies whenever the quadrature distribution exhibits narrower structures as in the ground state, even when it as a whole is much broader than the ground-state distribution.

Experimentally, the first-order-nonclassicality condition was first used for demonstrating the nonclassical character of a radiation quantum state

$$\hat{\varrho}_\eta = \eta|1\rangle\langle 1| + (1-\eta)|0\rangle\langle 0| \qquad (0 < \eta \leq 1) \tag{8.69}$$

[Lvovsky and Shapiro (2002)]. It is easy to verify that the Wigner function of such a (mixed) state may become negative, which is frequently considered as a (sufficient) condition for nonclassicality. However, negative values of the Wigner function do not appear when the contribution to the state (8.69) of the single-photon state $|1\rangle$ becomes too small, that is if $\eta \leq 0.5$. On the contrary, the condition (8.67) is fulfilled for any value of η.

8.3.3
Higher-order nonclassicality

Let us now return to the inequality (8.64) and analyze it for values of $n > 2$. For this purpose we first write the Bochner condition (8.62) in a matrix form, understanding $\Phi(\alpha_i - \alpha_j)$ as the $n \times n$ matrix $\Phi_{ij} \equiv \Phi(\alpha_i - \alpha_j)$:

$$\sum_{i,j=1}^{n} \Phi_{ij} \xi_i \xi_j^* \geq 0. \tag{8.70}$$

The symmetry property $\Phi(\alpha) = \Phi^*(-\alpha)$, which now reads as $\Phi_{ji} = \Phi_{ij}^*$, ensures that the left-hand side of the inequality represents a Hermitian form. Now we make use of the following theorem of linear algebra [Zhang (1999)]: An $n \times n$ complex matrix is positive semi-definite iff the determinant

$$D_k = \begin{vmatrix} 1 & \Phi_{12} & \cdots & \Phi_{1k} \\ \Phi_{12}^* & 1 & \cdots & \Phi_{2k} \\ \cdots & \cdots & \cdots & \cdots \\ \Phi_{1k}^* & \Phi_{2k}^* & \cdots & 1 \end{vmatrix} \tag{8.71}$$

of each of its principal submatrices is non-negative. Hence the inequality (8.70) is valid, iff for any order k ($k = 2, \ldots, n$) the condition

$$D_k \geq 0 \tag{8.72}$$

is fulfilled.

Recalling the inequality (8.64), we are now able to formulate the following necessary and sufficient nonclassicality condition: A quantum state is nonclassical iff there exist values α_i ($i = 1, \ldots, k$) for which the inequality

$$D_k < 0 \quad (k = 2, \ldots, \infty) \tag{8.73}$$

holds at least for one of the determinants D_k [Richter and Vogel (2002)]. In fact this determinant criterion implies an infinite hierarchy of conditions of nonclassicality. In practice, however, one would usually start with the condition for $k = 2$ (first-order nonclassicality). Whenever it is fulfilled, the higher-order conditions may be of little importance. Note that $D_2 < 0$ agrees with the condition (8.66) or its equivalent form (8.67).

If a quantum state does not show first-order nonclassicality, it can still be nonclassical with respect to higher orders. In particular, for $k = 3$, the condition (8.73) [together with Eq. (8.71)] reads

$$|\Phi(\alpha_1)|^2 + |\Phi(\alpha_2)|^2 + |\Phi(\alpha_1 + \alpha_2)|^2 - 2\text{Re}[\Phi(\alpha_1)\Phi(\alpha_2)\Phi^*(\alpha_1 + \alpha_2)] > 1, \tag{8.74}$$

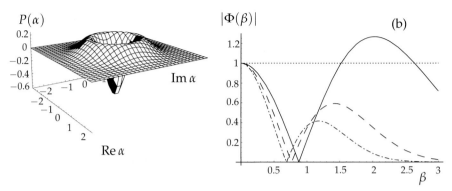

Fig. 8.5 (a) P function of $\hat{\varrho}_{\text{th}_+}$ for $\bar{n}_{\text{th}} = 0.7$. (b) Absolute value $|\Phi|$ of the characteristic function of the state $\hat{\varrho}_{\text{th}_+}$ for $\bar{n}_{\text{th}}=0.3$ (full line), 0.7 (dashed line), 1.1 (dash-dotted line). The dotted line marks the boundary between classical behavior and first-order nonclassical behavior according to the criterion (8.66). [After Shchukin, Richter and Vogel (2004).]

where the notation has been changed: $\alpha_1 - \alpha_2 \mapsto \alpha_1$, $\alpha_1 - \alpha_3 \mapsto \alpha_2$. By using Eq. (8.65) this condition can also be expressed in terms of the characteristic function of the quadrature distributions, $\Psi(y,\varphi)$. It is worth noting that in each order the condition (8.73) can be fully expressed in terms of $\Psi(y,\varphi)$, which can be inferred from the data measured by homodyne detection (Section 6.5.4).

An example of a state which does not obey the first-order nonclassicality criterion is the so-called one-photon added thermal state [Shchukin, Richter and Vogel (2004)]. It is defined by $\hat{\varrho}_{\text{th}_+} = \mathcal{N}\hat{a}^\dagger \hat{\varrho}_{\text{th}} \hat{a}$, where $\hat{\varrho}_{\text{th}}$ is a thermal state of mean photon number $\bar{n}_{\text{th}}$ ($\mathcal{N}$, normalization constant) and reads in the number basis as

$$\hat{\varrho}_{\text{th}_+} = \frac{1}{(1+\bar{n}_{\text{th}})^2} \sum_{m=1}^{\infty} \left(\frac{\bar{n}_{\text{th}}}{1+\bar{n}_{\text{th}}}\right)^{m-1} m|m\rangle\langle m|. \tag{8.75}$$

The characteristic function can be proved to be

$$\Phi(\beta) = L_1[(1+\bar{n}_{\text{th}})|\beta|^2]e^{-\bar{n}_{\text{th}}\beta^2} \tag{8.76}$$

[$L_i(x)$, Laguerre polynomial], from which it follows that the P function is well behaved:

$$P(\alpha) = -\frac{1}{\pi \bar{n}_{\text{th}}^2} L_1[(1+1/\bar{n}_{\text{th}})|\alpha|^2]e^{-|\alpha|^2/\bar{n}_{\text{th}}}. \tag{8.77}$$

It attains negative values around the origin of the phase space for any value of $\bar{n}_{\text{th}}$; see the example in Fig. 8.5(a). First-order nonclassicality, however, is only observed for sufficiently small values of $\bar{n}_{th}$, such as for example $\bar{n}_{th}=0.3$, as can be seen from Fig. 8.5(b). For larger values of $\bar{n}_{th}$ the nonclassicality is obviously of higher order [for details, see Shchukin, Richter and Vogel (2004)].

8.4
Nonclassical moments

From Section 8.2 we already know that observable moments play an important role in the study of nonclassical systems. It is therefore desirable to formulate very general nonclassicality criteria in terms of observable moments which are equivalent to the criteria introduced in Section 8.3 on the basis of the characteristic functions. This can again be achieved by means of the Bochner theorem, but in a somewhat modified form.

8.4.1
Reformulation of the Bochner condition

Let $\hat{f} = \hat{f}(\hat{a}, \hat{a}^\dagger)$ be an operator function that can be represented in normal order as

$$\hat{f} = \int d^2\alpha\, \underline{f}(\alpha) : \hat{D}(-\alpha) :, \tag{8.78}$$

where

$$\underline{f}(\alpha) = \frac{1}{\pi^2} \int d^2\beta\, e^{\alpha\beta^* - \alpha^*\beta} f(\beta), \tag{8.79}$$

with $f(\beta) \equiv f(\beta; s=1)$ being the c-number function associated with $\hat{f}$ in normal order [Eq. (4.78) together with Eqs. (4.51) and (4.77) for $s=1$], recall that $:\hat{D}(\alpha): = \hat{D}(\alpha; s=1) = e^{\alpha \hat{a}^\dagger} e^{-\alpha^* \hat{a}}$. Hence we may write, on recalling that $\Phi(\alpha) \equiv \Phi(\alpha; s=1) = \langle :\hat{D}(\alpha): \rangle$ [Eq. (4.90) for $s=1$], the expectation value $\langle :\hat{f}^\dagger \hat{f}: \rangle$ as

$$\langle :\hat{f}^\dagger \hat{f}: \rangle = \int d^2\alpha \int d^2\beta\, \Phi(\alpha - \beta) \underline{f}^*(\alpha) \underline{f}(\beta), \tag{8.80}$$

Now we can make contact with the Bochner theorem (8.62) in the continuous form [see, e.g., Kawata (1972)]: The function $\Phi(\alpha)$ is the characteristic function of a probability distribution on the complex plane iff for any smooth function $\underline{f}(\alpha)$ with compact support the inequality

$$\int d^2\alpha \int d^2\beta\, \Phi(\alpha - \beta) \underline{f}^*(\alpha) \underline{f}(\beta) \geq 0 \tag{8.81}$$

holds, which, under the assumptions made, is equivalent to

$$\langle :\hat{f}^\dagger \hat{f}: \rangle \geq 0. \tag{8.82}$$

From the above it is clear that a quantum state can be said to be nonclassical (in the sense that the P function fails to be a probability distribution) if the Bochner condition in the form (8.82) is violated, i.e., iff there exists an operator $\hat{f}$ with

$$\langle :\hat{f}^\dagger \hat{f}: \rangle < 0 \tag{8.83}$$

[Shchukin, Richter and Vogel (2005); Korbicz, Cirac, Wehr and Lewenstein (2005)]. Note that the class of operators $\hat{f}$ can be restricted to those whose normally ordered form (8.78) exists.

8.4.2
Criteria based on moments

As we will see, the condition (8.83) enables us to formulate nonclassicality conditions in terms of the normally ordered moments $\langle \hat{a}^{\dagger n} \hat{a}^m \rangle$ of the annihilation and creation operators. For this purpose let us first consider a (normally ordered) polynomial function

$$\hat{f}(\hat{a}, \hat{a}^\dagger) = \sum_{k=0}^{K} \sum_{l=0}^{L} c_{kl} \hat{a}^{\dagger k} \hat{a}^l \tag{8.84}$$

and require the conditions for the quadratic form

$$\langle :\hat{f}^\dagger \hat{f}: \rangle = \sum_{n,k=0}^{K} \sum_{m,l=0}^{L} c_{kl}^* c_{nm} \langle \hat{a}^{\dagger n+l} \hat{a}^{m+k} \rangle \tag{8.85}$$

to be non-negative (the coefficients c_{nm} are considered as independent variables). Introducing the matrix

$$\begin{pmatrix}
1 & \langle \hat{a} \rangle & \langle \hat{a}^\dagger \rangle & \langle \hat{a}^2 \rangle & \langle \hat{a}^\dagger \hat{a} \rangle & \langle \hat{a}^{\dagger 2} \rangle & \cdots \\
\langle \hat{a}^\dagger \rangle & \langle \hat{a}^\dagger \hat{a} \rangle & \langle \hat{a}^{\dagger 2} \rangle & \langle \hat{a}^\dagger \hat{a}^2 \rangle & \langle \hat{a}^{\dagger 2} \hat{a} \rangle & \langle \hat{a}^{\dagger 3} \rangle & \cdots \\
\langle \hat{a} \rangle & \langle \hat{a}^2 \rangle & \langle \hat{a}^\dagger \hat{a} \rangle & \langle \hat{a}^3 \rangle & \langle \hat{a}^\dagger \hat{a}^2 \rangle & \langle \hat{a}^{\dagger 2} \hat{a} \rangle & \cdots \\
\langle \hat{a}^{\dagger 2} \rangle & \langle \hat{a}^{\dagger 2} \hat{a} \rangle & \langle \hat{a}^{\dagger 3} \rangle & \langle \hat{a}^{\dagger 2} \hat{a}^2 \rangle & \langle \hat{a}^{\dagger 3} \hat{a} \rangle & \langle \hat{a}^{\dagger 4} \rangle & \cdots \\
\langle \hat{a}^\dagger \hat{a} \rangle & \langle \hat{a}^\dagger \hat{a}^2 \rangle & \langle \hat{a}^{\dagger 2} \hat{a} \rangle & \langle \hat{a}^\dagger \hat{a}^3 \rangle & \langle \hat{a}^{\dagger 2} \hat{a}^2 \rangle & \langle \hat{a}^{\dagger 3} \hat{a} \rangle & \cdots \\
\langle \hat{a}^2 \rangle & \langle \hat{a}^3 \rangle & \langle \hat{a}^\dagger \hat{a}^2 \rangle & \langle \hat{a}^4 \rangle & \langle \hat{a}^\dagger \hat{a}^3 \rangle & \langle \hat{a}^{\dagger 2} \hat{a}^2 \rangle & \cdots \\
\cdots & & & & & &
\end{pmatrix} \tag{8.86}$$

and employing Silverster's criterion, the required necessary and sufficient conditions are

$$d^{\mathbf{k}} \geq 0, \tag{8.87}$$

where $d^{\mathbf{k}}$, $\mathbf{k} = (k_1, \ldots, k_n)$, are the principal minors with rows and columns $k_1 < \cdots < k_n$. The leading principal minors are simply denoted by d_n.

The conditions (8.87) can be used analogously in the limiting case of infinite (convergent) series,

$$\hat{f} = \sum_{k,l=0}^{\infty} c_{kl} \hat{a}^{\dagger k} \hat{a}^l, \tag{8.88}$$

$$\langle :\hat{f}^\dagger \hat{f}: \rangle = \sum_{n,k,m,l=0}^{\infty} c_{kl}^* c_{nm} \langle \hat{a}^{\dagger n+l} \hat{a}^{m+k} \rangle, \tag{8.89}$$

where $\hat{f}$ belongs to the class of operators considered in Section 8.4.1. Hence the conditions (8.87) and (8.82) are equivalent to each other. On this basis we may formulate the following nonclassicality criterion:[10] A quantum state is nonclassical iff, at least for one **k**, the condition

$$d^{\mathbf{k}} < 0 \tag{8.90}$$

holds true [Shchukin and Vogel (2005a)]. The leading principal minor $n=2$ is excluded from consideration, since $d_2 = \langle \hat{a}^\dagger \hat{a} \rangle - \langle \hat{a}^\dagger \rangle \langle \hat{a} \rangle$ is simply the incoherent part of the photon number, which is always non-negative. Note that the moments in the matrix (8.86) can be determined experimentally (Sections 6.5.7 and 7.5).

The negativity of any principal minor $d^{\mathbf{k}}$ defines a sufficient condition for nonclassicality. To give an example, let us consider amplitude-squared squeezing [Hillery (1987)]:

$$\langle :(\Delta \hat{X}_\varphi)^2: \rangle < 0 \tag{8.91}$$

(for appropriately chosen phase φ), where

$$\hat{X}_\varphi = \hat{a}^2 e^{i\varphi} + \hat{a}^{\dagger 2} e^{-i\varphi}. \tag{8.92}$$

The condition (8.91) can be expressed in terms of a third-order principal minor:

$$d^{(1,4,6)} = \begin{vmatrix} 1 & \langle \hat{a}^2 \rangle & \langle \hat{a}^{\dagger 2} \rangle \\ \langle \hat{a}^{\dagger 2} \rangle & \langle \hat{a}^{\dagger 2} \hat{a}^2 \rangle & \langle \hat{a}^{\dagger 4} \rangle \\ \langle \hat{a}^2 \rangle & \langle \hat{a}^4 \rangle & \langle \hat{a}^{\dagger 2} \hat{a}^2 \rangle \end{vmatrix} < 0, \tag{8.93}$$

which more explicitly reads

$$d^{(1,4,6)} = \tfrac{1}{4} \langle :(\Delta \hat{X}_\varphi)^2: \rangle_{\min} \langle :(\Delta \hat{X}_\varphi)^2: \rangle_{\max} < 0. \tag{8.94}$$

---

10) For formulations of the criterion in terms of other moments, including normally ordered moments of two noncommuting quadratures, see Shchukin, Richter and Vogel (2005); Shchukin and Vogel (2005a). For some special cases see also Agarwal (1993); Klyshko (1996).

Here min (max) refers to the minimum (maximum) of the normally ordered variance with respect to its dependence on φ. Since the maximum cannot become negative, this condition directly analyzes the effect with respect to the optimized choice of the phase.[11]

8.5
Entanglement

After the experimental demonstration of the nonclassical effects considered in Section 8.2, attention has been focused on the realization of quantum states displaying observable quantum interference, with special emphasis on entanglement. Quantum interference effects occur when the quantum state of a physical system can be considered as being composed of two (or more) different states, e. g.,

$$|\Psi\rangle = \mathcal{N}(|\psi\rangle + |\chi\rangle) \tag{8.95}$$

($\mathcal{N}$, normalization constant; $|\psi\rangle \neq |\chi\rangle$). For any Hermitian operator $\hat{x}$ with eigenstates $|x\rangle$ the probability distribution $p_\Psi(x) = |\langle x|\Psi\rangle|^2$ of finding an eigenvalue x for the system being in the state $|\Psi\rangle$ reads

$$p_\Psi(x) = \tfrac{1}{2}\{p_\psi(x) + p_\chi(x) + 2\mathrm{Re}\,[\psi(x)\chi^*(x)]\}. \tag{8.96}$$

Besides the probability distributions for the system being in the states $|\psi\rangle$ and $|\chi\rangle$, $p_\psi(x) = |\langle x|\psi\rangle|^2$ and $p_\chi(x) = |\langle x|\chi\rangle|^2$, there occurs an interference term composed of the product of the two probability amplitudes $\psi(x) = \langle x|\psi\rangle$ and $\chi^*(x) = \langle \chi|x\rangle$ – quantities that are unknown in classical probability theory. Hence the interference term obviously represents a pure quantum effect, the strength of which increases with the separation of the states.

8.5.1
Separable and nonseparable quantum states

Let us consider a system that consists of two sub-systems, say the sub-systems 1 and 2. In the simplest case when the subsystems prepared in the pure states $|\psi_1\rangle$ and $|\chi_2\rangle$ are completely uncorrelated, then the composed system is prepared in the product state

$$|\Psi\rangle = |\psi_1\rangle \otimes |\chi_2\rangle, \tag{8.97}$$

which is an example of a separable state. In this case the expectation values of products of quantities of the two subsystems factorize. Now let $|\psi_1\rangle$ and $|\chi_2\rangle$

---

11) For further details the reader is referred to Shchukin and Vogel (2005a).

be superposition states,

$$|\psi_1\rangle = \mathcal{N}_1(|\psi_1^{(a)}\rangle + |\psi_1^{(b)}\rangle), \tag{8.98}$$

$$|\chi_2\rangle = \mathcal{N}_2(|\chi_2^{(c)}\rangle + |\chi_2^{(d)}\rangle) \tag{8.99}$$

$(|\psi_1^{(a)}\rangle \neq |\psi_1^{(b)}\rangle, |\chi_1^{(c)}\rangle \neq |\chi_1^{(d)}\rangle)$ and consider, e. g., the system state

$$|\Psi\rangle = \mathcal{N}(|\psi_1^{(a)}\rangle \otimes |\chi_2^{(c)}\rangle + |\psi_1^{(b)}\rangle \otimes |\chi_2^{(d)}\rangle), \tag{8.100}$$

which cannot be rewritten in a factorized (separable) form of the type (8.97). The state $|\Psi\rangle$, which describes a quantum-mechanically correlated bipartite system, is an example of an entangled state. Another typical example is the two-mode squeezed vacuum state (3.173).

The counterintuitive nature of entangled states was already realized in the early days of quantum mechanics. It becomes most evident when one tries to transfer the properties of an entangled state to macroscopic objects. The standard example is Schrödinger's cat [Schrödinger (1935)]. In his original Gedanken experiment Schrödinger considered a cat which is confined within a box together with a small amount of radioactive material. When a detector registers a radioactive decay it triggers a mechanism which kills the cat. Thus the system can be expected to be in an entangled state of the type

$$|\Psi_{\text{cat}}\rangle = \tfrac{1}{\sqrt{2}}(|\text{decay}\rangle|\text{cat dead}\rangle + |\text{no decay}\rangle|\text{cat alive}\rangle). \tag{8.101}$$

That is, if the detector has registered a decay the cat is dead, whereas the cat is still alive when the detector has not registered a decay. As long as we do not perform a measurement, say by opening the box, the cat is in a curious quantum superposition of being (simultaneously) dead and alive. Experimental realizations of mesoscopic versions of such entangled states have been achieved by using the atom–field interaction in a cavity (Chapter 12) and the vibronic coupling of trapped atoms (Chapter 13). Another example of the debate on the consequences of entanglement is the Einstein–Podolsky–Rosen paradox [Einstein, Podolsky and Rosen (1935)], which has stimulated a series of discussions on fundamental physical and philosophical aspects, including completeness of quantum theory, physical reality and locality.[12]

During the last years entangled states have received much attention in the framework of the rapidly developing field of quantum information. In particular, entanglement has been considered as a basic prerequisite for implementing quantum computation [see, e.g., Ekert and Josza (1996)] and quantum communication [see, e. g., Bennett and Wiesner (1992) and Bennett, Brassard, Crépeau, Jozsa, Peres and Wootters (1993)]. In this context, Shor's factoring

---

[12] For more details see, e. g., Bell (1997).

algorithm [Shor (1994)] and Grover's search algorithm [Grover (1997)] are the most prominent examples in quantum computation, were it is believed that the use of entangled states can lead to a substantial decrease in computational resources.

For a more detailed analysis of entanglement, we focus on bipartite systems – the simplest case in which entanglement may occur. In particular, we consider systems composed of subsystems of equal type, the dimension of the Hilbert space of each of them is allowed to be infinite. We are mainly interested in systems like two harmonic-oscillator modes, with special emphasis on two radiation modes or two modes of the quantized motion of trapped atoms (Chapter 13). Note that in the context of quantum information processing Hilbert spaces of finite dimensions are frequently considered, for example, two-dimensional spaces for qubits.

A generally accepted definition of entanglement must of course include mixed states, which cannot be described by state vectors [as in the example in Eq. (8.100)] but necessarily require a description in terms of density operators. Let us therefore consider a bipartite system prepared in an arbitrary state described by the density operator $\hat{\varrho}$. The state is called separable if it is a convex combination of factorizable states [Werner (1989)], i.e.,

$$\hat{\varrho} = \sum_{n=0}^{\infty} p_n \hat{\varrho}_1^{(n)} \otimes \hat{\varrho}_2^{(n)}, \tag{8.102}$$

where the density operators $\hat{\varrho}_1^{(n)}$ and $\hat{\varrho}_2^{(n)}$, respectively, characterize quantum states of the subsystems 1 and 2, and the p_n are ordinary probabilities:

$$p_n \geq 0, \quad \sum_{n=0}^{\infty} p_n = 1. \tag{8.103}$$

A state is called entangled if it is not separable.[13] Although the separability condition (8.102) looks rather simple, it turns out to be a complex problem to distinguish separable quantum states from inseparable ones. In particular, there has been no experimental implementation of the general separability condition.

8.5.2
Partial transposition and entanglement criteria

A partial answer to the problem of finding a tractable entanglement criterion is given by the Peres–Horodecki condition – a sufficient criterion applicable

---

13) At this point the question of a quantitative measure of the strength of entanglement may arise. There are different approaches to the problem [for a review, see, e. g., Plenio and Vedral (1998)]. In the case of pure states, the entropic measure $E = S_1 = S_2$ is generally accepted to be such a measure, where S_i is the von Neumann entropy of the ith subsystem.

to a broad class of states [Peres (1996); Horodecki, Horodecki and Horodecki (1996); Horodecki (1997)]. It is not difficult to verify that full transposition of an arbitrary density operator gives an operator that is again a density operator. In the case of a bipartite system one may also consider the partially transposed density operator, i. e., the operator which is obtained by partially transposing the density operator with respect to one of the subsystems. In contrast to full transposition, partial transposition does not necessarily map a quantum state onto a quantum state. However, it turns out that the partially transposed density operator of a separable state is again a density operator:[14]

$$\hat{\varrho}^{\text{PT}} = \sum_{n=0}^{\infty} p_n \hat{\varrho}_1^{(n)} \otimes \hat{\varrho}_2^{(n)\text{T}}. \tag{8.104}$$

This property of a separable quantum state allows one to use the failure of the partial transposition of the density operator to again have the properties of a density operator as a sufficient condition for entanglement.

This condition implies that a state is entangled when the partial transposition of the density operator fails to be non-negative. Before going into details, let us consider an arbitrary non-negative Hermitian operator $\hat{A}$, that is

$$\langle \psi | \hat{A} | \psi \rangle = \text{Tr}\left(\hat{A} | \psi \rangle \langle \psi |\right) \geq 0 \tag{8.105}$$

holds for any (normalizable) quantum state $|\psi\rangle$. The non-negative operator $|\psi\rangle\langle\psi|$ can be written as

$$|\psi\rangle\langle\psi| = \hat{f}^\dagger \hat{f}, \tag{8.106}$$

where $\hat{f}$ may be given in the form

$$\hat{f} = |\varphi\rangle\langle\psi|, \tag{8.107}$$

with $|\varphi\rangle$ being an arbitrary (normalizable) state of the system. Let $\hat{a}_1$ and $\hat{a}_2$ be the annihilation operators of the first and the second mode, respectively, of a two-mode bosonic system. In this case, $\hat{f}$ can be regarded as an operator-valued function of $\hat{a}_1, \hat{a}_1^\dagger$ and $\hat{a}_2, \hat{a}_2^\dagger$ [$\hat{f} = \hat{f}(\hat{a}_1, \hat{a}_1^\dagger, \hat{a}_2, \hat{a}_2^\dagger)$]. Since the expectation value of $|\varphi\rangle\langle\psi|$ for arbitrary coherent states exists, it is clear that the normally ordered form of $\hat{f}$ exists either. Hence we conclude that a Hermitian operator $\hat{A}$ is non-negative iff for any such operator $\hat{f}$ the inequality

$$\text{Tr}\left(\hat{A}\hat{f}^\dagger \hat{f}\right) \geq 0 \tag{8.108}$$

[14] Note that, since $\hat{\varrho}_2^{(n)\text{T}}$ is a density operator for all n, $\hat{\varrho}^{\text{PT}}$ is also a density operator, because of the expansion (8.104). Of course, this also remains valid if the transposition is performed with respect to the subsystem 1.

is fulfilled. Recalling the Peres–Horodecki condition, we may therefore say that, for a separable two-mode state, the inequality

$$\text{Tr}\,(\hat{\varrho}^{\text{PT}} \hat{f}^\dagger \hat{f}) \geq 0 \tag{8.109}$$

holds true for any (two-mode) operator $\hat{f}$ to be considered [Shchukin and Vogel (2005b)].

8.5.2.1 Negativity of the partial transposition in terms of moments

Let us expand the normally ordered form of $\hat{f} = \hat{f}(\hat{a}_1, \hat{a}_1^\dagger, \hat{a}_2, \hat{a}_2^\dagger)$,

$$\hat{f} = \sum_{n,m,k,l=0}^{\infty} c_{nmkl} \hat{a}_1^{\dagger n} \hat{a}_1^m \hat{a}_2^{\dagger k} \hat{a}_2^l, \tag{8.110}$$

so that the inequality (8.109) takes the form

$$\sum_{\substack{n,m,k,l=0 \\ p,q,r,s=0}}^{\infty} c_{pqrs}^* c_{nmkl} M_{pqrs,nmkl} \geq 0, \tag{8.111}$$

where

$$M_{pqrs,nmkl} = \text{Tr}\,(\hat{\varrho}^{\text{PT}} \hat{a}_1^{\dagger q} \hat{a}_1^p \hat{a}_1^{\dagger n} \hat{a}_1^m \hat{a}_2^{\dagger s} \hat{a}_2^r \hat{a}_2^{\dagger k} \hat{a}_2^l). \tag{8.112}$$

To relate the $M_{pqrs,nmkl}$ to measurable quantities, we first perform explicitly the partial transposition, leading to the moments

$$M_{pqrs,nmkl} = \langle \hat{a}_1^{\dagger q} \hat{a}_1^p \hat{a}_1^{\dagger n} \hat{a}_1^m \hat{a}_2^{\dagger l} \hat{a}_2^k \hat{a}_2^{\dagger r} \hat{a}_2^s \rangle. \tag{8.113}$$

These moments can then be expressed in terms of linear combinations of normally ordered moments by applying the relation (4.96) (for $s = -1$, $s' = 1$):

$$\hat{a}^p \hat{a}^{\dagger n} = \sum_{k=0}^{\min(p,n)} \frac{p!n!}{k!(p-k)!(n-k)!} \hat{a}^{\dagger n-k} \hat{a}^{p-k}. \tag{8.114}$$

The left hand side of the inequality (8.111) is a quadratic form with respect to the coefficients c_{nmkl}. Silvester's criterion states that this inequality is fulfilled for all c_{nmkl}, iff all the principal minors attributed to the quadratic form are non-negative. In this context it may be convenient to relate the multi-index combination (n, m, k, l) to a single index and number the $M_{pqrs,nmkl}$ by two indices like an ordinary matrix, $M_{pqrs,nmkl} \mapsto M_{ij}$. To define the ordinal number of a multi-index combination $\mathbf{u} = (n, m, k, l)$, we apply the following ordering prescription:

$$\mathbf{u} < \mathbf{u}' \leftrightarrow \begin{cases} |\mathbf{u}| < |\mathbf{u}'| \text{ or} \\ |\mathbf{u}| = |\mathbf{u}'| \text{ and } \mathbf{u} <' \mathbf{u}' \end{cases} \tag{8.115}$$

$[\mathbf{u}' = (n', m', k', l')]$, where $|\mathbf{u}| = n + m + k + l$ and $\mathbf{u} <' \mathbf{u}'$ means that the first nonzero difference $k' - k$, $l' - l$, $n' - n$, $m' - m$ is positive. According to this prescription, the ordered sequence of moments begins as follows:

$$1, \langle \hat{a}_1 \rangle, \langle \hat{a}_1^\dagger \rangle, \langle \hat{a}_2 \rangle, \langle \hat{a}_2^\dagger \rangle, \langle \hat{a}_1^2 \rangle, \langle \hat{a}_1^\dagger \hat{a}_1 \rangle, \langle \hat{a}_1^{\dagger 2} \rangle, \langle \hat{a}_1 \hat{a}_2 \rangle,$$
$$\langle \hat{a}_1^\dagger \hat{a}_2 \rangle, \langle \hat{a}_2^2 \rangle, \langle \hat{a}_1 \hat{a}_2^\dagger \rangle, \langle \hat{a}_1^\dagger \hat{a}_2^\dagger \rangle, \langle \hat{a}_2^\dagger \hat{a}_2 \rangle, \langle \hat{a}_2^{\dagger 2} \rangle, \dots \qquad (8.116)$$

Introducing the matrix

$$\begin{pmatrix} M_{11} & M_{12} & \dots & M_{1N} \\ M_{21} & M_{22} & \dots & M_{2N} \\ \dots & \dots & \dots & \dots \\ M_{N1} & M_{N2} & \dots & M_{NN} \end{pmatrix}$$

$$= \begin{pmatrix} 1 & \langle \hat{a}_1 \rangle & \langle \hat{a}_1^\dagger \rangle & \langle \hat{a}_2^\dagger \rangle & \langle \hat{a}_2 \rangle & \dots \\ \langle \hat{a}_1^\dagger \rangle & \langle \hat{a}_1^\dagger \hat{a}_1 \rangle & \langle \hat{a}_1^{\dagger 2} \rangle & \langle \hat{a}_1^\dagger \hat{a}_2^\dagger \rangle & \langle \hat{a}_1^\dagger \hat{a}_2 \rangle & \dots \\ \langle \hat{a}_1 \rangle & \langle \hat{a}_1^2 \rangle & \langle \hat{a}_1 \hat{a}_1^\dagger \rangle & \langle \hat{a}_1 \hat{a}_2^\dagger \rangle & \langle \hat{a}_1 \hat{a}_2 \rangle & \dots \\ \langle \hat{a}_2 \rangle & \langle \hat{a}_1 \hat{a}_2 \rangle & \langle \hat{a}_1^\dagger \hat{a}_2 \rangle & \langle \hat{a}_2^\dagger \hat{a}_2 \rangle & \langle \hat{a}_2^2 \rangle & \dots \\ \langle \hat{a}_2^\dagger \rangle & \langle \hat{a}_1 \hat{a}_2^\dagger \rangle & \langle \hat{a}_1^\dagger \hat{a}_2^\dagger \rangle & \langle \hat{a}_2^{\dagger 2} \rangle & \langle \hat{a}_2 \hat{a}_2^\dagger \rangle & \dots \\ \dots & \dots & \dots & \dots & \dots & \dots \end{pmatrix}, \qquad (8.117)$$

the following necessary and sufficient condition for the negativity of the partially transposed density operator of a two-mode bosonic system can be given in terms of the principal minors:

$$\exists \mathbf{k}: \quad d^\mathbf{k} < 0 \qquad (8.118)$$

[Shchukin and Vogel (2005b); Miranowicz and Piani (2006)]. This represents a very general, sufficient condition for entanglement in terms of measurable quantities.

8.5.2.2 Special entanglement conditions

From the above it follows that each individual inequality $d^\mathbf{k} < 0$ can be regarded as a special entanglement condition. Moreover, by appropriately identifying some of the coefficients c_{nmkl} in Eq. (8.110), the inequality (8.111) can be used in an analogous way to derive other types of entanglement conditions, which may be more useful than the condition (8.118). In this way, a number of sufficient entanglement conditions can be established.

For example, let us restrict our attention to moments up to the second order and consider the condition

$$d_5 < 0, \qquad (8.119)$$

which agrees with the one given by Simon (2000). Straightforward calculation shows that d_5 can be rewritten as

$$d_5 = \det A_1 \det A_2 + \left(\tfrac{1}{4} + \det C\right)^2 - \mathrm{Tr}(A_1 J C J A_2 J C^T J) \\ - \tfrac{1}{4}(\det A_1 + \det A_2), \tag{8.120}$$

where matrices A_i ($i=1,2$), C and J are defined by[15]

$$A_i = \begin{pmatrix} \langle(\Delta\hat{x}_i)^2\rangle & \tfrac{1}{2}\langle[\Delta\hat{x}_i,\Delta\hat{p}_i]_+\rangle \\ \tfrac{1}{2}\langle[\Delta\hat{x}_i,\Delta\hat{p}_i]_+\rangle & \langle(\Delta\hat{p}_i)^2\rangle \end{pmatrix}, \tag{8.121}$$

$$C = \begin{pmatrix} \langle\Delta\hat{x}_1\Delta\hat{x}_2\rangle & \langle\Delta\hat{x}_1\Delta\hat{p}_2\rangle \\ \langle\Delta\hat{p}_1\Delta\hat{x}_2\rangle & \langle\Delta\hat{p}_1\Delta\hat{p}_2\rangle \end{pmatrix}, \tag{8.122}$$

$$J = \begin{pmatrix} 0 & 1 \\ -1 & 0 \end{pmatrix}, \tag{8.123}$$

with $\hat{x}_i = (\hat{a}_i + \hat{a}_i^\dagger)/\sqrt{2}$ and $\hat{p}_i = i(\hat{a}_i^\dagger - \hat{a}_i)/\sqrt{2}$.

Another example[16] is the condition

$$d^{(1,2,4)} < 0, \tag{8.124}$$

with

$$d^{(1,2,4)} = \begin{vmatrix} 1 & \langle\hat{a}_1\rangle & \langle\hat{a}_2^\dagger\rangle \\ \langle\hat{a}_1^\dagger\rangle & \langle\hat{a}_1^\dagger\hat{a}_1\rangle & \langle\hat{a}_1^\dagger\hat{a}_2^\dagger\rangle \\ \langle\hat{a}_2\rangle & \langle\hat{a}_1\hat{a}_2\rangle & \langle\hat{a}_2^\dagger\hat{a}_2\rangle \end{vmatrix} \tag{8.125}$$

where the corresponding principal minor corresponds to the operator $\hat{f} = c_1 + c_2\hat{a}_1 + c_3\hat{a}_2$ in Eq. (8.110). The condition (8.124), which can be rewritten as

$$\langle\Delta\hat{a}_1^\dagger\Delta\hat{a}_1\rangle\langle\Delta\hat{a}_2^\dagger\Delta\hat{a}_2\rangle < |\langle\Delta\hat{a}_1\Delta\hat{a}_2\rangle|^2, \tag{8.126}$$

can be related to the following condition given by Duan, Giedke, Cirac and Zoller (2000):

$$\langle(\Delta\hat{u})^2\rangle + \langle(\Delta\hat{v})^2\rangle - (r^2 + r^{-2}) < 0, \tag{8.127}$$

where

$$\hat{u} = |r|\hat{x}_1 + r^{-1}\hat{x}_2, \quad \hat{v} = |r|\hat{p}_1 - r^{-1}\hat{p}_2, \tag{8.128}$$

15) $[\hat{A},\hat{B}]_+$ denotes the anti-commutator.
16) For further examples, see Shchukin and Vogel (2005b).

with r being an arbitrary (nonzero) real parameter. Expressing the operators $\hat{u}$ and $\hat{v}$ in terms of linear combinations of the creation and annihilation operators, substituting them into the inequality (8.127), and minimizing the left-hand side with respect to r, we arrive at

$$\langle \Delta \hat{a}_1^\dagger \Delta \hat{a}_1 \rangle \langle \Delta \hat{a}_2^\dagger \Delta \hat{a}_2 \rangle < (\operatorname{Re} \langle \Delta \hat{a}_1 \Delta \hat{a}_2 \rangle)^2. \tag{8.129}$$

We see that the condition (8.127) is a weaker form of the condition (8.126).

Of course, the method can also be used to formulate new entanglement conditions which are specific to particular problems. Choosing, for example,

$$\hat{f} = c_1 \hat{a}_1^{\dagger n} \hat{a}_1^m \hat{a}_2^{\dagger k} \hat{a}_2^l + c_2 \hat{a}_1^{\dagger p} \hat{a}_1^q \hat{a}_2^{\dagger r} \hat{a}_2^s \tag{8.130}$$

in Eq. (8.110), we may consider the condition

$$\langle \hat{a}_1^{\dagger m} \hat{a}_1^n \hat{a}_1^{\dagger n} \hat{a}_1^m \hat{a}_2^{\dagger l} \hat{a}_2^k \hat{a}_2^{\dagger k} \hat{a}_2^l \rangle \langle \hat{a}_1^{\dagger q} \hat{a}_1^p \hat{a}_1^{\dagger p} \hat{a}_1^q \hat{a}_2^{\dagger s} \hat{a}_2^r \hat{a}_2^{\dagger r} \hat{a}_2^s \rangle < |\langle \hat{a}_1^{\dagger m} \hat{a}_1^n \hat{a}_1^{\dagger p} \hat{a}_1^q \hat{a}_2^{\dagger s} \hat{a}_2^r \hat{a}_2^{\dagger k} \hat{a}_2^l \rangle|^2. \tag{8.131}$$

This type of condition offers a simple way of formulating conditions in terms of higher-order moments, which are of particular interest in characterizing non-Gaussian states.

References

Agarwal, G.S. (1993) *Opt. Commun.* **95**, 109.

Bell, J.S. (1997), *Speakable and Unspeakable in Quantum Mechanics* (Cambridge University Press, Cambridge).

Bennett, C.H. and S.J. Wiesner (1992) *Phys. Rev. Lett.* **69**, 2881.

Bennett, C.H., G. Brassard, C. Crépeau, R. Jozsa, A. Peres and W.K. Wootters (1993) *Phys. Rev. Lett.* **70**, 1895.

Bochner, S. (1933) *Math. Ann.* **108**, 378.

Carmichael, H.J. and D.F. Walls (1976) *J. Phys. B* **9**, L43.

Diedrich, F. and H. Walther (1987) *Phys. Rev. Lett.* **58**, 203.

Duan, L.-M., G. Giedke, J.I. Cirac and P. Zoller (2000) *Phys. Rev. Lett.* **84**, 2722.

Ekert, A. and R. Josza (1996) *Rev. Mod. Phys.* **68**, 733.

Einstein, A., B. Podolsky and N. Rosen (1935) *Phys. Rev.* **44**, 777.

Gardiner, C.W. (1983) *Handbook of Stochastic Methods* (Springer-Verlag, Berlin).

Grangier, P., R.E. Slusher, B. Yurke and A. La Porta (1987) *Phys. Rev. Lett.* **59**, 2153.

Grover, L.K. (1997) *Phys. Rev. Lett.* **79**, 325.

Hillery, M. (1985) *Phys. Lett.* **111A**, 409.

Hillery, M. (1987) *Phys. Rev. A* **36**, 3796.

Hong, C.K. and L. Mandel (1985) *Phys. Rev. Lett.* **54**, 323; *Phys. Rev. A* **32**, 974.

Hong, C.K. and L. Mandel (1986) *Phys. Rev. Lett.* **56**, 58.

Horodecki, M., P. Horodecki and R. Horodecki (1996) *Phys. Lett. A* **223**, 1.

Horodecki, P. (1997) *Phys. Lett. A* **232**, 333.

Johansen, L.M. (2004) *Phys. Rev. Lett.* **93**, 120402.

Johansen, L.M. (2004a) *Phys. Lett. A* **329**, 184.

Johansen, L.M. and A. Luis (2004) *Phys. Rev. A* **70**, 052115.

Kawata, T. (1972) *Fourier Analysis in Probability Theory* (Academic Press, New York).

Kimble, H.J. and L. Mandel (1976) *Phys. Rev. A* **13**, 2123.

Kimble, H.J., M. Dagenais and L. Mandel (1977) *Phys. Rev. Lett.* **39**, 691.

Klyshko, D.N. (1996) *Physics-Uspekhi* **39**, 573.

Korbicz, J.K., J.I. Cirac, J. Wehr and M. Lewenstein (2005) *Phys. Rev. Lett.* **94**, 153601.

Lvovsky, A.I. and J.H. Shapiro (2002) *Phys. Rev. A* **65**, 033830.

Machida, S., Y. Yamamoto and Y. Itaya (1987) *Phys. Rev. Lett.* **58**, 1000.

Machida, S. and Y. Yamamoto (1989) *Opt. Lett.* **14**, 1045.

Mandel, L. (1966) *Phys. Rev.* **152**, 438.

Mandel, L. (1986) *Phys. Scr. T* **12**, 34.

Miranowicz A. and M. Piani (2006) arXiv:quant-ph/0603239.

Peres, A. (1996) *Phys. Rev. Lett.* **77**, 1413.

Plenio, M.B. and V. Vedral (1998) *Contemp. Phys.* **39**, 431.

Polzik, E.S., J. Carri and H.J. Kimble (1992) *Phys. Rev. Lett.* **68**, 3020.

Richter, Th. and W. Vogel (2002) *Phys. Rev. Lett.* **89**, 283601.

Schrödinger, E. (1935) *Naturwiss.* **23**, 807.

Schubert, M., I. Siemers, R. Blatt, W. Neuhauser and P.E. Toschek (1992) *Phys. Rev. Lett.* **68**, 3016.

Shchukin, E., Th. Richter and W. Vogel (2004) *J. Opt. B: Quantum Semiclass. Opt.* **6**, S597.

Shchukin, E., Th. Richter and W. Vogel (2005) *Phys. Rev. A* **71**, 011802(R).

Shchukin, E., and W. Vogel (2005a) *Phys. Rev. A* **72**, 043808.

Shchukin, E., and W. Vogel (2005b) *Phys. Rev. Lett.* **95**, 230502.

Shor, P.W. (1994) in *Proceedings of the 35th Annual Symposium on the Foundations of Computer Science*, ed. by S. Goldwasser (IEEE Computer Society, Los Alamitos CA) p. 124.

Short, R. and L. Mandel (1983) *Phys. Rev. Lett.* **51**, 384.

Slusher, R.E., L.W. Hollberg, B. Yurke, J.C. Mertz and J.F. Valley (1985) *Phys. Rev. Lett.* **55**, 2409.

Slusher, R.E., B. Yurke, P. Grangier, A. La Porta, D.F. Walls and M. Reid (1987) *J. Opt. Soc. Am. B* **4**, 1453.

Simon, R. (2000) *Phys. Rev. Lett.* **84**, 2726.

Teich, M.C. and B.E.A. Saleh (1985) *J. Opt. Soc. Am. B* **2**, 275.

Titulaer, U.M. and R.J. Glauber (1965) *Phys. Rev.* **140**, B676.

van Kampen, N.G. (1981) *Stochastic Processes in Physics and Chemistry* (Elsevier, Amsterdam).

Vogel, W. (1990) *Phys. Rev. A* **42**, 5754.

Vogel, W. (2000) *Phys. Rev. Lett.* **84**, 1849; **85**, 2842.

Werner, R.F. (1989) *Phys. Rev. A* **40**, 4277.

Wu, L.-A., H.J. Kimble, J.L. Hall and H. Wu (1986) *Phys. Rev. Lett.* **57** 2520.

Wu, L.-A., M. Xiao and H.J. Kimble (1987) *J. Opt. Soc. Am. B* **4**, 1465.

Xiao, M., L.-A. Wu and H.J. Kimble (1987) *Phys. Rev. Lett.* **59**, 278.

Xiao, M., L.-A. Wu and H.J. Kimble (1988) *Opt. Lett.* **13**, 476.

Yuen, H.P. and J.H. Shapiro (1979) *Opt. Lett.* **4**, 334.

Zhang, F. (1999) *Matrix Theory* (Springer, New York).

9
Leaky optical cavities

Resonator-like devices play an important role in modern optics and rapid progress in the fields of laser physics and nonlinear and quantum optics[1] would be unthinkable without their use. Optical cavities are typically involved in the generation and/or amplification of light by atomic sources. The sources are placed inside the resonator-like cavity, which is in contact with the environment via appropriately chosen (fractionally transparent) walls, so that the radiation field inside the cavity (referred to as the internal field) may be partially transmitted to free space. In this manner, the field generated or amplified inside the cavity may contribute to the radiation field outside (referred to as the external field), which is the desired one in many applications. On the other hand, the dynamics of the light–matter interaction inside the resonator is influenced by that part of the external radiation field which passes through the fractionally transparent walls and therefore contributes to the internal radiation field. This is also true when the incoming radiation field is in the vacuum state. In this case the quantum fluctuations of the (incoming) vacuum act on the radiation field inside the cavity. In these and related situations a consistent quantum-physical description of the radiation field requires consideration of both the lossy resonator-like cavity and the active medium. In the hypothetical case of a lossless resonator the situation is quite simple. In practice, however, the losses of the resonator cannot be ignored. The radiation field of interest is usually the output radiation field, which simply represents the losses of the internal radiation field.

One approach to the problem of obtaining a quantum description of the action of a leaky optical cavity with output coupling is based on the familiar formalism of quantum noise theory (Chapter 5), in which the explicit nature of the input from a heat bath, and the output into it, is taken into account [see, e.g., Collett and Gardiner (1984); Collett and Walls (1985); Gardiner and Collett (1985); Yamamoto and Imoto (1986); Carmichael (1987); Gardiner (1989)]. The starting point is the conventional picture of a "small" system containing a

---

[1] Quantum optics in cavities [frequently called cavity quantum electrodynamics (cavity QED)] has been a very rapidly developing and growing field of quantum optics, see, e.g., the review by Meystre (1992).

Quantum Optics, Third, revised and extended edition. Werner Vogel and Dirk-Gunnar Welsch
Copyright © 2006 WILEY-VCH Verlag GmbH & Co. KGaA, Weinheim
ISBN: 3-527-40507-0

given radiation-field mode of the idealized (lossless) cavity, which is linearly coupled to a "large" (harmonic-oscillator) reservoir, and which may also underlie certain intra-cavity interactions. Physically, the reservoir is regarded as representing the radiation-field modes outside the resonator-like cavity, which are assumed to be coupled to the internal radiation-field mode owing to the fractional transparency of one or both mirrors. Accordingly, the input from the reservoir and the output into it are identified with the incoming and outgoing radiation fields, respectively. By assuming the reservoir spectrum to be flat and the reservoir–system coupling constant to be independent of frequency, the behavior of the internal-mode operators may then be calculated by quantum Langevin-equation methods. In the resulting Langevin equations the Langevin force operators are related to the input into the cavity from the reservoir. Analogously, time-reversed Langevin equations may be derived in which the Langevin forces are related to the output into the reservoir from the cavity. With the use of causality and boundary conditions, the relationship between correlation functions of the output and those of the system and the input may be developed. In this way, it becomes possible to calculate the output statistics, provided that the input statistics is known and that the system correlation functions can be evaluated. The advantage of this extended damping theory over the commonly used one is that, apart from the fact that it allows for the possibility that the input may be other than a vacuum or thermal input, it contains a prescription for calculating the properties of the light emitted from the cavity.

Clearly, the true situation is described by the existence of a (continuous) multi-mode radiation field, whose mode structure is more or less modified (in comparison with the free-space case) owing to the presence of the fractionally transparent (dielectric) walls of the cavity [Lang and Scully (1973); Ujihara (1975, 1977, 1978, 1979, 1984); Guedes, Penaforte and Baseia (1989); Guedes and Baseia (1990); Feng (1991); De Martini, Marocco, Mataloni, Crescentini and Loudon (1991); Khosravi and Loudon (1991); Knöll, Vogel and Welsch (1991a,b); Knöll and Welsch (1992); Dutra and Nienhuis (2000); Viviescas and Hackenbroich (2003)]. This multi-mode (space-time-dependent) radiation field interacts with certain kinds of atomic sources which usually form the active medium within the resonator. In performing an appropriate mode expansion of the radiation field, one can show that in some approximation the introduction of two kinds of photonic operators (namely system and reservoir operators) and the appropriate relations between them comes out naturally and need not be postulated. The equivalence of the two approaches can be used also to take into account the effect on the cavity field of unwanted losses such as absorption losses, by simply supplementing the Langevin equations with damping and noise terms that are attributed to the additional dissipative channels [Viviescas and Hackenbroich (2003); Khanbekyan, Knöll, Semenov, Vogel and Welsch (2004)].

Provided that absorption losses can be disregarded in the frequency interval under consideration, the formalism of field quantization in a dielectric with real, space-dependent refractive index (Section 2.4.1) applies [Knöll, Vogel and Welsch (1991b); Knöll and Welsch (1992)]. By starting from an expansion of the radiation field in terms of a continuous set of modes extended over the whole universe (Sections 9.1 and 9.2), the formalism naturally yields a description of the radiation field inside the cavity in terms of quantum Langevin equations (Section 9.3), which may be extended to include in the theory unwanted losses such as scattering and absorption losses (Section 9.7). In relating the radiation field outside the cavity to that inside it (Section 9.4), the time-dependent commutation relations (Section 9.5) are needed to handle the quantum Langevin equations and the input-output relations (Sections 9.6 and 9.8).

9.1
Radiation-field modes

Let us begin with the calculation of the mode continuum of the radiation field. For the sake of clarity we consider a one-dimensional cavity of length L, bounded by a perfectly reflecting plane mirror at $x=0$ and a fractionally transparent plane mirror at $x=L$ (Fig. 9.1). We further assume that the radiation field is polarized in the z direction.

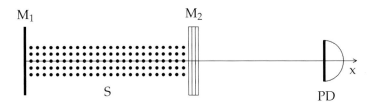

Fig. 9.1 Scheme of a one-dimensional cavity bounded by a perfectly reflecting mirror M_1 and a fractionally transparent (multi-slab dielectric) mirror M_2 (S, atomic sources; PD, photodetector).

9.1.1
Solution of the Helmholtz equation

To gain insight into the mode structure of the radiation field, let us first consider the simple case where the fractionally transparent mirror is a dielectric plate of thickness d and refractive index n, which for simplicity is assumed to be a positive constant number. In this case the mode functions $A(k,x)$ obey the Helmholtz equation

$$\frac{\partial^2 A(k,x)}{\partial x^2} + n^2(x)k^2 A(k,x) = 0, \qquad k^2 = \frac{\omega^2}{c^2}, \tag{9.1}$$

[cf. Eq. (2.195)], where

$$n(x) = \begin{cases} 1 & \text{if } 0 < x < L, \ L+d < x, \\ n & \text{if } L \leq x \leq L+d. \end{cases} \tag{9.2}$$

The boundary conditions at the surfaces of discontinuity are

$$A(k,0) = 0, \tag{9.3}$$

$$\lim_{\epsilon \to 0} [A(k, L+\epsilon) - A(k, L-\epsilon)] = 0, \tag{9.4}$$

$$\lim_{\epsilon \to 0} [A(k, L+d+\epsilon) - A(k, L+d-\epsilon)] = 0, \tag{9.5}$$

$$\lim_{\epsilon \to 0} \left[\frac{\partial A(k,x)}{\partial x} \bigg|_{x=L+\epsilon} - \frac{\partial A(k,x)}{\partial x} \bigg|_{x=L-\epsilon} \right] = 0, \tag{9.6}$$

$$\lim_{\epsilon \to 0} \left[\frac{\partial A(k,x)}{\partial x} \bigg|_{x=L+d+\epsilon} - \frac{\partial A(k,x)}{\partial x} \bigg|_{x=L+d-\epsilon} \right] = 0. \tag{9.7}$$

The solution of Eq. (9.1) together with Eqs (9.2)–(9.7) is elementary and straightforward. The result may be written as

$$A(k,x) = \left(\frac{\hbar}{4\pi \mathcal{A}\varepsilon_0 \omega} \right)^{1/2} \begin{cases} \underline{T}(\omega) \left(e^{ikx} - e^{-ikx} \right) & \text{if } 0 \leq x \leq L, \\ \dfrac{\underline{T}(\omega)}{\underline{T}^*(\omega)} e^{ikx} - e^{-ikx} & \text{if } L+d \leq x, \end{cases} \tag{9.8}$$

where $k > 0$ and $\mathcal{A}$ is the mirror area. Note that the functions $A(k,x)$ are normalized according to the condition (2.196) with $|c_\lambda|^2 = 2\omega_\lambda \varepsilon_0/\hbar$. The spectral response function of the cavity field $\underline{T}(\omega)$ is

$$\underline{T}(\omega) = \frac{\underline{t}(\omega)}{1 + \underline{r}(\omega)\exp(2il\omega/c)}, \quad l = L+d. \tag{9.9}$$

It is expressed in terms of the spectral transmission and reflection response functions of the right-hand mirror, $\underline{t}(\omega)$ and $\underline{r}(\omega)$, respectively. In the case of the simple dielectric plate considered here, $\underline{t}(\omega)$ and $\underline{r}(\omega)$ are calculated as [Eqs (6.130)–(6.132) with $\underline{T}(\omega) \mapsto \underline{t}(\omega)$ and $\underline{R}(\omega) \mapsto \underline{r}(\omega)$ therein]

$$\underline{t}(\omega) = \frac{(1-r^2)\exp[i(n-1)d\omega/c]}{1 - r^2 \exp(2ind\omega/c)} = \underline{t}^*(-\omega), \tag{9.10}$$

$$\underline{r}(\omega) = r\frac{\exp(2ind\omega/c) - 1}{1 - r^2 \exp(2ind\omega/c)} e^{-i\omega d/c} = \underline{r}^*(-\omega), \tag{9.11}$$

where

$$r^2 = \left(\frac{n-1}{n+1} \right)^2, \tag{9.12}$$

and $\underline{t}(\omega)$ and $\underline{r}(\omega)$ satisfy the conditions (6.133) and (6.134).

At this point we note that the validity of Eq. (9.9) is more general than that of Eqs (9.10) and (9.11), which only hold when the right-hand mirror is modeled by a dielectric plate of positive constant refractive index. Therefore, in what follows, we do not return to Eqs (9.10) and (9.11). Clearly, both the frequency dependences of $\underline{t}(\omega)$ and $\underline{r}(\omega)$ and the mode structure of the radiation field inside the right-hand wall are determined by the mirror actually used in a resonator device. To ensure high reflectivity and to sufficiently reduce internal losses, such a mirror is in general a complicated dielectric multi-slab configuration. Assuming the thickness of the mirror to be small compared with the length of the cavity, so that $L \approx L+d = l$, we may disregard the mode structure inside the mirror and regard Eq. (9.8) [together with Eq. (9.9) and appropriately chosen $\underline{t}(\omega)$] as complete [Ley and Loudon (1987)]. It should be noted that the following useful relations may be derived:

$$\underline{T}^*(\omega) = \underline{T}(-\omega), \tag{9.13}$$

$$\frac{\underline{T}(\omega)}{\underline{T}^*(\omega)} = \underline{t}(\omega)\underline{T}(\omega) - \underline{r}(\omega)\,e^{-2il\omega/c}. \tag{9.14}$$

9.1.2
Cavity-response function

In the case of a proper high-quality[2] resonator we may assume that (within the bandwidth of the radiation under study) the spectral transmission and reflection functions $\underline{t}(\omega)$ and $\underline{r}(\omega)$, respectively, are slowly varying with frequency. Let us (for a moment) suppose that they are constant, that is, $\underline{t}(\omega) \equiv \underline{t} = \text{const}$ and $\underline{r}(\omega) \equiv \underline{r} = \text{const}$. In this case, the poles of $\underline{T}(\omega)$ (as a function of ω in the whole complex plane) are determined by the solutions of

$$1 + \underline{r}\,e^{2il\Omega_n/c} = 0. \tag{9.15}$$

Writing $\underline{r}$ as

$$\underline{r} = |\underline{r}|\,e^{i\phi_r}, \tag{9.16}$$

from Eq. (9.15) we find

$$\Omega_m = \omega_m - \tfrac{1}{2}i\Gamma, \tag{9.17}$$

$$\omega_m = m\frac{\pi c}{l} + \delta\omega \quad (m \text{ integer}), \tag{9.18}$$

---

2) The quality of a resonator is usually described in terms of the quality factor $Q=\omega/\Gamma$, where ω is the mid-frequency of the line to which the resonator is tuned and Γ characterizes the width of the line.

$$\delta\omega = \frac{c}{2l}(\pi - \phi_r), \tag{9.19}$$

$$\Gamma = -\frac{c}{l}\ln|\underline{r}|. \tag{9.20}$$

In the case where the transmittance is sufficiently weak ($|\underline{t}|^2 \ll 1$) we may simplify Eq. (9.20) as follows:

$$\Gamma = -\frac{c}{l}\ln(1-|\underline{t}|^2)^{\frac{1}{2}} \simeq \frac{c}{2l}|\underline{t}|^2, \tag{9.21}$$

which allows a simple and intuitive interpretation. Imagine that a photon enters the cavity and leaves it after a time of flight

$$\tau_{\mathrm{fl}} = 2l/c \tag{9.22}$$

with probability $|\underline{t}|^2$. Hence, provided that times smaller than the time of flight are not resolved, $\Gamma \simeq |\underline{t}|^2 c/2l = |\underline{t}|^2/\tau_{\mathrm{fl}}$ is just the transition probability per unit time for the photon passing from inside to outside the cavity. As we shall see later, this interpretation is closely related to the damped-mode concept to be developed. Clearly, when $|\underline{t}|^2 \ll 1$, the "damping rate" Γ is small compared with the separation of frequency $\Delta\omega = \omega_{m+1} - \omega_m = \pi c/l$; that is, $\Gamma \ll \Delta\omega$. Note that the frequencies ω_m, apart from the frequency shift $\delta\omega$, coincide with the well-known eigenfrequencies of the radiation field inside the ideal-resonator cavity ($\underline{t}=0$). From Eqs (9.9) and (9.15) the behavior of $\underline{T}(\omega)$ in the vicinity of a given pole Ω_m can readily be derived. The result is

$$\underline{T}(\omega) \simeq \frac{ic}{2l}\frac{\underline{t}}{\omega - \Omega_m} = \frac{c}{2l}\frac{\underline{t}}{\frac{1}{2}\Gamma - i(\omega - \omega_m)}, \tag{9.23}$$

which reveals that for sufficiently small Γ ($\Gamma \ll \Delta\omega$) the spectral response function of the cavity field becomes effective in discriminating against values of ω not equal to ω_m. In what follows we confine ourselves to the case of small transmission coefficients ($|\underline{t}|^2 \ll 1$, or equivalently $\Gamma \ll \Delta\omega$).

We now specify the assumption that the spectral transmission response function $\underline{t}(\omega)$ is slowly varying with frequency. For this purpose, we assume that the variation of $\underline{t}(\omega)$ may be disregarded as long as the variation of frequency remains small compared with $\Gamma(\omega) = -c\ln|\underline{r}(\omega)|/l \simeq c|\underline{t}(\omega)|^2/2l$. In this case the (analytically continued) function $\underline{T}(\omega)$ may be expected to exhibit a pole structure similar to that discussed above:

$$\Omega_m = \omega_m - \tfrac{1}{2}i\Gamma_m, \tag{9.24}$$

$$\omega_m = m\frac{\pi c}{l} + \delta\omega_m \quad (m \text{ integer}), \tag{9.25}$$

$$\Gamma_m = -\frac{c}{l}\ln|\underline{r}(\Omega_m)| \simeq \frac{c}{2l}|\underline{t}(\Omega_m)|^2 \simeq \frac{c}{2l}|\underline{t}(\omega_m)|^2, \tag{9.26}$$

$$\delta\omega_m = \frac{c}{2l}[\pi - \phi_r(\Omega_m)], \tag{9.27}$$

$$\underline{T}(\omega) = \frac{ic}{2l}\frac{\underline{t}(\Omega_m)}{\omega - \Omega_m} \quad (\omega \approx \Omega_m). \tag{9.28}$$

Defining the Fourier transform of $\underline{T}(\omega)$ as

$$T(\tau) = \int \frac{d\omega}{2\pi}\underline{T}(\omega)\, e^{-i\omega\tau} \tag{9.29}$$

and using the pole structure of $\underline{T}(\omega)$, we may represent $T(\tau)$ in the form

$$T(\tau) = \Theta(\tau)\sum_m T_m\, e^{-i\Omega_m\tau}, \tag{9.30}$$

where

$$T_m = \frac{c}{2l}\underline{t}(\Omega_m) \simeq \frac{c}{2l}\underline{t}(\omega_m) \tag{9.31}$$

[$\Theta(\tau)$, unit step function]. Note that the relation

$$T(-\tau) = \int \frac{d\omega}{2\pi}\underline{T}^*(\omega)\, e^{-i\omega\tau} = \Theta(-\tau)\sum_m T_m^*\, e^{-i\Omega_m^*\tau} \tag{9.32}$$

holds, because of Eq. (9.13). Accordingly, we define the Fourier transform of the function

$$\underline{G}(\omega) = |\underline{T}(\omega)|^2 \tag{9.33}$$

as

$$G(\tau) = \int \frac{d\omega}{2\pi}\underline{G}(\omega)\, e^{-i\omega\tau} \tag{9.34}$$

leading to

$$G(\tau) = \frac{c}{2l}\sum_m \exp\left(-i\omega_m\tau - \tfrac{1}{2}\Gamma_m|\tau|\right), \tag{9.35}$$

where the approximation $\Gamma_m = c|\underline{t}(\omega_m)|^2/2l$ has been used [cf. Eq. (9.26)].

9.2
Source-quantity representation

Let us consider the operator of the canonical momentum of the radiation field, $\hat{\Pi}(x,t)$, which (in the Heisenberg picture) may be represented, by means of mode expansion according to Eq. (2.70), in the form

$$\hat{\Pi}(x,t) = \hat{\Pi}^{(+)}(x,t) + \hat{\Pi}^{(-)}(x,t), \tag{9.36}$$

$$\hat{\Pi}^{(+)}(x,t) = -\varepsilon_0 \int_0^\infty dk\, i\omega A(k,x)\hat{a}(k,t), \tag{9.37}$$

$$\hat{\Pi}^{(-)}(x,t) = [\hat{\Pi}^{(+)}(x,t)]^\dagger, \tag{9.38}$$

with the mode functions given by Eq. (9.8). To study the field outside the cavity (external field), it will be useful to decompose it into incoming and outgoing fields ($x>l$):

$$\hat{\Pi}^{(+)}(x,t) = \hat{\Pi}^{(+)}_{\text{in}}(x,t) + \hat{\Pi}^{(+)}_{\text{out}}(x,t), \tag{9.39}$$

where

$$\hat{\Pi}^{(+)}_{\text{in(out)}}(x,t) = -\varepsilon_0 \int_0^\infty dk\, i\omega A_{\text{in(out)}}(k,x)\hat{a}(k,t), \tag{9.40}$$

$$A_{\text{in}}(k,x) = -\left(\frac{\hbar}{4\pi\mathcal{A}\varepsilon_0\omega}\right)^{\frac{1}{2}} e^{-ikx}, \tag{9.41}$$

$$A_{\text{out}}(k,x) = \left(\frac{\hbar}{4\pi\mathcal{A}\varepsilon_0\omega}\right)^{\frac{1}{2}} \frac{\underline{T}(\omega)}{\underline{T}^*(\omega)} e^{ikx}. \tag{9.42}$$

Note that

$$A(k,x) = A_{\text{in}}(k,x) + A_{\text{out}}(k,x). \tag{9.43}$$

The information about the field incident on the cavity (incoming field) is contained in $\hat{\Pi}^{(\pm)}_{\text{in}}(x,t)$, whereas $\hat{\Pi}^{(\pm)}_{\text{out}}(x,t)$ describes the outgoing field available for further applications.

Recalling the pole structure of $\underline{T}(\omega)$, it is convenient to subdivide the k axis into intervals and introduce multi-mode fields with the aim of expressing the operator of the canonical momentum field as given in Eq. (9.37) in terms of (slowly varying) amplitude operators. The field inside the cavity (internal field) is written as ($0<x<L$)

$$\hat{\Pi}^{(+)}(x,t) = \sum_n \hat{\Pi}^{(+)}_n(x,t) e^{-i\omega_n t}, \tag{9.44}$$

$$\hat{\Pi}^{(+)}_n(x,t) = -\varepsilon_0 \int_{[n]} dk\, i\omega A(k,x) e^{i\omega_n t} \hat{a}(k,t). \tag{9.45}$$

In Eq. (9.45) the $[n]$ notation means integration over the nth interval k_n, $k_n + \Delta k_n$, where

$$k_n = [\omega_n - \tfrac{1}{2}(\omega_n - \omega_{n-1})]/c, \tag{9.46}$$

$$\Delta k_n = (\omega_{n+1} - \omega_{n-1})/(2c), \tag{9.47}$$

where the ω_n ($\omega_n > 0$) are the real parts of the zeros of the denominator of $\underline{T}(\omega)$ [cf. Eq. (9.25)]. When the radiation field outside the cavity is decomposed into an incoming and an outgoing part as in Eqs (9.39)–(9.42), the (slowly varying) amplitude operators can be defined as $(x>l)$

$$\hat{\Pi}^{(+)}_{n\,\text{in}}(x,t) = -\varepsilon_0 \int_{[n]} dk\, i\omega A_{\text{in}}(k,x)\, e^{i\omega_n(t+x/c)} \hat{a}(k,t), \tag{9.48}$$

$$\hat{\Pi}^{(+)}_{n\,\text{out}}(x,t) = -\varepsilon_0 \int_{[n]} dk\, i\omega A_{\text{out}}(k,x)\, e^{i\omega_n(t-x/c)} \hat{a}(k,t), \tag{9.49}$$

where $A_{\text{in}}(k,x)$ and $A_{\text{out}}(k,x)$ are given by Eqs (9.41) and (9.42), respectively. From Eqs (9.39) and (9.40), together with Eqs (9.48) and (9.49), we easily see that Eq. (9.39) may be rewritten in the form of

$$\hat{\Pi}^{(+)}(x,t) = \sum_n \left[\hat{\Pi}^{(+)}_{n\,\text{in}}(x,t)\, e^{-i\omega_n(t+x/c)} + \hat{\Pi}^{(+)}_{n\,\text{out}}(x,t)\, e^{-i\omega_n(t-x/c)} \right]. \tag{9.50}$$

Introducing the source-quantity representation according to Eqs (2.272)–(2.274), we may represent $\hat{\Pi}^{(+)}_n(x,t)$ as $(0<x<L)$

$$\hat{\Pi}^{(+)}_n(x,t) = \hat{\Pi}^{(+)}_{n\,\text{free}}(x,t) + \hat{\Pi}^{(+)}_{ns}(x,t), \tag{9.51}$$

where

$$-\frac{1}{\varepsilon_0} \hat{\Pi}^{(+)}_{n\,\text{free}}(x,t) = \int_{[n]} dk\, i\omega A(k,x)\, e^{i\omega_n t} \hat{a}_{\text{free}}(k,t), \tag{9.52}$$

$$-\frac{1}{\varepsilon_0} \hat{\Pi}^{(+)}_{ns}(x,t) = A \int dt' \int dx'\, \Theta(t-t') K^{(+)}_n(x,t;x',t')\, e^{i\omega_n t'} \hat{P}(x',t'), \tag{9.53}$$

and [cf. Eq. (2.277)]

$$K^{(+)}_n(x,t;x',t') = \frac{i}{\hbar} \int_{[n]} dk\, \omega^2 A(k,x) A^*(k,x')\, e^{-i(\omega-\omega_n)(t-t')}. \tag{9.54}$$

Accordingly, the source-quantity representation of $\hat{\Pi}^{(+)}_{n\,\text{in(out)}}(x,t)$ outside the cavity [Eqs (9.48) and (9.49)] is then $(x>l)$[3]

$$\hat{\Pi}^{(+)}_{n\,\text{in(out)}}(x,t) = \hat{\Pi}^{(+)}_{n\,\text{in(out) free}}(x,t) + \hat{\Pi}^{(+)}_{n\,\text{in(out) s}}(x,t), \tag{9.55}$$

where

$$-\frac{1}{\varepsilon_0} \hat{\Pi}^{(+)}_{n\,\text{in free}}(x,t) = \int_{[n]} dk\, i\omega A_{\text{in}}(k,x)\, e^{i\omega_n(t+x/c)} \hat{a}_{\text{free}}(k,t), \tag{9.56}$$

---

3) Note that outside the cavity the canonical momentum field can be regarded as being the electric field, when the (relevant) sources are inside the cavity.

$$-\frac{1}{\varepsilon_0}\hat{\Pi}^{(+)}_{n\,\text{out free}}(x,t) = \int_{[n]} dk\, i\omega\, A_{\text{out}}(k,x) e^{i\omega_n(t-x/c)} \hat{a}_{\text{free}}(k,t), \tag{9.57}$$

$$-\frac{1}{\varepsilon_0}\hat{\Pi}^{(+)}_{n\,\text{in(out)}\,s}(x,t)$$
$$= A \int dt' \int dx'\, \Theta(t-t') K^{(+)}_{n\,\text{in(out)}}(x,t;x',t') e^{i\omega_n t'} \hat{P}(x',t'), \tag{9.58}$$

$$K^{(+)}_{n\,\text{in}}(x,t;x',t')$$
$$= \frac{i}{\hbar}\int_{[n]} dk\, \omega^2 A_{in}(k,x) A(k,x')^* \exp[-i(\omega - \omega_n)(t-t') + i\omega_n x/c], \tag{9.59}$$

$$K^{(+)}_{n\,\text{out}}(x,t;x',t')$$
$$= \frac{i}{\hbar}\int_{[n]} dk\, \omega^2 A_{out}(k,x) A(k,x')^* \exp[-i(\omega - \omega_n)(t-t') - i\omega_n x/c]. \tag{9.60}$$

9.3
Internal field

Let us introduce some simplifications to the general expressions for the internal field. The main approximation consists in a coarse-grained averaging which implies that the time of flight of a photon through the cavity is assumed to be small compared with the time resolution of interest. On the basis of this concept, nonmonochromatic cavity modes can be introduced and the associated photon operators are found to obey quantum Langevin equations.

9.3.1
Coarse-grained averaging

We start from the free-field operator $\hat{\Pi}^{(+)}_{n\,\text{free}}(x,t)$ in the form (9.52). Using Eqs (9.8) and (9.29), and recalling the relation

$$\hat{a}_{\text{free}}(k,t) = e^{-i\omega(t-\tau)}\hat{a}_{\text{free}}(k,\tau), \tag{9.61}$$

we may rewrite Eq. (9.52) to obtain ($x < L \simeq l$)

$$-\frac{1}{\varepsilon_0}\hat{\Pi}^{(+)}_{n\,\text{free}}(x,t) = \int dt'\, T(t+x/c-t') e^{i\omega_n(t-t')} \hat{\mathcal{E}}^{(+)}_{n\,\text{in free}}(t')$$
$$- \int dt'\, T(t-x/c-t') e^{i\omega_n(t-t')} \hat{\mathcal{E}}^{(+)}_{n\,\text{in free}}(t'), \tag{9.62}$$

where

$$\hat{\mathcal{E}}^{(+)}_{n\text{ in free}}(t) = -i\left(\frac{\hbar}{4\pi A \varepsilon_0}\right)^{\frac{1}{2}} \int_{[n]} dk\, \sqrt{\omega}\, e^{i\omega_n t} \hat{a}_{\text{free}}(k, t) \tag{9.63}$$

is just the incoming free electric field (of center frequency ω_n), as can easily be seen by comparing Eq. (9.56) [together with Eq. (9.8) $(x > L + d = l)$] with Eq. (9.63):

$$-\varepsilon_0^{-1} \hat{\Pi}^{(+)}_{n\text{ in free}}(x, t) = \hat{\mathcal{E}}^{(+)}_{n\text{ in free}}(t + x/c) \quad (x > l). \tag{9.64}$$

Clearly, the incoming free field satisfies the homogeneous Maxwell equations and depends, as a function of t and x, only on the combination $t + x/c$.

In Eq. (9.62), which is exact, the first term on the right-hand side describes the transmitted part of the incoming field. The reflection at the totally reflecting (left-hand) mirror gives rise to the second term. Obviously there is a delay between these two parts of the radiation, the delay time being of the order of magnitude of the time of flight of a photon through the cavity. A global description of the dynamics of the radiation field inside the cavity (instead of the local description inherent in the Maxwell equations) can only be approximately valid, and of course implies some loss of information on the details of the radiation–matter dynamics inside the cavity.

We now specify the approximation scheme to be used below. The main assumptions may be summarized as follows.

1. We confine ourselves to resolving times $\Delta\tau$ large compared with the time of flight τ_{fl} of a photon through the cavity:

$$\tau_{\text{fl}} \ll \Delta\tau, \tag{9.65}$$

which means that, in the sense of some kind of coarse-grained averaging, time-of-flight effects due to the nonvanishing length of the cavity may be disregarded.

2. Further, we assume that (in the relevant spectral range) the transmission of the fractionally transparent (right-hand) mirror is sufficiently small:

$$|t(\omega)|^2 \ll 1. \tag{9.66}$$

From Eqs (9.22) and (9.26) we know that the inequality (9.66) is equivalent to

$$\tau_{\text{fl}} \ll \Gamma_m^{-1}. \tag{9.67}$$

Hence, since the resolving time $\Delta\tau$ may of course be small compared with Γ_m^{-1} (we assume that this condition is satisfied in order to resolve processes within a time scale Γ_m^{-1} with sufficient accuracy), the coarse-graining condition (9.65) and the condition of small transmission (9.66) [in the form of (9.67)] can be combined to give

$$\tau_{fl} \ll \Delta\tau \ll \Gamma_m^{-1} \tag{9.68}$$

or, taking into account that $\tau_{fl} \sim (\Delta\omega)^{-1} = (\omega_{m+1} - \omega_m)^{-1}$ [cf. Eqs (9.22) and (9.25)],

$$(\Delta\omega)^{-1} \ll \Delta\tau \ll \Gamma_m^{-1}. \tag{9.69}$$

These conditions enable us to ignore terms of order of magnitude $\tau_{fl}\Gamma_m$. Thus, using Eqs (9.30) and (9.31), we may rewrite Eq. (9.62) as ($0 < x < l$)

$$\begin{aligned}
-\frac{1}{\varepsilon_0}\hat{\Pi}^{(+)}_{n\text{ free}}(x,t) &= \int dt'\, T(t + x/c - t') e^{i\omega_n(t-t')} \hat{\mathcal{E}}^{(+)}_{n\text{ in free}}(t') \\
&\quad - \int dt'\, T(t - x/c - t') e^{i\omega_n(t-t')} \hat{\mathcal{E}}^{(+)}_{n\text{ in free}}(t') \\
&\simeq T_n \int dt'\, \Theta(t - t' + x/c) e^{-i\Omega_n(t-t'+x/c)} e^{i\omega_n(t-t')} \hat{\mathcal{E}}^{(+)}_{n\text{ in free}}(t') \\
&\quad - T_n \int dt'\, \Theta(t - t' - x/c) e^{-i\Omega_n(t-t'-x/c)} e^{i\omega_n(t-t')} \hat{\mathcal{E}}^{(+)}_{n\text{ in free}}(t') \\
&\simeq T_n \int dt'\, \Theta(t - t')\, e^{-i\omega_n x/c} e^{-\Gamma_n(t-t')/2} \hat{\mathcal{E}}^{(+)}_{n\text{ in free}}(t') \\
&\quad - T_n \int dt'\, \Theta(t - t') e^{(i\omega_n x/c)} e^{-\Gamma_n(t-t')/2} \hat{\mathcal{E}}^{(+)}_{n\text{ in free}}(t') \\
&= -i\frac{c}{l} t(\omega_n) \sin\left(\omega_n \frac{x}{c}\right) \int dt'\, \Theta(t - t')\, e^{-\Gamma_n(t-t')/2} \hat{\mathcal{E}}^{(+)}_{n\text{ in free}}(t').
\end{aligned} \tag{9.70}$$

Let us now consider the source-field part, defined by Eq. (9.53) together with Eq. (9.54). Combining Eqs (9.54) and (9.8), (9.33) and (9.34) yields ($0 < \{x, x'\} < L \simeq l$)

$$\begin{aligned}
K_n^{(+)}(x,t;x',t') &= \frac{i\omega_n}{2\mathcal{A}c\varepsilon_0} \int d\tau\, \Delta_n(\tau) e^{i\omega_n(\tau+t-t')} \{G\left[\tau + t - t' - (x - x')/c\right] \\
&\quad + G\left[\tau + t - t' + (x - x')/c\right] - G\left[\tau + t - t' - (x + x')/c\right] \\
&\quad - G\left[\tau + t - t' + (x + x')/c\right]\},
\end{aligned} \tag{9.71}$$

where

$$\Delta_n(\tau) = \int_{[n]} \frac{d\omega}{2\pi} \frac{\omega}{\omega_n} e^{i(\omega-\omega_n)\tau}, \tag{9.72}$$

which in our approximation scheme may be regarded as behaving like a δ function:

$$\Delta_n(\tau) \simeq \delta(\tau). \tag{9.73}$$

Substituting the results (9.71) and (9.73) into Eq. (9.53) for $K_n^{(+)}(x,t;x',t')$ and using Eq. (9.35), we derive $(0<x<l)$

$$-\frac{1}{\varepsilon_0}\hat{\Pi}_{ns}^{(+)}(x,t) = A\int dt'\int dx'\,\Theta(t-t')K_n^{(+)}(x,t;x',t')e^{i\omega_n t'}\hat{P}(x',t')$$

$$\simeq \frac{i\omega_n}{2c\varepsilon_0}\int dt'\int dx'\,\Theta(t-t')\{G[t-t'-(x-x')]+G[t-t'+(x-x')/c]$$
$$-G[t-t'-(x+x')/c]-G[t-t'+(x+x')/c]bigr\}e^{i\omega_n(t-t')}\hat{P}_n^{(+)}(x',t')$$

$$\simeq \frac{i\omega_n}{4l\varepsilon_0}\int dt'\int dx'\,\Theta(t-t')\bigl[e^{i\omega_n(x-x')/c}+e^{-i\omega_n(x-x')/c}$$
$$-e^{i\omega_n(x+x')/c}-e^{-i\omega_n(x+x')/c}\bigr]e^{-\Gamma_n|t-t'|/2}\hat{P}_n^{(+)}(x',t')$$

$$= \frac{i\omega_n}{l\varepsilon_0}\sin\left(\omega_n\frac{x}{c}\right)\int dt'\int dx'\,\Theta(t-t')\sin\left(\omega_n\frac{x'}{c}\right)e^{-\Gamma_n|t-t'|/2}\hat{P}_n^{(+)}(x',t'). \tag{9.74}$$

In this equation the notation

$$\hat{P}_n^{(+)}(x,t) = e^{i\omega_n t}\hat{P}(x,t) \tag{9.75}$$

is used, which indicates that $\hat{P}_n^{(+)}(x,t)$ effectively represents the slowly varying amplitude of the (positive-frequency part of the) dipole density.

9.3.2
Nonmonochromatic modes and Langevin equations

Introducing the standing-wave mode functions

$$A_n(x) = \left(\frac{\hbar}{lA\varepsilon_0\omega_n}\right)^{\frac{1}{2}}\sin(\omega_n x/c) \qquad (0<x<l) \tag{9.76}$$

and combining Eqs (9.51), (9.70) and (9.74) yields

$$-\varepsilon_0^{-1}\hat{\Pi}_n^{(+)}(x,t)e^{-i\omega_n t} = i\omega_n A_n(x)\hat{a}_n(t), \tag{9.77}$$

where

$$\hat{a}_n(t) = \frac{1}{\hbar}A\int dt'\int dx'\,\Theta(t-t')e^{-i\omega_n(t-t')}e^{-\Gamma_n(t-t')/2}\omega_n A_n(x')\hat{P}(x',t')$$
$$+ \left(\frac{c}{2l}\right)^{\frac{1}{2}}\underline{t}(\omega_n)\int dt'\,\Theta(t-t')e^{-i\omega_n(t-t')}e^{-\Gamma_n(t-t')/2}\hat{b}_n(t'), \tag{9.78}$$

$$\hat{b}_n(t) = -\left(\frac{2A\varepsilon_0 c}{\hbar\omega_n}\right)^{\frac{1}{2}}\hat{\mathcal{E}}_{n\text{ in free}}^{(+)}(t)e^{-i\omega_n t}. \tag{9.79}$$

From Eq. (9.78) the operators $\hat{a}_n$ are easily seen to obey the differential equations

$$\dot{\hat{a}}_n = -\left(i\omega_n + \tfrac{1}{2}\Gamma_n\right)\hat{a}_n + \frac{1}{\hbar}A\int dx'\, \omega_n A_n(x')\hat{P}(x') + \left(\frac{c}{2l}\right)^{\frac{1}{2}} \underline{t}(\omega_n)\hat{b}_n(t). \tag{9.80}$$

These equations of motion for the multi-mode photon annihilation and creation operators $\hat{a}_n$ and $\hat{a}_n^\dagger$, respectively, are of the type of quantum Langevin equations. The operators $\hat{f}_n(t) = \sqrt{c/2l}\,\underline{t}(\omega_n)\hat{b}_n(t)$ are the (random-operator) Langevin noise sources and the terms $-(\Gamma_n/2)\hat{a}_n$ give the drift motions (recall that $\sqrt{c/2l}|\underline{t}(\omega_n)| = \sqrt{\Gamma_n}$). The interpretation of $\hat{f}_n(t)$ as proper noise generators of course requires that $\langle \hat{f}_n(t)\rangle = 0$. Clearly, Eq. (9.80) is also valid when $\langle \hat{f}_n(t)\rangle \neq 0$. Effectively, the approximate representation of the operators $\hat{\Pi}_n^{(+)}(x,t)$ in Eq. (9.77) together with Eqs (9.76) and (9.78) (or (9.80)) corresponds to a description of the intra-cavity field in terms of damped (standing-wave) modes, which (from the point of view of the multi-mode description used) may be regarded as nonmonochromatic cavity modes.

From Eq. (9.79) we know that the (random-operator) Langevin noise sources are determined by the free-field part of the radiation incident on the cavity. Intuitively, it is plausible (see also Section 9.4.1) that the sources inside the cavity do not contribute to the field incident from outside onto the cavity, so that the incoming field is represented by the free-field part only. In Eq. (9.79), $\hat{\mathcal{E}}_{n\,\text{in free}}^{(+)}(t)$ may therefore be replaced by the corresponding operator of the full incoming field. It should be emphasized that this substitution remains valid even when the incident (far-)field is attributed to sources.

As we will see in Section 9.5.1, the operators $\hat{a}_n$ and $\hat{a}_n^\dagger$ satisfy bosonic commutation relations. Thus, the interaction term in Eq. (9.80) may be replaced by the commutator $[\hat{a}_n, \hat{H}_{\text{int}}]/i\hbar$, where $\hat{H}_{\text{int}}$ is the standard interaction Hamiltonian according to Eq. (2.265) where the operator of the canonical momentum field is expressed in terms of the operators $\hat{a}_n$ and $\hat{a}_n^\dagger$, namely

$$\hat{\Pi}(x) = -\varepsilon_0 \sum_n i\omega_n A_n(x)\,\hat{a}_n + \text{H.c.} \tag{9.81}$$

[cf. Eqs (9.44) and (9.77)]. Since in the equations of motion for the atomic source-quantity operators of the active medium the interaction terms may also be expressed in terms of commutators of the corresponding atomic operators and $\hat{H}_{\text{int}}$, with $\hat{\Pi}$ in the form (9.81), a coupled system of (nonlinear) equations of motion for the photonic operators $\hat{a}_n$ and $\hat{a}_n^\dagger$ and the atomic source-quantity operators is obtained. Clearly, using photonic Langevin equations of the type given in Eq. (9.80) makes little sense if the radiation-matter interaction is strong enough so that the characteristic interaction times become comparable with the time of flight of a photon through the cavity, because in coarse-graining approximation these times are not resolvable.

In close analogy with the approach described above, time-reversed Langevin equations may be derived. Instead of using the retarded source-quantity representation for the intra-cavity field we can also use the advanced solution, which can be obtained from Eqs (9.51)–(9.54) replacing $\Theta(t-t')$ by $-\Theta(t'-t)$ in the time integral in Eq. (9.53). Further manipulations, very similar to those shown above, again yield Eqs (9.76)–(9.77), but now the operators $\hat{a}_n$ obey the advanced quantum Langevin equations

$$\dot{\hat{a}}_n = -\left(i\omega_n - \tfrac{1}{2}\Gamma_n\right)\hat{a}_n + \frac{1}{\hbar}\mathcal{A}\int dx'\, \omega_n A_n(x')\hat{P}(x') + \left(\frac{c}{2l}\right)^{\frac{1}{2}} \underline{t}^*(\omega_n)\hat{b}_n^{[a]}(t),$$
(9.82)

where

$$\hat{b}_n^{[a]}(t) = -\left(\frac{2\mathcal{A}\varepsilon_0 c}{\hbar \omega_n}\right)^{\frac{1}{2}} \hat{\mathcal{E}}_{n\,\text{out free}}^{(+)}(t) e^{-i\omega_n t},$$
(9.83)

$$\hat{\mathcal{E}}_{n\,\text{out free}}^{(+)}(t) = i\left(\frac{\hbar}{4\pi\mathcal{A}\varepsilon_0}\right)^{\frac{1}{2}} \int_{[n]} dk\, \omega^{1/2} e^{i\omega_n t}\, \frac{T(\omega)}{T^*(\omega)}\, \hat{a}_{\text{free}}(k, t).$$
(9.84)

Outside the cavity the field may again be written in the form (9.50). In particular, it may easily be proved that

$$-\varepsilon_0^{-1} \hat{\Pi}_{n\,\text{out free}}^{(+)}(x, t) = \hat{\mathcal{E}}_{n\,\text{out free}}^{(+)}(t - x/c) \quad (x > l).$$
(9.85)

Whereas in the usual (retarded) Langevin equations (9.80) the noise operators are determined by the incoming radiation field, in the time-reversed Langevin equations (9.82) the noise operators obviously result from the outgoing radiation field.

In order to handle the quantum Langevin equations (9.80) [or the advanced equations (9.82)], it remains to derive the commutation relations for the operators that appear. In particular, it must be shown that the (nonmonochromatic mode) operators $\hat{a}_n$ and $\hat{a}_n^\dagger$ satisfy, in the approximations made, the bosonic commutation relations for equal times, so that they really represent photon annihilation and creation operators. Before dealing with the commutation relations in detail (Section 9.5), let us first study the field outside the cavity with the aim of relating it to the intra-cavity field.

9.4 External field

To treat the field outside the cavity, we apply the approximation scheme used in Section 9.3. After deriving appropriate free-field and source-field operators, we formulate the input-output relations required to relate the outgoing field to be observed to both the internal and the incoming field.

9.4.1
Source-quantity representation

The free-field part of the incoming radiation is given by Eq. (9.64) together with Eq. (9.63). Using Eq. (9.14) and recalling Eq. (9.61), we can easily see that the free-field part of the outgoing radiation as defined by Eq. (9.64) [together with Eq. (9.42)] may be rewritten as $(x > l)$

$$-\frac{1}{\varepsilon_0} \hat{\Pi}^{(+)}_{n\,\text{out free}}(x,t) = i \left(\frac{\hbar}{4\pi A \varepsilon_0}\right)^{\frac{1}{2}}$$
$$\times \int_{[n]} dk\, \omega^{1/2} \left\{ \underline{t}(\omega)\underline{T}(\omega) \exp\left[i\omega_n\left(t - \frac{x}{c}\right)\right] \hat{a}_{\text{free}}\left(k, t - \frac{x}{c}\right) \right.$$
$$\left. - \underline{r}(\omega)e^{-2il\omega_n/c} \exp\left[i\omega_n\left(t - \frac{x-2l}{c}\right)\right] \hat{a}_{\text{free}}\left(k, t - \frac{x-2l}{c}\right) \right\}.$$
(9.86)

Since $\underline{t}(\omega)$ and $\underline{r}(\omega)$ have been assumed to be slowly varying with ω, in the ω integrals in Eq. (9.86), $\underline{t}(\omega)$ and $\underline{r}(\omega)$ may be taken at ω_n [$\underline{t}(\omega), \underline{r}(\omega) \mapsto \underline{t}(\omega_n), \underline{r}(\omega_n)$]. Further, expressing $\underline{T}(\omega)$ in the first term in Eq. (9.86) by its Fourier transform $T(\tau)$ [see Eq. (9.30)], we may represent $\hat{\Pi}^{(+)}_{n\,\text{out free}}(x,t)$ in the form

$$-\frac{1}{\varepsilon_0} \hat{\Pi}^{(+)}_{n\,\text{out free}}(x,t) = \underline{r}(\omega_n) e^{-2il\omega_n/c} \hat{\mathcal{E}}^{(+)}_{n\,\text{in free}}[t - (x-2l)/c]$$
$$- \underline{t}(\omega_n) \int dt'\, T(t - x/c - t') \exp[i\omega_n(t - x/c - t')] \hat{\mathcal{E}}^{(+)}_{n\,\text{in free}}(t'), \quad (9.87)$$

where $\hat{\mathcal{E}}^{(+)}_{n\,\text{in free}}(t)$ is given by Eq. (9.63) [cf. also Eq. (9.64)]. Applying Eqs (9.30) and (9.31) together with the approximation scheme outlined in Section 9.3, we may simplify Eq. (9.87) to obtain

$$-\frac{1}{\varepsilon_0} \hat{\Pi}^{(+)}_{n\,\text{out free}}(x,t) \simeq \frac{\underline{t}(\omega_n)}{\underline{t}^*(\omega_n)} \hat{\mathcal{E}}^{(+)}_{n\,\text{in free}}(t - x/c)$$
$$- \frac{c}{2l} \underline{t}^2(\omega_n) \int dt'\, \Theta(t - x/c - t') e^{-\Gamma_n(t - x/c - t')/2} \hat{\mathcal{E}}^{(+)}_{n\,\text{in free}}(t'). \quad (9.88)$$

Note that in Eq. (9.88) the relation

$$\underline{r}(\omega_n) e^{-2il\omega_n/c} \simeq \frac{\underline{t}(\omega_n)}{\underline{t}^*(\omega_n)} \tag{9.89}$$

has been used [cf. Eqs (6.134) and (9.15)]. As expected, the incoming (free) field is partly reflected at the input port of the resonator [first term in Eq. (9.88)] and

9.4 External field

partly transmitted to build up an intra-cavity (free) field, a fraction of which is of course transmitted to the outside of the cavity [second term in Eq. (9.88)].

To calculate the source-field part of the incoming radiation [Eq. (9.58) together with Eq. (9.59)], we note that, by means of Eqs (9.8), (9.41), (9.32) and (9.72), the propagation function $K^{(+)}_{n\,\text{in}}(x,t;x',t')$, Eq. (9.59), may be represented as ($x' < l < x$)

$$
\begin{aligned}
K^{(+)}_{n\,\text{in}}(x,t;x',t') &= -\frac{i\omega_n}{2Ac\varepsilon_0}\int d\tau\, \Delta_n(\tau)\, \exp[i\omega_n(\tau + t + x/c - t')] \\
&\quad \times \{T[-(\tau + t - t' + (x + x')/c)] - T[-(\tau + t - t' + (x - x')/c)]\} \\
&\simeq -\frac{i\omega_n}{2Ac\varepsilon_0}\exp[i\omega_n(t + x/c - t')] \\
&\quad \times \{T[-(t - t' + (x + x')/c)] - T[-(t - t' + (x - x')/c)]\}.
\end{aligned} \quad (9.90)
$$

Recalling that $T(t) \sim \Theta(t)$ [see Eq. (9.30)], we find from Eq. (9.90) that

$$
\Theta(t - t')\, K^{(+)}_{n\,\text{in}}(x,t;x',t') = 0. \quad (9.91)
$$

Hence combining Eqs (9.53) and (9.91), we arrive at the result that, as expected, $\hat{\Pi}^{(+)}_{n\,\text{in}\,s}(x,t)$ vanishes:

$$
\hat{\Pi}^{(+)}_{n\,\text{in}\,s}(x,t) = 0 \quad (x > l). \quad (9.92)
$$

From Eqs (9.64) and (9.92) we then find that $\hat{\Pi}^{(+)}_{n\,\text{in}}(x,t)$ [see Eq. (9.55)] may be represented in the form

$$
-\varepsilon_0^{-1}\hat{\Pi}^{(+)}_{n\,\text{in}}(x,t) = \hat{\mathcal{E}}^{(+)}_{n\,\text{in free}}(t + x/c) \quad (x > l). \quad (9.93)
$$

Finally, let us consider the source-field part of the outgoing radiation [Eq. (9.58) together with Eq. (9.60)]. We combine Eqs (9.60), (9.8) ($0 < x' < L \simeq l$), (9.42), (9.29) and (9.72) to represent $K^{(+)}_{n\,\text{out}}(x,t;x',t')$ in the form ($x' < l < x$)

$$
\begin{aligned}
K^{(+)}_{n\,\text{out}}(x,t;x',t') &= \frac{i\omega_n}{2Ac\varepsilon_0}\int d\tau\, \Delta_n(\tau)\, \exp[i\omega_n(\tau + t - x/c - t')] \\
&\quad \times \{T[\tau + t - t' - (x - x')/c] - T[\tau + t - t' - (x + x')/c]\} \\
&\simeq \frac{i\omega_n}{2Ac\varepsilon_0}\exp[i\omega_n(t - x/c - t')] \\
&\quad \times \{T[t - t' - (x - x')/c] - T[t - t' - (x + x')/c]\}.
\end{aligned} \quad (9.94)
$$

Substituting into Eq. (9.58) for $K^{(+)}_{n\,\text{out}}(x,t;x',t')$ the result (9.94), using Eqs (9.30) and (9.31) and applying the approximation scheme under consideration yields the following representation of $\hat{\Pi}^{(+)}_{n\,\text{out s}}(x,t)$ ($x>l$):

$$-\frac{1}{\varepsilon_0}\hat{\Pi}^{(+)}_{n\,\text{out s}}(x,t) = \frac{i\omega_n}{2c\varepsilon_0}\int dt'\int dx'\,\Theta(t-t')\exp[i\omega_n(t-x/c-t')]$$
$$\times\{T[t-t'-(x-x')/c] - T[t-t'-(x+x')/c]\}\hat{P}^{(+)}_n(x',t')$$
$$\simeq \frac{i\omega_n}{4l\varepsilon_0}t(\omega_n)\int dt'\int dx'\,\Theta(t-x/c-t')$$
$$\times\left(e^{-i\omega_n x'/c} - e^{i\omega_n x'/c}\right)\exp\left[-\tfrac{1}{2}\Gamma_n(t-x/c-t')\right]\hat{P}^{(+)}_n(x',t')$$
$$= \frac{\omega_n}{2l\varepsilon_0}t(\omega_n)\int dt'\int dx'\,\Theta(t-x/c-t')\sin(\omega_n x'/c)$$
$$\times\exp\left[-\tfrac{1}{2}\Gamma_n(t-x/c-t')\right]\hat{P}^{(+)}_n(x',t'). \tag{9.95}$$

9.4.2
Input-output relations

We now substitute into Eq. (9.55) for $\hat{\Pi}^{(+)}_{n\,\text{out free}}(x,t)$ and $\hat{\Pi}^{(+)}_{n\,\text{out s}}(x,t)$ the results of Eqs (9.88) and (9.95), respectively. Using Eq. (9.78) together with Eqs (9.76) and (9.79) enables us to express the sum of the two integrals in terms of the (nonmonochromatic) photon annihilation operator $\hat{a}_n(t-x/c)$. Introducing slowly varying operators, $\tilde{\hat{a}}_n(t) = \hat{a}_n(t)e^{i\omega_n t}$, we obtain the following input-output relations ($x>l$):

$$-\frac{1}{\varepsilon_0}\hat{\Pi}^{(+)}_{n\,\text{out}}(x,t) = \left(\frac{\hbar\omega_n}{4l\mathcal{A}\varepsilon_0}\right)^{\frac{1}{2}}t(\omega_n)\,\tilde{\hat{a}}_n(t-x/c)$$
$$+ \frac{t(\omega_n)}{t^*(\omega_n)}\hat{\mathcal{E}}^{(+)}_{n\,\text{in free}}(t-x/c). \tag{9.96}$$

Since the sources are assumed to be localized inside the cavity, $\hat{\Pi}^{(+)}_{n\,\text{in}}(x,t)$, Eq. (9.93), and $\hat{\Pi}^{(+)}_{n\,\text{out}}(x,t)$, Eq. (9.96), depend as functions of x and t on the combinations $t+x/c$ and $t-x/c$, respectively, because the field outside the cavity satisfies the homogeneous Maxwell equations. Combining Eqs (9.50), (9.93) and (9.96), we finally arrive at the result that, outside the cavity ($x>l$),

$$-\frac{1}{\varepsilon_0}\hat{\Pi}^{(+)}(x,t) = \sum_n\Big\{\hat{\mathcal{E}}^{(+)}_{n\,\text{in free}}(t+x/c)\exp[-i\omega_n(t+x/c)]$$
$$+\left[\left(\frac{\hbar\omega_n}{4l\mathcal{A}\varepsilon_0}\right)^{\frac{1}{2}}t(\omega_n)\hat{a}_n(t-x/c) + \frac{t(\omega_n)}{t^*(\omega_n)}\hat{\mathcal{E}}^{(+)}_{n\,\text{in free}}(t-x/c)\right]\exp[-i\omega_n(t-x/c)]\Big\}, \tag{9.97}$$

where the photon operators $\hat{a}_n$ ($\hat{a}_n^\dagger$) obey the quantum Langevin equations (9.80) [together with Eq. (9.79)].

We recall that starting from the advanced source-quantity representation of the field operators yields, for the intra-cavity-radiation field, the time-reversed quantum Langevin equations (9.82). With regard to the radiation field outside the cavity, we note that using the advanced source-quantity representation, in a similar way to that leading to Eqs (9.93) and (9.96), we now obtain ($x>l$)

$$-\varepsilon_0^{-1}\hat{\Pi}_{n\,\text{out}}^{(+)}(x,t) = \hat{\mathcal{E}}_{n\,\text{out free}}^{(+)}(t-x/c) \tag{9.98}$$

and

$$-\frac{1}{\varepsilon_0}\hat{\Pi}_{n\,\text{in}}^{(+)}(x,t) = -\left(\frac{\hbar\omega_n}{4l\mathcal{A}\varepsilon_0}\right)^{\frac{1}{2}} \underline{t}^*(\omega_n)\hat{a}_n(t+x/c) + \frac{\underline{t}^*(\omega_n)}{\underline{t}(\omega_n)}\hat{\mathcal{E}}_{n\,\text{out free}}^{(+)}(t+x/c). \tag{9.99}$$

It should be pointed out that the input-output relations (9.96) or (9.99) may also be obtained by combining the retarded and advanced Langevin equations. This is of interest with regard to quantum noise theories, where time reversal is a way to define, under certain circumstances, outgoing fields.[4] Subtracting the retarded and advanced Langevin equations (9.80) and (9.82) from each other and taking into account Eqs (9.26), (9.79), (9.93) and (9.83), (9.98) yields Eq. (9.96). Similarly, Eq. (9.99) can be found.

9.5 Commutation relations

The Langevin-equation approach to the determination of the radiation field inside and outside of a resonator-like cavity as developed in Sections 9.3 and 9.4 may also be used in classical optics. In this case all the quantities introduced are, of course, c numbers. Operating with the Langevin equations in quantum optics requires knowledge about a series of commutation relations. This knowledge is needed not only to solve the quantum Langevin equations, but also to calculate correlation functions of the field escaping from the cavity.

We shall first give some general commutation rules, which we then use to derive the commutators for the various kinds of field operators.

The basic commutation relations may be summarized as follows:

$$[\hat{a}(k,t),\hat{a}(k',t)] = 0 = [\hat{a}^\dagger(k,t),\hat{a}^\dagger(k',t)], \tag{9.100}$$

$$[\hat{a}(k,t),\hat{a}^\dagger(k',t)] = \delta(k-k'), \tag{9.101}$$

[4] Recall that, since propagation of radiation is included in a field-theoretical description, the definition of the incoming and outgoing field rests on propagation directions, without the need for a time-reversed formulation of the theory.

$$[\hat{a}_{\text{free}}(k,t), \hat{a}_{\text{free}}(k',t')] = 0 = [\hat{a}^\dagger_{\text{free}}(k,t), \hat{a}^\dagger_{\text{free}}(k',t')], \quad (9.102)$$

$$[\hat{a}_{\text{free}}(k,t), \hat{a}^\dagger_{\text{free}}(k',t')] = \delta(k-k')\, e^{-i\omega(t-t')}. \quad (9.103)$$

Moreover, if $\hat{F}^{(j)}(x,t)$ and $\hat{G}^{(j)}(x,t)$ $(j=+,-)$ are field operators of the form

$$\hat{F}^{(+)}(x,t) = \int dk\, F^{(+)}(k,x)\hat{a}(k,t), \quad \hat{F}^{(-)}(x,t) = [\hat{F}^{(+)}(x,t)]^\dagger \quad (9.104)$$

$[F^{(+)}(k,x) = F(k,x),\, F^{(-)}(k,x) = F^*(k,x)]$, then from Eq. (2.313), we know that the commutator of $\hat{F}^{(j_1)}(x_1,t_1)$ and $\hat{G}^{(j_2)}(x_2,t_2)$ at different times may be expressed in terms of free-field and source-quantity commutators as

$$[\hat{F}^{(j_1)}(x_1,t_1), \hat{G}^{(j_2)}(x_2,t_2)]$$
$$= [\hat{F}^{(j_1)}_{\text{free}}(x_1,t_1), \hat{G}^{(j_2)}_{\text{free}}(x_2,t_2)] + \hat{\Delta}^{(j_1,j_2)}_{(F,G)}(x_1,t_1;x_2,t_2), \quad (9.105)$$

where

$$[\hat{F}^{(j_1)}_{\text{free}}(x_1,t_1), \hat{G}^{(j_2)}_{\text{free}}(x_2,t_2)] = \frac{\hbar}{i} K^{(j_1,j_2)}_{(F,G)}(x_1,t_1;x_2,t_2), \quad (9.106)$$

$$\hat{\Delta}^{(j_1,j_2)}_{(F,G)}(x_1,t_1;x_2,t_2) = \hat{D}^{(j_1,j_2)}_{(F,G)}(x_1,t_1;x_2,t_2) - \hat{D}^{(j_2,j_1)}_{(G,F)}(x_2,t_2;x_1,t_1), \quad (9.107)$$

$$\hat{D}^{(j_1,j_2)}_{(F,G)}(x_1,t_1;x_2,t_2)$$
$$= -\int dt'_1 \int dx'_1 \int dt'_2 \int dx'_2\, K^{(j_1)}_{(F)}(x_1,t_1;x'_1,t'_1) K^{(j_2)}_{(G)}(x_2,t_2;x'_2,t'_2)$$
$$\times \Theta(t_2-t'_2)\Theta(t'_2-t'_1)\Theta(t'_1-t_1)[\hat{P}(x'_1,t'_1), \hat{P}(x'_2,t'_2)], \quad (9.108)$$

$$K^{(j_1,j_2)}_{(F,G)}(x_1,t_1;x_2,t_2)$$
$$= j_1\,(1-\delta_{j_1 j_2})\frac{i}{\hbar}\int dk\, F^{(j_1)}(x_1,k)G^{(j_2)}(x_2,k)\, e^{-j_1 i\omega(t_1-t_2)}. \quad (9.109)$$

9.5.1
Internal field

We first study the commutation relations for the operators $\hat{a}_n$ and $\hat{a}^\dagger_m$ at equal times. Using Eq. (9.77), we may write

$$\left.\begin{array}{l}\varepsilon_0^{-2}[\hat{\Pi}^{(+)}_n(x,t), \hat{\Pi}^{(+)}_m(x',t)]\\ \varepsilon_0^{-2}[\hat{\Pi}^{(+)}_n(x,t), \hat{\Pi}^{(-)}_m(x',t)]\end{array}\right\} = \omega_n\omega_m A_n(x) A_m(x')\begin{cases}-[\hat{a}_n(t), \hat{a}_m(t)],\\ [\hat{a}_n(t), \hat{a}^\dagger_m(t)].\end{cases} \quad (9.110)$$

Since the commutator $[\hat{\Pi}^{(+)}_n(x,t), \hat{\Pi}^{(+)}_m(x',t)]$ vanishes identically [cf. Eqs (9.45) and (9.100)], we can easily see that for any values of n and m the commutator $[\hat{a}_n(t), \hat{a}_m(t)]$ must vanish too. We therefore find that

$$[\hat{a}_n(t), \hat{a}_m(t)] = 0 = [\hat{a}^\dagger_n(t), \hat{a}^\dagger_m(t)]. \quad (9.111)$$

The commutator $[\hat{\Pi}_n^{(+)}(x,t), \hat{\Pi}_m^{(-)}(x',t)]$ in the second line on the left-hand side of Eq. (9.110) is calculated by means of Eqs (9.45) ($0 < \{x, x'\} < L \simeq l$) and (9.101):

$$\frac{1}{\varepsilon_0^2}[\hat{\Pi}_n^{(+)}(x,t), \hat{\Pi}_m^{(-)}(x',t)] = \frac{\hbar}{i\omega_n}\delta_{nm}\left(\omega_n + i\frac{\partial}{\partial t}\right)K_n^{(+)}(x,t;x',t')\bigg|_{t'=t}, \quad (9.112)$$

where $K_n^{(+)}(x,t;x',t')$ is defined in Eq. (9.54). Using $K_n^{(+)}(x,t;x',t')$ in the form[5]

$$K_n^{(+)}(x,t;x',t') \simeq \frac{i\omega_n^2}{\hbar}A_n(x)A_n(x')\exp\left(-\tfrac{1}{2}\Gamma_n|t-t'|\right), \quad (9.113)$$

we may simplify Eq. (9.112) to obtain

$$\varepsilon_0^{-2}[\hat{\Pi}_n^{(+)}(x,t), \hat{\Pi}_m^{(-)}(x',t)] \simeq \omega_n^2 A_n(x)A_n(x')\delta_{nm}. \quad (9.114)$$

Hence, comparing the second line of Eq. (9.110) and Eq. (9.114), we may approximately set

$$[\hat{a}_n(t), \hat{a}_m^\dagger(t)] = \delta_{nm}. \quad (9.115)$$

Equations (9.111) and (9.115) show that the operators $\hat{a}_n$ and $\hat{a}_n^\dagger$ are indeed (approximate) bosonic operators, which may be regarded as photon annihilation and creation operators in the usual sense.

Next, let us determine the commutation relations for the (noise) operators $\hat{b}_n$ and $\hat{b}_n^\dagger$ at different times. Since the operators $\hat{b}_{n_1}(t_1)$ and $\hat{b}_{n_2}^\dagger(t_2)$ are proportional to the incoming free-field operators $\hat{\mathcal{E}}_{n_1 \text{ in free}}^{(+)}(t_1)$ and $\hat{\mathcal{E}}_{n_2 \text{ in free}}^{(-)}(t_2)$, respectively [see Eqs (9.79) and (9.63)], we immediately see from Eq. (9.102) that

$$[\hat{b}_{n_1}(t_1), \hat{b}_{n_2}(t_2)] = 0 = [\hat{b}_{n_1}^\dagger(t_1), \hat{b}_{n_2}^\dagger(t_2)]. \quad (9.116)$$

To calculate $[\hat{b}_{n_1}(t_1), \hat{b}_{n_2}^\dagger(t_2)]$, we apply Eqs (9.79), (9.63), (9.103) and (9.72), which, in the approximation scheme used here, lead to [$\hat{\tilde{b}}_n(t) = \hat{b}_n(t)e^{i\omega_n t}$]

$$[\hat{\tilde{b}}_{n_1}(t_1), \hat{\tilde{b}}_{n_2}^\dagger(t_2)] = \delta_{n_1 n_2}\left(1 + \frac{i}{\omega_{n_1}}\frac{\partial}{\partial t_1}\right)\Delta_{n_1}^*(t_1 - t_2)$$

$$= \delta_{n_1 n_2}\delta(t_1 - t_2). \quad (9.117)$$

Going back to the incoming-field operators $\hat{\Pi}_{n\,\text{in}}^{(\pm)}(x,t)$ ($x > l$), and recalling that they are free-field operators [see Eq. (9.93)], from Eqs (9.116) and (9.117)

---

[5] This result is found by comparing the first and the last versions of Eq. (9.74) and applying Eq. (9.76).

together with Eq. (9.79) we obtain ($x_i > l$)

$$\frac{1}{\varepsilon_0^2}[\hat{\Pi}_{n_1\,\text{in}}^{(+)}(x_1,t_1),\hat{\Pi}_{n_2\,\text{in}}^{(\pm)}(x_2,t_2)] = \begin{cases} 0, \\ \dfrac{\hbar\omega_{n_1}}{2\varepsilon_0 c\mathcal{A}}\delta_{n_1 n_2}\delta[t_1-t_2+(x_1-x_2)/c]. \end{cases} \quad (9.118)$$

In order to obtain the commutation relations for the intra-cavity-field operators $\hat{a}_n$ and $\hat{a}_n^\dagger$ and the (noise) operators $\hat{b}_n$ and $\hat{b}_n^\dagger$ at different times, we note that combining Eqs (9.77), (9.93) and (9.79) yields ($x_1 < l < x_2$)

$$\frac{1}{\varepsilon_0^2}[\hat{\Pi}_{n_1}^{(\pm)}(x_1,t_1),\hat{\Pi}_{n_2\,\text{in}}^{(+)}(x_2,t_2)]$$

$$= \mp i\omega_{n_1} A_{n_1}(x_1)\left(\frac{\hbar\omega_{n_2}}{2\varepsilon_0 c\mathcal{A}}\right)^{\frac{1}{2}} \begin{cases} [\hat{a}_{n_1}(t_1),\hat{b}_{n_2}(t_2+x_2/c)], \\ [\hat{a}_{n_1}^\dagger(t_1),\hat{b}_{n_2}(t_2+x_2/c)]. \end{cases} \quad (9.119)$$

Using Eqs (9.45) and (9.48) and applying Eqs (9.104)–(9.109) [with the allocations $\hat{F}^{(\pm)}(x_1,t_1) \mapsto \hat{\Pi}_{n_1}^{(\pm)}(x_1,t_1)e^{\mp i\omega_{n_1}t_1}$, $\hat{G}^{(\pm)}(x_2,t_2) \mapsto \hat{\Pi}_{n_2\,\text{in}}^{(+)}(x_2,t_2)e^{i\omega_{n_2}(t_2+x_2/c)}$] we obtain the source-quantity representation of the commutator $[\hat{\Pi}_{n_1}^{(\pm)}(x_1,t_1),\hat{\Pi}_{n_2\,\text{in}}^{(+)}(x_2,t_2)]$ as

$$\frac{1}{\varepsilon_0^2}[\hat{\Pi}_{n_1}^{(\pm)}(x_1,t_1),\hat{\Pi}_{n_2\,\text{in}}^{(+)}(x_2,t_2)] = \frac{1}{\varepsilon_0^2}[\hat{\Pi}_{n_1\,\text{free}}^{(\pm)}(x_1,t_1),\hat{\Pi}_{n_2\,\text{in free}}^{(+)}(x_2,t_2)]$$

$$+ \hat{\Delta}_{n_1,n_2\,\text{in}}^{(\pm,+)}(x_1,t_1;x_2,t_2), \quad (9.120)$$

where ($x_1 < l < x_2$)

$$\frac{1}{\varepsilon_0^2}[\hat{\Pi}_{n_1\,\text{free}}^{(\pm)}(x_1,t_1),\hat{\Pi}_{n_2\,\text{in free}}^{(+)}(x_2,t_2)]$$

$$= \begin{cases} 0, \\ -\dfrac{\hbar}{i\omega_{n_1}}\delta_{n_1 n_2}\left(\omega_{n_1}+i\dfrac{\partial}{\partial t_2}\right)K_{n_1\,\text{in}}^{(+)}(x_2,t_2;x_1,t_1), \end{cases} \quad (9.121)$$

$$\hat{\Delta}_{n_1,n_2\,\text{in}}^{(\pm,+)}(x_1,t_1;x_2,t_2)$$
$$= -\int dt_1'\int dx_1'\int dt_2'\int dx_2'\,\exp(\pm i\omega_{n_1}t_1'+i\omega_{n_2}t_2')$$
$$\times K_{n_1}^{(\pm)}(x_1,t_1;x_1',t_1')K_{n_2\,\text{in}}^{(+)}(x_2,t_2;x_2',t_2')$$
$$\times \{\Theta(t_2-t_2')\Theta(t_2'-t_1')\Theta(t_1'-t_1)[\hat{P}(x_1',t_1'),\hat{P}(x_2',t_2')]$$
$$- \Theta(t_1-t_1')\Theta(t_1'-t_2')\Theta(t_2'-t_2)[\hat{P}(x_2',t_2'),\hat{P}(x_1',t_1')]\}, \quad (9.122)$$

the propagation functions $K_n^{(\pm)}(x,t;x',t')$ and $K_{n\,\text{in}}^{(+)}(x,t;x',t')$ being defined according to Eqs (9.54) and (9.59), respectively. Using $K_{n\,\text{in}}^{(+)}(x,t;x',t')$ $(x'<l<x)$, in the (approximate) form given in Eq. (9.90) together with Eq. (9.30), and taking into account that in the approximation scheme used here $K_{n\,\text{in}}^{(+)}(x,t;x',t')$ is proportional to $\Theta[-(t+x/c-t')]$, we readily see from Eq. (9.121) that

$$[\hat{\Pi}_{n_1\,\text{free}}^{(-)}(x_1,t_1), \hat{\Pi}_{n_2\,\text{in free}}^{(+)}(x_2,t_2)] = 0 \quad \text{if} \quad t_1 < t_2 + x_2/c. \tag{9.123}$$

Further, in the integral term of Eq. (9.122) the three Θ functions together with the Θ function $\Theta[-(t_2+x_2/c-t_2')]$ arising from the propagation function $K_{n_2\,\text{in}}^{(+)}(x_2,t_2;x_2',t_2')$ reveal that under the condition $t_1 < t_2 + x_2/c$ the time-delayed commutator terms $\hat{\Delta}_{n_1,n_2\,\text{in}}^{(\pm,+)}(x_1,t_1;x_2,t_2)$ also vanish; that is,

$$\hat{\Delta}_{n_1,n_2\,\text{in}}^{(\pm,+)}(x_1,t_1;x_2,t_2) = 0 \quad \text{if} \quad t_1 < t_2 + x_2/c. \tag{9.124}$$

Hence we arrive at the result that

$$[\hat{\Pi}_{n_1}^{(\pm)}(x_1,t_1), \hat{\Pi}_{n_2\,\text{in}}^{(+)}(x_2,t_2)] = 0 \quad \text{if} \quad t_1 < t_2 + x_2/c, \tag{9.125}$$

and from Eqs (9.119) and (9.125) we conclude that

$$[\hat{a}_{n_1}(t_1), \hat{b}_{n_2}(t_2)] = 0 = [\hat{a}_{n_1}^\dagger(t_1), \hat{b}_{n_2}(t_2)] \quad \text{if} \quad t_1 < t_2. \tag{9.126}$$

These commutation relations are, of course, consistent with the quantum Langevin equations [Eq. (9.80) together with Eq. (9.79)]. In the language of quantum noise theory, the "dynamic-system" operators $\hat{a}_n$ and $\hat{a}_n^\dagger$ are said not to be influenced by the reservoir operators $\hat{b}_n$ and $\hat{b}_n^\dagger$ in the future.

9.5.2
External field

We briefly outline the determination of the time-dependent commutation relations for the outgoing field. We again apply Eqs (9.104)–(9.109) [with the allocations $\hat{F}^{(\pm)}(x_1,t_1) \mapsto \hat{\Pi}_{n_1\,\text{out}}^{(\pm)}(x_1,t_1)e^{\mp i\omega_{n_1}(t_1-x_1/c)}$, $\hat{G}^{(\pm)}(x_2,t_2) \mapsto \hat{\Pi}_{n_2\,\text{out}}^{(+)}(x_2,t_2)e^{-i\omega_{n_2}(t_2-x_2/c)}]$ to express [analogously to Eq. (9.120)] the commutators $[\hat{\Pi}_{n_1\,\text{out}}^{(\pm)}(x_1,t_1), \hat{\Pi}_{n_2\,\text{out}}^{(+)}(x_2,t_2)]$ $(x_i > l)$ in terms of free-field and source-quantity commutators as

$$\frac{1}{\varepsilon_0^2}[\hat{\Pi}_{n_1\,\text{out}}^{(\pm)}(x_1,t_1), \hat{\Pi}_{n_2\,\text{out}}^{(+)}(x_2,t_2)] = \frac{1}{\varepsilon_0^2}[\hat{\Pi}_{n_1\,\text{out free}}^{(\pm)}(x_1,t_1), \hat{\Pi}_{n_2\,\text{out free}}^{(+)}(x_2,t_2)]$$
$$+ \hat{\Delta}_{n_1\,\text{out},n_2\,\text{out}}^{(\pm,+)}(x_1,t_1;x_2,t_2). \tag{9.127}$$

In Eq. (9.127) the time-delayed commutator terms $\hat{\Delta}_{n_1\,\text{out},n_2\,\text{out}}^{(\pm,+)}(x_1,t_1;x_2,t_2)$ may easily be found from Eq. (9.122) with the propagation functions $K_{n_1\,\text{out}}^{(\pm)}(x_1,t_1;x_1',t_1')$ and $K_{n_2\,\text{out}}^{(+)}(x_2,t_2;x_2',t_2')$ instead of $K_{n_1}^{(\pm)}(x_1,t_1;x_1',t_1')$ and $K_{n_2\,\text{in}}^{(+)}(x_2,t_2;x_2',t_2')$, respectively. Note that in the approximation scheme used here $K_{n_i\,\text{out}}^{(\pm)}(x_i,t_i;x_1',t_i') \sim \Theta(t_i - x_i/c - t_i')$, $i=1,2$ [cf. Eq. (9.94) together with Eq. (9.30)]. The two Θ functions $\Theta(t_1 - x_1/c - t_1')$ and $\Theta(t_2 - x_1/c - t_2')$ together with the three Θ functions shown explicitly in Eq. (9.122) give rise to time conditions that obviously cannot be satisfied simultaneously. Hence the time-delayed terms $\hat{\Delta}_{n_1\,\text{out},n_2\,\text{out}}^{(\pm,+)}(x_1,t_1;x_2,t_2)$ do not contribute to the commutators sought, and Eq. (9.127) simplifies to

$$[\hat{\Pi}_{n_1\,\text{out}}^{(\pm)}(x_1,t_1),\hat{\Pi}_{n_2\,\text{out}}^{(+)}(x_2,t_2)] = [\hat{\Pi}_{n_1\,\text{out free}}^{(\pm)}(x_1,t_1),\hat{\Pi}_{n_2\,\text{out free}}^{(+)}(x_2,t_2)]. \tag{9.128}$$

The calculation of the remaining free-field commutators is straightforward. Using Eqs (9.57) [together with Eq. (9.42)], (9.102) and (9.103) and making the approximation (9.117), we obtain, analogously to the result found for the incoming field [cf. Eq. (9.118)], $(x_i > l)$

$$\frac{1}{\varepsilon_0^2}[\hat{\Pi}_{n_1\,\text{out}}^{(+)}(x_1,t_1),\hat{\Pi}_{n_2\,\text{out}}^{(\pm)}(x_2,t_2)] = \begin{cases} 0, \\ \dfrac{\hbar\omega_{n_1}}{2\varepsilon_0 c\mathcal{A}} \delta_{n_1 n_2} \delta[t_1-t_2-(x_1-x_2)/c], \end{cases} \tag{9.129}$$

It is worth noting that although both the incoming and the outgoing radiation fields may be regarded as effectively free fields [they obey the homogeneous Maxwell equations and the free-field commutation relations (9.118) and (9.129)], the overall field outside the cavity (i.e., the sum of incoming and outgoing fields), which also obeys the homogeneous Maxwell equations, cannot be regarded, in general, as an effectively free field, because time-delayed terms can contribute to the commutators between incoming and outgoing fields at different times, $[\hat{\Pi}_{n_1\,\text{out}}^{(\pm)}(x_1,t_1),\hat{\Pi}_{n_2\,\text{in}}^{(+)}(x_2,t_2)]$. Clearly, a light signal can travel from one space point (outside the cavity) to another space (outside the cavity) through the sources if both the incoming and the outgoing fields are involved. From intuitive arguments the commutators $[\hat{\Pi}_{n_1\,\text{out}}^{(\pm)}(x_1,t_1),\hat{\Pi}_{n_2\,\text{in}}^{(+)}(x_2,t_2)]$ may be expected to vanish provided that $t_1 - x_1/c < t_2 + x_2/c$:

$$[\hat{\Pi}_{n_1\,\text{out}}^{(\pm)}(x_1,t_1),\hat{\Pi}_{n_2\,\text{in}}^{(+)}(x_2,t_2)] = 0 \quad \text{if} \quad t_1 - x_1/c < t_2 + x_2/c, \tag{9.130}$$

which may again be proved by applying Eqs (9.104)–(9.109) with the corresponding allocations. Since the calculations may be performed analogously to those made to obtain, for example, the commutation relations (9.125), we omit them here.

It should be pointed out that the results, in the approximations made, are closely related to the corresponding results obtained from quantum noise theory [see, e. g., Gardiner (1991)]. This correspondence includes the quantum Langevin equations given in Section 9.3, the input-output relations given in Section 9.4, and the time-dependent commutation relations (see also Section 9.7). Further manipulations of the quantum Langevin equations in practical applications may therefore be based on standard methods of quantum noise theory, such as the formalism of master equations and the quantum regression theorem (Chapter 5). However, when the conditions for the application of quantum noise theory (essentially consisting in a time coarse-grained assumption) are not satisfied, as in the case of low-Q cavities, the general field-theoretical concept (which does not require a coarse-graining approximation) can of course be applied.

9.6
Field correlation functions

In the following we confine our attention to the field outside the cavity, so that we may regard $-\hat{\Pi}(x)/\varepsilon_0$ as the operator of the electric field strength,

$$-\varepsilon_0^{-1}\hat{\Pi}(x) = \hat{E}(x) \qquad (x > l), \tag{9.131}$$

provided that the (relevant) sources are inside the cavity. The commutation relations for the radiation-field operators at different times now enable us to express normally and time-ordered observable field correlation functions of the type

$$G^{(m,n)}(\{x_i, t_i, x_j, t_j\}) = \left\langle \left[\mathcal{T}_- \prod_{i=1}^{m} \hat{E}^{(-)}(x_i, t_i) \right] \left[\mathcal{T}_+ \prod_{j=m+1}^{m+n} \hat{E}^{(+)}(x_j, t_j) \right] \right\rangle \tag{9.132}$$

($\{x_i, x_j\} > l$) in terms of correlation functions of the outgoing radiation field or in terms of the intra-cavity field and/or the incoming field.

We decompose $\hat{E}^{(+)}(x,t)$ into the incoming and outgoing parts

$$\hat{E}_{\text{in}}^{(+)}(x,t) = \sum_n \hat{E}_{n\,\text{in}}^{(+)}(x,t) \exp[-i\omega_n(t+x/c)] \tag{9.133}$$

and

$$\hat{E}_{\text{out}}^{(+)}(x,t) = \sum_n \hat{E}_{n\,\text{out}}^{(+)}(x,t) \exp[-i\omega_n(t-x/c)], \tag{9.134}$$

respectively [cf. Eqs (9.39) and (9.50)]. Because of the commutation relations (9.130), in the resulting time-ordered products of operators $\hat{E}_{n\,\text{in}}^{(+)}$ and $\hat{E}_{m\,\text{out}}^{(+)}$

($\hat{E}^{(-)}_{n\,\text{in}}$ and $\hat{E}^{(-)}_{m\,\text{out}}$) these operators may obviously be rearranged in such a way that $\hat{E}^{(+)}_{n\,\text{in}}$ ($\hat{E}^{(-)}_{n\,\text{in}}$) are on the right (left) of $\hat{E}^{(+)}_{m\,\text{out}}$ ($\hat{E}^{(-)}_{m\,\text{out}}$). Hence if at the observation points the incoming radiation field may be assumed to be in the vacuum state, that is

$$\langle \hat{E}^{(-)}_{n\,\text{in}}(x,t)\cdots\rangle = 0 = \langle\cdots \hat{E}^{(+)}_{n\,\text{in}}(x,t)\rangle, \tag{9.135}$$

$G^{(m,n)}(\{x_i, t_i, x_j, t_j\})$ may be expressed in terms of correlation functions of the outgoing field, such as

$$G^{(m,n)}_{\{n_i,n_j\}}(\{x_i, t_i, x_j, t_j\}) = \left\langle \left[\prod_{i=1}^{m} \hat{E}^{(-)}_{n_i\,\text{out}}(x_i, t_i)\right]\left[\prod_{j=m+1}^{m+n} \hat{E}^{(+)}_{n_j\,\text{out}}(x_j, t_j)\right]\right\rangle \tag{9.136}$$

($\{x_i, x_j\} > l$), via

$$G^{(m,n)}(\{x_i, t_i, x_j, t_j\}) = \sum_{\{n_i,n_j\}} G^{(m,n)}_{\{n_i,n_j\}}(\{x_i, t_i, x_j, t_j\})$$

$$\times \exp\left[\sum_{i=1}^{m} i\omega_{n_i}(t_i - x_i/c) - \sum_{j=m+1}^{m+n} i\omega_{n_j}(t_j - x_j/c)\right]. \tag{9.137}$$

Note that [in contrast to Eq. (9.132)] the time-ordering symbols $\mathcal{T}_\pm$ may be omitted in Eq. (9.136), because the operators $\hat{E}^{(+)}_{n\,\text{out}}$ (and the operators $\hat{E}^{(-)}_{n\,\text{out}}$) behave like the (commuting) free-field operators [cf. the commutation relations (9.129)].

The conditions (9.135) are of course satisfied if the incoming radiation field is in the vacuum state. In many situations of practical interest, resonator-like devices are used with the aim of light generation and/or amplification depending on various (real) input radiation fields, and the properties of the output radiation are desired to be observed. Clearly, in these cases the observational scheme should be prepared in such a way that the conditions (9.135) are satisfied at the observation points to avoid detection of the input field.

To indicate explicitly that in Eq. (9.136) the operators $\hat{E}^{(\pm)}_{n\,\text{out}}(x,t)$ depend on t and x in the combination $t - x/c$ [cf. Eq. (9.96)], it is convenient to introduce the notation

$$\hat{E}^{(\pm)}_{n\,\text{out}}(x,t) \equiv \hat{\mathcal{E}}^{(\pm)}_{n\,\text{out}}(t - x/c). \tag{9.138}$$

Since $\hat{E}^{(+)}_{n\,\text{out}}$ (and $\hat{E}^{(-)}_{n\,\text{out}}$) are commuting quantities [cf. the commutation relations (9.129)], the right-hand side of Eq. (9.136) remains unchanged if we substitute for the operator products the time-ordered products according to

$$G^{(m,n)}_{\{n_i,n_j\}}(\{x_i,t_i,x_j,t_j\}) = \left\langle \left[\mathcal{T}^{(r)}_- \prod_{i=1}^{m} \hat{\mathcal{E}}^{(-)}_{n_i\,\text{out}}(t_i^{(r)})\right]\left[\mathcal{T}^{(r)}_+ \prod_{j=m+1}^{m+n} \hat{\mathcal{E}}^{(+)}_{n_j\,\text{out}}(t_j^{(r)})\right]\right\rangle$$

(9.139)

($\{x_i, x_j\} > l$), where $\mathcal{T}^{(r)}_\pm$ indicates time ordering with regard to the retarded times $t^{(r)} \equiv t - x/c$.

The advantage of the field correlation function in the form (9.137) together with Eq. (9.139) becomes clear when we try to express the measured quantities in terms of correlation functions of internal and incoming fields. For this reason, we substitute for $\hat{\mathcal{E}}^{(+)}_{n\,\text{out}}(t^{(r)})$ the result (9.96) [together with Eq. (9.79)]:

$$\hat{\mathcal{E}}^{(+)}_{n\,\text{out}}(t^{(r)}) = \left(\frac{\hbar\omega_n}{4l\mathcal{A}\varepsilon_0}\right)^{\frac{1}{2}} \left[t(\omega_n)\hat{a}_n(t^{(r)}) - \frac{t(\omega_n)}{t^*(\omega_n)}\left(\frac{2l}{c}\right)^{1/2} \hat{b}_n(t^{(r)})\right]. \quad (9.140)$$

Taking into account that $[\hat{a}_{n_k}(t_k^{(r)}), \hat{b}_{n_l}(t_l^{(r)})] = 0$ if $t_k^{(r)} < t_l^{(r)}$ [see Eq. (9.126)], in the resulting time-ordered products of operators $\hat{a}_{n_k}$ and $\hat{b}_{n_l}$ ($\hat{a}^\dagger_{n_k}$ and $\hat{b}^\dagger_{n_l}$) these operators may be rearranged with $\hat{b}_{n_l}$ ($\hat{b}^\dagger_{n_l}$) to the right (left) of $\hat{a}_{n_k}$ ($\hat{a}^\dagger_{n_k}$). In this way, we derive

$$G^{(m,n)}_{\{n_i,n_j\}}(\{x_i,t_i,x_j,t_j\}) = \left\langle \left[\mathcal{O}^{(r)}_- \prod_{i=1}^{m} \hat{\mathcal{E}}^{(-)}_{n_i\,\text{out}}(t_i^{(r)})\right]\left[\mathcal{O}^{(r)}_+ \prod_{j=m+1}^{m+n} \hat{\mathcal{E}}^{(+)}_{n_j\,\text{out}}(t_j^{(r)})\right]\right\rangle$$

(9.141)

($\{x_i, x_j\} > l$), where (similar to the $\mathcal{O}_\pm$ symbols defined in Section 2.8) the $\mathcal{O}^{(r)}_\pm$ symbols indicate the following ordering prescriptions:

(i) decomposition of the operators $\hat{\mathcal{E}}^{(\pm)}_{n\,\text{out}}$ according to Eq. (9.140);

(ii) ordering of the (quantum noise) operators $\hat{b}_n$ ($\hat{b}^\dagger_n$) and the (intra-cavity-field photon) operators $\hat{a}_n$ ($\hat{a}^\dagger_n$), with $\hat{b}_n$ ($\hat{b}^\dagger_n$) to the right (left) of $\hat{a}_n$ ($\hat{a}^\dagger_n$);

(iii) $\mathcal{T}^{(r)}_+$ ($\mathcal{T}^{(r)}_-$) ordering of the operators $\hat{a}_n(t^{(r)})$ [$\hat{a}^\dagger_n(t^{(r)})$].[6]

The disentangled form (9.141) may be applied directly to practical calculations. In particular, in the case of incoming vacuum noise,

$$\langle \hat{b}^\dagger_n(t^{(r)}) \cdots \rangle = 0 = \langle \cdots \hat{b}_n(t^{(r)}) \rangle, \quad (9.142)$$

6) Note that $\hat{b}_n(t^{(r)})$ [$\hat{b}^\dagger_n(t^{(r)})$] are commuting quantities; see Eq. (9.116).

Eq. (9.141) is simply

$$
G^{(m,n)}_{\{n_i,n_j\}}(\{x_i,t_i,x_j,t_j\}) = \Bigg\langle \Bigg[\mathcal{T}_-^{(r)} \prod_{i=1}^{m} \left(\frac{\hbar\omega_{n_i}}{4l\mathcal{A}\varepsilon_0}\right)^{\frac{1}{2}} \underline{t}^*(\omega_{n_i})\hat{a}^\dagger_{n_i}(t_i^{(r)})\Bigg]
$$
$$
\times \Bigg[\mathcal{T}_+^{(r)} \prod_{j=m+1}^{m+n} \left(\frac{\hbar\omega_{n_j}}{4l\mathcal{A}\varepsilon_0}\right)^{\frac{1}{2}} \underline{t}(\omega_{n_j})\hat{a}_{n_j}(t_j^{(r)})\Bigg] \Bigg\rangle
$$
(9.143)

$(\{x_i, x_j\} > l)$. Equation (9.143) reveals that the correlation functions of the electric-field strength of the radiation field outside the cavity (i.e., behind the fractionally transparent right-hand mirror) may be directly related to the correlation functions of the intra-cavity-field photon operators, the original time ordering now applying, with regard to the retarded times, to the intra-cavity-field photon operators.

If the incoming radiation field may be regarded as being, for example, in a coherent state,

$$\langle \hat{b}^\dagger_n(t^{(r)}) \cdots \rangle = \tilde{\beta}^*_n(t^{(r)})\langle \cdots \rangle, \quad \langle \cdots \hat{b}_n(t^{(r)}) \rangle = \langle \cdots \rangle \tilde{\beta}_n(t^{(r)}),$$
(9.144)

with the $\beta_n(t^{(r)})$ being c-number functions, Eq. (9.141) becomes ($\{x_i, x_j\} > l$)

$$G^{(m,n)}_{\{n_i,n_j\}}(\{x_i,t_i,x_j,t_j\})$$
$$= \Bigg\langle \Bigg[\mathcal{T}_-^{(r)} \prod_{i=1}^{m} \left(\frac{\hbar\omega_{n_i}}{4l\mathcal{A}\varepsilon_0}\right)^{\frac{1}{2}} \left(\underline{t}^*(\omega_{n_i})\hat{a}^\dagger_{n_i}(t_i^{(r)}) - \frac{\underline{t}^*(\omega_{n_i})}{\underline{t}(\omega_{n_i})}\left(\frac{2l}{c}\right)^{\frac{1}{2}}\tilde{\beta}^*_{n_i}(t_i^{(r)})\right)\Bigg]$$
$$\times \Bigg[\mathcal{T}_+^{(r)} \prod_{j=m+1}^{m+n} \left(\frac{\hbar\omega_{n_j}}{4l\mathcal{A}\varepsilon_0}\right)^{\frac{1}{2}} \left(\underline{t}(\omega_{n_j})\hat{a}_{n_j}(t_j^{(r)}) - \frac{\underline{t}(\omega_{n_j})}{\underline{t}^*(\omega_{n_j})}\left(\frac{2l}{c}\right)^{\frac{1}{2}}\tilde{\beta}_{n_j}(t_j^{(r)})\right)\Bigg]\Bigg\rangle.$$
(9.145)

The situation for an arbitrary state of the incoming field can also be considered using Eq. (9.141). In this case there appear, in addition to the correlation functions of the intra-cavity-field operators, correlation functions of the incoming-field operators.

It should be pointed out that in Eq. (9.140) the second term (proportional to $\underline{t}/\underline{t}^*$) arises from the reflected part of the incoming radiation field. If the incoming radiation field is not in the vacuum state and if it is required that the reflected part of this incoming (real) field should not contribute to the photodetection signal, the detection scheme must be chosen in such a way that the reflected radiation does not fall on the photodetectors. Clearly, in this case the conditions (9.142) are also satisfied and Eq. (9.143) applies, independently of whether or not the incoming radiation field represents the vacuum field.

Therefore, the properties solely of the radiation field generated and/or amplified inside the resonator-like cavity are observed. An example is a cavity with two fractionally transparent mirrors. In this case[7] one of the mirrors may be used as input port (for real input radiation) and the other as output port (with incoming vacuum). With regard to the output port, Eq. (9.143) applies. Obviously, the internal-field Langevin equations must then be complemented by the fluctuation and drift-motion terms related to the second fractionally transparent mirror.

9.7
Unwanted losses

In the derivation of the quantum Langevin equations (9.80) it is assumed that the cavity field only suffers from the required (radiative) losses due to the input-output coupling. There are of course also unwanted losses such as scattering and absorption losses, which are unavoidably connected with any material system. Though in the case of a proper (high-Q) cavity both the required and the unwanted losses are small, the unwanted losses can be of the same order of magnitude as the required ones. Since the main obstacle to the processing of nonclassical quantum states are decoherence effects associated with unwanted losses, the greatest care should be given to them.

As we know, the mode decomposition used in Sections 9.1 and 9.2 fails in the case of absorbing matter, and instead the Green function formalism developed in Section 2.4.2 may be applied. A first answer to the question of the effect of unwanted losses can be obtained from quantum noise theories. According to Section 5.3.1, quantum Langevin equations of the type (9.80) can be obtained within the framework of Markovian damping theory, by starting from the Hamiltonian

$$\hat{H} = \hat{H}_0 + \hat{H}_{\text{int}} + \hat{H}'_{\text{int}}, \qquad (9.146)$$

where

$$\hat{H}_0 = \sum_n \hbar \omega_n \hat{a}_n^\dagger \hat{a}_n + \sum_n \int_{[n]} d\omega \, \hbar \omega \, \hat{b}^\dagger(\omega) \hat{b}(\omega) \qquad (9.147)$$

---

[7] The extension of the theory to a one-dimensional resonator with two fractionally transparent mirrors is straightforward. The extension of the theory to three-dimensional cavities is more extensive, because of the calculation of three-dimensional mode structures [for the mode structure for certain cases of micro-droplets and micro-cavities, see, e. g., Lai, Leung and Young (1990); De Martini, Marocco, Mataloni, Crescentini and Loudon (1991)].

is the Hamiltonian of the cavity field and the field outside the cavity, which, for a moment, are thought of as being isolated from each other,

$$\hat{H}_{\text{int}} = i \sum_n \int dx\, A_n(x) \hat{P}(x) \hat{a}_n + \text{H.c.} \tag{9.148}$$

is the interaction energy between the cavity field and the active sources inside the cavity, and

$$\hat{H}'_{\text{int}} = \hbar \left(\frac{c}{2l}\right) \sum_n \int_{[n]} d\omega\, \underline{t}^*(\omega) \hat{b}^\dagger(\omega) \hat{a}_n + \text{H.c.} \tag{9.149}$$

is the interaction energy between the cavity field and the field outside the cavity, which is regarded as a dissipative system.

The equivalence between the quantum-noise theoretical approach and – within the approximation scheme used – the quantum-field theoretical approach to the problem of a leaky optical cavity may suggest that the effect of unwanted losses can be described by complementing the Hamiltonian (9.146) with additional interaction energies of the type

$$\hat{H}''_{\text{int}} = \hbar \left(\frac{c}{2l}\right) \sum_\sigma \sum_n \int_{[n]} d\omega\, \lambda^*_\sigma(\omega) \hat{c}^\dagger_\sigma(\omega) \hat{a}_n + \text{H.c.}, \tag{9.150}$$

where σ labels the additional dissipative channels. It is straightforward to prove that the application of Markovian damping theory leads to the quantum Langevin equations

$$\dot{\hat{a}}_n = -\left(i\omega_n + \tfrac{1}{2}\gamma_n\right)\hat{a}_n + \frac{1}{\hbar}A\int dx'\, w_n A_n(x') \hat{P}(x')$$
$$+ \left(\frac{c}{2l}\right)^{\frac{1}{2}} \underline{t}(\omega_n)\hat{b}_n(t) + \left(\frac{c}{2l}\right)^{\frac{1}{2}} \sum_\sigma \lambda_\sigma(\omega_n)\hat{c}_\sigma(t), \tag{9.151}$$

where

$$\gamma_n = \Gamma_n + \Gamma'_n. \tag{9.152}$$

Here the damping rate Γ_n is attributed to the input-output coupling and defined according to Eq. (9.26), and

$$\Gamma'_n = \sum_\sigma \Gamma_{n\sigma} \tag{9.153}$$

with

$$\Gamma_{n\sigma} = \frac{c}{2l}|\lambda_\sigma(\omega_n)|^2 \tag{9.154}$$

is the total damping rate that describes the unwanted losses. Hence the quantum Langevin equations (9.80) must be simply complemented with additional damping and noise terms to take into account the effect on the cavity field of unwanted losses.

It is also not difficult to prove that the input-output relations (9.96) remain unchanged, i. e., $[2\mathcal{A}c\varepsilon_0(\hbar\omega_n)^{-1}\hat{\mathcal{E}}^{(+)}_{n\,\text{in free}} \mapsto \hat{b}_n, -2\mathcal{A}c(\hbar\omega_n)^{-1}\hat{\Pi}^{(+)}_{n\,\text{out}} \mapsto \hat{b}_{n\,\text{out}}]$

$$\hat{b}_{n\,\text{out}}(t) = \left(\frac{c}{2l}\right)^{\frac{1}{2}} \underline{t}(\omega_n)\hat{a}_n(t) + \underline{r}(\omega_n)\hat{b}_n(t), \tag{9.155}$$

where

$$\underline{r}(\omega_n) = -\frac{\underline{t}(\omega_n)}{\underline{t}^*(\omega_n)}. \tag{9.156}$$

Clearly, the interaction energy (9.150) implies that only unwanted losses which primarily result from those unwanted losses, from which the cavity modes suffer, are considered. In particular, the effect of absorption losses inside the coupling mirror on the outgoing field, via reflection of the incoming field, is disregarded. As long as the input port is unused, this drawback does not play a role and may be ignored. In this case all the relations between the correlation function as given in Section 9.6 remain correspondingly valid even if the intracavity field is allowed to suffer from unwanted losses and the generalized quantum Langevin equations (9.151) are used. If, in cases of the input port being used, unwanted losses of the incoming field in the coupling mirror must be taken into account with respect to the reflected field, additional considerations are necessary.

9.8
Quantum-state extraction

From Section 9.6 we know that if the operator input-output relations are known, the correlation functions of the outgoing field can be expressed in terms of correlation functions of the cavity field and the incoming field. In this context the question of the calculation of the outgoing-field quantum state, as a whole, arises. Let us suppose, e. g., that during the passage of atoms through the cavity, the nth cavity mode is prepared in some quantum state and assume that the preparation time is sufficiently short compared with the decay time Γ_n^{-1}, so that the two time scales are clearly distinguishable. In this case we may assume that at some time t_0 (when the atoms leave the cavity) the cavity mode is prepared in a given quantum state and its evolution in the further course of time (i. e., for times $t \geq t_0$) can be treated as free-field evolution.

To calculate the quantum state of the outgoing field in the frequency interval $[n]$, we begin with its characteristic functional

$$\Phi_{n\,\text{out}}[\beta(\omega),t] = \left\langle \exp\left[\int_{[n]} d\omega\, \beta(\omega)\hat{b}^\dagger_{n\,\text{out}}(\omega,t) - \text{H.c.}\right]\right\rangle \tag{9.157}$$

in symmetric order [cf. Eq. (4.90) for $s=0$]. Assuming that the input port is unused, we may apply Eq. (9.155) [together with Eq. (9.156)], which implies that, on the relevant time scale $\Delta t \gg \Delta \omega^{-1}$,

$$\hat{b}_{n\,\text{out}}(\omega, t) = \left(\frac{c}{2l}\right)^{\frac{1}{2}} \underline{t}(\omega_n) \frac{1}{\sqrt{2\pi}} \int_{t_0}^{t+\Delta t} dt'\, e^{-i\omega(t-t')} \hat{a}_n(t')$$
$$+ \underline{r}(\omega_n) \hat{b}_n(\omega, t_0) e^{-i\omega(t-t_0)}. \tag{9.158}$$

Indeed, substitution of $\hat{b}_{n\,\text{out}}(\omega, t)$ into

$$\hat{b}_{n\,\text{out}}(t) = \int_{[n]} d\omega\, \hat{b}_{n\,\text{out}}(\omega, t) \tag{9.159}$$

just leads to Eq. (9.155). For times $t \geq t_0$, from Eq. (9.151) it follows that

$$\hat{a}_n(t) = e^{-(i\omega_n + \frac{1}{2}\Gamma_n)(t-t_0)} \hat{a}_n(t_0)$$
$$+ \left(\frac{c}{2l}\right)^{\frac{1}{2}} \int_{t_0}^{t} dt'\, e^{-(i\omega_n + \frac{1}{2}\Gamma_n)(t-t')} \left[\underline{t}(\omega_n) \hat{b}_n(t') + \sum_\sigma \lambda_\sigma(\omega_n) \hat{c}_\sigma(t')\right],$$
$$\tag{9.160}$$

and the combination of Eqs (9.158) and (9.160) yields

$$\hat{b}_{n\text{out}}(\omega, t) = F_n^*(\omega, t) \hat{a}_n(t_0) + \hat{B}_n(\omega, t), \tag{9.161}$$

where ($\Omega_n = \omega_n - i\Gamma_n/2$)

$$F_n(\omega, t) = \frac{i}{\sqrt{2\pi}} \left(\frac{c}{2l}\right)^{\frac{1}{2}} \underline{t}^*(\omega_n) e^{i\omega(t-t_0)} \frac{\exp\left[-i(\omega - \Omega_n^*)(t+\Delta t - t_0)\right] - 1}{\omega - \Omega_n^*}, \tag{9.162}$$

$$\hat{B}_n(\omega, t) = \int_{[n]} d\omega' \left[G_n^*(\omega, \omega', t) \hat{b}_n(\omega', t_0) + \sum_\sigma G_{n\sigma}^*(\omega, \omega', t) \hat{c}_\sigma(\omega', t_0)\right], \tag{9.163}$$

$$G_n(\omega, \omega', t) = \underline{t}^*(\omega_n) \underline{t}^*(\omega_n) v_n(\omega, \omega', t) + \underline{r}^*(\omega_n) e^{i\omega'(t-t_0)} \delta(\omega - \omega'), \tag{9.164}$$

$$G_{n\sigma}(\omega, \omega', t) = \underline{t}^*(\omega_n) \lambda_\sigma^*(\omega_n) v_n(\omega, \omega', t), \tag{9.165}$$

$$v_n(\omega, \omega', t) = \frac{1}{2\pi} \frac{c}{2l} \frac{e^{-i\omega\Delta t}}{\omega' - \omega}$$
$$\times \left[\frac{e^{i\omega(t+\Delta t - t_0)} - e^{i\Omega_n^*(t+\Delta t - t_0)}}{\omega - \Omega_n^*} - \frac{e^{i\omega'(t+\Delta t - t_0)} - e^{i\Omega_n^*(t+\Delta t - t_0)}}{\omega' - \Omega_n^*}\right]. \tag{9.166}$$

To further evaluate the characteristic functional (9.157), it is convenient to introduce a unitary, explicitly time-dependent transformation according to

$$\hat{b}_{n\,\text{out}}(\omega, t) = \sum_i \phi_i^*(\omega, t) \hat{b}_{n\,\text{out}}^{(i)}(t), \tag{9.167}$$

$$\hat{b}_{n\,\text{out}}^{(i)}(t) = \int_{[n]} d\omega\, \phi_i(\omega, t) \hat{b}_{n\,\text{out}}(\omega, t), \tag{9.168}$$

where, for chosen t, the nonmonochromatic mode functions $\phi_i(\omega, t)$ are a complete set of square integrable orthonormal functions:

$$\int_{[n]} d\omega\, \phi_i(\omega, t) \phi_j^*(\omega, t) = \delta_{ij}, \tag{9.169}$$

$$\sum_i \phi_i(\omega, t) \phi_i^*(\omega', t) = \delta(\omega - \omega'). \tag{9.170}$$

Needless to say that the commutation relation

$$[\hat{b}_{n\,\text{out}}^{(i)}(t), \hat{b}_{n\,\text{out}}^{(j)\dagger}(t)] = \delta_{ij} \tag{9.171}$$

holds.

Let $\hat{b}_{n\,\text{out}}^{(1)}(t)$ be the operator associated with the (nonmonochromatic) outgoing mode which is attributed to the (nonmonochromatic) cavity mode. Recalling Eq. (9.161), it is obvious to set, within the approximation scheme used,

$$\phi_1(\omega, t) = \frac{F_n(\omega, t)}{\sqrt{\eta_n(t)}}, \tag{9.172}$$

where

$$\eta_n(t) = \int_{[n]} d\omega\, |F_n(\omega, t)|^2, \tag{9.173}$$

so that, for chosen n, Eq. (9.168) takes the form

$$\hat{b}_{n\,\text{out}}^{(i)}(t) = \begin{cases} \sqrt{\eta_n(t)}\, \hat{a}_n(t_0) + \hat{B}_n^{(i)}(t) & \text{if } i = 1, \\ \hat{B}_n^{(i)}(t) & \text{otherwise,} \end{cases} \tag{9.174}$$

where

$$\hat{B}_n^{(i)}(t) = \int_{[n]} d\omega\, \phi_i(\omega, t) \hat{B}_n(\omega, t). \tag{9.175}$$

Introducing the operators $\hat{b}_{n\,\text{out}}^{(i)}(t)$ according to Eq. (9.167) in the characteristic functional (9.157) and taking into account the commutation relation (9.171), we see that the operator exponential factorizes as

$$\exp\left[\int_{[n]} d\omega\, \beta(\omega) \hat{b}_{n\,\text{out}}^\dagger(\omega, t) - \text{H.c.}\right] = \prod_i \exp[\beta_i \hat{b}_{n\,\text{out}}^{(i)\dagger}(t) - \text{H.c.}], \tag{9.176}$$

where

$$\beta_i = \beta_i(t) = \int_{[n]} d\omega \, \phi_i(\omega, t) \beta(\omega). \tag{9.177}$$

Since only $\hat{b}^{(1)}_{n\,\text{out}}(t)$ is attributed to the cavity mode [see Eq. (9.174)], we may assume that the characteristic function factorizes as well, with

$$\Phi^{(1)}_{n\,\text{out}}(\beta_1, t) = \left\langle \exp\left[\beta_1 \hat{b}^{(1)\dagger}_{n\,\text{out}}(t) - \text{H.c.}\right] \right\rangle \tag{9.178}$$

being the characteristic function of the relevant outgoing mode. Using Eq. (9.174) and noting that the commutation relation

$$\left[\hat{a}_n(t_0), \hat{B}^{(1)\dagger}_n(t)\right] = 0 \tag{9.179}$$

holds, we may rewrite Eq. (9.178) as

$$\Phi^{(1)}_{n\,\text{out}}(\beta_1, t) = \left\langle \exp\left[\beta_1 \sqrt{\eta_n(t)} \, \hat{a}^\dagger_n(t_0) - \text{H.c.}\right] \exp\left[\beta_1 \hat{B}^{(1)\dagger}_n(t) - \text{H.c.}\right] \right\rangle. \tag{9.180}$$

Noting that, according to Eq. (9.175) together with Eq. (9.163), $\hat{B}^{(1)}_n(t)$ is a functional of $\hat{b}_n(\omega, t_0)$ and $\hat{c}_\sigma(\omega, t_0)$ and assuming that the density operator (at the initial time t_0) factorizes with respect to the cavity field and the residual system, we obtain

$$\Phi^{(1)}_{n\,\text{out}}(\beta_1, t) = \left\langle \exp\left[\beta_1 \sqrt{\eta_n(t)} \, \hat{a}^\dagger_n(t_0) - \text{H.c.}\right] \right\rangle \left\langle \exp\left[\beta_1 \hat{B}^{(1)\dagger}_n(t) - \text{H.c.}\right] \right\rangle. \tag{9.181}$$

Inserting Eq. (9.163) into Eq. (9.175), we may rewrite $\hat{B}^{(1)}_n(t)$ as

$$\hat{B}^{(1)}_n(t) = \sqrt{\zeta^{(1)}_n(t)} \, \hat{b}^{(1)}_n(t) + \sum_\sigma \sqrt{\zeta^{(1)}_\sigma(t)} \, \hat{c}^{(1)}_\sigma(t), \tag{9.182}$$

where the operators $\hat{b}^{(1)}_n(t)$ and $\hat{c}^{(1)}_\sigma(t)$ are defined by

$$\hat{b}^{(1)}_n(t) = \int_{[n]} d\omega \, \frac{\chi^{(1)}_n(\omega, t)}{\sqrt{\zeta^{(1)}_n(t)}} \, \hat{b}_n(\omega, t_0), \tag{9.183}$$

$$\hat{c}^{(1)}_\sigma(t) = \int_{[n]} d\omega \, \frac{\chi^{(1)}_\sigma(\omega, t)}{\sqrt{\zeta^{(1)}_\sigma(t)}} \, \hat{c}_\sigma(\omega, t_0), \tag{9.184}$$

and the functions $\zeta^{(1)}_\mu(t)$ and $\chi^{(1)}_\mu(t)$ [$\mu = (n, \sigma), \sigma = 1, 2, \ldots$] read

$$\zeta^{(1)}_\mu(t) = \int_{[n]} d\omega \, |\chi^{(1)}_\mu(\omega, t)|^2, \tag{9.185}$$

$$\chi_\mu^{(1)}(\omega,t) = \int_{[n]} d\omega'\, \phi_1(\omega',t) G_\mu^*(\omega',\omega,t). \tag{9.186}$$

Inserting Eq. (9.182) into Eq. (9.181) and recalling Eq. (C.27), we may express the characteristic function $\Phi_{n\,\text{out}}(\beta,t;s) \equiv \Phi_{n\,\text{out}}^{(1)}(\beta_1,t;s)$ in s order of the quantum state of the relevant outgoing field [cf. Eq. (4.91)] in terms of the characteristic functions $\Phi_{n\,\text{cav}}(\beta';s')$, and $\Phi_\mu(\beta_\mu;s_\sigma)$ of the quantum states of the cavity field, the incoming field ($\mu=n$) and the unwanted dissipative channels ($\mu=\sigma$, $\sigma=1,2,\ldots$), respectively, as

$$\Phi_{n\,\text{out}}(\beta,t;s) = e^{-\xi_n(t)|\beta|^2/2}\Phi_{n\,\text{cav}}\!\left[\sqrt{\eta_n(t)}\,\beta;s'\right] \prod_\mu \Phi_\mu\!\left[\sqrt{\zeta_\mu(t)}\,\beta;s_\mu\right], \tag{9.187}$$

where

$$\xi_n(t) = \eta_n(t)s' + \sum_\mu \zeta_\mu(t)s_\mu - s. \tag{9.188}$$

From Eq. (9.187) the phase-space function in s order can be derived to be [cf. Eq. (4.93)]

$$P_{n\,\text{out}}(\alpha,t;s) = \frac{2}{\pi}\frac{1}{\xi_n(t)} \int d^2\alpha'\, P_{n\,\text{cav}}(\alpha';s') \prod_\mu \int d^2\alpha_\mu\, P_\mu(\alpha_\mu;s_\mu)$$

$$\times \exp\!\left[-\frac{2}{\xi_n(t)}\left|\sqrt{\eta_n(t)}\,\alpha' + \sum_{\mu'}\sqrt{\zeta_{\mu'}(t)}\,\alpha_{\mu'} - \alpha\right|^2\right], \tag{9.189}$$

provided that

$$\xi_n(t) \geq 0, \tag{9.190}$$

where the equality sign must be understood as a limiting process. In particular, the Wigner function $W_{n\,\text{out}}(\alpha,t)$ is obtained by setting $s=0$, that is $W_{n\,\text{out}}(\alpha,t) = P_{n\,\text{out}}(\alpha,t;s)|_{s=0}$.

In the case where both the incoming field and the dissipative channels are in the vacuum states with Wigner functions

$$W_\mu(\alpha) = P_\mu(\alpha;s)|_{s=0} = \frac{2}{\pi}e^{-2|\alpha|^2}, \tag{9.191}$$

from Eq. (9.189) (for $s=0$) it follows that the Wigner function of the quantum state of the relevant outgoing field reads

$$W_{n\,\text{out}}(\alpha,t) = \frac{2}{\pi}\frac{1}{1-\eta_n(t)} \int d^2\alpha'\, \exp\!\left[-\frac{2|\sqrt{\eta_n(t)}\,\alpha' - \alpha|^2}{1-\eta_n(t)}\right] W_{n\,\text{cav}}(\alpha'), \tag{9.192}$$

where, according to Eq. (9.173) [together with Eq. (9.162)],

$$\eta_n(t) = \frac{\Gamma_n}{\Gamma_n + \Gamma'_n}\left[1 - e^{-(\Gamma_n + \Gamma'_n)(t + \Delta t - t_0)}\right], \qquad (9.193)$$

with the damping rates Γ_n and Γ'_n, respectively, being defined by Eqs (9.26) and (9.153) [together with Eq. (9.154)]. Equation (9.192) reveals that, for almost perfect extraction of a quantum state from a high-Q cavity, the condition

$$\frac{\eta_n(t)}{1 - \eta_n(t)} \gg 1 \qquad (9.194)$$

must be satisfied, i. e., the value of the extraction efficiency $\eta_n(t)$ must be sufficiently close to unity. Note that for $(\Gamma_n + \Gamma'_n)(t - t_0) \gg 1$

$$\eta_n(t) \simeq \frac{\Gamma_n}{\Gamma_n + \Gamma'_n}. \qquad (9.195)$$

The actual required efficiency for almost perfect quantum-state extraction sensitively depends on the quantum state that is desired to be extracted. To illustrate this, let us consider a k-photon number state,

$$W^{(k)}_{n\,\text{cav}}(\alpha) = \frac{2}{\pi}(-1)^k e^{-2|\alpha|^2} L_k(4|\alpha|^2) \qquad (9.196)$$

[see Eq. (4.119)]. Substituting Eq. (9.196) into Eq. (9.192), we obtain the Wigner function of the outgoing field as

$$W^{(k)}_{n\,\text{out}}(\alpha, t) = \frac{2}{\pi}(-1)^k e^{-2|\alpha|^2}[2\eta_n(t) - 1]^k L_k\left[\frac{4\eta_n(t)}{2\eta_n(t) - 1}|\alpha|^2\right]. \qquad (9.197)$$

From Eq. (9.197) it is not difficult to see that the condition

$$\eta_n(t) > 1 - \frac{1}{2k} \qquad (9.198)$$

must be satisfied to guarantee that the k-photon number state prevails in the mixed output quantum state. In the simplest case of a one-photon number state, $k=1$, the condition reduces to $\eta_n(t) > 0.5$. That is to say, the weight of the one-photon number state exceeds the weight of the vacuum state in the mixed state of the outgoing field,

$$W^{(1)}_{n\,\text{out}}(\alpha, t) = [1 - \eta_n(t)]W^{(0)}_{n\,\text{out}}(\alpha) + \eta_n(t)W^{(1)}_{n\,\text{out}}(\alpha), \qquad (9.199)$$

only if the extraction efficiency exceeds 50%. The condition (9.198) clearly shows that with increasing value of k the required extraction efficiency rapidly approaches 100%.

The dependence on the extraction efficiency of the quantum state of the outgoing field is illustrated in Fig. 9.2 for the case in which a single-photon

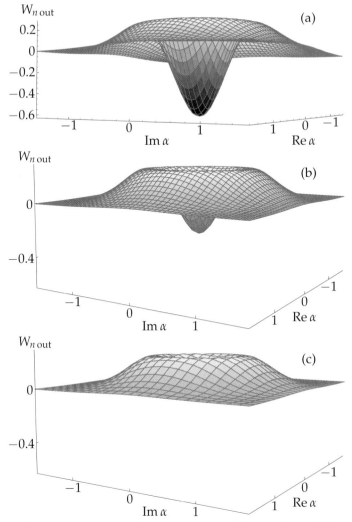

Fig. 9.2 Wigner function of the quantum state of the outgoing mode in the case of the cavity mode being (initially) prepared in a single-photon number state. (a) $\eta_n(t)=0.99$; (b) $\eta_n(t)=0.71$; (c) $\eta_n(t)=0.5$. [After Khanbekyan, Knöll, Semenov, Vogel and Welsch (2004).]

number state is required to be extracted. Figure 9.2(a) reveals that nearly perfect extraction requires an extraction efficiency that should be no smaller than $\eta_n(t)=0.99$, which for $t\to\infty$ corresponds to the requirement that $\Gamma'_n/\Gamma_n \lesssim 0.01$. As long as $\eta_n(t)>0.5$, the single-photon number state is the dominant state in the mixed output state, as can be seen from Fig. 9.2(b) [$\eta_n(t)=0.71$, i.e., $\Gamma'_n/\Gamma_n=0.429$ ($t\to\infty$)]. For $\eta_n(t)\leq 0.5$, i.e., $\Gamma'_n/\Gamma_n \geq 1$ ($t\to\infty$)], the features typical of a single-photon number state are lost, Fig. 9.2(c).

References

Carmichael, H.J. (1987) *J. Opt. Soc. Am. B* **4**, 1588.

Collett, M.J. and C.W. Gardiner (1984) *Phys. Rev. A* **30**, 1386.

Collett, M.J. and D.F. Walls (1985) *Phys. Rev. A* **32**, 2887.

De Martini, F., M. Marocco, P. Mataloni, L. Crescentini and R. Loudon (1991) *Phys. Rev. A* **43**, 2480.

Dutra, S.M. and G. Nienhuis (2000) *Phys. Rev. A* **62**, 063805.

Feng, X.-P. (1991) *Opt. Commun.* **83**, 162.

Gardiner, C.W. (1989) in *Springer Proceedings in Physics*, Vol. 41; *Quantum Optics*, eds V.J.D. Harvey and D.F. Walls (Springer-Verlag, Berlin) p. 98.

Gardiner, C.W. (1991) *Quantum Noise* (Springer-Verlag, Berlin).

Gardiner, C.W. and M.J. Collett (1985) *Phys. Rev. A* **31**, 3761.

Guedes, J. and B. Baseia (1990) *Phys. Rev. A* **42**, 6858.

Guedes, J., J.C. Penaforte and B. Baseia (1989) *Phys. Rev. A* **40**, 2463.

Khanbekyan, M, L. Knöll, A.A. Semenov, W. Vogel and D.-G. Welsch (2004) *Phys. Rev. A* **69**, 043807.

Khosravi, H. and R. Loudon (1991) *Proc. Roy. Soc. London A* **433**, 337.

Knöll, L., W. Vogel and D.-G. Welsch (1991a) *J. Mod. Opt.* **38**, 55.

Knöll, L., W. Vogel and D.-G. Welsch (1991b) *Phys. Rev. A* **43**, 543.

Knöll, L. and D.-G. Welsch (1992) *Progr. Quant. Electron.* **16**, 135.

Lai, H.M., P.T. Leung and K. Young (1990) *Phys. Rev. A* **41**, 5187; *ibid.* **41**, 5199.

Lang, R. and M.O. Scully (1973) *Opt. Commun.* **9**, 331.

Ley, M. and R. Loudon (1987) *J. Mod. Opt.* **34**, 227.

Meystre, P. (1992) In *Progress in Optics*, Vol. 30, ed. E. Wolf (North-Holland, Amsterdam), p. 261.

Ujihara, K. (1975) *Phys. Rev. A* **12**, 148.

Ujihara, K. (1977) *Phys. Rev. A* **16**, 652.

Ujihara, K. (1978) *Phys. Rev. A* **18**, 659.

Ujihara, K. (1979) *Phys. Rev. A* **20**, 1096.

Ujihara, K. (1984) *Phys. Rev. A* **29**, 3253.

Viviescas, C. and G. Hackenbroich (2003) *Phys. Rev. A* **67**, 013805.

Yamamoto, Y. and I. Imoto (1986) *IEEE J. Quant. Electron.* **22**, 2032.

10
Medium-assisted electromagnetic vacuum effects

The classical electromagnetic vacuum is simply the state in which all moments of the electric and induction fields identically vanish, and thus the fields themselves identically vanish. Hence, in classical electrodynamics the interaction of matter with the electromagnetic field – including field-assisted interaction between matter systems – always requires excited (source-attributed) fields, which, as is known, can be described in terms of positive semi-definite probability distribution functions in phase space. In quantum electrodynamics the situation is quite different, because the noncommutativity of canonical conjugate field quantities necessarily implies nonvanishing moments. As we know, the quantum electromagnetic vacuum can be regarded as the state in which all normally ordered field moments identically vanish. Clearly, the anti-normally ordered field moments cannot do so due to virtual photon creation and destruction – an effect in which the noise of the quantum vacuum becomes manifest.

Since the electromagnetic vacuum cannot be switched off, its interaction with atomic systems cannot be switched off either, thereby giving rise to a number of observable effects such as spontaneous emission, the Lamb shift, intermolecular energy transfer and the van der Waals force. Both virtual and real photons can be involved in the atom–field interaction. Whereas the interaction of ground-state atoms with the electromagnetic vacuum processes via virtual photon creation and destruction, the creation of real photons always requires excited atoms. A typical example of the first case is the van der Waals force between two ground-state atoms, whereas the spontaneous decay of an excited atomic state typically represents the second case.

The presence of linear media in the form of macroscopic bodies changes the structure of the electromagnetic field compared to that in the free space and in consequence the electromagnetic vacuum felt by an atom is changed. In a broader sense the effect is called the Casimir effect. It offers the possibility of controlling the interaction of atomic systems with the medium-assisted electromagnetic vacuum, with applications that range from cavity QED to integrated atom optics and electronics. On the basis of the quantization scheme for the electromagnetic field in dispersing and absorbing media (Section 2.4),

---

Quantum Optics, Third, revised and extended edition. Werner Vogel and Dirk-Gunnar Welsch
Copyright © 2006 WILEY-VCH Verlag GmbH & Co. KGaA, Weinheim
ISBN: 3-527-40507-0

the effect of the presence of macroscopic bodies on the spontaneous emission of a single atom is studied in Section 10.1, and Section 10.2 provides a unified approach to the problem of van der Waals and Casimir forces.

10.1
Spontaneous emission

Spontaneous emission is not only one of the most familiar quantum phenomena but it is also the basic process for generating light. Although a number of properties of the spontaneously emitted radiation can be described classically, spontaneous emission is a pure quantum effect the understanding of which requires quantization of the electromagnetic field. A dynamical theory of the spontaneous emission of a single (two-level) atom in free space was first given by Weisskopf and Wigner (1930).

Let us consider a single atomic system such as an atom or a molecule – briefly referred to as an atom in the following – (position $\mathbf{r}_A$, energy eigenvalues $E_n = \hbar \omega_n$) which in the presence of arbitrary linear dielectric bodies[1] interacts with the electromagnetic field via electric-dipole transitions, so that the multipolar-coupling Hamiltonian (2.229) can be given in the form of

$$\hat{H} = \hat{H}_C + \hat{H}_F + \hat{H}_{\text{int}}, \tag{10.1}$$

where

$$\hat{H}_C = \sum_n \hbar \omega_n \hat{A}_{nn} \tag{10.2}$$

is the atomic Hamiltonian,

$$\hat{H}_F = \int d^3 r \int_0^\infty d\omega\, \hbar \omega\, \hat{\mathbf{f}}^\dagger(\mathbf{r},\omega)\hat{\mathbf{f}}(\mathbf{r},\omega) \tag{10.3}$$

is the Hamiltonian of the electromagnetic field and the medium forming the bodies, and, according to Eq. (2.243),

$$\hat{H}_{\text{int}} = -\sum_{n,m} \mathbf{d}_{nm} \hat{\mathbf{E}}^{(+)}(\mathbf{r}_A)\, \hat{A}_{nm} + \text{H.c.}$$

$$= -i\sqrt{\frac{\hbar}{\pi\varepsilon_0}}\sum_{n,m}\int_0^\infty d\omega\, \frac{\omega^2}{c^2}\int d^3r'\, \mathbf{d}_{nm} G(\mathbf{r}_A,\mathbf{r}',\omega)\hat{\mathbf{f}}(\mathbf{r}',\omega)\hat{A}_{nm} + \text{H.c.} \tag{10.4}$$

---

1) For an extension to magnetodielectric bodies, see Ho, Buhmann, Knöll, Welsch, Scheel and Kästel (2003). It is worth noting that all formulas in this chapter which do not explicitly contain the permittivity, but are solely expressed in terms of the Green tensor (or related quantities), are also valid for magnetodielectric bodies, with the Green tensor being determined from the full Maxwell equations containing both the medium polarization and magnetization.

is the atom–field interaction energy in the electric-dipole approximation ($\hat{A}_{nm} = |n\rangle\langle m|$, $\hat{H}_C|n\rangle = \hbar\omega_n|n\rangle$). The typical features of spontaneous decay can already be understood on the basis of a two-state model of the atom, i.e.,

$$\hat{H}_C = \hbar(\omega_1 \hat{A}_{11} + \omega_2 \hat{A}_{22}) = \hbar\omega_1 + \hbar\omega_{21}\hat{A}_{22} \mapsto \hat{H}_C = \hbar\omega_{21}\hat{A}_{22} \quad (10.5)$$

($\omega_{21} = (E_2 - E_1)/\hbar > 0$; note that $\hat{A}_{11} + \hat{A}_{22} = 1$), and within the rotating-wave approximation, i.e.,[2]

$$\hat{H}_{\text{int}} = -i\sqrt{\frac{\hbar}{\pi\varepsilon_0}} \int_0^\infty d\omega \frac{\omega^2}{c^2} \int d^3r'\, \mathbf{d}_{21} \mathbf{G}(\mathbf{r}_A, \mathbf{r}', \omega) \hat{\mathbf{f}}(\mathbf{r}', \omega) \hat{A}_{21} + \text{H.c.}. \quad (10.6)$$

It should be stressed that the relevant information on the bodies is fully included in the Green tensor $\mathbf{G}(\mathbf{r}, \mathbf{r}', \omega)$ of the macroscopic Maxwell equations.

From Eq. (10.6) it is seen that when the atom is in the upper state $|2\rangle$ and the rest of the system is in the vacuum state $|\{0\}\rangle$,

$$\hat{\mathbf{f}}(\mathbf{r}, \omega)|\{0\}\rangle = 0, \quad (10.7)$$

then a single-quantum state of the combined field–body system,

$$|1(\mathbf{r}, \omega)\rangle = \hat{\mathbf{f}}^\dagger(\mathbf{r}, \omega)|\{0\}\rangle, \quad (10.8)$$

can be created owing to a transition of the atom into the lower state $|1\rangle$. Hence for the overall-system state vector at time t the ansatz

$$|\psi(t)\rangle = C_2(t) e^{-i\omega_{21}t} |\{0\}\rangle |2\rangle + \int d^3r \int_0^\infty d\omega\, e^{-i\omega t} \mathbf{C}_1(\mathbf{r}, \omega, t) |1(\mathbf{r}, \omega)\rangle |1\rangle \quad (10.9)$$

can be made. To determine the (slowly varying) expansion coefficients $C_2(t)$ and $\mathbf{C}_1(\mathbf{r}, \omega, t)$, we insert $|\psi(t)\rangle$ into the Schrödinger equation and obtain the following system of coupled differential equations:

$$\dot{C}_2(t) = -\frac{1}{\sqrt{\pi\varepsilon_0\hbar}} \int_0^\infty d\omega \frac{\omega^2}{c^2} e^{-i(\omega-\omega_{21})t}$$
$$\times \int d^3r\, \sqrt{\text{Im}\,\varepsilon(\mathbf{r}, \omega)}\, \mathbf{d}_{21} \mathbf{G}(\mathbf{r}_A, \mathbf{r}, \omega) \mathbf{C}_1(\mathbf{r}, \omega, t), \quad (10.10)$$

$$\dot{\mathbf{C}}_1(\mathbf{r}, \omega, t) = \frac{1}{\sqrt{\pi\varepsilon_0\hbar}} \frac{\omega^2}{c^2} \sqrt{\text{Im}\,\varepsilon(\mathbf{r}, \omega)}\, e^{i(\omega-\omega_{21})t} \mathbf{d}_{12} \mathbf{G}^*(\mathbf{r}_A, \mathbf{r}, \omega)\, C_2(t). \quad (10.11)$$

[2] Note that here, in addition to the transverse electric field considered in Eq. (2.244), the interaction of the atom with the longitudinal electric field which is attributed to the medium is also treated in the rotating-wave approximation.

To solve it under the initial conditions $C_2(t)|_{t=0}=1$, $\mathbf{C}_1(\mathbf{r},\omega,t)|_{t=0}=0$, we formally integrate Eq. (10.11) and insert the result in Eq. (10.10). Making use of the relation (A.3), after some algebra we derive the integro-differential equation

$$\dot{C}_2(t) = \int_0^t dt'\, K(t-t') C_2(t'), \tag{10.12}$$

where the kernel function reads

$$K(t) = -\frac{1}{\hbar\pi\varepsilon_0 c^2} \int_0^\infty d\omega\, \omega^2 e^{-i(\omega-\omega_{21})t} \mathbf{d}_{21} \operatorname{Im} \mathbf{G}(\mathbf{r}_A,\mathbf{r}_A,\omega) \mathbf{d}_{12}. \tag{10.13}$$

It is not difficult to see that Eq. (10.12) can be converted into the integral equation

$$C_2(t) = \int_0^t dt'\, K'(t-t') C_2(t') + 1, \tag{10.14}$$

where

$$K'(t) = \frac{1}{\hbar\pi\varepsilon_0 c^2} \int_0^\infty d\omega\, \omega^2 \frac{e^{-i(\omega-\omega_{21})t}-1}{i(\omega-\omega_{21})} \mathbf{d}_{21} \operatorname{Im} \mathbf{G}(\mathbf{r}_A,\mathbf{r}_A,\omega) \mathbf{d}_{12}. \tag{10.15}$$

Next let us calculate the intensity of the emitted radiation. From Eq. (2.295) it follows that the Heisenberg operator of the source-field part of the electric field reads

$$\hat{\mathbf{E}}_s(\mathbf{r},t) = \hat{\mathbf{E}}_s^{(+)}(\mathbf{r},t) + \text{H.c.}, \tag{10.16}$$

where, in the rotating-wave approximation,

$$\hat{\mathbf{E}}_s^{(+)}(\mathbf{r},t) = \int_0^t dt'\, \mathbf{K}_{(E)}^{(+)}(\mathbf{r},t;\mathbf{r}_A,t') \mathbf{d}_{12} \hat{A}_{12}(t'), \tag{10.17}$$

with $\mathbf{K}_{(E)}^{(+)}(\mathbf{r},t;\mathbf{r}_A,t')$ being expressed in terms of the Green tensor according to Eq. (2.296).[3] Since the free-field part is in the vacuum state, the intensity of the emitted radiation is fully determined by the source-field part,

$$I(\mathbf{r},t) = \langle \hat{\mathbf{E}}^{(-)}(\mathbf{r},t) \hat{\mathbf{E}}^{(+)}(\mathbf{r},t) \rangle = \langle \hat{\mathbf{E}}_s^{(-)}(\mathbf{r},t) \hat{\mathbf{E}}_s^{(+)}(\mathbf{r},t) \rangle, \tag{10.18}$$

and can therefore be expressed in terms of $C_2(t)$ as

$$I(\mathbf{r},t) = \left| \int_0^t dt'\, \mathbf{K}_{(E)}^{(+)}(\mathbf{r},t;\mathbf{r}_A,t') \mathbf{d}_{12} C_2(t') e^{-i\omega_{21}t'} \right|^2. \tag{10.19}$$

Note that only the far field contributes to the actually emitted radiation.

---

[3] Note that for a two-level atom the polarization $\hat{\mathbf{P}}(\mathbf{r})$ reduces to $\hat{\mathbf{P}}(\mathbf{r}) = \hat{\mathbf{d}}\delta(\mathbf{r}-\mathbf{r}_A) = (\mathbf{d}_{21}\hat{A}_{21} + \mathbf{d}_{12}\hat{A}_{12})\delta(\mathbf{r}-\mathbf{r}_A)$, and that, with respect to the emitted radiation, the difference between multipolar coupling and minimal coupling becomes meaningless.

From the above it is seen that all the relevant quantities can be expressed in terms of $C_2(t)$, with the presence of macroscopic bodies being fully included in the Green tensor of the system. The calculation of $C_2(t)$ requires the solution of Eq. (10.12) [or Eq. (10.14)] – a problem, which in general must be solved numerically. However, there are two limiting cases for which Eq. (10.13) [or Eq. (10.14)] can be further evaluated analytically without explicitly making use of the actual structure of the Green tensor – namely the cases of weak and strong atom–field coupling. In the analytical calculations it may be convenient to set

$$C_2(t) = \tilde{C}_2(t) e^{-i\delta\omega_{21} t} \tag{10.20}$$

in order to determine the shifted atomic transition frequency

$$\tilde{\omega}_{21} = \omega_{21} + \delta\omega_{21} \tag{10.21}$$

in a self-consistent way. Equation (10.12) then changes to

$$\dot{\tilde{C}}_2(t) = i\delta\omega_{21}\tilde{C}_2(t) + \int_0^t dt'\, \tilde{K}(t-t')\tilde{C}_2(t'), \tag{10.22}$$

where

$$\tilde{K}(t) = K(t) e^{i\delta\omega_{21} t}. \tag{10.23}$$

10.1.1
Weak atom–field coupling

We begin with the case of weak atom–field coupling, which is typical of the spontaneous emission observed, e.g., in free space. Let τ_c be the characteristic (correlation) time that defines the time interval in which $K(t)$ [Eq. (10.13)] and also $\tilde{K}(t)$ [Eq. (10.23)] are significantly different from zero. When $\tilde{C}_2(t)$ is slowly varying in this time interval, then in the time integral in Eq. (10.22) $\tilde{C}_2(t')$ may be replaced with $\tilde{C}_2(t)$ and in the remaining integral the upper limit t may be extended to infinity. In this approximation – known as the Markov approximation (cf. Section 5.1.3) – memory effects are disregarded, i.e., the temporal variation of $\tilde{C}_2(t)$ at any chosen time is solely determined by $\tilde{C}_2(t)$ at that time. In this way, the integro-differential equation (10.22) approximates to the simple differential equation

$$\dot{\tilde{C}}_2(t) = i\delta\omega_{21}\tilde{C}_2(t) + \tilde{C}_2(t) \lim_{t\to\infty} \int_0^t dt'\, \tilde{K}(t-t') = -\tfrac{1}{2}\Gamma \tilde{C}_2(t), \tag{10.24}$$

leading to an exponential decay of the upper atomic state,

$$\tilde{C}_2(t) = e^{-\tfrac{1}{2}\Gamma t}. \tag{10.25}$$

Inserting Eq. (10.23) [together with Eq. (10.13)] in the decomposition

$$\lim_{t\to\infty} \int_0^t d\tau\, \tilde{K}(\tau) = -i\delta\omega_{21} - \tfrac{1}{2}\Gamma \tag{10.26}$$

used in Eq. (10.24) and recalling the definition of the ζ function [Eq. (5.85)], we can easily see that the decay rate is given by[4]

$$\Gamma = \frac{2\tilde{\omega}_{21}^2}{\hbar\varepsilon_0 c^2} \mathbf{d}_{21} \mathrm{Im}\, \mathbf{G}(\mathbf{r}_A, \mathbf{r}_A, \tilde{\omega}_{21})\mathbf{d}_{12} \tag{10.27}$$

and for the shift of the transition frequency follows

$$\delta\omega_{21} = \frac{P}{\pi\hbar\varepsilon_0 c^2} \int_0^\infty d\omega\, \omega^2 \frac{\mathbf{d}_{21}\mathrm{Im}\,\mathbf{G}(\mathbf{r}_A,\mathbf{r}_A,\omega)\mathbf{d}_{12}}{\tilde{\omega}_{21} - \omega}. \tag{10.28}$$

Note that Eq. (10.28) does not explicitly determine $\delta\omega_{21}$, because it also appears via $\tilde{\omega}_{21}$ in the frequency integral on the right-hand side of the equation. If $\delta\omega_{21}$ is set equal to zero in the frequency integral, then Eq. (10.28) becomes an explicit expression for $\delta\omega_{21}$ which is valid in lowest (nonvanishing) order of perturbation theory – an approximation widely used in practical calculations.

10.1.1.1 Decay rate and quantum yield

For an atom in free space, the Green tensor $\mathbf{G}(\mathbf{r},\mathbf{r}',\omega)$ reduces to the simple free-space Green tensor $\mathbf{G}_0(\mathbf{r},\mathbf{r}',\omega)$ given by Eq. (2.298), from which $\mathrm{Im}\,\mathbf{G}_0(\mathbf{r},\mathbf{r},\omega) = \mathbf{I}\omega/(6\pi c)$ follows, leading to the well-known formula for the spontaneous-emission rate in free space:

$$\Gamma_0 = \frac{\tilde{\omega}_{21}^3 |\mathbf{d}_{21}|^2}{3\hbar\pi\varepsilon_0 c^3}. \tag{10.29}$$

When there are bodies in the neighborhood of the free-space region where the atom is located, then the Green tensor for the, now inhomogeneous, system can be given, in the free-space region, in the form

$$\mathbf{G}(\mathbf{r},\mathbf{r}',\omega) = \mathbf{G}_0(\mathbf{r},\mathbf{r}',\omega) + \mathbf{G}_S(\mathbf{r},\mathbf{r}',\omega), \tag{10.30}$$

where $\mathbf{G}_S(\mathbf{r},\mathbf{r}',\omega)$ is the more or less complicated scattering part, which typically describes the effect of reflection at the (surfaces of discontinuity of the) bodies. Combining Eq. (10.27) with Eqs (10.29) and (10.30), we can write the

---

[4] Γ corresponds to $\Gamma^{(r)}$ in Eq. (5.166) and the equations following. For notational convenience we drop the superscript here.

decay rate as[5]

$$\Gamma = \Gamma_0 + \frac{2\tilde{\omega}_{21}^2}{\hbar\varepsilon_0 c^2} \mathbf{d}_{21} \operatorname{Im} \mathbf{G}_S(\mathbf{r}_A, \mathbf{r}_A, \tilde{\omega}_{21}) \mathbf{d}_{21}. \tag{10.31}$$

In contrast to the homogeneous and isotropic free space, the decay rate now becomes a function of the atomic position and the orientation of the transition dipole moment. In other words, the homogeneous and isotropic vacuum fluctuations felt by an atom in the strictly free space are inhomogeneously and anisotropically changed by the presence of the bodies, in general. The nearer to a body that the atom is located, the stronger the effect to be expected.

Another effect of the bodies is that the spontaneous decay is not necessarily accompanied by the emission of a really observable photon, but instead a matter quantum can be created, because of material absorption (described in terms of the imaginary part of the permittivity). To quantify the effect, let us consider the emitted radiation energy W, which can be obtained by integration of $I(\mathbf{r},t)$ [Eq. (10.19)] with respect to time and integration over the surface of a sphere whose radius is much larger than the extension of the system under consideration,

$$W = 2c\varepsilon_0 \lim_{\rho \to \infty} \int_0^\infty dt \int_0^{2\pi} d\phi \int_0^\pi d\theta\, \rho^2 \sin\theta\, I(\mathbf{r},t) \tag{10.32}$$

($\rho = |\mathbf{r} - \mathbf{r}_A|$). The ratio W/W_0 ($W_0 = \hbar\tilde{\omega}_{21}$), which is obviously a measure of the fraction of the emitted (radiation) energy, on average, can be regarded as the quantum yield on the time scale $\sim \Gamma^{-1}$. Accordingly, $1 - W/W_0$ measures the fraction of the energy which is effectively absorbed by the bodies.

According to Eq. (10.19), together with Eqs (2.296) and (10.25), the intensity of the emitted radiation, $I(\mathbf{r},t)$, is given by

$$I(\mathbf{r},t) = \left| \int_0^t dt'\, \mathbf{K}_{(E)}^{(+)}(\mathbf{r},t;\mathbf{r}_A,t') \mathbf{d}_{12} e^{-(i\tilde{\omega}_{21}+\Gamma/2)t'} \right|^2$$

$$= \left| \frac{i}{\pi\varepsilon_0 c^2} \int_0^\infty d\omega\, \frac{e^{-(i\tilde{\omega}_{21}+\Gamma/2)t} - e^{-i\omega t}}{i\omega - (i\tilde{\omega}_{21}+\Gamma/2)} \omega^2 \operatorname{Im} \mathbf{G}(\mathbf{r},\mathbf{r}_A,\omega) \right|^2. \tag{10.33}$$

In the free-space case, where $\mathbf{G}(\mathbf{r},\mathbf{r}',\omega) = \mathbf{G}_0(\mathbf{r},\mathbf{r}',\omega)$ is valid, with $\mathbf{G}_0(\mathbf{r},\mathbf{r}',\omega)$ being given by Eq. (2.298), straightforward calculation of the integral in Eq. (10.33) leads to

$$I_0(\mathbf{r},t) = \left(\frac{\tilde{\omega}_{21}^2 d_{21} \sin\theta}{4\pi\varepsilon_0 c^2 \rho} \right)^2 e^{-\Gamma_0(t-\rho/c)} \Theta(t-\rho/c) + O(\rho^{-3}), \tag{10.34}$$

---

5) In fact, the imaginary part of the Green tensor at equal positions in Eq. (10.27) is singular for any realistic bulk material. Physically, this singularity is fictitious, because an atom, though surrounded by matter, should always be localized in a (more or less small) free-space region.

where the first term represents the relevant far-field intensity. It is straightforward to prove that Eq. (10.32) together with Eq. (10.34) leads to $W_0 = \hbar\tilde{\omega}_{21}$.

The effect of the body-induced change of the electromagnetic vacuum fluctuation can be used to control, to some extent, the process of spontaneous emission. So, spontaneous emission can be either enhanced [Purcell (1946)] or (almost) inhibited [Kleppner (1981)] compared with that in the free space. To be more specific, knowledge of the Green tensor for the body configuration of interest is required. It is worth noting that the Green tensor has been available for a large variety of configurations such as planar, spherically and cylindrically multi-layered ones [see, e. g., Tai (1994); Chew (1995)].

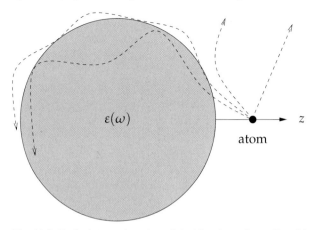

Fig. 10.1 Excited atom close to a dielectric microsphere. Possible ways of spontaneously emitting a photon are sketched by dashed arrows.

To give an impression of what can be observed, let us consider a two-level atom close to a dielectric microsphere (Fig. 10.1) whose permittivity is of the Drude–Lorentz type,

$$\epsilon(\omega) = 1 + \frac{\omega_P^2}{\omega_T^2 - \omega^2 - i\omega\gamma}. \tag{10.35}$$

Here ω_P corresponds to the coupling constant, and ω_T and γ are respectively the medium (transverse) oscillation frequency and the absorption linewidth. Note that the permittivity features a band gap between $\omega = \omega_T$ and $\omega = \omega_L = (\omega_T^2 + \omega_P^2)^{1/2}$. Figure 10.2 illustrates the dependence on the transition frequency of the decay rate. From the figure it is clearly seen that, when the atomic transition frequency agrees with (or is close to) the frequency of a field excitation of either whispering gallery type below the band gap ($\tilde{\omega}_{21} < \omega_T$) or surface-guided type inside the band gap ($\omega_T \leq \tilde{\omega}_{21} < \omega_L$),[6] the spontaneous

---

6) The orthogonal modes obtained by solving the homogeneous Helmholtz equation for a dielectric sphere of constant and real permittivity are commonly called whispering gallery modes (surface-guided modes) in the case of positive (negative) permittivity.

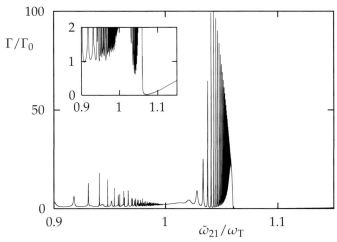

Fig. 10.2 The rate of spontaneous decay Γ of a two-level atom near a dielectric microsphere is shown as a function of the transition frequency $\tilde{\omega}_{21}$ for a radially oriented transition dipole moment, Γ_0 being the decay rate in free space [Eq. (10.29)]. The parameters in the permittivity (10.35) are chosen to be $\omega_P/\omega_T = 0.5$ and $\gamma/\omega_T = 10^{-4}$. The radius of the sphere is $R = 2\lambda_T$, and the distance between the atom and the surface of the sphere is $z_A = 0.1\lambda_T$ ($\lambda_T = 2\pi c/\omega_T$). [After Ho, Knöll and Welsch (2001).]

decay can be strongly enhanced. Figure 10.3 illustrates the fraction of the spontaneously emitted radiation energy, on average. The minima at the field resonance frequencies below the band gap indicate that, although the decay can be noticeably enhanced, the probability of emission of a really observable photon can be substantially reduced compared to the case of spontaneous emission in the free space. Obviously, a photon emitted at such a frequency is typically captured inside the microsphere for some time,[7] and hence the probability of photon absorption is increased. For transition frequencies inside the band gap, two regions can be distinguished. In the low-frequency region, where surface-guided waves are typically excited, radiative decay dominates, i. e., the atomic transition is accompanied by the emission of a photon escaping from the system. Here, the radiation penetration depth into the sphere is small and the probability of a photon being absorbed is also small. With increasing atomic transition frequency the penetration depth increases and the chance of a photon to escape drastically diminishes. As a result, photon absorption dominates.

[7] This time, which is inversely proportional to the width of the respective field resonance line, is of course small compared with the decay time Γ^{-1}. Otherwise the Markov approximation fails (see Section 10.1.2).

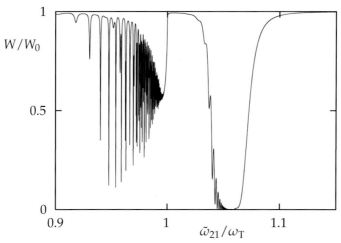

Fig. 10.3 The fraction W/W_0 of spontaneously emitted radiation of a two-level atom near a dielectric microsphere is shown as a function of the transition frequency $\tilde{\omega}_{21}$ for a radially oriented transition dipole moment. The other parameters are the same as in Fig. 10.2. [After Ho, Knöll and Welsch (2001).]

In particular, when the distance z_A between the atom and the surface of the sphere tends to zero,[8] then the decay rates $\Gamma^\perp$ and $\Gamma^\|$, respectively, for radially and tangentially oriented transition dipole moments approach the asymptotic values [Ho, Knöll and Welsch (2001)]

$$\Gamma^\perp = \frac{3\Gamma_0 c^3}{4\tilde{\omega}_{21}^3} \frac{\operatorname{Im}\varepsilon(\tilde{\omega}_{21})}{|\varepsilon(\tilde{\omega}_{21}) + 1|^2} \frac{1}{z_A^3}, \qquad \Gamma^\| = \tfrac{1}{2}\Gamma^\perp, \tag{10.36}$$

which solely result from absorption, as it is proportional to $\operatorname{Im}\varepsilon(\tilde{\omega}_{21})$. In other words, an effectively nonradiative decay is observed. The result reveals that, in the case of strong material absorption, the decay rate rises drastically as the atom approaches the surface of the microsphere, because of near-field assisted energy transfer from the atom to the medium – an effect that is typically observed for metals [see, e.g., Drexhage (1974)]. It should be pointed out that Eq. (10.36) also applies to the case of the atom being in front of a semi-infinite half space [Yeung and Gustafson (1996)], because in the short-distance limit the atom effectively regards the surface of the sphere as a plane.

10.1.1.2 Level shift

Let us return to the shift of the atomic transition frequency $\delta\omega_{21}$, Eq. (10.28). It reflects the fact that even in the case of an atom in the otherwise empty space,

[8] Recall that z must not be smaller than typical interatomic distances in the sphere. Otherwise a microscopic treatment is required.

the unperturbed atomic energy levels are not the observed ones, because the interaction of the atom with the always present electromagnetic vacuum gives rise to level shifts. The effect usually called the Lamb shift[9] was first demonstrated experimentally by Lamb and Retherford (1947) [for the first calculation of the Lamb shift, see Bethe (1947)].

Substituting the decomposition according to Eq. (10.30) for the Green tensor in Eq. (10.28), one can see that the term arising from the free-space Green tensor $G_0(\mathbf{r}_A, \mathbf{r}_A, \omega)$ is divergent and a refined description (including regularization) is required to adequately treat the ($\mathbf{r}_A$-independent) level shift caused by the interaction of the atom with the electromagnetic vacuum in free space.[10] To study the body-induced ($\mathbf{r}_A$-dependent) level shift, the shift observed in the free space may be thought of as being already included in the atomic transition frequency ω_{21}, so that ω_{21} is not the bare transition frequency but the transition frequency that is really observed in the free space, and thus $\delta\omega_{21}$ can be regarded as being determined by the scattering part of the Green tensor $G_S(\mathbf{r}_A, \mathbf{r}_A, \omega)$, leading to

$$\delta\omega_{21} = \frac{P}{\pi\hbar\varepsilon_0 c^2} \int_0^\infty d\omega\, \omega^2 \frac{\mathbf{d}_{21} \operatorname{Im} G_S(\mathbf{r}_A, \mathbf{r}_A, \omega) \mathbf{d}_{12}}{\tilde\omega_{21} - \omega}. \tag{10.37}$$

The integral in Eq. (10.37) can be further evaluated by means of contour integral techniques to obtain

$$\delta\omega_{21} = \delta\omega_{21}^{(1)} + \delta\omega_{21}^{(2)}, \tag{10.38}$$

where

$$\delta\omega_{21}^{(1)} = -\frac{\tilde\omega_{21}^2}{\hbar\varepsilon_0 c^2} \mathbf{d}_{21} \operatorname{Re} G_S(\mathbf{r}_A, \mathbf{r}_A, \tilde\omega_{21}) \mathbf{d}_{12} \tag{10.39}$$

and

$$\delta\omega_{21}^{(2)} = -\frac{\tilde\omega_{21}}{\pi\hbar\varepsilon_0 c^2} \int_0^\infty du\, u^2 \frac{\mathbf{d}_{21} G_S(\mathbf{r}_A, \mathbf{r}_A, iu) \mathbf{d}_{12}}{\tilde\omega_{21}^2 + u^2}. \tag{10.40}$$

In fact, the off-resonant contribution $\delta\omega_{21}^{(2)}$ to the shift of the transition frequency is not complete, because – apart from the two-state model – the underlying rotating-wave approximation does not take into account the purely off-resonant lower-state level shift [cf. Section 10.2, Eqs (10.70)–(10.72)]. However, even if $\delta\omega_{21}^{(2)}$ is complemented by the missing terms, it would typically remain

---

9) More generally, the effect of level shifting observed when a dynamical system interacts with a dissipative system (Chapter 5) is also called Lamb shift.
10) For the problem of the level shift in free space, which has widely been studied, we refer the reader to the literature [e. g., Milonni (1994)].

small compared with the resonant contribution $\delta\omega_{21}^{(1)}$ and may be therefore disregarded in many cases. In this approximation, Eq. (10.38) reduces to

$$\delta\omega_{21} = -\frac{\tilde{\omega}_{21}^2}{\hbar\varepsilon_0 c^2}\,\mathbf{d}_{21}\,\mathrm{Re}\,\mathbf{G}_S(\mathbf{r}_A,\mathbf{r}_A,\tilde{\omega}_{21})\mathbf{d}_{12}\,. \tag{10.41}$$

Whereas the effect of the bodies on the decay rate is determined by the imaginary part of the scattering part of the Green tensor, the real part is responsible for the shift of the transition frequency, as a comparison of Eq. (10.31) with Eq. (10.41) shows.

As in the case of the decay rate, further evaluation of Eq. (10.41) requires knowledge of the Green tensor for the actual body configuration. Again, the dependence of $\delta\omega_{21}$ on the distance z_A between an atom and the surface of a body in front of which the atom is situated becomes independent of the actual form of the body, if z_A is sufficiently small, leading to the short-distance law

$$\delta\omega_{21}^{\perp} = -\frac{3\Gamma_0 c^3}{16\tilde{\omega}_{21}^3}\frac{|\varepsilon(\tilde{\omega}_{21})|^2 - 1}{|\varepsilon(\tilde{\omega}_{21}) + 1|^2}\frac{1}{z_A^3}\,, \qquad \delta\omega_{21}^{\parallel} = \tfrac{1}{2}\delta\omega_{21}^{\perp} \tag{10.42}$$

[see, e. g., Ho, Knöll and Welsch (2001)]. This effect can be employed in scanning near-field optical microscopy to detect surface corrugation or impurities via the changes in the line shift of the radiation emitted by a probe atom, unless material absorption dominates the transition [see, e. g., Henkel and Sandoghdar (1998)]. The high sensitivity of the method results from the cubic dependence of the line shift on the inverse distance of the probe atom from the surface. As we will see in Section 10.2.1, body-induced level shifts are not only of spectroscopic relevance, but they are also closely related to the van der Waals force acting on an atom located near to macroscopic bodies.

10.1.2
Strong atom–field coupling

When a resonator-like arrangement of one or more macroscopic bodies features sharply peaked electromagnetic field resonances (such as the surface-guided waves or the whispering gallery waves in the case of a microsphere considered in Section 10.1.1, or the intra-cavity waves considered in Chapter 9) and the atomic transition frequency approaches the (mid-)frequency of such a resonator-assisted resonance line, then the strength of the atom–field coupling can drastically increase. More precisely, the correlation time, which in this case is determined by the inverse width of the resonator-assisted resonance line, can become much longer than the characteristic time scale on which the atomic-state population noticeably changes. As a consequence, the Markov approximation can fail and the temporal evolution of the occupation probability $|C_2(t)|^2$ of the upper atomic state can become nonexponential.

In order to gain insight into such a non-Markovian regime typical of strong atom–field coupling, let us consider a resonator-like equipment, referred to as cavity in the following, and assume that only one line of the cavity field, say the νth of (mid-)frequency ω_ν, is involved in the strong atom–field coupling – a case which requires the lines to be sufficiently well separated from each other in the relevant frequency interval. Accordingly we may decompose the integral kernel $\tilde{K}(t)$ as given by Eq. (10.23) together with Eq. (10.13) into two parts,

$$\tilde{K}(t) = \tilde{K}^{(1)}(t) + \tilde{K}^{(2)}(t), \tag{10.43}$$

where

$$\tilde{K}^{(1)}(t) = -\frac{1}{\hbar\pi\varepsilon_0 c^2} \int_{\omega_\nu - \frac{1}{2}\Delta\omega}^{\omega_\nu + \frac{1}{2}\Delta\omega} d\omega\, \omega^2 e^{-i(\omega-\tilde{\omega}_{21})t} \mathbf{d}_{21} \mathrm{Im}\, \mathbf{G}(\mathbf{r}_A, \mathbf{r}_A, \omega) \mathbf{d}_{12}, \tag{10.44}$$

$\Delta\omega$ being a measure of the separation of two adjacent lines, is related to the cavity-resonance line under consideration, and $\tilde{K}^{(2)}(t)$ which is related to the residual cavity field may be regarded as being responsible for the shift of the atomic transition frequency, similar to Eq. (10.28). To further evaluate $\tilde{K}^{(1)}(t)$, let us assume that the cavity-resonance line can be approximated by a Lorentzian,

$$\tilde{K}^{(1)}(t) = -\frac{\omega_\nu^2 e^{i\delta_\nu t}}{\hbar\pi\varepsilon_0 c^2} \mathbf{d}_{21} \mathrm{Im}\, \mathbf{G}(\mathbf{r}_A, \mathbf{r}_A, \omega_\nu) \mathbf{d}_{12} \gamma_\nu^2 \int_{\omega_\nu - \frac{1}{2}\Delta\omega}^{\omega_\nu + \frac{1}{2}\Delta\omega} d\omega\, \frac{e^{-i(\omega-\omega_\nu)t}}{(\omega-\omega_\nu)^2 + \gamma_\nu^2}, \tag{10.45}$$

where

$$\delta_\nu = \tilde{\omega}_{21} - \omega_\nu. \tag{10.46}$$

If the linewidth is sufficiently small compared with the line separation, i.e., $\gamma_\nu \ll \Delta\omega$, the upper (lower) limit of the integral may be extended to infinity (minus infinity), leading to

$$\tilde{K}^{(1)}(t) = -\tfrac{1}{2}\Gamma_\nu \gamma_\nu e^{i\delta_\nu t} e^{-\gamma_\nu |t|}, \tag{10.47}$$

where Γ_ν is defined according to Eq. (10.27), with $\tilde{\omega}_{21}$ being replaced with ω_ν,

$$\Gamma_\nu = \frac{2\omega_\nu^2}{\hbar\varepsilon_0 c^2} \mathbf{d}_{21} \mathrm{Im}\, \mathbf{G}(\mathbf{r}_A, \mathbf{r}_A, \omega_\nu) \mathbf{d}_{12}. \tag{10.48}$$

Hence the integro-differential equation (10.22) approximates to

$$\dot{\tilde{C}}_2(t) = \int_0^t dt'\, \tilde{K}^{(1)}(t-t') \tilde{C}_2(t')$$

$$= -\tfrac{1}{2}\Gamma_\nu \gamma_\nu \int_0^t dt'\, e^{(i\delta_\nu - \gamma_\nu)(t-t')} \tilde{C}_2(t'), \tag{10.49}$$

which corresponds to the differential equation

$$\ddot{\tilde{C}}_2(t) - (i\delta_\nu - \gamma_\nu)\dot{\tilde{C}}_2(t) + \tfrac{1}{4}\Omega_\nu^2 \tilde{C}_2(t) = 0 \tag{10.50}$$

$[\tilde{C}_2(0)=1, \dot{\tilde{C}}_2(0)=0]$, where

$$\Omega_\nu = \sqrt{2\Gamma_\nu \gamma_\nu} \tag{10.51}$$

is the vacuum Rabi frequency with respect to the νth resonance line of the cavity field.

Equation (10.50) can be solved by means of the standard ansatz $\tilde{C}_2(t) \sim e^{\lambda t}$, leading to

$$\lambda^2 - (i\delta_\nu - \gamma_\nu)\lambda + \tfrac{1}{4}\Omega_\nu^2 = 0, \tag{10.52}$$

from which

$$\lambda = \tfrac{1}{2}(i\delta_\nu - \gamma_\nu) \pm \tfrac{1}{2}\sqrt{(i\delta_\nu - \gamma_\nu)^2 - \Omega_\nu^2} \tag{10.53}$$

follows. Strong atom–field coupling requires that the inequalities[11]

$$|\delta_\nu| \ll \Omega_\nu, \quad \gamma_\nu \ll \Omega_\nu \tag{10.54}$$

are fulfilled. Hence the square root on the right-hand side of Eq. (10.53) may be expanded to linear order in δ_ν and γ_ν to approximately obtain

$$\lambda = \tfrac{1}{2}(\pm\Omega_\nu + \delta_\nu)i - \tfrac{1}{2}\gamma_\nu, \tag{10.55}$$

leading to

$$\tilde{C}_2(t) = e^{\frac{1}{2}(i\delta_\nu - \gamma_\nu)t}\left[\cos\left(\tfrac{1}{2}\Omega_\nu t\right) - \frac{i\delta_\nu - \gamma_\nu}{\Omega_\nu}\sin\left(\tfrac{1}{2}\Omega_\nu t\right)\right]$$

$$\simeq e^{\frac{1}{2}(i\delta_\nu - \gamma_\nu)t}\cos\left(\tfrac{1}{2}\Omega_\nu t\right). \tag{10.56}$$

In contrast to the irreversible exponential decay of the upper-state occupation probability which is typical of weak atom–field coupling, damped (vacuum) Rabi oscillations are typically observed in the case of strong atom–field coupling. In particular, on the time scale $\sim \Omega_\nu^{-1}$ the atom and the cavity field periodically exchange excitation – a process which is not disturbed by dissipation so that it becomes reversible on this time scale:

$$|\tilde{C}_2(t)|^2 = |C_2(t)|^2 = 1 - |C_1(t)|^2 = \cos^2\left(\tfrac{1}{2}\Omega_\nu t\right)$$
$$= \tfrac{1}{2}[1 + \cos(\Omega_\nu t)]. \tag{10.57}$$

---

11) Note that the second condition implies a sufficiently high-Q cavity with respect of the νth line, in general ($Q = \omega_\nu/\gamma_\nu \gg 1$).

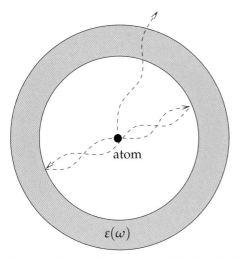

Fig. 10.4 Atom at the center of a dielectric spherical-shell cavity. Possible waves which an excited atom can spontaneously emit are sketched by dashed arrows.

Physically this means that a photon spontaneously emitted by the atom in the upper quantum state will be reabsorbed by the atom in the further course of time, thereby again exciting the atom, and the cycle of photon emission and reabsorption begins anew. The process can be obviously described by resonantly coupling the two-level atom to a single mode of an ideal cavity ($Q \to \infty$). Though only valid on a time scale that is sufficiently short compared with the inverse of the small, but always finite, width of the cavity line with which the atom strongly interacts, one advantage of such a model of strong atom–field coupling – known as the Jaynes–Cummings model (Section 12.1) – is the fact that the effect of arbitrarily excited states of the cavity field can be included in the theory more easily than would be the case in the exact description.

To illustrate the effect of strong atom–field coupling, let us consider an excited two-level atom located at the center of a dielectric spherical-shell cavity (Fig. 10.4). Typical examples of the temporal evolution of the occupation probability $|C_2(t)|^2$ of the upper atomic state are shown in Fig. 10.5. They were obtained by numerical integration of the exact integral equation (10.14), with the permittivity being of Drude–Lorentz type according to Eq. (10.35), by assuming the atomic transition being tuned to a cavity-resonance line in the middle of the band gap. The figure reveals that, with decreasing value of γ_ν (increasing Q value), the Rabi oscillations become more and more pronounced, in full agreement with Eq. (10.56).

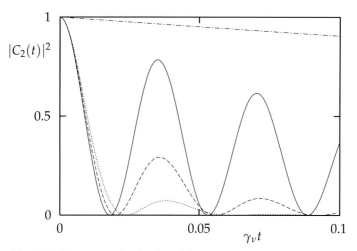

Fig. 10.5 The temporal evaluation of the occupation probability $|C_2(t)|^2$ of the upper atomic state is shown for an atom at the center of a spherical-shell cavity of inner radius $R = 30\,\lambda_T$ and thickness $d = \lambda_T$ [$\omega_P/\omega_T = 0.5$, $\tilde{\omega}_{21}/\omega_T\,\tilde{\omega}_v/\omega_T = 1.046448$, $\Gamma_0\lambda_T/(2c) = 10^{-6}$; $\gamma_v/\omega_T = 10^{-4}$ (solid line), $\gamma_v/\omega_T = 5 \times 10^{-4}$ (dashed line), $\gamma_v/\omega_T = 10^{-3}$ (dotted line). For comparison, the exponential decay in free space (dash-dotted line) is shown. [After Ho, Knöll and Welsch (2000).]

10.2
Vacuum forces

It is well known that there is an attractive force between electrically neutral, unpolarized ground-state atoms or molecules. This force, also called the van der Waals force, represents a pure quantum effect. Restricting his attention to two-level atoms and employing fourth-order time-independent perturbation theory, London (1930) derived the potential associated with the force to be

$$U(r) = -\frac{3\hbar\omega_{21}\alpha_{\text{stat}}^2}{4r^6} \tag{10.58}$$

(r, distance between the atoms; α_{stat}, static atomic polarizability). Verwey and Overbeek (1948) had already pointed out that the r^{-6} potential does not apply in the retarded limit, i.e., when the separation of the atoms is large compared with the atomic transition wavelength. A consistent quantum mechanical theory which closed this loophole was then given by Casimir and Polder (1948).[12] In particular, they found that, for large interatomic separations, the van der Waals potential varies as r^{-7} due to retardation. Moreover, the theory clearly showed that the origin of the force must be seen in the interaction of the atoms with the fluctuating electromagnetic quantum vacuum.

12) In this context, van der Waals forces are also referred to as Casimir–Polder forces.

10.2 Vacuum forces

Forces of van der Waals type are not only observed on a microscopic level but also on a macroscopic level. Typical examples are the force (also referred to as the van der Waals force) to which an atom is subject in the presence of macroscopic bodies, or the force (referred to as the Casimir force) between macroscopic bodies – dense media that may be thought of as consisting of a huge number of interacting atoms. Both types of force can be regarded as being macroscopic manifestations of microscopic van der Waals forces. In the following we restrict our attention to dielectric bodies, noting that the theory can also be extended to other materials.

10.2.1
Force on an atom

From Eq. (2.10) we know that the classical Lorentz force density $\mathbf{f}_L(\mathbf{r})$ acting on a charge density $\rho(\mathbf{r})$ and a current density $\mathbf{j}(\mathbf{r})$ in an electric field $\mathbf{E}(\mathbf{r})$ and an induction field $\mathbf{B}(\mathbf{r})$ reads

$$\mathbf{f}_L(\mathbf{r}) = \rho(\mathbf{r})\mathbf{E}(\mathbf{r}) + \mathbf{j}(\mathbf{r}) \times \mathbf{B}(\mathbf{r}), \tag{10.59}$$

and the total Lorentz force acting on the matter contained inside some space region (of volume V) can be obtained according to

$$\mathbf{F} = \int_V d^3 r \, \mathbf{f}_L(\mathbf{r}). \tag{10.60}$$

In the case of neutral matter which is electrically polarizable, the charge and current densities can be regarded as being the polarization charge and current densities $\rho_P(\mathbf{r})$ and $\mathbf{j}_P(\mathbf{r})$, respectively, so that in Eq. (10.59) we may set

$$\rho(\mathbf{r}) = \rho_P(\mathbf{r}) = -\nabla \mathbf{P}(\mathbf{r}) \tag{10.61}$$

and

$$\mathbf{j}(\mathbf{r}) = \mathbf{j}_P(\mathbf{r}) = \dot{\mathbf{P}}(\mathbf{r}). \tag{10.62}$$

Let us consider a neutral, electrically polarizable atom at position $\mathbf{r}_A$ and restrict our attention to the electric-dipole approximation, i.e.,

$$\mathbf{P}(\mathbf{r}) \equiv \mathbf{P}_A(\mathbf{r}) = \mathbf{d}\delta(\mathbf{r} - \mathbf{r}_A). \tag{10.63}$$

Combining Eqs (10.59)–(10.63), we find that the Lorentz force acting on the atom can be given in the form

$$\mathbf{F} = \mathbf{d}\nabla_A \otimes \mathbf{E}(\mathbf{r}_A) + \dot{\mathbf{d}} \times \mathbf{B}(\mathbf{r}_A)$$
$$= \nabla \otimes \mathbf{d}\mathbf{E}(\mathbf{r})\Big|_{\mathbf{r}=\mathbf{r}_A} + \frac{\partial [\mathbf{d} \times \mathbf{B}(\mathbf{r})]}{\partial t}\Big|_{\mathbf{r}=\mathbf{r}_A} \tag{10.64}$$

($\nabla_A \hat{=} \partial/\partial x_{Ak}$), where the second line follows from the first one by using the Maxwell equation (2.2). Equation (10.64), which as an operator-valued equation is also valid in quantum theory, may serve as a starting point for calculating the radiation force acting (in the electric-dipole approximation) on a neutral atom, by taking the expectation value with respect to the internal atomic quantum state (i.e., the electronic quantum state in the case of an atom) and the quantum state of the electromagnetic field,

$$\mathbf{F} = \nabla \langle \hat{\mathbf{d}} \hat{\mathbf{E}}(\mathbf{r}) \rangle |_{\mathbf{r}=\mathbf{r}_A} + \frac{\partial \langle \hat{\mathbf{d}} \times \hat{\mathbf{B}}(\mathbf{r}) \rangle}{\partial t} \bigg|_{\mathbf{r}=\mathbf{r}_A}. \tag{10.65}$$

This equation can be interpreted in several ways. So it can be regarded as giving the force in the Newtonian equation of motion for the center-of-mass coordinate, which is further evaluated within the frame of quantum mechanics or, if possible, also within the frame of classical mechanics. Clearly, the center-of-mass motion should be sufficiently slow, so that it (approximately) decouples from the internal motion in the spirit of a Born–Oppenheimer approximation. Equation (10.65) can be also regarded as determining the force that must be compensated for in the case when the center-of-mass coordinate may be considered as a given (classical) parameter controlled externally. Since there is no need here to distinguish between the possible interpretations, we do not use the operator hat on the center-of-mass coordinate $\mathbf{r}_A$ in Eq. (10.65). In particular, when the atom–field system is prepared in an energy eigenstate, then the expectation value of the magnetic part of the force obviously vanishes, so that only the expectation value of the electric part needs to be considered, i.e.,

$$\mathbf{F} = \nabla \langle \hat{\mathbf{d}} \hat{\mathbf{E}}(\mathbf{r}) \rangle |_{\mathbf{r}=\mathbf{r}_A}. \tag{10.66}$$

To calculate the van der Waals force which is observed in the case when the atom is subject to a body-assisted electromagnetic vacuum, we restrict our attention to energy eigenstates. In this case Eq. (10.66) applies, and a comparison with the (multipolar-coupling) interaction energy (2.239)[13] may suggest, at first glance, that the interaction energy plays the role of the potential of the force. However, the exact state vector with which the expectation value is calculated introduces an additional $\mathbf{r}_A$ dependence, which prevents this idea from being true in general. Fortunately, the expectation value calculated with the state vector in the lowest (leading) order of perturbation theory is an exception, so that in this most commonly considered case, the force can be simply obtained by taking the negative gradient of the (position-dependent) part

---

13) Note that Eq. (10.64) also holds with **E** being replaced by the transformed **E**′, so that in the further calculations the prime can be omitted for notational convenience.

of the body-induced shift of the corresponding energy, as we will do in the following.

10.2.1.1 Lowest-order perturbation theory

Let the body-assisted electromagnetic field be in the ground state and the atom in the nth energy eigenstate. The interaction Hamiltonian (2.239) in the form of Eq. (2.243) implies that the lowest-order energy shift δE_n is obtained in the second order of the perturbation series of the energy, i.e.,

$$\delta E_n = -\frac{1}{\hbar} \sum_k \mathcal{P} \int_0^\infty d\omega \int d^3r \, \frac{|\langle n|\langle\{0\}|\hat{H}_{\text{int}}|\{1(\mathbf{r},\omega)\}\rangle|k\rangle|^2}{\omega_{kn} + \omega} \tag{10.67}$$

$[\omega_{kn} = (E_k - E_n)/\hbar$ are the unperturbed transition frequencies], where [recall Eqs (10.7) and (10.8)]

$$\langle n|\langle\{0\}|\hat{H}_{\text{int}}|\{1(\mathbf{r},\omega)\}\rangle|k\rangle = -\langle n|\langle\{0\}|\hat{\mathbf{d}}\hat{\mathbf{E}}(\mathbf{r}_A)|\{1(\mathbf{r},\omega)\}\rangle|k\rangle$$

$$= -i\sqrt{\frac{\hbar}{\pi\varepsilon_0}} \frac{\omega^2}{c^2} \sqrt{\text{Im}\,\varepsilon(\mathbf{r},\omega)}\, \mathbf{d}_{nk} \mathbf{G}(\mathbf{r}_A, \mathbf{r}, \omega). \tag{10.68}$$

Combining Eqs (10.67) and (10.68) yields

$$\delta E_n = -\frac{1}{\pi\varepsilon_0 c^4} \sum_k \mathcal{P} \int_0^\infty d\omega \left[\frac{\omega^4}{\omega_{kn} + \omega} \right.$$

$$\left. \times \int d^3r\, \text{Im}\,\varepsilon(\mathbf{r},\omega) \mathbf{d}_{nk} \mathbf{G}(\mathbf{r}_A, \mathbf{r}, \omega) \mathbf{G}^*(\mathbf{r}, \mathbf{r}_A, \omega) \mathbf{d}_{kn} \right], \tag{10.69}$$

which, with the help of the relation (A.3), can be performed to obtain

$$\delta E_n = -\frac{1}{\pi\varepsilon_0 c^2} \sum_k \mathcal{P} \int_0^\infty d\omega \, \frac{\omega^2 \mathbf{d}_{nk} \text{Im}\, \mathbf{G}(\mathbf{r}_A, \mathbf{r}_A, \omega) \mathbf{d}_{kn}}{\omega_{kn} + \omega}. \tag{10.70}$$

Before proceeding let us briefly make contact with the frequency shift $\delta\omega_{21}$ as given by Eq. (10.28). We first note that from Eq. (10.70) the shifted transition frequencies $\tilde{\omega}_{nm}$ can be calculated according to

$$\tilde{\omega}_{nm} = \hbar^{-1}[E_n + \delta E_n - (E_m + \delta E_m)] = \omega_{nm} + \delta\omega_{nm}, \tag{10.71}$$

where the frequency shifts are given by

$$\delta\omega_{nm} = \hbar^{-1}(\delta E_n - \delta E_m). \tag{10.72}$$

Comparing $\delta\omega_{21}$ calculated from Eq. (10.72) [together with Eq. (10.70)] with $\delta\omega_{21}$ from Eq. (10.28), we see, on identifying $\tilde{\omega}_{21}$ in Eq. (10.28) with the unperturbed transition frequency ω_{21}, that in the rotating-wave approximation used to derive Eq. (10.28), the off-resonant lower-state level shift is completely ignored. Further, in Eq. (10.28) only the term with $k=1$ in the sum for δE_2 in Eq. (10.70) is taken into account, which is of course a consequence of the two-level approximation.

As already mentioned in Section 10.1.1, the body-induced level shifts in which we are interested are determined by the scattering part of the Green tensor [recall the decomposition (10.30) of the Green tensor]. Hence the potential $U_n(\mathbf{r}_A)$ for the van der Waals force

$$\mathbf{F}_n(\mathbf{r}_A) = -\nabla_A U_n(\mathbf{r}_A) \tag{10.73}$$

can be obtained from Eq. (10.70) by replacing therein the Green tensor by its scattering part:

$$U_n(\mathbf{r}_A) = -\frac{1}{\pi\varepsilon_0 c^2}\sum_k \mathcal{P}\int_0^\infty d\omega\, \frac{\omega^2 \mathbf{d}_{nk}\mathrm{Im}\,\mathbf{G}_S(\mathbf{r}_A,\mathbf{r}_A,\omega)\mathbf{d}_{kn}}{\omega_{kn}+\omega}. \tag{10.74}$$

Applying contour integral techniques, we may decompose the van der Waals potential $U_n(\mathbf{r}_A)$ into two parts [cf. Eqs (10.37)–(10.40)],

$$U_n(\mathbf{r}_A) = U_n^{(1)}(\mathbf{r}_A) + U_n^{(2)}(\mathbf{r}_A), \tag{10.75}$$

where

$$U_n^{(1)}(\mathbf{r}_A) = -\frac{1}{\varepsilon_0 c^2}\sum_k \Theta(\omega_{nk})\omega_{nk}^2 \mathbf{d}_{nk}\mathrm{Re}\,\mathbf{G}_S(\mathbf{r}_A,\mathbf{r}_A,\omega_{nk})\mathbf{d}_{kn} \tag{10.76}$$

is the resonant part and

$$U_n^{(2)}(\mathbf{r}_A) = \frac{1}{\pi\varepsilon_0 c^2}\sum_k \int_0^\infty du\, u^2\, \frac{\omega_{kn}\mathbf{d}_{nk}\mathbf{G}_S(\mathbf{r}_A,\mathbf{r}_A,iu)\mathbf{d}_{kn}}{\omega_{kn}^2+u^2} \tag{10.77}$$

the off-resonant part. In order to bring Eq. (10.77) into a more compact form, it may be convenient to introduce the lowest-order polarizability tensor[14] attributed to the atom in the nth excited state, viz.

$$\boldsymbol{\alpha}_n(\omega) \equiv \boldsymbol{\alpha}_n^{(0)}(\omega) = \lim_{\epsilon\to 0}\frac{2}{\hbar}\sum_k \frac{\omega_{kn}\mathbf{d}_{nk}\otimes\mathbf{d}_{kn}}{\omega_{kn}^2-\omega^2-i\omega\epsilon}, \tag{10.78}$$

leading to

$$U_n^{(2)}(\mathbf{r}_A) = \frac{\hbar}{2\pi\varepsilon_0 c^2}\int_0^\infty du\, u^2 \mathrm{Tr}\!\left[\boldsymbol{\alpha}_n(iu)\mathbf{G}^{(1)}(\mathbf{r}_A,\mathbf{r}_A,iu)\right]. \tag{10.79}$$

14) See, e. g., Fain and Khanin (1969).

In particular, for an atom in a spherically symmetric state, Eqs (10.76) and (10.79) simplify to

$$U_n^{(1)}(\mathbf{r}_A) = -\frac{1}{3\varepsilon_0 c^2} \sum_k \Theta(\omega_{nk})\omega_{nk}^2 |\mathbf{d}_{nk}|^2 \mathrm{Tr}\left[\mathrm{Re}\, \mathbf{G}_S(\mathbf{r}_A,\mathbf{r}_A,\omega_{nk})\right] \quad (10.80)$$

and

$$U_n^{(2)}(\mathbf{r}_A) = \frac{\hbar}{2\pi\varepsilon_0 c^2} \int_0^\infty du\, u^2 \alpha_n(iu)\, \mathrm{Tr}\, \mathbf{G}_S(\mathbf{r}_A,\mathbf{r}_A,iu). \quad (10.81)$$

To obtain Eq. (10.81) from Eq. (10.79), we have used the relation $\boldsymbol{\alpha}_n(\omega) = \alpha_n(\omega)\mathbf{I}$, where

$$\alpha_n(\omega) = \lim_{\epsilon\to 0} \frac{2}{3\hbar} \sum_k \frac{\omega_{kn}|\mathbf{d}_{nk}|^2}{\omega_{kn}^2 - \omega^2 - i\omega\epsilon}. \quad (10.82)$$

Equation (10.75), together with Eqs (10.76) and (10.77) [or, equivalently, (10.79)], apply to arbitrary causal dielectric bodies which linearly and locally respond to the electric field;[15] all relevant information on the bodies is contained in the scattering Green tensor. From Eq. (10.76) it is seen that [because of $\Theta(\omega_{nk})$] $U_n^{(1)}(\mathbf{r}_A)$ can only contribute to $U_n(\mathbf{r}_A)$ if the atom is excited. In this case $U_n^{(1)}(\mathbf{r}_A)$ can be expected to be the dominant contribution in general. It should be pointed out that since an excited state decays in the further course of time, the force acting on an initially excited atom varies with time until the atom has arrived back at the ground state – an effect that requires a dynamic description rather than the static one considered here [for a dynamic description, see Buhmann, Knöll, Welsch and Ho (2004)]. Note that the dipole matrix elements which enter the spontaneous decay rate attributed to an excited state also enter the excited-state van der Waals potential.

Let us study the van der Waals potential $U(\mathbf{r}_A) \equiv U_0(\mathbf{r}_A) = U_0^{(2)}(\mathbf{r}_A)$ of a ground-state atom in more detail. According to Eq. (10.81) it reads

$$U(\mathbf{r}_A) = \frac{\hbar}{2\pi\varepsilon_0 c^2} \int_0^\infty du\, u^2 \alpha(iu)\, \mathrm{Tr}\, \mathbf{G}_S(\mathbf{r}_A,\mathbf{r}_A,iu). \quad (10.83)$$

[$\alpha(\omega) \equiv \alpha_0(\omega)$]. From Eq. (10.83) it is clearly seen that the van der Waals force acting on a nonexcited, electrically neutral, polarizable particle, represents a quantum effect, which would vanish if $\hbar$ were set equal to zero in Eq. (10.83).[16] It is basically a pure vacuum effect, because the overall system is in the ground

---

15) It can be shown that Eq. (10.75) together with Eqs (10.76) and (10.77) [or, equivalently, (10.79)] also apply to causally, linearly and locally responding magnetodielectric bodies [Buhmann, Knöll, Welsch and Ho (2004)].
16) Note that the (scattering part of the) Green tensor is a classical quantity and the polarizability can also be introduced classically.

state, and hence a fully quantum theoretical treatment is necessary. By contrast, if the atom is in an excited state, the resonant contribution $U_n^{(1)}(\mathbf{r}_A)$ as given by Eq. (10.76) can be understood, in a sense, on the basis of a quasi-classical description, by considering the interaction of an oscillating classical dipole with the field scattered by the bodies.

10.2.1.2 Atom in front of a planar body

To give a typical example of the van der Waals potential of a ground-state atom, let us consider an atom in front of a planar body. It can be calculated in a straightforward way by inserting in Eq. (10.83) the well-known (scattering part of the) Green tensor for a multi-layer dielectric plate of infinite lateral extension.[16] Since the calculation is somewhat lengthy, we renounce it here and present the result at once:

$$U(z_A) = \frac{\hbar \mu_0}{8\pi^2} \int_0^\infty du\, u^2 \alpha(iu) \int_0^\infty dq\, \frac{q}{\kappa} e^{-2\kappa z_A} \left[r_{s-} - \left(1 + 2\frac{q^2 c^2}{u^2}\right) r_{p-} \right]. \tag{10.84}$$

Here, the atom is on the right of the plate, with z_A being the (positive) distance between the surface of the plate and the atom (see Fig. 10.6). Further, q is the absolute value of the transverse (lateral) component $\mathbf{q}$ of the wave vector $\mathbf{k}$, and $r_{\sigma-}$ ($\sigma = s, p$) are the (generalized) reflection coefficients for s-polarized and p-polarized waves[17] with respect to the surface (at $z=0$) which faces the atom, and

$$\kappa = \kappa(iu, q) = \sqrt{q^2 - \frac{(iu)^2}{c^2}}. \tag{10.85}$$

Since, depending on the actual layer structure of the plate, the reflection coefficients $r_{\sigma-} = r_{\sigma-}(iu, q)$ can be more or less complicated functions of u and

---

16) For a suitable representation of the Green tensor for a multi-layer dielectric structure of infinite lateral extension, see, e.g., Tomaš (1995). Since the transverse projection $\mathbf{q}$ of the wave vector is conserved and the polarizations $\sigma = s, p$ decouple, the scattering part of the Green tensor within each layer can be expressed in terms of reflection coefficients $r_{\sigma\pm} = r_{\sigma\pm}(\omega, q)$ referring to reflection of waves at the right (+) and left (−) wall (formed by the respective layers), as seen from the layer under consideration. Explicit (recurrence) expressions for the reflection coefficients are available if the walls are multi-slab magnetodielectrics like Bragg mirrors. For continuous wall profiles, Riccati-type equations have to be solved [Chew (1995)].

17) The corresponding polarization vectors are $\mathbf{e}_{s-} = \mathbf{q}/q \times \mathbf{e}_z$ and $\mathbf{e}_{p-} = -(iq\mathbf{e}_z - \kappa \mathbf{q}/q)/k$.

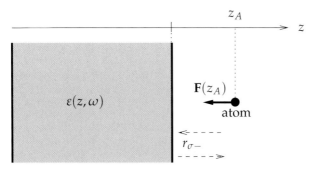

Fig. 10.6 A ground-state atom near to a (multi-layer) dielectric plate is subject to an attractive force $\mathbf{F}(z_A)$, Eq. (10.84), which, for chosen atomic position z_A, is determined by the (zeroth-order) atomic polarizability and the (generalized) reflection coefficients $r_{\sigma-}$ ($\sigma = s, p$). Note that only virtual photons are involved in the scattering at the plate.

q, further evaluation of the integrals in Eq. (10.84) requires application of numerical methods, in general.

In the limiting case of a perfectly reflecting plate such that

$$r_{p-} = -r_{s-} = 1, \tag{10.86}$$

the second integral in Eq. (10.84) can be easily calculated, leading to the attractive potential

$$U(z_A) = -\frac{\hbar}{16\pi^2 \varepsilon_0 z_A^3} \int_0^\infty du\, \alpha(iu) e^{-2uz_A/c} \left[1 + 2\left(\frac{uz_A}{c}\right) + 2\left(\frac{uz_A}{c}\right)^2\right], \tag{10.87}$$

which is exactly the formula first derived by Casimir and Polder (1948) for the potential of a ground-state atom in front of a perfectly conducting plate. In the short-distance (i. e., nonretarded) limit we may approximately set $e^{-2uz_A/c} = 1$ in the integral in Eq. (10.87) and neglect the second and third terms in the square brackets to recover, on using Eq. (10.82), the result of Lennard-Jones (1932):

$$U(z_A) = -\frac{1}{48\pi\varepsilon_0}\frac{1}{z_A^3}\sum_k |\mathbf{d}_{0k}|^2 = -\frac{\langle 0|\hat{\mathbf{d}}^2|0\rangle}{48\pi\varepsilon_0}\frac{1}{z_A^3}. \tag{10.88}$$

In the long-distance (i. e., retarded) limit the atomic polarizability $\alpha(iu)$ may be approximately replaced by its static value $\alpha(0)$ and put in front of the integral, leading to

$$U(z_A) = -\frac{3\hbar c\alpha(0)}{32\pi^2\varepsilon_0}\frac{1}{z_A^4}. \tag{10.89}$$

As already mentioned, the general formulas that do not explicitly make use of the material properties are also valid for other materials than dielectrics. Hence relations other than the ones given in Eq. (10.86) might be attributed to a perfectly reflecting plate. In particular, when $r_{p-} = -r_{s-} = -1$ is set, then the expression in the square brackets in Eq. (10.84) changes the sign; hence $U(z_A)$ changes to $-U(z_A)$ and as a result a repulsive force is observed. This case of a perfectly reflecting plate would correspond to an infinitely permeable magnetic plate – a case which is, of course, far from reality. Nevertheless, it reveals a very general aspect. The fact that Maxwell's equations in the absence of (free) charges and currents are invariant under a duality transformation between electric and magnetic fields can be exploited to extend the notion of forces acting on electrically polarizable objects to magnetically polarizable objects. Thus, knowing the attractive van der Waals force between two electrically polarizable particles (e. g., atoms), one can infer the existence of an analogous attractive force between two magnetically polarizable particles, which may be obtained from the former by replacing the electric polarizabilities by the corresponding magnetic ones. In contrast, the force between two polarizable particles of opposite type is repulsive [Feinberg and Sucher (1970)], which implies that an atom in front of a magnetic plate is subject to a repulsive force.

The van der Waals potential $U_n(\mathbf{r}_A)$ as given by Eq. (10.75) together with Eqs (10.76) and (10.77) [or (10.79)] is the body-induced shift of the (unperturbed) atomic energy level E_n. This offers the possibility of measuring it by means of spectroscopic methods. In particular, the powerful methods of laser spectroscopy used in cavity QED to study fundamental quantum phenomena can be employed also to perform direct and precise measurements of the van der Waals coupling between an atom and cavity walls, in which the interaction is quantitatively studied as a function of controlled separation and of the electronic state of the atom [Sandoghdar, Sukenik, Hinds and Haroche (1992)].

10.2.2
The Casimir force

In classical electrodynamics, electrically neutral material bodies at zero temperature which do not carry a permanent polarization (and/or magnetization) are not subject to a Lorentz force in the absence of external electromagnetic fields. The situation changes in quantum electrodynamics, since the body-assisted vacuum fluctuations of the electromagnetic field can give rise to a nonvanishing Lorentz force – the Casimir force. Let us assume that the macroscopic bodies consist of distinguishable, polarizable microconstituents commonly called atoms or molecules within the framework of molecular optics. From Section 10.2.1 it is clear that in the case of a large collection of (ground-state) atoms forming a macroscopic body, a van der Waals interaction of the body with other bodies should be observed. Clearly, the resulting Casimir force between macroscopic bodies is in general not simply the sum of the van der Waals forces acting on single (ground-state) atoms, because of many-particle interactions [see, e.g, Buhmann and Welsch (2006)]. Fortunately, the interaction of electromagnetic fields with linear magnetodielectric matter can be expressed, via the permittivity and the permeability, in terms of the Green tensor which a priori takes into account many-particle interactions. Hence, the Casimir force expressed in terms of the Green tensor takes into account many-particle interactions as well.

10.2.2.1 Basic equations

Let us again restrict our attention to dielectric bodies[19] and begin with the classical Lorentz force as given by Eq. (10.60) together with Eqs (10.59), (10.61) and (10.62). In contrast to Eq. (10.63), $\mathbf{P}(\mathbf{r})$ is now the macroscopic polarization field associated with the dielectric medium. Recalling Section 2.4.2, we may it write in the form of

$$\mathbf{P}(\mathbf{r}) = \int_0^\infty d\omega\, \underline{\mathbf{P}}(\mathbf{r}, \omega) + \text{c.c.}, \tag{10.90}$$

where

$$\underline{\mathbf{P}}(\mathbf{r}, \omega) = \varepsilon_0 [\varepsilon(\mathbf{r}, \omega) - 1] \underline{\mathbf{E}}(\mathbf{r}, \omega) + \underline{\mathbf{P}}_N(\mathbf{r}, \omega). \tag{10.91}$$

Inserting Eqs (10.61) and (10.62) into Eq. (10.59) and making use of Eq. (2.2), we find that the Lorentz force density can be rewritten as

$$\mathbf{f}_L(\mathbf{r}) = \nabla' \otimes \mathbf{P}(\mathbf{r}) \mathbf{E}(\mathbf{r}')\big|_{\mathbf{r}'=\mathbf{r}} + \frac{\partial [\mathbf{P}(\mathbf{r}) \times \mathbf{B}(\mathbf{r})]}{\partial t} + \nabla [\mathbf{P}(\mathbf{r}) \otimes \mathbf{E}(\mathbf{r})] \tag{10.92}$$

($\nabla' \hat{=} \partial/\partial x'_k$). Hence the total Lorentz force acting on the matter which fills some space region of volume V, Eq. (10.60), can be represented in the form

$$\mathbf{F} = \int_V d^3r\, \nabla' \otimes \mathbf{P}(\mathbf{r}) \mathbf{E}(\mathbf{r}')\big|_{\mathbf{r}'=\mathbf{r}} + \frac{d}{dt} \int_V d^3r\, \mathbf{P}(\mathbf{r}) \times \mathbf{B}(\mathbf{r}) + \int_{\partial V} d\mathbf{a} \mathbf{P}(\mathbf{r}) \otimes \mathbf{E}(\mathbf{r}). \tag{10.93}$$

In particular in the case of a body which is not embedded in a medium the surface integral taken with respect to the "outer" values of the integrand vanishes and Eq. (10.93) simplifies to

$$\mathbf{F} = \int_V d^3r\, \nabla' \otimes \mathbf{P}(\mathbf{r}) \mathbf{E}(\mathbf{r}')\big|_{\mathbf{r}'=\mathbf{r}} + \frac{d}{dt} \int_V d^3r\, \mathbf{P}(\mathbf{r}) \times \mathbf{B}(\mathbf{r}), \tag{10.94}$$

---

[19] For an extension to magnetodielectric bodies, see Raabe and Welsch (2005).

which corresponds to the single-atom equation (10.64). Note that for **P**(**r**) from Eq. (10.63), Eq. (10.94) reduces to Eq. (10.64).

Equivalently, by recalling the local momentum balance as given by Eq. (2.7), we see that the total Lorentz force (acting on the matter in a space region of volume V) can be expressed in terms of the electric and induction fields as

$$\mathbf{F} = \int_{\partial V} d\mathbf{a}\, T(\mathbf{r}) - \varepsilon_0 \frac{d}{dt}\int_V d^3r\, \mathbf{E}(\mathbf{r}) \times \mathbf{B}(\mathbf{r}), \tag{10.95}$$

where $T(\mathbf{r})$ is the ordinary stress tensor as given by Eq. (2.11), i.e.,

$$T(\mathbf{r}) = \varepsilon_0 \mathbf{E}(\mathbf{r}) \otimes \mathbf{E}(\mathbf{r}) + \mu_0^{-1} \mathbf{B}(\mathbf{r}) \otimes \mathbf{B}(\mathbf{r}) - \tfrac{1}{2}[\varepsilon_0 \mathbf{E}^2(\mathbf{r}) + \mu_0^{-1} \mathbf{B}^2(\mathbf{r})]I. \tag{10.96}$$

If the volume integral on the right-hand side of Eq. (10.95) can be regarded as being time-independent, then the total force is solely determined by the surface integral

$$\mathbf{F} = \int_{\partial V} d\mathbf{F}, \tag{10.97}$$

where

$$d\mathbf{F} = d\mathbf{a}\, T(\mathbf{r}) = T(\mathbf{r})\, d\mathbf{a} \tag{10.98}$$

can be regarded as the infinitesimal force element acting on the infinitesimal surface element $d\mathbf{a}$.

Equation (10.93) [or (10.94)] and the equivalent equation (10.95) [together with Eq. (10.96)] are respectively basic formulas for calculating radiation forces on macroscopic bodies. Note that Eq. (10.95) is more general than Eq. (10.93), because it is also valid for other materials than dielectrics. The formulas can be analogously used in quantum theory as well, by regarding them as operator-valued ones and taking the expectation values. Recall that the noise polarization $\hat{\mathbf{P}}_N(\mathbf{r}, \omega)$ is given by Eq. (2.210). In particular, Eq. (10.94) then reads

$$\mathbf{F} = \int_V d^3r\, \nabla'\langle\hat{\mathbf{P}}(\mathbf{r})\hat{\mathbf{E}}(\mathbf{r}')\rangle|_{\mathbf{r}'=\mathbf{r}} + \frac{d}{dt}\int_V d^3r\, \langle\hat{\mathbf{P}}(\mathbf{r}) \times \hat{\mathbf{B}}(\mathbf{r})\rangle, \tag{10.99}$$

which obviously corresponds to the single-atom equation (10.65). In a steady-state regime the second term on the right-hand side in this equation vanishes and the force formula reduces to

$$\mathbf{F} = \int_V d^3r\, \nabla'\langle\hat{\mathbf{P}}(\mathbf{r})\hat{\mathbf{E}}(\mathbf{r}')\rangle|_{\mathbf{r}'=\mathbf{r}'} \tag{10.100}$$

which corresponds to Eq. (10.66).

To calculate the Casimir force as the ground-state Lorentz force, we follow the line suggested by classical electrodynamics to derive Eq. (10.97) together with Eqs (10.98) and (10.96), noting that the (expectation value of the) second term on the right-hand side in Eq. (10.95) vanishes. Hence

$$\mathbf{F} = \int_{\partial V} d\mathbf{a}\, T(\mathbf{r}), \qquad (10.101)$$

where the (time-independent) Casimir stress tensor $T(\mathbf{r})$ can be obtained, in agreement with the classical equation (10.96), from the quantum-mechanical ground-state expectation value

$$T(\mathbf{r}, \mathbf{r}') = \varepsilon_0 \langle\{0\}|\hat{\mathbf{E}}(\mathbf{r}) \otimes \hat{\mathbf{E}}(\mathbf{r}')|\{0\}\rangle + \mu_0^{-1} \langle\{0\}|\hat{\mathbf{B}}(\mathbf{r}) \otimes \hat{\mathbf{B}}(\mathbf{r}')|\{0\}\rangle \\ - \tfrac{1}{2} I \left[\varepsilon_0 \langle\{0\}|\hat{\mathbf{E}}(\mathbf{r})\hat{\mathbf{E}}(\mathbf{r}')|\{0\}\rangle + \mu_0^{-1} \langle\{0\}|\hat{\mathbf{B}}(\mathbf{r})\hat{\mathbf{B}}(\mathbf{r}')|\{0\}\rangle \right] \quad (10.102)$$

in the coincidence limit,

$$T(\mathbf{r}) = \lim_{\mathbf{r}' \to \mathbf{r}} T(\mathbf{r}, \mathbf{r}'), \qquad (10.103)$$

where divergent bulk contributions are to be removed before taking the limit. This is always possible if the body under study is embedded in a material environment which is homogeneous at least in the vicinity of the body. If this is not the case, special care and additional considerations may be necessary. Note that in the calculation of the surface integral in Eq. (10.101) the "outer" values of the integrand should be used if ∂V is the interface between an inhomogeneous body and a near-surface homogeneous medium in which the body is embedded.

Recalling the commutation relations (2.208) and (2.209) and making use of the field representation as given by Eqs (2.211)–(2.214), we can calculate the field correlation functions in Eq. (10.102) in a straightforward way. By means of the commutation relations (2.208) and (2.209) it is not difficult to see that [recall Eq. (10.7)]

$$\langle\{0\}|\hat{\mathbf{f}}(\mathbf{r}, \omega) \otimes \hat{\mathbf{f}}^\dagger(\mathbf{r}', \omega')|\{0\}\rangle = \delta(\omega - \omega')\delta(\mathbf{r} - \mathbf{r}'), \qquad (10.104)$$

$$\langle\{0\}|\hat{\mathbf{f}}(\mathbf{r}, \omega) \otimes \hat{\mathbf{f}}(\mathbf{r}', \omega')|\{0\}\rangle = 0. \qquad (10.105)$$

Using Eqs (2.211)–(2.214), together with Eqs (10.104) and (10.105), we then derive, on employing the relation (A.3),

$$\langle\{0\}|\hat{\mathbf{E}}(\mathbf{r}) \otimes \hat{\mathbf{E}}(\mathbf{r}')|\{0\}\rangle = \frac{\hbar\mu_0}{\pi} \int_0^\infty d\omega\, \omega^2 \mathrm{Im}\, G(\mathbf{r}, \mathbf{r}', \omega), \qquad (10.106)$$

$$\langle\{0\}|\hat{\mathbf{B}}(\mathbf{r}) \otimes \hat{\mathbf{B}}(\mathbf{r}')|\{0\}\rangle = -\frac{\hbar\mu_0}{\pi} \int_0^\infty d\omega\, \nabla \times \mathrm{Im}\, G(\mathbf{r}, \mathbf{r}', \omega) \times \overleftarrow{\nabla}'. \qquad (10.107)$$

Combination of Eqs (10.102), (10.106) and (10.107) eventually yields

$$T(\mathbf{r}, \mathbf{r}') = \theta(\mathbf{r}, \mathbf{r}') - \tfrac{1}{2} I \,\mathrm{Tr}\, \theta(\mathbf{r}, \mathbf{r}'), \qquad (10.108)$$

where

$$\theta(\mathbf{r},\mathbf{r}') = \frac{\hbar}{\pi} \int_0^\infty d\omega \left[\frac{\omega^2}{c^2} \operatorname{Im} \mathbf{G}(\mathbf{r},\mathbf{r}',\omega) - \boldsymbol{\nabla} \times \operatorname{Im} \mathbf{G}(\mathbf{r},\mathbf{r}',\omega) \times \overleftarrow{\boldsymbol{\nabla}}' \right]. \tag{10.109}$$

Note that the permittivity $\varepsilon(\mathbf{r},\omega)$ does not appear explicitly in Eq. (10.109), but only via the Green tensor $\mathbf{G}(\mathbf{r},\mathbf{r}',\omega)$.[20] Having removed divergent bulk contributions [by replacing the Green tensor $\mathbf{G}(\mathbf{r},\mathbf{r}',\omega)$ with its scattering part $\mathbf{G}_S(\mathbf{r},\mathbf{r}',\omega)$, cf. Eq. (10.30)], we may take the imaginary part of the whole integral instead of the integrand in Eq. (10.109) and rotate the integration contour in the usual way toward the imaginary frequency axis, on which the Green tensor is real. In this way we arrive at

$$\mathbf{T}(\mathbf{r}) = \boldsymbol{\theta}_S(\mathbf{r}) - \tfrac{1}{2} \mathbf{I} \operatorname{Tr} \boldsymbol{\theta}_S(\mathbf{r}), \tag{10.110}$$

where

$$\boldsymbol{\theta}_S(\mathbf{r}) = -\frac{\hbar}{\pi} \int_0^\infty du \left[\frac{u^2}{c^2} \mathbf{G}_S(\mathbf{r},\mathbf{r},iu) + \boldsymbol{\nabla} \times \mathbf{G}_S(\mathbf{r},\mathbf{r},iu) \times \overleftarrow{\boldsymbol{\nabla}}' \right]. \tag{10.111}$$

Now we insert Eq. (10.110) into Eq. (10.101) to obtain the following expression for the Casimir force:

$$\mathbf{F} = \int_{\partial V} d\mathbf{a} \left[\boldsymbol{\theta}_S(\mathbf{r}) - \tfrac{1}{2} \mathbf{I} \operatorname{Tr} \boldsymbol{\theta}_S(\mathbf{r}) \right]. \tag{10.112}$$

10.2.2.2 Planar structures

Let us apply the theory to a planar dielectric structure defined according to

$$\varepsilon(\mathbf{r},\omega) = \begin{cases} \varepsilon_-(z,\omega) & z < 0, \\ \varepsilon(\omega) & 0 < z < d, \\ \varepsilon_+(z,\omega) & z > d \end{cases} \tag{10.113}$$

(Fig. 10.7). To determine the Casimir stress in the interspace $0 < z < d$, we need, according to Eqs (10.110) and (10.111), the (scattering part of the) Green tensor for both spatial arguments within the interspace ($0 < z = z' < d$). As in the example studied in Section 10.2.1.2, the calculation can be performed on the basis of the well-known Green tensor for a multi-layer dielectric structure of infinite lateral extension. We again renounce the rather lengthy but straightforward calculation and present the final result. For symmetry reasons it is clear that the stress tensor effectively reduces to the T_{zz} component, which can be given in the form of [Raabe and Welsch (2005)]

$$T_{zz}(z) = -\frac{\hbar}{8\pi^2} \int_0^\infty du \int_0^\infty dq \, \frac{q}{\kappa(iu,q)} g(z,iu,q), \tag{10.114}$$

---

[20] Equation (10.109) is also valid for magnetodielectrics, if $\mathbf{G}(\mathbf{r},\mathbf{r}',\omega)$ is understood as the Green tensor of the inhomogeneous Helmholtz equation with $\varepsilon(\mathbf{r},\omega)$ and $\mu(\mathbf{r},\omega)$ [Raabe and Welsch (2005)].

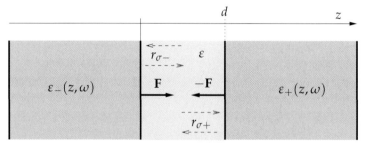

Fig. 10.7 Two (multi-layer) dielectric plates are subject to a mutual attractive force, whose absolute value (per unit area) $F = T_{zz}(d)$, Eq. (10.114), is, for chosen spacing d, determined by the coefficients of (virtual-)photon reflection $r_{\sigma-}$ ($r_{\sigma+}$) at the left (right) plate ($\sigma = s, p$).

where

$$\kappa(iu, q) = \sqrt{q^2 - \varepsilon(iu)\frac{(iu)^2}{c^2}} \tag{10.115}$$

and

$$\begin{aligned} g(z, iu, q) = &- 2[\kappa^2(1+\varepsilon^{-1}) + q^2(1-\varepsilon^{-1})]D_s^{-1} r_{s+} r_{s-} e^{-2\kappa d} \\ &- 2[\kappa^2(1+\varepsilon^{-1}) - q^2(1-\varepsilon^{-1})]D_p^{-1} r_{p+} r_{p-} e^{-2\kappa d} \\ &+ (\kappa^2 - q^2)(1-\varepsilon^{-1})D_s^{-1}\left[r_{s-} e^{-2\kappa z} + r_{s+} e^{-2\kappa(d-z)}\right] \\ &- (\kappa^2 - q^2)(1-\varepsilon^{-1})D_p^{-1}\left[r_{p-} e^{-2\kappa z} + r_{p+} e^{-2\kappa(d-z)}\right] \end{aligned} \tag{10.116}$$

with

$$D_\sigma = D_\sigma(iu, q) = 1 - r_{\sigma+} r_{\sigma-} e^{-2\kappa d} \tag{10.117}$$

[$r_{\sigma\pm}$ are the reflection coefficients referring to reflection of waves at the right (+) and left (−) wall, as seen from the interspace, cf. Fig. 10.7]. Note that in Eqs (10.116) and (10.117) $\varepsilon = \varepsilon(iu)$, $\kappa = \kappa(iu, q)$, and $r_{\sigma\pm} = r_{\sigma\pm}(iu, q)$.

To further (numerically) evaluate Eq. (10.114), knowledge of the dependence of the reflection coefficients on u and q is required. Let us here restrict our attention to (i) the retarded limit and (ii) the limit of perfectly reflecting plates. That is to say, we (i) assume that the distance d between the plates is not too small so that the permittivities can be replaced by their static values, and (ii) we set

$$r_{p\pm} = -r_{s\pm} = 1. \tag{10.118}$$

It is then not difficult to calculate the simplified integrals in Eq. (10.114) analytically to obtain the Casimir force per unit area as [$\varepsilon \equiv \varepsilon(0)$]

$$F = T_{zz}(d) = \frac{\hbar c \pi^2}{240} \frac{1}{\sqrt{\varepsilon}} \left(\frac{2}{3} + \frac{1}{3\varepsilon}\right) \frac{1}{d^4}, \tag{10.119}$$

which reduces to Casimir's and Polder's famous formula

$$F = \frac{\hbar c \pi^2}{240} \frac{1}{d^4} \tag{10.120}$$

[Casimir and Polder (1948)] in the case when the interspace between the plates is empty ($\varepsilon=1$).

References

Bethe, H.A. (1947) *Phys. Rev.* **72**, 339.

Buhmann, S.Y, L. Knöll, D.-G. Welsch and T.D. Ho (2004) *Phys. Rev. A* **70**, 052117.

Buhmann, S.Y and D.-G. Welsch (2006) *Appl. Phys. B* **82**, 189.

Casimir, H.B.G. and D. Polder (1948) *Phys. Rev.* **73**, 360.

Chew, W.C. (1995) *Waves and Fields in Inhomogeneous Media* (IEEE Press, New York).

Drexhage, K.H. (1974) in *Progress in Optics*, Vol. XXII, ed. E. Wolf (Elsevier, Amsterdam), p. 165.

Fain, V.M. and Ya.I. Khanin (1969) *Quantum Electronics* (MIT Press, Cambridge MA).

Feinberg, G. and J. Sucher (1970) *Phys. Rev. A* **2**, 2395 (1970).

Henkel, C. and V. Sandoghdar (1998) *Opt. Commun.* **158**, 250.

Ho, T.D, L. Knöll and D.-G.Welsch (2000) *Phys. Rev. A* **62**, 053804.

Ho, T.D, L. Knöll and D.-G.Welsch (2001) *Phys. Rev. A* **64**, 013804.

Ho, T.D., S.Y. Buhmann, L.Knöll, D.-G. Welsch, S. Scheel and J.Kästel (2003) *Phys. Rev. A* **68**, 043816.

Kleppner, D. (1981) *Phys. Rev. Lett.* **47**, 233.

Lamb, W.E. and R.C. Retherford (1947) *Phys. Rev.* **72**, 241.

Lennard-Jones, J.E. (1932) *Trans. Farad. Soc.* **28**, 333.

London, F. (1930) *Z. Physik* **63**, 245.

Milonni, P.W. (1994) *The Quantum Vacuum* (Academic Press, San Diego).

Purcell, E.M. (1946) *Phys. Rev.* **69**, 681.

Raabe, C. and D.-G. Welsch (2005) *Phys. Rev. A* **71**, 013814.

Sandoghdar, V., C.I. Sukenik, E.A. Hinds and S. Haroche (1992) *Phys. Rev. Lett.* **68**, 3432.

Tai, C.-T. (1994) *Dyadic Green Functions in Electromagnetic Theory* (IEEE Press, New York).

Tomaš, M.S. (1995) *Phys. Rev. A* **51**, 2545.

Verwey, E.J.W. and J.T.G. Overbeek (1948) *Theory of the Stability of Lyophobic Colloids* (Elsevier, Amsterdam).

Weisskopf, V. and E. Wigner (1930) *Z. Phys.* **63**, 54.

Yeung, M.S. and T.K. Gustafson (1996) *Phys. Rev. A* **54**, 5227.

11
Resonance fluorescence

Among the various light–matter interaction processes, resonant light scattering from microscopic objects (single atoms, ions etc.) are of particular interest, because they make it possible to study a series of typical quantum effects. An illustrative example is the resonance fluorescence from a single atom. The light emitted by the atom shows various nonclassical features. Even under resonance conditions where the frequency of the incident light is close to an atomic transition frequency, the effect of a single atom on the radiation field is of course small, and the observation requires refined experimental techniques.

In resonance-fluorescence experiments, the atom is usually placed in free space and irradiated by laser light tuned to an atomic transition, so that the atom can truly be excited into an upper quantum state. From Section 10.1 we know that, owing to the interaction of the atom with the electromagnetic vacuum, an excited atomic state can spontaneously decay. In the free space this decay is exponential and accompanied by the emission of a photon. Resonance fluorescence may therefore be regarded as being an interplay between the competing effects of (coherent) driving of an atomic transition and its (incoherent) decay – an interplay which is expected to give rise to interesting quantum-statistical features of the scattered radiation.

11.1
Basic equations

Let us consider N atoms situated at positions $\mathbf{r}_A$, $A=1,2,3,\ldots,N$, and assume that the atom–light interaction may be treated in the electric-dipole and rotating-wave approximations. From Eq. (2.295) [together with Eq. (2.266)] it then follows that the source part $\hat{\mathbf{E}}_s^{(+)}(\mathbf{r},t)$ of the positive-frequency part of the electric field[1]

$$\hat{\mathbf{E}}^{(+)}(\mathbf{r},t) = \hat{\mathbf{E}}_{\text{free}}^{(+)}(\mathbf{r},t) + \hat{\mathbf{E}}_s^{(+)}(\mathbf{r},t) \tag{11.1}$$

1) Recall that in the free space $\hat{\mathbf{E}}^{(+)}(\mathbf{r}) = \hat{\mathbf{E}}^{\perp(+)}(\mathbf{r})$.

Quantum Optics, Third, revised and extended edition. Werner Vogel and Dirk-Gunnar Welsch
Copyright © 2006 WILEY-VCH Verlag GmbH & Co. KGaA, Weinheim
ISBN: 3-527-40507-0

can be written in the form of

$$\hat{\mathbf{E}}_s^{(+)}(\mathbf{r},t) = \sum_A \hat{\mathbf{E}}_{sA}^{(+)}(\mathbf{r},t), \qquad (11.2)$$

where

$$\hat{\mathbf{E}}_{sA}^{(+)}(\mathbf{r},t) = \sum_{a,b}{}' \int dt'\Theta(t-t')\mathbf{K}_{(E)}^{(+)}(\mathbf{r},t,\mathbf{r}_A,t')\mathbf{d}_{Aab}\hat{A}_{Aab}(t'), \qquad (11.3)$$

with $\mathbf{K}_{(E)}^{(+)}(\mathbf{r},t,\mathbf{r}_A,t')$ being given by Eq. (2.296), and $\mathbf{d}_{Aab}$ and $\hat{A}_{Aab}=|a\rangle_{AA}\langle b|$, respectively, being the transition dipole matrix elements and the associated flip operators of the Ath atom. The primed sum in Eq. (11.6) indicates that irrelevant off-resonant transitions are to be excluded from consideration. In the following we will assume that the interaction of the atom with the electromagnetic vacuum is weak, which is always the case when the atom is in the free space, so that the Markov approximation applies. That is to say, we introduce slowly varying atomic operators $\tilde{A}_{Aab}(t)$,[2]

$$\hat{A}_{Aab}(t) = e^{-i\tilde{\omega}_{Aba}t}\tilde{A}_{Aab}(t), \qquad (11.4)$$

put them at an appropriately chosen retarded time $t - l_A/c$ in front of the integral in Eq. (11.6) and extend the time t in the remaining integral to ∞ to obtain

$$\hat{\mathbf{E}}_{sA}(\mathbf{r},t) = \sum_{ba}{}' \mathbf{g}_{ba}(\mathbf{r},\mathbf{r}_A)\tilde{A}_{Aab}(t-l_A/c), \qquad (11.5)$$

where

$$\mathbf{g}_{ab}(\mathbf{r},\mathbf{r}_A) = \lim_{t\to\infty} \int dt'\Theta(t-t')e^{i\tilde{\omega}_{Aba}(t-l_A/c-t')}\mathbf{K}_{(E)}^{(+)}(\mathbf{r},t,\mathbf{r}_A,t')\mathbf{d}_{Aab}. \qquad (11.6)$$

Further evaluation of Eq. (11.6) requires knowledge of the dyadic kernel function $\mathbf{K}_{(E)}^{(+)}(\mathbf{r},t,\mathbf{r}_A,t')$, which itself is determined, according to Eq. (2.296), by the Green tensor of the system. In particular if the atoms are in free space, we can combine Eq. (2.296) with Eq. (2.298) to derive, on restricting our attention to the far-field region relevant to the emitted radiation,

$$\mathbf{g}_{ab}(\mathbf{r},\mathbf{r}_A) = \frac{\tilde{\omega}_{Aba}^2}{4\pi\varepsilon_0 c^2}\left[\frac{\mathbf{d}_{Aab}}{|\mathbf{r}-\mathbf{r}_A|} - \frac{\mathbf{d}_{Aab}(\mathbf{r}-\mathbf{r}_A)\otimes(\mathbf{r}-\mathbf{r}_A)}{|\mathbf{r}-\mathbf{r}_A|^3}\right], \qquad (11.7)$$

and the simple relation $l_A = |\mathbf{r}-\mathbf{r}_A|$ is valid.

Let us consider normally and time-ordered field correlation functions of the general form

$$G_{\{k_ik_j\}}^{(m,n)}(\{\mathbf{r}_i,t_i,\mathbf{r}_j,t_j\}) = \left\langle {}_\circ^\circ \prod_{i=1}^{m}\prod_{j=m+1}^{m+n} \hat{E}_{k_i}^{(-)}(\mathbf{r}_i,t_i)\hat{E}_{k_j}^{(+)}(\mathbf{r}_j,t_j) {}_\circ^\circ \right\rangle. \qquad (11.8)$$

---

2) $\tilde{\omega}_{Aba}$ is thought of as containing the Lamb shift.

11.1 Basic equations

Applying Eq. (2.320),

$$G^{(m,n)}_{\{k_i k_j\}}(\{\mathbf{r}_i, t_i, \mathbf{r}_j, t_j\}) = \left\langle : \prod_{i=1}^{m} \prod_{j=m+1}^{m+n} \hat{E}^{(-)}_{sk_i}(\mathbf{r}_i, t_i) \hat{E}^{(+)}_{sk_j}(\mathbf{r}_j, t_j) : \right\rangle, \quad (11.9)$$

and using Eq. (11.5), we may express the $G^{(m,n)}_{\{k_i k_j\}}$ in terms of atomic correlation functions as

$$G^{(m,n)}_{\{k_i k_j\}}(\{\mathbf{r}_i, t_i, \mathbf{r}_j, t_j\}) = \left\langle \left\{ \mathcal{T}_{-} \prod_{i=1}^{m} \left[\sum_{A} \sum_{a,b}' g^*_{abk_i}(\mathbf{r}_i, \mathbf{r}_A) \hat{A}_{Aba}(t_i - l_{iA}/c) \right] \right\} \right.$$

$$\left. \times \left\{ \mathcal{T}_{+} \prod_{j=m+1}^{m+n} \left[\sum_{A} \sum_{a,b}' g_{abk_j}(\mathbf{r}_j, \mathbf{r}_A) \hat{A}_{Aab}(t_j - l_{jA}/c) \right] \right\} \right\rangle. \quad (11.10)$$

In Eqs (11.9) and (11.10) the spatial range of light observation is assumed to be outside the exciting beam, so that only the scattered light is observed, the free field being in the vacuum state. Note that in Eq. (11.10) the $\mathcal{T}_{\pm}$ time orderings involve the atomic operators, whose time arguments are the retarded times $t_i - l_{iA}/c$.

To study spectral properties, let us consider the case where the scattered light is detected after it has passed through a spectral filter, and apply the results given in Section 6.4:

$$G^{(m,n)}_{\{k_i k_j\}}(\{\mathbf{r}_i, t_i, \mathbf{r}_j, t_j\}) = \int dt'_1 \, T^*_f(t_1 - t'_1) \cdots \int dt'_{m+n} \, T_f(t_{m+n} - t'_{m+n})$$

$$\times \left\langle : \prod_{i=1}^{m} \hat{E}^{(-)}_{sk_i}(\mathbf{r}_f, t'_i - |\mathbf{r}_i - \mathbf{r}_f|/c) \prod_{j=m+1}^{m+n} \hat{E}^{(+)}_{sk_j}(\mathbf{r}_f, t'_j - |\mathbf{r}_j - \mathbf{r}_f|/c) : \right\rangle \quad (11.11)$$

($\mathbf{r}_f$, position vector of the entrance plane), where the response function $T_f(t)$ of the spectral filter may be assumed to be

$$T_f(t) = \tfrac{1}{2}\Gamma_f \Theta(t - \Delta t) \exp\left[-\left(i\omega + \tfrac{1}{2}\Gamma_f\right)(t - \Delta t)\right] \quad (11.12)$$

[ω, setting frequency; Γ_f, passband width; $c\Delta t$, difference between the optical and geometrical paths through the spectral apparatus; cf. Eqs (6.140) and (6.150)]. Instead of Eq. (11.10), we now obtain, using Eqs (11.2) and (11.5),[3]

$$G^{(m,n)}_{\{k_i k_j\}} = \int dt'_1 \, T^*_f(t_1 - t'_1) \cdots \int dt'_{m+n} \, T_f(t_{m+n} - t'_{m+n})$$

$$\times \left\langle \left\{ \mathcal{T}_{-} \prod_{i=1}^{m} \left[\sum_{A} \sum_{a,b}' g^*_{abk_i}(\mathbf{r}_f, \mathbf{r}_A) \hat{A}_{Aba}(t'_i - l_{ia}/c) \right] \right\} \right.$$

$$\left. \times \left\{ \mathcal{T}_{+} \prod_{j=m+1}^{m+n} \left[\sum_{a,b}' g_{abk_j}(\mathbf{r}_f, \mathbf{r}_A) \hat{A}_{Aab}(t'_j - l_{ja}/c) \right] \right\} \right\rangle. \quad (11.13)$$

---

[3] If correlations of different frequency components are required to be observed, spectral filters with different setting frequencies are needed. In this case the T_f in Eqs (11.11) and (11.13) may differ in the setting frequency (and passband width): $T_f \mapsto T_{f_i} \neq T_{f_j}$.

11.2
Two-level systems

In the case of resonance fluorescence from a single atom (at position $\mathbf{r}_A$) with a single two-level resonant transition involved in the scattering process, the model of an atomic two-level system applies, and Eq. (11.2), together with Eq. (11.5), reduces to

$$\hat{\mathbf{E}}_s^{(+)}(\mathbf{r},t) = \mathbf{g}(\mathbf{r},\mathbf{r}_A)\hat{A}_{12}(t-l_A/c) \tag{11.14}$$

$[\mathbf{g}(\mathbf{r},\mathbf{r}_A) \equiv \mathbf{g}_{12}(\mathbf{r},\mathbf{r}_A), \hat{A}_{12}(t) \equiv \hat{A}_{A12}(t)]$, Eq. (11.10) takes the form

$$G_{\{k_ik_j\}}^{(m,n)} = g_{\{k_ik_j\}}^{(m,n)} \left\langle \left[\mathcal{T}_-\prod_{i=1}^{m}\hat{A}_{21}(t_{ri})\right]\left[\mathcal{T}_+\prod_{j=m+1}^{m+n}\hat{A}_{12}(t_{rj})\right]\right\rangle, \tag{11.15}$$

and in the case of spectral filtering, Eq. (11.13) is

$$G_{\{k_ik_j\}}^{(m,n)} = g_{\{k_ik_j\}}^{(m,n)} \int dt_1'\, T_f^*(t_1 - t_1') \cdots \int dt_{m+n}'\, T_f(t_{m+n} - t_{m+n}')$$

$$\times \left\langle \left[\mathcal{T}_-\prod_{i=1}^{m}\hat{A}_{21}(t_{ri}')\right]\left[\mathcal{T}_+\prod_{j=m+1}^{m+n}\hat{A}_{12}(t_{rj}')\right]\right\rangle \tag{11.16}$$

(see also footnote 3). Here the abbreviated notations $t_r = t - l_A/c$ and

$$g_{\{k_ik_j\}}^{(m,n)} = \prod_{i=1}^{m}\prod_{j=m+1}^{m+n} g_{k_i}^* g_{k_j} \tag{11.17}$$

are used for the retarded times and the overall geometry factors, respectively.

Treating atomic relaxations such as spontaneous emission in the Markov approximation, we know that the problem of calculating the multi-time correlation functions appearing in Eq. (11.15) may be reduced to the problem of calculating the one-time averages $\langle \hat{A}_{ab}(t)\rangle$, by applying the quantum regression theorem (Section 5.5). Since these one-time averages are related to the atomic density-matrix elements (in the Schrödinger picture) by $\langle \hat{A}_{ab}(t)\rangle = \sigma_{ba}(t)$, the problem effectively reduces to solving the atomic density-matrix equations of motion, which in the case of the two-level atom considered are just the (optical) Bloch equations (5.177)–(5.180). Let us consider the case where the atom is driven by monochromatic laser light of frequency ω_L ($\omega_L \approx \tilde{\omega}_{21}$) and assume that the laser light may be treated as classical. In this semi-classical description, in Eqs (5.177)–(5.180) we may let

$$F_{21}^{(r)}(t) \mapsto -\frac{1}{\hbar}|\mathbf{d}_{21}\mathbf{E}_L|e^{-i(\omega_L t+\varphi_L)} \tag{11.18}$$

[cf. Eq. (5.174)], where $\mathbf{E}_L$ is the amplitude of the laser wave. Further, introducing slowly varying off-diagonal density-matrix elements according to

$$\tilde{\sigma}_{12}(t) = \sigma_{12}(t)e^{-i(\omega_L t+\varphi_L)}, \tag{11.19}$$

disregarding the excited-state filling rate $w_{12}^{(nr)}$, and using the abbreviated notations

$$\Gamma_1 = \Gamma^{(r)} + w_{21}^{(nr)}, \qquad \Gamma_2 = \tfrac{1}{2}\Gamma_1 + \gamma', \tag{11.20}$$

we may write Eqs (5.177)–(5.180) as

$$\dot{\sigma}_{22} = -\Gamma_1 \sigma_{22} - \tfrac{1}{2}i\Omega_R \tilde{\sigma}_{21} + \tfrac{1}{2}i\Omega_R \tilde{\sigma}_{12}, \tag{11.21}$$

$$\dot{\sigma}_{11} = \Gamma_1 \sigma_{22} + \tfrac{1}{2}i\Omega_R \tilde{\sigma}_{21} - \tfrac{1}{2}i\Omega_R \tilde{\sigma}_{12}, \tag{11.22}$$

$$\dot{\tilde{\sigma}}_{21} = (-i\delta\omega - \Gamma_2)\tilde{\sigma}_{21} + \tfrac{1}{2}i\Omega_R(\sigma_{11} - \sigma_{22}), \tag{11.23}$$

$$\dot{\tilde{\sigma}}_{12} = (i\delta\omega - \Gamma_2)\tilde{\sigma}_{12} - \tfrac{1}{2}i\Omega_R(\sigma_{11} - \sigma_{22}), \tag{11.24}$$

where Ω_R and $\delta\omega$ are the Rabi frequency and the detuning respectively:

$$\Omega_R = 2\hbar^{-1}|\mathbf{d}_{21}\mathbf{E}_L|, \tag{11.25}$$

$$\delta\omega = \tilde{\omega}_{21} - \omega_L. \tag{11.26}$$

Note that $\tilde{\sigma}_{12} = \tilde{\sigma}_{21}^*$.[4] Combining Eqs (11.21) and (11.22) yields

$$\frac{d}{dt}(\sigma_{11} + \sigma_{22}) = 0, \tag{11.27}$$

so that the two-level completeness relation

$$\sigma_{22} + \sigma_{11} = 1 \tag{11.28}$$

can be satisfied. Note that, in the case of spontaneous emission in free space, the Bloch equations (11.21)–(11.24) reduce to

$$\dot{\sigma}_{22} = -\Gamma_1 \sigma_{22} = -\dot{\sigma}_{11}, \quad \dot{\tilde{\sigma}}_{21} = -\tfrac{1}{2}\Gamma_1 \tilde{\sigma}_{21}, \quad \dot{\tilde{\sigma}}_{12} = \dot{\tilde{\sigma}}_{21}, \tag{11.29}$$

leading, in agreement with Eq. (10.25), to the exponential decay law $\sigma_{22}(t) = e^{-\Gamma_1 t}$ for an initially excited atom [$\sigma_{22}(0)=1$]. Recall that for a single (isolated) atom, which only interacts with the electromagnetic field, radiationless relaxation described by $w_{21}^{(nr)}$ and γ' does not occur and hence the relations $\Gamma_1 = \Gamma^{(r)}$ and $\Gamma_2 = \Gamma^{(r)}/2$ are valid, where $\Gamma^{(r)}$ is the spontaneous decay rate Γ as given by Eq. (10.27).

Since the Bloch equations can also be used to study the dynamics of two-level systems embedded in (dense) matter, where radiationless relaxation can become significant, we will include them in further considerations for the sake of generality. In particular, fast radiationless dephasing may be observed: $\Gamma_2 \gg \Gamma_1$. In this case the density-matrix equations of motion (11.21)–(11.24)

---

[4] This relation and Eq. (11.28) may be violated when the Bloch equations are used in an application of the quantum regression theorem.

may be simplified by adiabatically eliminating the off-diagonal elements. On a time scale Γ_1^{-1} it is sufficient to substitute into Eqs (11.21) and (11.22) for $\tilde{\sigma}_{21}$ and $\tilde{\sigma}_{12}$ the steady-state solutions of Eqs (11.23) and (11.24) respectively, which are

$$\tilde{\sigma}_{21} = \frac{\Omega_R}{2(\delta\omega - i\Gamma_2)}(\sigma_{11} - \sigma_{22}), \quad \tilde{\sigma}_{12} = (\tilde{\sigma}_{21})^*. \tag{11.30}$$

On the chosen time scale, the off-diagonal density-matrix elements adiabatically follow the inversion. Combining Eqs (11.21), (11.22) and (11.30), we arrive at

$$\dot{\sigma}_{22} = -\frac{\Omega_R^2 \Gamma_2}{2[(\delta\omega)^2 + \Gamma_2^2]}(\sigma_{22} - \sigma_{11}) - \Gamma_1\sigma_{22}, \tag{11.31}$$

$$\dot{\sigma}_{11} = \frac{\Omega_R^2 \Gamma_2}{2[(\delta\omega)^2 + \Gamma_2^2]}(\sigma_{22} - \sigma_{11}) + \Gamma_1\sigma_{22}. \tag{11.32}$$

We see that in the case of fast radiationless dephasing, balance equations for the atomic-state occupation probabilities may be derived. The method of balance equations is widely used (e. g., in laser theory and related fields) in order to simplify complicated multi-level density-matrix equations of motion and make them physically transparent. If we let $\dot{\sigma}_{11} = \dot{\sigma}_{22} = 0$ in Eqs (11.31) and (11.32), after some algebra we obtain, using the relation $\sigma_{11} + \sigma_{22} = 1$, the steady-state values of σ_{11} and σ_{22} as

$$\sigma_{11}(\infty) = \frac{1}{2}\frac{\Omega_R^2\Gamma_2 + 2\Gamma_1[(\delta\omega)^2 + \Gamma_2^2]}{\Omega_R^2\Gamma_2 + \Gamma_1[(\delta\omega)^2 + \Gamma_2^2]}, \tag{11.33}$$

$$\sigma_{22}(\infty) = \frac{1}{2}\frac{\Omega_R^2\Gamma_2}{\Omega_R^2\Gamma_2 + \Gamma_1[(\delta\omega)^2 + \Gamma_2^2]}. \tag{11.34}$$

Combining Eqs (11.30), (11.33) and (11.34) yields the steady-state value of $\tilde{\sigma}_{21}$ as

$$\tilde{\sigma}_{21}(\infty) = \frac{1}{2}\frac{\Omega_R\Gamma_1(\delta\omega + i\Gamma_2)}{\Omega_R^2\Gamma_2 + \Gamma_1[(\delta\omega)^2 + \Gamma_2^2]} \tag{11.35}$$

[with $\tilde{\sigma}_{12}(\infty) = \tilde{\sigma}_{21}^*(\infty)$].

11.2.1
Intensity

To determine the intensity of the scattered light observed by means of a broadband, point-like photodetector situated at position **r**, we have to calculate

$$I(t) \equiv G_{kk}^{(1,1)}(\mathbf{r},t,\mathbf{r},t) = \langle \hat{E}_k^{(-)}(\mathbf{r},t)\hat{E}_k^{(+)}(\mathbf{r},t)\rangle. \tag{11.36}$$

According to Eq. (11.15), we may write

$$I(t+r/c) = |\mathbf{g}|^2 \langle \hat{A}_{22}(t) \rangle = |\mathbf{g}|^2 \sigma_{22}(t) \tag{11.37}$$

($\mathbf{r}_A = 0$); recall that

$$\hat{A}_{ab}\hat{A}_{a'b'} = \delta_{ba'}\hat{A}_{ab'}, \quad \langle \hat{A}_{ab} \rangle = \sigma_{ba}. \tag{11.38}$$

Hence the intensity of the scattered light is determined by the atomic excited-state occupation probability $\sigma_{22}(t)$.[5] The remaining problem thus consists in solving the Bloch equations (11.21)–(11.24) in order to calculate $\sigma_{22}(t)$.

Making the standard ansatz

$$\sigma_{ab}(t) \sim e^{\lambda t}, \quad a,b = 1,2, \tag{11.39}$$

yields a fourth-order algebraic characteristic equation for determining λ. The roots λ_i, $i=1,2,3,4$, imply the following structure of the general solution:

$$\sigma_{ab}(t) = \sum_{i=1}^{4} c_{ab}^{(i)} e^{\lambda_i t}. \tag{11.40}$$

In view of Eq. (11.27), we may let $\lambda_4 = 0$, and the characteristic equation effectively reduces to a third-order algebraic equation, which may be solved by means of Cardan's formula to yield λ_i, $i=1,2,3$. Finally, the constants $c_{ab}^{(i)}$ are determined from the initial conditions. We leave the details of the calculation to the reader.

Since the atom is usually in the lower quantum state before the interaction with the laser is switched on, we may choose the initial conditions to be $\sigma_{ab}(0) = \delta_{a1}\delta_{b1}$. If we further assume that the laser is tuned to the transition frequency ($\delta\omega = 0$), the procedure outlined above yields the following expression for the atomic excited-state occupation probability:

$$\sigma_{22}(t) = \frac{\Omega_R^2}{2(\Omega_R^2 + \Gamma_1\Gamma_2)} \left(1 + \frac{\lambda_2}{\lambda_1 - \lambda_2} e^{\lambda_1 t} + \frac{\lambda_1}{\lambda_2 - \lambda_1} e^{\lambda_2 t} \right), \tag{11.41}$$

where

$$\lambda_{1,2} = -\tfrac{1}{2}(\Gamma_1 + \Gamma_2) \pm \sqrt{\tfrac{1}{4}(\Gamma_1 - \Gamma_2)^2 - \Omega_R^2}. \tag{11.42}$$

Note that when the driven atom, with regard to relaxations, only undergoes radiative damping through spontaneous emission, Eq. (11.42) may be written as

$$\lambda_{1,2} = -\tfrac{3}{2}\gamma \pm \sqrt{\tfrac{1}{4}\gamma^2 - \Omega_R^2}, \tag{11.43}$$

---

[5] It is not difficult to see that application of Eq. (11.37) together with Eq. (11.7) to the spontaneous emission of a two-level atom leads exactly to Eq. (10.34).

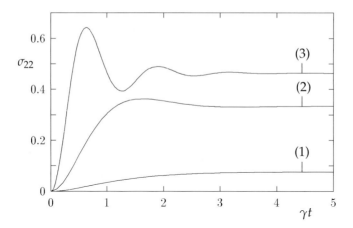

Fig. 11.1 The time evolution of the excited-state occupation probability $\sigma_{22}(t)$ of a two-level atom undergoing radiative damping ($\Gamma_1 = 2\Gamma_2 = 2\gamma$) and resonantly driven by a monochromatic laser beam, is shown for various values of the Rabi frequency: $\Omega_R/\gamma = 0.6$ (1); $\Omega_R/\gamma = 2$ (2); $\Omega_R/\gamma = 5$ (3).

where

$$\gamma = \tfrac{1}{2}\Gamma^{(r)} = \Gamma_2, \qquad \Gamma_1 = 2\gamma. \tag{11.44}$$

Typical examples of $\sigma_{22}(t)$ are shown in Fig. 11.1.

From Eq. (11.37) together with Eq. (11.34) the steady-state intensity of the resonance fluorescence ($t \to \infty$, $\delta\omega = 0$) is given by

$$I(\infty) = |g|^2 \sigma_{22}(\infty) = |g|^2 \frac{\Omega_R^2}{2(\Omega_R^2 + \Gamma_1\Gamma_2)}. \tag{11.45}$$

This result reveals that the intensity of the scattered light is, in general, a nonlinear function of the pump laser intensity [recall that $\Omega_R^2 \sim E_L^2$; cf. Eq. (11.25)]. Only in the weak-driving-field limit is resonance fluorescence a phenomenon of linear optics:

$$I(\infty) = |g|^2 \frac{\Omega_R^2}{2\Gamma_1\Gamma_2}, \qquad \Omega_R^2 \ll \Gamma_1\Gamma_2, \tag{11.46}$$

whereas in the high-driving-field limit, saturation may be observed:

$$I(\infty) = \tfrac{1}{2}|g|^2, \qquad \Omega_R^2 \gg \Gamma_1\Gamma_2 \tag{11.47}$$

[$\sigma_{11}(\infty) = \sigma_{22}(\infty) = 1/2$]. It should be pointed out that, in the weak-driving-field limit, when the atom undergoes radiative damping through spontaneous emission and the relations (11.44) are valid, from Eqs (11.34) and (11.35) the relation $|\tilde{\sigma}_{21}|^2 = \sigma_{22}$ is seen to hold, which implies coherent light scattering. In the high-driving-field limit $\tilde{\sigma}_{21}$ generally vanishes and the scattering is incoherent.

11.2.1.1 Limiting cases

To briefly illustrate the main features of the time evolution of the fluorescence intensity as given from Eqs (11.37) and (11.41) (cf. Fig. 11.1), let us again consider the two instructive cases of weak and strong pumping:

(i) *Weak-driving-field limit:* $\Omega_R^2 \ll \Gamma_1 \Gamma_2$. In this case Eq. (11.42) reduces to

$$\lambda_{1,2} = -\Gamma_{2,1}, \tag{11.48}$$

and from Eqs (11.37) and (11.41) we deduce that

$$I(t+r/c) = |g|^2 \frac{\Omega_R^2}{\Gamma_1 \Gamma_2}\left(1 - \frac{\Gamma_1}{\Gamma_1 - \Gamma_2} e^{-\Gamma_2 t} + \frac{\Gamma_2}{\Gamma_1 - \Gamma_2} e^{-\Gamma_1 t}\right). \tag{11.49}$$

The intensity is seen to monotonically approach the stationary value, and the characteristic time scale is given by the atomic relaxation times Γ_1^{-1} and Γ_2^{-1}, which for an isolated atom are just related to the time of radiation damping due to spontaneous emission. Note that the result (11.49) may also be derived from a perturbative solution of the Bloch equations.

(ii) *High-driving-field limit:* $\Omega_R^2 \gg \Gamma_1 \Gamma_2$. In this case from Eq. (11.42) we derive λ_1 and λ_2 as

$$\lambda_{1,2} = -\tfrac{1}{2}(\Gamma_1 + \Gamma_2) \pm i\Omega_R. \tag{11.50}$$

Thus, combining Eqs (11.37), (11.41) and (11.50) yields

$$I(t+r/c) = \tfrac{1}{2}|g|^2 \left\{1 - \exp[-\tfrac{1}{2}(\Gamma_1 + \Gamma_2)t]\cos(\Omega_R t)\right\}. \tag{11.51}$$

The intensity oscillates with the Rabi frequency Ω_R, the Rabi oscillations being damped owing to the atomic relaxation processes. The time scale of damping again corresponds to the time scale on which the intensity approaches the stationary value.

11.2.2
Intensity correlation and photon anti-bunching

To study the effect of photon anti-bunching in resonance fluorescence from a single atom [Carmichael and Walls (1976); Kimble and Mandel (1976)], we now consider the normally and time-ordered intensity correlation function of the scattered light,

$$G^{(2,2)}(\mathbf{r}, t+\tau, \mathbf{r}, t) = \langle \hat{E}_{k_1}^{(-)}(\mathbf{r},t) \hat{E}_{k_2}^{(-)}(\mathbf{r},t+\tau) \hat{E}_{k_2}^{(+)}(\mathbf{r},t+\tau) \hat{E}_{k_1}^{(+)}(\mathbf{r},t)\rangle, \tag{11.52}$$

which may be observed by measuring the correlation of counts in photodetection correlation experiments (Sections 6.1 and 8.2.1). Recall that the joint probability of observing an event during the time interval $t, t+\Delta t$ and an event during the time interval $t+\tau, t+\tau+\Delta t$ is proportional to $G^{(2,2)}(\mathbf{r}, t+\tau, \mathbf{r}, t)$. Introducing the atomic correlation functions[6]

$$G^{(II)}_{ab}(t+\tau, t) = \langle \hat{A}_{21}(t) \hat{A}_{ba}(t+\tau) \hat{A}_{12}(t) \rangle, \quad a, b = 1, 2, \tag{11.53}$$

and applying Eq. (11.15) together with the relations (11.38), we can easily find that the normally and time-ordered intensity correlation function $G^{(2,2)}(t+\tau, t) \equiv G^{(2,2)}_{k_1 k_2 k_2 k_1}(\mathbf{r}, t+\tau, \mathbf{r}, t)$ may be expressed in terms of $G^{(II)}_{22}(t+\tau, t)$ as

$$G^{(2,2)}(t+r/c+\tau, t+r/c) = |\mathbf{g}|^4 G^{(II)}_{22}(t+\tau, t). \tag{11.54}$$

In particular, for equal times

$$G^{(2,2)}(t+r/c, t+r/c) = |\mathbf{g}|^4 \langle \hat{A}_{21}(t) \hat{A}_{22}(t) \hat{A}_{12}(t) \rangle. \tag{11.55}$$

Taking into account the properties of the atomic flip operators [cf. the relations (11.38)], we can easily see that

$$G^{(2,2)}(t+r/c, t+r/c) = 0. \tag{11.56}$$

The result that the normally and time-ordered intensity correlation function at equal times is exactly zero, indicates perfect photon anti-bunching. In this case the joint probability of simultaneously detecting two events vanishes, and the anti-bunching condition (8.24) is satisfied. That is, the steady-state intensity correlation function

$$G^{(2)}(\tau) \equiv \lim_{t \to \infty} G^{(2,2)}(t+r/c+\tau, t+r/c) \tag{11.57}$$

necessarily has a positive initial slope:

$$G^{(2)}(\tau) > G^{(2)}(0) = 0. \tag{11.58}$$

Recall that photon anti-bunching, which may be regarded as a proof of the photon nature of light, is a pure quantum effect that cannot be explained on the basis of classical optics (Section 8.2.1).

To physically understand the effect of ideal photon anti-bunching in resonance fluorescence from a single atom, one may advance the following intuitive arguments. When the atom emits a photon to be detected, it undergoes

---
6) The superscript (II) indicates that the atomic correlation function is related to a field correlation function of second order with respect to the intensity of light.

a quantum jump from the upper to the lower quantum state. Clearly, in the lower quantum state (ground state) the atom cannot emit a second photon. After performing a pump-laser-induced transition from the lower to the upper quantum state, the atom is ready to emit a second photon to be detected. The probability of emitting the first photon at time t is proportional to the atomic excited-state occupation probability $\sigma_{22}(t)$ (cf. Section 11.2.1). Since, owing to the emission of the photon, the atom should be in the lower quantum state, the probability of emitting a photon at later time $t+\tau$ is expected to be proportional to the atomic excited-state occupation probability at time $t+\tau$ under the condition that the atom at time t is in the lower quantum state: $\sigma_{22}(t+\tau)|_{\sigma_{ab}(t)=\delta_{a1}\delta_{b1}}$. Hence the probability of emitting a photon at time t and a photon at time $t+\tau$ is expected to be equal to the product of the two excited-state occupation probabilities mentioned, so that

$$G_{22}^{(II)}(t+\tau,t) = \sigma_{22}(t)\,\sigma_{22}(t+\tau)|_{\sigma_{ab}(t)=\delta_{a1}\delta_{b1}}. \tag{11.59}$$

We prove Eq. (11.59) as follows. First, we introduce $\sigma_{ab}(t+\tau)|_{\sigma_{a'b'}(t)=\delta_{a'1}\delta_{b'1}}$ as the density-matrix elements $\sigma_{ab}(t+\tau)$ which solve, with regard to the time argument τ, the equations of motion (11.21)–(11.24) together with the initial conditions $\sigma_{ab}(t+\tau)|_{\tau=0}=\delta_{a1}\delta_{b1}$. Secondly, to calculate the atomic correlation function $G_{22}^{(II)}(t+\tau,t)$, we use the quantum regression theorem (Section 5.5). Applying, for example, Eq. (5.195) [$\langle \hat{C}_2(t+\tau)\hat{C}_1(t)\rangle \mapsto \langle \hat{A}_{21}(t)\hat{A}_{ba}(t+\tau)\hat{A}_{12}(t)\rangle$] and taking the relations (11.38) into account, we see that the $G_{ab}^{(II)}(t+\tau,t)$ defined in Eq. (11.53) obey, with regard to the time argument τ ($\tau \geq 0^+$), the density-matrix equations of motion (11.21)–(11.24):

$$\frac{dG_{22}^{(II)}}{d\tau} = -\Gamma_1 G_{22}^{(II)} - \tfrac{1}{2}i\Omega_R \tilde{G}_{21}^{(II)} + \tfrac{1}{2}i\Omega_R \tilde{G}_{12}^{(II)}, \tag{11.60}$$

$$\frac{dG_{11}^{(II)}}{d\tau} = \Gamma_1 G_{22}^{(II)} + \tfrac{1}{2}i\Omega_R \tilde{G}_{21}^{(II)} - \tfrac{1}{2}i\Omega_R \tilde{G}_{12}^{(II)}, \tag{11.61}$$

$$\frac{d\tilde{G}_{21}^{(II)}}{d\tau} = (-i\delta\omega - \Gamma_2)\tilde{G}_{21}^{(II)} + \tfrac{1}{2}i\Omega_R(G_{11}^{(II)} - G_{22}^{(II)}), \tag{11.62}$$

$$\frac{d\tilde{G}_{12}^{(II)}}{d\tau} = (i\delta\omega - \Gamma_2)\tilde{G}_{21}^{(II)} - \tfrac{1}{2}i\Omega_R(G_{11}^{(II)} - G_{22}^{(II)}), \tag{11.63}$$

where, in accordance with Eq. (11.19), slowly varying correlation functions $\tilde{G}_{12}^{(II)}$ ($=\tilde{G}_{21}^{(II)*}$) have been introduced,

$$\tilde{G}_{12}^{(II)}(t+\tau,t) = G_{12}^{(II)}(t+\tau,t)\exp\{-i[\omega_L(t+\tau)+\varphi_L]\}. \tag{11.64}$$

The initial conditions required for solving the above equations of motion are determined from the equal-time correlation functions $G_{ab}^{(II)}(t,t)$. Using the relations (11.38), we readily deduce that

$$G^{(\mathrm{II})}_{ab}(t+\tau,t)\big|_{\tau=0} = \langle \hat{A}_{21}(t)\hat{A}_{ba}(t)\hat{A}_{12}(t)\rangle = \delta_{a1}\delta_{b1}\sigma_{22}(t). \tag{11.65}$$

We see that, with regard to the time argument τ, the $G^{(\mathrm{II})}_{ab}(t+\tau,t)/\sigma_{22}(t)$ and the $\sigma_{ab}(t+\tau)\big|_{\sigma_{a'b'}(t)=\delta_{a'1}\delta_{b'1}}$ are determined from the same equations of motion and initial conditions. Hence they are equal,

$$\frac{G^{(\mathrm{II})}_{ab}(t+\tau,t)}{\sigma_{22}(t)} = \sigma_{ab}(t+\tau)\big|_{\sigma_{a'b'}(t)=\delta_{a'1}\delta_{b'1}}, \tag{11.66}$$

from which, in particular, Eq. (11.59) is seen to be valid ($a=b=2$). Note that in the considered case of the Rabi frequency being time-independent and the atom being initially in the lower quantum state Eq. (11.59) reduces to[7]

$$G^{(\mathrm{II})}_{22}(t+\tau,t) = \sigma_{22}(\tau)\sigma_{22}(t). \tag{11.67}$$

Combining Eqs (11.37), (11.54) and (11.59) we find that the normally and time-ordered intensity correlation function of the scattered light may be factored into two intensities:

$$G^{(2,2)}(t+r/c+\tau,t+r/c) = I(t+r/c+\tau)\big|_{\sigma_{ab}(t)=\delta_{a1}\delta_{b1}} I(t+r/c). \tag{11.68}$$

Here $I(t+r/c+\tau)\big|_{\sigma_{ab}(t)=\delta_{a1}\delta_{b1}}$ is the intensity observed at space point **r** and time $t+r/c+\tau$ under the condition that, at time t, the state of the atom has been reduced to the lower state. In particular, when Eq. (11.67) holds, Eq. (11.68) can be simplified to obtain

$$G^{(2,2)}(t+r/c+\tau,t+r/c) = I(\tau+r/c)\,I(t+r/c)$$
$$= |g|^4 \sigma_{22}(\tau)\sigma_{22}(t), \tag{11.69}$$

so that the results of Section 11.2.1 apply directly. It should be pointed out that when multi-level systems are involved in light scattering, the situation may be much more complicated.

In the study of intensity correlations the normalized intensity correlation function $\gamma^{(22)}(t+\tau,t)$ is frequently introduced,

$$\gamma^{(22)}(t+\tau,t) = \frac{G^{(2,2)}(t+\tau,t)}{I(t+\tau)\,I(t)}, \tag{11.70}$$

which, using Eq. (11.69), becomes

$$\gamma^{(22)}(t+r/c+\tau,t+r/c) = \frac{I(\tau+r/c)}{I(t+r/c+\tau)}. \tag{11.71}$$

---
7) Both $\sigma_{22}(\tau)$ and $\sigma_{22}(t)$ are now understood to be solutions of the optical Bloch equations with the initial conditions $\sigma_{ab}(0)=\delta_{1a}\delta_{b1}$.

In particular, for steady-state observation conditions

$$\gamma^{(22)}(\tau) = \lim_{t\to\infty} \gamma^{22}(t+r/c+\tau, t+r/c) = \frac{I(\tau+r/c)}{I(\infty)} = \frac{\sigma_{22}(\tau)}{\sigma_{22}(\infty)}. \quad (11.72)$$

In this case the normalized intensity correlation function obviously exhibits the same behavior as the normalized intensity [see Eqs (11.37) and (11.41) and Fig. 11.1], and it is clear that $\gamma^{22}(\tau)$ has a positive initial slope, which just reflects the effect of photon anti-bunching. It is worth noting that, since

$$\gamma^{22}(0) < 1, \quad (11.73)$$

the fluorescence light gives rise to a (nonclassical) sub-Poissonian photocounting statistics [Cook (1981)]; see also Section 8.2.2. Both the photon anti-bunching nature and the sub-Poissonian statistics of the resonance fluorescence from a two-level atom have been demonstrated experimentally (Sections 8.2.1 and 8.2.2).

11.2.3
Squeezing

To study squeezing in resonance fluorescence [Walls and Zoller (1981)], let us consider the normally ordered variance of the electric field strength of the radiation field:

$$\begin{aligned}\langle :[\Delta\hat{\mathbf{E}}(\mathbf{r},t)]^2: \rangle &= 2\langle \hat{\mathbf{E}}^{(-)}(\mathbf{r},t)\hat{\mathbf{E}}^{(+)}(\mathbf{r},t)\rangle + \{\langle [\hat{\mathbf{E}}^{(+)}(\mathbf{r},t)]^2\rangle + \text{c.c.}\} \\ &\quad - [\langle \hat{\mathbf{E}}^{(+)}(\mathbf{r},t)\rangle + \text{c.c.}]^2 \\ &= 2G_{kk}^{(1,1)}(\mathbf{r},t,\mathbf{r},t) + [G_{kk}^{(0,2)}(\mathbf{r},t,\mathbf{r},t) + \text{c.c.}] \\ &\quad - [G_{kk}^{(0,1)}(\mathbf{r},t,\mathbf{r},t) + \text{c.c.}]^2. \end{aligned} \quad (11.74)$$

Recall that light is squeezed when the normally ordered variance $\langle :[\Delta\hat{\mathbf{E}}(\mathbf{r},t)]^2: \rangle$ attains negative values (for appropriately chosen phases):

$$\langle :[\Delta\hat{\mathbf{E}}(\mathbf{r},t)]^2: \rangle < 0 \quad (11.75)$$

(cf. Sections 3.3 and 8.2.3). In the case of resonance fluorescence from a single two-level atom, on applying Eq. (11.15) and using the relations (11.38), we obtain

$$\langle :[\Delta\hat{\mathbf{E}}(\mathbf{r},t)]^2: \rangle = 2|\mathbf{g}|^2\sigma_{22}(t-r/c) - [\mathbf{g}\,\sigma_{21}(t-r/c) + \text{c.c.}]^2; \quad (11.76)$$

note that

$$\langle [\hat{\mathbf{E}}^{(+)}(\mathbf{r},t)]^2\rangle = |\mathbf{g}|^2 \langle \hat{A}_{12}(t-r/c)\hat{A}_{12}(t-r/c)\rangle = 0. \quad (11.77)$$

Introducing the slowly varying off-diagonal density-matrix element $\tilde{\sigma}_{21}$, Eq. (11.19), and separating from the product $\mathbf{g}\tilde{\sigma}_{21}(t-r/c)$ the slowly varying exponential phase factor $\exp[-i\tilde{\varphi}_{21}(t-r/c)]$, we may rewrite Eq. (11.76) as

$$\langle :[\Delta\hat{\mathbf{E}}(\mathbf{r},t)]^2:\rangle|\mathbf{g}|^{-2} = 2\sigma_{22}(t-r/c) \\ - 4|\tilde{\sigma}_{21}(t-r/c)|^2 \cos^2[\omega_L(t-r/c)+\phi(t-r/c)], \tag{11.78}$$

where

$$\phi(t-r/c) = \tilde{\varphi}_{21}(t-r/c) + \varphi_L. \tag{11.79}$$

If $\mathbf{g}$ is real, $\tilde{\varphi}_{21}(t)$ is simply the phase of $\tilde{\sigma}_{21}(t)$.

In particular, when the atom is driven by a single-mode cw laser, in the steady-state regime ($t\to\infty$) we may take the stationary values $\sigma_{22}(\infty)$ and $\tilde{\sigma}_{21}(\infty)$ as given in Eqs (11.34) and (11.35) to obtain

$$\langle :[\Delta\hat{\mathbf{E}}(\mathbf{r},t)]^2:\rangle|\mathbf{g}|^{-2} = -\frac{\Gamma_2\Omega_R^2}{\Gamma_1[\Gamma_2^2+(\delta\omega)^2+(\Gamma_2/\Gamma_1)\Omega_R^2]^2} \\ \times \left\{[\Gamma_2^2+(\delta\omega)^2]\left[\frac{\Gamma_1}{2\Gamma_2}\cos[2\omega_L(t-r/c)+2\phi(\infty)]-1+\frac{\Gamma_1}{2\Gamma_2}\right]-\frac{\Gamma_2}{\Gamma_1}\Omega_R^2\right\}, \tag{11.80}$$

which for exact resonance ($\delta\omega=0$) and radiative damping [$\Gamma_1=2\Gamma_2=2\gamma$; cf. Eq. (11.44)] reduces to (Fig. 11.2)

$$\langle :[\Delta\hat{\mathbf{E}}(\mathbf{r},t)]^2:\rangle|\mathbf{g}|^{-2} = -\frac{1}{2}\Omega_R^2\frac{\gamma^2\cos[2\omega_L(t-r/c)+2\phi(\infty)]-\frac{1}{2}\Omega_R^2}{(\gamma^2+\frac{1}{2}\Omega_R^2)^2}. \tag{11.81}$$

From Eq. (11.81) we see that, to obtain squeezed light, the condition

$$\Omega_R < \sqrt{2}\gamma \tag{11.82}$$

must be satisfied. The optimum phase condition is $\omega_L(t-r/c)+\phi(\infty)=n\pi$, $n=0,1,2,\ldots$, which, for a given value of Ω_R/γ, determines the minimum noise. Accordingly, the phase condition for maximum noise is $\omega_L(t-r/c)+\phi(\infty)=(2n+1)\pi/2$, $n=0,1,2,\ldots$. When the driving laser beam becomes too intense, so that in the fluorescence, the coherent Rayleigh component ($\sim\sigma_{21}$) is suppressed, the squeezing effect vanishes; cf. Fig. 11.2. Note that $\langle :[\Delta\hat{\mathbf{E}}(\mathbf{r},t)]^2:\rangle/|\mathbf{g}|^2$ tends to unity as γ/Ω_R goes to zero (high-driving-field limit). From Eqs (11.74)–(11.77) it can be seen that only a sufficiently large coherent part of the field can compensate the positive intensity term in order

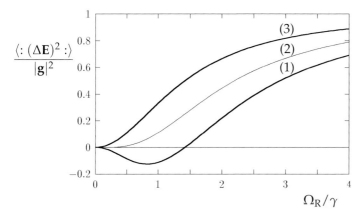

Fig. 11.2 The normally ordered variance $\langle :(\Delta \hat{\mathbf{E}})^2: \rangle$ of stationary resonance fluorescence from a two-level atom undergoing radiative damping ($\Gamma_1 = 2\Gamma_2 = 2\gamma$) and resonantly driven by a monochromatic laser beam is shown as function of the Rabi frequency Ω_R and for various values of the phase $\omega_L(t-r/c)+\phi(\infty)$: $n\pi$ (1); $(2n+1)\pi/4$ (2); $(2n+1)\pi/2$ (3); $n = 1,2,3,\ldots$.

to produce squeezing. Maximum squeezing, $\langle :(\Delta \hat{\mathbf{E}})^2: \rangle/|\mathbf{g}|^2 = -1/8$, may be attained for $\Omega_R = \sqrt{2/3}\gamma$ ($\delta\omega = 0$).

The results given above may be extended to the case of a sample of N identical two-level atoms being quasi-resonantly driven by a monochromatic laser beam, whose wavelength λ_L is small compared with the distance d between neighboring atoms. In this case ($\lambda_L \ll d$) we may ignore collective effects,[8] which implies that the source fields from different atoms are uncorrelated:

$$\langle \hat{A}_{Aab}(t) \hat{A}_{A'a'b'}(t') \rangle = \langle \hat{A}_{Aab}(t) \rangle \langle \hat{A}_{A'a'b'}(t') \rangle, \quad A \neq A'. \tag{11.83}$$

Applying Eq. (11.10) and using Eq. (11.83), after some algebra we obtain the generalized version of Eq. (11.80) as

$$\langle :[\Delta \hat{\mathbf{E}}(\mathbf{r},t)]^2: \rangle |\mathbf{g}|^{-2} = -N \frac{\Gamma_2 \Omega_R^2}{\Gamma_1 [\Gamma_2^2 + (\delta\omega)^2 + (\Gamma_2/\Gamma_1)\Omega_R^2]^2} \Big\{ [\Gamma_2^2 + (\delta\omega)^2]$$
$$\times \left[\frac{\Gamma_1}{2\Gamma_2} |\mathcal{C}^{(N)}| \cos[2\omega_L(t-r/c) + 2\phi^{(N)}(\infty)] - 1 + \frac{\Gamma_1}{2\Gamma_2} \right] - \frac{\Gamma_2}{\Gamma_1} \Omega_R^2 \Big\}, \tag{11.84}$$

where

$$\mathcal{C}^{(N)} = |\mathcal{C}^{(N)}| e^{2i\varphi_C} = \frac{1}{N} \sum_{A=1}^{N} \exp[2i(\mathbf{k}-\mathbf{k}')\mathbf{r}_A], \tag{11.85}$$

[8] A typical example of a collective effect is the so-called superradiance [Dicke (1954, 1964)]. If the distance of the atoms is small compared with the wavelength of the light, the intensity of the emitted light is proportional to N^2 in place of N.

$$\phi^{(N)}(\infty) = \phi(\infty) - \varphi_c. \tag{11.86}$$

In Eq. (11.85) the $\mathbf{r}_A$ are the position vectors of the atoms, and the vectors $\mathbf{k} \equiv \mathbf{k}_L$ and $\mathbf{k}' \equiv (\omega_L/c)\mathbf{r}/|\mathbf{r}|$ are respectively the wavenumber vectors of the incident laser light and the scattered light in the $\mathbf{r}$ direction, where $\mathbf{r}$ is the distance vector from the center of the scattering volume (origin of coordinates) to the chosen point of observation; note that $|\mathbf{r}_A| \ll |\mathbf{r}|$.

Comparing Eq. (11.84) with Eq. (11.80), we see that, when N atoms are involved in the resonant light scattering, the effect of squeezing will be enhanced by a factor N when the absolute value of $\mathcal{C}^{(N)}$ is held equal to unity. Distributing the atoms at regular positions is a way to enhance squeezing in resonance fluorescence [Vogel and Welsch (1985)]. For example, let us consider a linear chain of atoms, the distance vector between two neighboring atoms being $\mathbf{d}$. In this case, we can easily deduce from Eq. (11.85) that

$$|\mathcal{C}^{(N)}| = \frac{1}{N} \left| \frac{\sin[N(\mathbf{k} - \mathbf{k}')\mathbf{d}]}{\sin[(\mathbf{k} - \mathbf{k}')\mathbf{d}]} \right|, \tag{11.87}$$

from which we see that the scattered light may indeed exhibit enhanced squeezing in $\mathbf{k}'$ directions for which

$$(\mathbf{k} - \mathbf{k}')\mathbf{d} = n\pi, \quad n = 0, \pm 1, \pm 2, \ldots. \tag{11.88}$$

For comparison, the directions of the maxima of the coherent diffraction pattern are determined by the condition that

$$(\mathbf{k} - \mathbf{k}')\mathbf{d} = 2n\pi, \quad n = 0, \pm 1, \pm 2, \ldots. \tag{11.89}$$

Hence the maxima of the squeezing pattern correspond to both the maxima and the minima of the ordinary diffraction pattern, so that, depending upon the direction of observation, squeezed coherent and squeezed vacuum light may be observed.

Clearly, if the atoms are distributed at random positions, from Eq. (11.85) the value of $|\mathcal{C}^{(N)}|$ is seen to tend (for $\mathbf{k} \neq \mathbf{k}'$) to zero as the number of atoms, N, is sufficiently increased, and the squeezing effect is removed. An exception is the case of forward scattering ($\mathbf{k} = \mathbf{k}'$) [Heidmann and Reynaud (1985)]. From Eq. (11.85) it is clear that, for $\mathbf{k}$ close to $\mathbf{k}'$, the value of $\mathcal{C}^{(N)}$ may be close to unity.

Squeezing may be observed by means of a homodyne detection scheme (Section 6.5.3), in which a signal field to be detected is combined, through a lossless beam splitter, with a perfectly stable local-oscillator field on a photodetector. In particular, when the strength of the local-oscillator field greatly exceeds the signal field, the measured photocounting statistics is sub-Poissonian, provided that the signal field is squeezed. However, in single-atom light scattering the sub-Poissonian effect is expected to be extremely small due to a very small overall quantum efficiency,[9] so that it seems hopeless to measure it. For a small quantum efficiency, as in the case of resonantly scattered light from a trapped and cooled ion, a homodyne correlation measurement with a weak local oscillator is suited to detect the squeezing effect [Vogel (1991a)], for details of the method see Section 6.5.6.

11.2.4
Spectral properties

Let us now consider the power spectrum of the resonance fluorescence; that is, the spectrally resolved intensity $I(t;\omega,\Gamma_f) \equiv G_{kk}^{(1,1)}(\mathbf{r},t,\mathbf{r},t)$ with $G_{kk}^{(1,1)}(\mathbf{r},t,\mathbf{r},t)$ from Eq. (11.16) ($\mathbf{r}_A = 0$):

$$I(t+r/c;\omega,\Gamma_f) = |g|^2 \int dt_1' \int dt_2'\, T_f^*(t-t_1') T_f(t-t_2') G_{12}^{(I)}(t_1',t_2'), \quad (11.90)$$

where the atomic correlation function $G_{12}^{(I)}(t_1,t_2)$ is defined by

$$G_{12}^{(I)}(t_1,t_2) = \langle \hat{A}_{21}(t_1)\hat{A}_{12}(t_2)\rangle. \quad (11.91)$$

We substitute the result (11.12) into Eq. (11.90) for T_f and let $t - t_i' = \tau_i$, which yields

$$I(t+r_{op}/c;\omega,\Gamma_f) = |g|^2 (\tfrac{1}{2}\Gamma_f)^2 \int d\tau_1 \int d\tau_2\, \Theta(\tau_1)\Theta(\tau_2)$$
$$\times \exp[-\tfrac{1}{2}\Gamma_f(\tau_1+\tau_2) - i\omega(\tau_2-\tau_1)] G_{12}^{(I)}(t-\tau_1, t-\tau_2), \quad (11.92)$$

where $r_{op} = r + c\Delta t$ is the optical bath from the source (through the spectral filter) to the point of observation.

In particular, under steady-state conditions

$$S(\omega,\Gamma_f) = \frac{2}{\pi\Gamma_f |g|^2} \lim_{t\to\infty} I(t;\omega,\Gamma_f)$$
$$= \frac{\Gamma_f}{2\pi} \int d\tau_1 \int d\tau_2\, \Theta(\tau_1)\Theta(\tau_2) \exp[-\tfrac{1}{2}\Gamma_f(\tau_1+\tau_2) - i\omega(\tau_2-\tau_1)] G_{12}^{(I)}(\tau_2-\tau_1),$$
$$(11.93)$$

where

$$G_{12}^{(I)}(\tau) \equiv \lim_{t\to\infty} G_{12}^{(I)}(t+\tau,t). \quad (11.94)$$

We now let $\tau_2 - \tau_1 = \tau$ and $\tau_1 + \tau_2 = \tau'$ and perform the τ' integration to obtain

$$S(\omega,\Gamma_f) = \frac{1}{2\pi}\int_0^\infty d\tau\, \exp[-(i\omega + \tfrac{1}{2}\Gamma_f)\tau] G_{12}^{(I)}(\tau) + \text{c.c.}, \quad (11.95)$$

[9] In particular, the fluorescence light can be collected only within a small angular range.

from which we find that $S(\omega, \Gamma_f)$ may be rewritten as

$$S(\omega, \Gamma_f) = \frac{1}{2\pi} \int d\omega' \frac{\Gamma_f}{(\omega' - \omega)^2 + (\frac{1}{2}\Gamma_f)^2} S(\omega'), \qquad (11.96)$$

where the Wiener–Khintchine spectrum [cf. Eq. (6.167)]

$$S(\omega) = \frac{1}{2\pi} \int_0^\infty d\tau\, e^{-i\omega\tau} G_{12}^{(I)}(\tau) + \text{c.c.} \qquad (11.97)$$

corresponds to a (steady-state) Fourier analysis of the scattered light. In the limit as $\Gamma_f \to 0$ the "physical" spectrum $S(\omega, \Gamma_f)$ is equal to the Wiener–Khintchine spectrum $S(\omega)$:

$$\lim_{\Gamma_f \to 0} S(\omega, \Gamma_f) = S(\omega). \qquad (11.98)$$

Note that from Eq. (11.97) [together with Eqs (11.91) and (11.94)] the relation

$$G_{12}^{(I)}(0) = \sigma_{22}(\infty) = \int d\omega\, S(\omega) \qquad (11.99)$$

may be deduced. Since $\sigma_{22}(\infty)$ determines the overall steady-state intensity of the scattered light [cf. Eq. (11.45)], $S(\omega)\,d\omega$ corresponds to the contribution arising from frequencies within the interval $\omega, \omega + d\omega$.

Before going into the details of the further calculation, let us briefly outline the concept of dressed-atom states [Cohen-Tannoudji and Reynaud (1977)], which may be helpful in understanding the line structure to be expected. For this purpose, we view the exciting radiation field as a quantized single-mode field and describe the near-resonant interaction of the two-level atom with this field by a Hamiltonian $\hat{H}$ of the form Eqs (12.1)–(12.3) in Section 12.1. If $|a\rangle$, $a = 1, 2$, are the eigenstates of the unperturbed atomic Hamiltonian and $|n\rangle$ the eigenstates (photon-number states) of the unperturbed radiation-field Hamiltonian, the product states $|a\rangle|n\rangle$ are of course eigenstates of the Hamiltonian $\hat{H}_0$ of the uncoupled light–matter system with eigenvalues $E_{a,n}^{(0)} = \hbar(\omega_a + n\omega)$, Eq. (12.4):[10]

$$\hat{H}_0 |a\rangle|n\rangle = E_{a,n}^{(0)} |a\rangle|n\rangle, \qquad E_{a,n}^{(0)} = \hbar(\omega_a + n\omega) \qquad (11.100)$$

(ω, frequency of the radiation-field mode), which implies that, in the case of exact resonance, $\tilde{\omega}_{21} = \omega$, the eigenvalues of $\hat{H}_0$, except the ground state value, are twofold-degenerate, because the states $|1\rangle|n+1\rangle$ and $|2\rangle|n\rangle$ belong to equal energies:

$$E_{2,n}^{(0)} = E_{1,n+1}^{(0)}. \qquad (11.101)$$

10) Here the level shifts are thought of as being already included in the unperturbed atomic energies $\hbar\omega_a$, thus $\tilde{\omega}_{ab} = \hbar(\omega_a - \omega_b)/\hbar$.

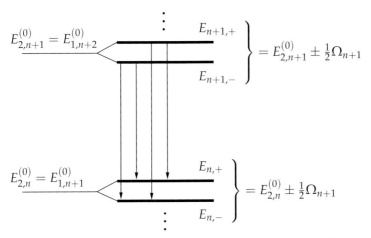

Fig. 11.3 The unperturbed, degenerate energy levels and the dressed-state energy levels of a two-level atom resonantly interacting with a quantized single-mode (driving) radiation field are shown schematically, and the dipole-allowed radiative transitions between the dressed states are indicated.

This degeneracy is removed by the interaction between atom and radiation–field mode, as can be seen by diagonalizing the full Hamiltonian $\hat{H}$ [Section 12.1, Eqs (12.5)–(12.16)], which for exact resonance yields (cf. Fig. 11.3)

$$\hat{H}|n,\pm\rangle = E_{n,\pm}|n,\pm\rangle, \tag{11.102}$$

$$E_{n,\pm} = \hbar(\omega_2 + n\omega \pm \tfrac{1}{2}\Omega_n), \tag{11.103}$$

$$|n,\pm\rangle = \sqrt{\tfrac{1}{2}}(|1\rangle|n+1\rangle \pm |2\rangle|n\rangle). \tag{11.104}$$

In Eq. (11.103) the n-photon Rabi frequency Ω_n defined in Eq. (12.11) may be written as

$$\Omega_n = 2\hbar^{-1}|\mathbf{d}_{21}\mathbf{E}(\mathbf{r}_A)|\sqrt{n+1}, \tag{11.105}$$

where $\mathbf{E}(\mathbf{r})$ is the electric-field mode function. The states $|n,\pm\rangle$, which are usually called dressed-atom states, describe the atom "dressed" by the interaction with the radiation field. From Eq. (11.103) the originally degenerate (excited) energy levels, $E_{2,n}^{(0)} = E_{1,n+1}^{(0)}$, are seen to be split into two levels separated from each other by $\hbar\Omega_n$ (Fig. 11.3).

If we now suppose that the Hamiltonian is complemented by the interaction energy between the two-level atom and the free-space mode continuum of the radiation field to allow for resonant light scattering (by spontaneous emission), we see from Eq. (11.104) that there are the following (dipole) transitions between the dressed states:

$$\langle n, \pm | \hat{d} | n+1, \pm \rangle = \pm \tfrac{1}{2} \langle 1 | \hat{d} | 2 \rangle, \tag{11.106}$$

$$\langle n, \mp | \hat{d} | n+1, \pm \rangle = \pm \tfrac{1}{2} \langle 1 | \hat{d} | 2 \rangle, \tag{11.107}$$

and from Eq. (11.103) we obtain the corresponding transition frequencies as (cf. Fig. 11.3)

$$\hbar^{-1}(E_{n+1,\pm} - E_{n,\pm}) = \omega_{21} \pm \tfrac{1}{2}(\Omega_{n+1} - \Omega_n), \tag{11.108}$$

$$\hbar^{-1}(E_{n+1,\pm} - E_{n,\mp}) = \omega_{21} \pm \tfrac{1}{2}(\Omega_{n+1} + \Omega_n). \tag{11.109}$$

Clearly, if the photon-number distribution of the field is sharply peaked at a sufficiently large value of the photon number n, so that it may be viewed as a classical driving field, we may let

$$\Omega_{n+1} \approx \Omega_n \approx \Omega_R. \tag{11.110}$$

In this case the four transition frequencies, Eqs (11.108) and (11.109), reduce to the three transition frequencies ω_{21} and $\omega_{21} \pm \Omega_R$. From the dressed-state picture the spectrum of the resonance fluorescence is therefore expected to exhibit a triplet structure, provided that the Rabi frequency (line separation) substantially exceeds the line broadening arising from dampings, such as spontaneous emission. Moreover, in this case the ratio of the (frequency-integrated) intensities of the center line and a side-band line is expected to be 2:1, because of the two dressed-state transitions contributing to the center line [cf. Eqs (11.106)–(11.109)].

It is worth noting that the dressed-state picture may also be useful for obtaining some insight into the line structure of the spectra in cases where more than two atomic levels are involved in the (resonant) light–matter interaction. Assume, for example, that there are additional atomic states between the atomic ground and excited states. If the corresponding transitions to these additional states are off-resonant, with regard to the driving field, they cannot give rise to dressed-state level splittings. Thus emission from the exited state into these lower-lying states can only yield doublet spectra.

So far relaxations, which typically determine transient and line-broadening effects, have been ignored. In the following we shall include them in the theory by applying the Bloch equations. It is worth noting that they may also be included in the theory within the framework of the dressed-state concept.

We confine attention to the steady-state regime and return to Eq. (11.95).[11] Applying the quantum regression theorem (Section 5.5), the desired atomic correlation function $G_{12}^{(I)}(t+\tau, t)$ $(t \to \infty)$ may be calculated in analogy with the

---

[11] For transient effects, the more general equation (11.92) applies [see, e.g., Herrmann, Süsse and Welsch (1973)].

lines shown in Section 11.2.2 for $G_{ab}^{(II)}(t+\tau,t)$. Introducing the set of atomic correlation functions

$$G_{ab}^{(I)}(t+\tau,t) = \langle \hat{A}_{ba}(t+\tau)\hat{A}_{12}(t)\rangle, \quad a,b = 1,2, \tag{11.111}$$

and the slowly varying quantities

$$\tilde{G}_{12}^{(I)}(t+\tau,t) = G_{12}^{(I)}(t+\tau,t)\exp(-i\omega_L\tau), \tag{11.112}$$

$$\tilde{G}_{21}^{(I)}(t+\tau,t) = G_{21}^{(I)}(t+\tau,t)\exp[i(\omega_L\tau)]\exp[2i(\omega_L t + \varphi_L)], \tag{11.113}$$

$$\tilde{G}_{aa}^{(I)}(t+\tau,t) = G_{aa}^{(I)}(t+\tau,t)\exp[i(\omega_L t + \varphi_L)], \tag{11.114}$$

the $\tilde{G}_{ab}^{(I)}(t+\tau,t)$ are easily proved to obey, with regard to τ ($\tau \geq 0^+$), the Bloch equations (11.21)–(11.24) [or Eqs (11.60)–(11.63)],[12] where the initial conditions are determined from Eq. (11.111) [together with Eqs (11.112)–(11.114)] as

$$G_{ab}^{(I)}(t+\tau,t)\big|_{\tau=0} = \langle \hat{A}_{ba}(t)\hat{A}_{12}(t)\rangle = \delta_{a1}\sigma_{2b}(t) \tag{11.115}$$

[cf. the relations (11.38)]. In particular, we see that

$$\lim_{t\to\infty}\tilde{G}_{11}^{(I)}(t+\tau,t)\big|_{\tau=0} = \tilde{\sigma}_{21}(\infty), \tag{11.116}$$

$$\lim_{t\to\infty}\tilde{G}_{12}^{(I)}(t+\tau,t)\big|_{\tau=0} = \sigma_{22}(\infty), \tag{11.117}$$

where the steady-state density-matrix elements $\sigma_{22}(\infty)$ and $\tilde{\sigma}_{21}(\infty)$ may be taken from Eqs (11.34) and (11.35), respectively.

Introducing the Laplace transforms

$$\tilde{S}_{ab}(s) = \int_0^\infty d\tau\, e^{-s\tau}\,\tilde{G}_{ab}^{(I)}(\tau), \tag{11.118}$$

where

$$\tilde{G}_{ab}^{(I)}(\tau) \equiv \lim_{t\to\infty}\tilde{G}_{ab}^{(I)}(t+\tau,t), \tag{11.119}$$

we find, on taking Eq. (11.112)–(11.114) into account, that the "physical" spectrum $S(\omega,\Gamma_f)$, Eq. (11.95), and the Wiener–Khintchine spectrum $S(\omega)$, Eq. (11.97), are related to $\tilde{S}_{12}(s)$ by

$$S(\omega,\Gamma_f) = \pi^{-1}\mathrm{Re}\{\tilde{S}_{12}[i(\omega-\omega_L) + \tfrac{1}{2}\Gamma_f]\}, \tag{11.120}$$

$$S(\omega) = \lim_{\Gamma_f\to 0}S(\omega,\Gamma_f) = \pi^{-1}\mathrm{Re}\{\tilde{S}_{12}[i(\omega-\omega_L)]\}. \tag{11.121}$$

Thus to calculate the spectrum it is sufficient to determine $\tilde{S}_{12}(s)$, Eq. (11.118),

12) Note that $\tilde{G}_{ab}^{(I)} \neq \tilde{G}_{ba}^{(I)*}$.

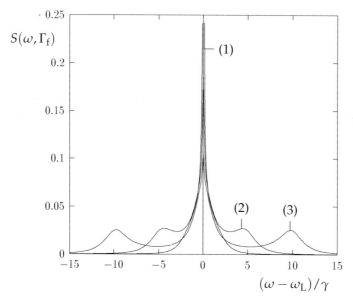

Fig. 11.4 The steady-state resonance fluorescence spectrum $S(\omega, \Gamma_f)$ of a two-level atom undergoing radiative damping ($\Gamma_1 = 2\Gamma_2 = 2\gamma$) and resonantly driven by a monochromatic laser beam of frequency ω_L is shown for various values of the Rabi frequency Ω_R/γ: 2 (1); 5 (2); 10 (3); the passband width of the spectral apparatus is chosen to be $\Gamma_f = 0.1\gamma$.

from the solution of the Laplace-transformed Bloch equations [with the initial conditions from Eqs (11.116) and (11.117)]:

$$s\tilde{S}_{22}(s) = -\Gamma_1 \tilde{S}_{22}(s) - \tfrac{1}{2}i\Omega_R \tilde{S}_{21}(s) + \tfrac{1}{2}i\Omega_R \tilde{S}_{12}(s), \tag{11.122}$$

$$s\tilde{S}_{11}(s) = \tilde{\sigma}_{21}(\infty) + \Gamma_1 \tilde{S}_{22}(s) + \tfrac{1}{2}i\Omega_R \tilde{S}_{21}(s) - \tfrac{1}{2}i\Omega_R \tilde{S}_{12}(s), \tag{11.123}$$

$$s\tilde{S}_{21}(s) = (-i\delta\omega - \Gamma_2)\tilde{S}_{21}(s) + \tfrac{1}{2}i\Omega_R [\tilde{S}_{11}(s) - \tilde{S}_{22}(s)], \tag{11.124}$$

$$s\tilde{S}_{12}(s) = \sigma_{22}(\infty) + (i\delta\omega - \Gamma_2)\tilde{S}_{12}(s) - \tfrac{1}{2}i\Omega_R [\tilde{S}_{11}(s) - \tilde{S}_{22}(s)]. \tag{11.125}$$

In this way, the determination of the spectrum is reduced to a straightforward algebraic problem. In particular, for exact resonance ($\delta\omega = 0$) we readily deduce that

$$\tilde{S}_{12}(s) = \frac{\sigma_{22}(\infty)}{s+\Gamma_2} - \frac{i}{2}\Omega_R \left[\frac{\tilde{\sigma}_{21}(\infty)}{s(s+\Gamma_2)} - \frac{is\Omega_R \sigma_{22}(\infty) + \Omega_R^2 \tilde{\sigma}_{21}(\infty)}{s(s+\Gamma_2)[(s+\Gamma_1)(s+\Gamma_2) + \Omega_R^2]} \right], \tag{11.126}$$

where

$$\sigma_{22}(\infty) = \frac{1}{2} \frac{\Omega_R^2}{\Gamma_1 \Gamma_2 + \Omega_R^2}, \tag{11.127}$$

$$\tilde{\sigma}_{21}(\infty) = \frac{i}{2} \frac{\Gamma_1 \Omega_R}{\Gamma_1 \Gamma_2 + \Omega_R^2} \tag{11.128}$$

[Eqs (11.34) and (11.35) for $\delta\omega=0$].

As expected, the spectrum $S(\omega, \Gamma_f)$ [Eq. (11.120) together with Eq. (11.126)] depends sensitively on the intensity of the driving field (Fig. 11.4).

11.2.4.1 Limiting cases

Let us particularly consider the two cases of weak and strong pumping.

(i) *Weak-driving-field limit*: $\Omega_R^2 \ll \Gamma_1 \Gamma_2$. In this limit the lowest order of perturbation theory applies. Keeping only terms up to second order in the Rabi frequency in Eq. (11.126) reduces Eq. (11.120) to

$$S(\omega, \Gamma_f) = S_{\text{coh}}(\omega, \Gamma_f) + S_{\text{incoh}}(\omega, \Gamma_f), \tag{11.129}$$

where

$$S_{\text{coh}}(\omega, \Gamma_f) = \frac{1}{\pi} \frac{\frac{1}{2}\Gamma_f |\tilde{\sigma}_{21}(\infty)|^2}{(\omega - \omega_L)^2 + \left(\frac{1}{2}\Gamma_f\right)^2}, \tag{11.130}$$

$$S_{\text{incoh}}(\omega, \Gamma_f) = \frac{1}{\pi} \frac{\left(\frac{1}{2}\Gamma_f + \Gamma_2\right)[\sigma_{22}(\infty) - |\tilde{\sigma}_{21}(\infty)|^2]}{(\omega - \omega_L)^2 + \left(\frac{1}{2}\Gamma_f + \Gamma_2\right)^2}, \tag{11.131}$$

and

$$|\tilde{\sigma}_{21}(\infty)|^2 = \frac{\Gamma_1}{2\Gamma_2} \sigma_{22}(\infty), \qquad \sigma_{22}(\infty) = \frac{1}{2} \frac{\Omega_R^2}{\Gamma_1 \Gamma_2}. \tag{11.132}$$

We see that the resulting line at $\omega=\omega_L$ consists of two parts. $S_{\text{coh}}(\omega, \Gamma_f)$ (Rayleigh component) is the part that obviously results from the coherently emitted light (whose intensity is proportional to $|\tilde{\sigma}_{21}|^2$). Accordingly, $S_{\text{incoh}}(\omega, \Gamma_f)$ is related to the incoherently emitted light. It is worth noting that in the case of radiative damping [$\Gamma_2 = \gamma$ and $\Gamma_1 = 2\gamma$; cf. Eq. (11.44)] the intensity of the incoherently emitted light vanishes ($\sigma_{22} = |\tilde{\sigma}_{21}|^2$), so that

$$S(\omega, \Gamma_f) = S_{\text{coh}}(\omega, \Gamma_f). \tag{11.133}$$

Thus in the high-resolution limit ($\Gamma_f \to 0$) the spectrum simply consists of the sharply peaked Rayleigh line:

$$S(\omega) = S_{\text{coh}}(\omega) = |\sigma_{21}(\infty)|^2 \delta(\omega - \omega_L) = \left(\frac{\Omega_R}{2\gamma}\right)^2 \delta(\omega - \omega_L). \tag{11.134}$$

(ii) *High-driving-field limit*: $\Omega_R^2 \gg \Gamma_1 \Gamma_2$. In this case Eqs (11.128) and (11.127) approximate to

$$\tilde{\sigma}_{21}(\infty) = 0, \quad \sigma_{22}(\infty) = \tfrac{1}{2}, \tag{11.135}$$

and the spectrum arises solely from incoherently scattered light. After some minor algebra, from Eq. (11.120) [together with Eqs (11.126) and (11.135)] we obtain the following result:

$$S(\omega, \Gamma_f) = S_{incoh}(\omega, \Gamma_f)$$
$$= \frac{1}{4\pi}[S_0(\omega, \Gamma_f) + S_+(\omega, \Gamma_f) + S_-(\omega, \Gamma_f)], \quad (11.136)$$

where

$$S_0(\omega, \Gamma_f) = \frac{\Gamma_2 + \frac{1}{2}\Gamma_f}{(\omega - \omega_L)^2 + (\Gamma_2 + \frac{1}{2}\Gamma_f)^2}, \quad (11.137)$$

$$S_\pm(\omega, \Gamma_f) = \frac{1}{2} \frac{\frac{1}{2}(\Gamma_1 + \Gamma_2 + \Gamma_f)}{[\omega - (\omega_L \pm \Omega_R)]^2 + [\frac{1}{2}(\Gamma_1 + \Gamma_2 + \Gamma_f)]^2}. \quad (11.138)$$

In particular, for radiative damping ($\Gamma_2 = \gamma$, $\Gamma_1 = 2\gamma$) and high resolution ($\Gamma_f \to 0$) Eqs (11.137) and (11.138) are

$$S_0(\omega) = \frac{\gamma}{(\omega - \omega_L)^2 + \gamma^2}, \quad (11.139)$$

$$S_\pm(\omega) = \frac{1}{2} \frac{\frac{3}{2}\gamma}{[\omega - (\omega_L \pm \Omega_R)]^2 + (\frac{3}{2}\gamma)^2}. \quad (11.140)$$

In accordance with the dressed-state concept, the spectrum is found to be a triplet, frequently called the Mollow triplet [Mollow (1969); for measurements see Schuda, Stroud and Hercher (1974); Wu, Grove and Ezekiel (1975, 1977) and Hartig, Rasmussen, Schieder and Walther (1976)]. The line separation is just given by the Rabi frequency. Note that the side-band width (3γ) is larger than the width of the central line (2γ); the central line exceeds the side bands in height by a factor of 3:

$$\frac{S_0(\omega_L)}{S_\pm(\omega_L \pm \Omega_R)} = 3. \quad (11.141)$$

11.2.4.2 Higher-order spectral properties

We recall that the spectral intensity considered so far is determined by the correlation function $G_{kk}^{(1,1)}$, which contains both the operator $\hat{E}^{(+)}$ and the operator $\hat{E}^{(-)}$ at first order, so that the $\mathcal{T}_\pm$ time orderings of the atomic source operators [cf. Eq. (11.16)] become superfluous. In the study of higher-order spectral properties, where $\hat{E}^{(+)}$ and/or $\hat{E}^{(-)}$ appear at higher than first order, these time orderings must be performed very carefully.

Spectral squeezing is a typical example. We know that the criterion for squeezing is that the normally ordered variance $\langle :(\Delta\hat{E})^2: \rangle$ may become negative. Decomposing $\hat{E}$ into $\hat{E}^{(+)}$ and $\hat{E}^{(-)}$, we easily see that $\langle :(\Delta\hat{E})^2: \rangle$ may be expressed in terms of the (equal-time) correlation functions $G^{(0,1)}$, $G^{(1,0)}$, $G^{(1,1)}$, $G^{(0,2)}$ and $G^{(2,0)}$ [cf. Eq. (11.74)]. In the case of spectral squeezing in resonance fluorescence [Collett, Walls and Zoller (1984)] all of these correlation functions may be calculated by appropriately applying Eq. (11.16). It is worth noting that the required time orderings in $G^{(0,2)}$ and $G^{(2,0)}$ prevent the observable (steady-state) squeezing spectrum from being related to a squeezing spectrum defined by a (steady-state) Fourier analysis of the scattered light [Knöll, Vogel and Welsch (1986, 1990)].[13] Another illustrative example of a higher-order spectral property is the spectrally resolved intensity correlation determined, according to Eq. (11.16), by the correlation function $G^{(2,2)}_{k_1 k_2 k_2 k_1}(\mathbf{r}, t+\tau, \mathbf{r}, t)$ [Knöll and Weber (1986); Cresser (1987)]. In particular, to answer the question of how the intensities of different lines of the Mollow triplet are correlated with each other, a spectral resolution which allows separation of the lines from each other but leaves the line shapes unresolved, is sufficient ($\Omega_R > \Gamma_f > \Gamma_{1,2}$). In this case the time orderings in Eq. (11.16) play a minor role, because the response functions of the (two) spectral filters are rapidly varying on the time scale of the decay of the relevant atomic correlation functions [for experimental results see Aspect, Roger, Reynaud, Dalibard and Cohen-Tannoudji (1980)].

11.3
Multi-level effects

If more than two atomic quantum states are involved in the resonant light–matter interaction, atomic excitation redistributions together with the discrete nature of resonant transitions (quantum jumps) may give rise to a series of new effects, such as dark resonances and intermittent fluorescence. In general, the dynamics of a driven multi-level atomic system depends sensitively on the (multi-level) damping parameters, the Rabi frequencies and the detunings of the driving fields, so that separating multi-level effects from each other may often be difficult. In the following we shall therefore restrict attention to a few illustrative examples and briefly discuss the features that may typically be observed under certain limiting conditions.

11.3.1
Dark resonances

Let us consider a three-level system as shown in Fig. 11.5 (the so-called Λ configuration). The transition from the (ground) state $|1\rangle$ to the excited state $|3\rangle$ is pumped by a laser of frequency ω_g (say a green laser). A second (red) laser

---

[13] The two kinds of squeezing spectra differ in time-delayed commutator terms, which do not appear in the measured spectrum.

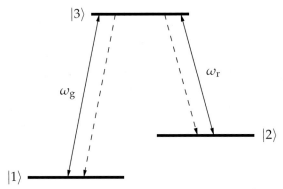

Fig. 11.5 Scheme of a three-level system of Λ configuration driven by two laser fields of frequencies ω_r and ω_g.

of frequency ω_r is used to pump the transition from a state $|2\rangle$ to the state $|3\rangle$.[14] To describe the dynamics of this system, the corresponding three-level density-matrix equations of motion may be deduced by straightforward extension of the derivation of the (two-level) optical Bloch equations to a three-level system (cf. Section 5.4.2):

$$\dot{\sigma}_{11} = w_{21}\sigma_{22} + w_{31}\sigma_{33} - \tfrac{1}{2}i\Omega_g\tilde{\sigma}_{13} + \tfrac{1}{2}i\Omega_g\tilde{\sigma}_{31}, \tag{11.142}$$

$$\dot{\sigma}_{22} = -w_{21}\sigma_{22} + w_{32}\sigma_{33} - \tfrac{1}{2}i\Omega_r\tilde{\sigma}_{23} + \tfrac{1}{2}i\Omega_r\tilde{\sigma}_{32}, \tag{11.143}$$

$$\dot{\sigma}_{33} = -(\dot{\sigma}_{22} + \dot{\sigma}_{11}), \tag{11.144}$$

$$\dot{\tilde{\sigma}}_{12} = [i(\delta\omega_g - \delta\omega_r) - \Gamma_{12}]\tilde{\sigma}_{12} + \tfrac{1}{2}i\Omega_g\tilde{\sigma}_{32} - \tfrac{1}{2}i\Omega_r\tilde{\sigma}_{13}, \tag{11.145}$$

$$\dot{\tilde{\sigma}}_{13} = (i\delta\omega_g - \Gamma_{13})\tilde{\sigma}_{13} - \tfrac{1}{2}i\Omega_r\tilde{\sigma}_{12} - \tfrac{1}{2}i\Omega_g(\sigma_{11} - \sigma_{33}), \tag{11.146}$$

$$\dot{\tilde{\sigma}}_{23} = (i\delta\omega_r - \Gamma_{23})\tilde{\sigma}_{23} - \tfrac{1}{2}i\Omega_g\tilde{\sigma}_{21} - \tfrac{1}{2}i\Omega_r(\sigma_{22} - \sigma_{33}), \tag{11.147}$$

$$\dot{\tilde{\sigma}}_{ba} = (\dot{\tilde{\sigma}}_{ab})^*, \tag{11.148}$$

where $\delta\omega_g = \omega_{31} - \omega_g$, $\delta\omega_r = \omega_{32} - \omega_r$, Ω_g and Ω_r are the Rabi frequencies of the green and red lasers respectively, the values of which may be comparable. Note that $\tilde{\sigma}_{13}$, $\tilde{\sigma}_{23}$ and $\tilde{\sigma}_{12}$ are slowly varying off-diagonal density-matrix elements:

$$\tilde{\sigma}_{13} = \sigma_{13}\exp[-i(\omega_g t + \varphi_g)], \tag{11.149}$$

$$\tilde{\sigma}_{23} = \sigma_{23}\exp[-i(\omega_r t + \varphi_r)], \tag{11.150}$$

$$\tilde{\sigma}_{12} = \sigma_{12}\exp[-i(\omega_g - \omega_r)t - i(\varphi_g - \varphi_r)] \tag{11.151}$$

[cf. Eq. (11.19)].

Although the transitions $|1\rangle \leftrightarrow |3\rangle$ and $|2\rangle \leftrightarrow |3\rangle$ are comparably driven, so that, at first glance, a trapping state would not be expected to appear, the fluo-

---

14) The state $|2\rangle$ is a long-lived (metastable) state, so that the transition $|1\rangle \leftrightarrow |2\rangle$ is weak.

rescence intensity may break down (dark resonance) under certain excitation conditions. The principal mechanism for this effect may be seen to be effective pumping of a superposition of the states $|1\rangle$ and $|2\rangle$ [Orriols (1979)]. To clarify this point, let us consider the temporal evolution of the intensity of the scattered light, which is, as we know, determined by the occupation probability of state $|3\rangle$, i.e., $\langle \hat{E}^{(-)}\hat{E}^{(+)}\rangle \sim \sigma_{33}$, and introduce a new set of (time-dependent) atomic states $|i'\rangle$ ($i=1,2,3$):

$$|1'\rangle = |1\rangle, \tag{11.152}$$
$$|2'\rangle = \exp[-i(\omega_g - \omega_r)t - i(\varphi_g - \varphi_r)]|2\rangle, \tag{11.153}$$
$$|3'\rangle = \exp[-i(\omega_g t + \varphi_g)]|3\rangle. \tag{11.154}$$

Note that the slowly varying off-diagonal density-matrix elements (11.149)–(11.151) are the density-matrix elements in the basis $|i'\rangle$, $\sigma_{i'j'} \equiv \langle i'|\hat{\sigma}|j'\rangle = \tilde{\sigma}_{ij}$, and $\sigma_{i'i'} = \sigma_{ii}$. Defining the superposition states

$$|\pm\rangle = \tfrac{1}{\sqrt{2}}(|1'\rangle \pm |2'\rangle) \tag{11.155}$$

and expressing $\sigma_{3'\pm} \equiv \langle 3'|\hat{\sigma}|\pm\rangle$, $\sigma_{\pm\pm} \equiv \langle \pm|\hat{\sigma}|\pm\rangle$ and $\sigma_{\mp\pm} \equiv \langle \mp|\hat{\sigma}|\pm\rangle$ in terms of the original density-matrix elements, we can easily deduce that

$$\sigma_{3'\pm} = \tilde{\sigma}_{31} \pm \tilde{\sigma}_{32}, \tag{11.156}$$
$$\sigma_{\pm\pm} = \tfrac{1}{2}(\sigma_{11} + \sigma_{22}) \pm \operatorname{Re} \tilde{\sigma}_{12}, \tag{11.157}$$
$$\sigma_{\mp\pm} = \tfrac{1}{2}(\sigma_{11} - \sigma_{22}) \pm i \operatorname{Im} \tilde{\sigma}_{12}. \tag{11.158}$$

Note that the (new) occupation probabilities will, of course, satisfy the condition

$$\sigma_{--} + \sigma_{++} + \sigma_{33} = 1. \tag{11.159}$$

Now it is a straightforward procedure to rewrite Eqs (11.142)–(11.148) to obtain the equations of motion obeyed by the above density-matrix elements. In particular, we deduce that[15]

$$\dot{\sigma}_{++} = \tfrac{1}{2}(w_{31} + w_{32})\sigma_{33} - \tfrac{1}{2}\Gamma_{12}(\sigma_{++} - \sigma_{--})$$
$$- \tfrac{1}{2}(\delta\omega_g - \delta\omega_r)(\sigma_{-+} + \sigma_{+-}) + \tfrac{1}{\sqrt{2}}(\Omega_g + \Omega_r)\operatorname{Im}\sigma_{+3'}, \tag{11.160}$$

$$\dot{\sigma}_{--} = \tfrac{1}{2}(w_{31} + w_{32})\sigma_{33} + \tfrac{1}{2}\Gamma_{12}(\sigma_{++} - \sigma_{--})$$
$$+ \tfrac{1}{2}(\delta\omega_g - \delta\omega_r)(\sigma_{-+} + \sigma_{+-}) + \tfrac{1}{\sqrt{2}}(\Omega_g - \Omega_r)\operatorname{Re}\sigma_{3'-}. \tag{11.161}$$

From inspection of Eqs (11.160) and (11.161) we can see that σ_{++} and σ_{--} are coupled to the two laser fields by $\Omega_g + \Omega_r$ and $\Omega_g - \Omega_r$ respectively. Thus when

$$\Omega \equiv \Omega_g = \Omega_r, \tag{11.162}$$

[15] For the full set of the equations of motion, see Vogel and Blatt (1992).

then the occupation probability σ_{--} is not influenced by the laser fields. Moreover, when the lasers are exactly tuned to the Raman resonance, that is, when the difference of the two laser frequencies $\omega_g - \omega_r$ is equal to the atomic transition frequency ω_{21}, so that

$$\delta\omega_g = \delta\omega_r, \tag{11.163}$$

then Eqs (11.160) and (11.161) simplify to

$$\dot{\sigma}_{++} = \tfrac{1}{2}(w_{31} + w_{32})\sigma_{33} - \tfrac{1}{2}\Gamma_{12}(\sigma_{++} - \sigma_{--}) + \sqrt{2}\,\Omega\,\mathrm{Im}\,\sigma_{+3'}, \tag{11.164}$$

$$\dot{\sigma}_{--} = \tfrac{1}{2}(w_{31} + w_{32})\sigma_{33} + \tfrac{1}{2}\Gamma_{12}(\sigma_{++} - \sigma_{--}). \tag{11.165}$$

From Eq. (11.165) we see that, in general, the process of occupying the state $|-\rangle$ by spontaneous emission from state $|3\rangle$ (first term) is, according to the term $-(\Gamma_{12}/2)\sigma_{--}$, in competition with a depopulation process. When the atom undergoes radiative damping and is driven by lasers with sufficiently small linewidths, then the depopulation rate $\Gamma_{12}/2$ in Eq. (11.165) is very small, so that the inequality

$$\Gamma_{12} \ll w_{31}, w_{32} \tag{11.166}$$

may assumed to be satisfied (cf. footnote 14). In the regime described, the first term in Eq. (11.165) is therefore the leading one. This implies that on a time scale of $2(w_{31}+w_{32})^{-1}$ the occupation probability σ_{--} is increased at the expense of σ_{33} (and σ_{++}) [cf. Eq. (11.159)]. Note that in the limit as $\Gamma_{12} \to 0$ we find from Eq. (11.165) that $\sigma_{33}(\infty) \to 0$, the optically active electron being trapped in the superposition state $|-\rangle$.

Thus the fluorescence intensity I ($I \sim \sigma_{33}$) observed at the beginning of the interaction of the atom with the laser fields, breaks down in the further course of time, which gives rise to the above-mentioned dark resonance. It should be noted that the superposition states $|+\rangle$ and $|-\rangle$ can be exchanged by a π change of the laser difference phase $\varphi_g - \varphi_r$ [cf. Eq. (11.155) together with Eqs (11.152)–(11.154)]. Hence the fluorescence can be switched on by appropriately switching the laser phase difference [Vogel and Blatt (1992)], which reveals that the considered dark resonance is a phase-sensitive (coherent) effect.

11.3.2
Intermittent fluorescence

Let us now consider a three-level system of the so-called V configuration (Fig. 11.6). The transition from the (ground) state $|1\rangle$ to the excited state $|3\rangle$ is pumped by a laser of frequency ω_g (green laser). A second (red) laser of frequency ω_r is used to pump the transition from state $|1\rangle$ to state $|2\rangle$. We assume that the transition $|1\rangle \leftrightarrow |3\rangle$ is dipole allowed, whereas the other transitions are

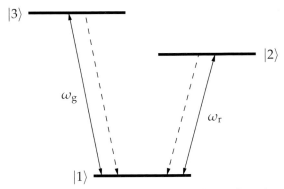

Fig. 11.6 Scheme of a three-level system of V configuration driven by two laser fields of frequencies ω_r and ω_g.

dipole forbidden. Moreover, we assume that the transition $|3\rangle \leftrightarrow |2\rangle$ is weak compared with the (pumped) transition $|1\rangle \leftrightarrow |2\rangle$.[16] The corresponding three-level density-matrix equations of motion are similar to Eqs (11.142)–(11.148), however with the transition $|1\rangle \leftrightarrow |2\rangle$ pumped by the red laser (in place of the transition $|2\rangle \leftrightarrow |3\rangle$).

If the atom, which is irradiated by the two laser beams, is at a certain time t in the (ground) state $|1\rangle$, during a relatively long period of time the strong (green) transition $|1\rangle \leftrightarrow |3\rangle$ is expected to dominate the atomic dynamics, which implies the generation of green fluorescence. During this period the atom undergoes many quantum jumps between the states $|3\rangle$ and $|1\rangle$. With a small but finite probability, it may also undergo a quantum jump from the state $|1\rangle$ to the state $|2\rangle$ by absorbing a red laser photon. Such an event is expected to switch off the transition $|1\rangle \leftrightarrow |3\rangle$ and the green fluorescence as long as the atom is in the long-lived state $|2\rangle$. The atom is expected to become invisible until it emits a red photon and therefore undergoes a quantum jump from state $|2\rangle$ to $|1\rangle$ owing to the weak transition $|1\rangle \leftrightarrow |2\rangle$ driven by the red laser. Now, the emission of green photons and the excitation of the transition $|1\rangle \leftrightarrow |3\rangle$ by the green laser may be viewed as the most probable process until again a dark period occurs, as described. In this way, the beginnings and ends of the dark periods may be regarded as indicating the instants of the quantum jumps $|1\rangle \to |2\rangle$ and $|2\rangle \to |1\rangle$ respectively.

The above intuitive arguments may be proved to be correct by calculating the intensity correlations of green and red photons, using the general results given in Section 11.1 and applying (as in Section 11.2.2) the quantum regression theorem together with the density-matrix equations of motion. From

---

16) For example, $|1\rangle \leftrightarrow |2\rangle$ may be a quadrupole transition.

Eq. (11.5) the source-field part of the scattered light is found to be

$$\hat{\mathbf{E}}_s^{(+)}(\mathbf{r},t) = \hat{\mathbf{E}}_{gs}^{(+)}(\mathbf{r},t) + \hat{\mathbf{E}}_{rs}^{(+)}(\mathbf{r},t), \qquad (11.167)$$

where

$$\hat{\mathbf{E}}_{gs}^{(+)}(\mathbf{r},t+r/c) \sim d_{13}\hat{A}_{13}(t), \qquad (11.168)$$

$$\hat{\mathbf{E}}_{rs}^{(+)}(\mathbf{r},t+r/c) \sim q_{12}\hat{A}_{12}(t), \qquad (11.169)$$

where d_{13} and q_{12} are respectively the (large) dipole and (small) quadrupole matrix elements. The four intensity correlation functions (green-green, red-red, green-red and red-green) are[17]

$$G_{gg}^{(2,2)} \sim |d_{13}|^4 \langle \hat{A}_{31}(t)\hat{A}_{33}(t+\tau)\hat{A}_{13}(t) \rangle, \qquad (11.170)$$

$$G_{rr}^{(2,2)} \sim |q_{12}|^4 \langle \hat{A}_{21}(t)\hat{A}_{22}(t+\tau)\hat{A}_{12}(t) \rangle, \qquad (11.171)$$

$$G_{gr}^{(2,2)} \sim |d_{13}|^2|q_{12}|^2 \langle \hat{A}_{21}(t)\hat{A}_{33}(t+\tau)\hat{A}_{12}(t) \rangle, \qquad (11.172)$$

$$G_{rg}^{(2,2)} \sim |q_{12}|^2|d_{13}|^2 \langle \hat{A}_{31}(t)\hat{A}_{22}(t+\tau)\hat{A}_{13}(t) \rangle. \qquad (11.173)$$

Recall that the joint probability of observing a green (respectively, red) photon at time $t+\tau$ and at the earlier time t a green (respectively, red) photon is proportional to $G_{gg}^{(2,2)}$ (respectively, $G_{rr}^{(2,2)}$). Accordingly, the joint probability of observing at time $t+\tau$ a green (respectively, red) photon and at time t a red (respectively, green) photon is proportional to $G_{gr}^{(2,2)}$ (respectively, $G_{rg}^{(2,2)}$). In Eqs (11.170)–(11.173) the atomic correlation functions $\langle \hat{A}_{a1}(t)\hat{A}_{bb}(t+\tau)\hat{A}_{1a}(t) \rangle$ ($a,b=2,3$) may be calculated in a similar way to Section 11.2.2:

$$G_{gg}^{(2,2)} \sim |d_{13}|^4 \sigma_{33}(t)\sigma_{33}(t+\tau)\big|_{\sigma_{ab}(t)=\delta_{a1}\delta_{b1}}, \qquad (11.174)$$

$$G_{gr}^{(2,2)} \sim |d_{13}|^2|q_{12}|^2 \sigma_{22}(t)\sigma_{33}(t+\tau)\big|_{\sigma_{ab}(t)=\delta_{a1}\delta_{b1}}, \qquad (11.175)$$

$$G_{rg}^{(2,2)} \sim |q_{12}|^2|d_{13}|^2 \sigma_{33}(t)\sigma_{22}(t+\tau)\big|_{\sigma_{ab}(t)=\delta_{a1}\delta_{b1}}, \qquad (11.176)$$

where $\sigma_{ab}(t+\tau)\big|_{\sigma_{a'b'}(t)=\delta_{a'1}\delta_{b'1}}$ are the solutions of the density-matrix equations of motion with initial conditions $\sigma_{ab}(t+\tau)\big|_{\tau=0}=\delta_{a1}\delta_{b1}$ (atom in state $|1\rangle$ at

---

[17] Tuning a filter to the green (respectively, red) light may suppress the red (respectively, green) light, without resolving the line shape of the green (or red) light. Note that the line shapes are determined by the damping parameters w_{ab}, Γ_{ab} and (for high driving fields) by the Rabi frequencies Ω_g and Ω_r (cf. Section 11.2.4). With regard to each light component (green, red), the filter response function T_f, (11.12), therefore reduces to a δ function, and both the intensities of the green and red light and the corresponding intensity correlations may be found by appropriate application of Eqs (11.9) and (11.10).

time t). Note that $G_{rr}^{(2,2)} \sim |q_{12}|^4$ is very small and may be disregarded for the following.

Equation (11.174) reveals that the (steady-state) conditional probability $p_{g|g}(\tau)$ of observing a green photon at time τ when at time $\tau=0$ a green photon has been observed, is

$$p_{g|g}(\tau) \sim |d_{13}|^2 \sigma_{33}(\tau)\big|_{\sigma_{ab}(0)=\delta_{a1}\delta_{b1}}. \tag{11.177}$$

Similarly, from Eqs (11.175) and (11.176) we derive the probability $p_{g|r}(\tau)$ (respectively, $p_{r|g}(\tau)$) of observing a green (respectively, red) photon at time τ when, at time $\tau=0$, a red (respectively, green) photon has been observed:

$$p_{g|r}(\tau) \sim |d_{13}|^2 \sigma_{33}(\tau)\big|_{\sigma_{ab}(0)=\delta_{a1}\delta_{b1}}, \tag{11.178}$$

$$p_{r|g}(\tau) \sim |q_{12}|^2 \sigma_{22}(\tau)\big|_{\sigma_{ab}(0)=\delta_{a1}\delta_{b1}}. \tag{11.179}$$

Note that the probability for detecting a green (respectively, red) photon is proportional to the green (respectively, red) intensity $I_g \sim \langle \hat{E}_{gs}^{(-)} \hat{E}_{gs}^{(+)} \rangle$ (respectively, $I_r \sim \langle \hat{E}_{rs}^{(-)} \hat{E}_{rs}^{(+)} \rangle$), which is proportional to $|d_{13}|^2 \sigma_{33}$ (respectively, $|q_{12}|^2 \sigma_{22}$); cf. Eqs (11.168) and (11.169).

We see that $p_{g|g}(0) = p_{g|r}(0) = p_{r|g}(0) = 0$, which reflects the fact that, at time $\tau=0$, owing to the emission of either a green or red photon, the atom is in state $|1\rangle$, and the (simultaneous) emission of a second photon is impossible. Further, if at a certain time a green photon is observed, the observation of a green photon at later time is much more probable than the observation of a red photon, because from Eqs (11.177) and (11.179) we can readily prove that, under the assumptions made, the inequality

$$p_{g|g}(\tau) \gg p_{r|g}(\tau) \tag{11.180}$$

is valid. Thus in most of the recorded events, green photons are observed to be followed by more green ones, which gives rise to the dominant green fluorescence. Nevertheless, from time to time a green photon is observed to be followed by a red one, according to the small probability $p_{r|g}(\tau)$. As can be seen from Eq. (11.179), this conditional probability is slowly rising since it rests on the excitation of the weakly coupled transition $|1\rangle \leftrightarrow |2\rangle$; that is, when this transition is excited due to the absorption of a red laser photon, the electron is trapped for a long time in the metastable state $|2\rangle$. When a red photon is emitted, according to Eq. (11.178), the green fluorescence is switched on, the time scale now being determined by the strong dipole transition $|1\rangle \leftrightarrow |3\rangle$.

Another, rather direct, explanation for the dark periods may be given by calculating the probability for the emission of the next photon at time τ after the emission of a photon at $\tau=0$ [Zoller, Marte and Walls (1987)]. There is a large body of work on the problem of intermittent fluorescence, and configurations

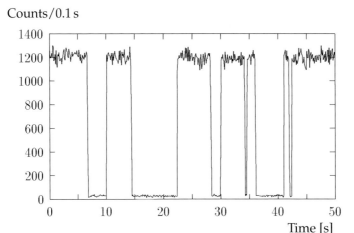

Fig. 11.7 Intermittent fluorescence from a single Ba$^+$ ion. [The measured data are used by kind permission by P.A. Appasamy, I. Siemers and P.E. Toschek, University of Hamburg.]

other than V, such as ladder and Λ, have been studied.[18] The effect has also been demonstrated experimentally using single trapped ions [Nagourney, Sandberg and Dehmelt (1986); Sauter, Blatt, Neuhauser and Toschek (1986); Bergquist, Hulet, Itano and Wineland (1986)]. A typical example of the behavior of the (green) fluorescence of a trapped Ba$^+$ ion intermitted by quantum jumps, is shown in Fig. 11.7. In the case of two trapped ions, quantum jumps have also been observed. In particular, the probability of simultaneous quantum jumps of both atoms has been found to be much larger than expected for the case when the atoms emit photons independently of each other [Sauter, Neuhauser, Blatt and Toschek (1986)].

11.3.3
Vibronic coupling

Let us consider a vibronic system, such as a molecule, and assume that the (harmonic) potential-energy surfaces, which govern the vibrational motion in the two electronic quantum states involved in the resonant light–matter interaction, are shifted as well as distorted with respect to each other (Fig. 11.8). The Hamiltonian of the free vibronic system may be written as

18) For further reading, see, e. g., Cook and Kimble (1985); Arecchi, Schenzle, DeVoe, Jungmann and Brewer (1986); Schenzle and Brewer (1986); Pegg, Loudon and Knight (1986); Cohen-Tannoudji and Dalibard (1986); Nienhuis (1987).

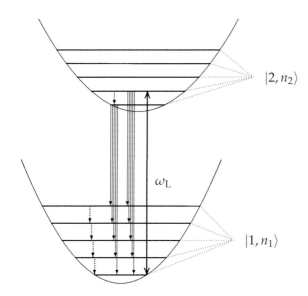

Fig. 11.8 Scheme of a vibronic system resonantly driven by a laser beam of frequency ω_L, the (harmonic) potential-energy surfaces for the vibrational motion in the two electronic quantum states being displaced and distorted.

$$\hat{H}_{vib} = \sum_{i=1}^{2} \sum_{n_i} \hbar \omega_{i,n_i} |i, n_i\rangle \langle i, n_i|, \tag{11.181}$$

where

$$\omega_{i,n_i} = \omega_i + \Omega_v^{(i)}\left(n_i + \tfrac{1}{2}\right), \quad i = 1, 2, \quad n_i = 0, 1, 2, \ldots, \tag{11.182}$$

and the vibronic quantum states

$$|i, n_i\rangle = |i\rangle |n_i\rangle \tag{11.183}$$

are closely related to the Born–Oppenheimer states, the two electronic quantum states and the associated vibrational states are enumerated by i and n_i respectively, and $\hbar \omega_i$ is the energy of the ith electronic quantum state, while $\Omega_v^{(i)}$ is the vibrational frequency in this state.

In particular, let us suppose that the distortion-assisted difference between the vibrational frequencies in the two electronic states, $\delta \Omega_v \equiv |\Omega_v^{(2)} - \Omega_v^{(1)}|$, is large compared with the linewidths of the driving laser field and the vibronic transitions. In this case a single vibronic transition may be driven resonantly by tuning the laser frequency to the corresponding transition frequency.[19]

---

19) When the difference in the two vibrational frequencies cannot be resolved, a large manifold of vibronic transitions may simultaneously be driven by the laser field [Vogel, Welsch and Kühn (1988)].

Clearly, in the course of time, a series of levels below the primarily excited one may also be excited to modify the light-scattering process, because of the various kinds and strengths of vibronic and vibrational relaxations. To illustrate some features which typically arise from the manifold of vibrational states in the lower electronic state, let us briefly study the case where the vibrationless transition $|1,0\rangle \leftrightarrow |2,0\rangle$ is resonantly driven [Vogel and Ullmann (1986); Vogel (1991b)].[20]

The various correlation functions of the scattered light may again be calculated by using the formulae given in Section 11.1 [the level label a now corresponds to (i, n_i)] and applying the quantum regression theorem. In particular, in close analogy with the derivation of Eq. (11.37), the intensity of the scattered light is determined from Eq. (11.10) by the occupation probability of the state $|2,0\rangle$ as

$$I(t + r/c) = |\mathbf{g}|^2 \sigma_{22}^{00}(t). \tag{11.184}$$

Here and in the following the notation $\sigma_{ij}^{mn} \equiv \langle i, n_i | \hat{\sigma} | j, n_j \rangle |_{n_i = m, n_j = n}$ is used.

The multi-level density-matrix equations of motion may be found by combining the (electronic) two-level Bloch equations [cf. Eqs (11.21)–(11.24) and Section 5.4.2] and the harmonic-oscillator master equations (cf. Section 5.3.2):

$$\dot{\sigma}_{22}^{00} = -\Gamma_1 \sigma_{22}^{00} - \tfrac{1}{2} i \Omega_R \tilde{\sigma}_{21}^{00} + \tfrac{1}{2} i \Omega_R \tilde{\sigma}_{12}^{00}, \tag{11.185}$$

$$\dot{\tilde{\sigma}}_{21}^{00} = (-i\delta\omega - \Gamma_2)\tilde{\sigma}_{21}^{00} + \tfrac{1}{2} i \Omega_R (\sigma_{11}^{00} - \sigma_{22}^{00}), \tag{11.186}$$

$$\dot{\tilde{\sigma}}_{12}^{00} = (i\delta\omega - \Gamma_2)\tilde{\sigma}_{12}^{00} - \tfrac{1}{2} i \Omega_R (\sigma_{11}^{00} - \sigma_{22}^{00}), \tag{11.187}$$

$$\dot{\sigma}_{11}^{00} = p_0 \Gamma_1 \sigma_{22}^{00} + \Gamma_v \sigma_{11}^{11} + \tfrac{1}{2} i \Omega_R \tilde{\sigma}_{21}^{00} - \tfrac{1}{2} i \Omega_R \tilde{\sigma}_{12}^{00}, \tag{11.188}$$

$$\dot{\sigma}_{11}^{nn} = p_n \Gamma_1 \sigma_{22}^{00} - n \Gamma_v \sigma_{11}^{nn} + (n+1) \Gamma_v \sigma_{11}^{n+1 n+1} \quad (n \geq 1), \tag{11.189}$$

where, according to Eq. (11.19), slowly varying off-diagonal density-matrix elements $\tilde{\sigma}_{12}^{00}$ ($= \tilde{\sigma}_{21}^{00*}$) have been introduced, and the detuning $\delta\omega$ is defined according to Eq. (11.26).[21] The rates Γ_1 and Γ_2 are respectively the electronic energy- and phase-relaxation rates, Γ_v is the vibrational-energy relaxation rate and Ω_R is the Rabi frequency of the driven transition $|1,0\rangle \leftrightarrow |2,0\rangle$. Further, interaction-matrix elements are factored into an electronic matrix element and a vibrational overlap integral $\langle n_i | n_j \rangle$, so that the vibronic coupling may be described in terms of vibrational overlap integrals. In this way, the filling

20) If the driving laser is tuned to a transition $|1,0\rangle \leftrightarrow |2,n\rangle$, $n > 0$, hot luminescence (i. e., spontaneous emission during the process of vibrational relaxation in the excited electronic state) may additionally appear [Kühn, Vogel and Welsch (1989)].

21) The equations of motion for off-diagonal density-matrix elements of the types σ_{12}^{n0} ($= \sigma_{21}^{0n*}$) and σ_{11}^{nm} ($= \sigma_{11}^{mn*}$), $n \neq m$, are omitted, because these elements are assumed not to be prepared initially, so that they vanish for all times.

rates of the states $|1,n\rangle$, $n=0,1,2,\ldots$, may be written as $p_n\Gamma_1$, where

$$p_n = |\langle n_2|n_1\rangle|^2_{n_2=0,n_1=n} \tag{11.190}$$

(note that $\sum_n p_n = 1$), and the Rabi frequency is

$$\Omega_R = \frac{2}{\hbar}\sqrt{p_0}\,|\mathbf{d}_{21}\mathbf{E}_L| \tag{11.191}$$

[cf. Eq. (11.25)]. The strength of vibrational overlap depends on the shifts with respect to each other of both the equilibrium positions and the frequencies of the vibrations in the two electronic quantum states. In particular, when the frequency shift is small compared with the vibrational frequencies ($|\Omega_v^{(1)}-\Omega_v^{(2)}|\ll\Omega_v^{(1)},\Omega_v^{(2)}$), the p_n, which may easily be calculated using the well-known energy eigenfunctions of a harmonic oscillator, simply represent a Poissonian distribution:

$$p_n = \frac{V^n}{n!}e^{-V}, \tag{11.192}$$

where the vibronic coupling strength V is related to the displacement of the corresponding normal coordinate, δQ_0, by

$$V = \frac{m\Omega_v^{(1)}}{2\hbar}(\delta Q_0)^2, \tag{11.193}$$

with m being the reduced mass.

To solve the multi-level master equations (11.185)–(11.189) and determine $\sigma_{22}^{00}(t)$ with initial conditions $\sigma_{ij}^{nm}(t)|_{t=0}=\delta_{i1}\delta_{j1}\delta_{n0}\delta_{m0}$, we note that these equations can be put into forms resembling the equations of motion for an (electronic) two-level system undergoing a non-Markovian dephasing [Vogel (1991b)]. Introducing the quantities

$$u = \tilde{\sigma}_{21}^{00} + \tilde{\sigma}_{12}^{00}, \tag{11.194}$$

$$v = i(\tilde{\sigma}_{21}^{00} - \tilde{\sigma}_{12}^{00}), \tag{11.195}$$

$$w = \sigma_{22}^{00} - \sigma_{11}, \tag{11.196}$$

where

$$\sigma_{11} = \sum_{n=0}^{\infty} \sigma_{11}^{nn}, \tag{11.197}$$

after some algebra, we deduce from Eqs (11.185)–(11.189) the following equations of motion for u, v and w:

$$\dot{w} = -\Gamma_1(w+1) - \Omega_R v, \tag{11.198}$$

$$\dot{v} = \delta w u - \Gamma_2 v - \int_0^t d\tau\, M(t-\tau)v(\tau) + \Omega_R w, \tag{11.199}$$

$$\dot{u} = -\delta w v - \Gamma_2 u. \tag{11.200}$$

In Eq. (11.199) the field-induced memory function

$$M(t) = \tfrac{1}{2}\Omega_R^2 \Gamma_1 \int_0^t d\tau\, K(t-\tau)\, e^{-\Gamma_1 \tau} \tag{11.201}$$

is introduced, where

$$K(t) = 1 - \sum_{n=0}^{\infty} p_n[1 - \exp(-\Gamma_v t)]^n = 1 - \exp\left(-V e^{-\Gamma_v t}\right). \tag{11.202}$$

The memory function $M(t)$ obviously reflects the dynamics of the vibrational motion in the lower electronic quantum state, where the two competing processes of vibrational-state population and depopulation are governed by the strength of vibronic coupling, V, and the rate of vibrational relaxation, Γ_v, respectively. Clearly, in the limiting cases of sufficiently weak vibronic coupling ($V \to 0$) or extremely fast vibrational relaxation ($\Gamma_v^{-1} \to 0$) the memory function vanishes ($M(t) \to 0$), and Eqs (11.198)–(11.200) just correspond to the familiar (two-level) Bloch equations. For a given strength of vibronic coupling the effect of $M(t)$ on the electronic-state dynamics is most pronounced in the case of slow vibrational relaxation ($\Gamma_v \ll \Gamma_1$). In this case Eq. (11.201) [together with Eq. (11.202)] may be simplified to obtain the memory function as

$$M(t) \simeq \tfrac{1}{2}\Omega_R^2 \left[1 - \exp\left(-V e^{-\Gamma_v t}\right)\right]\left(1 - e^{-\Gamma_1 t}\right). \tag{11.203}$$

We see that $M(t)$ is built up on a time scale of Γ_1^{-1} to attain a quasi-stationary value

$$M = \tfrac{1}{2}\Omega_R^2 \left(1 - e^{-V}\right), \qquad \Gamma_1^{-1} \ll t \ll \Gamma_v^{-1}, \tag{11.204}$$

which, in the further course of time, decays to zero,

$$M(t) = \tfrac{1}{2}\Omega_R^2 \left[1 - \exp\left(-V e^{-\Gamma_v t}\right)\right], \tag{11.205}$$

the time scale being given by Γ_v^{-1}.

This behavior of $M(t)$ typically influences the temporal evolution of the intensity of the scattered light, as can be seen from Fig. 11.9. When the driving field is sufficiently strong, Rabi oscillations are observed. However, with increasing time, the intensity substantially decreases and the Rabi oscillations are smoothed out. Eventually, in the long-time limit the intensity slightly increases to attain a steady-state value. Note that the Rabi oscillations may survive somewhat longer than in the case of a two-level system.

From the point of view of the multi-level description (11.185)–(11.189), pumping of the transition $|1,0\rangle \leftrightarrow |2,0\rangle$ is followed by spontaneous emissions $|2,0\rangle \to |1,n\rangle$, $n=0,1,2,\ldots$, to also excite higher vibrational quantum states in the electronic ground state ($n>0$). These states cannot serve as starting states

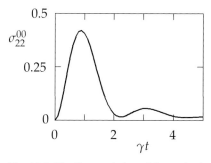
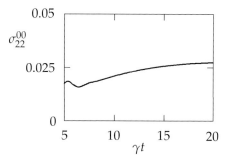

Fig. 11.9 The time evolution of the excited-state occupation probability $\sigma_{22}^{00}(t)$ is shown for a vibronic system undergoing radiative damping ($\Gamma_1 = 2\Gamma_2 = 2\gamma$), the parameters being $\Omega_R/\gamma = 3$, $\Gamma_v/\gamma = 0.1$ and $V = 3$.

for further pumping, because the corresponding vibronic transitions are out of resonance with the driving laser field. Moreover, slow vibrational relaxation reduces the repopulation of the ground state $|1,0\rangle$ to a low level. In this way, the optically active electron is effectively trapped and the intensity of the scattered light nearly breaks down. On a long-time scale of Γ_v^{-1} the intensity slightly increases, because of the nonexponential filling of the state $|1,0\rangle$ due to vibrational relaxation [cf. the double-exponential structure of $M(t)$ as given in Eq. (11.205)].[22] From the point of view of the two-level description (11.198)–(11.200), one would say that the decrease in the intensity results from a field-induced dephasing, which is responsible for detuning the two-level transition so that it goes out of resonance with the driving laser field.

It should be noted that, with regard to the intensity correlation of the scattered light, the results (11.54) and (11.59) for the two-level case may easily be extended to the case of the multi-level vibronic system, considered above, to obtain

$$G^{(2,2)}(t+r/c+\tau, t+r/c) = I(t+r/c+\tau)\big|_{\sigma_{ij}^{nm}(t)=\delta_{i1}\delta_{j1}\delta_{nm}p_n} I(t+r/c)$$

$$= |\mathbf{g}|^4 \sigma_{22}^{00}(t+\tau)\big|_{\sigma_{ij}^{nm}(t)=\delta_{i1}\delta_{j1}\delta_{nm}p_n} \sigma_{22}^{00}(t). \quad (11.206)$$

Here $\sigma_{22}^{00}(t+\tau)\big|_{\sigma_{ij}^{nm}(t)=\delta_{i1}\delta_{j1}\delta_{nm}p_n}$ is determined from the solution of the equations of motion as given in Eqs (11.185)–(11.189), the initial conditions being $\sigma_{ij}^{nm}(t+\tau)\big|_{\tau=0} = \delta_{i1}\delta_{j1}\delta_{nm}p_n$, where p_n is given in Eq. (11.192). This initial preparation reflects the fact that after the emission of a photon (at time t) the system is, with probability p_n, in the nth vibrational state of the electronic ground state. (Note that the probabilities for the transitions $|2,0\rangle \to |1,n\rangle$ are just proportional to the p_n.) The difference in the initial conditions needed

---

[22] When the ground state is filled exponentially, a long-time increase of the intensity does not appear; see Vogel and Ullmann (1986).

to calculate the intensity and the intensity correlation may lead to different short-time behavior of the intensity (as a function of t) and the (normalized) intensity correlation (as a function of τ), whereas the long-time behavior (time scale Γ_v^{-1}) of the two functions is the same.

By extending the result (11.95), the steady-state power spectrum of the scattered light is given as [Vogel and Welsch (1986)]:[23]

$$S(\omega,\Gamma_f) = \sum_n p_n S_n(\omega,\Gamma_f), \tag{11.207}$$

$$S_n(\omega,\Gamma_f) = \frac{1}{2\pi} \int_0^\infty d\tau \, \exp\left[-(i\omega + \tfrac{1}{2}\Gamma_f)\tau\right] G_{12}^{n0(\mathrm{I})}(\tau) + \text{c.c.}, \tag{11.208}$$

where

$$G_{12}^{n0(\mathrm{I})}(\tau) \equiv \lim_{t\to\infty} G_{12}^{n0(\mathrm{I})}(t+\tau,t), \tag{11.209}$$

$$G_{12}^{n0(\mathrm{I})}(t+\tau,t) = \langle \hat{A}_{21}^{0n}(t+\tau) \hat{A}_{12}^{n0}(t) \rangle \tag{11.210}$$

[cf. Eqs (11.94) and (11.91)]. Application of the quantum regression theorem for determining the $G_{ij}^{nm(\mathrm{I})}(t+\tau,t)$ requires that the density-matrix equations of motion (11.185)–(11.189) are complemented by the equations of motion for the off-diagonal density-matrix elements σ_{12}^{n0} and σ_{11}^{nm}, because the nonvanishing initial values of the $G_{ij}^{nm(\mathrm{I})}(t+\tau,t)$ are

$$G_{11}^{n0(\mathrm{I})}(t+\tau,t)\big|_{\tau=0} = \tilde{\sigma}_{21}^{00}(\infty) e^{-i\omega_{21}t}, \tag{11.211}$$

$$G_{12}^{n0(\mathrm{I})}(t+\tau,t)\big|_{\tau=0} = \sigma_{22}(\infty). \tag{11.212}$$

We omit a detailed analysis here and refer the interested reader to the literature.

From a dressed-state approach one may see, analogously to the discussion in Section 11.2.4, that the line which corresponds to the driven (vibrationless) transition and is related to the Fourier transform of $G_{12}^{00(\mathrm{I})}(\tau)$ shows a triplet structure, provided that the Rabi frequency is sufficiently large. A detailed analysis reveals that the side peaks of the triplet may become narrower and higher as in the case of a two-level atom, which is consistent with the above-mentioned fact that the Rabi oscillations may survive over a longer time. The Raman lines, which are just related to the Fourier transforms of the $G_{12}^{n0(\mathrm{I})}(\tau)$, $n>0$, split into doublets, since the higher ($n>0$) vibrational energy levels of the electronic ground state cannot give rise to level splittings under the excitation conditions considered.

We finally note that a trapped ion undergoing quantized center-of-mass motion (of vibration type) in a trap potential may also be regarded as a vibronic

---

[23] Here $\bar{n}\Omega_v^{(1)} \ll \omega_{21}$, where $\bar{n} = \sum_n p_n = V$.

system, whose Hamiltonian of course differs from that of the molecule-like vibronic system considered above; for details see Chapter 13. Compared with the situation in a molecule, the trap potential is externally given and thus it does not depend on the electronic state of the atom. The vibronic coupling arises from the kick effects due to the emission and absorption of photons, which in momentum space play a similar role to the (position) displacement of the potentials of the molecule. The effects of quantized motion in the resonance fluorescence of a trapped ion have been studied, for example, in connection with the spectrum [Cirac, Blatt, Parkins and Zoller (1993)] and the squeezing [de Matos Filho and Vogel (1994)] of the fluorescence radiation.

References

Arecchi, F.T., A. Schenzle, R.G. DeVoe, K. Jungmann and R.G. Brewer (1986) *Phys. Rev. A* **33**, 2124.

Aspect, A., G. Roger, S. Reynaud, J. Dalibard and C. Cohen-Tannoudji (1980) *Phys. Rev. Lett.* **45**, 617.

Bergquist, J.C., R.G. Hulet, W.M. Itano and D.J. Wineland (1986) *Phys. Rev. Lett.* **57**, 1699.

Carmichael, H.J. and D.F. Walls (1976) *J. Phys. B* **9**, L43.

Cirac, J.I., R. Blatt, A.S. Parkins and P. Zoller (1993) *Phys. Rev. A* **48**, 2169.

Cohen-Tannoudji, C. and J. Dalibard (1986) *Europhys. Lett.* **1**, 441.

Cohen-Tannoudji, C. and S. Reynaud (1977) *J. Phys. B* **10**, 345, 365.

Collett, M.J., D.F. Walls and P. Zoller (1984) *Opt. Commun.* **52**, 145.

Cook, R.J. (1981) *Phys. Rev. A* **23**, 1243.

Cook, R.J. and H.J. Kimble (1985) *Phys. Rev. Lett.* **54**, 1023.

Cresser, J.D. (1987) *J. Phys. B* **20**, 4915.

Dicke, R.H. (1954) *Phys. Rev.* **93**, 99.

Dicke, R.H. (1964) in *Quantum Electronics*, eds N. Bloembergen and P. Grivet (Dunod, Paris), p. 35.

Hartig, W., W. Rasmussen, R. Schieder and H. Walther (1976) *Z. Phys. A* **278**, 205.

Heidmann, A. and S. Reynaud (1985) *J. Physique* **45**, 1937.

Herrmann, J., K.-E. Süsse and D.-G. Welsch (1973) *Ann. Phys. (Leipzig)* **30**, 37.

Kimble, H.J. and L. Mandel (1976) *Phys. Rev. A* **13**, 2123.

Knöll, L., W. Vogel and D.-G. Welsch (1986) *J. Opt. Soc. Am. B* **3**, 1315.

Knöll, L., W. Vogel and D.-G. Welsch (1990) *Phys. Rev. A* **42**, 503.

Knöll, L. and G. Weber (1986) *J. Phys. B* **19**, 2817.

Kühn, H., W. Vogel and D.-G. Welsch (1989) *Chem. Phys. Lett.* **158**, 233.

de Matos Filho, R.L. and W. Vogel (1994) *Phys. Rev. A* **49**, 2812.

Mollow, B.R. (1969) *Phys. Rev.* **188**, 1969.

Nagourney, W., J. Sandberg and H.G. Dehmelt (1986) *Phys. Rev. Lett.* **56**, 2797.

Nienhuis, G. (1987) *Phys. Rev. A* **35**, 4639.

Orriols, G. (1979) *Nuovo Cim.* **53B**, 1.

Pegg, D.T., R. Loudon and P.L. Knight (1986) *Phys. Rev. A* **33**, 4085.

Sauter, Th., R. Blatt, W. Neuhauser and P.E. Toschek (1986) *Opt. Commun.* **60**, 287.

Sauter, Th., W. Neuhauser, R. Blatt and P.E. Toschek (1986) *Phys. Rev. Lett.* **57**, 1696.

Schenzle, A. and R.G. Brewer (1986) *Phys. Rev. A* **34**, 3127.

Schuda, F., C.R. Stroud Jr and M. Hercher (1974) *J. Phys. B* **1**, L198.

Vogel, W. (1991a) *Phys. Rev. Lett.* **67**, 2451.

Vogel, W. (1991b) *J. Opt. Soc. Am. B* **8**, 129.

Vogel, W. and R. Blatt (1992) *Phys. Rev. A* **45**, 3319.

Vogel, W. and Th. Ullmann (1986) *J. Opt. Soc. Am. B* **3**, 441.

Vogel, W. and D.-G. Welsch (1985) *Phys. Rev. Lett.* **54**, 1802.

Vogel, W. and D.-G. Welsch (1986) *J. Opt. Soc. Am. B* **3**, 1692.

Vogel, W., D.-G. Welsch and H. Kühn (1988) *J. Opt. Soc. Am. B* **5**, 67.

Walls, D.F. and P. Zoller (1981) *Phys. Rev. Lett.* **47**, 709.

Wu, F.Y., R.E. Grove and S. Ezekiel (1975) *Phys. Rev. Lett.* **35**, 1426.

Wu, F.Y., R.E. Grove and S. Ezekiel (1977) *Phys. Rev. A* **15**, 227.

Zoller, P., M. Marte and D.F. Walls (1987) *Phys. Rev. A* **35**, 198.

12
A single atom in a high-Q cavity

From the study of spontaneous emission (Section 10.1) we know that, in the presence of macroscopic bodies, the strength of the interaction between atoms and the radiation field can drastically change compared with the case when the atoms are in free space. In particular, if the bodies form a resonator-like equipment – referred to as a cavity – giving rise to a well-pronounced line spectrum of the field, a noticeably enhanced atom–field coupling is observed for atomic transitions tuned to the lines of the cavity field. As a consequence, a photon emitted by an excited atom does not escape at once but is captured by the cavity for some time in general and can thus be reabsorbed by the atom. Moreover, subsequent induced emission together with external pumping may lead to the well-known lasing effects. In fact cavity-induced modifications of radiation–matter interaction processes depend sensitively on the geometrical and optical properties of the cavity, in particular on its Q value.

As long as Q is small and allows one to regard the cavity as being a moderate disturbance of free space, so that the weak coupling regime effectively continues to hold, the effect of the cavity on the atomic motion and the emitted light may be included in the theory by appropriately modifying the relaxation rates appearing in the optical Bloch equations for the electronic-state dynamics of the atom [Rice and Carmichael (1988)]. Further, non-Markovian relaxation theory may be applied to explain the measured modifications in the Mollow spectrum of the light emitted from an atom inside a cavity [Lezama, Zhu, Morin and Mossberg (1989)].

When Q becomes sufficiently large, so that the coupling between the atom and the cavity field becomes strong, the situation may be changed drastically, since the back-action of the radiation field on the atom now plays a dominant role. As already demonstrated in Section 10.1.2, strong atom–field coupling requires sufficiently narrow lines of the cavity field, with some of which being tuned to specific atomic transitions. When only one atomic transition is involved in the strong coupling regime, then, on a time scale small compared with the inverse width of the corresponding line of the cavity field, the problem effectively reduces to a problem of Jaynes–Cummings type – the near-resonant interaction of a two-level atom with a quantized mode of a perfect cavity ($Q \to \infty$). When the Jaynes–Cummings model was proposed [Jaynes and Cummings (1963); Paul (1963)], its predictions seemed to be somewhat artificial and far from practical relevance. Owing to progress in experimental techniques, however, the situation has now changed. Experiments with the micromaser, in which a beam of long-living Rydberg atoms is injected into a cooled single-mode high-Q cavity at such a low rate that, at most, one atom at a time is present inside the cavity, have made it possible to experimentally prove a series of predictions derived from the Jaynes–Cummings model [see, e. g., Haroche (1984); Haroche and Raimond (1985); Meschede, Walther and Müller (1985); Rempe, Walther and Klein (1987)]. In the optical domain the regime of strong coupling between an atom and a cavity field can also be demonstrated using a cavity with mirrors of extremely high reflectivity [Thompson, Rempe and Kimble (1992)].

The further improvement of the available cavities has created exciting perspectives for realizing fundamental Gedanken experiments from the early days of quantum mechanics. In this context new types of nonclassical states have been realized, such as Schrödinger-cat states [Brune, Hagley, Dreyer, Maitre, Maali, Wunderlich, Raimond and Haroche (1996)], Einstein–Podolsky–Rosen pairs of atoms [Hagley, Maitre, Nogues, Wunderlich, Brune, Raimond and Haroche (1997)], and trapping states of the micromaser [Weidinger, Varcoe, Heerlein and Walther (1999)]. The feasibility of performing quantum nondemolition measurements of the cavity field has been demonstrated [Nogues, Rauschenbeutel, Osnaghi, Brune, Raimond and Haroche (1999)] and elementary quantum logic operations have been realized [Rauschenbeutel, Nogues, Osnaghi, Bertet, Brune, Raimond and Haroche (1999)].

12.1
The Jaynes–Cummings model

In the Jaynes–Cummings model the near-resonant interaction of a two-level atom with a quantized single-mode radiation field is described by the Hamiltonian

$$\hat{H} = \hat{H}_0 + \hat{H}_{\text{int}}, \tag{12.1}$$

where the Hamiltonian $\hat{H}_0$ governs the free motion of the atom and the radiation-field mode,[1]

$$\hat{H}_0 = \hbar\omega_1 \hat{A}_{11} + \hbar\omega_2 \hat{A}_{22} + \hbar\omega \hat{a}^\dagger \hat{a}, \tag{12.2}$$

---

1) Since only one mode is considered, the mode label is omitted for notational convenience. Further, the atomic energies $\hbar\omega_1$ and $\hbar\omega_2$ are thought of as being the shifted ones, so that $\omega_{21} = \omega_2 - \omega_1$ is the shifted transition frequency and the tilde notation used in Chapter 10 can be omitted.

and $\hat{H}_{\text{int}}$ describes their coupling to each other,

$$\hat{H}_{\text{int}} = -\hbar\lambda(\hat{a}^\dagger \hat{A}_{12} + \hat{A}_{21}\hat{a}). \tag{12.3}$$

The coupling parameter λ is the electric-dipole matrix element of the atomic transition multiplied by the cavity mode function at the location of the atom. Without loss of generality, we may assume that λ is real-valued and $\lambda > 0$. The Hamiltonian given in Eqs (12.1)–(12.3) corresponds to the Hamiltonian already used in Section 11.2.4 to explain the line structure of the spectrum of resonance fluorescence from a single atom in free space. Whereas in Section 11.2.4 the radiation-field mode simply represents the field driving the atomic transition, here and in the following it represents both the driving field and the field radiated by the atom in the cavity.

The eigenvalue equation for $\hat{H}_0$ may be written as

$$\hat{H}_0 |a, n\rangle = \hbar(\omega_a + n\omega)|a, n\rangle, \tag{12.4}$$

where a labels the two electronic states of the atom ($a = 1, 2$) and n is the photon number. It is straightforward to diagonalize the full Jaynes–Cummings Hamiltonian $\hat{H}$ by making the ansatz that the eigenstates of $\hat{H}$ are superpositions of the degenerate (or in the case of nonvanishing detuning, near-degenerate) eigenstates of $\hat{H}_0$ as follows:

$$|n, \pm\rangle = c_\pm(|1, n+1\rangle + \alpha_\pm |2, n\rangle), \tag{12.5}$$

so that

$$\hat{H}|n, \pm\rangle = E_{n,\pm}|n, \pm\rangle \equiv \hbar\omega_{n,\pm}|n, \pm\rangle. \tag{12.6}$$

Applying in Eq. (12.6) the Hamiltonian $\hat{H}$ [Eqs (12.1)–(12.3)] to $|n, \pm\rangle$, as given in Eq. (12.5), and comparing the coefficients of $|1, n+1\rangle$ and $|2, n\rangle$ on both sides of the resulting equation, we obtain two equations for determining $\omega_{n,\pm}$ and $\alpha_\pm$. The $c_\pm$ are determined from the normalization condition $\langle n, \pm | n, \pm\rangle = 1$. After some algebra, we arrive at

$$\omega_{n,\pm} = \tfrac{1}{2}[\omega_2 + \omega_1 + (2n+1)\omega \pm \Delta_n], \tag{12.7}$$

$$\alpha_\pm = -\frac{1}{\Omega_n}(\omega_{21} - \omega \pm \Delta_n), \tag{12.8}$$

$$c_\pm = \frac{1}{\sqrt{1 + \alpha_\pm^2}}, \tag{12.9}$$

where

$$\Delta_n = \sqrt{\delta^2 + \Omega_n^2}, \tag{12.10}$$

$$\Omega_n = 2\lambda\sqrt{n+1}, \tag{12.11}$$

$$\delta = \omega_{21} - \omega. \tag{12.12}$$

From Eqs (12.8)–(12.12) we may rewrite Eq. (12.5) to represent the eigenstates $|n, \pm\rangle$ in the form

$$|n, +\rangle = \cos \Theta_n |1, n+1\rangle - \sin \Theta_n |2, n\rangle, \tag{12.13}$$

$$|n, -\rangle = \sin \Theta_n |1, n+1\rangle + \cos \Theta_n |2, n\rangle, \tag{12.14}$$

where

$$\sin \Theta_n = \frac{\Omega_n}{\sqrt{(\Delta_n - \delta)^2 + \Omega_n^2}}, \tag{12.15}$$

$$\cos \Theta_n = \frac{\Delta_n - \delta}{\sqrt{(\Delta_n - \delta)^2 + \Omega_n^2}}. \tag{12.16}$$

The states $|n, \pm\rangle$ are usually called dressed-atom states [Cohen-Tannoudji and Reynaud (1977)], because they may be regarded as describing the atom "dressed" by the interaction with the radiation mode rather than the free atom. In particular, in the case of exact resonance ($\delta = 0$) the eigenstates $|n, \pm\rangle$ take the simple form

$$|n, \pm\rangle = \frac{1}{\sqrt{2}}(|1, n+1\rangle \mp |2, n\rangle), \tag{12.17}$$

and the corresponding eigenfrequencies $\omega_{n,\pm}$ are

$$\omega_{n,\pm} = (\omega_2 + n\omega) \pm \tfrac{1}{2}\Omega_n. \tag{12.18}$$

We see that the light–matter interaction considered gives rise to level splittings $\hbar\Omega_n$ (dynamic Stark effect) where Ω_n is called the *n*-photon Rabi frequency. It should be pointed out that, even in the case of an excited atom interacting with the photon vacuum ($n=0$), a splitting occurs, the so-called vacuum Rabi splitting [for an experimental demonstration see Thompson, Rempe and Kimble (1992)]. Note that the vacuum Rabi frequency $\Omega_{n=0}$ defined by Eq. (12.11) corresponds exactly to Ω_ν introduced by Eq. (10.51) in the limit $\gamma_\nu \to 0$ but finite $\Gamma_\nu \gamma_\nu$.

Clearly, the states $|n, \sigma\rangle \equiv |n, \pm\rangle$ given in Eqs (12.13) and (12.14) are the excited eigenstates of the Hamiltonian $\hat{H}$. To make them complete requires addition of the ground state $|1, 0\rangle$, which is an eigenstate of both $\hat{H}_0$ and $\hat{H}$,

$$\hat{H}|1, 0\rangle = \hat{H}_0|1, 0\rangle = \hbar\omega_1|1, 0\rangle, \tag{12.19}$$

because in this state there is no resonant coupling of the atom to the radiation-field mode. The completeness relation is then

$$|1, 0\rangle\langle 1, 0| + \sum_{\sigma=\pm} \sum_{n=0}^{\infty} |n, \sigma\rangle\langle n, \sigma| = \hat{I}. \tag{12.20}$$

12.1 The Jaynes–Cummings model

Let us represent the unitary time-evolution operator $\hat{U}(t) = \exp(-i\hat{H}t/\hbar)$ needed to study the dynamics of the coupled radiation–matter system when it is initially prepared in a state that is not an eigenstate of $\hat{H}$. In the basis of the dressed states (eigenstates of $\hat{H}$) we simply have

$$\hat{U}(t) = e^{-i\omega_1 t}|1,0\rangle\langle 1,0| + \sum_{\sigma=\pm}\sum_{n=0}^{\infty} e^{-i\omega_{n,\sigma}t}|n,\sigma\rangle\langle n,\sigma|, \tag{12.21}$$

which follows directly from application of the completeness relation (12.20). In Eq. (12.21), expressing the dressed states $|n,\pm\rangle$ in terms of the unperturbed states $|a,n\rangle$, cf. Eqs (12.13)–(12.16), we may represent $\hat{U}$ in the basis of the eigenstates of the unperturbed Hamiltonian $\hat{H}_0$. After some algebra, we obtain

$$\hat{U}(t) = e^{-i\omega_1 t}|1,0\rangle\langle 1,0| + \sum_{n=0}^{\infty} e^{-\frac{1}{2}i[\omega_1+\omega_2+\omega(2n+1)]t}$$

$$\times \left\{ \left[\cos\left(\tfrac{1}{2}\Delta_n t\right) + i\frac{\delta}{\Delta_n}\sin\left(\tfrac{1}{2}\Delta_n t\right)\right]|1,n+1\rangle\langle 1,n+1| \right.$$

$$+ \left[\cos\left(\tfrac{1}{2}\Delta_n t\right) - i\frac{\delta}{\Delta_n}\sin\left(\tfrac{1}{2}\Delta_n t\right)\right]|2,n\rangle\langle 2,n|$$

$$\left. + i\frac{\Omega_n}{\Delta_n}\sin\left(\tfrac{1}{2}\Delta_n t\right)(|1,n+1\rangle\langle 2,n| + |2,n\rangle\langle 1,n+1|)\right\}, \tag{12.22}$$

from which the matrix elements of $\hat{U}$ in the basis of the $|a,n\rangle$,

$$U_{an,bm}(t) = \langle a,n|\hat{U}(t)|b,m\rangle \tag{12.23}$$

($\{a,b\}=1,2$, $\{n,m\}=0,1,2,\ldots$), can easily be obtained. Note that the nonvanishing matrix elements are $U_{1n,1n}$, $U_{2n,2n}$, $U_{1(n+1),2n}$ and $U_{2n,1(n+1)}$, all others are zero. Taking into account that the density operators of the system at the two times t and t' are related by

$$\hat{\varrho}(t) = \hat{U}(t-t')\hat{\varrho}(t')\hat{U}^\dagger(t-t'), \tag{12.24}$$

we may express the matrix elements of the density operator at time t,

$$\varrho_{an,bm}(t) = \langle a,n|\hat{\varrho}(t)|b,m\rangle, \tag{12.25}$$

in terms of the matrix elements of the density operator $\hat{\varrho}(t')$ and the time evolution operator $\hat{U}(t-t')$:

$$\varrho_{an,bm}(t) = \sum_{a'n'}\sum_{b'm'} U_{an,b'n'}(t-t') U^*_{bm,b'm'}(t-t') \varrho_{a'n',b'm'}(t'), \tag{12.26}$$

which represents the general solution of the density-matrix equations of motion for the Jaynes–Cummings problem.

The above formulae apply directly to the study of the time evolution of atomic and photonic quantities, without any further approximations. The Jaynes–Cummings model has been used successfully to study various problems in micromaser and optical-cavity experiments in the very-high-Q regime. It also plays an important role in the study of fundamental quantum features in the light–matter interaction on the basis of an exact solution of coupled light–matter equations of motion. In this context, it has been extended to allow for more complicated light–matter interactions, such as the near-resonant interaction of two quantized light modes with a three-level atomic system [see, e. g., Yoo and Eberly (1985)].

In particular, the extension of the above results to the case of the so-called multi-photon Jaynes–Cummings model, is straightforward. Assuming that the atomic transition is in near-resonance with a k-photon transition ($k = 2, 3, \ldots$) of the cavity field, we may describe this form of interaction between an atom and the quantized cavity mode by substituting into Eq. (12.1) for $\hat{H}_{\text{int}}$ an effective interaction Hamiltonian $\hat{H}_{\text{int}}^{(k)}$ as follows:[2]

$$\hat{H}_{\text{int}} \mapsto \hat{H}_{\text{int}}^{(k)} = -\hbar \lambda^{(k)} \left(\hat{a}^{\dagger k} \hat{A}_{12} + \hat{A}_{21} \hat{a}^k \right). \tag{12.27}$$

The determination of the eigenstates of the k-photon Jaynes–Cummings Hamiltonian $\hat{H}_0 + \hat{H}_{\text{int}}^{(k)}$ and of the time-evolution operator, may be performed in close analogy to the approach outlined above. It can easily be proved that the dressed states are now

$$|n, +\rangle^{(k)} = \cos \Theta_n^{(k)} |1, n+k\rangle - \sin \Theta_n^{(k)} |2, n\rangle, \tag{12.28}$$

$$|n, -\rangle^{(k)} = \sin \Theta_n^{(k)} |1, n+k\rangle + \cos \Theta_n^{(k)} |2, n\rangle, \tag{12.29}$$

[cf. Eqs (12.13) and (12.14)], where $\sin \Theta_n^{(k)}$ and $\cos \Theta_n^{(k)}$ may be calculated using Eqs (12.15) and (12.16), respectively, [together with Eq. (12.10)] and substituting, for the one-photon quantities δ and Ω_n, the corresponding k-photon quantities:

$$\delta \mapsto \delta^{(k)} = \omega_{21} - k\omega, \tag{12.30}$$

$$\Omega_n \mapsto \Omega_n^{(k)} = 2\lambda^{(k)} \sqrt{\frac{(n+k)!}{n!}}. \tag{12.31}$$

Accordingly, the time-evolution operator (12.21) may be expressed in terms of the dressed states $|n, \pm\rangle^{(k)}$ or the unperturbed states $|i, n\rangle$ [Vogel and Welsch (1989)]. Since the states $|1, q\rangle$ for $q = 0, \ldots, k-1$ are not affected by the atom–field interaction, in place of Eq. (12.21) the unitary time-evolution operator

---

[2] For the derivation of effective multi-photon interaction operators, we refer to Section 2.5.3.

now reads as

$$\hat{U}^{(k)}(t) = \sum_{q=0}^{k-1} e^{-i(\omega_1+q\omega)t}|1,q\rangle\langle 1,q| + \sum_{\sigma=\pm}\sum_{n=0}^{\infty} e^{-i\omega_{n,\sigma}t}|n,\sigma\rangle^{(k)(k)}\langle n,\sigma|. \quad (12.32)$$

Substituting in this expression the explicit form of the multi-photon dressed states, Eqs (12.28) and (12.29), one may obtain an expression for the time-evolution operator in terms of the bare states in analogy to that given in Eq. (12.22) for the one-photon model. It is noteworthy that the k-quantum Jaynes–Cummings model plays an important role in describing the vibronic coupling of an appropriately laser-driven trapped atom (Section 13.3.1).

12.2
Electronic-state dynamics

To measure the time-dependent occupation probabilities of the electronic states of an atom interacting with a cavity field, one may transmit equally prepared single atoms with different velocities through the cavity and measure their final preparation by electronic-state sensitive ionization as a function of the time of flight through the cavity. The interaction time between the atom and the cavity field corresponds to the time of flight which is controlled by the velocity of the atoms.[3]

12.2.1
Reduced density matrix

To calculate the temporal evolution of atomic quantities, we start from the general solution for the atom–field density matrix [Eq. (12.26) together with Eqs (12.22) and (12.23)] and take the trace with respect to the field to obtain the reduced density matrix σ_{ab} of the atomic (electronic) subsystem:

$$\sigma_{ab}(t) = \sum_{n=0}^{\infty} \varrho_{an,bn}(t). \quad (12.33)$$

When we identify the time t' with the onset of the interaction between the atom and the cavity field (e.g., when the atom enters the cavity), we may write the initial density matrix in factored form as

$$\varrho_{an,bm}(t') = \rho_{nm}(t')\sigma_{ab}(t'). \quad (12.34)$$

Here and in the following the (reduced) density operator of the radiation-field mode is denoted by $\hat{\rho}$, with ρ_{nm} being the matrix elements in the photon-number basis. Since $\sigma_{22} + \sigma_{11} = 1$ and $\sigma_{21} = \sigma_{12}^*$, it is sufficient to perform the

---

[3] Such a measurement scheme is typically used in micromaser experiments, cf. Section 12.4.

calculations for two density-matrix elements of the atomic system. Let us consider the excited-state occupation probability σ_{22} measured in the above mentioned detection scheme. Combining Eqs (12.22)–(12.26) and (12.33), (12.34), we deduce that ($t'=0$)

$$\sigma_{22}(t) = \sigma_{22}^{\text{inc}}(t) + \sigma_{22}^{\text{coh}}(t), \tag{12.35}$$

where

$$\sigma_{22}^{\text{inc}}(t) = \sum_{n=0}^{\infty} \left\{ \frac{2\delta^2 + \Omega_n^2[1+\cos(\Delta_n t)]}{2\Delta_n^2} \sigma_{22}(0)\rho_{nn}(0) \right.$$

$$\left. + \frac{\Omega_n^2[1-\cos(\Delta_n t)]}{2\Delta_n^2} \sigma_{11}(0)\rho_{n+1 n+1}(0) \right\} \tag{12.36}$$

is the part of the atomic excited-state occupation probability arising from the initially prepared diagonal density-matrix elements (incoherent preparation), whereas the term

$$\sigma_{22}^{\text{coh}}(t) = \text{Re}\left\{ \sum_{n=0}^{\infty} \left[i\frac{\Omega_n}{\Delta_n}\sin(\Delta_n t) - \frac{\delta\Omega_n}{\Delta_n^2}[1-\cos(\Delta_n t)]\right] \sigma_{12}(0)\rho_{n+1 n}(0) \right\} \tag{12.37}$$

only appears when there is also a coherent initial preparation characterized by the corresponding off-diagonal density-matrix elements.

In particular, for exact resonance ($\delta=0$) Eqs (12.36) and (12.37) simplify to

$$\sigma_{22}^{\text{inc}}(t) = \frac{1}{2} \sum_{n=0}^{\infty} \{[1 + \cos(\Omega_n t)]\sigma_{22}(0)\rho_{nn}(0)$$

$$+ [1 - \cos(\Omega_n t)]\sigma_{11}(0)\rho_{n+1 n+1}(0)\}, \tag{12.38}$$

$$\sigma_{22}^{\text{coh}}(t) = \text{Re}\left[\sum_{n=0}^{\infty} i\sin(\Omega_n t)\sigma_{12}(0)\rho_{n+1 n}(0)\right]. \tag{12.39}$$

Let us assume that the atom is initially prepared incoherently, so that the atomic excited-state occupation probability $\sigma_{22}(t)$ is given by $\sigma_{22}^{\text{inc}}(t)$. Further, assuming that the atom is initially in the excited state [$\sigma_{ab}(0) = \delta_{ab}\delta_{a2}$], Eqs (12.35), (12.38) and (12.39) reduce to

$$\sigma_{22}(t) = \sigma_{22}^{\text{inc}}(t) = \frac{1}{2}\left[1 + \sum_{n=0}^{\infty} \rho_{nn}(0)\cos(\Omega_n t)\right]. \tag{12.40}$$

If the initial field is prepared in a photon-number state $|k\rangle$ [$\rho_{nn}(0) = \delta_{nk}$], the excited-state occupation probability oscillates with the corresponding Rabi frequency Ω_k:

$$\sigma_{22}(t) = \frac{1}{2}[1 + \cos(\Omega_k t)]. \tag{12.41}$$

In particular, if the cavity field is initially in the vacuum state ($k=0$), the atomic occupation probability oscillates with the vacuum Rabi frequency Ω_0. In this case Eq. (12.41) corresponds to Eq. (10.57) and describes the effect of vacuum Rabi oscillations in spontaneous emission in an ideal cavity ($Q \to \infty$).

12.2.2
Collapse and revival

In the case of arbitrary initial preparation of the field, Eq. (12.41) must be averaged over the photon-number distribution $\rho_{kk}(0)$ to obtain the result (12.40), which represents a superposition of various Rabi oscillations. For example, let us first assume that the cavity field is initially in a coherent state $|\alpha\rangle$. Recalling that the photon-number distribution of a coherent state is Poissonian [cf. Eq. (3.60)], we obtain from Eq. (12.40)

$$\sigma_{22}(t) = \tfrac{1}{2}\left[1 + \sum_{n=0}^{\infty} \frac{\langle \hat{n}(0)\rangle^n}{n!} e^{-\langle \hat{n}(0)\rangle} \cos(\Omega_n t)\right]. \qquad (12.42)$$

Note that $\langle \hat{n}(0)\rangle = |\alpha|^2$ is the mean photon number in the initial coherent state $|\alpha\rangle$. The dependence of the Rabi frequency on the photon number, $\Omega_n = 2\lambda\sqrt{n+1}$, Eq. (12.11), prevents, in general, the result of averaging from being obtained in closed form. The problem may be solved approximately if the mean number of photons is sufficiently large, $\langle \hat{n}(0)\rangle \gg 1$, so that the relative variance of the photon number becomes small:

$$\frac{\langle [\Delta \hat{n}(0)]^2\rangle}{\langle \hat{n}(0)\rangle^2} \simeq \frac{1}{\langle \hat{n}(0)\rangle}. \qquad (12.43)$$

In this case in Eq. (12.42) we may approximate $\Omega_n = 2\lambda\sqrt{n+1}$ as

$$\Omega_n = 2\lambda\sqrt{\langle \hat{n}(0)\rangle + 1}\left\{1 + \frac{1}{2}\frac{n - \langle \hat{n}(0)\rangle}{\langle \hat{n}(0)\rangle + 1} - \frac{1}{8}\left[\frac{n - \langle \hat{n}(0)\rangle}{\langle \hat{n}(0)\rangle + 1}\right]^2 + \dots\right\}$$

$$\simeq 2\lambda\sqrt{\langle \hat{n}(0)\rangle}\left[1 + \frac{n - \langle \hat{n}(0)\rangle}{2\langle \hat{n}(0)\rangle}\right]. \qquad (12.44)$$

The n summation in Eq. (12.42) may now be carried out to obtain

$$\sigma_{22}(t) \simeq \tfrac{1}{2}\mathrm{Re}\left\{1 + e^{-\langle \hat{n}(0)\rangle}\exp\left[2i\lambda\sqrt{\langle \hat{n}(0)\rangle}\,t\right]\exp\left[-i\lambda\sqrt{\langle \hat{n}(0)\rangle}\,t\right]\right.$$

$$\left. \times \exp\left[\langle \hat{n}(0)\rangle \exp\left(\frac{i\lambda t}{\sqrt{\langle \hat{n}(0)\rangle}}\right)\right]\right\}. \qquad (12.45)$$

With regard to the double exponential in Eq. (12.45), we note that, for times t satisfying the condition

$$\lambda t \ll \sqrt{\langle \hat{n}(0)\rangle}, \qquad (12.46)$$

the inner exponential may be expanded up to terms of second order [recall that $\langle \hat{n}(0) \rangle \gg 1$]. In this time regime we may therefore rewrite Eq. (12.45) as

$$\sigma_{22}(t) \simeq \tfrac{1}{2}\left\{1 + \cos\left[2\lambda\sqrt{\langle \hat{n}(0) \rangle}\, t\right] \exp\left(-\tfrac{1}{2}\lambda^2 t^2\right)\right\}. \tag{12.47}$$

This equation reveals that for $\langle \hat{n}(0) \rangle \gg 1, \lambda^2 t^2$ the excited-state occupation probability oscillates with the effective Rabi frequency

$$\Omega_{\text{eff}} = 2\lambda\sqrt{\langle \hat{n}(0) \rangle} \tag{12.48}$$

and undergoes a collapse [Cummings (1965)] with a characteristic time τ_c independent of the mean photon number:

$$\tau_c = \frac{\sqrt{2}}{\lambda}. \tag{12.49}$$

The collapse of the excited-state occupation probability is caused by destructive interference of the quantum Rabi oscillations at different frequencies, Eq. (12.42). It has nothing to do with dissipative processes. This becomes clear when we follow up the excited-state occupation probability in the further course of time $[\lambda t \ll \sqrt{\langle \hat{n}(0) \rangle}]$. The periodicity of the (slowly varying) double exponential with the period

$$\tau_r = \frac{2\pi}{\lambda}\sqrt{\langle \hat{n}(0) \rangle}, \tag{12.50}$$

Eq. (12.45), suggests that the collapse of the excited-state occupation probability is followed by a series of revivals at times

$$\tau_r^{(k)} = k\tau_r, \quad k = 1, 2, 3, \ldots. \tag{12.51}$$

It should be pointed out that from Eq. (12.44) [together with Eq. (12.42)] the application of Eq. (12.45) becomes questionable for times t with $\lambda t \geq \sqrt{\langle \hat{n}(0) \rangle}$. Thus it is already invalid in the vicinity of the first revival.

The appearance of revivals may be established as follows. Considering the dominant oscillations in the vicinity of the maximum of the Poisson distribution,

$$n \approx \langle \hat{n}(0) \rangle + j, \quad j \ll \langle \hat{n}(0) \rangle, \tag{12.52}$$

from Eq. (12.44) we may estimate the difference between two neighboring Rabi frequencies to be

$$\Omega_n - \Omega_{n-1} \approx \frac{\lambda}{\sqrt{\langle \hat{n}(0) \rangle}}, \tag{12.53}$$

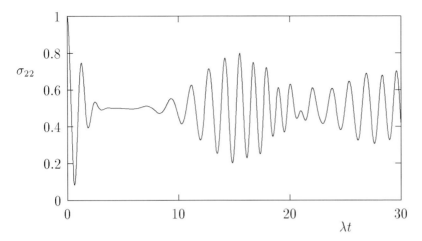

Fig. 12.1 Time evolution of the atomic excited-state occupation probability of an atom initially prepared in the upper state interacting with a field initially prepared in a coherent state $|\alpha\rangle$ of mean photon number $\langle\hat{n}(0)\rangle = 5$.

from which we can see that the value of the relative phase of the two Rabi oscillations at time $t + \tau_r^{(k)}$, with $\tau_r^{(k)}$ from Eq. (12.51), differs from the value at time t by $2\pi k$:

$$(\Omega_n - \Omega_{n-1})\tau_r^{(k)} \approx 2\pi k, \tag{12.54}$$

and hence

$$(\Omega_n - \Omega_{n-j})\tau_r^{(k)} \approx 2\pi k j \tag{12.55}$$

[recall the conditions (12.52)]. Thus superimposing the oscillations in the vicinity of the maximum of the Poisson distribution at a time $t + \tau_r^{(k)}$ is expected to approximately yield the result of interference at the earlier time t. Apart from the ignored effect of the absolute phase value, the phase matching cannot be achieved for all (relevant) Rabi oscillations, because of the square-root dependence of Ω_n on n, and partially destructive interferences prevent the revivals from being complete [Eberly, Narozhny and Sanchez-Mondragon (1980)]. An example of the collapse-revival behavior of the excited-state occupation probability is shown in Fig. 12.1.

Let us consider the revivals in the Jaynes–Cummings model in more detail. With the help of the Poisson summation formula

$$\sum_{n=0}^{\infty} f_n = \sum_{k=-\infty}^{\infty} \int_0^{\infty} dn\, f(n) e^{2\pi i k n} + \tfrac{1}{2} f_0 \tag{12.56}$$

the sum over n in Eq. (12.40) can be converted into an infinite sum of integrals:

$$\sigma_{22}(t) = \frac{1}{2}\left[1 + \sum_{k=-\infty}^{\infty} w_k(t) + \frac{1}{2}p(0)\cos(2\lambda t)\right], \tag{12.57}$$

$$w_k(t) = \text{Re}\left\{\int_{-\infty}^{\infty} dn\, p(n)\exp[2iS_k(n)]\right\}, \tag{12.58}$$

where $p(n) \equiv \rho_{nn}(0)$ for $n \geq 0$ and $p(n) \equiv 0$ for $n < 0$, and

$$S_k(n) = \pi k n - \lambda t\sqrt{n+1}. \tag{12.59}$$

When $p(n)$ is slowly varying compared with $\exp[2iS_k(n)]$, the integration over n, Eq. (12.58), can (approximately) be performed by expanding the phase $S_k(n)$ around the point of stationary phase n_k [Fleischhauer and Schleich (1993)]:

$$S_k(n) \approx S_k(n_k) + \frac{1}{2}\left.\frac{d^2 S_k(n)}{dn^2}\right|_{n=n_k}(n-n_k)^2, \tag{12.60}$$

where n_k is defined by

$$\left.\frac{dS_k(n)}{dn}\right|_{n=n_k} = \pi k - \frac{\lambda t}{2\sqrt{n_k+1}} = 0, \tag{12.61}$$

which implies that

$$\sqrt{n_k + 1} = \frac{\lambda t}{2\pi k}. \tag{12.62}$$

For positive values of t, positive values of k provide a point of stationary phase, and we have, on using Eq. (12.59),

$$S_k(n_k) = -\pi k - \frac{\lambda^2 t^2}{4\pi k}, \quad \left.\frac{d^2 S_k(n)}{dn^2}\right|_{n=n_k} = \frac{2\pi^3 k^3}{\lambda^2 t^2}. \tag{12.63}$$

Note that if $k=0$, a point of stationary phase is only found for $t=0$, and a separate consideration is required.[4] From the above, we may use Eq. (12.57) in the approximate form ($t \geq 0$)

$$\sigma_{22}(t) = \frac{1}{2}\left[1 + w_0(t) + \sum_{k=1}^{\infty} w_k(t) + \frac{1}{2}p(0)\cos(2\lambda t)\right], \tag{12.64}$$

where $w_k(t)$ ($k > 0$) is obtained from Eq. (12.58) together with Eqs (12.60), (12.62) and (12.63) as

$$w_k(t) = \text{Re}\left\{p(n_k)\exp\left[-i\frac{\lambda^2 t^2}{2\pi k}\right]\int_{-\infty}^{\infty} dn\, \exp\left[2i\frac{\pi^3 k^3}{\lambda^2 t^2}n^2\right]\right\}$$

$$= p\left(\frac{\lambda^2 t^2}{4\pi^2 k^2} - 1\right)\frac{\lambda t}{\sqrt{2\pi^2 k^3}}\cos\left(\frac{\lambda^2 t^2}{2\pi k} - \frac{\pi}{4}\right). \tag{12.65}$$

4) See, e.g., Eq. (12.47).

This result reveals that the photon-number statistics of the initial field essentially determine the shape of the revivals as a function of time. In particular, if the photon-number distribution $p(n)$ is (approximately) centered at the mean number of photons, $\langle \hat{n}(0) \rangle$, the envelope $p[(\lambda^2 t^2)/(4\pi^2 k^2) - 1]$ of $w_k(t)$ is (approximately) centered at time

$$\tau_r^{(k)} = \frac{2\pi k}{\lambda} \sqrt{\langle \hat{n}(0) \rangle + 1}, \tag{12.66}$$

which for $\langle \hat{n}(0) \rangle \gg 1$ reduces to Eq. (12.51) (together with Eq. (12.50)). Note that Eq. (12.65) implies that the photon number and the time are related according to

$$n \leftrightarrow \frac{\lambda^2 t^2}{4\pi^2 k^2} - 1. \tag{12.67}$$

In the two-photon Jaynes–Cummings model, instead of Eq. (12.42), we have

$$\sigma_{22}^{(2)}(t) = \frac{1}{2}\left[1 + \sum_{n=0}^{\infty} \frac{\langle \hat{n}(0) \rangle^n}{n!} e^{-\langle \hat{n}(0) \rangle} \cos(\Omega_n^{(2)} t)\right], \tag{12.68}$$

where $\Omega_n^{(2)}$ is now given by Eq. (12.31). Provided that the mean number of photons is large $[\langle \hat{n}(0) \rangle \gg 1]$, in Eq. (12.68) we may expand $\Omega_n^{(2)}$ and disregard terms of the order of magnitude of $\langle \hat{n}(0) \rangle^{-1}$ or smaller:

$$\Omega_n^{(2)} = 2\lambda^{(2)}\left(n + \frac{3}{2} - \frac{1}{8n} \pm \ldots\right) \simeq 2\lambda^{(2)} n + 3\lambda^{(2)}. \tag{12.69}$$

In this case, carrying out the n summation, yields

$$\sigma_{22}^{(2)}(t) \simeq \frac{1}{2}\mathrm{Re}\{1 + e^{-\langle \hat{n}(0) \rangle} \exp(i3\lambda^{(2)} t) \exp[\langle \hat{n}(0) \rangle \exp(i2\lambda^{(2)} t)]\}. \tag{12.70}$$

Although the structure of this result is similar to that of Eq. (12.45), there are substantial differences. By expanding the inner exponential in the double exponential in Eq. (12.70) up to second order in time, we can easily see that the characteristic collapse time depends on the mean number of photons:

$$\tau_c^{(2)} = \frac{1}{\lambda^{(2)}\sqrt{2\langle \hat{n}(0) \rangle}}, \tag{12.71}$$

and the effective Rabi frequency is

$$\Omega_{\mathrm{eff}}^{(2)} = 2\lambda^{(2)} \langle \hat{n}(0) \rangle. \tag{12.72}$$

From the periodicity of the double exponential, the kth revival time is found to be independent of the mean number of photons:

$$\tau_r^{(2)(k)} = k\tau_r^{(2)}, \quad \tau_r^{(2)} = \frac{\pi}{\lambda^{(2)}}. \tag{12.73}$$

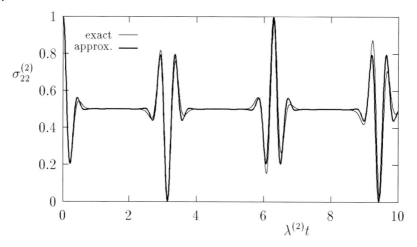

Fig. 12.2 Two-photon Jaynes–Cummings model: time evolution of the atomic excited-state occupation probability of an atom initially prepared in the upper state interacting with a field initially prepared in a coherent state $|\alpha\rangle$ of mean photon number $\langle\hat{n}(0)\rangle = 5$.

Further, from Eq. (12.69) [together with Eq. (12.68)], the error in Eq. (12.70) is expected to be small as long as the time obeys the condition that

$$t \ll \frac{4\langle\hat{n}(0)\rangle}{\lambda^{(2)}}. \tag{12.74}$$

Hence Eq. (12.70) may be used to follow the time evolution of the excited-state occupation probability over a few collapse-revival cycles, provided that the mean number of photons is large enough. In this context, Eq. (12.70) may be approximately rewritten as[5]

$$\sigma_{22}^{(2)}(t) = \tfrac{1}{2}\left\{1 + \sum_{k\geq 0}(-1)^k \cos[\Omega_{\text{eff}}^{(2)}(t - k\tau_r^{(2)})]\exp\left[-\left(\frac{t - k\tau_r^{(2)}}{\tau_c^{(2)}}\right)^2\right]\right\}, \tag{12.75}$$

which may be regarded as an expansion with respect to the revival number k. In Fig. 12.2 the collapse-revival behavior of the atomic excited-state occupation probability in the two-photon Jaynes–Cummings model is shown for the case where the cavity field is initially in a coherent state $|\alpha\rangle$, and the exact result (12.68) is compared with the approximation (12.75). Note that (apart from the alternating sign) the revivals are almost completely due to the quasi-linear dependence on n of the two-photon Rabi frequency, cf. Eq. (12.69).

---

5) The kth-order term is obtained by expanding $\sigma_{22}^{(2)}(t)$ around the kth revival.

12.2.3
Quantum nature of the revivals

It is worth noting that the revivals are a true quantum effect, which results from the discreteness of the photon-number states. If it is ignored [$\langle\hat{n}(0)\rangle \to \infty$], the difference between neighboring Rabi frequencies is effectively reduced to zero, and hence, according to Eq. (12.54), revivals cannot occur for finite times $\tau_r^{(k)}$.

To illustrate this, let us return to Eq. (12.42). When $\langle\hat{n}(0)\rangle \gg 1$ the Poissonian photon-number distribution can by replaced approximately by a Gaussian distribution:

$$\rho_{nn}(0) = \frac{\langle\hat{n}(0)\rangle^n}{n!} e^{-\langle\hat{n}(0)\rangle} \simeq \frac{1}{\sqrt{2\pi\langle\hat{n}(0)\rangle}} \exp\left[-\frac{(n-\langle\hat{n}(0)\rangle)^2}{2\langle\hat{n}(0)\rangle}\right], \quad (12.76)$$

where [recall Eq. (12.43)]

$$n - \langle\hat{n}(0)\rangle = \left[\sqrt{n} - \sqrt{\langle\hat{n}(0)\rangle}\right]\left[\sqrt{n} + \sqrt{\langle\hat{n}(0)\rangle}\right]$$

$$\simeq 2\sqrt{\langle\hat{n}(0)\rangle}\left[\sqrt{n} - \sqrt{\langle\hat{n}(0)\rangle}\right]. \quad (12.77)$$

Using Eqs (12.76) and (12.77), Eq. (12.42) takes the approximate form ($\Omega_n \simeq 2\lambda\sqrt{n}$)

$$\sigma_{22}(t) = \frac{1}{2}\left[1 + \sum_{n=0}^{\infty} \frac{1}{\sqrt{2\pi\langle\hat{n}(0)\rangle}} \exp\left[-2\left(\sqrt{n}-\sqrt{\langle\hat{n}(0)\rangle}\right)^2\right] \cos(2\lambda\sqrt{n}\,t)\right], \quad (12.78)$$

which still allows for the revivals in good agreement with the exact formula.[6]

From inspection of Eq. (12.78) one might expect that for sufficiently large $\langle\hat{n}(0)\rangle$ the summation over n can be performed in the sense of an integration:

$$\sigma_{22}(t) = \frac{1}{2}\left[1 + \int_0^\infty \frac{dn}{\sqrt{2\pi\langle\hat{n}(0)\rangle}} \exp\left[-2\left(\sqrt{n}-\sqrt{\langle\hat{n}(0)\rangle}\right)^2\right] \cos(2\lambda\sqrt{n}\,t)\right], \quad (12.79)$$

from which we obtain

$$\sigma_{22}(t) = \frac{1}{2}\left[1 + \sqrt{\frac{2}{\pi}} \int_0^\infty dx \frac{x}{\sqrt{\langle\hat{n}(0)\rangle}} \exp\left[-2\left(x-\sqrt{\langle\hat{n}(0)\rangle}\right)^2\right] \cos(2\lambda xt)\right]$$

$$\approx \frac{1}{2}\left[1 + \sqrt{\frac{2}{\pi}} \int_{-\infty}^\infty dx \exp\left[-2\left(x-\sqrt{\langle\hat{n}(0)\rangle}\right)^2\right] \cos(2\lambda xt)\right]$$

$$= \frac{1}{2}\left[1 + \cos\left(2\lambda\sqrt{\langle\hat{n}(0)\rangle}\,t\right) \exp(-\tfrac{1}{2}\lambda^2 t^2)\right]; \quad (12.80)$$

6) Note that Eq. (12.78) can also be handled with the help of the Poisson summation formula (12.56), so that Eqs (12.64) and (12.65) apply, with a Gaussian in place of the Poissonian envelope.

that is, we reproduce the approximate result (12.47). We can see that, if the (discrete) photon nature of light is "smoothed" out, the revivals are indeed lost and only the decay is preserved.

It is worth noting that during the collapse-revival cycles the quantum correlations between atom and field may be changed substantially [Phoenix and Knight (1988); Gea-Banacloche (1990)]. If the atom–field system evolving under the Jaynes–Cummings Hamiltonian (12.1) [together with Eqs (12.2) and (12.3)] is initially prepared in a pure state $|\Psi(0)\rangle$, it remains, of course, in a pure state $|\Psi(t)\rangle = \hat{U}(t)|\Psi(0)\rangle$ for all times, which implies that

$$|\Psi(t)\rangle = c^{(1)}(t)\,|\Psi_f^{(1)}(t)\rangle \otimes |\Psi_a^{(1)}(t)\rangle + c^{(2)}(t)\,|\Psi_f^{(2)}(t)\rangle \otimes |\Psi_a^{(2)}(t)\rangle, \quad (12.81)$$

where the subscripts a and f indicate the atomic and field states, respectively. Using the time evolution operator $\hat{U}$ in the form (12.22), the states $|\Psi_a^{(i)}\rangle$ and $|\Psi_f^{(i)}\rangle$ and the coefficients $c^{(i)}$ ($i=1,2$) can be calculated in a straightforward way. We therefore omit the calculations here and refer the reader to the literature [see, e.g., Phoenix and Knight (1991)]. In the general case the resulting state $|\Psi(t)\rangle$ cannot be factored into a product of atom and field states, thus the two subsystems develop due to the interaction into an entangled quantum state (Section 8.5). Clearly, when one of the coefficients $c^{(i)}$ vanishes, the atom and the field are not correlated and both subsystems are in pure states. The calculations show that during subsequent collapses the atom and the field may approach pure states at the times $t=(2k+1)\tau_r/2$ ($k=0,1,2,\ldots$), although the approach becomes progressively less perfect. These times correspond to half of the times between subsequent revival peaks.

12.2.4
Coherent preparation

So far we have considered atomic motion for an incoherent initial preparation. To study effects that have their origin in a coherent initial preparation [Vogel, Welsch and Leine (1987)], let us assume that the atom is initially prepared in a coherent superposition of the two quantum states as follows:

$$\begin{aligned}&\sigma_{22}(0) = \sin^2\chi, \qquad &&\sigma_{11}(0) = \cos^2\chi,\\ &\sigma_{12}(0) = \tfrac{1}{2}e^{-i\varphi}\sin(2\chi), \quad &&\sigma_{21}(0) = \sigma_{12}^*(0),\end{aligned} \quad (12.82)$$

which can be achieved by pre-pumping the atom with a classical light field, the parameters χ and φ being controlled respectively by the product of the Rabi frequency and the interaction time and by the phase of the pre-pumping field. With regard to the cavity mode, we assume that it is initially in a coherent state $|\alpha\rangle$, so that the complete matrix elements of the initial cavity-field

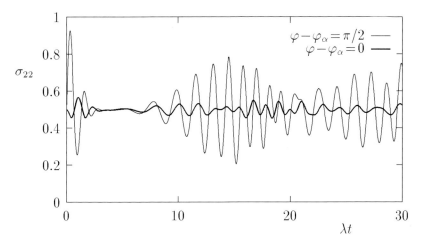

Fig. 12.3 Time evolution of the atomic excited-state occupation probability of an atom initially prepared in a coherent superposition of two states interacting with a field initially prepared in a coherent state $|\alpha\rangle$ of mean photon number $\langle \hat{n}(0) \rangle = 5$, for $\chi = \pi/4$, and $\varphi - \varphi_\alpha = 0$ and $\pi/2$.

density operator $\hat{\rho}(0) = |\alpha\rangle\langle\alpha|$ are

$$\rho_{nm}(0) = \frac{\alpha^n (\alpha^*)^m}{\sqrt{n!\,m!}} e^{-|\alpha|^2} = \frac{\sqrt{\langle \hat{n}(0) \rangle^{n+m}}}{\sqrt{n!\,m!}} e^{i(n-m)\varphi_\alpha} e^{-\langle \hat{n}(0) \rangle} \qquad (12.83)$$

[cf. Eq. (3.59)].

For example, let us choose $\chi = \pi/4$, which is best suited, according to the initial preparation of the atom, (12.82), to producing substantial coherence effects. Note that, in this case, both the ground and excited states of the atom are occupied with probability one-half, and $|\sigma_{12}(0)|$ attains its maximum value of one-half. From Eqs (12.35), (12.38) and (12.39), together with the initial conditions (12.82) and (12.83), we obtain the result that

$$\sigma_{22}(t) = \sigma_{22}^{\mathrm{inc}}(t) + \sigma_{22}^{\mathrm{coh}}(t), \qquad (12.84)$$

$$\sigma_{22}^{\mathrm{inc}}(t) = \frac{1}{4}\left\{2 - e^{-\bar{n}} + \sum_{n=0}^{\infty} \frac{\langle \hat{n}(0) \rangle^n [n+1-\langle \hat{n}(0) \rangle]}{(n+1)!} e^{-\langle \hat{n}(0) \rangle} \cos(\Omega_n t)\right\}, \qquad (12.85)$$

$$\sigma_{22}^{\mathrm{coh}}(t) = \frac{1}{2}\sin(\varphi - \varphi_\alpha) \sum_{n=0}^{\infty} \frac{\langle \hat{n}(0) \rangle^n}{n!} e^{-\langle \hat{n}(0) \rangle} \sqrt{\frac{\langle \hat{n}(0) \rangle}{n+1}} \sin(\Omega_n t). \qquad (12.86)$$

From Eq. (12.86) the contribution of $\sigma_{22}^{\mathrm{coh}}(t)$ to $\sigma_{22}(t)$ is seen to depend sensitively on the atomic initial phase φ relative to the phase φ_α of the initial coherent state $|\alpha\rangle$, cf. Fig. 12.3. If $\varphi - \varphi_\alpha = (k+1/2)\pi$ ($k = 0, 1, 2, \ldots$), $\sigma_{22}^{\mathrm{coh}}(t)$ contributes with maximum weight to $\sigma_{22}(t)$, whereas $\sigma_{22}^{\mathrm{coh}}(t)$ vanishes for

$\varphi - \varphi_\alpha = k\pi$. That is, coherently prepared atoms "distinguish" sensitively between a coherent initial cavity field [$\rho_{nm}(0) \neq 0$ for $n \neq m$] and an incoherent initial cavity field [$\rho_{nm}(0) = 0$ for $n \neq m$] which have equal photon-number distributions $\rho_{nn}(0)$.

12.3
Field dynamics

We have seen that the electronic-state dynamics typically consists of a collapse of the Rabi-oscillating occupation probabilities, which is followed by a series of revivals. Clearly, the dynamics of the electronic subsystem is unavoidably connected with changes in the properties of the radiation-field mode.

12.3.1
Reduced density matrix

To study the time evolution of the field mode, we recall the general solution for the atom–field density matrix given in Eq. (12.26), together with Eqs (12.22) and (12.23). From these results we can easily derive the reduced density matrix ρ_{nm} for the field mode by taking the trace with respect to the atomic degrees of freedom:

$$\begin{aligned}
\rho_{nm}(t) &= \varrho_{1n,1m}(t) + \varrho_{2n,2m}(t) \\
&= U_{1n,1n}(t-t')[U_{1m,1m}(t-t')]^* \varrho_{1n,1m}(t') \\
&\quad + U_{2n,1(n+1)}(t-t')[U_{2m,1(m+1)}(t-t')]^* \varrho_{1(n+1),1(m+1)}(t') \\
&\quad + U_{2n,2n}(t-t')[U_{2m,2m}(t-t')]^* \varrho_{2n,2m}(t') \\
&\quad + U_{1n,2(n-1)}(t-t')[U_{1m,2(m-1)}(t-t')]^* \varrho_{2(n-1),2(m-1)}(t') \\
&\quad + U_{1n,1n}(t-t')[U_{1m,2(m-1)}(t-t')]^* \varrho_{1n,2(m-1)}(t') \\
&\quad + U_{2n,1(n+1)}(t-t')[U_{2m,2m}(t-t')]^* \varrho_{1(n+1),2m}(t') \\
&\quad + U_{2n,2n}(t-t')[U_{2m,1(m+1)}(t-t')]^* \varrho_{2n,1(m+1)}(t') \\
&\quad + U_{1n,2(n-1)}(t-t')[U_{1m,1m}(t-t')]^* \varrho_{2(n-1),1m}(t'),
\end{aligned} \qquad (12.87)$$

where the initial density-matrix elements $\varrho_{in,jm}(t')$ may again be written in the factored form (12.34). Explicit expressions for the matrix elements of the time-evolution operator may be taken from Eq. (12.22). Equation (12.87) then enables us to study the time evolution of the quantum-statistical properties of the cavity field for an arbitrary initial atomic-state preparation [Vogel and Welsch (1989)].

Let us consider the case where the atom is initially in the upper quantum state $[\sigma_{ab}(t')=\delta_{a2}\delta_{b2}]$, so that Eq. (12.87) reduces to

$$\rho_{nm}(t) = U_{2n,2n}(t-t')[U_{2m,2m}(t-t')]^*\rho_{nm}(t')$$
$$+ U_{1n,2(n-1)}(t-t')[U_{1m,2(m-1)}(t-t')]^*\rho_{n-1\,m-1}(t'), \qquad (12.88)$$

and from Eq. (12.22) we deduce the result ($t'=0$)

$$\rho_{nm}(t) = e^{-i\omega(n-m)t}$$
$$\times \left\{ \left[\cos\left(\tfrac{1}{2}\Delta_n t\right) - i\frac{\delta}{\Delta_n}\sin\left(\tfrac{1}{2}\Delta_n t\right)\right]\left[\cos\left(\tfrac{1}{2}\Delta_m t\right) + i\frac{\delta}{\Delta_m}\sin\left(\tfrac{1}{2}\Delta_m t\right)\right]\rho_{nm}(0) \right.$$
$$\left. + \frac{\Omega_{n-1}}{\Delta_{n-1}}\sin\left(\tfrac{1}{2}\Delta_{n-1}t\right)\frac{\Omega_{m-1}}{\Delta_{m-1}}\sin\left(\tfrac{1}{2}\Delta_{m-1}t\right)\rho_{n-1\,m-1}(0) \right\}. \qquad (12.89)$$

12.3.2
Photon statistics

Let us first consider the temporal evolution of the diagonal density-matrix elements $\rho_{nn}(t)$ determining the photon-number statistics. From Eq. (12.89) we obtain

$$\rho_{nn}(t) = \left[\cos^2\left(\tfrac{1}{2}\Delta_n t\right) + \left(\frac{\delta}{\Delta_n}\right)^2 \sin^2\left(\tfrac{1}{2}\Delta_n t\right)\right]\rho_{nn}(0)$$
$$+ \left(\frac{\Omega_{n-1}}{\Delta_{n-1}}\right)^2 \sin^2\left(\tfrac{1}{2}\Delta_{n-1}t\right)\rho_{n-1\,n-1}(0), \qquad (12.90)$$

which in the case of exact resonance ($\delta=0$) becomes

$$\rho_{nn}(t) = \cos^2\left(\tfrac{1}{2}\Omega_n t\right)\rho_{nn}(0) + \sin^2\left(\tfrac{1}{2}\Omega_{n-1}t\right)\rho_{n-1\,n-1}(0). \qquad (12.91)$$

Typical examples of the photon-number distribution at various times are shown in Figs 12.4 and 12.5. For comparison with the atomic motion (Fig. 12.1), the atom initially prepared in the upper quantum state is assumed to be resonantly interacting with a cavity field initially prepared in a coherent state $|\alpha\rangle$ of mean photon number $\langle\hat{n}(0)\rangle = 5$. In Fig. 12.4 the chosen times are in the interval during which the (Rabi-oscillating) atomic excited-state occupation probability collapses. We see that, in the beginning, the initially prepared Poissonian photon-number distribution tends to a nonclassical, sub-Poissonian distribution $[\langle[\Delta\hat{n}(t)]^2\rangle/\langle\hat{n}(t)\rangle < 1, t > 0]$. In the further course of time, particularly near the end of the collapse, the photon-number distribution becomes more and more structured and the sub-Poisson effect decreases. The behavior of the photon-number distribution for times after the collapse and before the first revival, is shown in Fig. 12.5. Although during this time interval the atomic excited-state occupation probability $\sigma_{22}(t)$ is nearly constant (cf. Fig. 12.1), from Fig. 12.5 the photon-number distribution is found to change substantially. These changes, which mainly concern redistributions,

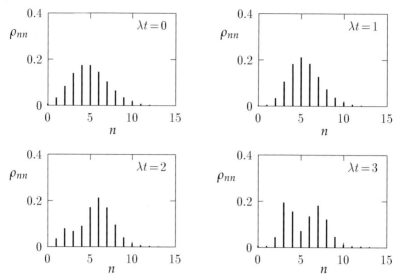

Fig. 12.4 Time evolution of the photon-number distribution during the collapse of the atomic excited-state occupation probability. The atom is initially in the upper state and the initial field is in a coherent state $|\alpha\rangle$ of mean photon number $\langle \hat{n}(0)\rangle = 5$. The values of $\langle[\Delta\hat{n}(t)]^2\rangle/\langle\hat{n}(t)\rangle$ are 1 ($\lambda t=0$), 0.717 (1), 0.887 (2) and 0.9715 (3).

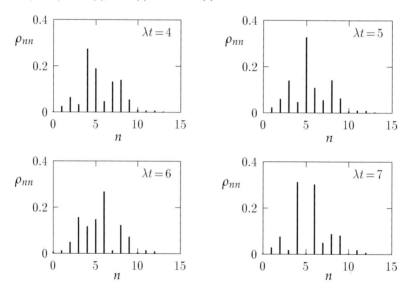

Fig. 12.5 Time evolution of the photon-number distribution is shown after the collapse of the atomic excited-state occupation probability and before the first revival. The atom is initially in the upper state and the initial field is in a coherent state $|\alpha\rangle$ of mean photon number $\langle\hat{n}(0)\rangle=5$. The values of $\langle[\Delta\hat{n}(t)]^2\rangle/\langle\hat{n}(t)\rangle$ are 0.959 ($\lambda t=4$), 0.953 (5), 0.949 (6) and 0.975 (7).

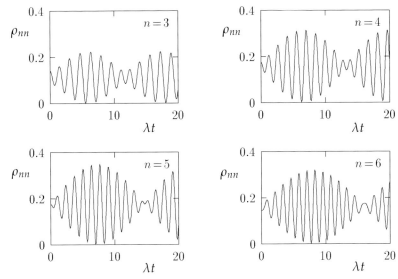

Fig. 12.6 Time evolution of leading photonic density matrix elements $\rho_{nn}(t)$. The atom is initially in the upper state and the initial field is in a coherent state $|\alpha\rangle$ of mean photon number $\langle\hat{n}(0)\rangle=5$.

leave both the mean number of photons and the variance of the number of photons almost constant. This is consistent with the 0.5:0.5 occupation probabilities of the two atomic quantum states during the considered time interval, which suggests that the atom gives, on average, half of its excitation energy to the cavity field. In the case shown in Fig. 12.5 the value of the mean number of photons is close to 5.5, according to the assumed initial value of the mean number of photons, $\langle\hat{n}(0)\rangle=5$. The given values of $\langle[\Delta\hat{n}(t)]^2\rangle/\langle\hat{n}(t)\rangle$ indicate slightly sub-Poissonian radiation.

For deeper insight into the time evolution of the photon-number distribution, it may be helpful to consider the full time dependence of the leading photonic density-matrix elements $\rho_{nn}(t)$ [$n \approx \langle\hat{n}(0)\rangle$]. As can be seen from Eq. (12.91), the time evolution of each density-matrix element ρ_{nn} is determined by a superposition of two oscillations with neighboring Rabi frequencies Ω_n and Ω_{n-1}. The difference frequency may be regarded as a beat frequency, provided that n is sufficiently large. In this case the motion of the leading density-matrix elements ρ_{nn} effectively consists of Rabi oscillations superimposed by a beating, as is illustrated in Fig. 12.6. From Eq. (12.54) the beat period is seen to correspond to the revival time τ_r. Hence when the atomic excited-state occupation probability changes only slightly, the photon-number distribution may change substantially and vice versa, cf. Figs 12.1 and 12.6.

12.4
The Micromaser

As previously mentioned, the Jaynes–Cummings model has been successfully applied to the study of the micromaser also called the single-atom Rydberg maser [Meschede, Walther and Müller (1985); Rempe, Walther and Klein (1987); Walther (1992)], in which a beam of long-living Rydberg atoms is injected into a cooled, single-mode high-Q cavity at such a low rate that, at most, one atom at a time is present inside the cavity, see the scheme in Fig. 12.7. In order to obtain a well-defined time of interaction of the individual atoms with the cavity field, the atomic beam is transmitted through a velocity selector before it is injected into the cavity. Further, before entering the cavity, the atoms are pre-pumped by a laser to a highly excited Rydberg state, and the cooled high-Q microwave cavity is tuned to resonance with a single atomic Rydberg transition. Since the natural lifetime of a Rydberg state is very long [Haroche (1984); Haroche and Raimond (1985)], in the time interval during which an atom interacts with the cavity field, the atom can be regarded as a nearly undamped two-level system. In this way, one effectively deals with a single atom interacting with a single cavity mode, as described by the Jaynes–Cummings model. After the atom has passed the cavity, its state can be measured by ionization, from which information on the properties of the cavity field may also be obtained. A direct field measurement would require coupling out some part of the cavity field, and as a consequence, the desired high quality of the cavity would be lost.

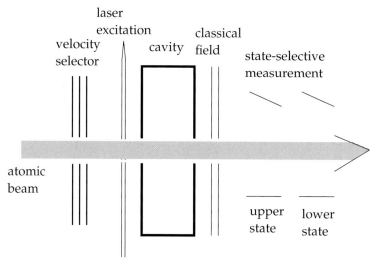

Fig. 12.7 Draft scheme of the elements of a micromaser, including detection through state-selective field ionization. [After Wagner, Brecha, Schenzle and Walther (1992).]

Experiments with a two-photon Rydberg maser have also been performed [Brune, Raimond, Goy, Davidovich and Haroche (1987)]. Since attempts are usually made to enhance the atomic two-photon transition probability by an intermediate level (detuned from the single-photon resonance), the two-photon Jaynes–Cummings model as outlined in Sections 12.1 and 12.2 does not apply directly. In this case the model of a three-level atom interacting with a single-mode cavity field [Yoo and Eberly (1985)] seems to be more appropriate.

A theory of the micromaser may be developed as follows [Filipowicz, Javanainen and Meystre (1986)]. From the scheme described, one may distinguish three characteristic time scales. The first is determined by the interaction time τ_{int}, which corresponds to the time of flight of a single atom through the microwave cavity. This time varies only slightly from atom to atom, because of the velocity selector employed. After the atom has left the cavity, the field evolves freely up to the entrance into the cavity of the next atom in the atomic beam. The corresponding time interval τ_{free} determines the second time scale. Because of the atom-number statistics (which are usually Poissonian), this time interval may be regarded as a random variable. The requirement that, at most, one atom interacts with the cavity field, implies that τ_{int} should be small compared with τ_{free}: $\tau_{int} \ll \tau_{free}$. With regard to the assumed high-Q value of the cavity, τ_{int} should also be small compared with the lifetime τ_{cav} of a cavity photon (third time scale): $\tau_{int} \ll \tau_{free} < \tau_{cav}$, so that in zeroth approximation, effects of damping of the cavity mode during the time of flight of an individual atom through the cavity may be ignored.

Let us assume that, at time t_k, the kth atom prepared in the upper quantum state,

$$\sigma_{ab}(t_k) = \delta_{ab}\delta_{a2}, \tag{12.92}$$

enters the cavity. Under the given conditions, the time evolution of the density matrix of the cavity field, $\rho_{nm}(t)$, during the time interval t_k, $t_k + \tau_{int}$ may be described within the framework of the Jaynes–Cummings model using Eq. (12.89) ($\delta = 0$):

$$\rho_{nm}(t_k+\tau_{int}) = \exp[-i\omega(n-m)\tau_{int}] [\cos(\tfrac{1}{2}\Omega_n\tau_{int})\cos(\tfrac{1}{2}\Omega_m\tau_{int})\rho_{nm}(t_k) \\ + \sin(\tfrac{1}{2}\Omega_{n-1}\tau_{int})\sin(\tfrac{1}{2}\Omega_{m-1}\tau_{int})\rho_{n-1\,m-1}(t_k)]. \tag{12.93}$$

In particular, the diagonal density-matrix elements are

$$\rho_{nn}(t_k+\tau_{int}) = \cos^2(\tfrac{1}{2}\Omega_n\tau_{int})\rho_{nn}(t_k) + \sin^2(\tfrac{1}{2}\Omega_{n-1}\tau_{int})\rho_{n-1\,n-1}(t_k). \tag{12.94}$$

Note that if the photonic density matrix is diagonal at time t_k, it is also diagonal at time $t_k+\tau_{int}$, so that it is sufficient to consider only the diagonal matrix elements. This is the case when the initial density matrix of the cavity field is diagonal:

$$\rho_{nm}(t_1) = \rho_{nn}(t_1)\delta_{nm}, \qquad (12.95)$$

which typically describes the initial situation where the cavity field is in thermal equilibrium (at temperature T):

$$\rho_{nn}(t_1) = (\rho_{\text{th}})_{nn} = \frac{\exp\left(-\frac{\hbar\omega}{k_B T}n\right)}{\text{Tr}\left[\exp\left(-\frac{\hbar\omega}{k_B T}\hat{n}\right)\right]}. \qquad (12.96)$$

If $\tau_{\text{free}} \ll \tau_{\text{cav}}$ then $\rho_{nn}(t)$ may be regarded as constant during the time interval $t_k + \tau_{\text{int}}$, $t_k + \tau_{\text{int}} + \tau_{\text{free}}$, so that the $(k+1)$th atom that enters the cavity finds a cavity field with (diagonal) density matrix $\rho_{nn}(t_{k+1}) = \rho_{nn}(t_k + \tau_{\text{int}})$. In this case, Eq. (12.94) represents a recurrence relation, which may be solved in a straightforward way to obtain the (diagonal) density-matrix elements of the cavity field as functions of the number of atoms transmitted through the cavity.

Let us suppose that the interaction time τ_{int} is chosen such that for a given photon number $n = p$

$$\tau_{\text{int}} = m\frac{2\pi}{\Omega_p} \qquad (12.97)$$

($m > 0$, integer). When the cavity field is initially a low-temperature thermal field, so that $\rho_{nn}(t_1) \approx 0$ for $n \geq p$, from inspection of Eq. (12.94) together with Eq. (12.97) it can be seen that the photon-number states $|p+l\rangle$, $l = 1, 2, 3, \ldots$ cannot be built up. For $n = p + 1$, Eq. (12.94) is

$$\rho_{p+1\,p+1}(t_k + \tau_{\text{int}}) = \cos^2\left(\tfrac{1}{2}\Omega_{p+1}\tau_{\text{int}}\right)\rho_{p+1\,p+1}(t_k), \qquad (12.98)$$

which, under the initial condition that $\rho_{p+1\,p+1}(t_1) = 0$, is solved by

$$\rho_{p+1\,p+1}(t) = 0 \qquad (12.99)$$

for all times t. Hence for $n = p + l$, $l = 1, 2, 3, \ldots$, Eq. (12.94) is solved by

$$\rho_{p+l\,p+l}(t) = 0, \qquad (12.100)$$

and for the maximum value of n, $n = p$, Eq. (12.94) takes the form

$$\rho_{pp}(t_k + \tau_{\text{int}}) = \rho_{pp}(t_k) + \sin^2\left(\tfrac{1}{2}\Omega_{p-1}\tau_{\text{int}}\right)\rho_{p-1\,p-1}(t_k). \qquad (12.101)$$

From these results one may expect that, after transmitting a sufficiently large number of atoms through the cavity, so that the second term in

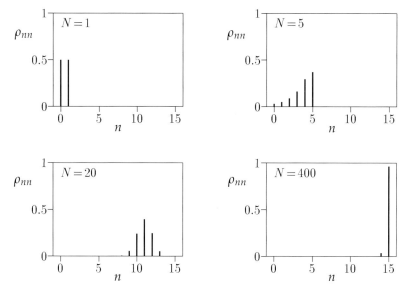

Fig. 12.8 The photon-number statistics of a lossless micromaser for $\Omega_{15}\tau_{\text{int}} = 2\pi$ is shown for various numbers N of atoms transmitted through the cavity.

Eq. (12.101) is approximately zero (and all matrix elements ρ_{nn} with $n < p$ are approximately zero), the cavity field tends to be in the photon-number state $|p\rangle$.[7] Typical examples of the photon-number statistics of the idealized micromaser considered are shown in Fig. 12.8 for $\Omega_{15}\tau_{\text{int}} = 2\pi$ and various values of the number N of atoms that have passed through the cavity. In this figure the zero-temperature limit is considered, so that the initial state of the cavity field is the vacuum state. With increasing number of atoms transmitted through the cavity, the photon-number distribution is seen to be shifted towards larger values of the number of photons. After $N = 400$ atoms have passed through the cavity, the state of the field is seen to be very close to the number state $|n-15\rangle$. When the target Fock state of the cavity field has been reached in this manner the idealized micromaser would be trapped in this state, which has also been called the trapping state.

Clearly, in a micromaser operating under more realistic conditions, a pure photon-number state is hardly achievable. Cavity losses, together with the effects of finite cavity temperature, give rise to smoothings, which lead, for example, to violation of Eq. (12.100). However, it was possible to demon-

---

7) Applying Eq. (12.94) ($n < p$) together with (12.100) iteratively, in the matrix elements with $n < p$, trigonometric functions are multiplied. Since their absolute values are less than unity, after a sufficiently large number of iterations (corresponding to the number of atoms that have passed through the cavity) only the matrix element ρ_{pp} survives.

strate the sub-Poissonian character of the photon-number distribution of the maser field experimentally [Rempe, Schmidt-Kaler and Walther (1990)]. Later the parameters were significantly improved [Weidinger, Varcoe, Heerlein and Walther (1999)]. A cavity Q factor of about 3×10^{10} could be realized, which corresponds to an average lifetime of a photon in the cavity of 0.2 s. This renders it possible to clearly demonstrate the signatures of trapping states in micromaser experiments. To do this, the electronic states of the atoms have been measured after the interaction with the cavity field. Two different ways have been used to demonstrate the existence of trapping states. The first possibility is based on the detection of the electronic-state inversion as a function of the interaction time τ_{int}. When the interaction time is close to the trapping-state condition (12.97), the initially excited atoms cannot leave the cavity in the electronic ground state, which would be accompanied by depositing more photons to the cavity. The second possibility consists of the determination of the variance of the emergent atoms in the lower state. For the micromaser in the trapping state, one expects a sub-Poissonian statistics of the atoms in the ground state, which is in agreement with experimental results.

Cavity losses must be taken into account if, for example, the time between succeeding injections of atoms into the cavity is not small compared with the decay time of the cavity field: $\tau_{\text{free}} \ll \tau_{\text{cav}}$. In this case the time evolution of the density matrix of the cavity field during the time interval $t_k + \tau_{\text{int}}$, $t_k + \tau_{\text{int}} + \tau_{\text{free}}$ may be described within the framework of damping theory (Chapters 5 and 9), by applying a quantum Langevin equation of the type (9.151). In this case, the density-matrix elements of the cavity field obey the following equations of motion:

$$\dot{\rho}_{nn} = w_\downarrow [(n+1)\rho_{n+1\,n+1} - n\rho_{nn}] + w_\uparrow [n\rho_{n-1\,n-1} - (n+1)\rho_{nn}] \quad (12.102)$$

[cf. Eq. (5.110)], where the relaxation rates $w_\downarrow$ and $w_\uparrow$ may be represented as

$$w_\downarrow = \Gamma_{\text{cav}}(n_{\text{th}} + 1), \quad w_\uparrow = \Gamma_{\text{cav}} n_{\text{th}} \quad (12.103)$$

[cf. Eqs (5.88) and (5.89)]. Here, Γ_{cav} corresponds to γ_n in Eq. (9.152), and n_{th} is the mean number of excitations of a bosonic reservoir in thermal equilibrium. Provided that the cavity is cooled to a very low temperature ($n_{\text{th}} \approx 0$), Eq. (12.102) reduces to

$$\dot{\rho}_{nn} = \Gamma_{\text{cav}}[(n+1)\rho_{n+1\,n+1} - n\rho_{nn}]. \quad (12.104)$$

The time evolution of the photon number distribution of the micromaser field may then be calculated by iterative application of both Eq. (12.94) and Eq. (12.102), and averaging over the ensemble of times between successive entrances of atoms into the cavity [for details, see Filipowicz, Javanainen and Meystre (1986); Meystre and Sargent III (1990)].

12.5
Quantum-state preparation

In the context of the micromaser we have considered the demonstration of trapping states, which is closely related to the realization of number states of the cavity field, or a sub-Poissonian photon statistics. Atom–field interactions in high-Q cavities are also suited for the preparation of other interesting quantum states. In this section we will illustrate some of these possibilities by considering the creation of entangled states of the Schrödinger-cat type and of Einstein–Podolsky–Rosen (EPR) pairs of atoms.

12.5.1
Schrödinger-cat states

We begin with the possibilities of generating entangled quantum states of a type of Schrödinger-cat state (8.101) [Brune, Haroche, Raimond, Davidovich and Zagury (1992)]. In the experimental implementation by Brune, Hagley, Dreyer, Maitre, Maali, Wunderlich, Raimond and Haroche (1996), the entanglement is realized between the cavity field and the electronic quantum states of an atom passing through the cavity. A Rb atom is initially prepared in a superposition of two (long-living) circular Rydberg states $|1\rangle$ and $|2\rangle$, by transmitting it through a Ramsey zone[8] consisting of a low-Q cavity in which the atom undergoes a resonant $\pi/2$ pulse interaction with a microwave field. Next, the atom enters the high-Q cavity which is tuned slightly off resonance with respect to the atomic $|1\rangle \leftrightarrow |2\rangle$ transition so that the atom and the field cannot exchange energy. The cavity field is prepared by a pulsed source in a coherent state $|\alpha\rangle$, the mean photon number $|\alpha|^2$ can be varied from 1 to 10. The atom–field coupling produces a phase shift of the field, by the single-atom dispersion effect, which depends on the electronic state. In this manner the phase of the coherent field in the cavity is entangled with the electronic state of the atom after the interaction. Another low-Q Ramsey cavity is used to apply again a $\pi/2$ pulse and subsequently the atoms are counted in the states $|1\rangle$ and $|2\rangle$ by two field ionization detectors. The two low-Q Ramsey cavities are fed by the same source whose frequency is swept across the atomic transition frequency. The measured signal is the probability of finding the atom in state $|1\rangle$ as a function of the frequency of the source.

Let us consider the preparation of the Schrödinger-cat state in more detail. For this purpose we need the unitary time evolution operator (12.22) in the limit of large detuning, $\delta \gg \Omega_n$. Taking into account the leading terms therein, we arrive at

$$\hat{U}(t) = \exp\left(-\frac{i}{\hbar}\hat{H}_0 t\right)\hat{U}_{\text{int}}(t), \qquad (12.105)$$

---

[8] See Ramsey (1985).

where the interaction part of the time evolution operator reads as

$$\hat{U}_{int}(t) = \exp\left(i\frac{\lambda^2}{\delta}\hat{n}t\right)|1\rangle\langle 1| + \exp\left(-i\frac{\lambda^2}{\delta}(\hat{n}+1)t\right)|2\rangle\langle 2|. \tag{12.106}$$

The initial preparation of the atom–field system is described by the quantum state

$$|\Psi(0)\rangle = \frac{1}{\sqrt{2}}(|1\rangle + |2\rangle)|\alpha\rangle. \tag{12.107}$$

Due to the effect of the dispersive interaction the resulting time-evolved state in the interaction picture, $|\Psi(t)\rangle = \hat{U}_{int}(t)|\Psi(0)\rangle$, is of the form

$$|\Psi(0)\rangle = \frac{1}{\sqrt{2}}(|1\rangle|\alpha e^{i\Phi}\rangle + e^{-i\Phi}|2\rangle|\alpha e^{-i\Phi}\rangle). \tag{12.108}$$

The phase shift caused by the dispersive interaction is given by

$$\Phi = \frac{\lambda^2}{\delta}t. \tag{12.109}$$

The resulting state $|\Psi(t)\rangle$ is clearly entangled, the phases of the coherent states being different in the two electronic states. It can be interpreted in the spirit of Schrödinger's Gedanken experiment (Section 8.5), where the two coherent states in Eq. (12.108) replace – on a mesoscopic rather than a macroscopic level – the states of the cat being dead and alive in Eq. (8.101).

12.5.2
Einstein–Podolsky–Rosen pairs of atoms

Another interesting possibility consists of the use of the cavity for preparing an entangled state of two atoms, an Einstein–Podolsky–Rosen pair of atoms [Hagley, Maitre, Nogues, Wunderlich, Brune, Raimond and Haroche (1997)]. The two spatially separated atoms enter the cavity successively. Now the cavity is tuned on resonance with the atomic transition. Initially the system is prepared in the uncorrelated state

$$|\Psi(0)\rangle = |2\rangle_1|1\rangle_2|0\rangle. \tag{12.110}$$

The first atom enters the initially empty cavity (state $|0\rangle$) in the excited state $|2\rangle_1$ and the second atom enters the cavity in the ground state, $|1\rangle_2$. Let us consider the effect of the interaction of the first atom on the quantum state in the interaction picture. For exact resonance, the interaction part of the unitary evolution operator (12.22) reads as

$$\hat{U}_{int}(t) = |1,0\rangle\langle 1,0| + \sum_{n=0}^{\infty}[\cos(\tfrac{1}{2}\Omega_n t)(|1,n+1\rangle\langle 1,n+1| + |2,n\rangle\langle 2,n|)$$
$$+ i\sin(\tfrac{1}{2}\Omega_n t)(|1,n+1\rangle\langle 2,n| + |2,n\rangle\langle 1,n+1|)]. \tag{12.111}$$

Applying this evolution to the interaction between the first atom and the cavity field for an interaction time t_1 obeying the condition $\Omega_0 t_1 = \pi/2$, the state vector is given by

$$|\Psi(t_1)\rangle = \tfrac{1}{\sqrt{2}}(|2\rangle_1|1\rangle_2|0\rangle + i|1\rangle_1|1\rangle_2|1\rangle). \tag{12.112}$$

It represents a superposition of atom 1 leaving the empty cavity in the excited state and depositing a photon in the cavity and being in the ground state. In any case, atom 2 is in the ground state.

Now the second atom interacts with the cavity for an interaction time $t_2 = 2t_1$. It can either absorb a photon from the cavity and leaves the cavity in the excited state or it remains in the ground state when the cavity is empty. The state at the time $\tau > t_1 + t_2$ at which the two interactions are completed, resulting from the action of the unitary evolution operator on the state $|\Psi(t_1)\rangle$, is given by

$$|\Psi(\tau)\rangle = \tfrac{1}{\sqrt{2}}(|2\rangle_1|1\rangle_2 - |1\rangle_1|2\rangle_2)|0\rangle. \tag{12.113}$$

The result of these two interactions consists of a quantum state in which the cavity is empty and decorrelated from the atoms. The atoms, however, leave the cavity in the maximally entangled EPR state

$$|\Psi_{EPR}\rangle = \tfrac{1}{\sqrt{2}}(|2\rangle_1|1\rangle_2 - |1\rangle_1|2\rangle_2). \tag{12.114}$$

The entangled atoms prepared in the experiments are spatially separated by distances of the order of centimeters.

It is worth noting that the method of preparing EPR pairs of atoms can be extended to create entanglement between more than two atoms. Experiments of this type have also been performed [Rauschenbeutel. Nogues, Osnaghi, Bertet, Brune, Raimond and Haroche (2000)].

12.6
Measurements of the cavity field

Unfortunately, a high-Q cavity field cannot be detected directly. First, any coupling out of photons would decrease the Q value of the cavity. Second, the quantum statistical properties of the output field from a cavity may significantly differ from those of the internal field, cf. also Section 9. Thus to gain insight into the properties of the intra-cavity field one needs indirect methods. For this purpose, appropriately prepared atoms are transmitted through the cavity and their electronic quantum states are measured by state-sensitive ionization. By using a Ramsey zone in front of the high-Q cavity and a second one between the cavity and the detector, one may prepare and analyze coherent superpositions of electronic states. In the following we will describe some measurements schemes and consider methods that allow one to obtain insight into the full information on the quantum state of the cavity field.

12.6.1
Quantum state endoscopy

Let us consider a two-level (test) atom that resonantly interacts with the cavity field mode according to the Jaynes–Cummings Hamiltonian (12.3). The atom is initially prepared in an electronic superposition state, $|\pm\rangle = (|1\rangle \pm e^{-i\psi}|2\rangle)/\sqrt{2}$, and the occupation of the excited electronic state after the interaction with the field is measured by state-selective ionization. Repeating the procedure by using a sequence of equally prepared atoms and equal preparation of the cavity mode, one obtains the occupation probabilities $P_2^\pm(t)$ of the electronic state $|2\rangle$ as functions of the interaction time t. Performing these measurements with two different initial preparations of the atoms in the states $|\pm\rangle$, the recorded difference signal is of the form [Vogel, Welsch and Leine (1987)]

$$P_2^+(t) - P_2^-(t) = 2\sum_{n=0}^{\infty} a_n \sin(\Omega_n t), \tag{12.115}$$

where

$$a_n = \frac{e^{i\psi}}{2i}\rho_{n\,n+1} + \text{c.c.}. \tag{12.116}$$

The off-diagonal density-matrix elements $\rho_{n\,n+1}$ can be directly obtained from the coefficients a_n for two phases ψ, such as $\psi=0$ and $\psi=\pi/2$. Provided that the interaction time t can be varied in a sufficiently large interval $(0,T)$, the Fourier transform of $P_2^+(t) - P_1^-(t)$ consists of sharp peaks, whose values yield the sought coefficients a_n as[9]

$$a_n = \frac{2}{T}\int_0^T dt\, \sin(\Omega_n t)[P_2^+(t) - P_2^-(t)] \tag{12.117}$$

($T \to \infty$). To measure the diagonal density-matrix elements ρ_{nn}, it is sufficient to prepare the atom in the excited state, $P_2(t)|_{t=0}=1$, and observe the atomic-state inversion $\Delta P = P_2 - P_1 = 2P_2 - 1$,

$$\Delta P(t) = \sum_{n=0}^{\infty} \rho_{nn} \cos(\Omega_n t). \tag{12.118}$$

[9] If T is not large enough, then the peaks in the Fourier integral contain non-negligible contributions of the tails of the corresponding sinc functions. In this case the coefficients a_n can be calculated from a set of linear equations obtained from Eq. (12.115) for different times.

The photon statistics of the intra-cavity field, ρ_{nn}, can be obtained by Fourier transforming the recorded data $\Delta P(t)$.

The described method directly yields insight in the photon statistics of the cavity field and in the off-diagonal density matrix elements $\rho_{n\,n+1}$ of the field density matrix. In general, the full information on the quantum state would additionally require the elements $\varrho_{n\,n+k}$ for $k>1$. One can overcome the lack of information in special cases where the quantum state of the cavity field is a priori known to be a pure state, $|\psi\rangle = \sum_n c_n |n\rangle$, such that ϱ_{mn} is given by $\varrho_{mn} = c_m c_n^*$ with $c_m c_{m+1}^* \neq 0\ \forall m$.[10] If this condition is fulfilled the determination of the coefficients a_n in Eq. (12.116) renders it possible to determine the expansion coefficients c_n of the pure quantum state, the corresponding method has been called quantum state endoscopy [Bardroff, Mayr and Schleich (1995); Bardroff, Mayr, Schleich, Domokos, Brune, Raimond and Haroche (1996)]. In this case, Eq. (12.115) [together with Eq. (12.116)] can be taken at a sufficiently large number of time points (and at least at two phases) in order to obtain, after truncating the state at a sufficiently large photon number $n_{\max}$, a system of equations for the expansion coefficients c_m, which can be solved numerically.

Important parts of the described technique have already been realized in experiments. In particular, precise measurements of the electronic-state occupations as a function of the time of flight of the atom through the cavity have been performed [Brune, Schmidt-Kaler, Maali, Dreyer, Hagley, Raimond and Haroche (1996)], the corresponding atomic-state inversion being of the type as given in Eq. (12.118). It has been demonstrated that the measured data may indeed be Fourier-analyzed in order to determine the photon number distribution.

12.6.2
QND measurement of the photon number

Another method of measuring the photon-number statistics is based on a quantum-nondemolition (QND) approach [Brune, Haroche, Lefevre, Raimond and Zagury (1990); Brune, Haroche, Raimond, Davidovich and Zagury (1992)]. This can be realized by using three-level Rydberg atoms with states $|a\rangle$, $a = 0, 1, 2$ (with $E_2 > E_1 > E_0$). The transition $|1\rangle \leftrightarrow |2\rangle$ is coupled to the cavity field and the auxiliary $|0\rangle \leftrightarrow |1\rangle$ transition is far off resonance and does not affect the cavity field. Moreover, the cavity mode is sufficiently detuned from the electronic transition frequency ω_{21}, so that the atom–field interaction is of the dispersive type given in Eq. (12.106). Consequently, the interaction in-

---

[10] If $c_m c_{m+1}^*$ is zero for some values of m, one may coherently displace the quantum state to be determined in order to obtain the expansion coefficient of the displaced state and finally one can transform back to the original quantum-state coefficients.

troduces a phase shift of the electronic state $|1\rangle$ relative to $|0\rangle$, which depends on the number of photons in the cavity. This phase shift can be measure by a Ramsey technique. It consists of coherent manipulations of the auxiliary transition $|0\rangle \leftrightarrow |1\rangle$ before and after the interaction of the atom with the cavity field together with a subsequent state-selective ionization. The interference fringes observed in this manner, by using a velocity selected atom beam, depend on the photon-number statistics of the cavity field and the latter can be determined from the measured data.[11]

The method can be more easily realized experimentally when it is used for detecting the presence of a single photon in the cavity. In this case one may replace the (weak) dispersive atom–field coupling of the $|1\rangle \leftrightarrow |2\rangle$ transition with the (stronger) resonant one. Let us consider the situation for the atom initially prepared in the lower state and the cavity containing one photon, $|1,1\rangle$. In this case the resonant interaction according to Eq. (12.111) represents a coherent oscillation,

$$\hat{U}_{\text{int}}(t)|1,1\rangle = \cos\left(\tfrac{1}{2}\Omega_0 t\right)|1,1\rangle + i\sin\left(\tfrac{1}{2}\Omega_0 t\right)|2,0\rangle, \qquad (12.119)$$

between the initial state and the atom being in the upper state, with the cavity being in the vacuum state, $|2,0\rangle$. By fixing the interaction time according to $\tau = 2\pi/\Omega_0$, the atom leaves the cavity in the ground state and the photon remains in the cavity. However, Eq. (12.119) reveals that the phase of the state undergoes a shift by π

$$\hat{U}_{\text{int}}(\tau)|1,1\rangle = e^{i\pi}|1,1\rangle. \qquad (12.120)$$

On the other hand, when the atom is initially in the ground state and the field in the vacuum state, $|1,0\rangle$, the resonant interaction leaves the initial state unchanged. The phase shift caused by the presence of a photon in the cavity is again observed by using the auxiliary $|0\rangle \leftrightarrow |1\rangle$ transition in a Ramsey-type ionization measurement. This resonant interaction scheme has been experimentally realized and it has been possible to see a single photon inside a high-Q cavity without destroying it [Nogues, Rauschenbeutel, Osnaghi, Brune, Raimond and Haroche (1999)].

12.6.3
Determining arbitrary quantum states

To determine arbitrary quantum states, a two-mode nonlinear atomic homodyne detection scheme could be used [Wilkens and Meystre (1993)], in which

---

[11] This method of measuring the photon number statistics could also be extended in order to determine the quantum state of the cavity field in terms of the Wigner function. This requires one to coherently displace the intra-cavity field before detecting the photon number statistics; see Section 12.6.3.

the signal cavity mode is mixed with a local-oscillator cavity mode according to the interaction Hamiltonian

$$\hat{H}_{int} = \hbar\kappa[\hat{\sigma}_+(\hat{a}+\hat{a}_L) + (\hat{a}^\dagger + \hat{a}_L^\dagger)\hat{\sigma}_-]. \tag{12.121}$$

Let assume that the local oscillator mode can be treated classically so that one may replace the operator $\hat{a}_L$ by a c number α_L, $\hat{a}_L \mapsto \alpha_L$. This corresponds to a coherent displacement of the initial state of the cavity mode. In particular, when $|\alpha_L|$ is sufficiently large, then the atomic-state inversion $\Delta P(t)$ after the interaction can be rewritten as

$$\Delta P(t) = \tfrac{1}{2}\left[e^{i2\kappa t|\alpha_L|}\Phi(ie^{i\varphi_L}\kappa t) + \text{c.c.}\right]. \tag{12.122}$$

For $|\alpha_L|\to\infty$ the atomic occupation probabilities $P_{2(1)}(t)$ can be directly related to the characteristic function $\Phi(\beta)$ of the Wigner function $W(\beta)$ of the cavity mode. Varying the interaction time and the phase φ_L of α_L, the whole function $\Phi(\beta)$ can be scanned, in principle. Knowing $\Phi(\beta)$, the Wigner function can then be obtained by Fourier transformation. The scheme was also analyzed by taking into account the quantized nature of the local oscillator. It was found that the classical treatment of the local oscillator restricts the time scale to times less than a vacuum Rabi period [Zaugg, Wilkens and Meystre (1993); Dutra, Knight and Moya-Cessa (1993)].

To avoid this problem of the two-mode-scheme, one can also coherently displace the quantum state of the cavity field before performing the measurements. Thus the initial state of the cavity field $\hat{\rho}$ is replaced by $\hat{D}^\dagger(\alpha)\hat{\rho}\hat{D}(\alpha)$. The Jaynes–Cummings interaction with the (single-mode) cavity field, in place of Eq. (12.118), now yields for the electronic-state inversion

$$\Delta P(t) = \sum_{n=0}^{\infty} \rho_{nn}(-\alpha)\cos(\Omega_n t), \tag{12.123}$$

where $\rho_{nn}(-\alpha) = \langle n|\hat{D}^\dagger(\alpha)\hat{\rho}\hat{D}(\alpha)|n\rangle$. The displaced diagonal matrix elements $\rho_{nn}(-\alpha)$ can again be obtained from $\Delta P(t)$ by Fourier transformation and from $\rho_{nn}(-\alpha)$ the quantum state of the cavity mode can be obtained by applying the methods described in Section 7.3.2. Alternatively, the quantum state can also be reconstructed when the interaction time is left fixed and $\alpha = |\alpha|e^{i\varphi}$ is varied [Bodendorf, Antensberger, Kim and Walther (1998)].

It is interesting that one can even perform a direct measurement of the Wigner function of the cavity field [Lutterbach and Davidovich (1997)]. A velocity-selected atomic beam interacts resonantly with two Ramsey zones placed in front of and behind the high-Q cavity. The frequency of the high-Q cavity is sufficiently detuned from the atomic resonance in order to avoid electronic transitions, the corresponding time evolution operator is given in Eq. (12.106). By applying a microwave generator to the cavity, the field density operator $\hat{\rho}$ of interest is coherently displaced, $\hat{\rho} \mapsto \hat{D}^\dagger(\alpha)\hat{\rho}\hat{D}(\alpha)$. After preparing a superposition of the two electronic states in the first Ramsey zone, during the transmission of the atoms through the displaced cavity field they undergo phase shifts depending on the electronic states and the intra-cavity photon statistics. In the second Ramsey zone the electronic states are again transformed coherently. Eventually, electronic-state sensitive ionization can be used to determine the electronic-state inversion. By appropriately fixing the phases in the Ramsey zones and the interaction time of the atoms in the cavity, the inversion is given by

$$\Delta P = \text{Tr}\left[\hat{D}^\dagger(\alpha)\hat{\rho}\hat{D}(\alpha)(-1)^{\hat{n}}\right] = \tfrac{1}{2}\pi W(\alpha). \tag{12.124}$$

In this manner the measured electronic inversion directly represents the Wigner function of the quantum state of the cavity field in a phase-space point that is chosen by the coherent displacement of the field.

References

Bardroff, P.J., E. Mayr and W.P. Schleich (1995) *Phys. Rev. A* **51**, 4963.

Bardroff, P.J., E. Mayr, W.P. Schleich, P. Domokos, M. Brune, J.M. Raimond and S. Haroche (1996) *Phys. Rev. A* **53**, 2736.

Bodendorf, C.T., G. Antensberger, M.S. Kim and H. Walther (1998) *Phys. Rev. A* **57**, 1371.

Brune, M., E. Hagley, J. Dreyer, X. Maitre, A. Maali, C. Wunderlich, J.M. Raimond and S. Haroche (1996) *Phys. Rev. Lett.* **77**, 4887.

Brune, M., S. Haroche, V. Lefevre, J.M. Raimond and N. Zagury (1990) *Phys. Rev. Lett.* **65**, 976.

Brune, M., S. Haroche, J.M. Raimond, L. Davidovich and N. Zagury (1992) *Phys. Rev. A* **45**, 5193.

Brune, M., J.M. Raimond, P. Goy, L. Davidovich and S. Haroche (1987) *Phys. Rev. Lett.* **59**, 1899.

Brune, M., F. Schmidt-Kaler, A. Maali, J. Dreyer, E. Hagley, J.M. Raimond and S. Haroche (1996) *Phys. Rev. Lett.* **76**, 1800.

Cohen-Tannoudji, C. and S. Reynaud (1977) *J. Phys. B* **10**, 345; 365.

Cummings, F.W. (1965) *Phys. Rev.* **140**, A1051.

Dutra, S.M., P.L. Knight and H. Moya-Cessa (1993) *Phys. Rev. A* **48**, 3168.

Eberly, J.H., N.B. Narozhny and J.J. Sanchez-Mondragon (1980) *Phys. Rev. Lett.* **44**, 1323.

Filipowicz, P., J. Javanainen and P. Meystre (1986) *Phys. Rev. A* **34**, 3077.

Fleischhauer, M. and W. Schleich (1993) *Phys. Rev. A* **47**, 4258.

Gea-Banacloche, J. (1990) *Phys. Rev. Lett.* **65**, 3385.

Hagley, E., X. Maitre, G. Nogues, C. Wunderlich, M. Brune, J.M. Raimond and S. Haroche (1997) *Phys. Rev. Lett.* **79**, 1.

Haroche, S. (1984) in *New Trends in Atomic Physics*, eds G. Grynberg and R. Stora (Elsevier, New York), p. 195.

Haroche, S. and J.M. Raimond (1985) *Adv. At. Mol. Phys.* **20**, 347.

Jaynes, E.T. and F.W. Cummings (1963) *Proc. IEEE* **51**, 89.

Lezama, A., Y. Zhu, S. Morin and T.W. Mossberg (1989) *Phys. Rev. A* **39**, 2754.

Lutterbach L.G. and L. Davidovich (1997) *Phys. Rev. Lett.* **78**, 2547.

Meschede, D., H. Walther and G. Müller (1985) *Phys. Rev. Lett.* **54**, 551.

Meystre, P. and M. Sargent III (1990) *Elements of Quantum Optics* (Springer-Verlag, Berlin).

Nogues, G., A. Rauschenbeutel, S. Osnaghi, M. Brune, J.M. Raimond and S. Haroche (1999) *Nature* **400**, 239.

Paul, H. (1963) *Ann. Phys.* (Leipzig) **11**, 411.

Phoenix, S.J.D. and P.L. Knight (1988) *Ann. Phys.* **186**, 381.

Phoenix, S.J.D. and P.L. Knight (1991) *Phys. Rev. A* **44**, 6023.

Ramsey, N.F. (1985) *Molecular Beams* (Oxford University Press, Oxford), ch. 5.

Rauschenbeutel, A., G. Nogues, S. Osnaghi, P. Bertet, M. Brune, J.M. Raimond and S. Haroche (1999) *Phys. Rev. Lett.* **83**, 5166.

Rauschenbeutel, A., G. Nogues, S. Osnaghi, P. Bertet, M. Brune, J.M. Raimond and S. Haroche (2000) *Science* **288**, 2024.

Rempe, G., F. Schmidt-Kaler and H. Walther (1990) *Phys. Rev. Lett.* **64**, 2783.

Rempe, G., H. Walther and N. Klein (1987) *Phys. Rev. Lett.* **58**, 353.

Rice, P.R. and H.J. Carmichael (1988) *IEEE J. Quant. Electron.* **24**, 1351.

Thompson R.J., G. Rempe and H.J. Kimble (1992) *Phys. Rev. Lett.* **68**, 1132.

Vogel, W. and D.-G. Welsch (1989) *Phys. Rev. A* **40**, 7113.

Vogel, W., D.-G. Welsch and L. Leine (1987) *J. Opt. Soc. Am. B* **4**, 1633.

Wagner, C., R.J. Brecha, A. Schenzle and H. Walther (1992) *Phys. Rev. A* **46**, R5350.

Walther, H. (1992) *Phys. Rep.* **219**, 263.

Weidinger, M., B.T.H. Varcoe, R. Heerlein and H. Walther (1999) *Phys. Rev. Lett.* **82**, 3795.

Wilkens, M. and P. Meystre (1993) *Phys. Rev. A* **43**, 3832.

Yoo, H.-I. and J.H. Eberly (1985) *Phys. Rep.* **118**, 239.

Zaugg, T., M. Wilkens and P. Meystre (1993) *Found. Phys.* **23**, 857.

13
Laser-driven quantized motion of a trapped atom

In many cases of atom–radiation interaction processes the atomic positions can be considered as classically controllable parameters. However, there are also cases where their center-of-mass motion must be treated quantum mechanically. Due to absorption and emission of photons an atom undergoes kicks and thus its motional quantum state is changed by the corresponding momentum transfers onto the atom. These quantum mechanical effects in the interaction of atoms and light can be discarded, if the temperature is too high for their observation. As an example one may estimate the de Broglie wavelength, $\lambda = \hbar/mv$, for atoms of mass m propagating with thermal velocity $v = (3k_B T/m)^{1/2}$. For H atoms at room temperature the resulting λ value is smaller than the atomic Bohr radius. Consequently, the observation of quantum effects of atomic motion requires a regime of extremely low temperatures, which nowadays can be achieved by methods of laser cooling. This has opened exciting new developments in modern physics and new areas of research have been established, such as the field of atom optics.

Among the manifold studies on the quantized motion of ultra-cold atoms we will consider the laser-induced dynamics of a single ion in a Paul trap. Since trapping and observation of a single ion became possible [Neuhauser, Hohenstatt, Toschek and Dehmelt (1980)], the further development of the experimental techniques has created interesting possibilities for preparing and measuring quantum states [see, e. g., Monroe, Meekhof, King and Wineland (1996); Leibfried, Meekhof, King, Monroe, Itano and Wineland (1996)]. The phenomena under consideration show some resemblance to those studied in the preceding chapter for the interaction of a single atom with a quantized cavity field, as the quantized center-of-mass motion of an atom in a trap potential can play a similar role to the cavity field before. However, the laser-induced coupling of electronic and motional degrees of freedom of a trapped atom will exhibit new types of nonlinear effect, which have no counterpart in the interaction of an atom with photons in a cavity.

Quantum Optics, Third, revised and extended edition. Werner Vogel and Dirk-Gunnar Welsch
Copyright © 2006 WILEY-VCH Verlag GmbH & Co. KGaA, Weinheim
ISBN: 3-527-40507-0

13.1
Quantized motion of an ion in a Paul trap

To give an example of a trap, let us consider a quadrupole trap, also known as a Paul trap [Paul, Osberghaus and Fischer (1958)], which is suitable for studying the interaction of a single trapped ion with light. It typically consists of a ring electrode and two end-cap electrodes as shown in Fig. 13.1. Between the ring and end-cap electrodes a direct-current (dc) voltage and a radio-frequency (rf) voltage of frequency ω_{rf} are applied so that an ion of charge Q in the center of the trap experiences the time dependent potential[1]

$$V(x_1, x_2, x_3, t) = Q[Q_{11}(t)x_1^2 + Q_{22}(t)x_2^2 + Q_{33}(t)x_3^2]. \tag{13.1}$$

The diagonal elements $Q_{kk}(t)$ of the traceless quadrupole tensor,

$$\sum_{k=1}^{3} Q_{kk}(t) = 0, \tag{13.2}$$

are given as

$$Q_{kk}(t) = \bar{Q}_{kk}[\zeta_{dc} + \zeta_{rf}\cos(\omega_{rf}t)], \tag{13.3}$$

with ζ_{dc} and ζ_{rf} being the contributions to the potential due to the dc and rf voltage, respectively. It is noteworthy that static trapping of a charged particle in a three-dimensional quadrupole field is impossible due to the validity of Eq. (13.2), which is a direct consequence of the Laplace equation for the scalar potential.

The nonrelativistic equations of the center-of-mass motion for an atom in a time-dependent potential of the type (13.1) are obtained in the form of three decoupled differential equations of the Mathieu type,

$$\ddot{x}_k + \frac{2Q\bar{Q}_{kk}}{m}[\zeta_{dc} + \zeta_{rf}\cos(\omega_{rf}t)]x_k = 0. \tag{13.4}$$

The Mathieu equation has stable and unstable solutions depending on the parameters ζ_{dc} and ζ_{rf} [for stability diagrams see, e. g., Ghosh (1995)]. For a stable solution, the motion described by Eq. (13.4) consists of an oscillation with the frequency ω_{rf}, which is called micromotion, and a usually much slower oscillation with secular frequencies ν_k. Due to the different time scales of these oscillations the micromotion can be neglected in an equation of motion that is averaged over a period of the rf frequency. The resulting averaged equation of motion then describes only the harmonic, secular motion, and the effective potential reads

$$V(x_1, x_2, x_3) = \sum_{k=1}^{3} \tfrac{1}{2} m \nu_k^2 x_k^2, \tag{13.5}$$

---

[1] For simplicity we assume that the principal axes x_k ($k=1,2,3$) of the trap correspond to the Cartesian coordinates.

13.1 Quantized motion of an ion in a Paul trap

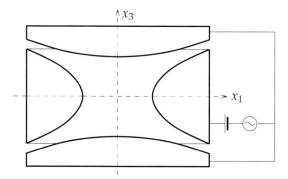

Fig. 13.1 Draft scheme of a Paul trap: Between the end-cap and ring electrodes dc and rf voltages are applied.

where the frequencies ν_k of the secular motion are given by

$$\nu_k = \sqrt{2\left[\left(\frac{QQ_{kk}\tilde{\zeta}_{rf}}{m\omega_{rf}}\right)^2 - \frac{QQ_{kk}\tilde{\zeta}_{dc}}{m}\right]}. \tag{13.6}$$

Stable solutions are obtained for $\tilde{\zeta}_{dc} \ll \tilde{\zeta}_{rf}$, so that the dc term in Eq. (13.6) can be ignored. Therefore, due to Eq. (13.2) the frequencies fulfill the condition

$$\nu_3 = \nu_1 + \nu_2 \tag{13.7}$$

in the typical case when $|Q_{33}|=|Q_{11}+Q_{22}|$. Moreover, in the case of rotational symmetry of the trap with respect to the x_3 axis two frequencies become equal, $\nu_1 = \nu_2$.

The quantization of the motion of an atom in the (effective) potential (13.5) is straightforward. The canonical (center-of-mass) positions and momenta, x_k and p_k, are replaced by the corresponding Hermitian operators, $\hat{x}_k$ and $\hat{p}_k$, respectively, which obey the familiar commutation relations

$$[\hat{x}_k, \hat{p}_{k'}] = i\hbar\delta_{kk'}, \quad [\hat{x}_k, \hat{x}_{k'}] = [\hat{p}_k, \hat{p}_{k'}] = 0. \tag{13.8}$$

Annihilation and creation operators of vibrational quanta, $\hat{a}_k$ and $\hat{a}_k^\dagger$, can be introduced in the usual way,

$$\hat{\mathbf{r}} = \sum_{k=1}^{3} \Delta x_k\, \mathbf{e}_k (\hat{a}_k^\dagger + \hat{a}_k), \quad \hat{\mathbf{p}} = \sum_{k=1}^{3} i\Delta p_k\, \mathbf{e}_k (\hat{a}_k^\dagger - \hat{a}_k), \tag{13.9}$$

where $\mathbf{e}_k$ are the unit vectors in the directions of the principal axes of the trap, and the widths of the ground-state wave functions, Δx_k and Δp_k, in position and momentum basis, respectively, are given by

$$\Delta x_k = \sqrt{\frac{\hbar}{2m\nu_k}}, \quad \Delta p_k = \sqrt{\frac{\hbar m \nu_k}{2}}. \tag{13.10}$$

The Hamiltonian of the free motion of the atomic center-of-mass in the effective potential then results in the standard form of the three-dimensional harmonic oscillator:

$$\hat{H}_{cm} = \sum_{k=1}^{3} \hbar v_k \left(\hat{a}_k^\dagger \hat{a}_k + \tfrac{1}{2} \right). \tag{13.11}$$

13.2
Interaction of a moving atom with light

The quantization of atoms interacting with light has been introduced in Section 2.3. We start from the minimal-coupling Hamiltonian for a hydrogen-type ion, with Q_n (Q_e) and m_n (m_e), respectively, being the charge and the mass of the nucleus (electron). With respect to the resonant interactions considered in the following, we may disregard the $\hat{A}^2$ term in Eq. (2.121) (for details, see Section 2.5) and write

$$\hat{H}_{int} = -\sum_{a=n,e} \frac{Q_a}{m_a} \hat{\mathbf{p}}_a \hat{\mathbf{A}}(\hat{\mathbf{r}}_a). \tag{13.12}$$

In order to distinguish between interactions that rely on electronic transitions and those that act directly on the center-of-mass of the atom, we introduce the center-of-mass coordinate and momentum,

$$\hat{\mathbf{r}} = \frac{m_e \hat{\mathbf{r}}_e + m_n \hat{\mathbf{r}}_n}{m}, \quad \hat{\mathbf{p}} = \hat{\mathbf{p}}_e + \hat{\mathbf{p}}_n, \tag{13.13}$$

and the relative coordinate and momentum,

$$\hat{\mathbf{r}}_{rel} = \hat{\mathbf{r}}_e - \hat{\mathbf{r}}_n, \quad \hat{\mathbf{p}}_{rel} = \frac{m_n \hat{\mathbf{p}}_e - m_e \hat{\mathbf{p}}_n}{m}, \tag{13.14}$$

where $m = m_n + m_e$ is the total mass of the atom. Clearly, the usual commutation relations hold for these pairs of canonical coordinates, and the center-of-mass and relative coordinates commute.

With the help of center-of-mass and relative coordinates the Hamiltonian (13.12) can be rewritten as a sum of two terms,

$$\hat{H}_{int} = \hat{H}_{int}^{(cm)} + \hat{H}_{int}^{(el)}, \tag{13.15}$$

where the Hamiltonian $\hat{H}_{int}^{(cm)}$ contains only the center-of-mass momentum,

$$\hat{H}_{int}^{(cm)} = -\frac{Q_n}{m} \hat{\mathbf{p}} \hat{\mathbf{A}}(\hat{\mathbf{r}}_n) - \frac{Q_e}{m} \hat{\mathbf{p}} \hat{\mathbf{A}}(\hat{\mathbf{r}}_e), \tag{13.16}$$

whereas $\hat{H}_{int}^{(el)}$ contains the relative momentum,

$$\hat{H}_{int}^{(el)} = \frac{Q_n}{m_n} \hat{\mathbf{p}}_{rel} \hat{\mathbf{A}}(\hat{\mathbf{r}}_n) - \frac{Q_e}{m_e} \hat{\mathbf{p}}_{rel} \hat{\mathbf{A}}(\hat{\mathbf{r}}_e). \tag{13.17}$$

The two terms, Eq. (13.16) and (13.17), are responsible for different types of interactions of the atom with radiation. Whereas the former describes the interaction of the charged center-of-mass of the atom with radiation, the latter describes the internal, electronic transitions that are typically induced by (optical) radiation.

13.2.1
Radio-frequency radiation

Let us first consider the interaction Hamiltonian (13.16). Since an ion has a nonvanishing total charge of $Q = Q_n + Q_e$, its center-of-mass motion in the trapping potential can be resonantly driven by use of radio-frequency fields with frequencies of the order of the vibrational trap frequencies, i.e., 10–100 MHz. The wavelength of such a radio-frequency field is typically very large compared with the spatial extension of the atomic center-of-mass wave function in the trap potential. Therefore, the spatial dependence of the vector potential in Eq. (13.16) can be neglected altogether and we obtain

$$\hat{H}_{\text{int}}^{(\text{cm})} = -\frac{Q}{m}\hat{\mathbf{p}}\hat{\mathbf{A}}, \tag{13.18}$$

where $\hat{\mathbf{A}} \equiv \hat{\mathbf{A}}(0)$ is the vector potential taken at the origin of the trap potential. Performing a rotating-wave approximation (cf. Section 2.5.2) with respect to the vibrational frequencies ν_k, we arrive at

$$\hat{H}_{\text{int}}^{(\text{cm})} = -\sum_{k=1}^{3} \mathbf{d}_k \hat{\mathbf{E}}^{(+)} \hat{a}_k^\dagger + \text{H.c.}. \tag{13.19}$$

Here the positive-frequency part $\hat{\mathbf{E}}^{(+)}$ of the electric-field strength of the rf field is typically considered as a classical field, $\hat{\mathbf{E}}^{(+)} \mapsto \mathbf{E}^{(+)}(t)$. The direct coupling of the field to a trapped ion appears in the form of a dipole interaction. The effective electric dipole moment of the ion in the vibrational ground state is defined by

$$\mathbf{d}_k = Q\Delta x_k \mathbf{e}_k. \tag{13.20}$$

It is the electric dipole moment produced by the vibrational ground-state fluctuations of the ion. It is noteworthy that the corresponding unitary time evolution governed by the Hamiltonian (13.19) may represent a coherent displacement of the atomic center-of-mass wave packet, provided that a monochromatic microwave field is used and tuned on resonance with the motional frequency under consideration.

Clearly, such a dipole-type coupling to radiation in the radio-frequency regime will also lead to a damping of the vibrational motion by spontaneous emission of quanta of radio-frequency radiation. However, the characteristic damping rate corresponding to this process,

$$\Gamma_k^{(\text{cm})} = \frac{v_k^3 d_k^2}{3\hbar\pi c^3 \epsilon_0} = \frac{v_k^2 Q^2}{6\pi m c^3 \epsilon_0} \tag{13.21}$$

[Eq. (10.29) together with Eqs (13.20) and (13.10)], is negligibly small for such frequencies.

13.2.2
Optical radiation

The interaction of an atom with optical radiation is described by (the second part of) the interaction Hamiltonian (13.17). Using a traveling-mode expansion of the vector potential according to Eq. (2.87),

$$\hat{\mathbf{A}}(\hat{\mathbf{r}}_a) = \sum_{l,\sigma} \mathbf{A}_{l,\sigma}(\hat{\mathbf{r}}_a) \hat{b}_{l,\sigma} + \text{H.c.}, \tag{13.22}$$

with $\hat{b}_{l,\sigma}$ being the photon annihilation operators and the mode functions given by

$$\mathbf{A}_{l,\sigma}(\hat{\mathbf{r}}_a) = \sqrt{\frac{\hbar}{2\epsilon_0 c k_l V}} \, \mathbf{e}_{l,\sigma} e^{i\mathbf{k}_l \hat{\mathbf{r}}_a}, \tag{13.23}$$

we may rewrite the interaction Hamiltonian (13.17) in the form

$$\hat{H}_{\text{int}}^{(\text{el})} = \sum_{l,\sigma} \mathbf{A}_{l,\sigma}(\hat{\mathbf{r}}) \hat{\mathbf{M}}_l \hat{b}_{l,\sigma} + \text{H.c..} \tag{13.24}$$

Here the (vector) operators $\hat{\mathbf{M}}_l$ are defined by

$$\hat{\mathbf{M}}_l = \left[\frac{Q_n}{m_n} \exp\left(-i\mathbf{k}_l \hat{\mathbf{r}}_{\text{rel}} \frac{m_e}{m}\right) - \frac{Q_e}{m_e} \exp\left(i\mathbf{k}_l \hat{\mathbf{r}}_{\text{rel}} \frac{m_n}{m}\right) \right] \hat{\mathbf{p}}_{\text{rel}}. \tag{13.25}$$

Representing the interaction Hamiltonian (13.24) in the basis of the internal (electronic) energy eigenstates $|i\rangle$ of the trapped atom yields ($\hat{A}_{ij} = |i\rangle\langle j|$)

$$\hat{H}_{\text{int}}^{(\text{el})} = \sum_{ij} \sum_{l,\sigma} \mathbf{A}_{l,\sigma}(\hat{\mathbf{r}}) \langle i|\hat{\mathbf{M}}_l|j\rangle \hat{A}_{ij} \hat{b}_{l,\sigma} + \text{H.c..} \tag{13.26}$$

Clearly, Eq. (13.26) also holds for neutral atoms that are magneto-optically trapped.

In the electric-dipole approximation the exponential operators in Eq. (13.25) can be set equal to unity, leading to the familiar result

$$\langle i|\hat{\mathbf{M}}_l|j\rangle = -i\omega_{ij}\langle i|\hat{\mathbf{d}}|j\rangle \tag{13.27}$$

(cf. Section 2.5.1), where $\omega_{ij}=\omega_i-\omega_j$ is the transition frequency between the electronic states $|i\rangle$ and $|j\rangle$. In particular, for classically describable laser radiation which is near resonant to an electronic transition $|1\rangle \leftrightarrow |2\rangle$ and whose electric field has a positive-frequency amplitude $E_L^{(+)}(t)$ and wave vector $\mathbf{k}_L$ [that is, $\hat{\mathbf{E}}^{(+)}(\hat{\mathbf{r}}) \mapsto \mathbf{E}_L^{(+)}(t)e^{i\mathbf{k}_L\hat{\mathbf{r}}}$], the interaction Hamiltonian (13.26) then reduces, in rotating-wave approximation with respect to the electronic transition, to the familiar dipole coupling term

$$\hat{H}_{\text{int}}^{(\text{el})} = -d_{21}E_L^{(+)}(t)e^{i\mathbf{k}_L\hat{\mathbf{r}}}\hat{A}_{21} + \text{H.c.}, \tag{13.28}$$

where d_{21} is the projection of the electric-dipole matrix element in the direction of the electric-field amplitude, and the exponentials $e^{i\mathbf{k}_L\hat{\mathbf{r}}}$ are responsible for the momentum recoil of the atom during absorption and emission of photons of momentum $\hbar\mathbf{k}_L$. Note that the basic structure of the interaction Hamiltonian also holds for electric-dipole forbidden transitions, such as magnetic dipole or electric quadrupole transitions or other multipole transition. Then only the electric dipole moment has to be replaced by the appropriate expression for the type of transition under consideration.

13.3
Dynamics in the resolved sideband regime

For the realization of resonant interactions of a single trapped atom with laser fields, the so-called resolved sideband regime is required – a regime able to resolve the energy levels of the vibrational levels of the trapped atom. Besides laser-induced electronic transitions, vibronic transitions, that is, transitions of both the vibrational and electronic states, and pure vibrational transitions can also occur, as will be shown in the following.

13.3.1
Nonlinear Jaynes–Cummings model

The interaction Hamiltonian (13.26) describes rather complex coupling effects of quantized light, quantized atomic motion and electronic states, in general. When the atom is driven by a strong, quasi-monochromatic laser field, then the interaction Hamiltonian can be further simplified. First, the field operators can be replaced by c-numbers, as done when arriving at the Hamiltonian (13.28). Second, if the laser linewidth is small enough, one may address individual vibronic transitions. Such interactions in the resolved (vibrational) sideband regime allow one to further simplify the Hamiltonian by performing a vibrational rotation-wave approximation, resulting in a nonlinear Jaynes–Cummings model [Vogel and de Matos Filho (1995)]. It provides an exactly solvable model for various types of vibronic interaction in a strongly nonlinear regime. In particular, in the Lamb–Dicke limit, when the quantized atomic motion is only weakly affected by momentum transfer due to the emission and absorption of photons, the model reduces to the standard Jaynes–Cummings model [Blockley, Walls and Risken (1992)] used in Chapter 12 to describe the atom–field interaction in a cavity.

Let us consider the resonant interaction of an electronic transition of the trapped atom with a strong laser and focus on the electric-dipole approximation. In this case the Hamiltonian reads

$$\hat{H} = \hat{H}_0 + \hat{H}_{\text{int}}, \tag{13.29}$$

where $\hat{H}_0$ describes the free internal (electronic) and external (center-of-mass) motion according to Eq. (13.11), and $\hat{H}_{\text{int}} \equiv \hat{H}_{\text{int}}^{(\text{el})}$, with $\hat{H}_{\text{int}}^{(\text{el})}$ being given by Eq. (13.28). The laser field is assumed to be monochromatic and its frequency ω_L is quasi-resonant with the transition frequency ω_{21} of the electronic transition $|1\rangle \leftrightarrow |2\rangle$, i.e., $\omega_L \approx \omega_{21}$. We further assume that the wave vector $\mathbf{k}_L$ of the driving laser has only a component along one principal axis of the trap, say the x axis, so that only one motional degree of freedom appears in the interaction Hamiltonian. In this case the interaction Hamiltonian simplifies to[3]

$$\hat{H}_{\text{int}} = \kappa \hat{A}_{21} E_L e^{-i\omega_L t} \hat{g}(\hat{x}) + \text{H.c.}, \tag{13.30}$$

where $\kappa = -d_{21}$ and

$$\hat{g}(\hat{x}) = e^{ik_L \hat{x}}. \tag{13.31}$$

For a harmonic trap potential, the position operator $\hat{x}$ can be expressed, according to Eqs (13.9) and (13.10), in terms of the annihilation and creation operators in the x direction, $\hat{a}$ and $\hat{a}^\dagger$ respectively, as

$$k_L \hat{x} = \eta(\hat{a} + \hat{a}^\dagger), \tag{13.32}$$

where the Lamb–Dicke parameter

$$\eta = k_L \Delta x = \frac{\hbar k_L}{\Delta p} \tag{13.33}$$

is a measure of the spread in position, Δx, of the center-of-mass wave function of the atom in the ground state of the trap potential relative to the wavelength

---

[3] Note that this interaction Hamiltonian can also be used to describe the Raman excitation of an electric-dipole-forbidden electronic transition [Toschek (1985); Lindberg and Javanainen (1986); Heinzen and Wineland (1990)]. In this case κ is the Raman coupling strength, the field E_L is replaced by the product $E_1 E_2^*$ of the complex amplitudes of the two Raman lasers, and ω_L and k_L are the frequency and wave number, respectively, of the beat node of the two lasers.

$\lambda_L = 2\pi/k_L$ of the driving laser. Alternatively, η can be viewed as the ratio of the photon momentum to the spread Δp of the atomic momentum distribution in the motional ground state. That is, in the Lamb–Dicke regime, where η is small, the effect of the momentum transfer on the atomic wave packet due to absorption or emission of a single photon is also small. By using the Baker–Campbell–Hausdorff formula (C.27) we may rewrite Eq. (13.31), together with Eq. (13.32), in the form

$$\hat{g}(\hat{x}) = e^{-\eta^2/2} e^{i\eta \hat{a}^\dagger} e^{i\eta \hat{a}} = e^{-\eta^2/2} \sum_{l,m=0}^{\infty} \frac{(i\eta)^{l+m}}{l!m!} \hat{a}^{\dagger l} \hat{a}^m. \quad (13.34)$$

Inserting this result into Eq. (13.30) and transforming to the interaction picture, $\hat{H}_{\text{int}} = \hat{U}_0^\dagger(t) \hat{H}_{\text{int}} \hat{U}_0(t)$, yields

$$\hat{H}_{\text{int}} = \tfrac{1}{2}\hbar \Omega_L \hat{A}_{21} e^{-\eta^2/2} \sum_{l,m=0}^{\infty} \frac{(i\eta)^{l+m}}{l!m!} \hat{a}^{\dagger l} \hat{a}^m e^{-i[\omega_L - \omega_{21} + (m-l)\nu]t} + \text{H.c.}, \quad (13.35)$$

where $\hat{U}_0(t) = \exp(-i\hat{H}_0 t/\hbar)$, and $\Omega_L = 2\kappa E_L/\hbar$ describes the interaction of the laser with the electronic transition $|1\rangle \leftrightarrow |2\rangle$.[3]

The interaction Hamiltonian (13.35) consists of a variety of contributions, which oscillate at multiples of the vibrational frequency ν. Let us now assume that the laser is resonant with the kth vibrational sideband,

$$\omega_L = \omega_{21} - k\nu, \quad (13.36)$$

and we are in the resolved sideband regime, $\nu \gg \Omega_L, \Gamma$, with Γ being a representative measure of the linewidths of the vibronic transitions which result from the natural linewidths and the linewidth of the irradiating laser. This condition implies that one can address a series of vibronic transitions of equal transition frequencies, in the chosen case the $|1,n\rangle \leftrightarrow |2,n-k\rangle$ transitions, independently of all other transitions. Consequently, in the resolved-sideband regime one can perform a vibrational rotating-wave approximation by neglecting, in the interaction Hamiltonian, all those terms oscillating with the vibrational frequency ν and multiples of it. Thus we obtain from Eq. (13.35), by use of the condition (13.36) and choosing $k \geq 0$, the approximate interaction Hamiltonian

$$\hat{H}_{\text{int}} = \tfrac{1}{2}\hbar\Omega_L \hat{A}_{21} \hat{f}_k(\hat{n};\eta) \hat{a}^k + \text{H.c.}, \quad (13.37)$$

where the operator-valued function $\hat{f}_k(\hat{n};\eta)$ reads

$$\hat{f}_k(\hat{n};\eta) = e^{-\eta^2/2} \sum_{l=0}^{\infty} \frac{(i\eta)^{2l+k}}{l!(l+k)!} :\hat{n}^l: . \quad (13.38)$$

[3] Note that $|\Omega_L|$ corresponds to the Rabi frequency Ω_R as defined by Eq. (11.25).

The interaction Hamiltonian (13.37) [together with Eq. (13.38)] represents a nonlinear k-quantum Jaynes–Cummings model [Vogel and de Matos Filho (1995)]. Since it describes the mutual coupling of a closed set of vibronic states allowing for transitions $|1, n\rangle \leftrightarrow |2, n-k\rangle$, it can be exactly diagonalized in a similar way to the multi-photon Jaynes–Cummings model [see Section 12.1, Eqs (12.27)–(12.32)]. Whereas in cavity QED the model describes the emissions and absorptions of photons due to electronic transitions, here it describes emission and absorption of k vibrational quanta when the atom undergoes electronic transitions.

Besides describing multi-quantum emissions/absorptions, the nonlinear k-quantum Jaynes–Cummings model depends in a nonlinear manner on the vibrational excitation. Clearly, in order to observe the nonlinear effects, the Lamb–Dicke parameter must not be too small. Let us consider a situation where the Lamb–Dicke parameter is large enough so that the spatial extension of the motional wave function is not small compared with the wavelength of the driving laser field. For sufficiently large motional quantum numbers the (rather extended) motional wave function therefore overlaps with a substantially large fraction of the wavelength of the laser wave and it may simultaneously experience the positive and negative values of the electric field strength. As a consequence, the Jaynes–Cummings-type interaction may, for appropriately chosen vibrational excitations of the atom in the trap potential, effectively average to zero. This behavior periodically occurs as a function of the motional quantum number of the atom.

So far we have used the resonance condition (13.36) for the case $k \geq 0$, which includes the irradiation of the carrier on the kth red sideband. For tuning the laser at resonance with the kth blue sideband, $k < 0$, the interaction Hamiltonian (13.35) together with the resonance condition (13.36) in the vibrational rotating-wave approximation leads to

$$\hat{H}_{\text{int}} = \tfrac{1}{2} \hbar \Omega_L \hat{A}_{21} \hat{a}^{\dagger |k|} \hat{f}_{|k|}(\hat{n}; \eta) + \text{H.c.}, \tag{13.39}$$

which also describes the mutual coupling of the vibronic states of a closed set of states: $|1, n\rangle \leftrightarrow |2, n+|k|\rangle$. Thus it can easily be diagonalized in the same manner to the interaction Hamiltonian (13.37).

Let us consider the excitation of the first red vibrational sideband by choosing $k=1$ in Eq. (13.37),

$$\hat{H}_{\text{int}} = \tfrac{1}{2} \hbar \Omega_L \hat{A}_{21} \hat{f}_1(\hat{n}; \eta) \hat{a} + \text{H.c.}. \tag{13.40}$$

This interaction Hamiltonian describes the coupling of the transitions $|2, n\rangle \leftrightarrow |1, n+1\rangle$, similar to the standard Jaynes–Cummings model. It can be diagonalized in the way as described in Section 12.1, provided that the detuning δ of the laser is small compared with the motional frequency, $\delta \ll \nu$.

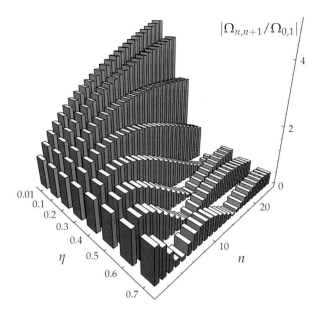

Fig. 13.2 Absolute values of the Rabi frequencies $\Omega_{n,n+1}$ of the nonlinear one-quantum transition for various values of the Lamb–Dicke parameter η. [After Vogel and de Matos Filho (1995).]

Moreover, the Rabi frequencies $\Omega_n = 2\lambda\sqrt{n+1}$ are replaced with the vibronic Rabi frequencies (assuming Ω_L to be real)

$$\Omega_{n,n+1} = \frac{2}{i\hbar}\langle 2,n|\hat{H}_{\text{int}}^{(1)}|1,n+1\rangle = \Omega_L \frac{\eta}{\sqrt{n+1}} L_n^1(\eta^2) e^{-\eta^2/2}. \quad (13.41)$$

The dependence on n of $|\Omega_{n,n+1}|$ is illustrated in Fig. 13.2. In the Lamb–Dicke regime, where the atomic center-of-mass wave function is well localized with respect to the wavelength of the driving laser field, i.e., $\eta\sqrt{n+1} \ll 1$, Eq. (13.41) simplifies to $\Omega_{n,n+1} \simeq \eta\Omega_L\sqrt{n+1}$, which corresponds to the standard Jaynes–Cummings model. The conditions for the Lamb–Dicke regime are well fulfilled for the curve shown for $\eta = 0.01$ in the figure. For somewhat larger Lamb–Dicke parameters the Rabi frequencies are increasing (as functions of n) more slowly than in the Lamb–Dicke regime. When the Lamb–Dicke parameter is further increased, see, e.g., the curve for $\eta = 0.5$, the Rabi frequencies show an oscillating behavior. It represents the nonlinear effects arising from the spatial extension of the atomic wave function.

The occupation probability of the excited electronic state, $\sigma_{22}(t)$, for the laser-driven atom can be obtained from the results derived for the standard Jaynes–Cummings model in Section 12.2 [by replacing again Ω_n with the nonlinear Rabi frequencies $\Omega_{n,n+1}$, as given in Eq. (13.41)]. One may also observe collapse and revival phenomena. However, the nonlinear Rabi frequencies allow one to realize various kinds of revivals. For example, for $\eta = 0.5$ and around $n = 10$ the function $\Omega_{n,n+1}$ is almost linear in n. As is shown in Section 12.2.2, the almost linear dependence on n of the Rabi frequency in the two-photon Jaynes–Cummings model implies almost complete revivals. A similar behavior is expected in the nonlinear (one-quantum) model, e. g., by choosing $\eta = 0.5$ and preparing the motion initially in a coherent state $|\alpha\rangle$ of mean motional quantum number $\langle \hat{n} \rangle = |\alpha|^2 \approx 10$ [Vogel and de Matos Filho (1995)]. This example shows that the nonlinearities in the trapped-ion model can be used to considerably modify the dynamics known from the Jaynes–Cummings model in cavity QED.

13.3.2
Decoherence effects

The nonlinear Jaynes–Cummings model was first realized in a trapped-ion experiment [Meekhof, Monroe, King, Itano and Wineland (1996)], with the Jaynes–Cummings transition being driven in a Raman configuration as illustrated in Fig. 13.3. After a (variably) chosen atom–field interaction time the electronic ground-state occupation is measured by testing an auxiliary transition for the occurrence of fluorescence. To be more specific, the ion is driven by a bichromatic laser field,

$$E_L^{(+)}(\hat{x}, t) \mapsto \mathcal{E}_L e^{i(k_L \hat{x} - \omega_L t)} + \mathcal{E}_L' e^{i(k_L' \hat{x} - \omega_L' t)}, \tag{13.42}$$

where $\mathcal{E}_L, \mathcal{E}_L'$ are the field amplitudes, k_L, k_L' are the x components of the wave vectors, and ω_L, ω_L' are the frequencies of the two fields. For a sufficiently large detuning Δ of the two fields from the electronic state $|3\rangle$, the latter can be eliminated. The resonance condition in the case when the Raman lasers drive the first blue motional sideband is $\omega_L - \omega_L' = \omega_{21} + \nu$. The interaction Hamiltonian is of the form of Eq. (13.39) for $k = -1$, with Ω_L representing now the coupling strength of the Raman interaction. The transitions $|1\rangle \leftrightarrow |2\rangle$ and $|1\rangle \leftrightarrow |3\rangle$ are (electric) dipole transitions with the spontaneous decay rates γ and γ', respectively. Note that the auxiliary state $|3\rangle$ is used to enhance the coupling strength of the Raman scheme.

Even when the detuning Δ is quite large, such as approximately 500 natural linewidths in the experiment, from time to time an excitation of the auxiliary state may occur. Due to the short lifetime of this state, compared with the time scale of the Raman interaction, the excited atom will immediately undergo a spontaneous decay towards either the state $|1\rangle$ or $|2\rangle$. Assuming equally strong decay channels, $\gamma \approx \gamma'$, the probabilities of the two transitions become equal. Let us consider the case where the trapped atom is initially cooled to the motional ground state in the electronic state $|1\rangle$, that is to the vibronic state $|1, 0\rangle$. As long as no excitation of the auxiliary electronic state occurs, the

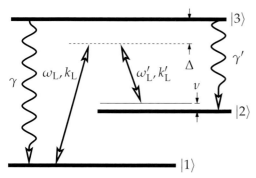

Fig. 13.3 Raman excitation scheme of the $|1\rangle \leftrightarrow |2\rangle$ transition.

Raman interaction leads to coherent oscillations of the vibronic $|1,0\rangle \leftrightarrow |2,1\rangle$ transition.

To consider the effect of an excitation of the electronic state $|3\rangle$, let assume that the Lamb–Dicke parameter η is sufficiently small. In the experiment counter-propagating laser beams with an effective value of η of about 0.2 were used. For spontaneous emission this leads to values of about 0.1 for the Lamb–Dicke parameters. In this case the leading transitions approximately preserve the motional quantum state. Hence among the dominant vibronic transitions including the auxiliary state $|3\rangle$ are either $|1,0\rangle \to |3,0\rangle \to |2,0\rangle$ or $|2,1\rangle \to |3,1\rangle \to |1,1\rangle$. These two possibilities are of particular interest since they switch the phase of the coherent oscillation of the electronic transition $|1\rangle \leftrightarrow |2\rangle$ by the value of π, which leads to a substantial decoherence effect via dephasing. In addition, in the case of the first type of transition the system is trapped in the motional ground state of the second electronic state, $|2,0\rangle$. As can be seen from Fig. 13.3, when the system is in this state it is no longer resonantly coupled by the Raman interaction to the electronic ground state $|1\rangle$. The theoretical results found by quantum trajectory simulations [Di Fidio and Vogel (2000)] are in agreement with the experimental ones,[4] as can be seen from Fig. 13.4. Both the damping of the coherent oscillation and the asymmetry in the decay behavior are sufficiently well reproduced. The results show that the discontinuous interruption of the coherent dynamics, by quantum jumps between the laser-driven Jaynes–Cummings transition and the auxiliary electronic state, can be regarded as the main mechanism for the observed decoherence. It turns out that a very small number of quantum jumps can lead to significant damping effects.

---

4) For a more detailed experimental study of the decoherence effects, see Ozeri, Langer, Jost, DeMarco, Ben-Kish, Blakestad, Britton, Chiaverini, Itano, Hume, Leibfried, Rosenband, Schmidt and Wineland (2005).

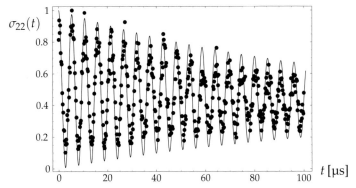

Fig. 13.4 Damped dynamics of the Raman-driven trapped atom initially prepared in the vibronic state $|1,0\rangle$. The full lines represent the theory. [After Di Fidio and Vogel (2000).] The dots are experimental data. [After Meekhof, Monroe, King, Itano and Wineland (1996).]

Nevertheless there are open questions, particularly when the atom is initially in an excited vibrational state. In the experiment an increase in the decoherence rate with increasing excitation of the initial center-of-mass motion of the trapped atom was observed. The measured dependence of the decoherence rate Γ on the motional quantum number n was inferred from the experimental data to be

$$\Gamma_n \simeq \Gamma_0 (n+1)^{0.7}, \tag{13.43}$$

where Γ_0 is the damping rate for the atom initially in the motional ground state. The value of Γ_0 can be easily explained, together with the mentioned asymmetric decay, by the mechanism of quantum jumps. One could try to explain the relation (13.43) by combining the quantum jumps with additional classical stochastic effects.[5] However, any such noise may be expected to further increase the value of the rate Γ_0, contrary to the experimental result. Altogether this situation reflects the difficulties in precisely understanding the mechanisms of decoherence to an extent needed for applications of such systems in quantum information processing, even in the seemingly simple case of a single trapped atom. Hence the study of mechanisms of decoherence is a subject that requires further research.

13.3.3
Nonlinear motional dynamics

Let us consider a scheme that renders it possible to drive nonlinearly the motional state of an atom in a trap potential, without significantly affecting the electronic state. Two lasers of wave vectors $\mathbf{k}_L$, $\mathbf{k}'_L$ and frequencies

---

5) For such an attempt, see Budini, de Matos and Zagury (2002).

ω_L, $\omega_L' = \omega_L + \delta\omega$ ($\delta\omega \ll \omega_L$), drive the atom in a Raman scheme as shown in Fig. 13.5. The two lasers are assumed to be sufficiently far detuned by $\Delta = \omega_{21} - \omega_L$ from the closest neighboring dipole transition $|1\rangle \leftrightarrow |2\rangle$ in order to avoid a population of the excited state $|2\rangle$. On the other hand, they should be close enough, $\Delta/\omega_{21} \ll 1$, to get a sufficiently strong Raman coupling that is dominated by this transition.

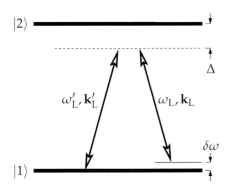

Fig. 13.5 Raman excitation scheme for driving the quantized motion in a nonlinear manner.

Under these conditions one may adiabatically eliminate the excited electronic state from the equations of motion. Since the atom is assumed to be initially in the electronic ground state, one may eventually trace over the electronic states, similar to the case of effective nonlinear interaction Hamiltonians considered in Section 2.5.3. This yields an effective interaction Hamiltonian which contains only the position operator $\hat{\mathbf{r}}$ of the motional subsystem of the atom,

$$\hat{H}_{\text{int}} = \tfrac{1}{2}\hbar\Omega e^{-i(\delta\omega t - \delta\mathbf{k}\hat{\mathbf{r}})} + \text{H.c.}, \qquad (13.44)$$

where $\delta\mathbf{k} = \mathbf{k}_L - \mathbf{k}_L'$. The effective two-photon coupling strength is given in terms of the single-photon coupling strengths Ω_L, $\Omega_{L'}$ of the two lasers as

$$\Omega = \frac{\Omega_L \Omega_L'^*}{2\Delta}. \qquad (13.45)$$

The phase φ of $\Omega = |\Omega|e^{i\varphi}$ is determined by the phase difference of the two lasers, so that it can be held very stable.

The further procedure is similar to that in Section 13.3.1, but now we deal with three motional directions. By using Eq. (13.9), the term $\delta\mathbf{k}\hat{\mathbf{r}} = \sum_{i=1}^{3} \eta_i(\hat{a}_i + \hat{a}_i^\dagger)$ can be decomposed into three terms corresponding to the motion along the three principal axes of the trap, with η_i being the Lamb–Dicke parameters in these directions. The corresponding exponentials in Eq. (13.44) can be factorized for each degree of freedom according to

Eqs (13.31) and (13.34). This yields an interaction Hamiltonian in the normally ordered form:

$$\hat{H}_{int} = \tfrac{1}{2}\hbar\Omega\, e^{-i\delta\omega t}\, e^{-(\eta_1^2+\eta_2^2+\eta_3^2)/2}$$

$$\times \sum_{mm'}\sum_{nn'}\sum_{ll'} \frac{(i\eta_1)^{m+m'}(i\eta_2)^{n+n'}(i\eta_3)^{l+l'}}{m!\,m'!\,n!\,n'!\,l!\,l'!}\, \hat{a}_1^{\dagger m}\hat{a}_2^{\dagger n}\hat{a}_3^{\dagger l}\,\hat{a}_1^{m'}\hat{a}_2^{n'}\hat{a}_3^{l'} + \text{H.c.}. \quad (13.46)$$

It reveals that manifold nonlinear motional couplings can be induced by the Raman scheme under study. It is worth noting that, compared to the situation in nonlinear optics, here the strengths of the nonlinearities can be controlled more easily by the values of the Lamb–Dicke parameters. This allows one to realize rather strong nonlinearities.

Choosing the laser beat frequencies to be multiples of the three vibrational frequencies,[6]

$$\delta\omega = s_1 \nu_1 + s_2 \nu_2, \quad s_{1,2} = 0, \pm 1, \pm 2, \dots, \quad (13.47)$$

we may apply the vibrational rotating-wave approximation. Further, we focus on the nondegenerate case in which the ratio ν_2/ν_1 cannot be expressed in terms of a rational number p/q ($p,q=1,2,3,\dots$).[7] Transforming the interaction Hamiltonian (13.46) into the interaction picture, we then obtain, on discarding those contributions that are oscillating with the vibrational frequencies,

$$\hat{H}_{int} = \tfrac{1}{2}\hbar\Omega \sum_n \hat{g}_{n-s_1}(\hat{a}_1^\dagger,\hat{a}_1;\eta_1)\, \hat{g}_{n-s_2}(\hat{a}_2^\dagger,\hat{a}_2;\eta_2)\, \hat{g}_n(\hat{a}_3^\dagger,\hat{a}_3;\eta_3) + \text{H.c.}, \quad (13.48)$$

where the operator-valued function $\hat{g}_k(\hat{a}^\dagger,\hat{a};\eta)$ is defined by

$$\hat{g}_k(\hat{a}^\dagger,\hat{a};\eta) = \begin{cases} \hat{a}^{\dagger|k|}\,\hat{f}_{|k|}(\hat{n};\eta) & \text{if } k \geq 0, \\ \hat{f}_{|k|}(\hat{n};\eta)\,\hat{a}^{|k|} & \text{if } k < 0, \end{cases} \quad (13.49)$$

with $\hat{f}_k(\hat{n};\eta)$ from Eq. (13.38) [Wallentowitz and Vogel (1997)]. The interaction Hamiltonian (13.48) governs, similar to nonlinear optics, a variety of nonlinear phenomena. They are controlled by the resonance condition (13.47) and by the Lamb–Dicke parameters.

Let us consider, for example, the quantum mechanical counterpart of an optical parametric oscillator. The Lamb–Dicke parameter η_3 can be set equal to zero by using a laser-beam geometry in which the projection of the difference wave vector $\delta\mathbf{k}$ on the x_3 axis becomes zero. In this case only the x_1 and x_2 motional components couple to each other. By choosing the detuning [Eq. (13.47)]

---

6) This choice is the most general one since for a quadrupole potential the relation $\nu_3 = \nu_1 + \nu_2$ is valid, cf. Eq. (13.7).
7) As long as the Lamb–Dicke parameters are sufficiently small this assumption must be fulfilled for small numbers p,q only.

as $\delta\omega = 2\nu_1 - \nu_2$ ($s_1 = 2$, $s_2 = -1$), the interaction Hamiltonian (13.48) reduces to

$$\hat{H}_{int} = \tfrac{1}{2}\hbar\Omega \hat{f}_2(\hat{n}_1;\eta_1)\hat{a}_1^2\hat{a}_2^\dagger \hat{f}_1(\hat{n}_2;\eta_2) + \text{H.c.}. \tag{13.50}$$

In contrast to an ordinary optical parametric oscillator, here the coupling constant is given in terms of the excitation-dependent operator functions $\hat{f}_1$ and $\hat{f}_2$. For sufficiently small Lamb–Dicke parameters, $\eta_{1,2} \ll 1$, Eq. (13.50) can be further simplified to obtain

$$\hat{H}_{int} = -\tfrac{i}{2}\hbar\eta_1^2\eta_2\Omega \hat{a}_1^2\hat{a}_2^\dagger + \text{H.c.} \tag{13.51}$$

[Agarwal and Banerji (1997)], which corresponds to the standard form of the interaction Hamiltonian of an optical parametric oscillator.

An optical parametric oscillator usually requires a strong pump mode in order to get a sufficiently strong coupling. In this case one may replace, in the so-called parametric approximation, the pump-mode operator $\hat{a}_2^\dagger$ with a c number α_2^*. The unitary time evolution then corresponds to the action of the squeeze operator (3.102). On the other hand, the interaction Hamiltonian (13.51) for a trapped atom allows one to realize the parametric dynamics in the quantum regime. However, even for small values of the Lamb–Dicke parameters the applicability of this interaction Hamiltonian is limited to sufficiently low excitations of the involved vibrational modes. In fact, the validity of Eq. (13.51) requires that $\eta_i^2 N_i \ll 1$ ($i = 1, 2$), where N_i is an upper limit for the excitation of the ith motional mode. Whenever higher excited motional states become relevant, the dynamics must be described by the interaction Hamiltonian (13.50) containing the nonlinear excitation-dependence of the coupling strength.

To illustrate the effect of the nonlinear excitation-dependence of the coupling strength as given in the interaction Hamiltonian (13.48), let us assume that the difference wave vector $\delta \mathbf{k}$ points in the x_1 direction so that $\eta_2 = \eta_3 = 0$. Furthermore, assuming the Raman detuning of the two lasers is chosen such that a k quantum transition is directly driven, $\delta\omega = k\nu$ ($k \geq 0$), we may approximate Eq. (13.48) by

$$\hat{H}_{int} = \tfrac{1}{2}\hbar\Omega \hat{f}_k(\hat{n};\eta)\hat{a}^k + \text{H.c.}. \tag{13.52}$$

In particular, when the first vibrational sideband ($k = 1$) is excited, then this interaction Hamiltonian further simplifies to

$$\hat{H}_{int} = \tfrac{1}{2}\hbar\Omega \hat{f}_1(\hat{n};\eta)\hat{a} + \text{H.c.}, \tag{13.53}$$

which in the Lamb–Dicke regime reduces to

$$\hat{H}_{int} = \tfrac{1}{2}i\hbar\Omega\eta\hat{a} + \text{H.c.}. \tag{13.54}$$

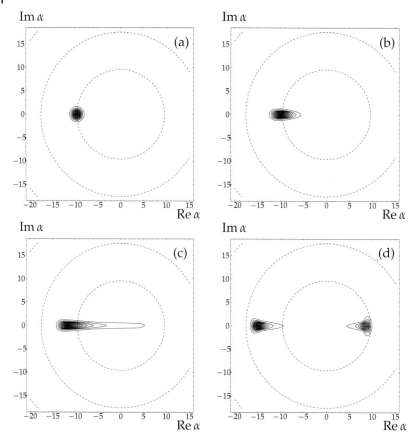

Fig. 13.6 Time evolution of a coherent state which is initially positioned on the boundary between two phase-space zones. The Lamb–Dicke parameter is $\eta = 0.2$ and the dimensionless times $\eta|\Omega|t$ are 0 (a), 4 (b), 8 (c), and 16 (d). The contours represent the Q function and the dashed lines indicate the partitioning of the phase space. [After Wallentowitz and Vogel (1997).]

In this case the corresponding unitary time evolution operator

$$\hat{U}_{\text{int}}(t) = \hat{D}(-\tfrac{1}{2}\eta\Omega^* t) \tag{13.55}$$

leads to a coherent displacement of the motional quantum state [for an experimental implementation, see Monroe, Meekhof, King and Wineland (1996)].

The effect of the interaction Hamiltonian (13.53) can be thought of as a nonlinearly modified displacement [Wallentowitz and Vogel (1997)], the c-number displacement variable being replaced by an operator proportional to $\hat{f}_1(\hat{n};\eta)$. To get some insight into the modification, one may consider the expectation value $\langle \alpha | \hat{f}_1(\hat{n};\eta) | \alpha \rangle = f_1(|\alpha|^2;\eta)$ as a function of the phase-space amplitude $|\alpha|$, which typically oscillates with $|\alpha|$. As a result, the "nonlinear displacement" collapses at the amplitudes defining the zeros of $\hat{f}_1(|\alpha|^2; \eta)$. Effectively this leads to a partitioning of the phase-space into adjacent zones. Due to the changing sign of the function f_1 the direction of the displacement changes when crossing the zeros of f_1. An example is shown in Fig. 13.6, where the initial coherent state is located on the boundary between two phase-space zones. Consequently, the "nonlinear displacement" acts in opposite directions in the two neighboring zones. This leads to a coherent splitting of the quantum state which is accompanied by quantum interference effects. Note that, in the shown Q function, the interference fringes are smoothed out.

There are various possibilities for using the nonlinear, dynamic phase-space partitioning for the generation of nonclassical effects in the atomic center-of-mass motion. For example, starting from the motional ground state, one can realize a displacement that eventually squeezes the quantum state onto the circle in phase space representing the first zero of the function $f_1(|\alpha|^2; \eta)$. In this manner one may generate a strongly amplitude-squeezed state [Wallentowitz and Vogel (1997)]. For a Lamb–Dicke parameter of $\eta = 0.25$ a noise reduction of $\langle \Delta \hat{n}^2 \rangle / \langle \hat{n} \rangle = 0.006$ can be obtained. On the basis of the interaction Hamiltonian (13.52) for a two-quantum excitation, $k=2$, the time-evolution operator represents a nonlinear generalization of the squeeze operator [Wallentowitz and Vogel (1998)]. For a three-quantum excitation, $k=3$, one may generate motional quantum states displaying star-like structures [Wallentowitz, Vogel and Knight (1999)].

13.4
Preparing motional quantum states

Trapped atoms offer exciting possibilities for preparing and manipulating quantum states, with special emphasis on highly nonclassical states. It is well known that nonclassical states very sensitively respond to any kind of disturbance, in general. Although the vibrational frequencies of an atom in a Paul trap are typically in the MHz range, so that direct radiative damping is extremely small, there are other decoherence mechanisms such as, for example, unwanted quantum jumps discussed in Section 13.3.2. As already mentioned therein, there are a number of open problems whose solution is a big challenge to research.

13.4.1
Sideband laser-cooling

In order to prepare a trapped atom in a desired motional quantum state, it should be initially prepared in the ground state, which can be achieved by laser cooling [Meekhof, Monroe, King, Itano and Wineland (1996); Monroe,

Meekhof, King and Wineland (1996)]. In a first step, the atom is usually pre-cooled via Doppler cooling.[8] The result typically consists of a thermal distribution of vibrational number states with a mean number of a few quanta. A dipole-allowed transition $|1\rangle \leftrightarrow |2\rangle$ of the atom is driven by a laser that is detuned by $\Delta = \omega_{21} - \omega_L$ to the red of the electronic resonance ($\Delta > 0$). Due to thermal motion the (initially hot) atom is moving forth and back in the trap potential. Whenever the atom is moving towards the incident laser beam, due to the Doppler effect, the laser is tuned closer to resonance. Consequently, the probability of absorbing a laser-photon is enhanced by the atomic motion. After absorbing a laser photon of frequency $\omega_L = \omega_{21} + \Delta$ the atom is in the excited state $|2\rangle$ and it is subsequently decaying with rate Γ. The frequency ω_{em} of the emitted photon is approximately given by $\omega_{em} \approx \omega_{21}$, within an uncertainty of the size of the natural linewidth Γ, and therefore a net loss of energy appears. During a series of such absorption-emission cycles the atomic motion is cooled down to a limit determined by the frequency spread of the emitted photons, $\Delta\omega_{em} \approx \Gamma/2$. The so-called Doppler limit for the temperature T of the atom's external motion is therefore given by $k_B T = \hbar\Gamma/2$.

This limit can be overcome by applying resolved sideband cooling in a next step [Wineland and Dehmelt (1975)]. For this purpose the decay rate Γ must be small compared with the vibrational frequency:

$$\Gamma \ll \nu. \tag{13.56}$$

In the Lamb–Dicke regime, the laser is incident in the direction of atomic motion to be cooled and it is tuned on resonance with the first red motional sideband, $\omega_L = \omega_{21} - \nu$. The absorption of a laser photon of frequency ω_L by the atom is accompanied by the annihilation of a vibrational energy quantum. The absorption is followed by the spontaneous emission of a photon of frequency ω_{21}. Hence, one vibrational energy quantum $\hbar\nu$ is absorbed per scattering event, on average. The cooling process proceeds until the mean vibrational quantum number attains its minimum value

$$\langle n \rangle_{min} = \left(\frac{\Gamma}{2\nu}\right)^2. \tag{13.57}$$

Under the condition (13.56) this implies that $\langle n \rangle_{min} \ll 1$. Thus the atom can effectively be cooled into its vibrational ground state. Since for typical trap frequencies in the MHz range the condition (13.56) can hardly be fulfilled by using a dipole transition, one can think of using a dipole-forbidden transition. In this case, however, Γ is too small in general to realize noticeable light scattering required for fast enough cooling. This problem can be overcome by

---

8) For details, see, e.g., Wineland, Drullinger and Walls (1978); Neuhauser, Hohenstatt, Toschek and Dehmelt (1978); for a review, see Stenholm (1986).

driving the first resolved sideband of a dipole-forbidden transition and shortening its lifetime via a coupling by radiation to a fast-decaying level.

Experimentally, resolved-sideband cooling was demonstrated for the first time by using a ^{198}Hg$^+$ ion [Diedrich, Bergquist, Itano and Wineland (1989)], where the trapped ion was prepared in the vibrational ground state with a probability of about 95%. Later on, resolved-sideband Raman cooling was performed on a trapped ^{9}Be$^+$ ion [Monroe, Meekhof, King, Jefferts, Itano and Wineland (1995)], where the quantized motion (in one dimension) could be cooled to the motional ground state with a probability of 98%. In the experiment the direct drive of the resolved sideband is replaced by a stimulated Raman drive, which allows one to use transitions between metastable levels (such as hyperfine or Zeeman electronic ground states). Optical frequency modulators render it possible to realize very small effective laser linewidths. The internal state of the atom is recycled by spontaneous Raman transitions.

13.4.2
Coherent, number and squeezed states

Let us first consider the preparation of coherent motional states and the coherent displacement of a given motional state. One way consists of a direct drive of the center-of-mass of the atom. Consider a classical, monochromatic microwave which resonantly drives a single motional mode, say mode 1. In the interaction Hamiltonian (13.19) we may thus make the replacement $\mathbf{d}_k \hat{\mathbf{E}}^{(-)} \mapsto \delta_{k1} d_1 E_0 e^{i(\nu_1 t - \varphi_0)}$, leading in the interaction picture to

$$\hat{H}_{\text{int}}^{(\text{cm})} = -d_1 E_0 e^{-i\varphi_0} \hat{a}_1 + \text{H.c..} \tag{13.58}$$

The corresponding unitary time evolution yields a coherent displacement, $\hat{U}_{\text{int}}^{(\text{cm})}(t) \equiv \hat{D}(\alpha)$, where $\alpha = id_1 E_0 e^{i\varphi_0} t/\hbar$ is the complex displacement amplitude. Its absolute value can be controlled by the interaction time and/or the field amplitude E_0, the phase is controlled by the phase φ_0 of the microwave. Experimentally the method can be realized by applying a sinusoidally varying potential at the trap oscillation frequency on one of the trap compensation electrodes for a fixed time, as was demonstrated by Meekhof, Monroe, King, Itano and Wineland (1996). The achieved displacement $\alpha \approx 1.7$ of the motional ground state shows that the method is useful for typical experiments in a microscopic regime.

Alternatively, coherent displacements can be realized by a Raman drive of the atomic motion, cf. Section 13.3.3. By resonantly driving the first motional sideband in the Lamb–Dicke regime, according to Eq. (13.55), the unitary time evolution operator represents a coherent displacement operator. In the experiments by Monroe, Meekhof, King and Wineland (1996) coherent displacement amplitudes up to $\alpha \approx 3$ were realized in this way. Note that in such a scheme, for larger displacement amplitudes, the nonlinear effects inherent in the interaction Hamiltonian (13.53) may become important.

Number states, whose generation is difficult for radiation-field modes, can be easily obtained for the quantized motion of a trapped atom [Meekhof, Monroe, King, Itano and Wineland (1996)]. The first step of the preparation is the cooling to the vibrational ground state. Suppose the atom is initially in the state $|1,0\rangle$, that is in both its electronic and vibrational ground state. Now one can successively apply two interactions of the nonlinear Jaynes–Cummings type, cf. Section 13.3.1. First, one applies a π pulse on the first blue sideband. This interaction is described by the interaction Hamiltonian (13.39) for $k=-1$ and it flips the atom into the first number state in the excited electronic state, $|2,1\rangle$. Second, one applies a π pulse on the first red sideband, cf. the interaction Hamiltonian (13.40). In this example one would prepare the second vibrational number state in the electronic ground state, $|1,2\rangle$. In a similar manner one can prepare higher vibrational number states by repeatedly applying appropriate sequences of laser pulses.

Motional squeezed states can be obtained by applying a Raman drive of the atomic motion, cf. Section 13.3.3. Consider a Raman excitation of a vibrational two-quantum resonance as described by the interaction Hamiltonian (13.52) for $k=2$. For a sufficiently small Lamb–Dicke parameter simplifies to

$$\hat{H}_{\text{int}} = -\tfrac{1}{4}\hbar\Omega\eta^2\,\hat{a}^2 + \text{H.c.}. \tag{13.59}$$

In this case the unitary time evolution operator agrees with the squeeze operator, $\hat{U}_{\text{int}}(t) \equiv \hat{S}(\xi)$, the absolute value of the squeeze parameter being $|\xi|=|\Omega|\eta^2 t/2$. Applying this interaction to an atom which was initially cooled to its vibrational ground state, one prepares a vibrational squeezed ground state [Meekhof, Monroe, King, Itano and Wineland (1996)]. Since highly excited number states contribute to the squeezed ground state whenever a significant squeezing effect occurs, one may quickly approach the limits of the approximate interaction Hamiltonian (13.59).[9]

13.4.3
Schrödinger-cat states

The combination of methods for preparing relatively elementary quantum states allows one also to prepare more complicated quantum states, such as entangled states (Section 8.5). Let us consider the preparation of a trapped atom in an entangled state of the Schrödinger-cat type, with the motional states being entangled with the electronic ones [Monroe, Meekhof, King and Wineland (1996)]. Of course, such a cat state is not truly macroscopic. The

---

[9] For the nonlinear effects which may occur under more general conditions, see Wallentowitz and Vogel (1998).

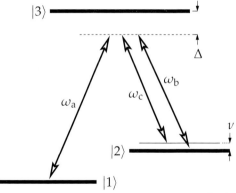

Fig. 13.7 The atomic energy level scheme of ^{9}Be$^+$ which is relevant for the preparation of entangled Schrödinger-cat states.

basic energy level scheme of the atom is shown in Fig. 13.7. The two states $|1\rangle$ and $|2\rangle$ are formed by two highly stable hyperfine ground states. The state $|3\rangle$ provides the Raman coupling[10] for the three lasers a, b and c. Let us assume that the Raman lasers b and c only affect the quantized motion in the state $|2\rangle$, which can be achieved by appropriate polarizations of the lasers.

The state preparation consists of the following steps. Initially the atom is in the vibrational and electronic ground state $|1,0\rangle = |1\rangle|0\rangle_m$.[11] In the first step the laser pair a,b is on and the interaction time is chosen to apply a $\pi/2$ rotation of the electronic states, resulting in

$$|\Psi_1\rangle = \tfrac{1}{\sqrt{2}}\left(|1\rangle|0\rangle_m - ie^{-i\mu}|2\rangle|0\rangle_m\right), \tag{13.60}$$

where μ is the phase difference of the laser pair. The second step is the coherent displacement of the motional state by the Raman drive of the type given in Eqs (13.54) and (13.55), acting in the excited electronic state. This interaction is driven by the laser pair b,c (of phase difference $-\phi/2$) and yields

$$|\Psi_2\rangle = \tfrac{1}{\sqrt{2}}\left(|1\rangle|0\rangle_m - ie^{-i\mu}|2\rangle|\alpha e^{-\tfrac{i}{2}\phi}\rangle_m\right). \tag{13.61}$$

The third step consists of a π pulse on the vibrationless transition, the carrier (lasers a,b, phase χ), which basically exchanges the motional states in the two electronic states,

$$|\Psi_3\rangle = \tfrac{1}{\sqrt{2}}\left(e^{i(\chi-\mu)}|1\rangle|\alpha e^{-\tfrac{i}{2}\phi}\rangle_m + ie^{-i\chi}|2\rangle|0\rangle_m\right). \tag{13.62}$$

10) The interaction Hamiltonians given in Sections 13.3.1 and 13.3.3 apply.
11) Here we distinguish motional states from electronic ones by the subscript m.

In the fourth step the motional state in the excited electronic state is again displaced (lasers b,c, phase $\phi/2$), resulting in the Schrödinger-cat-like state

$$|\Psi_4\rangle = \tfrac{1}{\sqrt{2}}\left(e^{i(\chi-\mu)}|1\rangle|\alpha e^{-\frac{i}{2}\phi}\rangle_m + ie^{-i\chi}|2\rangle|\alpha e^{\frac{i}{2}\phi}\rangle_m\right). \tag{13.63}$$

This is an entangled state with two independent coherent states being correlated to the electronic states. Needless to say that the state (13.63), which is of the same type as the radiation-field state (12.108) generated in a cavity, corresponds to a Schrödinger-cat state on a mesoscopic level rather than a macroscopic one.

One can proceed and perform a fifth step which consists of a $\pi/2$ pulse on the carrier (lasers a,b, phase 0) and yields, up to a normalization constant,

$$|\Psi_5\rangle \sim |1\rangle\left(|\alpha e^{-\frac{i}{2}\phi}\rangle_m - e^{i\delta}|\alpha e^{\frac{i}{2}\phi}\rangle_m\right) - i|2\rangle\left(|\alpha e^{-\frac{i}{2}\phi}\rangle_m + e^{i\delta}|\alpha e^{\frac{i}{2}\phi}\rangle_m\right), \tag{13.64}$$

where $\delta = \mu - 2\chi + \pi$. The state $|\Psi_5\rangle$ can be used to prepare a coherent superposition of two motional coherent states. This can be done in the sixth step by a reduction of the electronic state via conditional measurement. For example, one may measure the fluorescence by irradiating a transition from the state $|2\rangle$ to an auxiliary state $|4\rangle$. When no fluorescence is observed, the atom is reduced to the electronic state $|1\rangle$ and the resulting purely motional state $|\Psi_6\rangle_m$ is

$$|\Psi_6\rangle_m \sim \left(|\alpha e^{-\frac{i}{2}\phi}\rangle_m - e^{i\delta}|\alpha e^{\frac{i}{2}\phi}\rangle_m\right). \tag{13.65}$$

Special cases of these kinds of states[12] are the even and odd coherent states defined by Eq. (13.68).

13.4.4
Motional dark states

States which are produced by the methods in Sections 13.4.2 and 13.4.3 are not very long-lived in general. From Section 11.3.3, we know that a three-level system can be prepared in long-lived states – the dark states. As we will see, the basic idea of dark states can be transferred to trapped atoms, where it offers the possibility of preparing long-lived motional states, such as squeezed states [Cirac, Parkins, Blatt and Zoller (1993)], even and odd coherent states [de Matos Filho and Vogel (1996a)], nonlinear coherent states [de Matos Filho and Vogel (1996b)], pair coherent states [Gou, Steinbach and Knight (1996)], and others.

Let us consider a trapped atom, two electronic states of which are driven by one or several lasers in the resolved sideband regime. In the rotating wave approximation, each interaction may be described separately by a nonlinear

---
12) They are often also called states of Schrödinger-cat type.

Jaynes–Cummings interaction Hamiltonian, such as is given in Eqs (13.37) and (13.39).[13] The sum of these interactions eventually yields the total interaction Hamiltonian. For the preparation of dark states, the decay of the upper electronic state to the ground state by radiative damping is of great importance. These electronic transitions are accompanied by kicks due to the momentum transfer of the emitted photons on the atom, which act on the quantum state of the atomic center-of-mass motion. The master equation describing the vibronic dynamics of the trapped atom can be obtained by specifying the general master equation (5.59). Using for the system–reservoir coupling the interaction of the atom with a reservoir of radiation modes, Eq. (13.26), one can derive the following master equation in the interaction picture:

$$\frac{d\hat{\varrho}}{dt} = -\frac{i}{\hbar}[\hat{H}_{\text{int}}, \hat{\varrho}] + \tfrac{1}{2}\Gamma(2\hat{A}_{12}\hat{\varrho}_r\hat{A}_{21} - \hat{A}_{22}\hat{\varrho} - \hat{\varrho}\hat{A}_{22}), \qquad (13.66)$$

where

$$\hat{\varrho}_r = \tfrac{1}{2}\int_{-1}^{1} ds\, W(s) e^{i\eta s(\hat{a}+\hat{a}^\dagger)} \hat{\varrho}\, e^{-i\eta s(\hat{a}+\hat{a}^\dagger)}, \qquad (13.67)$$

accounts for the changes in the motional state due to the momentum recoil of the atom during the emission of photons, with the function $W(s)$ representing the angular distribution of the spontaneous emission. The density operator $\hat{\varrho}$ describes the combined vibronic quantum state including the center-of-mass motion in the direction under study. It is obtained by tracing over the motional degrees of freedom which are not of interest.

We begin with the generation of even and odd coherent states. Quite generally, they are defined as superpositions of two coherent states whose complex amplitudes point in opposite directions in the phase space:

$$|\alpha\rangle_\pm = \mathcal{N}_\pm(|\alpha\rangle \pm |-\alpha\rangle) \qquad (13.68)$$

($\mathcal{N}_\pm$, normalization constants) [Dodonov, Malkin and Man'ko (1974); Peřina (1984)]. Note that only even (odd) number states contribute to the number statistics of $|\alpha\rangle_+$ ($|\alpha\rangle_-$). The quantum interference caused by the superposition of the coherent states is of particular interest. For example, the quadrature distributions of $|\alpha\rangle_\pm$ which correspond to the position distributions are

$$p_\pm(x) = |\mathcal{N}_+|^2[|\langle x|\alpha\rangle|^2 + |\langle x|-\alpha\rangle|^2 \pm 2\mathrm{Re}(\langle \alpha|x\rangle\langle x|-\alpha\rangle)]. \qquad (13.69)$$

The first two terms are the position distributions of the two coherent states $|\pm\alpha\rangle$, whereas the third term represents the quantum interference between these states.

---
13) For simplicity we will consider, in the following, only dark states in one vibrational mode.

Motional even/odd coherent states can be obtained as dark states in the following manner [de Matos Filho and Vogel (1996a)]. A trapped atom is driven in the resolved sideband regime by two lasers, the first one is in resonance with the (red) second sideband and the second one resonantly drives the carrier. These interactions are described by the interaction Hamiltonian (13.37) for $k=0,2$. In particular, in the case of the Lamb–Dicke parameter being sufficiently small, the following simplifications can be made. First, the kicks due to spontaneous emission can be discarded, $\hat{\varrho}_r \simeq \hat{\varrho}$. Second, the interaction Hamiltonian can be simplified by taking into account only the leading terms in the series (13.38). In this case the master equation (13.66) approximates to

$$\frac{d\hat{\varrho}}{dt} = -\frac{i}{\hbar}[\hat{H}_{\text{int}}, \hat{\varrho}] + \tfrac{1}{2}\Gamma(2\hat{A}_{12}\hat{\varrho}\hat{A}_{21} - \hat{A}_{22}\hat{\varrho} - \hat{\varrho}\hat{A}_{22}), \tag{13.70}$$

and the effective interaction Hamiltonian is of the form

$$\hat{H}_{\text{int}} = -\tfrac{1}{2}\hbar\eta^2\Omega_2\hat{A}_{21}\left(\hat{a}^2 - \frac{2\Omega_0}{\eta^2\Omega_2}\right) + \text{H.c.}, \tag{13.71}$$

where Ω_k ($k=0,2$) is the driving-laser coupling strength with regard to the kth sideband. Now we are interested in a steady-state solution, $\hat{\varrho}_s$, of the type

$$\hat{\varrho}_s = |1\rangle|\psi\rangle\langle\psi|\langle 1|. \tag{13.72}$$

When such a stationary state exists, the atom ends in its electronic ground state and it stops to fluoresce despite the lasers continuously driving the system. Thus the atom is prepared in a motional dark state. It remains to derive the properties of the motional quantum state $|\psi\rangle$ under these conditions. In the steady-state regime in which we are interested, we may set the temporal derivative of the density operator equal to zero in Eq. (13.70). Inserting the ansatz (13.72) into the master equation, we can see that the relaxation term vanishes identically. Physically, this is obvious since an atom in the ground state cannot spontaneously decay. The remaining condition for a steady-state solution reduces to $[\hat{H}_{\text{int}}, \hat{\varrho}_s] = 0$, which is fulfilled if the motional quantum state obeys the eigenvalue equation

$$\hat{a}^2|\psi\rangle = \frac{2\Omega_0}{\eta^2\Omega_2}|\psi\rangle. \tag{13.73}$$

The complex eigenvalue can be well controlled by the relative amplitudes and the phase difference of the two lasers, which can be held very stable by using electro/acousto-optical modulation techniques to derive the two fields from the same laser.

It is not difficult to see that the recurrence relation resulting from Eq. (13.73) in the number-state basis couples between even or odd number states only.

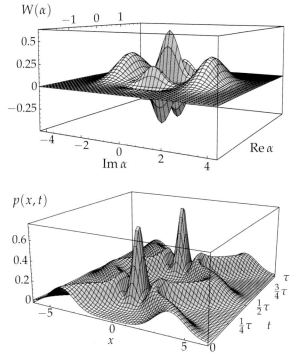

Fig. 13.8 Wigner function $W(\alpha)$ and position distribution $p(x,t)$ for an atom whose motional steady-state is close to an even coherent state. Initially the atom was prepared in the vibrational ground state, the parameters for the atom–field interaction are $\eta=0.05$, $2\Omega_0/\eta^2\Omega_2=4$, and $\eta^2\Omega_2=1.4\Gamma$ [$\tau=\nu/(2\pi)$]. [After de Matos Filho and Vogel (1996a).]

Further, the relaxation effects in the approximate master equation (13.70) do not alter the motional state. Last but not least, the interaction Hamiltonian (13.71) also couples only between even or odd number states. Thus we may imagine the following scenario. Let us assume that we have initially cooled the atom to its vibrational ground state. Subsequently we switch on the interaction with the two lasers according to the interaction Hamiltonian (13.71). For nonvanishing coupling strengths Ω_k the vibrational ground state does not fulfill the eigenvalue equation (13.73). Consequently, the atom cannot be in its steady state and the motional quantum state is modified by scattering photons until it fulfills Eq. (13.73). Since this process couples only even number states, the steady state is expected to be given by the even coherent state

$$|\psi\rangle = |\alpha\rangle_+, \qquad \alpha = \sqrt{\frac{2\Omega_0}{\eta^2\Omega_2}}. \qquad (13.74)$$

The example in Fig. 13.8 clearly shows that in this way motional dark states can be generated which are indeed close to even coherent states. Both the

Wigner function and the position distribution in the figure show the typical features of an even coherent state. Note that, if the two main peaks of the position distribution approach each other, quantum interference fringes are observed. It should be pointed out that odd coherent states can be prepared in a similar manner if one starts with the atom being initially excited in its first vibrational number state.

Another interesting possibility consists of the preparation of nonlinear coherent states [de Matos Filho and Vogel (1996b)] which are defined as the right-hand eigenstates of the non-Hermitian operator $\hat{f}(\hat{n})\hat{a}$,

$$\hat{f}(\hat{n})\hat{a}|\chi;f\rangle = \chi|\chi;f\rangle, \tag{13.75}$$

where $\hat{f}(\hat{n})$ is a function of the number operator. The notation $|\chi;f\rangle$ indicates that the properties of the states sensitively depend on the chosen function $\hat{f}(\hat{n})$. Note that for $\hat{f}(\hat{n})=1$ the ordinary coherent states are obtained. In general one may expect that, due to the function $\hat{f}(\hat{n})$, the states $|\chi;f\rangle$ feature coherence properties which are more or less modified in the case when higher number states contribute to them. It is straightforward to represent them in the number-state basis, since the definition (13.75) yields a simple recurrence relation connecting only two elements, $\langle n|\chi;f\rangle$ and $\langle n+1|\chi;f\rangle$. One finds that[14]

$$|\chi;f\rangle = \sum_n |n\rangle\langle n|\chi;f\rangle, \tag{13.76}$$

where

$$\langle n|\chi;f\rangle = \mathcal{N}\frac{g(n)}{\sqrt{n!}} \tag{13.77}$$

with

$$g(n) = \delta_{n,0} + (1-\delta_{n,0})\prod_{k=0}^{n-1}[f(k)]^{-1}. \tag{13.78}$$

To generate nonlinear coherent states as motional dark states of a trapped atom, the atom is simultaneously driven on the carrier and the first (red) vibrational sideband. Here the Lamb–Dicke parameter can become rather large, so that we use the interaction Hamiltonians of the type (13.37) for $k=0,1$. Their combination yields the total interaction Hamiltonian

$$\hat{H}_{\text{int}} = \tfrac{1}{2}\hbar\Omega_1 e^{-\eta^2/2}\hat{A}_{21}\left(\hat{F} + \frac{\Omega_0}{\Omega_1}\right) + \text{H.c.}, \tag{13.79}$$

---

[14] For more details on the properties of the nonlinear coherent states, see, e. g., Man'ko, Marmo, Sudarshan and Zaccaria (1997); Man'ko, Marmo, Porzio, Solimeno and Zaccaria (2000).

with the operator $\hat{F}$ being given by

$$\hat{F} = \sum_{l=0}^{\infty} \frac{(i\eta)^{2l+1}}{l!(l+1)!} \hat{a}^{\dagger l} \hat{a}^{l+1} + \frac{\Omega_0}{\Omega_1} \sum_{l=1}^{\infty} \frac{(i\eta)^{2l}}{(l!)^2} \hat{a}^{\dagger l} \hat{a}^{l}. \qquad (13.80)$$

Searching for a steady state of the form (13.72) that is solution to the full master equation (13.66) results in in the eigenvalue equation

$$\hat{F}|\psi\rangle = -\frac{\Omega_0}{\Omega_1} |\psi\rangle. \qquad (13.81)$$

Comparing Eqs (13.81) and (13.75), we conclude that

$$|\psi\rangle = |\chi; f\rangle, \qquad (13.82)$$

where

$$\chi = \frac{i\Omega_0}{\eta \Omega_1} \qquad (13.83)$$

and

$$\hat{f}(\hat{n}) = \sum_{n=0}^{\infty} \frac{L_n^1(\eta^2)}{(n+1)L_n^0(\eta^2)} |n\rangle\langle n|. \qquad (13.84)$$

Since the Laguerre polynomials $L_n^k(\eta^2)$ are oscillating functions of n (with different periodicities for $k=0$ and $k=1$), the expansion coefficients (13.77) are oscillating functions of n either, where the oscillation period decreases with increasing value of η. It turns out that, for moderate values of η, the state $|\psi\rangle$ can be localized within a single peak of the structured coefficients, which yields an amplitude-squeezed state displaying coherence properties. By changing the eigenvalue χ, which is well controlled by the driving lasers, one may prepare a coherent superposition of two sub-states which are placed in the same direction from the origin of the phase space.[15]

For weak interactions of the lasers with the atom, the master equation (13.66) can be simplified by adiabatically eliminating the electronic degrees of freedom, because the atom is almost always confined in the electronic ground state. As a result, the damping term in Eq. (13.66), i.e., the second term on the right-hand side, changes to a new one whose structure is also determined by the driving laser configuration. In this manner one can engineer the coupling of the motional subsystem with an effective reservoir [Poyatos, Cirac and Zoller (1996); for an experimental demonstration, see Myatt, King, Turchette, Sackett, Kielpinski, Itano, Monroe and Wineland (2000)].

15) For a realization of more general nonlinear coherent states, see Kis, Vogel and Davidovich (2001).

13.5
Measuring the quantum state

When one is able to prepare well-defined quantum states of trapped atoms, one also needs methods to measure them. Based on the concepts of quantum-state reconstruction developed in Chapter 7, in the following we will consider measurement schemes specific to reconstructing the quantum state of a trapped atom.

13.5.1
Tomographic methods

Let us begin with the reconstruction of the motional quantum state of a trapped atom by directly measuring the characteristic function of the phase-dependent quadrature distribution [Wallentowitz and Vogel (1995, 1996)]. For this purpose a weak electronic transition $|1\rangle \leftrightarrow |2\rangle$ of the atom is simultaneously driven in the resolved-sideband regime by two classical laser fields, as schematically shown in Fig. 13.9. The laser frequencies $\omega_{b/r}$ are tuned to the first blue/red motional sidebands, $\omega_{b/r} = \omega_{21} \pm \nu$. The two interactions are described by interaction Hamiltonians of the Jaynes–Cummings type, Eq. (13.39) for $|k|=1$ for the blue sideband and Eq. (13.40) for the red one. Assuming that the coupling strengths of the two driving laser fields are equal and that the system is driven in the Lamb–Dicke regime, the total interaction Hamiltonian is of the form

$$\hat{H}_{int} = \tfrac{1}{2}\hbar(\Omega \hat{A}_{12} + \Omega^* \hat{A}_{21})\hat{x}(\varphi), \qquad (13.85)$$

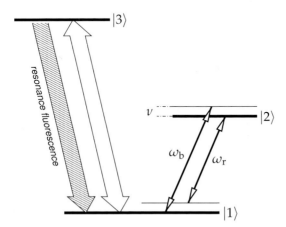

Fig. 13.9 Excitation scheme for the tomographic reconstruction of the motional quantum state of a trapped atom.

where Ω is the sum of the vibronic coupling strengths of the two driven transitions, and φ is the phase difference of the two lasers which can be controlled precisely. This interaction Hamiltonian is proportional to the phase-sensitive quadrature $\hat{x}(\varphi)$, which yields the information desired for the tomographic reconstruction of the motional quantum state. The coupling of this observable to the electronic transition $|1\rangle \leftrightarrow |2\rangle$ renders it possible to map the information onto the populations of the corresponding electronic states. To detect the electronic-state populations of the weak transition one usually measures the fluorescence on a strong auxiliary transition $|1\rangle \leftrightarrow |3\rangle$, which yields an overall quantum efficiency being very close to unity.[16] To be more specific, the interaction of the atom with the bichromatic laser field is switched on for an interaction time t. After this interaction the $|1\rangle \leftrightarrow |3\rangle$ transition is irradiated and the fluorescence is recorded. Repeating these measurements, the statistics of the occurrence of fluorescence give the occupation probability $\sigma_{11}(t) \equiv \sigma_{11}(t;\varphi)$ of the atom in the ground state $|1\rangle$ right after the interaction time t,

$$\sigma_{11}(t;\varphi) = \mathrm{Tr}[\hat{A}_{11}\hat{U}_{\mathrm{int}}(t;\varphi)\hat{\varrho}(0)\hat{U}_{\mathrm{int}}^{\dagger}(t;\varphi)], \tag{13.86}$$

where $\hat{U}_{\mathrm{int}}(t) \equiv \hat{U}_{\mathrm{int}}(t;\varphi)$ is the unitary time evolution due to the interaction (13.85) for chosen φ, and $\hat{\varrho}(0)$ is the initial density operator of the vibronic system, which is commonly assumed to be factorized into a product of the vibrational and electronic density operators $\hat{\rho}(0)$ and $\hat{\sigma}(0)$, respectively. For calculating the trace in Eq. (13.86) it is advantageous to represent $\hat{\rho}(0)$ in the basis of the eigenstates $|x,\varphi\rangle$ of the $\hat{x}(\varphi)$. When the atom is initially in the electronic ground state, $\hat{\sigma}(0) = \hat{A}_{11}$, then Eq. (13.86) leads to

$$\sigma_{11}^{(\mathrm{inc})}(t) = \tfrac{1}{2}[1 - \mathrm{Re}\Psi(\tau,\varphi)] \tag{13.87}$$

($\tau = |\Omega|t$), where $\Psi(\tau,\varphi)$ is the characteristic function of the quadrature distribution $p(x,\varphi)$ [cf. Eq. (7.10)] of the motional quantum state. Hence the recorded fluorescence statistics directly yield insight into the real part of the characteristic function of the quadrature distribution. To obtain the imaginary part of $\Psi(\tau,\varphi)$, one can prepare the atom initially in a coherent superposition of the two electronic states, such that $\sigma_{11}(0) = |\sigma_{12}(0)| = \tfrac{1}{2}$. This preparation requires the application of a $\pi/2$ pulse in the Lamb–Dicke regime onto an atom in the electronic ground state. By appropriately adjusting $\arg[\sigma_{12}(0)]$, the resulting signal reads

$$\sigma_{11}^{(\mathrm{coh})}(\tau;\varphi) = \tfrac{1}{2}[1 + \mathrm{Im}\Psi(\tau,\varphi)]. \tag{13.88}$$

---

16) For highly efficient measurements of this type, see Nagourney, Sandberg and Dehmelt (1986); Sauter, Neuhauser, Blatt and Toschek (1986); Bergquist, Hulet, Itano and Wineland (1986).

Combining the two types of measurement, we obtain the full characteristic function:

$$\Psi(\tau, \varphi) = 2[\sigma_{11}^{(\text{inc})}(\tau; \varphi) - \tfrac{1}{2}] + 2i[\sigma_{11}^{(\text{coh})}(\tau; \varphi) - \tfrac{1}{2}]. \tag{13.89}$$

Performing the measurements for sufficiently many values of the difference phase φ between the two driving lasers within a π interval, one can readily derive various forms of representations of the motional quantum state of the atom, see Chapter 7. For example, the reconstruction of the density matrix in a quadrature-state basis reduces to a single Fourier integral of the measured fluorescence signal. Alternatively, it is possible to directly sample the density matrix in the number basis from the measured signal.

A related method of tomographic reconstruction of the motional quantum state is based on pulsed atom–laser interactions in the Lamb–Dicke regime [D'Helon and Milburn (1996)]. Here, the $|1\rangle \leftrightarrow |2\rangle$ transition is driven by a standing-wave laser pulse tuned to ω_{21}. Its duration τ_P is much shorter than the vibrational period, $\nu\tau_P \ll 1$. Moreover, the center of the trap potential must coincide with a node of the standing wave. To derive the desired atom–field interaction, one transforms the interaction Hamiltonian (13.30) to the interaction picture.[17] In the Lamb–Dicke approximation the operator-valued (standing-wave) mode function therein can be approximated according to $\hat{g}(\hat{x}) = \sin(k_L\hat{x}) \simeq k_L\hat{x}$. The resulting interaction Hamiltonian is very similar to that in Eq. (13.85) and reads as

$$\hat{H}_{\text{int}} = \tfrac{1}{2}\hbar[\Omega(t)\hat{A}_{12} + \Omega^*(t)\hat{A}_{21}]\hat{x}(t), \tag{13.90}$$

where $\hat{x}(t)$ is the freely evolving position operator of the center-of-mass motion. The laser pulse shape is included in the time-dependent coupling strength $\Omega(t)$. Due to the shortness of the driving pulse its action at time t consists of a unitary kick,

$$\hat{K} = T\exp\left[-\frac{i}{\hbar}\int dt'\,\hat{H}_{\text{int}}(t')\right] \simeq \exp\left[-\frac{i}{2}(\theta\hat{A}_{12} + \theta^*\hat{A}_{21})\hat{x}(t)\right], \tag{13.91}$$

with $\theta = \int dt'\Omega(t')$ being the pulse area. Its effect on the state is formally the same as that of the unitary operator for the interaction (13.85). Thus the characteristic function $\Psi(\theta, \nu t)$ is obtained in a similar way to the previous scheme, cf. Eq. (13.89), by two types of measurement of $\sigma_{11}^{(\text{inc})}$ and $\sigma_{11}^{(\text{coh})}$ right after the application of the standing-wave pulse. Instead of controlling the scaled interaction time τ and the phase difference φ between two lasers, in the present scheme, control of the pulse area θ and of the (freely evolving) phase νt is required.

17) In this interaction Hamiltonian the vibrational rotating-wave approximation has not been performed. The pulse envelope is included via a time-dependence of E_L, $E_L \mapsto E_L(t)$.

An approximate quadrature measurement can be performed by using the fact that a squeezed coherent state approaches a quadrature eigenstate in the limit of infinitely strong squeezing,

$$|x, \varphi\rangle \sim \lim_{|\xi|\to\infty} \hat{D}(xe^{i\varphi}/\sqrt{2})\hat{S}(|\xi|e^{i2\varphi})|0\rangle, \tag{13.92}$$

see Section 3.4. The quadrature distribution $p(x, \varphi)$ can be asymptotically obtained as

$$p(x, \varphi) \sim \lim_{|\xi|\to\infty} \langle 0|\hat{\rho}'|0\rangle, \tag{13.93}$$

where $\hat{\rho}'$ is a coherently displaced and squeezed version of the density operator $\hat{\rho}\equiv\hat{\rho}(0)$ to be determined,

$$\hat{\rho}' = \hat{S}^\dagger(|\xi|e^{i2\varphi})\hat{D}^\dagger(xe^{i\varphi}/\sqrt{2})\hat{\rho}\hat{D}(xe^{i\varphi}/\sqrt{2})\hat{S}(|\xi|e^{i2\varphi}). \tag{13.94}$$

From Eqs (13.93) and (13.94) it can be seen that, provided the quantum state to be measured can be strongly squeezed and coherently displaced, the quadrature distributions could be obtained from the occupation probability of the vibrational ground state measured after these manipulations. The phase control of the quadrature is based on the free evolution of the vibrational system. One could think of the following scenario [Poyatos, Cirac and Zoller (1996)]. (i) Wait for a time t such that $\varphi = \nu t$. (ii) Perform a sudden displacement of the center of the trap to the right for a distance d such that $x = \sqrt{m\nu/(2\hbar)}\, d$. (iii) Change the trap frequency instantaneously from ν to a lower ν'.[18] (iv) Determine the population of the vibrational ground state.[19] After determining $p(x, \varphi)$ in this manner, standard methods for the tomographic reconstruction of the quantum state [Chapter 7] can be applied. Note that the accuracy of the method is limited by the achievable squeeze parameter $|\xi| = \ln(\nu/\nu')/2$.

13.5.2
Local methods

Motional quantum states of a trapped atom can also be determined by local methods as introduced in Section 7.4. For this purpose the state to be reconstructed is first coherently displaced, $\hat{\rho} \mapsto \hat{\rho}(-\alpha) = \hat{D}^\dagger(\alpha)\hat{\rho}\hat{D}(\alpha)$. Then the Jaynes–Cummings dynamics is measured, where the number statistics, $P_n = \rho_{nn}$, occurring in the expression for the measured electronic-state occupation, $\sigma_{11}(t)$, is replaced by the displaced number distribution,

---

[18] A sudden change in the trap frequency leads to squeezing with the squeeze parameter being $|\xi| = \ln(\nu/\nu')/2$ [Janszky and Yushin (1986)]. Here a strong change in the trap frequency is needed since the measurement scheme requires strong squeezing.

[19] It can be obtained from the measured Jaynes–Cummings revivals [Meekhof, Monroe, King, Itano and Wineland (1996)].

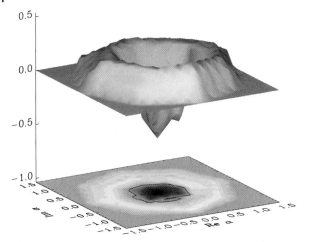

Fig. 13.10 Reconstructed Wigner function of the vibrational number state $|n=1\rangle$ for a trapped $^9\text{Be}^+$ ion. [After Leibfried, Meekhof, King, Monroe, Itano and Wineland (1996).]

$P_n(-\alpha) = \rho_{nn}(-\alpha) = \langle n, \alpha | \hat{\varrho} | n, \alpha \rangle$. Inverting the expression for $\sigma_{11}(t)$, which is given as a linear combination of the $P_n(-\alpha)$ for all relevant n, eventually yields the displaced number statistics. By changing α in a succession of ensemble measurements, the complete information on the original (unshifted) motional quantum state can be obtained from the measured data. In particular, applying the methods introduced in Section 7.4, a pointwise reconstruction of phase-space functions is possible. This was successfully demonstrated by reconstructing the Wigner function $W(\alpha)$ of a single $^9\text{Be}^+$ ion stored in a rf Paul trap [Leibfried, Meekhof, King, Monroe, Itano and Wineland (1996)]. In the experiment, the relevant oscillation frequency in the trap potential is $\nu/2\pi \approx 11.2\,\text{MHz}$, and the transition between the states $|1\rangle$ and $|2\rangle$ is a stimulated Raman transition between the hyperfine ground states $^2S_{1/2}$ ($F=2$, $m_F=-2$) and $^2S_{1/2}$ ($F=1$, $m_F=-1$), respectively, separated by about 1.25 GHz. The coherent displacements of the initially prepared motional quantum states are realized by applying classical, spatially uniform rf fields. The example in Fig. 13.10 shows the reconstructed Wigner function of the number state $|n=1\rangle$.

The method can be further simplified in order to measure the Wigner function directly [Lutterbach and Davidovich (1997)]. For this purpose the laser driving the Jaynes–Cummings dynamics is tuned in resonance with the electronic transition, so that the interaction Hamiltonian (13.37) for $k=0$ applies. Let us assume that the state of the system, after a well-defined coherent displacement of the motional state, is given by $\hat{\varrho} = \hat{\rho}(-\alpha) \otimes |2\rangle\langle 2|$. Now the resonant interaction is switched on for the interaction time t. After the interaction the electronic-state occupation is tested by a fluorescence measurement. The

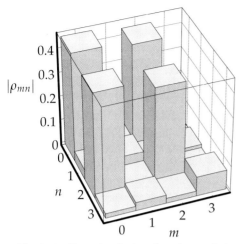

Fig. 13.11 From the displaced number statistics, reconstructed absolute values of density-matrix elements of an approximate $1/\sqrt{2}(|n=0\rangle - i|n=2\rangle)$ motional quantum state. The state was displaced by $|\alpha|=0.79$ for eight phases on a circle. [After Leibfried, Meekhof, King, Monroe, Itano and Wineland (1996).]

electronic inversion reads as

$$\sigma_{22}(t) - \sigma_{11}(t) = \sum_{n=0}^{\infty} \cos(|\Omega_{n,n}|t) P_n(-\alpha), \quad (13.95)$$

where

$$\Omega_{n,n} = \Omega_L e^{-\eta^2/2} L_n(\eta^2) = \Omega_L \left[1 - \eta^2 \left(n + \tfrac{1}{2}\right) + O(\eta^4 n^2)\right]. \quad (13.96)$$

For a sufficiently small Lamb–Dicke parameter, $\eta \ll \sqrt{n+1}$, the terms $O(\eta^4 n^2)$ can be disregarded. Fixing the interaction time by $t = \pi/(\eta^2 \Omega_L)$, from Eq. (13.95) we obtain

$$\sigma_{22}(t) - \sigma_{11}(t) = \sin\left(\frac{\pi}{\eta^2}\right) \sum_{n=0}^{\infty} (-1)^n P_n(-\alpha) = \frac{\pi}{2} \sin\left(\frac{\pi}{\eta^2}\right) W(\alpha). \quad (13.97)$$

We can see that the electronic-state inversion is directly proportional to the Wigner function of the initial state of the atomic center-of-mass motion in a given phase-space point, which is defined by the displacement amplitude α. The applicability of this method can be extended by engineering the vibronic interaction Hamiltonian [de Matos Filho and Vogel (1998)], which allows one to linearize the n dependence of the coupling strength under less restrictive conditions than in Eq. (13.96).

Finally, it should be mentioned that the density matrix in the number-state basis, ρ_{mn}, can be obtained from the displaced number statistics on a circle according to the theory given in Section 7.3.2. This was also demonstrated in the above-mentioned experiment [Leibfried, Meekhof, King, Monroe, Itano and Wineland (1996)]. An example is shown in Fig. 13.11.

13.5.3
Determination of entangled states

Up to now we have considered methods that allow one to determine the information on the quantum state of the motional subsystem of a trapped atom. As we have seen in Section 13.4.3, however, the system may also be prepared in an entangled vibronic state. A measurement of the full information on the combined quantum state of the vibronic system may be realized by employing the interaction Hamiltonian (13.37) for $k=0$ [Wallentowitz, de Matos Filho and Vogel (1997)]. In the first step, the initially prepared vibronic state $\hat{\varrho}$ is coherently displaced in the phase space of the motional subsystem, $\hat{\varrho} \mapsto \hat{\varrho}(-\alpha) = \hat{D}^\dagger(\alpha)\hat{\varrho}\hat{D}(\alpha)$. Subsequently, the driving laser on the carrier which realizes the interaction (13.37), is switched on for an interaction time t. This procedure is followed by probing the atom for fluorescence on a strong, auxiliary transition. Provided the atom is detected in the excited electronic state $|2\rangle$ (no fluorescence), the density operator $\hat{\varrho}(t)$ of the system is reduced to

$$\hat{\varrho}(t) = |2\rangle\langle 2| \otimes \hat{\rho}(t), \tag{13.98}$$

where $\hat{\rho}(t) \sim \langle 2|\hat{\varrho}(t)|2\rangle$ is the corresponding density operator of the motional subsystem. Its diagonal elements in the number-state basis, due to the interaction (13.37) for $k=0$, read as

$$\rho_{nn}(t) = \text{Im}\left[\varrho_{12}^{nn}(-\alpha)\right]\sin(\Omega_{nn}t) \\ + \varrho_{22}^{nn}(-\alpha)\cos^2\left(\tfrac{1}{2}\Omega_{nn}t\right) + \varrho_{11}^{nn}(-\alpha)\sin^2\left(\tfrac{1}{2}\Omega_{nn}t\right). \tag{13.99}$$

The result reveals that, for appropriately chosen interaction times t, one can map the displaced vibronic density-matrix elements $\varrho_{ab}^{nn}(-\alpha)$ onto the reduced vibrational number statistics $\rho_{nn}(t)$. The latter can be measured by employing the same interaction on the carrier as in the case of a quantum nondemolition measurement of the reduced motional number statistics [de Matos Filho and Vogel (1996c); Davidovich, Orszag and Zagury (1996)].[20]

Suppose that one has determined the displaced vibronic density-matrix elements $\varrho_{ij}^{nn}(-\alpha)$ ($i=1,2; j=1,2$) as functions of the displacement amplitude α. They contain the full information on the (initial) quantum state of the combined vibronic system. It may be advantageous to represent the vibronic

---

20) The interaction Hamiltonian (13.37) for $k=0$ commutes with the vibrational number operator.

quantum state in terms of the Wigner-function matrix $W_{ij}(\alpha)$ [Wallentowitz, de Matos Filho and Vogel (1997)],

$$W_{ij}(\alpha) = \frac{2}{\pi} \sum_{n=0}^{\infty} (-1)^n \varrho_{ij}^{nn}(-\alpha). \tag{13.100}$$

This Wigner-function matrix has the following properties. Its trace with respect to the electronic subsystem is the (reduced) motional Wigner function,

$$W(\alpha) = \sum_{i=1,2} W_{ii}(\alpha). \tag{13.101}$$

Moreover, integration of $W_{ij}(\alpha)$ with respect to the motional phase-space amplitude α yields the (reduced) electronic density matrix σ_{ij},

$$\sigma_{ij} = \int d^2\alpha\, W_{ij}(\alpha). \tag{13.102}$$

The method under consideration is suited to the determination of the Wigner-function matrix of entangled states of the Schrödinger-cat type, such as the state $|\Psi_4\rangle$ given in Eq. (13.63).

The method can be extended to determine the composed quantum state of the electronic subsystem and two vibrational modes [Wallentowitz, de Matos Filho, Gou and Vogel (1999)]. In particular, by taking the trace of the quantum state with respect to the electronic degrees of freedom, one readily obtains the complete information on entangled states of two vibrational modes. This renders it possible to apply the entanglement criteria considered in Section 8.5.2 to the data recorded in this manner.

References

Agarwal, G.S. and J. Banerji (1997) *Phys. Rev. A* **55**, 4007.

Bergquist, J.C., R.G. Hulet, W.M. Itano and D.J. Wineland (1986) *Phys. Rev. Lett.* **57**, 1699.

Blockley, C.A., D.F. Walls and H. Risken (1992) *Europhys. Lett.* **17**, 509.

Budini, A.A., R.L. de Matos and N. Zagury *Phys. Rev. A* **65**, 041402(R).

Cirac, J.I., A.S. Parkins, R. Blatt and P. Zoller (1993) *Phys. Rev. Lett.* **70**, 556.

Davidovich, L., M. Orszag and N. Zagury (1996) *Phys. Rev. A* **54**, 5118.

Di Fidio, C. and W. Vogel (2000) *Phys. Rev. A* **62**, 031802(R).

Diedrich, F., J.C. Bergquist, W.M. Itano and D.J. Wineland (1989) *Phys. Rev. Lett.* **62**, 403.

Dodonov, V.V., I.A. Malkin and V.I. Man'ko (1974), *Physica* **72**, 597.

Ghosh, P.K. (1995) *Ion Traps* (Clarendon Press, Oxford).

Gou, S.-C., J. Steinbach and P.L. Knight (1996) *Phys. Rev. A* **54**, R1014.

D'Helon C. and G.J. Milburn (1996) *Phys. Rev. A* **54**, R25.

Heinzen, D.J. and D.J. Wineland (1990) *Phys. Rev. A* **42**, 2977.

Janszky, J. and Y.Y. Yushin (1986) *Opt. Commun.* **59**, 151.

Kis, Z., W. Vogel and L. Davidovich (2001) *Phys. Rev. A* **64**, 033401.

Leibfried, D., D.M. Meekhof, B.E. King, C. Monroe, W.M. Itano and D. Wineland (1996) *Phys. Rev. Lett.* **77**, 4281.

Lindberg, M. and J. Javanainen (1986) *J. Opt. Soc. Am. B* **3**, 1008.

Lutterbach, L.G. and L. Davidovich (1997) *Phys. Rev. Lett.* **78**, 2547.

Man'ko, V.I., G. Marmo, E.C.G. Sudarshan and F. Zaccaria (1997) *Phys. Scr.* **55**, 528.

Man'ko, V.I., G. Marmo, A. Porzio, S. Solimeno and F. Zaccaria (2000) *Phys. Rev. A* **62**, 053407.

de Matos Filho, R.L. and W. Vogel (1996a) *Phys. Rev. Lett.* **76**, 608.

de Matos Filho, R.L. and W. Vogel (1996b) *Phys. Rev. A* **54**, 4560.

de Matos Filho, R.L. and W. Vogel (1996c) *Phys. Rev. Lett.* **76**, 4520.

de Matos Filho, R.L. and W. Vogel (1998) *Phys. Rev. A* **58**, R1661.

Meekhof, D.M., C. Monroe, B.E. King, W.M. Itano and D.J. Wineland (1996) *Phys. Rev. Lett.* **76**, 1796.

Monroe, C., D.M. Meekhof, B.E. King, S.R. Jefferts, W.M. Itano and D.J. Wineland (1995) *Phys. Rev. Lett.* **75**, 4011.

Monroe, C., D.M. Meekhof, B.E. King and D.J. Wineland (1996) *Science* **272**, 1131.

Myatt, C.J., B.E. King, Q.A. Turchette, C.A. Sackett, D. Kielpinski, W.M. Itano, C. Monroe and D.J. Wineland (2000) *Nature* **403**, 269.

Nagourney, W., J. Sandberg and H. Dehmelt (1986) *Phys. Rev. Lett.* **56**, 2797.

Neuhauser W., M. Hohenstatt, P.E. Toschek and H. Dehmelt (1978) *Phys. Rev. Lett.* **41**, 233.

Neuhauser, W., M. Hohenstatt, P.E. Toschek and H.G. Dehmelt (1980) *Phys. Rev. A* **22**, 1137.

Ozeri, R., C. Langer, J.D. Jost, B. DeMarco, A. Ben-Kish, B.R. Blakestad, J. Britton, J. Chiaverini, W.M. Itano, D.B. Hume, D. Leibfried, T. Rosenband, P.O. Schmidt and D.J. Wineland (2005) *Phys. Rev. Lett.* **95**, 030403.

Paul, W., O. Osberghaus and E. Fischer (1958) *Forschungsberichte des Wirtschafts- und Verkehrsministeriums Nordrhein-Westfalen Nr. 415* (Westdeutscher Verlag, Köln und Opladen).

Peřina, J. (1984) *Quantum Statistics of Linear and Nonlinear Optical Phenomena* (Reidel, Dordrecht, 1985).

Poyatos, J.F., J.I. Cirac and P. Zoller (1996) *Phys. Rev. Lett.* **77**, 4728.

Poyatos, J.F., R. Walser, J.I. Cirac, P. Zoller and R. Blatt (1996) *Phys. Rev. A* **53**, R1966.

Sauter, Th., W. Neuhauser, R. Blatt and P.E. Toschek (1986) *Phys. Rev. Lett.* **57**, 1696.

Stenholm, S. (1986) *Rev. Mod. Phys.* **58**, 699.

Toschek, P.E. (1985) *Ann. Phys. (Paris)* **10**, 761.

Vogel, W. and R.L. de Matos Filho (1995) *Phys. Rev. A* **52**, 4214.

Wallentowitz, S. and W. Vogel (1995) *Phys. Rev. Lett.* **75**, 2932.

Wallentowitz, S. and W. Vogel (1996) *Phys. Rev. A* **54**, 3322.

Wallentowitz, S. and W. Vogel (1997) *Phys. Rev. A* **55**, 4438.

Wallentowitz, S. and W. Vogel (1998) *Phys. Rev. A* **58**, 679.

Wallentowitz S., R.L. de Matos Filho, S.-C. Gou, and W. Vogel (1999) *Eur. Phys. J. D* **6**, 397 (1999).

Wallentowitz, S., R.L. de Matos Filho and W. Vogel (1997) *Phys. Rev. A* **56**, 1205.

Wallentowitz S., W. Vogel and P.L. Knight (1999) *Phys. Rev. A* **59**, 531 (1999).

Wineland, D.J. and H. Dehmelt (1975) *Bull. Am. Phys. Soc.* **20**, 637.

Wineland, D.J., R. Drullinger and D.F. Walls (1978) *Phys. Rev. Lett.* **40**, 1639.

A
The medium-assisted Green tensor

A.1
Basic relations

The Green tensor $G_{ij}(\mathbf{r},\mathbf{r}',\omega)$ in Eq. (2.207) has the properties

$$G_{ij}^*(\mathbf{r},\mathbf{r}',\omega) = G_{ij}(\mathbf{r},\mathbf{r}',-\omega^*), \tag{A.1}$$

$$G_{ji}(\mathbf{r}',\mathbf{r},\omega) = G_{ij}(\mathbf{r},\mathbf{r}',\omega), \tag{A.2}$$

$$\int d^3s \, \frac{\omega^2}{c^2} \, \text{Im}\,\varepsilon(\mathbf{s},\omega) \, G_{ik}(\mathbf{r},\mathbf{s},\omega) G_{jk}^*(\mathbf{r}',\mathbf{s},\omega) = \text{Im}\,G_{ij}(\mathbf{r},\mathbf{r}',\omega). \tag{A.3}$$

Obviously, Eq. (A.1) is a direct consequence of the corresponding relation (2.179) for the permittivity. To prove the reciprocity relation (A.2), we note that the Green tensor $\mathbf{G}(\mathbf{r},\mathbf{r}',\omega)$ as a function of $\mathbf{r}$ and $\mathbf{r}'$ can be regarded as being the matrix elements in a position basis of a (tensor-valued) Green operator $\hat{\mathbf{G}} = \hat{\mathbf{G}}(\omega)$ in an abstract 1-particle Hilbert space:

$$\mathbf{G}(\mathbf{r},\mathbf{r}',\omega) = \langle \mathbf{r}|\hat{\mathbf{G}}|\mathbf{r}'\rangle, \tag{A.4}$$

where the matrix elements of the position operator $\hat{\mathbf{r}}$ are given by

$$\langle \mathbf{r}|\hat{\mathbf{r}}|\mathbf{r}'\rangle = \mathbf{r}\delta(\mathbf{r}-\mathbf{r}'), \tag{A.5}$$

and the matrix elements of the associated momentum operator $\hat{\mathbf{p}}$ read

$$\langle \mathbf{r}|\hat{\mathbf{p}}|\mathbf{r}'\rangle = \frac{1}{i}\nabla\delta(\mathbf{r}-\mathbf{r}') \tag{A.6}$$

($[\hat{x}_i,\hat{p}_j]=i\delta_{ij}$). Let $\hat{\mathbf{H}} = \hat{\mathbf{H}}(\omega)$ be the tensor-valued operator

$$\hat{\mathbf{H}} = i\hat{\mathbf{p}} \times i\hat{\mathbf{p}} \times -\hat{q}^2\hat{\mathbf{I}} = \hat{\mathbf{p}}^2\hat{\mathbf{I}} - \hat{\mathbf{p}}\otimes\hat{\mathbf{p}} - \hat{q}^2\hat{\mathbf{I}} \tag{A.7}$$

($\hat{\mathbf{I}}$, unit operator), where

$$\hat{q}^2 = \frac{\omega^2}{c^2}\varepsilon(\hat{\mathbf{r}},\omega). \tag{A.8}$$

Quantum Optics, Third, revised and extended edition. Werner Vogel and Dirk-Gunnar Welsch
Copyright © 2006 WILEY-VCH Verlag GmbH & Co. KGaA, Weinheim
ISBN: 3-527-40507-0

Equation (2.207) then corresponds to the operator equation

$$\hat{H}\hat{G} = \hat{I}, \tag{A.9}$$

as can be easily seen. From Eq. (A.9) it then follows that the equation

$$\hat{G} = \hat{H}^{-1} \tag{A.10}$$

is valid, and thus we find, after multiplying it from the right by $\hat{H}$,

$$\hat{G}\hat{H} = \hat{I}. \tag{A.11}$$

Writing down Eqs (A.9) and (A.11) in the position basis, a comparison of the two equations immediately shows that the Green tensor obeys Eq. (A.2).

In order to prove Eq. (A.3), we introduce operators $\hat{O}^+$ defined by

$$(\hat{O}^+)_{ij} = (\hat{O}_{ji})^\dagger \equiv \hat{O}^\dagger_{ji}. \tag{A.12}$$

From Eq. (A.9) it then follows that

$$\hat{G}^+ \hat{H}^+ = \hat{I}. \tag{A.13}$$

Multiplying Eq. (A.9) from the left by $\hat{G}^+$ and Eq. (A.13) from the right by $\hat{G}$ and subtracting the resulting equations from each other, we find

$$\hat{G}^+(\hat{H} - \hat{H}^+)\hat{G} = \hat{G}^+ - \hat{G}. \tag{A.14}$$

From Eq. (A.7) it is easily seen that

$$(\hat{H} - \hat{H}^+) = \hat{I}(\hat{q}^{2\dagger} - \hat{q}^2). \tag{A.15}$$

Representing Eq. (A.14) [together with Eq. (A.15)] in the position basis and recalling the reciprocity relation (A.2), we eventually arrive at Eq. (A.3).

A.2
Asymptotic behavior

The Green tensor $G_{ij}(\mathbf{r}, \mathbf{r}', \omega)$ is holomorphic in the upper half-plane of complex ω, because of the holomorphic behavior of $\varepsilon(\mathbf{r}, \omega)$. In order to study the behavior of $G_{ij}(\mathbf{r}, \mathbf{r}', \omega)$ for $|\omega| \to \infty$ and $|\omega| \to 0$, we first introduce the tensor-valued projection operators

$$\hat{I}^\perp = \hat{I} - \hat{I}^\|, \quad \hat{I}^\| = \frac{\hat{\mathbf{p}} \otimes \hat{\mathbf{p}}}{\hat{\mathbf{p}}^2} \tag{A.16}$$

and decompose $\hat{H}$, Eq. (A.7), as

$$\hat{H} = (\hat{\mathbf{p}}^2 - \hat{q}^2)\hat{I}^\perp - \hat{q}^2 \hat{I}^\|. \tag{A.17}$$

Applying the Feshbach formula[1], we then may decompose the Green tensor operator $\hat{G} = \hat{H}^{-1}$, Eq. (A.10), as

$$\hat{G} = \hat{H}^{-1} = \hat{I}^{\|}(\hat{I}^{\|}\hat{H}\hat{I}^{\|})^{-1}\hat{I}^{\|}$$
$$+ [\hat{I}^{\perp} - \hat{I}^{\|}(\hat{I}^{\|}\hat{H}\hat{I}^{\|})^{-1}\hat{I}^{\|}\hat{H}\hat{I}^{\perp}]\hat{K}[\hat{I}^{\perp} - \hat{I}^{\perp}\hat{H}\hat{I}^{\|}(\hat{I}^{\|}\hat{H}\hat{I}^{\|})^{-1}\hat{I}^{\|}], \tag{A.18}$$

where

$$\hat{K} = [\hat{I}^{\perp}\hat{H}\hat{I}^{\perp} - \hat{I}^{\perp}\hat{H}\hat{I}^{\|}(\hat{I}^{\|}\hat{H}\hat{I}^{\|})^{-1}\hat{I}^{\|}\hat{H}\hat{I}^{\perp}]^{-1}. \tag{A.19}$$

It is not difficult to prove that

$$\hat{I}^{\|}\hat{H}\hat{I}^{\|} = -\hat{I}^{\|}\hat{q}^{2}\hat{I}^{\|}, \quad \hat{I}^{\|}\hat{H}\hat{I}^{\perp} = -\hat{I}^{\|}\hat{q}^{2}\hat{I}^{\perp}, \tag{A.20}$$

$$\hat{I}^{\perp}\hat{H}\hat{I}^{\|} = -\hat{I}^{\perp}\hat{q}^{2}\hat{I}^{\|}, \quad \hat{I}^{\perp}\hat{H}\hat{I}^{\perp} = \hat{I}^{\perp}(\hat{p}^{2} - \hat{q}^{2})\hat{I}^{\perp}. \tag{A.21}$$

Combining Eqs (A.18)–(A.21) and using Eq. (A.8), we obtain for $\hat{G}$, the expression

$$\hat{G} = -\frac{c^2}{\omega^2}\hat{I}^{\|}(\hat{I}^{\|}\hat{\varepsilon}\hat{I}^{\|})^{-1}\hat{I}^{\|}$$
$$+ [\hat{I}^{\perp} - \hat{I}^{\|}(\hat{I}^{\|}\hat{\varepsilon}\hat{I}^{\|})^{-1}\hat{I}^{\|}\hat{\varepsilon}\hat{I}^{\perp}]\hat{K}[\hat{I}^{\perp} - \hat{I}^{\perp}\hat{\varepsilon}\hat{I}^{\|}(\hat{I}^{\|}\hat{\varepsilon}\hat{I}^{\|})^{-1}\hat{I}^{\|}], \tag{A.22}$$

where $[\hat{\varepsilon} = \varepsilon(\hat{\mathbf{r}}, \omega)]$

$$\hat{K} = \left[\hat{I}^{\perp}\left(\hat{p}^2 - \frac{\omega^2}{c^2}\hat{\varepsilon}\right)\hat{I}^{\perp} + \frac{\omega^2}{c^2}\hat{I}^{\perp}\hat{\varepsilon}\hat{I}^{\|}(\hat{I}^{\|}\hat{\varepsilon}\hat{I}^{\|})^{-1}\hat{I}^{\|}\hat{\varepsilon}\hat{I}^{\perp}\right]^{-1}. \tag{A.23}$$

Now the desired limiting processes can be performed easily. For $|\omega| \to 0$ we find $[\hat{\varepsilon}^{(0)} = \varepsilon^{(0)}(\hat{\mathbf{r}}) \equiv \varepsilon(\hat{\mathbf{r}}, \omega = 0)]$

$$\lim_{|\omega| \to 0} \frac{\omega^2}{c^2}\hat{G} = -\hat{I}^{\|}(\hat{I}^{\|}\hat{\varepsilon}^{(0)}\hat{I}^{\|})^{-1}\hat{I}^{\|} \tag{A.24}$$

and

$$\lim_{|\omega| \to 0} \hat{q}^2\hat{G} = -\hat{\varepsilon}^{(0)}\hat{I}^{\|}(\hat{I}^{\|}\hat{\varepsilon}^{(0)}\hat{I}^{\|})^{-1}\hat{I}^{\|}. \tag{A.25}$$

For $|\omega| \to \infty$ we arrive at, on recalling that $\hat{I}^{\|}\hat{I}^{\perp} = \hat{I}^{\perp}\hat{I}^{\|} = 0$ and $\varepsilon(\mathbf{r}, \omega) \to 1$ if $|\omega| \to \infty$,

$$\lim_{|\omega| \to \infty} \frac{\omega^2}{c^2}\hat{G} = \lim_{|\omega| \to \infty} \hat{q}^2\hat{G} = -\hat{I}. \tag{A.26}$$

---

1) See, e. g., Newton, R.G. (1982) *Scattering Theory of Waves and Particles* (Springer, Berlin).

Obviously, the first term on the right-hand side of Eq. (A.22) is the singular part of $\hat{G}$ for $|\omega| \to 0$. Performing in that term the Taylor expansion

$$\varepsilon(\hat{\mathbf{r}}, \omega) = \varepsilon^{(0)}(\hat{\mathbf{r}}) + \omega\, \varepsilon^{(1)}(\hat{\mathbf{r}}) + \ldots, \tag{A.27}$$

where $\varepsilon^{(0)}(\hat{\mathbf{r}})$ is real and $\varepsilon^{(1)}(\hat{\mathbf{r}})$ is imaginary [see Eq. (2.179)], we find that

$$\operatorname{Re} G_{ij}(\mathbf{r}, \mathbf{r}', \omega) \sim \omega^{-2} \qquad (|\omega| \to 0), \tag{A.28}$$

$$\operatorname{Im} G_{ij}(\mathbf{r}, \mathbf{r}', \omega) \sim \omega^{-1} \qquad (|\omega| \to 0). \tag{A.29}$$

B
Equal-time commutation relations

In order to prove the familiar equal-time commutation relations for the medium-assisted electromagnetic field (Section 2.4.2), let us first consider the commutation relations of the electric-field strength and the vector potential. Using Eqs (2.213) and (2.218) together with Eqs (2.211) and (2.220), recalling the commutation relations (2.208) and (2.209), and applying the integral relation (A.3), after some algebra we derive

$$[\hat{E}_k(\mathbf{r}), \hat{E}_{k'}(\mathbf{r}')] = [\hat{A}_k(\mathbf{r}), \hat{A}_{k'}(\mathbf{r}')] = 0, \tag{B.1}$$

$$[\varepsilon_0 \hat{E}_k(\mathbf{r}), \hat{A}_{k'}(\mathbf{r}')] = \frac{2i\hbar}{\pi} \int d^3s \int_0^\infty d\omega \, \frac{\omega}{c^2} \, \text{Im}\, G_{kl}(\mathbf{r}, \mathbf{s}, \omega) \, \delta_{lk'}^\perp(\mathbf{s} - \mathbf{r}'). \tag{B.2}$$

In Eq. (B.2) the ω-integral can be performed by applying the rule

$$\int_0^\infty d\omega \ldots = \lim_{\epsilon \to 0} \int_\epsilon^\infty d\omega \ldots. \tag{B.3}$$

Recall that, according to Eq. (A.29), $\text{Im}\, G_{il}(\mathbf{r}, \mathbf{s}, \omega)$ behaves as ω^{-1} as ω approaches zero. Thus we may transform, on using Eqs (A.1), (A.4) and (A.16), the ω-integral into a principal-part ($\mathcal{P}$) integral, so that Eq.(B.2) reads

$$[\varepsilon_0 \hat{E}_k(\mathbf{r}), \hat{A}_l(\mathbf{r}')] = \frac{\hbar}{\pi} \int d^3s \, \mathcal{P} \int_{-\infty}^\infty \frac{d\omega}{\omega} \frac{\omega^2}{c^2} \, \langle \mathbf{r} | \hat{G}(\omega) | \mathbf{s} \rangle \langle \mathbf{s} | \hat{I}^\perp | \mathbf{r}' \rangle. \tag{B.4}$$

Note that $\langle \mathbf{r} | \hat{I}^{\perp(\|)} | \mathbf{r}' \rangle = \delta^{\perp(\|)}(\mathbf{r} - \mathbf{r}')$.

The evaluation of the ω-integral in Eq. (B.4) can be performed by means of contour-integral techniques. Since $G_{km}(\mathbf{r}, \mathbf{s}, \omega)$ is a holomorphic function of ω in the upper complex frequency half-plane with the asymptotic behavior according to Eq. (A.26), the ω-integrals can be calculated by contour integration along an infinitely small half-circle $|\omega| = \rho$, $\rho \to 0$, and an infinitely large half-circle $|\omega| = R$, $R \to \infty$, in the upper complex half-plane,

$$\mathcal{P} \int_{-\infty}^\infty d\omega \ldots = \lim_{\rho \to 0} \int_{\substack{|\omega|=\rho \\ \text{Im}\, \omega > 0}} d\omega \ldots - \lim_{R \to \infty} \int_{\substack{|\omega|=R \\ \text{Im}\, \omega > 0}} d\omega \ldots. \tag{B.5}$$

From Eq. (A.24) it follows that

$$\lim_{|\omega| \to 0} \frac{\omega^2}{c^2} \hat{G} \hat{I}^\perp = -\hat{I}^\| (\hat{I}^\| \hat{\varepsilon}^{(0)} \hat{I}^\|)^{-1} \hat{I}^\| \hat{I}^\perp = 0 \tag{B.6}$$

($\hat{I}^\| \hat{I}^\perp = 0$), and therefore the integral over the small half-circle vanishes. Finally, from Eq. (A.26) we see that

$$\lim_{|\omega|\to\infty} \frac{\omega^2}{c^2} \hat{G}\hat{I}^\perp = -\hat{I}^\perp. \tag{B.7}$$

Hence,

$$\mathcal{P}\int_{-\infty}^{\infty} \frac{d\omega}{\omega} \frac{\omega^2}{c^2} \langle \mathbf{r}|\hat{G}(\omega)\hat{I}^\perp|\mathbf{r}'\rangle = i\pi\delta^\perp(\mathbf{r}-\mathbf{r}'), \tag{B.8}$$

and the sought commutator reads

$$[\varepsilon_0 \hat{E}_k(\mathbf{r}), \hat{A}_l(\mathbf{r}')] = i\hbar \delta_{kl}^\perp(\mathbf{r}-\mathbf{r}'). \tag{B.9}$$

The corresponding commutation relations for the displacement field [Eq. (2.215) together with Eqs (2.216) and (2.211)] can be derived in a quite similar way. As expected, the result is

$$[\hat{D}_k(\mathbf{r}), \hat{D}_{k'}(\mathbf{r}')] = 0, \tag{B.10}$$

$$[\hat{D}_k(\mathbf{r}), \hat{A}_{k'}(\mathbf{r}')] = i\hbar \delta_{kk'}^\perp(\mathbf{r}-\mathbf{r}'), \tag{B.11}$$

because the polarization $\hat{\mathbf{P}}(\mathbf{r}) = \hat{\mathbf{D}}(\mathbf{r}) - \varepsilon_0 \hat{\mathbf{E}}(\mathbf{r})$ is related to the degrees of freedom of the matter and it should therefore commute with the radiation field operators. Recalling the relations $\nabla \times \hat{\mathbf{A}} = \hat{\mathbf{B}}$, $\hat{\mathbf{\Pi}} = -\varepsilon_0 \hat{\mathbf{E}}^\perp$, $-\nabla\hat{\varphi} = \hat{\mathbf{E}}^\|$, and $\hat{\mathbf{P}} = \hat{\mathbf{D}} - \varepsilon_0 \hat{\mathbf{E}}$ and using the commutation relations (B.1), (B.9)–(B.11), it is not difficult to derive further commutation relations, e.g.,

$$[\varepsilon_0 \hat{E}_k(\mathbf{r}), \hat{B}_l(\mathbf{r}')] = -i\hbar\, \epsilon_{klm}\, \partial_m^r \delta(\mathbf{r}-\mathbf{r}'), \tag{B.12}$$

$$[\hat{A}_k(\mathbf{r}), \hat{\Pi}_l(\mathbf{r}')] = i\hbar\, \delta_{kl}^\perp(\mathbf{r}-\mathbf{r}'), \tag{B.13}$$

$$[\hat{\varphi}(\mathbf{r}), \hat{\varphi}(\mathbf{r}')] = [\hat{\varphi}(\mathbf{r}), \hat{A}_k(\mathbf{r}')] = [\hat{\varphi}(\mathbf{r}), \hat{E}_k(\mathbf{r}')] = [\hat{\varphi}(\mathbf{r}), \hat{D}_k(\mathbf{r}')] = 0. \tag{B.14}$$

It should be pointed out that the commutation relations given above are also valid when additional charged particles are present. Needless to say, quantities of the medium-assisted electromagnetic field and quantities of the additional charged particles, commute.

C
Algebra of bosonic operators

C.1
Exponential-operator disentangling

Let us briefly summarize some basic rules of bosonic operator algebra.[1] For simplicity, we restrict our attention to operator functions $\hat{F}(\hat{a}, \hat{a}^\dagger)$ of a single harmonic oscillator,

$$[\hat{a}, \hat{a}^\dagger] = 1. \tag{C.1}$$

The extension to multi-mode systems is straightforward. Note that operator functions are understood in the sense of power-series expansions in the operators $\hat{a}$ and $\hat{a}^\dagger$. Let us first consider an expression of the type

$$\hat{G}(z; \hat{a}, \hat{a}^\dagger) = e^{\hat{a}z} \hat{F}(\hat{a}, \hat{a}^\dagger) e^{-\hat{a}z}, \tag{C.2}$$

where z is a c-number variable. Expanding $\hat{F}(\hat{a}, \hat{a}^\dagger)$ in a power series and performing the exponential-operator transformations step by step, by inserting the identity operator $\hat{I} = e^{\hat{a}z} e^{-\hat{a}z}$ into any pair of neighboring boson operators, we readily verify that

$$\hat{G}(z; \hat{a}, \hat{a}^\dagger) = \hat{F}(\hat{a}, e^{\hat{a}z} \hat{a}^\dagger e^{-\hat{a}z}). \tag{C.3}$$

We now differentiate $\hat{G}$ as given in Eq. (C.2) and use Eq. (C.3) to obtain

$$\frac{d\hat{G}(z; \hat{a}, \hat{a}^\dagger)}{dz} = e^{\hat{a}z}[\hat{a}, \hat{F}(\hat{a}, \hat{a}^\dagger)] e^{-\hat{a}z}$$
$$= [\hat{a}, \hat{G}(z; \hat{a}, \hat{a}^\dagger)] = [\hat{a}, \hat{F}(\hat{a}, e^{\hat{a}z} \hat{a}^\dagger e^{-\hat{a}z})]. \tag{C.4}$$

Equation (C.4) may be used to evaluate $e^{\hat{a}z} \hat{a}^\dagger e^{-\hat{a}z}$. For this purpose, we choose $\hat{F} = \hat{a}^\dagger$, so that

$$\hat{G} = e^{\hat{a}z} \hat{a}^\dagger e^{-\hat{a}z} \tag{C.5}$$

---

[1] For more details we refer the reader to standard text books on quantum mechanics.

Quantum Optics, Third, revised and extended edition. Werner Vogel and Dirk-Gunnar Welsch
Copyright © 2006 WILEY-VCH Verlag GmbH & Co. KGaA, Weinheim
ISBN: 3-527-40507-0

and

$$\frac{d\hat{G}}{dz} = e^{\hat{a}z}[\hat{a}, \hat{a}^\dagger]e^{-\hat{a}z} = 1. \tag{C.6}$$

Since $\hat{G}|_{z=0} = \hat{a}^\dagger$, from Eq. (C.6) we find that

$$\hat{G} = \hat{a}^\dagger + z. \tag{C.7}$$

Hence

$$e^{\hat{a}z}\hat{a}^\dagger e^{-\hat{a}z} = \hat{a}^\dagger + z, \tag{C.8}$$

$$e^{-\hat{a}^\dagger z^*}\hat{a}e^{\hat{a}^\dagger z^*} = \hat{a} + z^*. \tag{C.9}$$

Equation (C.8) [or (C.9)] is a special case of the Baker–Hausdorff lemma

$$e^{\hat{A}z}\hat{B}e^{-\hat{A}z} = \sum_{n=0}^{\infty} \frac{z^n}{n!}[\hat{A}, \hat{B}]_n, \tag{C.10}$$

where

$$[\hat{A}, \hat{B}]_n = [\hat{A}, [\hat{A}, \hat{B}]_{n-1}], \quad [\hat{A}, \hat{B}]_0 = \hat{B}, \tag{C.11}$$

with $\hat{A}$ and $\hat{B}$ being arbitrary operators. It can be proved straightforwardly by power-series expansion in z of the operator $e^{\hat{A}z}\hat{B}e^{-\hat{A}z}$.

We now return to Eq. (C.4). Using the relation (C.8), we may rewrite Eq. (C.4) as

$$\frac{d\hat{G}(z; \hat{a}, \hat{a}^\dagger)}{dz} = [\hat{a}, \hat{F}(\hat{a}, \hat{a}^\dagger + z)]. \tag{C.12}$$

From this it is evident that

$$\left.\frac{d\hat{G}}{dz}\right|_{z=0} = [\hat{a}, \hat{F}(\hat{a}, \hat{a}^\dagger)]. \tag{C.13}$$

On the other hand, from Eq. (C.3) together with Eq. (C.8), we find that

$$\hat{G}(z; \hat{a}, \hat{a}^\dagger) = \hat{F}(\hat{a}, \hat{a}^\dagger + z), \tag{C.14}$$

and thus

$$\left.\frac{d\hat{G}}{dz}\right|_{z=0} = \lim_{z\to 0}\frac{d\hat{F}(\hat{a}, \hat{a}^\dagger + z)}{dz} = \frac{\partial \hat{F}(\hat{a}, \hat{a}^\dagger)}{\partial \hat{a}^\dagger}. \tag{C.15}$$

Comparing Eqs (C.13) and (C.15), we obtain the relation

$$[\hat{a}, \hat{F}(\hat{a}, \hat{a}^\dagger)] = \frac{\partial \hat{F}(\hat{a}, \hat{a}^\dagger)}{\partial \hat{a}^\dagger}. \tag{C.16}$$

The relation

$$[\hat{F}(\hat{a},\hat{a}^\dagger),\hat{a}^\dagger] = \frac{\partial \hat{F}(\hat{a},\hat{a}^\dagger)}{\partial \hat{a}} \tag{C.17}$$

can be proved analogously.

Let us now consider an exponential operator of the form

$$\hat{G}(z;\hat{a},\hat{a}^\dagger) = \exp\{[\hat{F}_1(\hat{a},\hat{a}^\dagger) + \hat{F}_2(\hat{a},\hat{a}^\dagger)]z\} \tag{C.18}$$

and seek a representation of $\hat{G}$ in the form of a product of two operator exponentials. To find this disentangled form of $\hat{G}$, we make the ansatz

$$\hat{G}(z;\hat{a},\hat{a}^\dagger) = \hat{G}_1(z;\hat{a},\hat{a}^\dagger)\,\hat{G}_2(z;\hat{a},\hat{a}^\dagger), \tag{C.19}$$

$$\hat{G}_1(z;\hat{a},\hat{a}^\dagger) = \exp[\hat{F}_1(\hat{a},\hat{a}^\dagger)z]. \tag{C.20}$$

From Eqs (C.18) and (C.19) it follows that $\hat{G}_2$ obeys the differential equation

$$\frac{d\hat{G}_2}{dz} = \hat{F}'_2 \hat{G}_2, \qquad \hat{F}'_i = \hat{G}_1^{-1} \hat{F}_i \hat{G}_1 \tag{C.21}$$

($i=1,2$). As is well known, the solution of Eq. (C.21) (with the initial condition $\hat{G}_2|_{z=0}=1$) may be written in the form of the $\mathcal{Z}$-ordered exponential (z, real number)

$$\hat{G}_2(z;\hat{a},\hat{a}^\dagger) = \mathcal{Z} \exp\left[\int_0^z dz'\, \hat{F}'_2(z';\hat{a},\hat{a}^\dagger)\right], \tag{C.22}$$

where the $\mathcal{Z}$ ordering is defined by

$$\mathcal{Z}\hat{F}'_2(z_1;\hat{a},\hat{a}^\dagger)\hat{F}'_2(z_2;\hat{a},\hat{a}^\dagger) = \begin{cases} \hat{F}'_2(z_1;\hat{a},\hat{a}^\dagger)\hat{F}'_2(z_2;\hat{a},\hat{a}^\dagger) & \text{if } z_1 > z_2, \\ \hat{F}'_2(z_2;\hat{a},\hat{a}^\dagger)\hat{F}'_2(z_1;\hat{a},\hat{a}^\dagger) & \text{if } z_2 > z_1. \end{cases} \tag{C.23}$$

The operator $\hat{F}'_2$ obviously satisfies the differential equation

$$\frac{d\hat{F}'_2}{dz} = \hat{G}_1^{-1}[\hat{F}_2,\hat{F}_1]\hat{G}_1 = [\hat{F}'_2,\hat{F}'_1]. \tag{C.24}$$

If the commutator $[\hat{F}_2,\hat{F}_1]$ is a c number, that is, $[\hat{F}_2,\hat{F}_1]=\alpha$, from Eq. (C.24) it follows that $\hat{F}'_2 = \hat{F}_2 + \alpha z$ (note that the initial condition $\hat{F}'_2|_{z=0} = \hat{F}_2$ holds). Substituting this expression into Eq. (C.22), we can readily see that, in the present case, the $\mathcal{Z}$-ordering symbol can be omitted and the integration over z can be performed directly in the exponent. In this way we derive

$$\hat{G}_2 = \exp(\hat{F}_2 z + \tfrac{1}{2}\alpha z^2), \tag{C.25}$$

so that Eqs (C.18)–(C.20) lead to the disentangling prescription[2]

$$\exp(\hat{F}_1 z + \hat{F}_2 z) = \exp(\hat{F}_1 z) \exp(\hat{F}_2 z) \exp\left(\tfrac{1}{2}[\hat{F}_2, \hat{F}_1] z^2\right), \tag{C.26}$$

which for $z = 1$ yields the well-known Baker–Campbell–Hausdorff formula. Making the identifications $\hat{F}_1 = \alpha_1 \hat{a}^\dagger$ and $\hat{F}_2 = \alpha_2 \hat{a}$ (with α_1 and α_2 being c numbers), we can easily see that Eq. (C.26) leads to the relation ($z=1$)

$$e^{\alpha_1 \hat{a}^\dagger + \alpha_2 \hat{a}} = e^{\alpha_1 \hat{a}^\dagger} e^{\alpha_2 \hat{a}} e^{\tfrac{1}{2}\alpha_1 \alpha_2}. \tag{C.27}$$

C.2
Normal and anti-normal ordering

In quantum optics, various kinds of operator ordering play an important role. In this context it is often useful, and in certain cases necessary, to bring operator functions into a given order. Let us consider the normal ordering indicated by the $\mathcal{N}$ symbol or the $::$ notation, which means ordering of the operators $\hat{a}$ and $\hat{a}^\dagger$ with the (creation) operators $\hat{a}^\dagger$ to the left of the (annihilation) operators. To bring a given operator function $\hat{F}(\hat{a}, \hat{a}^\dagger)$ into its normally ordered form $\hat{F}^{(N)}(\hat{a}, \hat{a}^\dagger)$ [by means of the commutation relation (C.1)], the operators $\hat{a}$ and $\hat{a}^\dagger$ in $\hat{F}(\hat{a}, \hat{a}^\dagger)$ must be rearranged in such a way that

$$\hat{F}(\hat{a}, \hat{a}^\dagger) = \hat{F}^{(N)}(\hat{a}, \hat{a}^\dagger) = :\hat{F}^{(N)}(\hat{a}, \hat{a}^\dagger):. \tag{C.28}$$

An example is given in Eq. (C.27). To formulate a general rule, let us consider an operator function

$$\hat{F}(\hat{a}, \hat{a}^\dagger) = \hat{a}\, \hat{G}(\hat{a}, \hat{a}^\dagger) \tag{C.29}$$

and assume that the normally ordered form $\hat{G}^{(N)}$ of the operator $\hat{G}$ is known, so that

$$\hat{F}(\hat{a}, \hat{a}^\dagger) = \hat{a} \hat{G}^{(N)}(\hat{a}, \hat{a}^\dagger), \tag{C.30}$$

which may be rewritten as

$$\hat{F}(\hat{a}, \hat{a}^\dagger) = [\hat{a}, \hat{G}^{(N)}(\hat{a}, \hat{a}^\dagger)] + \hat{G}^{(N)}(\hat{a}, \hat{a}^\dagger)\, \hat{a}. \tag{C.31}$$

Using Eq. (C.16) yields

$$\hat{F}(\hat{a}, \hat{a}^\dagger) = \frac{\partial \hat{G}^{(N)}(\hat{a}, \hat{a}^\dagger)}{\partial \hat{a}^\dagger} + \hat{G}^{(N)}(\hat{a}, \hat{a}^\dagger)\, \hat{a}. \tag{C.32}$$

[2] For a large variety of examples of disentangling prescriptions, see Wilcox (1967).

C.2 Normal and anti-normal ordering

The expression on the right-hand side of Eq. (C.32) obviously represents the operator $\hat{F}$ in normal order. We may therefore write

$$\hat{F}(\hat{a}, \hat{a}^\dagger) = \frac{\partial \hat{G}^{(N)}(\hat{a}, \hat{a}^\dagger)}{\partial \hat{a}^\dagger} + \hat{G}^{(N)}(\hat{a}, \hat{a}^\dagger)\,\hat{a} = :\left(\hat{a} + \frac{\partial}{\partial \hat{a}^\dagger}\right)\hat{G}^{(N)}(\hat{a}, \hat{a}^\dagger):.$$

Clearly, since in $\hat{G}^{(N)}$ the operators $\hat{a}^\dagger$ are to the left of the operators $\hat{a}$, the latter may be replaced by $\hat{a} + \partial/\partial \hat{a}^\dagger$:

$$\hat{F}^{(N)}(\hat{a}, \hat{a}^\dagger) = :\left(\hat{a} + \frac{\partial}{\partial \hat{a}^\dagger}\right)\hat{G}^{(N)}\left(\hat{a} + \frac{\partial}{\partial \hat{a}^\dagger}, \hat{a}^\dagger\right):. \tag{C.33}$$

Recalling Eq. (C.30), we finally arrive at the result that

$$\hat{F}^{(N)}(\hat{a}, \hat{a}^\dagger) = :\hat{F}\left(\hat{a} + \frac{\partial}{\partial \hat{a}^\dagger}, \hat{a}^\dagger\right):. \tag{C.34}$$

It is worth noting that this derivation implies that Eq. (C.34) not only applies to the particular operator function in Eq. (C.29) but to any operator function allowing power-series expansion. In the derivation of Eq. (C.34) we started from an operator function of the form $\hat{F}(\hat{a}, \hat{a}^\dagger) = \hat{a}\hat{G}(\hat{a}, \hat{a}^\dagger)$. If we start from $\hat{F}(\hat{a}, \hat{a}^\dagger) = \hat{G}(\hat{a}, \hat{a}^\dagger)\hat{a}^\dagger$ then, on the basis of similar arguments, it may be shown that an operator function $\hat{F}$ can also be brought into normal order by applying the relation

$$\hat{F}^{(N)}(\hat{a}, \hat{a}^\dagger) = :\hat{F}\left(\hat{a}, \hat{a}^\dagger + \overleftarrow{\frac{\partial}{\partial \hat{a}}}\right):. \tag{C.35}$$

In contrast to Eq. (C.34), in Eq. (C.35) the operators $\hat{a}^\dagger$ are replaced by $\hat{a}^\dagger + \overleftarrow{\partial/\partial \hat{a}}$, where the arrow indicates that $\partial/\partial \hat{a}$ must be thought of as acting on the $\hat{a}$ operators to its left.

We finally note that the case of anti-normal order can be treated quite similarly. An operator function is said to be in anti-normal order when the operators $\hat{a}$ and $\hat{a}^\dagger$ are ordered with the operators $\hat{a}^\dagger$ to the right of the operators $\hat{a}$. We indicate this kind of ordering prescription by the $\mathcal{A}$ symbol or the $\ddagger\ddagger$ notation. To bring a given operator function $\hat{F}(\hat{a}, \hat{a}^\dagger)$ into the anti-normally ordered form $\hat{F}^{(A)}(\hat{a}, \hat{a}^\dagger)$,

$$\hat{F}(\hat{a}, \hat{a}^\dagger) = \hat{F}^{(A)}(\hat{a}, \hat{a}^\dagger) = \ddagger \hat{F}^{(A)}(\hat{a}, \hat{a}^\dagger)\ddagger, \tag{C.36}$$

we may use, in close analogy to Eqs (C.34) and (C.35), the relations

$$\hat{F}^{(A)}(\hat{a}, \hat{a}^\dagger) = \ddagger \hat{F}\left(\hat{a}, \hat{a}^\dagger - \frac{\partial}{\partial \hat{a}}\right)\ddagger, \tag{C.37}$$

$$\hat{F}^{(A)}(\hat{a}, \hat{a}^\dagger) = \ddagger \hat{F}\left(\hat{a} - \overleftarrow{\frac{\partial}{\partial \hat{a}^\dagger}}, \hat{a}^\dagger\right)\ddagger. \tag{C.38}$$

References

Wilcox, R.M. (1967) *J. Math. Phys.* **8**, 962.

D
Sampling function for the density matrix in the number basis

In order to show that the function $f_{mn}(x)$ defined by Eq. (7.70) solves the integral equation (7.68), we follow the derivation given by Leonhardt (1997). We first rewrite Eq. (7.69) as

$$\phi_n''(x) = [u(x) - n]\phi_n(x), \qquad (D.1)$$

where

$$u(x) = \tfrac{1}{4}x^2 - \tfrac{1}{2}. \qquad (D.2)$$

From Eq. (D.1) it follows that

$$(k+l)\phi_k(x)\phi_l(x) = 2u(x)\phi_k(x)\phi_l(x) + 2\phi_k'(x)\phi_l'(x) - [\phi_k(x)\phi_l(x)]''. \qquad (D.3)$$

Thus we may write

$$(k+l)\int dx\, \phi_k(x)\phi_l(x)[\phi_m(x)\phi_n(x)]' = -\int dx\, [\phi_k(x)\phi_l(x)]''[\phi_m(x)\phi_n(x)]'$$
$$+ \int dx\, [2u(x)\phi_k(x)\phi_l(x) + 2\phi_k'(x)\phi_l'(x)][\phi_m(x)\phi_n(x)]'. \qquad (D.4)$$

Differentiating Eq. (D.3) and using Eq. (D.1), we derive

$$[\phi_m(x)\phi_n(x)]''' = 2[2u(x) - m - n][\phi_m(x)\phi_n(x)]'$$
$$+ 2u'(x)\phi_m(x)\phi_n(x) - (m-n)W_{mn}(x), \qquad (D.5)$$

where

$$W_{mn}(x) = \phi_m(x)\phi_n'(x) - \phi_m'(x)\phi_n(x). \qquad (D.6)$$

In the first integral on the right-hand side of Eq. (D.4) we integrate by parts,[1]

$$\int dx\, [\phi_k(x)\phi_l(x)]''[\phi_m(x)\phi_n(x)]' = \int dx\, \phi_k(x)\phi_l(x)[\phi_m(x)\phi_n(x)]''', \qquad (D.7)$$

---

1) Here and in the following, we assume that the integrands which are to be taken at $x = \pm\infty$ vanish, which is of course the case for regular wave functions. The assumption also remains correct if one of the four wave functions is irregular.

and substitute into the resulting integral the expression (D.5). Equation (D.4) then takes the form of

$$(k+1)\int dx\, \phi_k(x)\phi_l(x)[\phi_m(x)\phi_n(x)]'$$
$$= \int dx\, \phi_k(x)\phi_l(x)\{2[m+n-u(x)][\phi_m(x)\phi_n(x)]' - 2u'(x)\phi_m(x)\phi_n(x)\}$$
$$+ \int dx\, \{2\phi'_k(x)\phi'_l(x)[\phi_m(x)\phi_n(x)]' + W'_{kl}(x)W_{mn}(x)\}, \tag{D.8}$$

where the relations

$$W'_{mn}(x) = (m-n)\phi_m(x)\phi_n(x) \tag{D.9}$$

and $m - n = k - l$ have been used. Equation (D.9) follows by differentiating Eq. (D.6) and applying Eq. (D.1). The second integral on the right-hand side of Eq. (D.8) can be rewritten, on integrating by parts, as

$$\int dx\, \{2\phi'_k(x)\phi'_l(x)[\phi_m(x)\phi_n(x)]' + W'_{kl}(x)W_{mn}(x)\}$$
$$= -\int dx\, \{2[\phi'_k(x)\phi'_l(x)]'\phi_m(x)\phi_n(x) + W_{kl}(x)W'_{mn}(x)\}. \tag{D.10}$$

Application of Eq. (D.1) [together with Eq. (D.6)] yields the relation

$$2[\phi'_k(x)\phi'_l(x)]' = [2u(x) - k - l][\phi_k(x)\phi_l(x)]' - (k-l)W_{kl}(x). \tag{D.11}$$

Substituting it into Eq. (D.10), we derive ($m - n = k - l$)

$$\int dx\, \{2\phi'_k(x)\phi'_l(x)[\phi_m(x)\phi_n(x)]' + W'_{kl}(x)W_{mn}(x)\}$$
$$= -\int dx\, [2u(x) - k - l][\phi_k(x)\phi_l(x)]'\phi_m(x)\phi_n(x)$$
$$= \int dx\, \phi_k(x)\phi_l(x)\{2u'(x)\phi_m(x)\phi_n(x) + [2u(x) - k - l][\phi_m(x)\phi_n(x)]'\}. \tag{D.12}$$

We now combine Eqs (D.8) and (D.12) to obtain

$$2(m+n-k-l)\int dx\, \phi_k(x)\phi_l(x)[\phi_m(x)\phi_n(x)]' = 0, \tag{D.13}$$

from which it follows that

$$\int dx\, \phi_k(x)\phi_l(x)[\phi_m(x)\phi_n(x)]' = C_{mn}\delta_{km}\delta_{ln} \qquad (k-l=m-n). \tag{D.14}$$

Obviously, $C_{mn} = 0$ if $\phi_m(x)$ and $\phi_n(x)$ are both regular wave functions, i.e., $\phi_m(x) = \psi_m(x)$ and $\phi_n(x) = \psi_n(x)$. Let us now assume that one of the wave functions is irregular, i.e., $\phi_m(x) = \psi_m(x)$ and $\phi_n(x) = \chi_n(x)$. Recalling Eq. (D.9) [together with Eq. (D.6)], we can easily see that the Wronskian

$$W_n = \psi_n(x)\chi'_n(x) - \psi'_n(x)\chi_n(x) \tag{D.15}$$

is a constant that must be nonvanishing, because of the linear independence of the functions $\psi_n(x)$ and $\chi_n(x)$. We thus derive

$$\int dx\, \psi_m(x)\psi_n(x)[\psi_m(x)\chi_n(x)]'$$
$$= \int dx\, \{\psi_m(x)\chi_n(x)[\psi_m(x)\psi_n(x)]' + \psi_m^2(x)W_n\}$$
$$= -\int dx\, \psi_m(x)\psi_n(x)[\psi_m(x)\chi_n(x)]' + W_n, \tag{D.16}$$

from which it follows that

$$\int dx\, \psi_m(x)\psi_n(x)[\psi_m(x)\chi_n(x)]' = \tfrac{1}{2}W_n \tag{D.17}$$

($C_{mn} = W_n/2$). Choosing $W_n = 2/\pi$, we just arrive at Eq. (7.68) together with Eq. (7.71).

References

Leonhardt, U. (1997) *Measuring the Quantum State of Light* (Cambridge University Press).

Index

a

absorption matrix 211
adiabatic elimination 372
Airy formula 199
angular momentum density 17
annihilation operator 24, 74
anti-bunched light 4, 5, 270
– condition 272
– intensity correlation 270, 271
– resonance fluorescence 272, 273, 375
– sub-Poissonian light 274
atom optics 443
atomic system
– V configuration 394, 398
– Λ configuration 391
– absorption 169
– balance equation 372
– Bloch equation 164, 168, 370, 388, 392
– bound state 51
– damped 161
– dephasing 163, 169
– depopulation rate 168
– dipole approximation 51, 353
– dipole transition 177, 338, 368, 448
– dressed state 384, 385
– electric-dipole operator 51
– energy relaxation 161, 163, 169
– filling rate 168
– flip operator 52
– induced emission 169
– ladder configuration 398
– Langevin equation 162, 163, 164, 165
– level shift 163, 342, 347
– Lorentz force 353
– master equation 163, 164, 166
– multipole transition 52, 449
– phase relaxation 169
– photodetector 174
– polarizability tensor 356
– population inversion 168
– quantum jump 395, 398
– radiation force 354
– radiationless dephasing 371
– radiationless relaxation 165
– radiative damping 165, 168
– resonance fluorescence 367
– saturation 374
– spontaneous emission 168, 338, 402
– three-level system 392
– transition rate 163
– transition-dipole moment 409
– trapped atom 443
– trapped ion 383, 398, 404
– two-level system 161, 339, 370, 408
– van der Waals force 354
– van der Waals potential 356
– vibronic coupling 398

b

Baker–Campbell–Hausdorff relation 490
Baker–Hausdorff lemma 488
balance equation 17
beam splitter 8, 9, 198, 205
– absorbing 210
– absorption matrix 211
– asymmetric 206
– coherent-state transformation 210, 212
– commutation relation 206
– dielectric plate 205
– homodyne detection 213
– input-output relation 206, 210, 211
– phase-space function 208, 212
– quantum-state transformation 208, 212
– reflectance 206
– SU(2) disentangling 210
– SU(2) group transformation 208
– SU(4) group transformation 211
– symmetric 206
– transformation matrix 211
– transmittance 206
– U(2) group transformation 206
– unitary transformation operator 208
Bernoulli transformation 193
– inverse 193
Bloch equation 7, 164, 168
– Laplace transformed 388
– resonance fluorescence 370, 388
– semi-classical 370

Quantum Optics, Third, revised and extended edition. Werner Vogel and Dirk-Gunnar Welsch
Copyright © 2006 WILEY-VCH Verlag GmbH & Co. KGaA, Weinheim
ISBN: 3-527-40507-0

– three-level system   392
Bochner theorem   282, 287
Born approximation   141
Born–Oppenheimer state   399
bosonic algebra   487
– exponential-operator disentangling   487
bosonic system
– coherent state   79
– displaced number state   87
– number state   73
– phase state   104
– quadrature eigenstate   102
– quantum state   73
– squeezed state   88
bunched light   3, 272

c

canonical equation
– electromagnetic field   22
canonical variable
– center-of-mass motion   445
– electromagnetic field   21, 23, 24, 31, 47
– matter   31
– mode expansion   23
Casimir effect   337
Casimir force   26, 353, 360
– Green tensor   364
– perfectly reflecting plate   365
– planar structure   364
– stress   363, 364
Casimir–Polder force, see van der Waals force
cavity   299
– advanced Langevin equation   313
– cavity QED   299
– characteristic function   329
– commutation relation   317, 318, 321
– correlation function   323, 324, 326
– damping   299, 328
– damping rate   304, 328
– detection of light   435
– dielectric plate   301
– dissipative channel   328
– eigenfrequency   304
– external field   313, 321, 323, 326
– extraction efficiency   334
– Hamiltonian   327
– incoming field   306, 314, 315
– input-output relation   316, 325, 329
– input/output   9
– internal field   308, 318, 326
– Langevin equation   308, 312, 328
– leaky   15, 135, 299
– microwave cavity   428
– mode function   301
– multi-mode field   300, 306
– nonmonochromatic mode   311, 331
– outgoing field   306, 314, 315
– phase-space function   333
– QND measurement   437
– quality   303, 350
– quantum-state extraction   329
– quantum-state preparation   431
– quantum-state reconstruction   435, 438
– reconstruction of the Wigner function   440
– response function   302
– source-quantity representation   307, 314, 318, 320, 321
– spectral response function   303
– spontaneous emission   349, 415
– unwanted losses   327, 328
– Wigner function   333
chaotic light, see thermal state
charge conservation   17
charge density
– noise   45
coarse-grained averaging   54, 142, 308, 313
coherence condition   85
coherent state   3, 79, 272
– P function   129
– electric-field variance   84, 86
– light pulse   87
– mean field   84, 86, 87
– mean photon number   82, 85
– multi-mode system   85
– nonlinear   470
– nonorthogonality   83
– number basis   82
– number distribution   82
– over-completeness   82, 84
– photon-number variance   82, 85
– Poisson counting statistics   181, 274
– resolution of the identity   83
– single-mode system   79
– squeezed   90
commutation relation
– annihilation operator   25, 28
– beam splitter   206
– bosonic   25, 28, 74, 190
– cavity   317, 318, 321
– center-of-mass motion   445
– creation operator   25, 28
– damped harmonic oscillator   136
– different times   6, 16, 65, 66, 68, 136, 154, 190, 318
– electromagnetic field   21, 24
– equal times   21, 24, 25, 27, 31, 36, 43, 90, 136, 190, 317, 318, 445, 485
– free-space electromagnetic field   66
– noise generator   136, 319
– system–reservoir   154
– time delay   6, 68

continuity equation   17, 45
correlation function
– cavity output   323, 324, 326
– classical counterpart   266
– damped system   169
– detected radiation   182, 186, 189, 266
– electromagnetic field   16, 69
– input-output relation   326
– intensity   184, 186, 203, 228, 266, 270, 375
– resonance fluorescence   368
– source quantity   16, 71
– spectral   202
Coulomb energy   20, 37, 43, 48, 49
Coulomb gauge   19, 22, 26, 47
– generalized   42
Coulomb potential   19
creation operator   24, 74
current density
– noise   45
– transverse   19

d

damping   3, 135
– (dynamic) system   135
– Born approximation   141
– center-of-mass motion   448
– coarse-grained averaging   142
– commutation relation   136, 154
– correlation time   142
– decay time   142
– density matrix   147
– expectation value   146
– Fokker–Planck equation   148, 156, 160
– harmonic oscillator   138, 151
– Heisenberg equation of motion   139
– Heisenberg picture   139, 170
– Langevin equation   135, 137, 142, 143, 145, 146
– leaky cavity   139, 328
– level shift   153, 160, 163
– line broadening   386
– Markov approximation   142, 145, 154, 169
– Markovian   136
– master equation   147, 155, 160, 163
– micromaser   432
– multi-level system   168
– multi-time correlation   169
– phase-space representation   148
– quantum regression theorem   169, 172
– radiationless dephasing   158
– rate   143, 154, 448
– reservoir   135
– Schrödinger picture   148, 170
– slowly varying amplitude operator   139
– system–reservoir Hamiltonian   137
– two-level system   161

dark resonance   391
decoherence   12
– nonlinear Jaynes–Cummings model   454
density matrix
– P function   267
– characteristic function   248
– coherent-state basis   116
– damping   147
– direct sampling   252, 255
– equation of motion   115, 175
– Jaynes–Cummings model   411
– master equation   147, 155, 160, 163
– phase-space representation   126
– properties   114
– reconstruction   10, 248, 254, 477
– reduced   116, 147
– relation to Q function   250
– relation to displaced number distribution   254
– relation to quadrature distribution   248, 250
– relation to Wigner function   249
– sampling function   493
– statistical operator   113
density operator, see density matrix
dielectric medium   15
– absorbing   44
– causality   40
– dispersing   44
– dissipation-fluctuation theorem   40
– frequency-dependent permittivity   15
– Kramers–Kronig relation   41
– multi-slab configuration   303
– noise charge density   45
– noise current density   45
– noise polarization   40
– nonabsorbing   41
– nondispersing   41
– permittivity   41
– susceptibility   40
dielectric plate   9
– beam splitter   205
– mode function   198
– response function   199, 201
dipole approximation   7, 51, 61, 176
– Lorentz force   353
– radiation force   354
– resonance fluorescence   367
– trapped atom   448
dipole-forbidden transition   52
dispersion relation   26, 44
displaced number distribution
– measurement   215, 439
– relation to density matrix   254
– relation to phase-space representation   256
displaced number state   87, 216

– completeness relation 88
– number basis 88
displacement operator 80
– s-order 121
– anti-normal order 120
– normal order 119
dissipative system, *see* reservoir
dressed-atom state 384, 385, 410
– level splitting 386
– transition 386
duality transformation 360
dynamic Stark effect 410

e

effective Hamiltonian 7, 56, 72
– four-wave mixing 278
– squeezed-light generation 278
– three-photon resonance 59
– two-photon absorption/emission 56
Einstein's hypothesis 2
Einstein–Podolsky–Rosen (EPR)
– paradox 291
– state 434
electromagnetic field 16
– balance equation 17
– canonical equation 22
– canonical variable 21, 24, 47
– commutation relation 6, 21, 24, 25, 27, 31, 36, 43, 66, 68, 190, 317, 318, 321, 485
– correlation function 69, 182, 184, 186, 202, 266, 270, 323, 368
– Coulomb gauge 19
– dielectric medium 39
– energy density 17
– free field 6, 20, 62, 65, 307, 314
– Green tensor representation 47
– Hamiltonian 21, 24, 25, 29, 32, 36
– input-output relation 316
– input/output 72
– Lagrangian 20
– longitudinal 19
– medium-assisted 44, 337
– mode expansion 22, 27, 28, 36, 44, 61, 86, 100, 305, 448
– momentum density 17
– monochromatic mode 22
– nonmonochromatic mode 28
– photoelectric detection 173
– photon 24
– propagation function 62, 65
– quantum-state reconstruction 237
– radiation 32
– reservoir 164
– resonance fluorescence 367
– source field 62, 65, 307, 315, 340, 367, 396
– source-quantity representation 60

– spontaneous emission 338
– time-dependent commutation relation 65
– transverse 19
– traveling plane wave 26, 27
– vacuum energy 25
– wave equation 22
electromagnetic field/matter 30
– canonical variable 31, 35
– Coulomb energy 20, 37, 48, 49
– coupling 32, 37, 50
– dielectric medium 39
– elementary excitation 46
– Hamiltonian 31, 32, 35, 43, 46, 47, 49
– Lagrangian 18, 20, 34, 42
– magnetization 33
– minimal coupling 31, 43, 47, 56
– multipolar coupling 33, 35, 49, 51, 56
– polarization 33, 40
– semi-classical 370
energy balance 17
energy density 17
energy transfer 55, 337
entanglement 6, 265, 290
– definition 292
– entropic measure 292
– EPR 291, 435
– moments-based criteria 295
– normally ordered moments 294
– partial transposition 293
– Peres–Horodecki condition 292
– special conditions 295
Euler–Lagrange equation 19, 20

f

Fabry–Perot 15, 198
Fano factor
– electronic 194
– photonic 194
Fock state, *see* number state
Fokker–Planck equation 148, 156, 160
– harmonic oscillator 156, 160
four-port device
– beam splitter 198
– dielectric plate 198
– spectral filter 198
four-wave mixing 278
free electromagnetic field, *see* electromagnetic field
frequency mixing 56
frequency shift, *see* level shift

g

gauge freedom 19
Glauber state, *see* coherent state
Green tensor 45, 481
– Casimir force 364

– field expansion   47
– free space   65, 342
– scattering part   342
– source-quantity representation   64
– spontaneous emission   341
– van der Waals force   357

h

Hamilton's principle   18, 20
Hamiltonian
– approximate   16, 50
– center-of-mass motion   446
– center-of-mass motion/light   446
– decomposition   32, 36
– dynamic/dissipative system   163
– effective   7, 56, 59, 72, 278
– electromagnetic field   21, 24, 25, 29, 32, 36
– electromagnetic field/matter   31, 32, 35, 36, 43, 49
– Jaynes–Cummings   408, 453
– matter   32, 36
– medium-assisted electromagn. field   46
– mode expansion   24
– nonlinear Jaynes–Cummings   452
– nonmonochromatic modes   29
– photoelectric detection of light   174
– system–reservoir   137, 151, 159, 161, 165
– trapped atom   446
– vibronic system   398
Hanbury Brown–Twiss experiment   3, 270
harmonic oscillator
– annihilation/creation operator   25
– balance equation   156, 158
– center-of-mass motion   446
– coherent state   79
– damped   135, 151
– dephasing   156, 158
– displaced number state   87
– energy relaxation   151, 156, 158
– Fokker–Planck equation   156, 158, 160
– irregular wave function   252
– Langevin equation   135, 151, 160
– master equation   155, 160
– number state   73
– phase relaxation   158
– phase state   104
– quadrature state   102
– quantum   24
– radiationless dephasing   158
– radiationless relaxation   159
– squeezed state   88
– transition rate   153, 156
heat bath, *see* reservoir
Heisenberg equation of motion
– damping   139
– electromagnetic field/matter   53, 61

– free electromagnetic field   22
– photonic   25, 61
– source quantity   61
– two-photon absorption/emission   56
Heisenberg picture   22, 60, 139, 146, 170, 305
Heisenberg's uncertainty principle   97, 135
Helmholtz equation   22, 44, 198, 301
higher-order harmonics   56
homodyne detection   10, 15, 205, 281
– Q function   10, 226
– balanced eight-port scheme   10, 223
– balanced four-port scheme   10, 217, 266
– beam splitter   205, 210, 213
– correlation measurement   228, 231, 266
– difference-count probability   10, 218
– displaced photon-number statistics   216
– four-port scheme   213
– higher-order moments   231
– joint-event probability   218
– local oscillator   205, 213, 219, 225
– mean number of counts   213
– photon-number fluctuation   214
– quadrature   213, 214, 217, 220
– quantum-state displacement   216
– shot-noise level   214
– signal field   205
– squeezed light   215, 280, 281
– unbalanced four-port scheme   213
– variance of counts   214
hot luminescence   400
Husimi function, *see* phase-space representation, Q function

i

idler field   59
input-output relation
– beam splitter   206, 210, 211
– cavity   316
– correlation function   326
– spectral filter   200
intensity correlation, *see* correlation function, intensity
interaction Hamiltonian
– approximate   50
– bound atomic state   51
– electric-dipole approximation   51, 61, 176
– many atoms   61
– minimal coupling   33, 51, 446
– multipolar coupling   37
– photoelectric detection of light   174, 176
– rotating-wave approximation   53, 176
– trapped atom   446, 448, 451
interaction picture   175, 451
intra-atomic electric field   51

inversion
– population   168

j
Jaynes–Cummings model   6, 7, 351, 408
– coherent preparation   422
– collapse   415, 416, 419
– decoherence effects   454
– density matrix   411
– eigenvalue problem   409
– electronic-state dynamics   413
– field dynamics   424
– Hamiltonian   408, 412
– level splitting   410
– micromaser   428
– multi-photon transition   412
– nonlinear   12, 449, 452
– reduced density matrix   413
– revival   415, 416, 417, 419
– spontaneous emission   415
– sub-Poissonian light   425
– time-evolution operator   411
– trapped atom   449
– two-photon transition   419
– vacuum Rabi oscillation   415

k
Kramers–Kronig relation   15, 41

l
Lagrangian
– Coulomb gauge   20
– electromagnetic field   20
– electromagnetic field/matter   18, 20, 34, 42
– generalized Coulomb gauge   42
Lamb shift, *see* level shift
Lamb–Dicke parameter   450
Lamb–Dicke regime   451
Lambert–Beer law   196
– extinction coefficient   197
Langevin equation   135, 137, 142, 151, 160, 162, 165, 300, 311
– advanced   313
– damping term   143
– drift motion   312
– harmonic oscillator   151, 160
– hierarchy   146
– level shift   143
– multiplicative noise   144
– noise generator   136, 143, 145, 312, 319
– two-level system   162, 165
Langevin force, *see* Langevin equation, noise generator
laser   2, 3
– single-mode   131
laser spectroscopy   1

level shift   143, 153, 160, 163, 337, 342, 347
light absorption   174, 197
light cone   66
local oscillator   15, 205, 213, 219, 225, 280
Lorentz force   17, 52, 353, 361
– dipole approximation   353
– stress tensor   362
loss compensation   253

m
Mach–Zehnder interferometer   281
magnetization   33
– current   34
Markov approximation   142, 145, 154, 169, 341, 368
master equation   147, 148
– harmonic oscillator   155, 160
– trapped atom   467
– two-level system   163, 164, 166
matter
– commutation relation   31, 36, 43, 445
– Hamiltonian   32, 36
– magnetization   33
– nonrelativistic   18
– polarization   33, 40
– trapped atom   443
Maxwell equation   16, 40, 44
– Green tensor   45
– stress tensor   17, 362
medium-assisted electromagnetic field   7, 44
– canonical variable   47
– dynamical variable   46
– Hamiltonian   46
– source-quantity
micromaser   408, 428
– damping   432
– density matrix   429
– photon-number state   431
– sub-Poissonian light   432
– time scaling   429
– trapping state   431
– two-photon transition   429
micromotion   444
minimal coupling   31, 47
– Hamiltonian   31, 43, 47
minimum-uncertainty state   97
mixture of states, *see* density matrix
mode
– canonical variable   23
– cavity   301
– completeness relation   23, 44
– Coulomb gauge   22, 26
– dielectric plate   198
– expansion   22, 28, 36, 44, 61, 86, 100, 305, 448

– expansion of the canonical momentum   23, 27
– expansion of the vector potential   23, 27
– function   22, 29
– monochromatic   22, 22, 44
– nonmonochromatic   28, 311, 331
– nonorthogonality   29
– operator   22
– ortho-normalization relation   23, 25, 44
– spatio-temporal   30
– traveling plane wave   26, 86, 100, 448
molecular optics   360
Mollow triplet   390
momentum balance   17, 362
momentum density   17
multi-mode system
– coherent state   85
– number state   78
– squeezed state   98
multi-photon absorption/emission   7, 56
multi-photon resonance   56, 59
multi-wave mixing   5, 7, 56
multipolar coupling   33, 49
– Hamiltonian   35, 49
– Lagrangian   34
multipole moment   35

n

negative-frequency part   54
Newtonian equation of motion   18
– trapped atom   444
nonclassical   5, 265
– characteristic function   281
– entanglement   6, 290
– normally ordered moments   288, 294
– quantum interference   290
– state   127, 131
nonclassical light   1, 5, 184
– P function   267
– anti-bunched light   5, 270, 272
– application   281
– condition   272, 274, 276
– generation   278
– squeezed light   5, 276, 278
– sub-Poisson counting statistics   184
– sub-Poissonian light   5, 273, 425, 432
nonclassicality
– Bochner condition   282, 288
– characteristic function criteria   281
– first order   283
– hierarchy of conditions   285
– higher order   285
– moments-based criteria   289, 295
– weak measurement   269
nonlinear optics   1, 7, 56
number operator   74
– eigenvalue problem   74

– photoelectric detection of light   190, 191
– photonic   25, 78
– QND measurement   437
number state   73
– P function   132
– Q function   133
– completeness relation   76
– displaced   87
– electric-field variance   78
– multi-mode system   78
– single-mode system   73
– Wigner function   134

o

off-resonant light–matter coupling   53
one-photon resonance   54
operator disentangling   490
operator ordering
– s-order   121
– anti-normal   80, 121, 127, 491
– anti-standard   121
– associated c-number function   117, 120, 122, 148, 219, 225
– cavity output   325
– normal   69, 75, 80, 117, 121, 127, 177, 182, 186, 189, 192, 203, 257, 266, 323, 490
– other than boson systems   148
– photoelectric detection of light   182, 183, 186, 189, 192, 203
– standard   121, 269
– symmetrical   121
– time   68, 69, 177, 182, 186, 266, 323
optical bistability   56
optical cavity, see cavity
optical homodyne tomography, see quantum-state reconstruction, tomography
optical parametric oscillator   7, 59, 281
optical path   199
optoelectronics   1

p

P function, see phase-space representation
parametric down-conversion   275
parity operator
– displaced   123
– Wigner function   123
passband width   200
passive instrument   8, 15, 65, 72
– absorbing beam splitter   210
– beam splitter   198
– cavity   7, 299
– four-port device   206
– linear macroscopic body   8
– nonabsorbing beam splitter   205
– spectral filter   7, 198
Paul trap   444

periodic boundary condition 26
permittivity 41
– absorption 41
– complex 41
– dispersion 41
– Drude–Lorentz type 344
– space-dependent 41
phase
– canonical 107, 260
– commutation relation 109, 111
– cosine/sine 109
– difference 111
– Dirac 104
– exponential 105
– exponential moment 260
– Hermitian 109, 111
– London 107
– operational 228
– operator 5, 104, 105, 109
– reconstruction 260
– relation to quadrature distribution 261
– state 5, 107, 108
– Susskind–Glogower 105
– truncated Hilbert space 108
– uncertainty 109
phase conjugation 56
phase state 104
– canonical 107
– cosine/sine 109
– resolution of the identity 107, 110
phase-rotated quadrature, see quadrature
phase-space function, see phase-space representation
phase-space representation 113
– P function 119, 166, 222, 267, 274, 276
– Q function 120, 124, 226, 228, 250
– s-order 121, 124, 156, 333
– s-ordered moment 128
– anti-normal order 120
– associated c-number function 118, 120, 123
– cavity 333
– characteristic function 128, 149, 281
– classical 117
– density operator 126
– Fokker–Planck equation 147, 151
– normal order 117
– operator expansion 124
– other than bosonic systems 148
– positive P representation 129
– reconstruction 226, 244, 246, 256
– relation to displaced number distribution 256
– relation to quadrature distribution 242, 244, 246
– symmetric order 120
– Wigner function 121, 123, 244, 440, 477

– Wigner function matrix 479
photoelectric detection of light 3, 9, 69, 173, 266
– Bernoulli transformation 193
– broad-band photodetector 178, 180
– correlation of counts 183, 184, 186
– counting probability 179, 181, 182, 191, 194
– counting statistics 9, 187
– detection operator 191
– efficiency 181, 190, 218, 228, 246, 253
– electric-field correlation function 182, 186, 189, 266
– factorial moments of counts 180
– Fano factor 194
– homodyne detection 205
– interferometric detection 198
– mean number of counts 183, 196
– mean number of photons 194
– measurement time 189
– moments of counts 181, 183, 186
– multiplication process 186
– nonperturbative corrections 195
– perfect detection 193
– photocounting rate 183
– photocurrent 173, 186
– photocurrent correlation 187
– photoelectron 174, 179
– photon-number statistics 187, 191, 193
– point-like detector 181
– probability of photoelectron emission 177
– renormalized efficiency 196
– spectral detection 197
– sub-Poisson counting statistics 184, 194, 274
– sub-Poisson photon statistics 194, 274
– super-Poisson counting statistics 184
– variance of counts 183, 194
– variance of photons 194
– volume of detection 191
– volume of quantization 191
photoelectric effect 2
– photoelectric detection of light 9
photon 3, 24
– annihilation/creation operator 24, 28, 74
– cavity 312
– coherent state 5
– micromaser 431
– nonmonochromatic 29
– number operator 25, 74, 78
– number state 5, 74
– number statistics 191
– number-density operator 28
– photoelectric detection of light 190, 191
– QND measurement 437

– time of flight   304, 308, 309, 313
photon anti-bunching, *see* anti-bunched light
photon bunching, *see* bunched light
Poisson bracket   21
Poissonian distribution   82, 184, 273, 401
polarizability tensor   356
polarization
– current   34
– matter   33, 40
– noise   40
– plane wave   27
positive operator valued measure (POVM)   192, 217, 224
positive-frequency part   54
potential
– Coulomb   19
– scalar   18, 19, 47
– trap   444
– vector   18, 47
potential energy surface
– distorted   398
– shifted   398
power spectrum, *see* Wiener–Khintchine spectrum
Power–Zienau transformation   38
principle of least action, *see* Hamilton's principle

q

Q function, *see* phase-space representation
quadrature   94
– eigenstate   102
quadrature distribution
– characteristic function   240
– coherent state   103
– joint   228, 244, 258
– measurement   220
– relation to canonical phase   260
– relation to density matrix   248, 250
– relation to moments   257
– relation to phase-space representation   242
– relation to quantum state   240
– relation to Wigner function   244
– sum   244
quadrature state   102
– coherent-state basis   103
– completeness relation   104
– number basis   103
quadrupole trap, *see* Paul trap
quantization
– canonical   15, 21, 31, 36, 43
– center-of-mass motion   445
– dielectric medium   39
– electromagnetic field/matter   6, 15, 31, 36, 43

– magnetodielectric medium   47
– medium-assisted electromagn. field   45
– volume   26, 191
quantized center-of-mass motion
– coherent displacement   460
– coherent state   463
– cold atom   11
– dark state   466, 468, 470
– entangled state   464, 478
– even/odd coherent state   468
– Hamiltonian   446
– nonlinear coherent state   470
– nonlinear coupling   458
– nonlinear displacement   461
– nonlinear dynamics   456
– nonlinear parametric interaction   459
– nonlinear squeezing   461
– number state   464
– Raman-driven   457
– reservoir engineering   471
– Schrödinger-cat state   464
– squeezed state   464
– trap potential   444
– trapped atom   443
quantum coherence   85
quantum communication
quantum computation   291
quantum efficiency, *see* photoelectric detection of light, efficiency
quantum electrodynamics   15
quantum information   291
quantum interference   290
– trapped atom   470
quantum Langevin equation, *see* Langevin equation
quantum noise theory, *see* damping
quantum nondemolition (QND) measurement   173, 437
quantum ruler   217
quantum state
– bosonic system   73
– coherent   79, 463
– coherent-state expansion   83
– dark state   466
– displaced number   87
– entangled   99, 291, 422, 433
– EPR   434
– even/odd coherent   283, 468
– motional   461
– nonlinear coherent   470
– number   73, 464
– phase   104
– preparation, *see* quantum-state preparation
– quadrature   102
– reconstruction, *see* quantum-state reconstruction

- Schrödinger cat 291, 433, 464
- squeezed 88, 464
- thermal 130
quantum-state determination, *see* quantum-state reconstruction
quantum-state measurement, *see* quantum-state reconstruction
quantum-state preparation 10
- cavity 431
- EPR state 434
- number state 275, 431, 464
- Schrödinger-cat state 433, 464
- single-photon state 275
- squeezed state 278, 464
- trapped atom 461
quantum-state reconstruction 10, 237
- Q function 226
- canonical phase 260
- cavity 435, 438
- density matrix 248, 250, 254, 477
- direct sampling 250, 255, 261
- displaced number distribution 215, 439
- endoscopy 436
- entangled state 478
- local 256, 475
- moments 257
- operational phase 228
- phase-space representation 226, 244, 256
- quadrature distribution 220
- tomographic 472
- tomography 239, 472
- trapped atom 472
- Wigner function 244, 256, 440, 477
quantum-state transformation 208, 212
quasi-probability distribution, *see* phase-space function
quorum 238

r

Rabi frequency 371, 385, 400, 410, 451
- n-photon state 385, 410
- photon vacuum 350, 415
Rabi oscillation 375, 402
- damped 350
- photon vacuum 350, 415
radiation force 354, 362
radiationless dephasing
- harmonic oscillator 158
- two-level system 169
Radon transformation 244
- inverse 244
Raman line 404
Raman resonance 394
Rayleigh line 389
reciprocity relation 481
regression theorem

- quantum 169, 172
- resonance fluorescence 377, 386, 400
relaxation, *see* damping
reservoir 135
resonance fluorescence 4, 6, 367
- V configuration 394, 398
- Λ configuration 391
- anti-bunched light 272, 375, 379
- Bloch equations 370, 388
- coherent 389
- correlation function 368
- dark periods 395
- dark resonance 391
- field-strength variance 379
- high-driving-field limit 375, 380, 389
- higher-order spectral properties 390
- hot luminescence 400
- incoherent 375, 389, 390
- intensity 372, 374, 400
- intensity correlation 375, 378, 403
- intermittent 394, 398
- ladder configuration 398
- line splitting 390, 404
- Markov approximation 368
- Mollow triplet 390
- multi-level system 391
- power spectrum 383, 404
- Raman line 404
- Raman resonance 394
- Rayleigh component 380, 389
- regression theorem 370, 377, 386, 400
- sample of atoms 381
- single atom 229, 370, 375, 379
- source field 367, 396
- spectral filtering 370
- spectral intensity correlation 391
- spectral properties 369
- spectral squeezing 390
- squeezed light 379
- squeezing pattern 382
- sub-Poissonian light 379
- trapped electron 394
- trapped ion 383, 398
- two-level system 370
- vibronic coupling 398
- weak-driving-field limit 375, 389
- Wiener–Khintchine spectrum 384, 387
resonant light–matter coupling 53
resonator, *see* cavity
response function
- cavity 302, 303
- dielectric plate 199, 201
- photoelectric detection of light 178
- reflection 199, 202, 302
- spectral 178, 199, 302, 303
- transmission 199, 201, 302
retarded solution 61

rotating-wave approximation 7, 53, 57, 139, 142, 176
– resonance fluorescence 367
– spontaneous emission 339
– trapped atom 447, 449

s

sampling 238
– density matrix in number-state basis 250, 255
– exponential phase moment 261
– normally ordered moment 258
scanning near-field optical spectroscopy 348
scattered light 71
Schrödinger equation 115, 251
Schrödinger picture 115, 147, 148, 170
Schwarz inequality 271
separation of variables 22
setting frequency 200
signal field 15, 59, 213
single-atom Rydberg maser, *see* micromaser
source-quantity representation 60
– cavity 307, 314
– commutation relation 66, 67
– correlation function 69, 183
– Green tensor 64
– medium-assisted electromagn. field 63
– operator ordering 70, 177, 183
– photoelectric detection of light 183
spectral filter 15
– correlation function 202
– Fabry–Perot 9, 198
– Fourier decomposition 204
– input-output relation 200
– intrinsic spectral property 202
– physical spectrum 202
– power spectrum 205
– resonance fluorescence 369
spontaneous emission 6, 338
– cavity 349, 415
– decay rate 165, 168, 342
– dielectric bodies 338
– enhancement 9, 344
– Green tensor 341
– inhibition 9, 344
– intensity 340
– level shift 342, 347
– magnetodielectric bodies 338
– Markov approximation 341
– microsphere 344
– nonradiative decay 346
– quantum yield 343
– Rabi oscillation 350
– Schrödinger equation 339
– source field 340
– spherical-shell cavity 351
– strong atom–field coupling 348
– two-level system 339
– vacuum Rabi frequency 350
– vibronic system 402
– weak atom–field coupling 341
squeeze operator
– multi-mode system 99
– single-mode system 88
– two-mode system 99
squeezed light 5, 276
– P function 276
– application 281
– condition 276
– four-wave mixing 279
– higher-order squeezing 277
– homodyne detection 215, 280, 281, 382
– intensity measurement 281
– Mach–Zehnder interferometer 281
– normally ordered field variance 98, 99
– optical parametric oscillator 281
– polarization interferometer 281
– resonance fluorescence 379, 390
– second-order squeezing 277
– spectroscopy 281
– squeeze operator 278
squeezed state 88, 276
– amplitude-squared 289
– coherent 90, 91
– coherent-state basis 92
– field correlation 101
– minimum uncertainty 97
– multi-mode system 98
– nonorthogonality 91
– normally ordered field variance 100
– number basis 92
– over-completeness 91
– quadrature squeezing 88
– quadrature variance 94
– resolution of the identity 91
– squeeze operator 88, 99, 278
– squeezing spectrum 101
– $SU(1,1)$ group transformation 89
– two-mode squeezed vacuum 99
– vacuum 90, 99
– white noise 102
– Wigner function 134
squeezing, *see* squeezed state and squeezed light
statistics 174
– Bernoulli's scheme 179
– characteristic function 128, 179, 181, 182
– joint probability 266
– marginal probability 267
– multiplication process 186
– stochastic process 169
sub-Poissonian light 5, 273

– P function 274
– anti-bunched light 274
– condition 274
– Franck–Hertz scheme 275
– Jaynes–Cummings model 425
– micromaser 432
– parametric down-conversion 275
– pump-noise-suppressed lasing 276
– resonance fluorescence 275, 379
– short-time measurement 274
– shot-noise level 273
– sub-Poisson counting statistics 274
– sub-Poisson photon statistics 274
super-radiance 381

t

thermal state 130
– P function 131
– chaotic light 130
– coherently displaced 131
– density operator 130
three-photon resonance 60
time resolution 54, 304, 308, 309
time-delay term 68
time-evolution operator 175, 278
transformation matrix 211
transition probability
– harmonic oscillator 153, 156
– multi-level atom 168
– photoelectric detection of light 174
transition rate, see transition probability
transverse current density 19
transverse vector function 19, 20
trapped atom 443
– Doppler cooling 462
– effective potential 444
– Lamb–Dicke parameter 450
– Lamb–Dicke regime 451
– laser-cooling 461
– master equation 467
– micromotion 444
– nonlinear Jaynes–Cummings model 449, 452
– Paul trap 444
– quantized motion 443, 445
– quantum-state preparation 461
– quantum-state reconstruction 472
– reconstruction of entangled state 478
– reservoir engineering 471
– resolved sideband cooling 462
– resolved sideband regime 449, 451
– secular motion 445
trapped ion 398

trapped ion, see trapped atom
two-mode squeezed vacuum 99
two-photon absorption/emission 56
two-photon coherent state, see squeezed state, coherent

v

vacuum
– Casimir effect 337
– energy 25
– Rabi splitting 410
– squeezed 90, 99
van der Waals force 55, 337, 348, 352
– body-assisted vacuum 354
– Green tensor 357
– ground-state atom 357
– long-distance limit 359
– perfectly reflecting plate 359
– planar structure 358
– potential 356
– short-distance limit 359
vibronic system 398
– Born–Oppenheimer state 399
– density-matrix equations of motion 400
– electronic matrix element 400
– electronic quantum state 399
– electronic relaxation 400
– hot luminescence 400
– non-Markovian dephasing 401
– normal coordinate 401
– population/depopulation 402
– potential energy surface 398
– Raman line 404
– relaxations 400
– resonance fluorescence 398
– spontaneous emission 402
– trapped ion 404
– vibrational frequency 399
– vibrational overlap integral 400
– vibrational quantum state 399
– vibrational relaxation 400
– vibronic coupling 400
volume of detection 191

w

wave equation 22
wave-number vector 26
white noise 102
– squeezed 102
Wiener–Khintchine spectrum 205
– resonance fluorescence 384, 387
Wigner function, see phase-space representation